T0337980

Oogenesis

Oogenesis: The Universal Process

Editors

Marie-Hélène Verlhac
CNRS/Université Pierre et Marie Curie, Paris, France

and

Anne Villeneuve
Stanford University School of Medicine, Stanford, CA, USA

WILEY-BLACKWELL
A John Wiley & Sons, Ltd., Publication

Wiley-Blackwell is an imprint of John Wiley & Sons, formed by the merger of Wiley's global Scientific, Technical and Medical business with Blackwell Publishing.

Registered office: John Wiley & Sons Ltd, The Atrium, Southern Gate, Chichester, West Sussex, PO19 8SQ, UK

Other Editorial Offices:
9600 Garsington Road, Oxford, OX4 2DQ, UK

111 River Street, Hoboken, NJ 07030-5774, USA

For details of our global editorial offices, for customer services and for information about how to apply for permission to reuse the copyright material in this book please see our website at www.wiley.com/wiley-blackwell

Library of Congress Cataloguing-in-Publication Data

Oogenesis : the universal process / [edited by] Marie-Hélène Verlhac and Anne Villeneuve.
p. cm.
Includes index.
ISBN 978-0-470-69682-8
1. Oogenesis. I. Verlhac, Marie-Hélène. II. Villeneuve, Anne.
QL965.O656 2010
571.8′45–dc22

2009051040

ISBN: 978-0-470-69682-8 (H/B)

A catalogue record for this book is available from the British Library.

Set in 10.5/12.5 Times by Thomson Digital, Noida, India.
Printed in Great Britain by CPI Antony Rowe, Chippenham, Wiltshire.

First Impression 2010

Contents

List of contributors

David Albertini, Department of Molecular and Integrative Physiology and Department of Anatomy and Cell Biology, University of Kansas Cancer Centre, Kansas City, KS 66160, USA; Marine Biological Laboratory, Woods Hole, MA 02543, USA

Aldine Amiel, UMR 7009, UPMC-CNRS, Developmental Biology Unit, Observatoire Océanologique, 06230 Villefranche sur mer, France

Susan L. Barrett, Department of Obstetrics and Gynecology, Feinberg School of Medicine and Center for Reproductive Science, Northwestern University, Evanston, IL 60208 USA; The Oncofertility Consortium, Chicago, IL 60611, USA

Frédéric Baudat, CNRS UPR 1142, Institut de Génétique Humaine, 141 rue de la Cardonille, 34396 Montpellier CEDEX 5, France

Claudia Baumann, Department of Clinical Studies, Center for Animal Transgenesis and Germ Cell Research, New Bolton Center, University of Pennsylvania, School of Veterinary Medicine, 382 West Street Road, Kennett Square, PA 19348 USA

Eulàlia Belloc, Gene Regulation Program, Centre for Genomic Regulation (CRG), C/Dr Aiguader, 88, 08003, Barcelona, Spain

John Bromfield, Department of Molecular and Integrative Physiology, University of Kansas Cancer Centre, Kansas City KS 66160, USA

Stéphane Brunet, Biologie du Développement – UMR 7622, UPMC-CNRS, 9 Quai St Bernard, 75252 Paris CEDEX 05, France

Julien Burger, Institut Jacques Monod, CNRS and Université Paris Diderot, Buffon Building, 15 rue Hélène Brion, 75205 Paris CEDEX 13, France

Anna Castro, CNRS UMR 5237, IFR 122, Universités Montpellier 2 et 1, Centre de Recherche de Biochimie Macromoléculaire, 1919 Route de Mende, 34293 Montpellier CEDEX 5, France

Patrick Chang, UMR 7009, UPMC-CNRS, Developmental Biology Unit, Observatoire Océanologique, 06230 Villefranche sur mer, France

Alexander Combes, Institute for Molecular Biosciences, The University of Queensland, Brisbane, QLD 4072, Australia

Rabindranath De La Fuente, Department of Clinical Studies, Center for Animal Transgenesis and Germ Cell Research, New Bolton Center, University of Pennsylvania, School of Veterinary Medicine, 382 West Street Road, Kennett Square PA 19348, USA

Julien Dumont, Desai Laboratory, Department of Cellular and Molecular Medecine, Ludwig Institute for Cancer Research, University of California, 9500 Gilman Drive, La Jolla, CA 92093-0660, USA

Carolina Eliscovich, Gene Regulation Program, Centre for Genomic Regulation (CRG), C/Dr Aiguader, 88, 08003, Barcelona, Spain

Ronald Ellis, Department of Molecular Biology, School of Osteopathic Medicine, B303 Science Center, The University of Medicine and Dentistry of New Jersey, 2 Medical Center Drive, Stratford NJ 08084, USA

Malcolm Faddy, School of Mathematical Sciences, Queensland University of Technology, Brisbane, QLD 4001, Australia

José-Eduardo Gomes, Institut Jacques Monod, CNRS and Université Paris Diderot, Buffon Building, 15 rue Hélène Brion, 75205 Paris CEDEX 13, France

Roger Gosden, Weill Medical College of Cornell University, 1305 York Avenue, New York, NY 10021, USA

Evelyn Houliston, UMR 7009, UPMC-CNRS, Developmental Biology Unit, Observatoire Océanologique, 06230 Villefranche sur mer, France

Laurinda A. Jaffe, Department of Cell Biology, L-5004, University of Connecticut Health Center, 263 Farmington Avenue, Farmington, CT 06032, USA

Catherine Jessus, Biologie du Développement – UMR 7622, UPMC-CNRS, 9 Quai St Bernard, 75252 Paris CEDEX 05, France

Eujin Kim, Weill Medical College of Cornell University, 1305 York Avenue, New York, NY 10021, USA

Takeo Kishimoto, Graduate School of Bioscience and Biotechnology, Laboratory of Cell and Developmental Biology, Tokyo Institute of Technology, Nagatsuta 4259, Midoriku, Yokohama 226-8501, Japan

Peter Koopman, Institute for Molecular Biosciences and ARC Centre of Excellence in Biotechnology and Development, The University of Queensland, Brisbane, QLD 4072, Australia

Bora Lee, Weill Medical College of Cornell University 1305 York Avenue, New York, NY 10021, USA

Thierry Lorca, CNRS UMR 5237, IFR 122, Universités Montpellier 2 et 1, Centre de Recherche de Biochimie Macromoléculaire, 1919 Route de Mende, 34293 Montpellier CEDEX 5, France

Katia Manova, Memorial Sloan-Kettering Cancer Center, New York, NY 10065, USA

Raúl Méndez, Gene Regulation Program, Centre for Genomic Regulation (CRG), C/Dr Aiguader, 88, 08003, Barcelona, Spain

Jorge Merlet, Institut Jacques Monod, CNRS and Université Paris Diderot Buffon Building, 15 rue Hélène Brion, 75205 Paris CEDEX 13, France

Susanna Mlynarczyk-Evans, Department of Developmental Biology, Stanford University School of Medicine, Beckman Center, B300, 279 Campus Drive, Stanford CA 94305-5329, USA

Tsuyoshi Momose, UMR 7009, UPMC-CNRS, Developmental Biology Unit, Observatoire Océanologique, 06230 Villefranche sur mer, France

Tomoko Nishiyama, Institute of Molecular Pathology, Dr. Bohr-Gasse 7, 1030 Vienna, Austria

Rachael P. Norris, Department of Cell Biology, L-5004, University of Connecticut Health Center, 263 Farmington Avenue, Farmington CT 06032, USA

Isabel Novoa, Gene Regulation Program, Centre for Genomic Regulation (CRG), C/Dr Aiguader, 88, 08003, Barcelona, Spain

Lionel Pintard, Institut Jacques Monod, CNRS and Université Paris Diderot, Buffon Building, 15 rue Hélène Brion, 75205 Paris CEDEX 13, France

Sabine Santucci-Darmanin, FRE 3086, CNRS, Faculté de Médecine, Université de Nice Sophia-Antipolis, Avenue de Valombrose, 06107 Nice CEDEX 2, France

Cassy Spiller, Institute for Molecular Biosciences, The University of Queensland and ARC Centre of Excellence in Biotechnology and Development, Brisbane QLD 4072, Australia

Kazunori Tachibana, Graduate School of Bioscience and Biotechnology, Laboratory of Cell and Developmental Biology, Tokyo Institute of Technology, Nagatsuta 4259, Midoriku, Yokohama 226-8501, Japan

M. Emilie Terret, Molecular Biology Program, Memorial Sloan-Kettering Cancer Center, Box # 97, 1275 York Avenue, New York, NY 10021, USA

Marie-Hélène Verlhac, Biologie du Développement – UMR 7622, UPMC-CNRS, 9 Quai St Bernard, 75252 Paris CEDEX 05, France

Anne Villeneuve, Department of Developmental Biology, Stanford University School of Medicine, Beckman Center, B300, 279 Campus Drive, Stanford, CA 94305-5329, USA

Maria M. Viveiros, Department of Animal Biology, Center for Animal Transgenesis and Germ Cell Research, New Bolton Center University of Pennsylvania, School of Veterinary Medicine, 382 West Street Road, Kennett Square, PA 19348, USA

Gary M. Wessel, Department of Molecular Biology, Cell Biology, and Biochemistry, Brown University, Box G-L173, 185 Meeting Street, Providence RI 02912, USA

Karen Wingman Lee, Biologie du Développement – UMR 7622, UPMC-CNRS 9 Quai St Bernard, 75252 Paris CEDEX 05, France

Julian L. Wong, Department of Molecular Biology, Cell Biology, and Biochemistry, Brown University, Providence, RI 02912, USA

Teresa K. Woodruff, Department of Obstetrics and Gynecology Feinberg School of Medicine Northwestern University and The Oncofertility Consortium, Chicago IL 60611, USA; Center for Reproductive Science, Northwestern University, Evanston IL 60208, USA

Feikun Yang, Department of Clinical Studies, Center for Animal Transgenesis and Germ Cell Research, New Bolton Center, University of Pennsylvania, School of Veterinary Medicine, 382 West Street Road, Kennett Square, PA 19348, USA

Foreword

August Weissman dedicated his book, 'The Germ-Plasm' (1892) to the memory of Charles Darwin. Weissman understood the urgent need for a proper theory of heredity, knew that Darwin's ideas on the subject were inadequate, and equally clearly recognized that, unlike "the perishable body of the individual" something —the "hereditary substance"—had to be passed from generation to generation in eggs and sperm and hence, "the continuity of the germ-plasm". It took another 10–15 years before Thomas Hunt Morgan accepted that the behaviour of chromosomes explained Mendel's laws (of which Weissman was unaware; indeed, neither 'chromosomes' nor 'nucleus' feature in the index of his book), and one might say that it took the structure of DNA, and the idea that "DNA makes RNA makes protein" to bring biology into the modern era. We don't think twice, these days, about the continuity of life on earth, and accept without question that cells only arise from pre-existing cells; this is all so integral to the biologist's world view that a number of great mysteries hardly ever come to light. Broadly speaking, these underlie the topic of this collection of essays about oogenesis. How does the germ-plasm manage to avoid the body's mortality?

Quite apart from deep questions of this kind, the details of how eggs come to be eggs are fascinating and instructive well beyond the relatively narrow field of reproductive biology. Likewise the events just before and after fertilization, when the egg meets the sperm and starts to become a new body. This book contains a series of essays, authoritative and fascinating reviews of all aspects of oogenesis.

The reviews follow a kind of chronological or developmental order from questions about sex determination in worms to assisted reproduction in humans. The simple-sounding decision of what sex to become is anything but, and we are reminded that it is quite possible to be a hermaphrodite and survive perfectly successfully. We discuss the setting-aside of germ cells from the soma early in development as well as the surprisingly complicated decision-making processes that lead to the differentation of eggs or sperm. Meiosis is a necessary common process for both kinds of gamete, and we have reviews of what is known about meiotic chromosome pairing and homologous recombination. In oocytes, the meiotic divisions often take place shortly before the cell becomes a fully-fledged, fertilizable egg, and is subject to some elaborate controls that are still far from completely understood.

The choice between becoming an egg or a sperm is one of the most complex of development, and it is made long before changes in cell morphology take place. This fate decision depends on sex chromosomes and depends on interactions between gonadal somatic cell lineages and the germ cells themselves. Indeed, metazoans have evolved a complex array of interactions between the soma and germ line that regulate reproductive success. During the growth period of oogenesis, meiotically-arrested oocytes accumulate large quantities of dormant maternal mRNAs. Meiotic resumption requires cascades of successive unmasking, translation, and discarding of these maternal mRNAs. Not only is the the timing of specific translation finely regulated during this period, but the embryonic axis and even the establishment of the next generation of germ cells are also defined through the localization of such dormant mRNAs within the oocyte. And of course, meiosis is an integral component of the oogenesis program, accomplishing the essential reduction of diploid chromosome number to a haploid complement in preparation for zygotic development. Crossovers between homologous chromosomes not only generate genetic diversity, but are actually required for the accurate segregation of homologous chromosomes in most organisms. At a fundamental level, the ability to reduce chromosome number two-fold requires the formation of correct pairwise associations between homologous chromosomes and further recombination. Chromosomes in the germ line exhibit unique structural and functional properties that are essential to coordinate the complex events of meiosis with subsequent changes leading towards nuclear and epigenetic maturation during gametogenesis.

Once meiosis is (almost) complete and sufficient growth has been achieved, the oocyte is ready to exit the prophase I arrest of meiosis and undergo the two meiotic divisions. Once again, communication between somatic cells and the oocyte are required to control this unique prophase-to-metaphase transition. The oocyte normally undergoes a highly asymmetric division that is critical to ensure the formation of a competent resource-rich egg, capable of generating a living euploid descendent after fertilization. In the last few years, our understanding of the principles of meiotic spindle assembly has significantly improved, due to the elucidation of common mitotic and meiotic principles as well as special features that apply to female meiosis and the generation of extreme asymmetry in the formation of polar bodies. There is great interest in the business of chromosome segregation from a medical standpoint, since chromosome non-disjunction produces all kinds of problems including developmental arrest, miscarriages, or severe birth defects such as Down's syndrome. The basis for these errors are still a matter of intense investigation, with a long-term view to prevention as well as diagnosis.

The regulation of the cell cycle during the life of an oocyte is extremely interesting, with multiple arrest points. Here, there is tremendous specificity and variability from organism to organism, bewildering to the unwary. In some species, it is the arrival of the sperm that reinitiates meiosis. In others, hormonal signals prepare the oocyte for fertilization, and elaborate mechanisms exist to ensure that the sperm hits the egg at the right phase of the cell cycle. So clams release oocytes into the sea and the arrival of the sperm initiates completion of meiosis; frogs and women lay eggs that are arrested in second meiotic metaphase waiting for the sperm to arrive, but sea urchins complete both meiotic divisions and arrest in a dormant G-zero state to await fertilization. Limpets and starfish eggs like to be fertilized while meiotic divisions are in progress; sometimes one marvels that there are any successful matings at all! Extensive studies have gradually revealed the core signalling components required for oocytes to wait for the sperm, and show how common components can be used and reused in different ways to achieve the same end by a variety of routes.

Fertilization marks the completion and culmination of oogenesis. It is a multi-step event that leads to the fusion of two complementary gametes. Compatibility of the particular egg with the correct sperm is determined before the gametes fuse in a variety of ways including the complex behaviour of courtship as well as gamete attraction and gamete molecular recognition and adhesion. The extracellular molecules on each gamete that participate in this species-selective process are thought to co-evolve within a species while diversifying from sister organisms so as to minimize cross-species interactions. But fertilization also initiates early development, and, germane to the oocyte to embryo transition, is the need to dispose of some maternal products. This is achieved via their specific and timely degradation, triggered by the arrival of the sperm.

The mammalian ovary is endowed with a fixed number of follicles because in the female, germline stem cells have been exhausted around the time of birth. The reserve population of potential oocytes, represented by primordial follicles, is gradually depleted by recruitment to the growing stages of oogenesis, but most of these would-be eggs undergo atresia by apoptosis. Over the course of the reproductive lifespan in human females, the total number of follicles declines from about a million to a threshold of around one thousand, below which ovulatory cycles are unsustainable and the menopause intervenes. Thus, ageing of the follicle population commences from the moment it has been established, and is irreversible, but the initial reserve is normally sufficient for fecundity until mid-life. Such basic knowledge of the journey of an oocyte has major implications for our understanding of the molecular mechanisms of aneuploidy as well as the design of clinical procedures to address infertility. Understanding ovarian follicle development is crucial for physicians interested to determine the best assisted reproductive technologies to use for women with fertility-threatening diseases and for scientists to develop experimental foeto-protective strategies.

The study of oocytes has made enormous contributions to the understanding of the molecular composition of the factors promoting M-phase entry. The power and complementarity of investigations into the mechanisms of maturing oocytes on the one hand and yeast genetic studies on the other, coupled with the revolution in molecular cloning allowed us to unravel the basis of cell cycle regulation. But although the heroic phase of the story of maturation promoting factor and points of no return may be over, the study of oocyte and oogenesis is still producing new seeds and comes up with interesting new model organisms that give evolutionary perspective to sexual reproduction. For example, the jellyfish *Clytia* offers a fresh perspective on regulation of oogenesis and its evolutionary history because of the phylogenetic position of the organism and by the simplicity, transparency and experimental accessibility of the female gonad. The development of diverse model systems will surely bring answers to this fascinating question of the evolutionary origins and advantages of sex.

Dr Tim Hunt
Cancer Research UK

Section I

Oocyte determination

1

The sperm/oocyte decision, a *C. elegans* perspective

Ronald Ellis

Department of Molecular Biology, School of Osteopathic Medicine, The University of Medicine and Dentistry of New Jersey, Stratford, NJ 08084, USA

> No trumpets sound when the important decisions of our life are made. Destiny is made known silently.
>
> *Agnes de Mille*

1.1 Introduction

The decision of germ cells to differentiate as spermatocytes or oocytes is dramatically different from other decisions made during development. First, the magnitude of the response is far greater than in most cell-fate decisions. For example, microarray analyses identified at least 250 oocyte-enriched genes and 650 spermatocyte-enriched genes in *Caenorhabditis elegans* (Reinke *et al.*, 2000). By contrast, touch-receptor cells are defined by only a few dozen genes (reviewed by Goodman, 2006; Bounoutas and Chalfie, 2007). Second, most cell-fate decisions occur in individual cells, or pairs of daughter cells that are being formed by division. However, germ cells retain cytoplasmic contacts with their neighbours during much of development. In *C. elegans,* for example, primordial germ cells begin spermatogenesis or oogenesis as part of a syncytium. Indeed, some cells connected to the syncytium undergo spermatogenesis while others are initiating oogenesis. Third, developing oocytes contain a variety of messenger RNAs and proteins that are needed for embryonic development, and some of these molecules must be prevented from influencing the sperm/oocyte decision itself. Thus, this regulatory decision is unique. Since sperm and oocytes are the most ancient sexually dimorphic cells (reviewed by White-Cooper,

Oogenesis: The Universal Process Marie-Hélène Verlhac and Anne Villeneuve

Doggett and Ellis, 2009), evolution has had a long time to shape solutions to these problems.

In most animals, primordial germ cells differentiate into spermatocytes in males or oocytes in females. However, hermaphrodites like *C. elegans* make both types of gametes in the same gonad, which simplifies the study of how these fates are controlled. In particular, hermaphrodite genetics makes it easy to identify and maintain sterile mutants. Furthermore, these animals are transparent, so developing germ cells can be observed in living worms. Finally, mutant hermaphrodites that make only sperm or only oocytes are easy to identify. Thus, research has been able to create a detailed picture of how the sperm/oocyte decision is regulated in *C. elegans*.

1.2 *C. elegans* hermaphrodites are modified females

Although most species of nematodes produce males and females, hermaphroditism has arisen independently on many occasions (Kiontke and Fitch, 2005). Even in the genus *Caenorhabditis,* two species appear to have acquired this trait independently (Cho *et al.*, 2004; Kiontke *et al.*, 2004). In these species, the *XX* hermaphrodites develop female bodies, but some of their germ cells undergo spermatogenesis late in larval development, producing a small supply of sperm that are stored in the spermatheca. Early in adulthood, hermaphrodites switch to the production of oocytes, which can be fertilized by their own sperm. This pattern of development shows that primordial germ cells have the ability to form either spermatocytes or oocytes, and analysis of *C. remanei* confirms that this capacity is found in related male/female species (Haag, Wang and Kimble, 2002).

Two traits make self-fertile hermaphrodites like *C. elegans* different from cross-fertile hermaphrodites, which are able to mate with each other. First, these nematodes produce sperm by altering germ cell fates in *XX* animals for a short period of time, prior to the onset of oogenesis. Thus, the number of self-sperm is limited by the duration of production. Second, self-fertile hermaphrodites have female gonads, so they provide an excellent model for oogenesis. By contrast, most cross-fertile hermaphrodites have male and female gonads.

1.3 The hermaphrodite gonad provides the normal environment for oogenesis

In many species, the female gonad is essential for germ cells to initiate and carry out oogenesis. This is not true for nematodes, since some mutations that alter the sperm/oocyte decision cause males to make oocytes (for examples, see Barton and Kimble, 1990; Ellis and Kimble, 1995). However, the hermaphrodite gonad does provide the normal setting for oogenesis in nematodes, and oocytes in males do not progress to fertilization. Furthermore, some experiments imply that cells in the somatic gonad directly influence the sperm/oocyte decision (McCarter *et al.*, 1997).

1.3.1 Structure of the hermaphrodite gonad

In *C. elegans,* the hermaphrodite gonad is composed of two symmetrical tubes that meet at a central uterus (Figure 1.1). Each tube contains a large ovotestis and a spermatheca, which adjoins the uterus. The entire process of germ cell differentiation takes place in the two ovotestes, which are each composed of a distal tip cell and five pairs of sheath cells (Figure 1.1; McCarter *et al.*, 1997; Hall *et al.*, 1999, and see www.wormatlas.org for a concise review). Each stage of oogenesis occurs in a separate region of the ovotestis.

The distal tip cells create a stem cell niche, where mitosis continues throughout the animal's life. In the area just beyond the distal tip cells (known as the transition zone), germ cells begin meiosis. This region is not ensheathed by cells of the somatic gonad, although it is covered by a basement membrane. Next, most developing oocytes arrest in the pachytene phase of prophase I while in contact with the large sheath cell 1 pair. Near the bend in the ovotestis, under the sheath cell 2 pair, most oocytes resume progression through meiosis, and some undergo apoptosis (Gumienny *et al.*, 1999). Finally, sheath

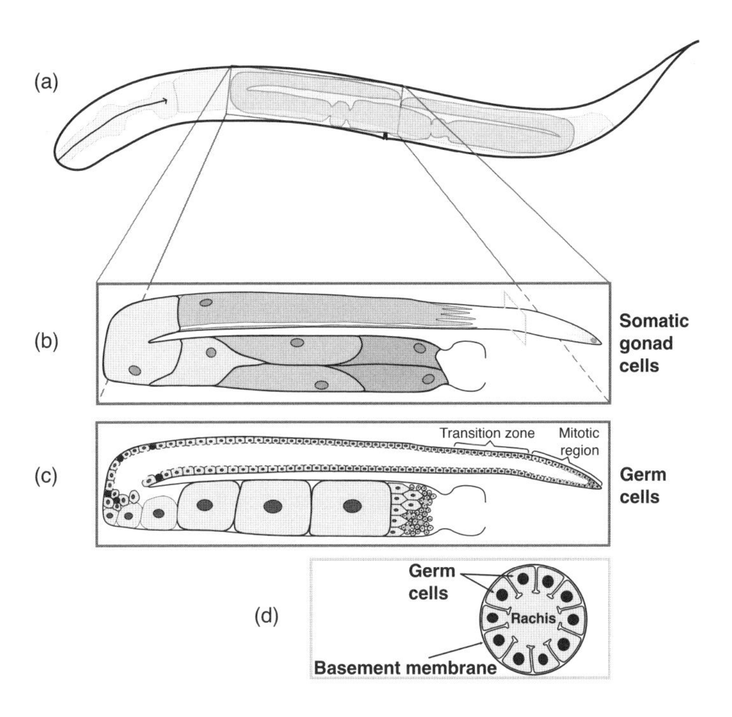

Figure 1.1 Structure of the hermaphrodite gonad. (a) Diagram of a young adult hermaphrodite, showing the digestive system in light green, and the gonad in grey. Anterior is to the left, and ventral is down. (b) Inset diagram of the anterior ovotestis, showing cells of the somatic gonad. The distal tip cell is yellow. Sheath cell 1 is dark blue, sheath cell 2 is light blue, and sheath cell 3 is tan. The second member of each pair is on the opposite side of the gonad, with only the edge of sheath cell 1 visible. Sheath cell pair 4 is peach, and sheath cell pair 5 is orange. (c) Inset diagram of the anterior ovotestis, showing the germ cells. Cells expressing female transcripts and proteins are pink, and those expressing male transcripts are blue. Cell corpses are black circles, and residual bodies are blue circles. (d) Cross-section of the gonad. A full colour version of this figure appears in the colour plate section.

cells 3, 4 and 5 contain extensive actin/myosin networks that support rapidly growing oocytes and control ovulation.

1.3.2 Interactions between gonad and germline

The somatic gonad is descended from two founder cells present in newly hatched larvae (Kimble and Hirsh, 1979). The simplicity of this lineage allows the elimination of groups of gonadal cells by killing their ancestors with a laser microbeam (Kimble and White, 1981; McCarter *et al.*, 1997). When a sheath/spermatheca (SS) precursor cell is killed, the ovotestis contains only a single member of each sheath cell pair, and often produces oocytes instead of sperm (McCarter *et al.*, 1997). Thus, the somatic gonad appears to influence the sperm/oocyte decision. However, killing germ cells sometimes causes animals to make oocytes instead of sperm, so it remains possible that the somatic gonad influences the sperm/oocyte decision indirectly, by promoting robust growth of the germline.

1.4 The core sex-determination pathway regulates somatic and germ cell fates

In *C. elegans,* the same genes regulate sexual fates in both the soma and germline. They act through a signal transduction pathway to control the master transcription factor TRA-1 (Figure 1.2).

1.4.1 The *X*: A ratio determines sex

In nematodes, sexual identity is specified by the ratio of *X* chromosomes to sets of autosomes (Madl and Herman, 1979). Signalling elements on these chromosomes regulate the activity of *xol-1,* a gene that promotes male development (reviewed by Wolff and Zarkower, 2008). In males, XOL-1 represses three *sdc* genes, allowing the expression of HER-1. In hermaphrodites, the absence of XOL-1 allows the SDC

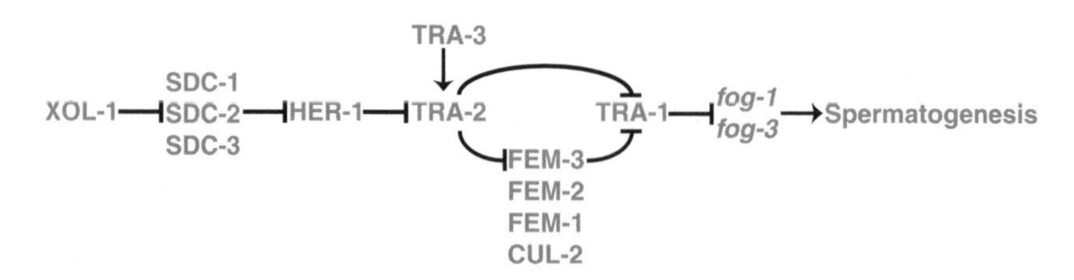

Figure 1.2 The core sex-determination pathway. Genes promoting male fates are blue, and those promoting female fates are pink. Arrows indicate positive interactions, and '—|' indicates negative interactions. Proteins are indicated by capital letters, and genes by lowercase italics. A full colour version of this figure appears in the colour plate section.

proteins to block the transcription of *her-1*. The SDC proteins also promote dosage compensation (reviewed by Wolff and Zarkower, 2008).

1.4.2 Sexual fates are coordinated by the secreted protein HER-1

HER-1 is a small, secreted protein that causes somatic cells to adopt male fates and germ cells to become sperm. Thus, it acts like a male sex hormone. In *XX* animals, ectopic expression of HER-1 is sufficient to cause spermatogenesis (Perry *et al.*, 1993). In *XO* animals, *her-1* mutations result in hermaphroditic development and the production of oocytes, so *her-1* is required to maintain spermatogenesis (Hodgkin, 1980). However, it is not needed for spermatogenesis *per se,* since null mutants make sperm before switching to oogenesis (Hodgkin, 1980). Although most cells secrete HER-1, mosaic analyses indicate that the germline is most strongly influenced by production from the intestine, which is the major site for protein production and secretion in the worm, and possibly by the somatic gonad as well (Hunter and Wood, 1992).

1.4.3 HER-1 inactivates the TRA-2 receptor

The only target of HER-1 is TRA-2. It produces a large transcript that encodes the transmembrane protein TRA-2A, and two small transcripts that encode the intracellular fragment TRA-2B (Okkema and Kimble, 1991). HER-1 binds the TRA-2A receptor (Okkema and Kimble, 1991; Kuwabara, Okkema and Kimble, 1992; Kuwabara and Kimble, 1995) at an interaction site defined by a dominant mutation in *tra-2* that transforms *XO* animals into hermaphrodites (Hodgkin and Albertson, 1995; Kuwabara, 1996). The complementary site on HER-1 was identified by mutations that block binding in HEK 293 cells (Hamaoka *et al.*, 2004). Although genetic analyses imply that HER-1 inactivates TRA-2A, how it works is unknown. However, *tra-3* behaves like a positive regulator of *tra-2* (Hodgkin, 1980). Since TRA-3 is a calpain protease (Barnes and Hodgkin, 1996) that cleaves TRA-2A *in vitro* (Sokol and Kuwabara, 2000), it might cleave TRA-2A *in vivo* to release an active, intracellular fragment. If so, perhaps the interaction between HER-1 and TRA-2A prevents cleavage.

1.4.4 TRA-2 prevents the FEM proteins from causing TRA-1 degradation

The pathway branches at TRA-2. First, TRA-2 negatively regulates three *fem* genes, which are needed for spermatogenesis and male development (Doniach and Hodgkin, 1984; Kimble, Edgar and Hirsh, 1984; Hodgkin, 1986). FEM-1 has ankyrin repeats (Spence, Coulson and Hodgkin, 1990), FEM-2 is a type 2C protein phosphatase (Pilgrim *et al.*, 1995), and FEM-3 is novel (Ahringer *et al.*, 1992). These proteins cooperate to lower the activity of TRA-1, a transcription factor that controls all sexual fates in the nematode (Hodgkin and Brenner, 1977; Zarkower and Hodgkin, 1992). To do this, FEM-1 binds to CUL-2, a member of the E3 ubiquitin ligase complex that promotes male fates (Starostina *et al.*, 2007), and these four proteins act together to target TRA-1 for ubiquitinylation and degradation. The net effect is that TRA-1 protein levels are low in males and high in hermaphrodites (Figure 1.3; Schvarzstein and Spence, 2006). Since TRA-2 binds to FEM-3 (Mehra *et al.*, 1999), it might work by inhibiting this FEM/CUL-2 complex and protecting TRA-1.

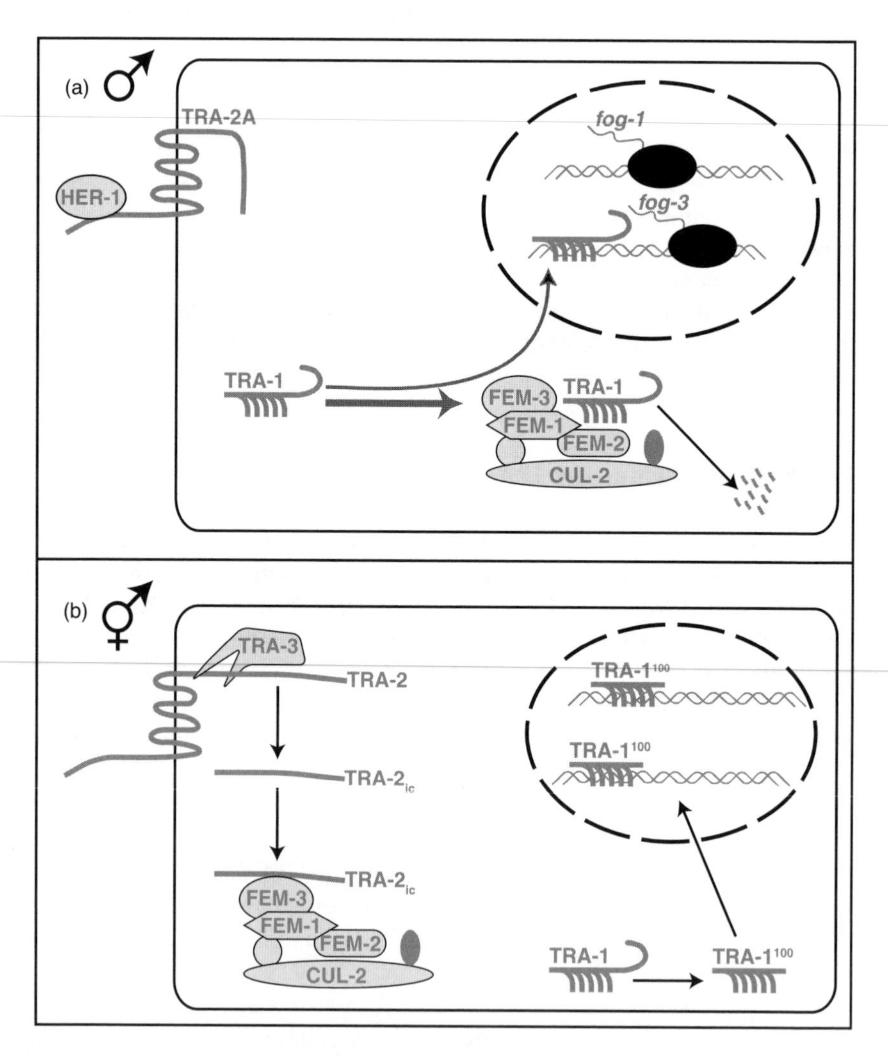

Figure 1.3 Model for the sperm/oocyte decision in adults. (a) In males, HER-1 binds to and represses the TRA-2A receptor; in this diagram, we do not depict cleavage of TRA-2A, but it has not yet been proven that HER-1 prevents this cleavage. The FEM/CUL-2 complex degrades full length TRA-1, which is needed to maintain spermatogenesis in older animals; thus, some TRA-1A is shown being degraded, and some entering the nucleus and regulating targets. The *fog-1* and *fog-3* genes are transcribed and promote spermatogenesis. In the figure, the black ellipses represent RNA polymerase, and the dark blue ellipsis represents ubiquitin. (b) In adult hermaphrodites, TRA-2 and TRA-3 are active, and prevent the FEM/CUL-2 complex from degrading TRA-1A. One possibility is that cleavage of TRA-2A by TRA-3 releases an intracellular fragment that inhibits the FEM complex by binding FEM-3. TRA-1 is cleaved to produce an aminoterminal fragment that represses transcription. A full colour version of this figure appears in the colour plate section.

1.4.5 TRA-2 also regulates TRA-1 directly

TRA-2 also regulates sexual fates through a second branch in the pathway, which involves direct contact with TRA-1 (Lum *et al.*, 2000; Wang and Kimble, 2001). The sites required for this interaction were identified by deletion studies in the yeast two-hybrid system, and are located on the intracellular portion of TRA-2A, a region also found in the smaller protein TRA-2B. Furthermore, several unusual *tra-2* mutations, often called mixomorphic alleles, disrupt TRA-2/TRA-1 binding. These alleles slightly decrease *tra-2* activity in somatic tissues, causing some cells to adopt male fates (Doniach, 1986; Schedl and Kimble, 1988). However, in the germline they are dominant and cause hermaphrodites to produce only oocytes, just like females. Thus, the interaction between TRA-2 and TRA-1 is necessary for hermaphrodites to make sperm, though it is not clear if this interaction regulates sexual fates in other tissues. An intracellular fragment of TRA-2 can be imported into the nucleus (Lum *et al.*, 2000), so it might interact with TRA-1 there *in vivo*. This fragment could be produced by cleavage of TRA-2A, or by translation of the smaller *tra-2* transcripts.

1.4.6 TRA-2, FEM-1 and FEM-3 stability is also regulated

Mutations in RPN-10, a component of the 26S proteasome, prevent hermaphrodite spermatogenesis and cause males to make yolk (Shimada *et al.*, 2006). In the intestine, these mutations increase the amount of TRA-2 protein in nuclei, so wild-type RPN-10 probably helps degrade TRA-2. Perhaps *rpn-10* mutations affect only the sperm/oocyte decision and yolk production, because these processes are more sensitive to changes in TRA-2 activity than other aspects of sex determination.

A similar but opposite effect involves *sel-10,* an F-box protein that regulates the levels of FEM-1 and FEM-3 (Jager *et al.*, 2004). Co-immunoprecipitation experiments show that SEL-10 binds both FEM-1 and FEM-3 and targets them for ubiquitinylation and degradation (Jager *et al.*, 2004), and yeast two-hybrid experiments indicate that SEL-10 also binds SKR-1, a component of the E3 ubiquitin ligase complex (Killian *et al.*, 2008). Mutations in *sel-10* alter some somatic fates and can suppress *tra-2(mixomorphic)* alleles in the germline.

1.5 Transcriptional control of germ cell fates

The two branches of the sex-determination pathway converge on TRA-1, a member of the Ci and Gli family of transcription factors (Zarkower and Hodgkin, 1992). Although *tra-1* produces two transcripts, only *tra-1A* has a known function, so its product is called TRA-1 below.

1.5.1 TRA-1 represses male genes in the germline and soma

Mutations that inactivate *tra-1* cause *XX* animals to develop male bodies (Hodgkin, 1987). Several somatic targets of TRA-1 have been identified, including: *egl-1,* a gene that regulates apoptosis (Conradt and Horvitz, 1998; Conradt and Horvitz, 1999); *mab-3,* a homologue of *Drosophila doublesex* that specifies many male cell fates (Shen and Hodgkin, 1988; Raymond *et al.*, 1998; Yi, Ross and Zarkower, 2000); *ceh-30,* a gene that prevents specific cell deaths in males (Peden *et al.*, 2007; Schwartz and Horvitz, 2007); and *dmd-3,* another *doublesex* homologue (Mason, Rabinowitz and Portman, 2008). So far, all of these somatic targets are male genes that are repressed by TRA-1 in *XX* animals.

Somatic targets of TRA-1 usually have a single binding site, either in the promoter, an intron, or an enhancer. By contrast, the major targets of TRA-1 in germ cells have multiple binding sites in their promoters, near the start of transcription (Chen and Ellis, 2000; Jin, Kimble and Ellis, 2001b). Both of these targets, *fog-1* and *fog-3,* are essential for spermatogenesis (Barton and Kimble, 1990; Ellis and Kimble, 1995). Mutations in either gene are epistatic to mutations in *tra-1,* and cause males to make oocytes. Furthermore, inactivation of *tra-1* increases *fog-3* expression (Chen and Ellis, 2000). Thus, TRA-1 controls germ cell fates by repressing transcription.

1.5.2 TRA-1 might also activate targets in the germline

If TRA-1 only worked by repressing *fog-1* and *fog-3*, then null alleles of *tra-1* should cause spermatogenesis. Instead, these mutations cause both *XX* and *XO* animals to produce sperm early in life, and then switch to oogenesis (Hodgkin, 1987; Schedl *et al.*, 1989). This result leads to two major conclusions. First, *tra-1* is not essential for either germ cell fate, since null mutants make both sperm and oocytes. And second, *tra-1* normally represses spermatogenesis in young animals, but promotes spermatogenesis in older males. One set of transgenic experiments is consistent with these observations: mutations in some of the *tra-1* binding sites of *fog-3* inactivate the transgene, implying that those sites mediate activation by TRA-1 (Chen and Ellis, 2000).

1.5.3 TRA-1 cleavage might be critical for oogenesis and female development

If TRA-1 indeed acts both as a repressor and an activator in the germline, how does it work? The Ci and Gli proteins also act as repressors in some contexts, and activators in others (Alexandre, Jacinto and Ingham, 1996; Ruiz i Altaba, 1999). The N-termini of these proteins contain five zinc fingers that are essential for repression, and the C-termini contain sequences required for activation. The full-length protein activates transcription of some targets, but cleavage releases an N-terminal fragment that represses transcription (reviewed by Jiang, 2002).

In *C. elegans,* TRA-1 is cleaved to produce a shorter product, called TRA-1^{100} (Schvarzstein and Spence, 2006). This product is abundant in adult hermaphrodites, which are producing oocytes. Furthermore, some *tra-1* nonsense mutations are dominant and cause oogenesis if the system for nonsense-mediated decay has also been disrupted. Since these mutants encode only the N-terminal half of TRA-1, the TRA-1^{100}

fragment must specify oogenesis. Although animals that lack a germline do not accumulate full-length TRA-1, they do make TRA-1^{100} in the soma, where it promotes female cell fates. By contrast, animals that are producing only sperm accumulate significant amounts of full-length TRA-1 (Schvarzstein and Spence, 2006). Thus, one simple model is that TRA-1^{100} promotes female development and oogenesis, whereas full-length TRA-1 promotes spermatogenesis (Figure 1.3).

1.5.4 Do other transcription factors cooperate with TRA-1 in germ cells?

In the soma, *tra-4* works with *tra-1* to repress transcription of male genes (Grote and Conradt, 2006). TRA-4 is a homologue of the transcriptional repressor PLZF, and appears to act in a complex with NASP-1, a histone chaperone, and HDA-1, a histone deacetylase. Thus, these proteins are likely to repress male genes by altering chromatin structure. So far, there is no evidence that members of this complex regulate the sperm/oocyte decision. However, the transcript levels of many genes that act during spermatogenesis are high in males and low in adult hermaphrodites (reviewed by L'Hernault, 2006), and transgenic experiments confirm that several genes active during spermatogenesis are regulated transcriptionally (Merritt *et al.*, 2008). Thus, it is likely that transcriptional control of germ cell fates occurs downstream of *tra-1*. Perhaps either TRA-4 or a group of germline genes regulates chromatin structure as part of the sperm/oocyte switch.

1.6 Translational regulation of the sperm/oocyte decision

Both *fog-1* and *fog-3* act at the end of the sex-determination pathway to control germ cell fates. If either gene is inactive, all germ cells differentiate as oocytes, so *fog-1* and *fog-3* are needed to specify spermatogenesis (Barton and Kimble, 1990; Ellis and Kimble, 1995).

1.6.1 FOG-1 is a cytoplasmic polyadenylation element binding protein

The *fog-1* gene makes two transcripts, but only the larger one has a known function. It encodes a CPEB protein with two RNA recognition motifs and a zinc finger (Luitjens *et al.*, 2000; Jin, Kimble and Ellis, 2001b). All of these RNA-binding domains are essential for activity, and FOG-1 interacts with its own 3′UTR (Jin *et al.*, 2001a), so it probably regulates translation like other CPEB proteins (reviewed by Richter, 2007). Antibody staining revealed that FOG-1 is expressed in germ cells long before a sperm-specific marker, which is consistent with models in which FOG-1 controls the sperm/oocyte decision (Figure 1.4c; Lamont and Kimble, 2007). Although *fog-1* itself, *fog-3*, and other genes have potential FOG-1 binding sites in their 3′UTRs, the steps that occur between FOG-1 activation and the expression of genes involved in spermatogenesis are not known.

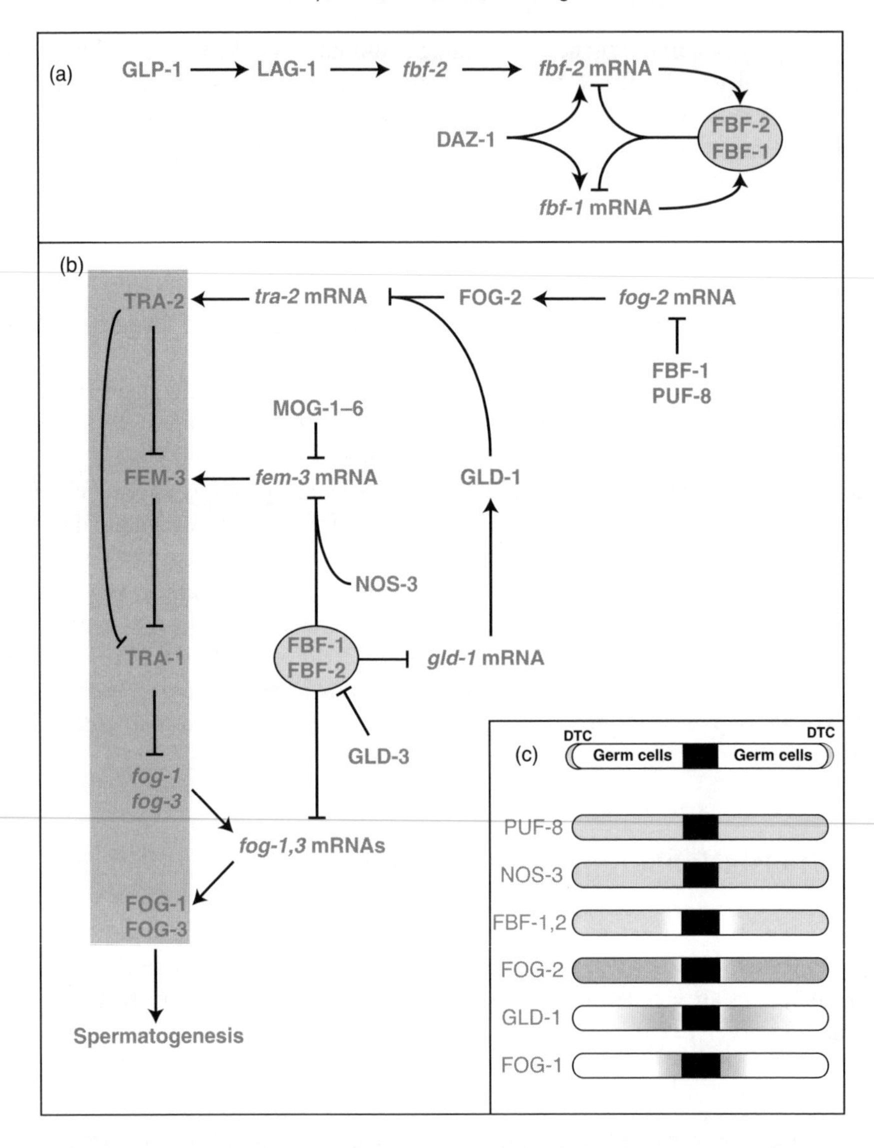

Figure 1.4 Translational regulation of germ cell fates. (a) The distal tip cell promotes FBF activity. In germ cells, the GLP-1 (Notch) receptor is activated by a signal from the distal tip cells (reviewed by Kimble and Crittenden, 2007). Working through the transcription factor LAG-1, it promotes transcription of *fbf-2*. The FBF proteins in turn promote mitotic proliferation or female germ cell fates. Through a feedback loop, they also inhibit their own translation; repression of *fbf-1* by FBF-2 and repression of *fbf-2* by FBF-1 have been demonstrated, and auto-repression is inferred. Proteins are shown in uppercase, and genes in lower case. Arrows indicate positive interactions, and '⊣' indicates negative interactions. (b) Modulation of the core sex-determination pathway by translational regulators (highlighted in grey; see text). The FBF proteins act at several points in the sex-determination pathway to prevent the translation of messenger RNAs that promote spermatogenesis. Similarly, GLD-1 acts with FOG-2 to prevent translation of *tra-2* messages, which normally promote oogenesis. GLD-1 also binds *tra-1* messages. All molecules that promote male fates are blue, and those that promote female fates are pink. (c) Expression of translational regulators in L3 hermaphrodites. A schematic of the L3 gonad is shown at top, with the distal tip cells (DTC, yellow) at either end, and

1.6.2 FOG-3 is a tob protein that might function with FOG-1

FOG-3 acts at the same step in the pathway as FOG-1, and both genes are essential for spermatogenesis. In fact, the only genetic distinction between them is that *fog-1* is very sensitive to changes in gene dose, whereas *fog-3* is not (Barton and Kimble, 1990; Ellis and Kimble, 1995). For example, *fog-1/+* males cannot sustain spermatogenesis, and eventually begin producing oocytes.

FOG-3 is the only nematode member of the large Tob and BTG family of proteins (Chen *et al.*, 2000). Other family members bind a diverse set of regulatory proteins, but in most cases their biochemical functions are not clear (reviewed by Jia and Meng, 2007). However, recent studies show that human Tob protein can promote the deadenylation of target messenger RNAs (Ezzeddine *et al.*, 2007). It does this by binding both the CCR4–CAF1 deadenylation complex and poly(A)-binding protein. If FOG-3 acts similarly, then both FOG proteins might control the translation of mRNAs by regulating their poly(A) tails. However, it remains possible that FOG-3 cooperates with unknown genes to do something else, like regulate transcription.

1.6.3 The three FEM proteins directly promote spermatogenesis

The primary function of the FEM proteins is to eliminate TRA-1. However, they have a second function in *C. elegans*, revealed by the fact that *tra-1; fem* double mutants make oocytes, even though they have male bodies (Hodgkin, 1986) and express high levels of *fog-3* (Chen and Ellis, 2000). How the FEM proteins promote spermatogenesis is not known. However, this activity seems to be a recent innovation, since it is not found in the related species *C. briggsae* (Hill *et al.*, 2006).

1.7 Other translational regulators specify hermaphrodite development

Male nematodes make sperm because HER-1 inactivates the TRA-2 receptor, allowing the FEM proteins to eliminate TRA-1 (Figures 1.2 and 1.3). Since hermaphrodites don't express HER-1, how do they produce sperm? Researchers have identified several translational regulators that modulate the activity of the sex-determination pathway to allow hermaphroditic development (Figure 1.4).

Figure 1.4 (*Continued*) other somatic cells (black) in the centre. Rough sketches of the protein levels of key translational regulators are shown below; since none of these studies compared different proteins in the same animals, the regions shown are only approximate. The PUF-8 expression pattern is based on a PUF-8::GFP transgene (Ariz, Mainpal and Subramaniam, 2009). NOS-3 is based on antibody staining (Kraemer *et al.*, 1999), as are FBF (Zhang *et al.*, 1997), FOG-2 (Clifford *et al.*, 2000), GLD-1 (Jones, Francis and Schedl, 1996) and FOG-1 (Lamont and Kimble, 2007). A full colour version of this figure appears in the colour plate section.

1.7.1 FOG-2 and GLD-1 repress translation of *tra-2* to allow spermatogenesis

Mutations in *fog-2* transform *XX* animals into true females, but do not affect males (Schedl and Kimble, 1988). Thus, *fog-2* alters the sperm/oocyte decision to allow hermaphroditic development. Mutations in *gld-1* affect many aspect of oogenesis, so *XX* animals are sterile rather than female (Francis *et al.*, 1995a). However, one of the phenotypes controlled by *gld-1* is hermaphrodite spermatogenesis; in null mutants all germ cells begin oogenesis instead of spermatogenesis, although they fail to complete it (Francis *et al.*, 1995a; Jones, Francis and Schedl, 1996). Genetic tests imply that both *fog-2* and *gld-1* act upstream of *tra-2* (Schedl and Kimble, 1988; Francis, Maine and Schedl, 1995b).

Cloning revealed that FOG-2 was created by a gene duplication event and co-opted into the sex-determination pathway to allow hermaphrodite development, and that it contains an F-box (Clifford *et al.*, 2000). Although many F-box proteins work as part of the E3 ubiquitin ligase complex to mark targets for degradation (reviewed by Kipreos and Pagano, 2000; Kipreos, 2005), FOG-2 associates with GLD-1 but does not destabilize it (Clifford *et al.*, 2000). This interaction with GLD-1 is mediated by the carboxyl terminus of FOG-2, which has been under positive selection during recent evolution (Nayak, Goree and Schedl, 2005).

GLD-1 is a translational regulator that contains a KH domain (Jones and Schedl, 1995) and appears to act as a dimer (Ryder *et al.*, 2004). It binds the 3′UTR of *tra-2* messenger RNAs, and can form a ternary complex that includes FOG-2 (Clifford *et al.*, 2000) and blocks translation (Jan *et al.*, 1999). The target site is defined by dominant mutations in two Direct Repeat Elements of the *tra-2* 3′UTR, which cause hermaphrodites to make oocytes rather than sperm (Doniach, 1986; Goodwin *et al.*, 1993); deletion of these repeats prevents GLD-1 binding (Jan *et al.*, 1999). Thus, FOG-2 and GLD-1 lower TRA-2 levels in young hermaphrodites to allow spermatogenesis. GLD-1 also regulates many other messages in the developing germline (Lee and Schedl, 2001; Marin and Evans, 2003; Mootz, Ho and Hunter, 2004; Schumacher *et al.*, 2005), including *tra-1* (Lakiza *et al.*, 2005), but none of these interactions appears to require FOG-2.

1.7.2 The FBF proteins repress translation of *fem-3* to allow oogenesis

Although FOG-2 and GLD-1 allow spermatogenesis to begin, hermaphrodites need to ensure that some germ cells eventually differentiate as oocytes. Mutations in several genes show that the level of FEM-3 is restricted so that this change can happen at the appropriate time.

As with *tra-2,* dominant mutations have been identified in the 3′UTR of *fem-3*, but they have the opposite effect, causing all germ cells to differentiate as sperm (Barton, Schedl and Kimble, 1987; Ahringer and Kimble, 1991; Ahringer *et al.*, 1992). These mutations disrupt a point mutation element (PME) that binds to and is regulated by FBF-1 and FBF-2 (Zhang *et al.*, 1997), two nematode members of the PUF family of translational regulatory proteins (reviewed by Wickens *et al.*, 2002). Since inactivation of both proteins causes constitutive spermatogenesis, just like the dominant mutations in the *fem-3* 3′UTR, FBF-1 and FBF-2 normally repress translation of *fem-3* messenger RNAs. Mutations in either *fbf-1* or *fbf-2* alone have more subtle but complex effects, which suggest that they also inhibit each other (Lamont *et al.*, 2004). Finally, the FBF proteins can also bind *fog-1* messages and repress their translation, and seem likely to act on *fog-3* transcripts as well, since they contain putative binding sites (Thompson *et al.*, 2005).

FBF-1 and FBF-2 are assisted by NOS-3, a homologue of the translational regulatory protein Nanos from *Drosophila* (Kraemer *et al.*, 1999). Furthermore, RNA interference shows that NOS-1 and NOS-2 act redundantly with NOS-3 to prevent spermatogenesis. Since only NOS-3 binds the FBF proteins in the yeast two-hybrid system, perhaps the other NOS proteins only form a complex with FBF-1 or FBF-2 when *fem-3* messages are present. The co-regulation of *fem-3* by the FBF proteins and NOS-3 parallels the regulation of hunchback by Pumilio and Nanos in *Drosophila,* suggesting that these translational regulatory networks are ancient.

1.7.3 Other translational regulators reinforce these decisions

The activities of the *fbf* genes are themselves tightly regulated (Figure 1.4a). First, the translational regulator DAZ-1 can bind *fbf* messenger RNAs and promote translation, thus favouring oogenesis (Karashima, Sugimoto and Yamamoto, 2000; Otori, Karashima and Yamamoto, 2006). Second, GLD-3, a homologue of bicaudal-C, can bind the FBF proteins and inhibit their interaction with the *fem-3* 3′UTR (Eckmann *et al.*, 2002). This inhibitory interaction is mutual, since the FBF proteins repress the expression of GLD-3 (Eckmann *et al.*, 2004). Third, the distal tip cell acts through the Notch pathway to promote the expression of *fbf-2* (Lamont *et al.*, 2004). Thus, FBF activity is controlled in part by translational regulation.

The activities of *fog-2* and *gld-1* are also under translational control. FBF-1 and PUF-8, a related protein, act redundantly to regulate FOG-2 proteins levels (Bachorik and Kimble, 2005). And the two FBF proteins regulate the translation of *gld-1* (Crittenden *et al.*, 2002).

1.7.4 Essential RNA-binding proteins also influence the sperm/oocyte switch

Several essential genes also regulate the expression of *fem-3*. Most of these genes were identified in general screens for mutations that caused hermaphrodites to produce sperm throughout their lives, and are named *mog-1* through *mog-6* (Graham and Kimble, 1993; Graham, Schedl and Kimble, 1993). Although mutations in these genes cause constitutive spermatogenesis, the mutants do not make as many sperm as *fem-3(gf)* mutants. Since the *mog* mutations are suppressed by mutations in the *fem* genes, but not by mutations in *fog-2*, they could act upstream of *fem-3*. Furthermore, mutations in the *mog* genes activate reporter constructs that have been fused to the *fem-3* 3′UTR, which implies that the MOG proteins regulate translation of *fem-3* (Gallegos *et al.*, 1998). For technical reasons, these experiments used transgenes that were only active in the soma, so it is unclear if the reporters were co-regulated by the FBF proteins, which are largely restricted to the germline (Zhang *et al.*, 1997). However, mutations in the *fem-3* PME did increase translation of the reporter constructs, so perhaps somatic members of the PUF family (Walser *et al.*, 2006) can work in concert with the MOG proteins to control their translation.

Molecular cloning revealed that MOG-1, MOG-4 and MOG-5 are DEAH helicases, a family that includes proteins that bind RNA (Puoti and Kimble, 1999; Puoti and Kimble, 2000). Since *mog; fem-3* double mutants make oocytes that give rise to dead eggs, these genes are also essential for embryonic development (Graham and Kimble, 1993; Graham, Schedl and Kimble, 1993). These three helicases interact with MEP-1, a zinc finger protein that regulates the expression of *fem-3* in germ cells (Belfiore *et al.*, 2002), and works with PIE-1 to block the expression of germline messages in the soma (Unhavaithaya *et al.*, 2002). Furthermore, MOG-6 is an unusual cyclophilin that also interacts with MEP-1 (Belfiore *et al.*, 2004). Thus, this large group of proteins appears essential for initiating oogenesis in hermaphrodites, and regulating gene expression in the early embryo.

Another essential gene influences the switch from spermatogenesis to oogenesis – *mag-1* (Li, Boswell and Wood, 2000). The *mag-1(RNAi)* animals resemble *mog* mutants in two ways: they make sperm constitutively, and *mag-1(RNAi); fem-3* double mutants make oocytes that give rise to dead eggs. But unlike *mog-1,* the *mag-1(RNAi); fog-2* double mutants make only oocytes, just like *fog-2* animals; so *mag-1* might act upstream of *fog-2.* Although epistasis experiments that involve RNA interference are not conclusive, this difference raises the possibility that MAG-1 is a positive regulator of *tra-2,* rather than a negative regulator of *fem-3.*

MAG-1 is likely to work with RNP-4, the homologue of yeast Y14, since *rnp-4 (RNAi)* animals have similar phenotypes and the two proteins co-immunoprecipitate (Kawano *et al.*, 2004). Both MAG-1 and RNP-4 are components of the exon-junction complex, which is formed during splicing. Since the mammalian homologues Magoh and Y14 remain associated with mRNAs following splicing and promote translation (Nott, Le Hir and Moore, 2004), perhaps MAG-1 and RNP-4 promote translation of a message needed for oogenesis, like *tra-2.*

ATX-2 also regulates sex determination in the germline (Ciosk, DePalma and Priess, 2004; Maine *et al.*, 2004), since RNA interference causes many hermaphrodites to produce sperm constitutively. Surprisingly, this phenotype is not completely suppressed by *fog-2(q71*null*)* mutations, but is suppressed by *tra-2(q122*gf*)* mutations. Thus, these mutations have distinct effects, even though both disrupt translational regulation of *tra-2.* How ATX-2 promotes oogenesis is not known.

Finally, the essential gene *laf-1* has the opposite effect; *laf-1/+* animals make oocytes instead of sperm, just like the *fog* mutants. Analysis of double mutants indicates that *laf-1* might regulate *tra-2* translation (Goodwin *et al.*, 1997; Jan *et al.*, 1997).

1.7.5 The relative activities of TRA-2 and FEM-3 determine germ cell fates

The existence of elaborate regulatory networks focused on *tra-2* and *fem-3* highlights the importance of these genes in the developing germline. In fact, several observations support the idea that the relative levels of TRA-2 and FEM-3 are the critical factor in the sperm/oocyte decision. First, the *tra-2(gf)* mutations increase the production of wild-type TRA-2 protein, causing oogenesis. Second, the *fem-3(gf)* mutations, which should increase the production of wild-type FEM-3, cause constitutive spermatogenesis. And third, these mutations compensate for each other, since *tra-2(gf); fem-3(gf)* double mutants are self-fertile hermaphrodites (Barton, Schedl and Kimble, 1987).

1.8 The sperm/oocyte decision is intimately linked to the initiation of meiosis

Many of the genes that modulate the sex-determination pathway in hermaphrodites also regulate the decision of germ cells to remain in mitosis or enter meiosis (reviewed by Kimble and Crittenden, 2007). For example, GLD-1 and NOS-3 work together to stop mitosis and promote meiosis, as do GLD-2 and GLD-3 and some of the MOG proteins (Belfiore *et al.*, 2004; Eckmann *et al.*, 2004; Hansen *et al.*, 2004). By contrast, FBF-1, FBF-2, FOG-1 and FOG-3 play redundant roles promoting germ cells to remain in mitosis (Crittenden *et al.*, 2002; Thompson *et al.*, 2005). In addition, PUF-8 acts redundantly with the translational regulator MEX-3 to keep germ cells in mitosis (Ariz, Mainpal and Subramaniam, 2009), and ATX-2 also promotes mitotic proliferation (Ciosk, DePalma and Priess, 2004; Maine *et al.*, 2004).

Some of the genes that regulate sex determination also act at later stages during meiosis. For example, GLD-1 is needed to maintain germ cells in oogenesis, since oocytes return to mitosis and form tumours in *gld-1(null)* mutants (Francis *et al.*, 1995a). PUF-8 plays an analogous role in the male germline, preventing spermatocytes from returning to mitosis and forming tumours (Subramaniam and Seydoux, 2003). And DAZ-1 is required for germ cells to progress beyond the pachytene phase of oogenesis (Karashima, Sugimoto and Yamamoto, 2000).

1.8.1 Translational regulators define zones within the germline syncytium

Why do so many translational regulators control both the sperm/oocyte decision, and the entry into meiosis? The germline is a long tube in which cell fates are arranged from a stem cell niche at the distal end to fully differentiated germ cells at the proximal end (Figure 1.1). Since much of the tube is a syncytium, perhaps translational regulators promote the localized production of target proteins, thus dividing the syncytium into zones, each with germ cells at a different stage of development.

A few observations support this model. The interactions between different translational regulators appear to set up zones of protein expression, with FBF activity high near the distal tip, and FOG-1 and GLD-1 high proximally (Figure 1.4). Furthermore, experiments with transgenes show that most genes in the germline are controlled in large part by their 3′UTRs, rather than their promoters (Merritt *et al.*, 2008), confirming the importance of translational regulation. By contrast, most proteins that mediate signal transduction do not play dual roles. For example, none of the proteins that form the HER-1 to TRA-1 signal transduction pathway regulate mitosis or meiosis. Instead, the only members of this pathway that influence the cell cycle are the translational regulator FOG-1 and its partner FOG-3. Similarly, the GLP-1 signal transduction pathway controls mitosis without influencing the sperm/oocyte decision, whereas many of the translational regulators it influences do both (reviewed by Kimble and Crittenden, 2007).

1.8.2 The sperm/oocyte decision is likely to occur near the entry into meiosis

Since primordial germ cells and germ cells in the early stages of meiosis look the same in both sexes, it has been hard to identify the point at which each cell decides between spermatogenesis and oogenesis. Several early markers of spermatogenesis are first detected in pachytene germ cells during prophase I (Jones, Francis and Schedl, 1996). By contrast, some early markers of oogenesis are found more distally along the tube, in cells that are making the transition to meiosis, and even in some mitotic cells. Since distal cells express female markers but more proximal ones do not (M.H. Lee and T. Schedl, shown by Ellis and Schedl, 2006), these transcripts might accurately reflect the sexual fates of individual nuclei in the germline syncytium. Thus, the sperm/oocyte decision probably occurs between late mitosis and the pachytene phase of meiosis I.

Although many translational regulators control both the sperm/oocyte decision and the entry into meiosis, there is no simple correlation between these two fates (Table 1.1). Some genes promote spermatogenesis and mitosis. Some promote spermatogenesis and meiosis. Some promote oogenesis and mitosis, and others promote oogenesis and meiosis. One simple model to explain this complex pattern is that the two decisions are made at almost the same point in the germline tube. Thus, each of these translational regulators might originally have been expressed in this region because of its role in either sex determination or the entry into meiosis, but was eventually recruited into the other pathway to tighten the control of target messages.

This hypothesis is supported by temperature-shift experiments conducted using *fog-1,* which specifies sexual fate, and *glp-1,* which promotes mitosis (Barton and Kimble, 1990). These studies indicate that *fog-1* is needed continually to promote spermatogenesis, and that temperature shifts that affect both *fog-1* and *glp-1* alter both decisions, as if the genes were acting on cell fates at roughly the same time.

Table 1.1 Pleiotropic genes regulate the sperm/oocyte decision

Gene	Sperm/oocyte decision	Mitosis/meiosis decision	Biochemical function
atx-2	Promotes oogenesis	Promotes mitosis	Ataxin family
fbf-1	Promotes oogenesis	Promotes mitosis	PUF translational regulator
fbf-2	Promotes oogenesis	Promotes mitosis	PUF translational regulator
puf-8	Promotes oogenesis	Promotes mitosis	PUF translational regulator
nos-3	Promotes oogenesis	Promotes meiosis	Nanos translational regulator
daz-1	Promotes oogenesis	Promotes meiosis	Translational regulator
gld-1	Promotes *XX* spermatogenesis	Promotes meiosis	KH translational regulator
gld-3	Promotes spermatogenesis	Promotes meiosis	bicaudal-C homologue
fog-1	Promotes spermatogenesis	Promotes mitosis	CPEB translational regulator
fog-3	Promotes spermatogenesis	Promotes mitosis	Tob protein

1.8.3 Some translational regulators are also essential for embryogenesis

Translational regulators might also be crucial for repressing transcripts that are essential for the future embryo, but which would be harmful in developing oocytes. For example, all three of the *fem* genes show maternal effects, which suggests that their transcripts are packaged into oocytes to help control the sex of the future embryo (Hodgkin, 1986). The phenotype of *fem-3(gf)* mutants implies that high levels of FEM-3 protein cause spermatogenesis, so translational inhibitors like the FBF proteins might play a critical role in preventing *fem-3* transcripts from blocking oogenesis. As discussed above, other translational regulators are essential, apparently because they control targets involved in sex determination, as well as targets needed for embryogenesis.

1.9 The future

Although we now understand a great deal about how the sperm/oocyte decision is made in *C. elegans*, this information has opened up a new set of questions for the future; questions that should dominate the next several years of research.

1.9.1 What are the primary targets controlled by the sperm/oocyte decision?

Although we know that *fog-1* and *fog-3* act at the end of the sex-determination pathway to promote spermatogenesis, we do not know what their targets are. Possible candidates include *fog-1* and *fog-3* themselves, and genes like *cpb-1,* that act early in spermatogenesis. Furthermore, we do not know what genes are activated early in oogenesis by the absence of *fog-1* and *fog-3* activity. Identifying these targets and working out how the action of *fog-1* and *fog-3* controls their activities is critical for understanding the sperm/oocyte decision. Over the next few years, our focus should move from studying the sex-determination process *per se* to elucidating the mechanics of cell fate determination in the germline.

1.9.2 How has the sperm/oocyte decision changed during evolution?

This question entails two very different lines of research. The first concerns whether there has been broad conservation of genes involved in the sperm/oocyte decision. Although *fog-1* and *fog-3* have homologues in all animals, some of which are expressed in germ cells, it is not known if any of these homologues regulates germ cell fates. Furthermore, the possible conservation of genes downstream of *fog-1* and *fog-3* remains a complete mystery.

The second line of enquiry concerns how the sperm/oocyte decision changes during evolution. Comparative analysis of nematode species is beginning to provide some answers to this question. Genes of the core pathway are conserved in structure and function amongst relatives of *C. elegans*, and most show only subtle differences between species. For example *fem-2* and *fem-3* mutations always cause oogenesis in *C. elegans*, but only cause oogenesis under some conditions in *C. briggsae* (Hill *et al.*, 2006). By contrast, genes that modulate the core pathway seem to be evolving rapidly. For example, *fog-2* and *gld-1* are needed for hermaphrodite spermatogenesis in *C. elegans,* but not in *C. briggsae,* which lacks a *fog-2* gene (Nayak, Goree and Schedl, 2005). However, *C. briggsae* has recruited a different member of the F-box family of proteins, SHE-1, to specify hermaphrodite development (Guo, Lang and Ellis, 2009). Finally, both *fog-2* and *she-1* control *tra-2* activity, and knocking down *tra-2* function in the male/female species *C. remanei* can help create self-fertile animals (Baldi, Cho and Ellis, 2009).

1.9.3 How does the somatic gonad influence the sperm/oocyte decision?

So far, we know that the distal tip cells signal to nearby germ cells to remain in mitosis, and that a variety of somatic cells in the male act through HER-1 to cause germ cells to adopt male fates and begin spermatogenesis. However, the selective killing of cells in the hermaphrodite gonad showed that there might be additional signals (above). Furthermore, a surprising genetic experiment supports this hypothesis: mutations in the *fshr-1* gene, which acts in the somatic gonad, promote spermatogenesis over oogenesis (Cho, Rogers and Fay, 2007). How *fshr-1* works, and what additional interactions occur between the soma and germline remain mysterious. Since the somatic gonad is critical for the development of germ cells in most animals, these studies could open up entirely new avenues for research.

Acknowledgements

I would like to thank members of my laboratory and Judith Kimble for valuable comments. This work was supported by NIH (National Institutes of Health) grant GM085282-01.

References

Ahringer, J. and Kimble, J. (1991) Control of the sperm-oocyte switch in *Caenorhabditis elegans* hermaphrodites by the *fem-3* 3′ untranslated region. *Nature*, **349**(6307), 346–348.

Ahringer, J., Rosenquist, T.A., Lawson, D.N. and Kimble, J. (1992) The Caenorhabditis elegans sex determining gene fem-3 is regulated post-transcriptionally. *EMBO J.*, **11**(6), 2303–2310.

Alexandre, C., Jacinto, A. and Ingham, P.W. (1996) Transcriptional activation of hedgehog target genes in *Drosophila* is mediated directly by the cubitus interruptus protein, a member of the GLI family of zinc finger DNA-binding proteins. *Genes. Dev.*, **10**(16), 2003–2013.

Ariz, M., Mainpal, R. and Subramaniam, K. (2009) *C. elegans* RNA-binding proteins PUF-8 and MEX-3 function redundantly to promote germline stem cell mitosis. *Dev. Biol.*, **326**(2), 295–304.

Bachorik, J.L. and Kimble, J. (2005) Redundant control of the *Caenorhabditis elegans* sperm/oocyte switch by PUF-8 and FBF-1, two distinct PUF RNA-binding proteins. *Proc. Natl. Acad. Sci. USA*, **102**(31), 10893–10897.

Baldi, C., Cho, S. and Ellis, R.E. (2009) Mutations in two independent pathways are sufficient to create hermaphroditic nematodes. *Science*, **326**,1002–5.

Barnes, T.M. and Hodgkin, J. (1996) The tra-3 sex determination gene of Caenorhabditis elegans encodes a member of the calpain regulatory protease family. *EMBO J.*, **15**(17), 4477–4484.

Barton, M.K. and Kimble, J. (1990) *fog-1,* a regulatory gene required for specification of spermatogenesis in the germ line of *Caenorhabditis elegans. Genetics*, **125**(1), 29–39.

Barton, M.K., Schedl, T.B. and Kimble, J. (1987) Gain-of-function mutations of *fem-3,* a sex-determination gene in *Caenorhabditis elegans. Genetics*, **115**(1), 107–119.

Belfiore, M., Mathies, L.D., Pugnale, P. *et al.* (2002) The MEP 1 zinc-finger protein acts with MOG DEAH box proteins to control gene expression via the fem-3 3′ untranslated region in Caenorhabditis elegans. *RNA*, **8**(6), 725–739.

Belfiore, M., Pugnale, P., Saudan, Z. and Puoti, A. (2004) Roles of the *C. elegans* cyclophilin-like protein MOG-6 in MEP-1 binding and germline fates. *Development*, **131**(12), 2935–2945.

Bounoutas, A. and Chalfie, M. (2007) Touch sensitivity in Caenorhabditis elegans. *Pflugers Arch.*, **454**(5), 691–702.

Chen, P.J. and Ellis, R.E. (2000) TRA-1A regulates transcription of *fog-3,* which controls germ cell fate in *C. elegans. Development*, **127**(14), 3119–3129.

Chen, P.J., Singal, A., Kimble, J. and Ellis, R.E. (2000) A novel member of the tob family of proteins controls sexual fate in *Caenorhabditis elegans* germ cells. *Dev. Biol.*, **217**(1), 77–90.

Cho, S., Jin, S.W., Cohen, A. and Ellis, R.E. (2004) A phylogeny of *Caenorhabditis* reveals frequent loss of introns during nematode evolution. *Genome Res.*, **14**(7), 1207–1220.

Cho, S., Rogers, K.W. and Fay, D.S. (2007) The *C. elegans* glycopeptide hormone receptor ortholog, FSHR-1, regulates germline differentiation and survival. *Curr. Biol.*, **17**(3), 203–212.

Ciosk, R., DePalma, M. and Priess, J.R. (2004) ATX-2, the *C. elegans* ortholog of ataxin 2, functions in translational regulation in the germline. *Development*, **131**(19), 4831–4841.

Clifford, R., Lee, M.H., Nayak, S. *et al.* (2000) FOG-2, a novel F-box containing protein, associates with the GLD-1 RNA binding protein and directs male sex determination in the *C. elegans* hermaphrodite germline. *Development*, **127**(24), 5265–5276.

Conradt, B. and Horvitz, H.R. (1998) The *C. elegans* protein EGL-1 is required for programmed cell death and interacts with the Bcl-2-like protein CED-9. *Cell*, **93**(4), 519–529.

Conradt, B. and Horvitz, H.R. (1999) The TRA-1A sex determination protein of *C. elegans* regulates sexually dimorphic cell deaths by repressing the *egl-1* cell death activator gene. *Cell*, **98**(3), 317–327.

Crittenden, S.L., Bernstein, D.S., Bachorik, J.L. *et al.* (2002) A conserved RNA-binding protein controls germline stem cells in *Caenorhabditis elegans. Nature*, **417**(6889), 660–663.

Doniach, T. (1986) Activity of the sex-determining gene *tra-2* is modulated to allow spermatogenesis in the *C. elegans* hermaphrodite. *Genetics*, **114**(1), 53–76.

Doniach, T. and Hodgkin, J. (1984) A sex-determining gene, *fem-1,* required for both male and hermaphrodite development in *Caenorhabditis elegans. Dev. Biol.*, **106**(1), 223–235.

Eckmann, C.R., Crittenden, S.L., Suh, N. and Kimble, J. (2004) GLD-3 and control of the mitosis/meiosis decision in the germline of *Caenorhabditis elegans. Genetics*, **168**(1), 147–160.

Eckmann, C.R., Kraemer, B., Wickens, M. and Kimble, J. (2002) GLD-3, a bicaudal-C homolog that inhibits FBF to control germline sex determination in *C. elegans. Dev. Cell*, **3**(5), 697–710.

Ellis, R.E. and Kimble, J. (1995) The *fog-3* gene and regulation of cell fate in the germ line of *Caenorhabditis elegans. Genetics*, **139**(2), 561–577.

Ellis, R.E. and Schedl, T. (April 4, 2006) Sex determination in the germ line, in *WormBook* (ed. The *C. elegans* Research Community). www.wormbook.org. doi: 10.1895/wormbook.1.82.2

Ezzeddine, N., Chang, T.C., Zhu, W. *et al.* (2007) Human TOB, an antiproliferative transcription factor, is a poly(A)-binding protein-dependent positive regulator of cytoplasmic mRNA deadenylation. *Mol. Cell Biol.*, **27**(22), 7791–7801.

Francis, R., Barton, M.K., Kimble, J. and Schedl, T. (1995a) *gld-1,* a tumor suppressor gene required for oocyte development in *Caenorhabditis elegans. Genetics*, **139**(2), 579–606.

Francis, R., Maine, E. and Schedl, T. (1995b) Analysis of the multiple roles of *gld-1* in germline development: interactions with the sex determination cascade and the *glp-1* signaling pathway. *Genetics*, **139**(2), 607–630.

Gallegos, M., Ahringer, J., Crittenden, S. and Kimble, J. (1998) Repression by the 3′ UTR of *fem-3,* a sex-determining gene, relies on a ubiquitous mog-dependent control in *Caenorhabditis elegans. EMBO J.*, **17**(21), 6337–6347.

Goodman, M.B.(January 6, 2006) Mechanosensation, in *WormBook* (ed. The *C. elegans* Research Community). www.wormbook.org. doi: 10.1895/wormbook.1.62.1

Goodwin, E.B., Hofstra, K., Hurney, C.A. *et al.* (1997) A genetic pathway for regulation of *tra-2* translation. *Development*, **124**(3), 749–758.

Goodwin, E.B., Okkema, P.G., Evans, T.C. and Kimble, J. (1993) Translational regulation of *tra-2* by its 3′ untranslated region controls sexual identity in *C. elegans. Cell*, **75**(2), 329–339.

Graham, P.L. and Kimble, J. (1993) The *mog-1* gene is required for the switch from spermatogenesis to oogenesis in *Caenorhabditis elegans. Genetics*, **133**(4), 919–931.

Graham, P.L., Schedl, T. and Kimble, J. (1993) More *mog* genes that influence the switch from spermatogenesis to oogenesis in the hermaphrodite germ line of *Caenorhabditis elegans. Dev. Genet.*, **14**(6), 471–484.

Grote, P. and Conradt, B. (2006) The PLZF-like protein TRA-4 cooperates with the Gli-like transcription factor TRA-1 to promote female development in *C. elegans. Dev. Cell*, **11**(4), 561–573.

Gumienny, T.L., Lambie, E., Hartwieg, E. *et al.* (1999) Genetic control of programmed cell death in the *Caenorhabditis elegans* hermaphrodite germline. *Development*, **126**(5), 1011–1022.

Guo, Y., Lang, S. and Ellis, R.E. (2009) Independent recruitment of F box genes to regulate hermaphrodite development during nematode evolution. *Curr. Biol.*, **19**, 1853–60.

Haag, E.S., Wang, S. and Kimble, J. (2002) Rapid coevolution of the nematode sex-determining genes *fem-3* and *tra-2. Curr. Biol.*, **12**(23), 2035–2041.

Hall, D.H., Winfrey, V.P., Blaeuer, G. *et al.* (1999) Ultrastructural features of the adult hermaphrodite gonad of *Caenorhabditis elegans:* relations between the germ line and soma. *Dev. Biol.*, **212**(1), 101–123.

Hamaoka, B.Y., Dann, C.E. 3rd, Geisbrecht, B.V. and Leahy, D.J. (2004) Crystal structure of Caenorhabditis elegans HER-1 and characterization of the interaction between HER-1 and TRA-2A. *Proc. Natl. Acad. Sci. USA*, **101**(32), 11673–11678.

Hansen, D., Wilson-Berry, L., Dang, T. and Schedl, T. (2004) Control of the proliferation versus meiotic development decision in the *C. elegans* germline through regulation of GLD-1 protein accumulation. *Development*, **131**(1), 93–104.

Hill, R.C., de Carvalho, C.E., Salogiannis, J. *et al.* (2006) Genetic flexibility in the convergent evolution of hermaphroditism in *Caenorhabditis* nematodes. *Dev. Cell*, **10**(4), 531–538.

Hodgkin, J. (1980) More sex-determination mutants of *Caenorhabditis elegans. Genetics*, **96**(3), 649–664.

Hodgkin, J. (1986) Sex determination in the nematode *C. elegans:* analysis of *tra-3* suppressors and characterization of *fem* genes. *Genetics*, **114**(1), 15–52.

Hodgkin, J. (1987) A genetic analysis of the sex-determining gene, *tra-1,* in the nematode *Caenorhabditis elegans. Genes. Dev.*, **1**(7), 731–745.

Hodgkin, J. and Albertson, D.G. (1995) Isolation of dominant *XO*-feminizing mutations in *Caenorhabditis elegans:* new regulatory *tra* alleles and an *X* chromosome duplication with implications for primary sex determination. *Genetics*, **141**(2), 527–542.

Hodgkin, J.A. and Brenner, S. (1977) Mutations causing transformation of sexual phenotype in the nematode Caenorhabditis elegans. *Genetics*, **86**(2 Pt. 1), 275–287.

Hunter, C.P. and Wood, W.B. (1992) Evidence from mosaic analysis of the masculinizing gene *her-1* for cell interactions in *C. elegans* sex determination. *Nature*, **355**(6360), 551–555.

Jager, S., Schwartz, H.T., Horvitz, H.R. and Conradt, B. (2004) The Caenorhabditis elegans F-box protein SEL-10 promotes female development and may target FEM-1 and FEM-3 for degradation by the proteasome. *Proc. Natl. Acad. Sci. USA*, **101**(34), 12549–12554.

Jan, E., Motzny, C.K., Graves, L.E. and Goodwin, E.B. (1999) The STAR protein, GLD-1, is a translational regulator of sexual identity in *Caenorhabditis elegans*. *EMBO J.*, **18**(1), 258–269.

Jan, E., Yoon, J.W., Walterhouse, D. *et al.* (1997) Conservation of the *C. elegans tra-2* 3′UTR translational control. *EMBO J.*, **16**(20), 6301–6313.

Jia, S. and Meng, A. (2007) Tob genes in development and homeostasis. *Dev. Dyn.*, **236**(4), 913–921.

Jiang, J. (2002) Degrading Ci: who is Cul-pable? *Genes. Dev.*, **16**(18), 2315–2321.

Jin, S.W., Arno, N., Cohen, A. *et al.* (2001a) In *Caenorhabditis elegans,* the RNA-binding domains of the cytoplasmic polyadenylation element binding protein FOG-1 are needed to regulate germ cell fates. *Genetics*, **159**(4), 1617–1630.

Jin, S.W., Kimble, J. and Ellis, R.E. (2001b) Regulation of cell fate in *Caenorhabditis elegans* by a novel cytoplasmic polyadenylation element binding protein. *Dev. Biol.*, **229**(2), 537–553.

Jones, A.R., Francis, R. and Schedl, T. (1996) GLD-1, a cytoplasmic protein essential for oocyte differentiation, shows stage- and sex-specific expression during *Caenorhabditis elegans* germline development. *Dev. Biol.*, **180**(1), 165–183.

Jones, A.R. and Schedl, T. (1995) Mutations in *gld-1,* a female germ cell-specific tumor suppressor gene in *Caenorhabditis elegans,* affect a conserved domain also found in Src-associated protein Sam68. *Genes. Dev.*, **9**(12), 1491–1504.

Karashima, T., Sugimoto, A. and Yamamoto, M. (2000) *Caenorhabditis elegans* homologue of the human azoospermia factor DAZ is required for oogenesis but not for spermatogenesis. *Development*, **127**(5), 1069–1079.

Kawano, T., Kataoka, N., Dreyfuss, G. and Sakamoto, H. (2004) Ce-Y14 and MAG-1, components of the exon-exon junction complex, are required for embryogenesis and germline sexual switching in *Caenorhabditis elegans. Mech. Dev.*, **121**(1), 27–35.

Killian, D.J., Harvey, E., Johnson, P. *et al.* (2008) SKR-1, a homolog of Skp1 and a member of the SCF (SEL-10) complex, regulates sex-determination and LIN-12/Notch signaling in *C. elegans. Dev. Biol.*, **322**(2), 322–331.

Kimble, J. and Crittenden, S.L. (2007) Controls of germline stem cells, entry into meiosis, and the sperm/oocyte decision in *Caenorhabditis elegans. Annu. Rev. Cell Dev. Biol.*, **23**,405–433.

Kimble, J., Edgar, L. and Hirsh, D. (1984) Specification of male development in *Caenorhabditis elegans:* the fem genes. *Dev. Biol.*, **105**(1), 234–239.

Kimble, J. and Hirsh, D. (1979) The postembryonic cell lineages of the hermaphrodite and male gonads in *Caenorhabditis elegans. Dev. Biol.*, **70**(2), 396–417.

Kimble, J.E. and White, J.G. (1981) On the control of germ cell development in *Caenorhabditis elegans. Dev. Biol.*, **81**(2), 208–219.

Kiontke, K. and Fitch, D.H.(August 11, 2005) The phylogenetic relationships of *Caenorhabditis* and other rhabditids, in *WormBook* (ed. The *C. elegans* Research Community). www.wormbook.org. doi: 10.1895/wormbook.1.11.1

Kiontke, K., Gavin, N.P., Raynes, Y. *et al.* (2004) Caenorhabditis phylogeny predicts convergence of hermaphroditism and extensive intron loss. *Proc. Natl. Acad. Sci. USA*, **101**(24), 9003–9008.

Kipreos, E.T.(December 1, 2005) Ubiquitin-mediated pathways in *C. elegans*, in *WormBook* (ed. The *C. elegans* Research Community). www.wormbook.org. doi: 10.1895/wormbook.1.36.1

Kipreos, E.T. and Pagano, M. (2000) The F-box protein family. *Genome Biol.*, **1**(5), reviews 3002.1–3002.7.

Kraemer, B., Crittenden, S., Gallegos, M. *et al.* (1999) NANOS-3 and FBF proteins physically interact to control the sperm-oocyte switch in *Caenorhabditis elegans. Curr. Biol.*, **9**(18), 1009–1018.

Kuwabara, P.E. (1996) A novel regulatory mutation in the *C. elegans* sex determination gene *tra-2* defines a candidate ligand/receptor interaction site. *Development*, **122**(7), 2089–2098.

Kuwabara, P.E. and Kimble, J. (1995) A predicted membrane protein, TRA-2A, directs hermaphrodite development in *Caenorhabditis elegans. Development*, **121**(9), 2995–3004.

Kuwabara, P.E., Okkema, P.G. and Kimble, J. (1992) *tra-2* encodes a membrane protein and may mediate cell communication in the *Caenorhabditis elegans* sex determination pathway. *Mol. Biol. Cell*, **3**(4), 461–473.

L'Hernault, S.W.(February 20, 2006) Spermatogenesis, in *WormBook* (ed. The *C. elegans* Research Community). www.wormbook.org. doi: 10.1895/wormbook.1.85.1

Lakiza, O., Frater, L., Yoo, Y. *et al.* (2005) STAR proteins quaking-6 and GLD-1 regulate translation of the homologues GLI1 and tra-1 through a conserved RNA 3′UTR-based mechanism. *Dev. Biol.*, **287**(1), 98–110.

Lamont, L.B., Crittenden, S.L., Bernstein, D. *et al.* (2004) FBF-1 and FBF-2 regulate the size of the mitotic region in the *C. elegans* germline. *Dev. Cell*, **7**(5), 697–707.

Lamont, L.B. and Kimble, J. (2007) Developmental expression of FOG-1/CPEB protein and its control in the *Caenorhabditis elegans* hermaphrodite germ line. *Dev. Dyn.*, **236**(3), 871–879.

Lee, M.H. and Schedl, T. (2001) Identification of in vivo mRNA targets of GLD-1, a maxi-KH motif containing protein required for *C. elegans* germ cell development. *Genes Dev.*, **15**(18), 2408–2420.

Li, W., Boswell, R. and Wood, W.B. (2000) *mag-1,* a homolog of *Drosophila mago nashi,* regulates hermaphrodite germ-line sex determination in *Caenorhabditis elegans. Dev. Biol.*, **218**(2), 172–182.

Luitjens, C., Gallegos, M., Kraemer, B. *et al.* (2000) CPEB proteins control two key steps in spermatogenesis in *C. elegans. Genes Dev.*, **14**(20), 2596–2609.

Lum, D.H., Kuwabara, P.E., Zarkower, D. and Spence, A.M. (2000) Direct protein-protein interaction between the intracellular domain of TRA-2 and the transcription factor TRA-1A modulates feminizing activity in *C. elegans. Genes Dev.*, **14**(24), 3153–3165.

Madl, J.E. and Herman, R.K. (1979) Polyploids and sex determination in *Caenorhabditis elegans. Genetics*, **93**(2), 393–402.

Maine, E.M., Hansen, D., Springer, D. and Vought, V.E. (2004) Caenorhabditis elegans atx-2 promotes germline proliferation and the oocyte fate. *Genetics*, **168**(2), 817–830.

Marin, V.A. and Evans, T.C. (2003) Translational repression of a *C. elegans* Notch mRNA by the STAR/KH domain protein GLD-1. *Development*, **130**(12), 2623–2632.

Mason, D.A., Rabinowitz, J.S. and Portman, D.S. (2008) *dmd-3,* a doublesex-related gene regulated by *tra-1,* governs sex-specific morphogenesis in *C. elegans. Development*, **135**(14), 2373–2382.

McCarter, J., Bartlett, B., Dang, T. and Schedl, T. (1997) Soma-germ cell interactions in *Caenorhabditis elegans:* multiple events of hermaphrodite germline development require the somatic sheath and spermathecal lineages. *Dev. Biol.*, **181**(2), 121–143.

Mehra, A., Gaudet, J., Heck, L. *et al.* (1999) Negative regulation of male development in *Caenorhabditis elegans* by a protein-protein interaction between TRA-2A and FEM-3. *Genes Dev.*, **13**(11), 1453–1463.

Merritt, C., Rasoloson, D., Ko, D. and Seydoux, G. (2008) 3′ UTRs are the primary regulators of gene expression in the *C. elegans* germline. *Curr. Biol.*, **18**(19), 1476–1482.

Mootz, D., Ho, D.M. and Hunter, C.P. (2004) The STAR/Maxi-KH domain protein GLD-1 mediates a developmental switch in the translational control of *C. elegans* PAL-1. *Development*, **131**(14), 3263–3272.

Nayak, S., Goree, J. and Schedl, T. (2005) *fog-2* and the evolution of self-fertile hermaphroditism in *Caenorhabditis. PLoS Biol.*, **3**(1), e6.

Nott, A., Le Hir, H. and Moore, M.J. (2004) Splicing enhances translation in mammalian cells: an additional function of the exon junction complex. *Genes. Dev.*, **18**(2), 210–222.

Okkema, P.G. and Kimble, J. (1991) Molecular analysis of *tra-2,* a sex determining gene in *C.elegans. EMBO J.*, **10**(1), 171–176.

Otori, M., Karashima, T. and Yamamoto, M. (2006) The *Caenorhabditis elegans* homologue of deleted in azoospermia is involved in the sperm/oocyte switch. *Mol. Biol. Cell*, **17**(7), 3147–3155.

Peden, E., Kimberly, E., Gengyo-Ando, K. *et al.* (2007) Control of sex-specific apoptosis in *C. elegans* by the BarH homeodomain protein CEH-30 and the transcriptional repressor UNC-37/Groucho. *Genes. Dev.*, **21**(23), 3195–3207.

Perry, M.D., Li, W., Trent, C. *et al.* (1993) Molecular characterization of the *her-1* gene suggests a direct role in cell signaling during *Caenorhabditis elegans* sex determination. *Genes Dev.*, **7**(2), 216–228.

Pilgrim, D., McGregor, A., Jackle, P. *et al.* (1995) The *C. elegans* sex-determining gene *fem-2* encodes a putative protein phosphatase. *Mol. Biol. Cell*, **6**(9), 1159–1171.

Puoti, A. and Kimble, J. (1999) The *Caenorhabditis elegans* sex determination gene *mog-1* encodes a member of the DEAH-Box protein family. *Mol. Cell Biol.*, **19**(3), 2189–2197.

Puoti, A. and Kimble, J. (2000) The hermaphrodite sperm/oocyte switch requires the Caenorhabditis elegans homologs of PRP2 and PRP22. *Proc. Natl. Acad. Sci. USA*, **97**(7), 3276–3281.

Raymond, C.S., Shamu, C.E., Shen, M.M. *et al.* (1998) Evidence for evolutionary conservation of sex-determining genes. *Nature*, **391**(6668), 691–695.

Reinke, V., Smith, H.E., Nance, J. *et al.* (2000) A global profile of germline gene expression in *C. elegans*. *Mol. Cell*, **6**(3), 605–616.

Richter, J.D. (2007) CPEB: a life in translation. *Trends Biochem. Sci.*, **32**(6), 279–285.

Ruiz i Altaba, A. (1999) Gli proteins encode context-dependent positive and negative functions: implications for development and disease. *Development*, **126**(14), 3205–3216.

Ryder, S.P., Frater, L.A., Abramovitz, D.L. *et al.* (2004) RNA target specificity of the STAR/GSG domain post-transcriptional regulatory protein GLD-1. *Nat. Struct. Mol. Biol.*, **11**(1), 20–28.

Schedl, T., Graham, P.L., Barton, M.K. and Kimble, J. (1989) Analysis of the role of *tra-1* in germline sex determination in the nematode *Caenorhabditis elegans*. *Genetics*, **123**(4), 755–769.

Schedl, T. and Kimble, J. (1988) *fog-2,* a germ-line-specific sex determination gene required for hermaphrodite spermatogenesis in *Caenorhabditis elegans*. *Genetics*, **119**(1), 43–61.

Schumacher, B., Hanazawa, M., Lee, M.H. *et al.* (2005) Translational repression of *C. elegans* p53 by GLD-1 regulates DNA damage-induced apoptosis. *Cell*, **120**(3), 357–368.

Schvarzstein, M. and Spence, A.M. (2006) The *C. elegans* sex-determining GLI protein TRA-1A is regulated by sex-specific proteolysis. *Dev. Cell*, **11**(5), 733–740.

Schwartz, H.T. and Horvitz, H.R. (2007) The *C. elegans* protein CEH-30 protects male-specific neurons from apoptosis independently of the Bcl-2 homolog CED-9. *Genes. Dev.*, **21**(23), 3181–3194.

Shen, M.M. and Hodgkin, J. (1988) *mab-3,* a gene required for sex-specific yolk protein expression and a male-specific lineage in *C. elegans*. *Cell*, **54**(7), 1019–1031.

Shimada, M., Kanematsu, K., Tanaka, K. *et al.* (2006) Proteasomal ubiquitin receptor RPN-10 controls sex determination in *Caenorhabditis elegans*. *Mol. Biol. Cell*, **17**(12), 5356–5371.

Sokol, S.B. and Kuwabara, P.E. (2000) Proteolysis in *Caenorhabditis elegans* sex determination: cleavage of TRA-2A by TRA-3. *Genes. Dev.*, **14**(8), 901–906.

Spence, A.M., Coulson, A. and Hodgkin, J. (1990) The product of *fem-1,* a nematode sex-determining gene, contains a motif found in cell cycle control proteins and receptors for cell-cell interactions. *Cell*, **60**(6), 981–990.

Starostina, N.G., Lim, J.M., Schvarzstein, M. *et al.* (2007) A CUL-2 ubiquitin ligase containing three FEM proteins degrades TRA-1 to regulate *C. elegans* sex determination. *Dev. Cell*, **13**(1), 127–139.

Subramaniam, K. and Seydoux, G. (2003) Dedifferentiation of primary spermatocytes into germ cell tumors in *C. elegans* lacking the Pumilio-like protein PUF-8. *Curr. Biol.*, **13**(2), 134–139.

Thompson, B.E., Bernstein, D.S., Bachorik, J.L. *et al.* (2005) Dose-dependent control of proliferation and sperm specification by FOG-1/CPEB. *Development*, **132**(15), 3471–3481.

Unhavaithaya, Y., Shin, T.H., Miliaras, N. *et al.* (2002) MEP-1 and a homolog of the NURD complex component Mi-2 act together to maintain germline-soma distinctions in *C. elegans*. *Cell*, **111**(7), 991–1002.

Walser, C.B., Battu, G., Hoier, E.F. and Hajnal, A. (2006) Distinct roles of the Pumilio and FBF translational repressors during *C. elegans* vulval development. *Development*, **133**(17), 3461–3471.

Wang, S. and Kimble, J. (2001) The TRA-1 transcription factor binds TRA-2 to regulate sexual fates in Caenorhabditis elegans. *EMBO J.*, **20**(6), 1363–1372.

White-Cooper, H., Doggett, K. and Ellis, R.E. (2009) The evolution of spermatogenesis, in *Sperm Biology: An Evolutionary Perspective* (eds T.R. Birkhead, D.J. Hosken and S. Pitnick), Academic Press, Burlington, MA, pp. 151–183.

Wickens, M., Bernstein, D.S., Kimble, J. and Parker, R. (2002) A PUF family portrait: 3′UTR regulation as a way of life. *Trends Genet.*, **18**(3), 150–157.

Wolff, J.R. and Zarkower, D. (2008) Somatic sexual differentiation in *Caenorhabditis elegans*. *Curr. Top. Dev. Biol.*, **83**,1–39.

Yi, W., Ross, J.M. and Zarkower, D. (2000) *mab-3* is a direct *tra-1* target gene regulating diverse aspects of *C. elegans* male sexual development and behavior. *Development*, **127**(20), 4469–4480.

Zarkower, D. and Hodgkin, J. (1992) Molecular analysis of the *C. elegans* sex-determining gene *tra-1:* a gene encoding two zinc finger proteins. *Cell*, **70**(2), 237–249.

Zhang, B., Gallegos, M., Puoti, A. *et al.* (1997) A conserved RNA-binding protein that regulates sexual fates in the *C. elegans* hermaphrodite germ line. *Nature*, **390**(6659), 477–484.

2

Sex determination and gonadal development

Alexander Combes,[1,3] Cassy Spiller[1,2,3] and Peter Koopman[1,2]

[1]*Institute for Molecular Biosciences, The University of Queensland, Brisbane, QLD 4072, Australia*
[2]*ARC Centre of Excellence in Biotechnology and Development, The University of Queensland, Brisbane, QLD 4072, Australia*
[3]*These authors contributed equally to this work.*

2.1 Introduction

In mammals, sex determination is a direct result of chromosomal constitution determined at fertilization, with females harbouring XX and males XY sex chromosomes. Despite the apparent simplicity of this system, the cascade of events that are triggered during embryogenesis to enforce gender is remarkably fragile. Correct sex determination relies on intricate genetic and cellular interactions to direct differentiation of the bipotential gonadal primordium into either a testis or an ovary. Once this is determined, hormones produced by the gonads reinforce gender in the form of secondary sexual characteristics, which define our emotional and physical behaviours and identities. Due to the complexity of this system, there are many stages at which aberrations can occur, giving rise to disorders of sexual development.

Whilst sex determination and gonadal development comprise a fascinating struggle for gender identity, the ultimate purpose of this process is to provide the correct environment to nurture the germ cells of the individual. These cells wholly represent the individual's ability to reproduce and bestow unique genetic information to future generations. Specification, migration and differentiation of germ cells is completely controlled by the somatic cell environment. Once sex differentiation has occurred, the germ cells are directed to develop into oocytes (female) or spermatozoa (male), both highly specialized cell types.

Our current understanding of these processes has been gleaned from over 50 years of research across many model organisms. The mouse model is the established system

Oogenesis: The Universal Process Marie-Hélène Verlhac and Anne Villeneuve

for studies relating to human sex determination, and therefore the majority of information discussed below pertains to mouse and human sex determination and gonadal development. This chapter will follow the timeline of sexual development, from specification of germ cells to differentiation into oocytes and spermatozoa, and cover key genetic and cellular events in the soma that give rise to the genital ridge and the gonads.

2.2 Early murine embryo and germ cell development

Primordial germ cells (PGCs) are specified very early in development. They begin as a small population of cells identifiable by specific pluripotent markers within an environment that is rapidly growing and differentiating into the multitude of cell types needed for a complete organism. Within this dynamic environment the germ cells migrate to the site of the forming genital ridges, where they begin differentiation down the male or female developmental pathway. At all of these stages the germ cells are responding to cues from the surrounding somatic cells, and so to fully appreciate their unique journey it is first useful to understand the environment in which they are specified.

2.2.1 Germ layers of the developing embryo

The development of the mouse embryo from initial gamete fusion to differentiation and morphogenesis requires 19–20 days of gestation. Following fertilization, the embryo divides slowly with little increase in mass until implantation into the wall of the uterus at 4.5 days *post coitum* (dpc). At this point the blastocyst consists of an epiblast, the primitive endoderm and polar trophectoderm. Following implantation, the embryo elongates, and begins to form the ectoplacental cone and trophoblast giant cells. At 6.5 dpc the primitive streak appears and gastrulation occurs, during which epiblast cells form the mesoderm and endoderm tissues. Along with the ectoderm, it is from these three primary germ (tissue) layers that the multitude of specialized tissues in the resulting embryo will be generated. This coordinated development relies on intricately timed cell movement in response to tightly regulated signalling and transcription factor activities (Tam and Loebel, 2007). This complex network ensures that the correct tissue progenitors are laid down for subsequent embryo development. Organ specification from these primary germ layers is complete at around 13–14 dpc, and the remaining period up until birth involves mainly foetal growth. For further reading on embryogenesis see Zernicka-Goetz (2002) and Tam *et al.* (2003).

2.2.2 Origin and specification of germ cells

It is amidst the myriad of morphological changes of gastrulation that the germ cells begin their fascinating developmental pathway. At this early stage, male and female germ cells display identical morphology and behaviour.

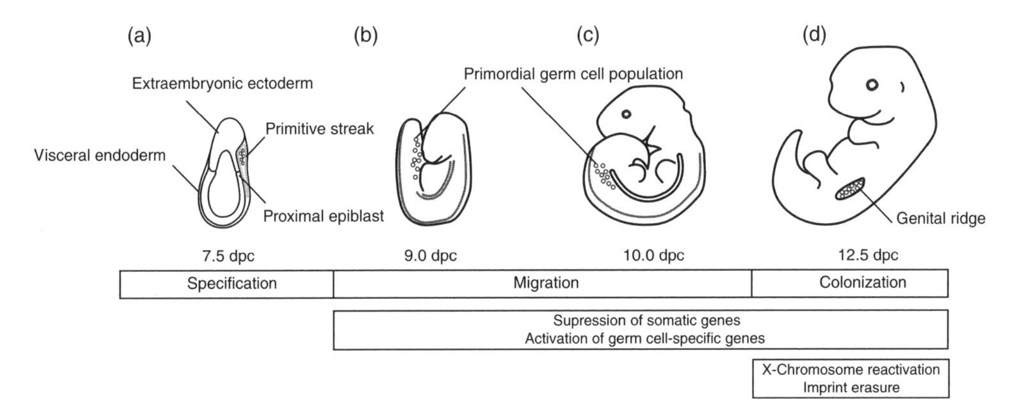

Figure 2.1 Germ cell specification and migration during early mouse development. The primordial germ cells are first identified at 7.25 dpc within the proximal epiblast (a). This population proliferates and migrates through the hindgut (b and c) to colonize the genital ridges by 11.0–12.5 dpc (d). Throughout this process, genetic regulation reinforces the germ cell lineage with suppression of somatic cell genes and upregulation of germ cell-specific genes. X-Chromosome reactivation occurs in female gonads prior to imprint erasure in both sexes. Cartoons for the mouse embryos were adapted from Sasaki and Matsui (2008) and Boldajipour and Raz (2007). A full colour version of this figure appears in the colour plate section.

PGC location

The germ cell precursors have been identified as a small cluster of cells in the proximal epiblast as early as 7.25 dpc in the developing embryo (Figure 2.1a) (McLaren, 1983a; Lawson and Hage, 1994; Parameswaran and Tam, 1995). Their fate as PGCs is sealed only when they have moved into the extraembryonic tissues at the proximal region of the allantois and, as a cluster of around 45 cells, begin to express germ cell-specific markers (Lawson and Hage, 1994; Ohinata *et al.*, 2005; Saitou, Barton and Surani, 2002; Tanaka and Matsui, 2002). It has been demonstrated through clonal lineage analysis (Lawson and Hage, 1994) and transplantation experiments (Tam and Zhou, 1996) that the founding epiblast cells have not been preprogrammed for PGC fate and can therefore give rise to both somatic cells and gametes (McLaren, 1983a; Tam and Zhou, 1996). Around the time of this specification, random X-chromosome inactivation takes place in female germ cells as it does in all somatic cells, which is important for modulating X-linked gene dosage (Tsang *et al.*, 2001; Monk and McLaren, 1981).

Signalling for PGC specification

The initial specification of PGCs from the proximal epiblast cells requires paracrine signals that originate from the surrounding somatic cells. Most notably, members of the bone morphogenetic protein (BMP) family, BMP4 and BMP8b, produced by the extraembryonic ectoderm, are responsible for cell reprogramming to produce the PGC lineage at approximately 6.0 dpc. BMPs signal through homologues of *Caenorhabditis elegans* SMA protein and *Drosophila* Mothers against decapentaplegic (SMAD) signal transducers to induce upregulation of PGC-specific genes. The PGC population is dramatically reduced in *Bmp4* and *Smad1/5* loss-of-function models, highlighting their requirement for this purpose (Hayashi *et al.*, 2002; Tremblay, Dunn and Robertson, 2001; Arnold *et al.*, 2006). BMP signalling induces expression of the gene transcript interferon-inducible transmembrane protein (*Ifitm3/Fragilis/mil-1*) in the prospective PGC population by 6.25 dpc (Tanaka and Matsui, 2002; Saitou, Barton and Surani, 2002). A cell adhesion molecule, E-cadherin (E-CAD), is also expressed by the cluster of PGC precursors, suggesting an important role for cell–cell contact for correct PGC determination (Bendel-Stenzel *et al.*, 2000; Di Carlo and De Felici, 2000). The total subset of cells that go on to comprise the founding PGC population then express the SET domain and zinc finger binding protein encoded by B-lymphocyte-induced maturation protein 1 (*Blimp1/Prdm1*) by 6.5 dpc (Ohinata *et al.*, 2005). By 7.25 dpc the specified PGCs now express the chromosome organizational and RNA processing protein encoded by developmental pluripotency associated 3 (*Dppa/Stella/PGC7*) (Saitou, Barton and Surani, 2002; Sato *et al.*, 2002).

PGC pluripotency and gene markers

In order to remain capable of generating a new organism, the newly specified germ cells must actively suppress somatic differentiation whilst maintaining expression of various pluripotent and lineage-specific genes (Table 2.1) (Seydoux and Braun, 2006; Ohinata *et al.*, 2005). For this purpose, a unique set of transcription factors and signalling molecules have been identified within the PGC population. *Blimp1* (Ohinata *et al.*, 2005; Vincent *et al.*, 2005) and the transcription factor *Smad1* (Chang *et al.*, 2001a; Hayashi *et al.*, 2002) are thought to be responsible for suppression of some somatic lineage genes. Downregulated somatic genes include *Hoxb1, Fgf8* and *Snail* (Ancelin *et al.*, 2006; Hayashi, de Sousa Lopes and Surani, 2007). Pluripotent genes maintained in PGCs include SRY-box containing gene 2 (*Sox2*) (Yabuta *et al.*, 2006; Ohinata *et al.*, 2005), nanog homeobox (*Nanog*) (Yamaguchi *et al.*, 2005; Chambers *et al.*, 2007) and germline transcription factor Octamer-4 (*Oct4*) (Scholer *et al.*, 1990).

Other genes expressed by PGCs following their specification include tissue non-specific alkaline phosphatase (*Tnap*) (Chiquoine, 1954; Ginsburg, Snow and McLaren, 1990; Tam and Zhou, 1996; MacGregor, Zambrowicz and Soriano, 1995), stage-specific embryonic antigen 1 (*Ssea1*) and PR domain containing14 (*Prdm14*) (related to *Blimp1*). This unique gene expression profile has allowed PCGs to be distinguished from surrounding somatic tissues and has been exploited for experimental purposes (see Table 2.1). For example, high levels of TNAP activity were first observed by Chiquoine (1954), and have since been used to identify the founder PGC population and its subsequent development during embryogenesis (Ginsburg, Snow and McLaren, 1990).

2.2.3 Germ cell migration and proliferation

Once PGCs have been specified and express appropriate gene markers, they begin their journey to the primitive gonad, where they later differentiate into functional

Table 2.1 Genes and proteins expressed by primordial germ cells during their specification, migration and colonization of the developing murine gonad

Developmental process	Gene/protein	Gene information	Reference
Specification	*Fragilis/Ifitm3*	Interferon-inducible transmembrane protein involved in cell adhesion	Tanaka and Matsui, 2002; Saitou, Barton and Surani, 2002
	E-cadherin	Cell adhesion molecule	Di Carlo and De Felici, 2000; Bendel-Stenzel *et al.*, 2000
	Blimp1/Prdm1	SET domain and zinc finger-containing protein	Ohinata *et al.*, 2005
	Stella/PGC7	Protein involved in RNA processing and chromosomal organization	Sato *et al.*, 2002; Saitou, Barton and Surani, 2002
	Smad 1	Receptor-regulated transcription regulators	Tremblay, Dunn and Robertson, 2001; Hayashi *et al.*, 2002; Chang, Lau and Matzuk, 2001b
	TNAP	Tissue non-specific alkaline phosphatase	MacGregor, Zambrowicz and Soriano, 1995; Tam and Zhou, 1996; Ginsburg, Snow and McLaren, 1990
Pluripotency	*Nanog*	Homeodomain-bearing transcription factor	Chambers *et al.*, 2007; Yamaguchi *et al.*, 2005
	Oct4/Pou5f1	Transcription factor	Scholer *et al.*, 1990
	Sox2	Transcription factor	Yabuta *et al.*, 2006; Ohinata *et al.*, 2005
Migration	CXCR4 ((C-X-C motif) receptor 4)	Chemokine receptor for stromal derived growth factor-1 (SDF1)	Stebler *et al.*, 2004; Molyneaux *et al.*, 2003
	Steel/c-kit	Receptor for stem cell factor	Buehr *et al.*, 1993b; Matsui *et al.*, 1991; Godin and Wylie, 1991
	Gp130	Receptor for leukaemia inhibitory factor (LIF)	Matsui *et al.*, 1991
	Pin1 (peptidyl-prolyl isomerase)	Phosphoprotein modifier	Atchison, Capel and Means, 2003
	Wt1 (Wilms' tumour suppressor)	Zinc finger protein	Natoli *et al.*, 2004
	Ssea1 (Stage-specific embryonic antigen-1)	Trisaccharide of the form galactose [β1–4]*N*-acetylglucosamine[α1–3] fucose	Fox *et al.*, 1981
	Nanos3	Zinc finger protein with putative RNA-binding activity	Jaruzelska *et al.*, 2003; Tsuda *et al.*, 2003

(continued)

Table 2.1 *(Continued)*

Developmental process	Gene/protein	Gene information	Reference
Colonization	ADAM	Cell–cell contacts	Rosselot *et al.*, 2003
	B1 integrins	Heterodimeric receptors involved in cell–cell contact	Anderson *et al.*, 1999; Rosselot *et al.*, 2003
	Mvh/ddx4 (Mouse vasa homologue)	RNA helicase, DEAD box polypeptide 4	Toyooka *et al.*, 2000; Noce, Okamoto-Ito and Tsunekawa, 2001
	Gcna1 (germ cell nuclear antigen 1)	Unknown	Enders and May, 1994
	Gcl (germ cell-less)	Unknown – nuclear envelope component	Kimura *et al.*, 1999; Masuhara *et al.*, 2003
	Dazl (deleted in azoospermia like)	PABP-binding transcription factor	Saunders *et al.*, 2003; Collier *et al.*, 2005
Epigenetic reprogramming	DNMT3A	DNA methyltransferase	Hata *et al.*, 2002; Kaneda *et al.*, 2004; Bourc'his *et al.*, 2001
	DNMT3L	DNA methyltransferase	Hata *et al.*, 2002; Kaneda *et al.*, 2004; Bourc'his *et al.*, 2001
	DNMT3B	DNA methyltransferase	Hata *et al.*, 2002; Kaneda *et al.*, 2004; Webster *et al.*, 2005; Bourc'his and Bestor, 2004

gametes. The migratory pathway of germ cells from their primary colony in the posterior primitive streak to the primitive gonads has been tracked using TNAP expression (Chiquoine, 1954; Mintz and Russell, 1957). The journey begins within 24 hours of specification with initial passive incorporation of PGCs into the developing hindgut (Clark and Eddy, 1975). Between 8.5 dpc and 9.5 dpc PGCs move anteriorly through the hindgut wall by active migration (Figure 2.1b and c). Leaving the hindgut via the gut mesentery, PGCs migrate into the nascent genital ridges on either side of the posterior dorsal aorta (Lawson and Hage, 1994; Anderson *et al.*, 2000; Molyneaux *et al.*, 2001).

Interestingly, germ cells isolated in culture 24 hours prior to migration (8.5 dpc) have been shown to be incapable of active locomotion (Godin, Wylie and Heasman, 1990; Godin and Wylie, 1991), suggesting that they require some signal to begin migration. Once initiated, successful migration has been shown to rely on germ cell–germ cell interactions, and by 10.5 dpc the PGCs are networked by long processes, contrary to earlier belief that migration occurred independently for each PGC (Gomperts *et al.*, 1994). Gomperts *et al.* (1994) have postulated that, following initial PGC colonization of the genital ridges, the remaining cells arrive by virtue of the long processes between these cells both drawing and directing their migration.

Signals for migration

It has been established that during the journey to the genital ridge, PGCs are receptive to various signals originating from other PGCs and the surrounding tissues (Wylie, 1993). Importantly, it is believed that the somatic cell environment determines the migratory pathway rather than being a cell autonomous response (Wylie, 1999). It is also of interest to note that, when the PGCs begin their migration, neither the dorsal mesentery nor genital ridges have developed (Clark and Eddy, 1975). The origins of the signals that trigger migration are therefore unknown. Several hours after PGC migration is initiated the gonadal structures develop, at which point they exert a chemoattractive effect on the germ cells. In culture, explanted gonadal primordium increases the number of germ cells and promotes migration (Godin, Wylie and Heasman, 1990).

Throughout migration, PGCs rapidly proliferate in response to numerous extracellular growth factors. Some of these factors and receptors identified to date include stromal derived factor-1 (SDF1) and its receptor chemokine (C-X-C motif) receptor 4 (Molyneaux *et al.*, 2003; Stebler *et al.*, 2004), stem cell factor and its receptor c-KIT (Godin and Wylie, 1991; Matsui *et al.*, 1991) and LIF and receptor interleukin 6 signal transducer (Matsui *et al.*, 1991; De Felici, 2000). Required growth factors include fibroblast growth factors (FGF)-2, -4 and -8 (Matsui, 1992; Resnick *et al.*, 1992), interleukin 4 (Cooke, Heasman and Wylie, 1996) and genes such as peptidyl-prolyl isomerase (*Pin1*) (Atchison, Capel and Means, 2003) and Wilms' tumour suppressor gene (*Wt1*) (Natoli *et al.*, 2004).

Despite the migratory guides produced by the soma, a number of PGCs depart from the pathway and end up in ectopic locations (Gobel *et al.*, 2000). In these instances the PGCs undergo meiosis, characteristic of the female developmental pathway, regardless of their genetic sex, and ultimately undergo apoptosis (this response will be discussed in further detail in later sections) (McLaren, 1983a; Molyneaux *et al.*, 2001; Boldajipour and Raz, 2007; Upadhyay and Zamboni, 1982; McLaren, 1983b).

2.3 Genital ridge colonization

Almost 24 hours after PGCs receive the signal to begin migration, the gonads themselves begin to develop. Derived from the urogenital ridges as paired swellings parallel to the neural tube from 9.0 dpc, they arise largely as a result of cell proliferation at the coelomic epithelium (Schmahl *et al.*, 2000; Karl and Capel, 1998; Byskov, 1986). As the PGCs begin colonization of the newly formed genital ridges (Figure 2.1d) they are proliferating with a cell cycle time of 12–14 hours (Tam and Snow, 1981). This proliferation occurs for a further 1–2 days such that the original population of founder cells (around 45) has increased to 25 000–30 000 cells by 13.5 dpc (Donovan *et al.*, 1986; Tam and Snow, 1981). Resident germ cells are now referred to as gonocytes, and undergo several changes independent of sex differentiation. These include alterations to their cellular morphology as they take on a conspicuous large and rounded shape (Donovan *et al.*, 1986) and become less motile (De Felici, Dolci and Pesce, 1992; Garcia-Castro *et al.*, 1997). In addition, cell adhesion molecules such as integrins and members of the ADAM family are expressed to presumably facilitate new cell–cell adhesions with the local somatic cell environment (Rosselot *et al.*, 2003).

Pluripotency versus differentiation

Following their specification, germ cells actively suppress somatic genes in order to retain a pluripotent state. This is important, as germ cells are unique in their ability to maintain a differentiated yet pluripotent nature and, throughout all developmental stages, this state must be tightly controlled. However, once they reach the genital ridge, this genetic programme changes such that gonocytes have a decreased ability to form pluripotent stem cells (Matsui, 1992; Resnick *et al.*, 1992; McLaren, 1984), and a different set of gene markers are expressed. Previously expressed genes such as *Tnap* and *Sseal* become downregulated, whereas mouse vasa homologue (*Mvh*) (McLaren, 1984; Toyooka *et al.*, 2000), germ cell nuclear antigen 1 (*Gcnal*) (Enders and May, 1994), deleted in azoospermia like (*Dazl*) (Saunders *et al.*, 2003; Noce, Okamoto-Ito and Tsunekawa, 2001) and germ cell-less (*Gcl*) (Kimura *et al.*, 1999) are upregulated. Despite this apparent loss of pluripotency, the germ cells are still considered to be multipotent, primarily because they are capable of forming teratomas, which comprise various types of somatic tissues, in both ovaries and testes (Kanatsu-Shinohara and Shinohara, 2006). Once sex differentiation has taken place, pluripotent markers of female germ cells including *Oct4* are downregulated as the germ cells enter meiosis at 13.5 dpc. These same markers persist in male germ cells as they enter G1/G0 arrest, but are extinguished by birth.

X-chromosome reactivation, epigenetic imprint erasure and re-methylation

PGC arrival at the genital ridge signals the reactivation of the silent X-chromosome for female germ cells, which occurs from 11.5 to 13.5 dpc and is dependent on interactions with the XX genital ridge (Tam, Zhou and Tan, 1994). Germ cells that fail to colonize the gonad never achieve X-chromosome reactivation, despite entering into meiosis before eventual degeneration (Tsang *et al.*, 2001; Upadhyay and Zamboni, 1982; McLaren and Monk, 1981). Interestingly, female germ cells that find themselves in a testis will also reactivate their X-chromosome, suggesting that the signal for reactivation is not ovarian specific (McLaren and Monk, 1981; Jamieson *et al.*, 1998).

Epigenetic reprogramming is also initiated in the gonocytes located in the genital ridge. Up until this point the germ cells have carried parent-of-origin-specific imprinting marks and so exhibit monoallelic expression of many genes (Maatouk *et al.*, 2006). Erasure of methylation from these regions and chromatin restructuring is required for the gonocytes to give rise to totipotent cells of a new embryo, with appropriate methylation according to the sex of the embryo (Allegrucci *et al.*, 2005; McLaren, 2003). This erasure is thought to begin as early as 10.5 dpc, with the majority of genes studied displaying hypomethylation by 12.5 dpc (Szabo and Mann, 1995; Hajkova *et al.*, 2002). Several germ cell-specific genes that become hypomethylated (and therefore expressed) at this time include *Mvh*, *Dazl* and synaptonemal complex protein 3 (*Sycp3*) (Maatouk *et al.*, 2006). The gonocytes maintain this state of DNA demethylation until the next stage of epigenetic reprogramming that comprises re-establishment of a new methylation status that occurs in a sex-specific manner. To date there are approximately 100 genes known to be regulated by this imprinting, the majority occurring in the female gametes (Jue, Bestor and Trasler, 1995; Ueda *et al.*, 2000). Maternal imprinting in the female germline occurs after birth while oocytes are arrested in meiotic prophase I (Ueda *et al.*, 2000). In the male germ line, just three loci have been identified to undergo paternal imprinting. This occurs following sex determination from 14.5 dpc to after birth, but prior to meiosis (Jue, Bestor and Trasler, 1995; Ueda *et al.*, 2000; Li *et al.*, 2004; Davis *et al.*, 2000).

Having made the journey to the genital ridges and undertaken the various gene expression patterns and epigenetic modifications, the gonocytes are now waiting for sex determination of the soma to occur before they are directed to one of two fates: oogenesis or spermatogenesis.

2.4 Sex determination

The sex chromosome complement of an ovum is invariably X, and so contribution of an X or a Y chromosome from the spermatozoa determines the genetic sex of an individual. However, the development of a normal sexual phenotype by inheritance of an X or a Y chromosome is far from a foregone conclusion. Following fertilization, the sex programme lays dormant while the fertilized ovum progresses through development to the early embryo. The first manifestation of sexual dimorphism in the embryo occurs around 7 weeks of development in humans or 10.5 days in mice, with the activation of genetic pathways in the gonadal primordium that promote development of either a testis or an ovary.

One unique feature of gonadal development is that the testis and ovary both develop from the genital ridge. As such the genital ridge contains populations of bipotential cell types that adopt corresponding roles in testis or ovarian fate. These are: the supporting cells, which differentiate into either Sertoli or granulosa cells; vascular precursors, which adopt different identities and structure dependant on their environment; and mesenchymal interstitial cells which give rise to steroidogenic Leydig or theca cells and other cell types that are less defined. Molecular pathways for both fates are present and receptive to activation to regulate this dimorphic system.

2.4.1 Male

In the bipotential environment of the early gonad, the presence of a Y chromosome initiates testis development through the expression of a single gene in the somatic cell lineage, Sex-determining region on the Y chromosome (*Sry*). *Sry* was identified by gene mapping within a chromosomal region that caused human sex reversal when deleted in males or ectopically present in females (Sinclair *et al.*, 1990). Further, it was shown to be the only gene necessary to direct male development, through genetic experiments in which chromosomally female mice transgenic for *Sry* developed as males (Koopman *et al.*, 1991). Multiple cases of human sex reversal have been reported to result from mutations in the DNA binding and bending high mobility group (HMG) domain of SRY (Harley *et al.*, 1992; Jager *et al.*, 1992; Mitchell and Harley, 2002; Pontiggia *et al.*, 1994; Schmitt-Ney *et al.*, 1995).

Soon after *Sry* expression, a related gene, Sry-like HMG box containing gene 9 (*Sox9*), is also upregulated in the early testis (Morais da Silva *et al.*, 1996; Kent *et al.*, 1996). *Sox9* is critical to testis determination, as *Sox9* knockout mice display male-to-female sex reversal (Barrionuevo *et al.*, 2006; Chaboissier *et al.*, 2004). Furthermore, *Sox9* can functionally substitute for *Sry*, as ectopic expression is sufficient to initiate testis development in XX gonads (Vidal *et al.*, 2001). In humans, mutations in *SOX9* give rise to campomelic dysplasia, a syndrome characterized by skeletal abnormalities and often associated with XY sex reversal (Foster *et al.*, 1994; Wagner *et al.*, 1994). Thus, both *Sry* and *Sox9* are necessary and sufficient for male sex determination.

2.4.2 Female

For many years the molecular regulation of ovarian development remained a mystery while the discovery of *Sry* fuelled an intense focus on testis-specific genes. The dominant action of *Sry* led to a view that ovarian development was a default pathway. However, this view was confounded by cases of human XX sex reversal in which genetically female individuals developed as phenotypic males. These findings led to the hypothesis that *Sry* was required for repressing a hypothetical factor 'Z' that normally repressed testis development in the female (McElreavey *et al.*, 1993). *Sry*-negative cases of XX sex reversal were then explained, theoretically, by disruption of the testis-repressing factor Z resulting in activation of the male pathway in females.

While no definitive Z factor has been found to date, genes involved in the Wingless type MMTV integration site (Wnt) signalling pathway and a forkhead transcription factor (FOXL2) have been reported to repress aspects of testis development (Kim *et al.*, 2006b; Ottolenghi *et al.*, 2007; Chassot *et al.*, 2008; Tomizuka *et al.*, 2008; Liu *et al.*, 2009; Maatouk *et al.*, 2008). The germ cell lineage also plays a critical role in maintaining ovarian fate, as loss of this cell type results in partial sex reversal after birth (to be discussed in detail later). Recent characterization of genes involved in testis and ovarian development has revealed increasing evidence of an active antagonism between the male and female molecular pathways during sex determination.

2.4.3 Antagonism between the pathways

During the initial stages of sex determination, genes promoting both male and female pathways are expressed in mutually exclusive domains in XY and XX gonads, respectively (Kim *et al.*, 2006b). The expression of members of the Wnt signalling pathway *Wnt4* and β-Catenin at this stage promotes ovarian development while opposing testis development (Vainio *et al.*, 1999; Kim *et al.*, 2006b; Chassot *et al.*, 2008; Maatouk *et al.*, 2008). In a complementary domain, expression of *Fgf9* promotes testis development while repressing ovarian fate (Kim *et al.*, 2006b; Figure 2.2). The expression of *Sry* in XY gonads tips the balance of these two opposing signals towards testis development by upregulating *Sox9*. *Sox9* expression results in upregulation of paracrine signals Fgf9 (Kim *et al.*, 2006b) and prostaglandin D2 (Wilhelm *et al.*, 2007), which repress *Wnt4* expression and/or promote *Sox9* expression in undifferentiated somatic cells. In an XX environment, the Wnt signal prevails to repress *Fgf9* expression and promote ovarian development (Kim *et al.*, 2006b). Disruption or delay in expression of genes in either pathway at this early stage leads to partial or full development of the opposing fate, resulting in sex reversal. Sex reversal can also occur at later stages but is repressed in the XX gonad by the presence of meiotic germ cells (Yao, DiNapoli and Capel, 2003).

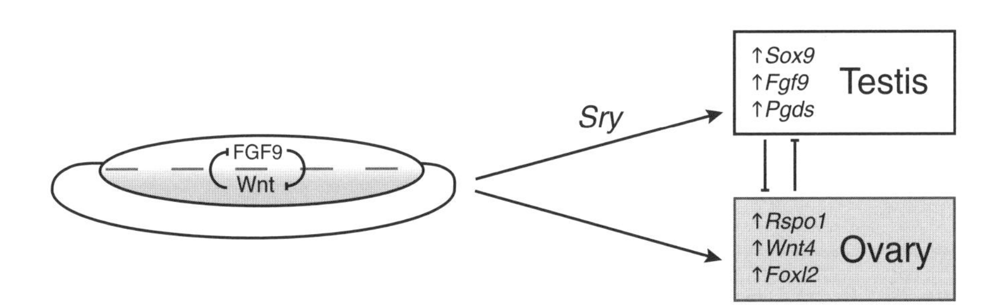

Figure 2.2 Sex determination. Mutually antagonistic signals promoting testis (*Fgf9*) and ovary (Wnt signalling) development are expressed in XX and XY gonadal primordia. Male-specific expression of *Sry* upregulates the expression of *Sox9*, which initiates testis development. *Sox9* expression is propagated and maintained through FGF and prostaglandin signalling. In the absence of *Sry*, Wnt signalling represses the testis pathway and initiates ovarian development

Consistent with this, loss of germ cells results in postnatal transdifferentiation of granulosa cells into Sertoli cells. Thus gonadal sex is determined through a molecular struggle during the early stages of sex determination (Figure 2.2). Downstream of sex determination, the molecular cues involved in establishing sex enact differentiation programs to construct the functional and morphological differences that distinguish the testis and ovary.

2.5 Ovary development

Structural development of the ovary progresses gradually from the time of sex determination to maturation (Figure 2.3). Germ cells display the first morphological

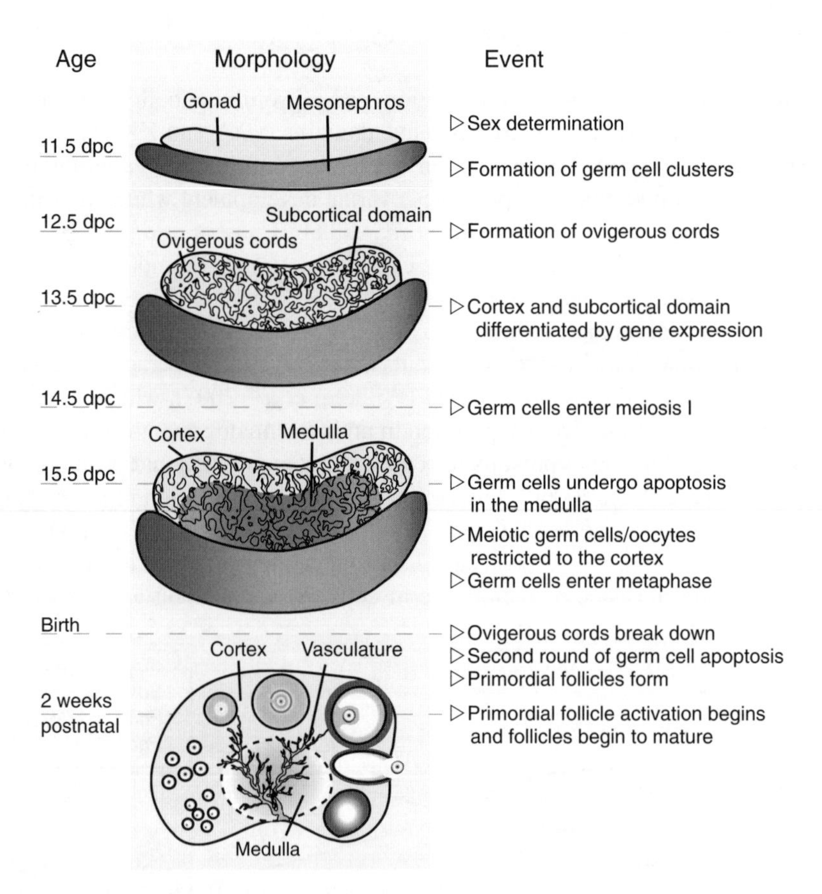

Figure 2.3 Ovarian development in mouse. Ovarian development progresses gradually with the formation of germ cell clusters and ovigerous cords. The ovary is segmented into cortical and subcortical domains by gene expression before these areas are morphologically distinguished. Germ cell entry into meiosis precedes a round of germ cell apoptosis in the medulla, leaving the remaining population (now called oocytes) restricted to the cortex. At birth, ovigerous cords break down coincident with a second major round of germ cell apoptosis. The remaining oocytes form a pool of primordial follicles, subsets of which are activated in a multiphase process throughout the reproductive lifespan of the individual

changes in the ovary by forming clusters from as early as 11.5 dpc, either by aggregation (Gomperts *et al.*, 1994) or through multiple divisions of a single progenitor (Pepling and Spradling, 1998). Germ cell clusters interact with somatic cells in the formation of ovigerous cords, where they become enclosed by pregranulosa cells and delineated by a basement membrane (Konishi *et al.*, 1986; Odor and Blandau, 1969). External to the ovigerous cords is the interstitium, which is composed of mesenchymal cells and the developing ovarian vascular system. Ovigerous cords fill the early ovary, which is soon segmented into two areas: cortex and medulla. The cortex occupies the outer portion of the ovary and encloses the central subcortical or medulla region. These regions are identified by differential gene expression as early as 12.5 dpc in the mouse, with morphological differences developing further with time. By 13.5 dpc the cortex is marked by expression of *Bmp2* (Yao *et al.*, 2004), and the medulla is marked by expression of *Wnt4*, Follistatin (*Fst*) (Yao *et al.*, 2004) and transgenic markers for pregranulosa cells (Albrecht and Eicher, 2001). At this stage, germ cells within the ovigerous cords enter and arrest in prophase I of meiosis, and are referred to as oocytes. Following several rounds of apoptosis, the remaining oocytes become surrounded by a single layer of granulosa cells and delineated by a basement membrane to form primordial follicles (Hirshfield, 1991). Follicular development progresses after the first oocytes reach diplotene stage (Byskov, 1986; Byskov and Lintern-Moore, 1973), and is a continuous process with regular activation of a subset of primordial follicles until the pool of follicles is depleted (Kezele, Nilsson and Skinner, 2002). For a comprehensive review see: Eppig, 2001; McGee and Hsueh, 2000. The genetic regulation of early follicle development and detail on germ–soma interactions will be covered in later chapters.

In the following section we will review recent findings that are building the framework for understanding the genetics of early ovarian differentiation. Furthermore, we will explore the interdependence of the maintenance of ovarian identity and feminized germ cells, as the two seem inextricably linked.

2.5.1 Genetic factors in determining and maintaining ovarian fate

Current knowledge on the genetic regulation of ovarian fate has arisen from analysis of XX sex-reversal conditions in humans, goats, and the mouse model. Multiple reports have identified a vigorous ovary-specific programme of gene expression from 11.5 dpc, identifying players in the genetic regulation of ovarian differentiation (Cederroth *et al.*, 2007). As a result, new ovarian-specific expression profiles are generated as part of concerted efforts to characterize the molecular profile of urogenital development (Beverdam and Koopman, 2006; Cory *et al.*, 2007; Little *et al.*, 2007; Nef *et al.*, 2005; www.gudmap.org). In addition, advances are being made through exploration of genetic pathways implicated in ovarian development. Two main regulators have been identified that are involved in determining and maintaining ovarian fate: Wnt signalling and *Foxl2*. Wnt signalling has been shown to promote germ cell survival, and *Foxl2* has been shown to regulate granulosa cell development. Disruption of either of these pathways results in postnatal sex reversal, possibly due to loss of germ cells. However, current data suggests that these pathways also have a role in regulating ovarian sex determination.

Wnt4

Wnt4 is expressed at 10 dpc in both XX and XY gonads, but is downregulated in XY gonads at 11.5 dpc (Vainio *et al.*, 1999). In the ovary, *Wnt4* expression is downregulated from 12.5 dpc and remains low in primordial follicles (Hsieh *et al.*, 2002). Analysis of XX gonads lacking *Wnt4* revealed rounded testis-like morphology at birth and a reduced number of oocytes (Vainio *et al.*, 1999). *Wnt4* is also required for suppressing endothelial cell migration: in the absence of *Wnt4*, testis-like vascular patterns are established in XX gonads as marked by the formation of the coelomic vessel – a prominent testis-specific artery positioned under the ventromedial surface of the gonad (Jeays-Ward *et al.*, 2003).

Fst and Bmp2

Elements of the *Wnt4*$^{-/-}$ phenotype, including coelomic vessel formation, were also seen in mice lacking *Fst*, implicating *Fst* as an effector of *Wnt4* action in the ovary (Yao *et al.*, 2004). *Fst* is expressed in an XX-specific pattern from 11.5 dpc, increasing at 12.5 dpc in wild-type mice (Yao *et al.*, 2004), but is absent in *Wnt4*$^{-/-}$ gonads, indicating that it is genetically downstream of *Wnt4*. Germ cells in the ovarian cortex are almost completely lost in both *Wnt4*- and *Fst*-null gonads before birth, complicating the partial sex reversal observed at later stages, since it is known that XX germ cells are required for maintaining ovarian fate (Behringer *et al.*, 1990; Guigon *et al.*, 2005). Nevertheless, *Wnt4* appears to act through *Fst* to repress endothelial cell migration and promote germ cell survival (Yao *et al.*, 2004). *Bmp2* is another gene expressed specifically in the XX gonad from 11.5 dpc. At 12.5 dpc, *Bmp2* expression is restricted to the coelomic domain of the ovary (Yao *et al.*, 2004). *Bmp2* expression is dependent on *Wnt4,* but not *Fst,* as it is absent in *Wnt4*$^{-/-}$ gonads but persists in *Fst*$^{-/-}$. *Bmp2*-null mice die before gonadogenesis (Zhang and Bradley, 1996), thus a conditional allele will need to be generated for analysis of the ovarian function of this gene.

R-spondin and β-Catenin

Since characterization of the *Wnt4*-null mice, other members of the Wnt signalling pathway have been implicated in ovarian development. In particular, R-spondin1 (*Rspo1*) activates the Wnt signalling pathway in the ovary, complementing the role of *Wnt4*. Wnt signalling is modulated through β-Catenin and regulates multiple processes including cell growth and development. Canonical Wnt signalling involves binding of a secreted Wnt protein ligand to a receptor complex involving a frizzled receptor and a co-receptor, lymphoid enhancer-binding factor (Kim *et al.*, 2006a). The signal from the bound ligand/receptor complex results in activation of β-Catenin, which is subsequently translocated into the nucleus to regulate gene expression in cooperation with the transcription factor T-cell transcription factor (Jho *et al.*, 2002).

Recent findings involving RSPO1 have underscored the importance of Wnt signalling in ovarian development. A point mutation in human *RSPO1* was identified to underlie a case of familial XX sex reversal. *RSPO1* was confirmed to be associated with XX sex reversal in a genetically independent individual harbouring a deletion including exon 4 of the coding sequence (Parma *et al.*, 2006). An independent study identified a homozygous point mutation in a splice donor site in *RSPO1* that appears to be causative in an XX individual with both testicular and ovarian gonadal tissue (Tomaselli *et al.*, 2008). While loss of *RSPO1* appears to be sufficient to cause XX sex reversal in humans, recent generation and characterization of *RSPO1* knockout mice indicates that this is not the only factor required for ovary development in all mammals (Chassot *et al.*, 2008; Tomizuka *et al.*, 2008).

In mice, XX individuals lacking *Rspo1* developed an ectopic coelomic vessel (Tomizuka *et al.*, 2008; Chassot *et al.*, 2008) and external genitalia were masculinized (Chassot *et al.*, 2008). Sex-specific duct development was also abnormal with both Wölffian (male) and Müllerian (female) ducts persisting in various stages of development (Tomizuka *et al.*, 2008; Chassot *et al.*, 2008). Wnt signalling was compromised in the absence of *Rspo1*, with the most severe outcome being the presence of both ovarian and testicular tissue in some XX gonads by 18.5 dpc (Tomizuka *et al.*, 2008; Chassot *et al.*, 2008). Absence of *Rspo1* in the XX gonads led to an increase in germ cell apoptosis, which may have led to transdifferentiation of Sertoli cells in these models.

Use of a reporter line responsive to β-Catenin-mediated transcriptional activity indicated that β-Catenin activity is primarily localized to somatic cells in the ovary and in the Wölffian and Müllerian ducts in both sexes. No activity was observed in the XY gonad (Chassot *et al.*, 2008). Levels of reporter activity were greatly reduced, but not abolished, in gonads lacking *Wnt4*; yet *Rspo1* levels remained unchanged, indicating that *Rspo1* is upstream of *Wnt4* (Chassot *et al.*, 2008). Conversely, reporter activity and *Wnt4* expression were lost in XX gonads lacking *Rspo1* (Chassot *et al.*, 2008). In the absence of β-Catenin, *Rspo1* levels remain unchanged, but *Wnt4* and *Fst* are down-regulated, indicating that β-Catenin acts as a mediator between Rspo1 and Wnt4 signals (Liu *et al.*, 2009; Figure 2.4). Ectopic expression of *Sox9* in XX gonads inhibited expression of the reporter, providing evidence that *Sox9* inhibits β-Catenin-mediated transcription (Chassot *et al.*, 2008).

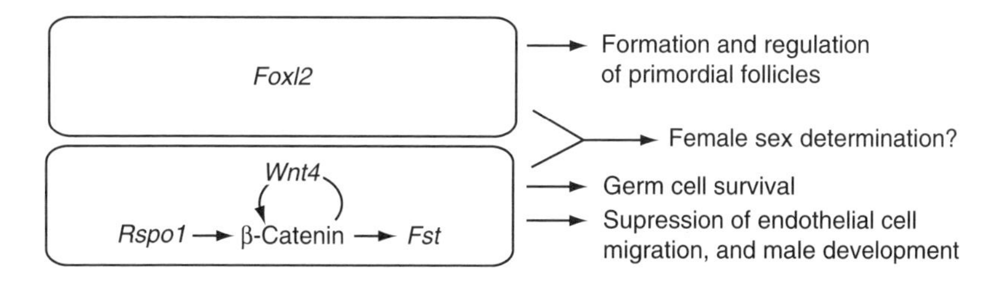

Figure 2.4 *Foxl2* and Wnt signalling control sex determination, follicle formation, and germ cell survival in the ovary. Loss of *Foxl2* results in defects in formation and activation of primordial follicles. Wnt signalling is mediated by *Rspo1* and *Wnt4*, which act through β-Catenin to upregulate *Fst*, repress endothelial cell migration, and ensure germ cell survival. Disruption in either pathway leads to loss of germ cells which causes transdifferentiation of Sertoli cells around birth. However, combined loss of both pathways leads to primary sex reversal in some cells in the ovary, indicating a role in female sex determination

Recent experiments directed at the sex-specific roles of β-Catenin have demonstrated that deletion of β-Catenin from the supporting cell lineage is not sufficient to induce primary female-to-male sex reversal (Liu *et al.*, 2009). It is possible that the expression of β-Catenin outside the supporting cell lineage was able to maintain ovarian identity in this case, as stabilization of β-Catenin in the same cell lineage in the testis causes repression of Sertoli cell identity and activation of the female pathway (Maatouk *et al.*, 2008).

These studies demonstrate a critical role for Wnt signalling in ovary development mediated by *Wnt4* and *Rspo1,* through β-Catenin. Activation of the Wnt signalling pathway in the testis triggers ovarian development, but loss of Wnt signalling in the mouse ovary is not sufficient to upregulate the testis pathway.

Dax1

Differences in ovarian phenotypes between mouse and human, such as those resulting from loss of *Rspo1*, are not without precedent. Regional duplications in the X chromosome containing nuclear receptor subfamily 0, group B, member 1 (*DAX1/ NROB1*) cause dosage-sensitive male-to-female sex reversal in humans (Bardoni *et al.*, 1994; Phelan and McCabe, 2001). However, overexpression of additional copies of *Dax1* in XY mice does not cause male-to-female sex reversal, but only shows delayed testis development on a wild-type background (Swain *et al.*, 1998).

Dax1 is expressed from early stages in the genital ridge in mouse (Ikeda *et al.*, 1996) and is maintained in the ovary until 14.5 dpc, at which time expression decreases (Ikeda *et al.*, 2001). However, Dax1 is not ovary specific; it is also expressed in various cell types of the testis at different times in mouse (Ikeda *et al.*, 2001), and is maintained at similar levels in developing testis and ovaries in human embryos (Hanley *et al.*, 2000). The molecular mechanism of *Dax1* action remains unclear; however, it has been shown to play roles in both testicular and ovarian development (Bardoni *et al.*, 1994; Swain *et al.*, 1998; Ludbrook and Harley, 2004). Male-to-female sex reversal was achieved by crossing the *Dax1* overexpressing mouse line with male mice harbouring a 'weak' *Sry* allele, indicating that this gene can induce female development, and has a conserved function between mouse and human (Swain *et al.*, 1998). Cross-species analysis has identified evolution of multiple mechanisms of sex determination. Thus roles for central genes such as *Dax1* and *Rspo1* may have different weighting in sex determination even between mouse and human, which appear to utilize the same molecular pathways. However, comparative analysis remains a powerful approach for gene discovery, and was responsible for uncovering another gene critical to human ovarian development from analysis in the goat.

Foxl2

The characterization of sex reversal conditions in goat and humans led to the identification of *Foxl2* as a candidate ovary-determining gene (Crisponi *et al.*, 2001; Pailhoux *et al.*, 2001). Mutations in *Foxl2* have been shown to underlie blepharophimosis, ptosis and epicanthus inversus syndrome (BPES) in human (Crisponi *et al.*, 2001; De Baere *et al.*, 2001). BPES is characterized by premature ovarian failure and defects in eyelid formation (BPES [MIM 110100], www.ncbi.nlm.nih.gov/entrez/dispomim.cgi?id=110100). In the mouse, *Foxl2* is expressed in XX gonads by 12.5 dpc in mesenchymal pregranulosa cells, and maintained in granulosa cells of early follicles, but declines at later stages of folliculogenesis (Loffler, Zarkower and Koopman, 2003; Schmidt *et al.*, 2004). *Foxl2*$^{-/-}$ mice were generated and analysed with respect to the presumptive role in ovarian development (Uda *et al.*, 2004; Schmidt *et al.*, 2004). XX *Foxl2*-null mice displayed premature ovarian failure (in this case, follicle depletion) due to defects in granulosa cell development, which did not complete the squamous-to-cuboidal morphological transition normally associated with follicle development (Schmidt *et al.*, 2004; Uda *et al.*, 2004). Despite this, oocyte differentiation was only partially affected, with levels of oocyte regulators Growth differentiation factor-9 (*Gdf9*), *c-kit,* and Folliculogenesis specific basic helix-loop-helix (*Figla*) comparable to wild-type controls (Uda *et al.*, 2004). Insight into the mechanism of ovarian failure was gained by assessing regulation of follicle activation. During the first three days of postnatal life there was no significant difference between numbers of primordial follicles/oocytes in wild-type and mutant embryos. However at eight weeks of postnatal development, all primary follicles in the *Foxl2*^{LacZ} homozygous mice were activated due to ectopic upregulation of *Gdf9* in all oocytes, triggering unrestrained follicle activation. Activated follicles in mutant mice underwent apoptosis due to the absence of functional granulosa cells, resulting in follicle depletion (Schmidt *et al.*, 2004), and postnatal transdifferentiation of granulosa cells into Sertoli cells (Ottolenghi *et al.*, 2005).

Combined loss of Foxl2 and Wnt4

Analysis of *Foxl2*$^{-/-}$/*Wnt4*$^{-/-}$ mice has given some insight into the separate and cumulative effects of these two factors in regulating ovarian identity. As was predicted, the ablation of both genes amplified the partial sex reversal observed in single mutants, and resulted in formation of testicular and ovarian tissue in the *Foxl2*$^{-/-}$/*Wnt4*$^{-/-}$ XX gonad, which extended to the presence of both male and female germ cells. The timing of the differentiation of Sertoli cells and male germ cells in these XX gonads has not been identified, but because male germ cells differentiated, it was assumed that some transdifferentiation occurred before 16.5 dpc, when all female germ cells have entered meiosis (Ottolenghi *et al.*, 2007).

Wnt signalling and *Foxl2* cooperate to establish and maintain ovarian development. Both pathways exhibit anti-testis activity in addition to regulating distinct functions in ovarian development (Ottolenghi *et al.*, 2007; Kim *et al.*, 2006b; Chassot *et al.*, 2008; Maatouk *et al.*, 2008). Wnt signalling acts to promote germ cell survival through *Rspo1*, *Wnt4*, and *Fst*. These factors appear to be active earlier in ovarian development, from 11.5 to 13.5 dpc in the mouse. Sex reversal from individual loss of these factors appears to result from an absence of a feminizing influence exerted by female germ cells when they are depleted. However, a role for Wnt signalling in regulating ovarian sex determination is also likely. *Foxl2* is expressed from at least 12.5 dpc, but it appears to have a direct role in granulosa cell development. Germ cells survive in the absence of *Foxl2*; however the primordial follicles formed at birth are morphologically abnormal. In gonads lacking *Foxl2*, follicle activation, and the genes that control it, are dysregulated (Ottolenghi *et al.*, 2005). All follicles are activated and soon degenerate, again resulting in sex reversal due to loss of germ cells (Ottolenghi *et al.*, 2005).

Analysis of ovarian development in the absence of both *Foxl2* and *Wnt4* indicates a primary sex reversal. This model is yet to be fully analysed, but the presence of both spermatogonia and oocytes suggests that disruptions in both of these pathways is sufficient to cause primary sex reversal on a cellular level (Ottolenghi *et al.*, 2007). However in this model, the sex-reversal phenotype was not fully penetrant throughout the gonad, with some ovarian tissue and oocytes remaining. Future combination of *Rspo1* and *Foxl2* deletions in the one mouse will give further insight into the combined effect of these pathways (Figure 2.4).

2.5.2 Female germ cells are required for correct ovarian development

Both naturally occurring and genetically modified models of germ cell depletion have highlighted the necessity for female germ cells in the somatic cell differentiation of a functional ovary. Unlike the situation in the testis, loss of female germ cells results in varied effects that are dependent on the developmental stage at which germ cell depletion occurs.

Loss of mitotic oogonia

The loss of resident mitotic oogonia and early meiotic oogonia during early development does not affect either the formation of ovigerous cords or the correct differentiation of the interstitium (Merchant-Larios and Centeno, 1981; Merchant, 1975). However, within ovigerous cords, the supporting soma remains characteristic of pregranulosa cells, and this structure will never break down into follicles but rather endures for many weeks/months before regressing (Merchant, 1975; Mazaud *et al.*, 2002; Merchant-Larios and Centeno, 1981). These data suggest that germ cells are not required for the initial differentiation of the gonad or the formation of differentiated ovigerous cords, but rather for subsequent follicle histogenesis and epithelial differentiation into granulosa cells. To date two genes have been implicated in these roles in mitotic oogonia, *Figla* (Soyal, Amleh and Dean, 2000) and OG2 homeobox (*Og2x/Nobox*) (Rajkovic *et al.*, 2004). Disruption of these germ cell-specific genes results in failure, and a delay of ovigerous cord breakdown, respectively, and both are accompanied by extensive oocyte death after birth. Additionally, ablation of *Dazl*, in which germ cells are lost early during meiosis, also results in sterile ovigerous cords. Whether these genes function to regulate these processes, or are simply required to maintain germ cell survival at these times is not known.

Loss of meiotic oogonia

Germ cell loss at later developmental stages gives rise to a more severe somatic phenotype. In the instances where meiotic germ cells are depleted, the pregranulosa cells can be observed to transdifferentiate into Sertoli-like cells to give rise to seminiferous-like cords (SLCs) (Charpentier and Magre, 1990; Vigier *et al.*, 1988). Several models of this ovarian sex reversal have been identified, including mice overexpressing anti-Müllerian hormone (*Amh*) (Lyet *et al.*, 1995; Behringer *et al.*, 1990), deficiency of *Wnt4* (Yao *et al.*, 2004; Vainio *et al.*, 1999) and deficiency of *Rspo1* (Tomizuka *et al.*, 2008). In these situations the SLCs express AMH, display the specific junctional complexes of Sertoli cells and express the testis-specific gene *Sox9* (Taketo-Hosotani *et al.*, 1985; Taketo *et al.*, 1993; Vigier *et al.*, 1984). Rarely are these effects observed when germ cells are lost at the mitotic stage (Whitworth, Shaw and Renfree, 1996) but rather only in models in which germ cells are lost around the time of follicle formation (Merchant-Larios and Centeno, 1981; Mazaud *et al.*, 2002). This suggests that granulosa cells must be at a certain stage of maturation before they acquire the potential to transdifferentiate, and that this maturation is dependent on oocyte presence (Guigon and Magre, 2006). Interestingly, it has also been observed that oogonia possess the ability to inhibit differentiation of seminiferous cords in male testes when cocultured with reassociated testis somatic cells (Yao, DiNapoli and Capel, 2003). These data suggest that the oocytes are simultaneously antagonistic to the testis differentiation pathway, whilst also required for attainment of pregranulosa cell potential for transdifferentiation into Sertoli cells. This phenomenon highlights the intricate relationship between the germ cells and the somatic cells of the ovary that changes as development progresses.

Loss of preovulatory follicles

In addition to promoting granulosa cell differentiation and follicle histogenesis, oocytes are also required at later stages of follicular development. Here the oocytes have been shown to be responsible for signalling to thecal cells and preventing premature leutinization of granulosa cells. Surgical removal of oocytes from follicles (oocytectomy) has been shown to result in premature differentiation of granulosa cells into luteal cells (Nekola and Nalbandov, 1971), although no specific factor has been implicated in this transformation to date. In the same way, Vanderhyden *et al.* (1992) observed a decrease of granulosa cell proliferation following oocytectomy. Two growth factors have been implicated in this oocyte-dependent stimulation: GDNF9 (Elvin *et al.*, 1999; Dong *et al.*, 1996) and BMP15 (Galloway *et al.*, 2000), where ablation of these genes results in a similar phenotype to the oocytectomy.

Together these experiments have highlighted the necessity of premeiotic germ cells for the differentiation of pregranulosa cells into granulosa cells, and ovigerous cord breakdown after birth. Meiotic germ cells are needed for the attainment of transdifferentiation potential for granulosa cells to form SLCs. And finally, primary follicles direct both the timing of luteinization of granulosa cells and the recruitment of thecal cells for preovulatory development.

2.5.3 Vascularization of the ovary

One of the most striking examples of the divergence of ovarian and testis fates is the development of sex-specific vasculature. Vascular systems in both organs begin to develop by proliferation of vasculogenic precursors and endothelial cells in the bipotential gonad (Brennan, Karl and Capel, 2002). Following sex determination, endothelial cells from the mesonephros are induced to migrate into the testis, and form a major contribution to the forming vascular system (Brennan, Karl and Capel, 2002). In the ovary this migration is actively repressed by Wnt signalling, *Fst* being the most downstream effector identified (Yao *et al.*, 2004; Tomizuka *et al.*, 2008; Jeays-Ward *et al.*, 2003; Chassot *et al.*, 2008). Ovarian vasculature develops through rapid proliferation of pre-existing endothelial cells (Brennan, Karl and Capel, 2002). Early ovarian vasculature expresses both arterial and venous markers (Brennan, Karl and Capel, 2002). The vascular network permeates the developing ovary and is closely apposed to germ cell clusters, then ovigerous cords (Bullejos, Bowles and Koopman, 2002). Vasculature also plays a critical role in the cycle of follicle development and is integrated with the layer of steroid-producing theca cells surrounding each follicle (Fraser, 2006).

2.5.4 Theca cell development

Theca cells are ovarian endocrine cells that regulate follicle development, ovulation and pregnancy. They produce androgens that are used as a substrate for the synthesis of oestrogen by granulosa cells. Theca cells differentiate from fibroblastic cells in the ovarian mesenchyme in response to signals secreted by developing follicles. The first theca cells differentiate within a week after birth in mice, thus do not appear to play a role in embryonic development of the ovary. However, the induction and function of this cell type plays a critical role in regulating follicle development (Erickson *et al.*, 1985; Magoffin, 2005).

2.6 Testis development

In stark contrast to the gradual development of the ovary, the testis is promptly and thoroughly reorganized after sex determination (Figure 2.5). Within a 48 hour period in the mouse, or 4–5 weeks in humans (Ostrer *et al.*, 2007), the testis undergoes cell proliferation, differentiation, vascularization, and structural reorganization to form a functioning embryonic organ (Brennan and Capel, 2004). During this process the testis is divided into two structural compartments: the testis cords and the interstitium. The testis cords are tubule-like structures that grow to occupy the majority of the gonad, containing Sertoli cells which support the germ cell lineage and form the basis of the reproductive function of the testis. Cords are surrounded by peritubular myoid cells that cooperate with Sertoli cells to produce a basement membrane around the cords (Skinner, Tung and Fritz, 1985). The interstitium surrounding the testis cords is home to the steroidogenic cells of the testis. The production and export of hormones such as testosterone serves to masculinize the embryo and regulate development of secondary sex structures, including male genitalia, to establish the male phenotype. Disruptions in testis development lead to

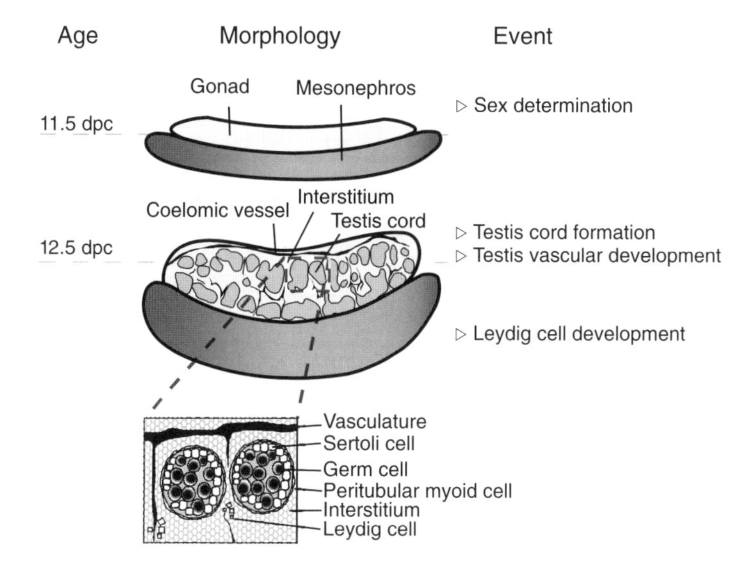

Figure 2.5 Testis development in the mouse. Following sex determination, the XY gonad is rapidly organized into a structured embryonic organ with functional testis cords and interstitial compartments serviced by a prominent vascular system. Cords are composed of a core of germ cells, surrounded by Sertoli cells, which are encased by peritubular myoid cells. External to the cords is the interstitium, which contains steroid-producing Leydig cells and a testis-specific vascular system

a spectrum of human conditions from hypospadias and malformed gonads to female development of a genetically male individual (sex reversal). Most of these disorders are accompanied by infertility. We will explore what is known of the regulation of testis development, focusing on the differentiation of key cell lineages and downstream cellular events involved in establishing testis structure.

2.6.1 Sertoli cell differentiation

Sertoli cell differentiation is the single most important event in testis development, as Sertoli cells trigger testis development. Sertoli cells differentiate from the somatic cell lineage in response to expression of *Sry*. *Sry* expression is tightly regulated in a spatial and temporal manner, whereby a wave of expression is initiated at 10.5 dpc in the centre of the mouse gonad, peaking at 11.5 dpc throughout the gonad, and ending at 12.5 dpc in the rostral, then the caudal pole (Bullejos and Koopman, 2001; Jeske *et al.*, 1995; Wilhelm *et al.*, 2005). Technically, expression of *Sry* defines pre-Sertoli cells, which then differentiate into Sertoli cells with the upregulation of *Sox9* and the formation of testis cords (Sekido *et al.*, 2004; Wilhelm *et al.*, 2005).

Sertoli cell recruitment

The upregulation of *Sox9* is intrinsically linked to the SRY protein, but the details of this interaction remained elusive until recently. *In vivo* characterization of the *Sox9* promoter identified a 1.4 kb testis-specific enhancer containing multiple binding sites for SRY and another transcription factor, steroidogenic factor 1 (SF1) (Sekido and Lovell-Badge, 2008). These sites act synergistically to regulate the expression of *Sox9*, resolving a long-standing question by demonstrating that SRY can directly activate expression of *Sox9* (Sekido and Lovell-Badge, 2008).

In addition to direct genetic activation, somatic cells can be recruited to Sertoli cell fate by paracrine signalling. This was alluded to in XX–XY gonadal chimera experiments where approximately 10% of Sertoli cells were found to be XX in origin (Palmer and Burgoyne, 1991a). This finding demonstrated that the requirement for a Y chromosome (and therefore *Sry* expression), was not absolute for differentiation of Sertoli cells (Palmer and Burgoyne, 1991a). Non-cell autonomous induction of Sertoli cell fate has since been shown to involve both prostaglandin and fibroblast growth factor signalling (Kim *et al.*, 2006b; Wilhelm *et al.*, 2005; Wilhelm *et al.*, 2007). A threshold number of Sertoli cells are required to complete testis development. Reduction in the numbers of *Sry*-expressing cells, or a delay in *Sry* expression, can lead to defective testis development and result in sex reversal (Albrecht *et al.*, 2003; Bullejos and Koopman, 2005; Palmer and Burgoyne, 1991a; Schmahl *et al.*, 2003).

2.6.2 Cellular events downstream of sertoli cell differentiation

Cell proliferation

Following *Sry* expression, the male gonad undergoes rapid growth to soon outsize a female gonad of comparable age (Hunt and Mittwoch, 1987; Mittwoch, Delhanty and Beck, 1969; Mittwoch and Mahadevaiah, 1980). This growth was characterized using 5′-bromo-2′-deoxyuridine incorporation to label dividing cells in the genital ridge (Schmahl *et al.*, 2000). The size increase correlated to an increase in somatic cell proliferation in XY and not XX gonads. Proliferation was observed to contribute to two subpopulations of cells. Proliferation up to 11.5 dpc was detected at the coelomic epithelium of the XY gonad in SF1-positive cells that subsequently contribute to Sertoli and interstitial cell types (Karl and Capel, 1998). From 11.75 dpc, proliferation continued at and near the coelomic epithelium in the XY gonad; however the proliferating cells were SF1 negative, giving rise to endothelial and interstitial cell types. Proliferation in XX gonads at comparable stages occurred at much lower levels. At 12.5 dpc, proliferating cells were observed throughout gonads of both sexes, though by this time XY gonads were twice the size of an XX gonad. Male-like proliferation is observed in XX gonads transgenic for *Sry*, indicating that this process is reliant on Sry expression (Schmahl *et al.*, 2000).

As may be expected, growth factors have a significant role in promoting proliferation in the early gonad. Insulin signalling and *Fgf9* are required for Sertoli cell proliferation, their absence resulting in XY sex reversal (Colvin *et al.*, 2001; Nef *et al.*, 2003). Platelet-derived growth factor receptor α (*Pdgfr*α) also contributes to proliferation, as evidenced by reduced levels in mice deficient for this receptor (Brennan *et al.*, 2003). Cell proliferation appears to influence testis development through expansion of the pre-Sertoli cell lineage. As threshold levels of pre-Sertoli and Sertoli cells are required to initiate testis formation, defects in cell proliferation can restrict the number of cells available to meet this threshold.

Cell migration

Early testis culture experiments demonstrated that normal testis development relied on cell migration from the mesonephros. Testis explants cultured without the mesonephros, or with the mesonephros separated from the testis by a permeable membrane, failed to form testis cords (Buehr, Gu and McLaren, 1993a; Merchant-Larios, Moreno-Mendoza and Buehr, 1993; Tilmann and Capel, 1999). Importantly, although cell migration from the mesonephros is required for testis cord formation, absence of migration does not hinder Sertoli or Leydig cell development (Merchant-Larios, Moreno-Mendoza and Buehr, 1993). Early lineage-tracing experiments reported the presence of multiple cell types in the migrating population (Buehr, Gu and McLaren, 1993a; Merchant-Larios, Moreno-Mendoza and Buehr, 1993; Martineau *et al.*, 1997; Nishino *et al.*, 2001). However, recent analysis has clarified that the migrating population is almost exclusively composed of endothelial cells (Cool *et al.*, 2008; Combes *et al.*, 2009).

Formation of testis vasculature

While initial vasculature of XX and XY genital ridges appears the same, by 12.5 dpc, sexual dimorphism in gonadal vasculature is clearly evident (Byskov, 1986; Nagamine and Carlisle, 1996; Pelliniemi, 1975). The most prominent feature of this system is the coelomic vessel (Brennan, Karl and Capel, 2002). The formation of XY-specific vasculature occurs via cell migration and is concurrent with testis cord development (Brennan, Karl and Capel, 2002; Coveney *et al.*, 2008).

Studies of knockout mice have revealed dependence of testis-specific vascular formation on *Fgf9* and *Pdgfrα*. Mice deficient in these genes exhibit disrupted vascular formation in the testis due to defects in cell proliferation, endothelial cell migration, and organization (Brennan *et al.*, 2003; Colvin *et al.*, 2001). Conversely, formation of the coelomic vessel is observed in XX gonads of mice deficient for *Rspo1*, *Wnt4*, and *Fst*, though this vessel does not branch into the gonad as in XY conditions (Chassot *et al.*, 2008; Jeays-Ward *et al.*, 2003; Tomizuka *et al.*, 2008; Yao *et al.*, 2004). Therefore, vascularization by cell migration is promoted in an XY environment and repressed by *Wnt4* expression in XX gonads.

Formation of testis cords

Cord formation is the final stage in development for the embryonic testis. Previous events of Sertoli cell differentiation, cell proliferation and vascular development converge in the formation of the testis cords. Cords divide the testis into two functional compartments to enable the dual functions of hormone production in the interstitium, and sperm maturation and export from the cords.

Cord formation is initiated as *Sry*-expressing pre-Sertoli cells differentiate into *Sox9*-expressing Sertoli cells. By 11 dpc, pre-Sertoli cells are defined along the length of the gonad (Wilhelm *et al.*, 2005) and appear to be evenly distributed amongst germ cells and other cell types in the interstitium (Combes *et al.*, 2009). As they mature, pre-Sertoli cells increase production of extracellular matrix proteins. The production of cytokeratins marks the beginning of pre-Sertoli cell differentiation to an epithelized phenotype of a mature Sertoli cell (Frojdman *et al.*, 1992). Sertoli cells become polarized through secretion of extracellular matrix proteins towards one side of the cell (Frojdman *et al.*, 1992). Testis cord formation is marked by an increase in extracellular matrix proteins surrounding the cords. These include: collagen type II (Paranko, 1987), IV and V, laminin, fibronectin, heparin sulfate proteoglycan (Pelliniemi *et al.*, 1984), cytokeratin and vimentin (Frojdman *et al.*, 1989; Paranko, 1987).

On a cellular level, cord formation occurs through Sertoli cell self-association and intercellular interactions. Sertoli cells have the capacity to self-associate (Hadley *et al.*, 1985), but this capacity alone does not lead to cord formation as testis cords do not assemble when deprived of input from migrating cells from the mesonephros (Buehr, Gu and McLaren, 1993a; Merchant-Larios, Moreno-Mendoza and Buehr, 1993; Tilmann and Capel, 1999). Migrating endothelial cells are required to partition the field of Sertoli and germ cells into testis cords as they traverse the gonad (Combes *et al.*, 2009). Other cell types involved in cord formation include the germ cells and peritubular myoid cells. Germ cells form the core of testis cords but are not required for cord formation (Buehr, Gu and McLaren, 1993a). On the other hand, disruptions in peritubular myoid differentiation are correlated with defects in cord development (Brennan *et al.*, 2003; Yao and Capel, 2002).

2.6.3 Leydig cell development

Foetal Leydig cells differentiate from the steroidogenic lineage in the interstitium from 12.5 dpc. Leydig cells produce and export testosterone, which controls development of the male reproductive tract and exerts long-range effects on embryonic organs and tissue such as the brain and developing muscles. Leydig cell development is induced from a pool of progenitor cells in a process regulated by Notch signalling (Tang *et al.*, 2008). To date, two signalling molecules produced by Sertoli cells have been implicated in regulating Leydig cell differentiation: desert hedgehog (DHH) and platelet-derived growth factor A (PDGFA). mRNA for the *Dhh* gene is expressed by Sertoli cells, with the receptor Patched1 expressed in the cytoplasm of peritubular myoid cells, in a speckled pattern in what is thought to be Leydig cells, and in endothelial cells (Bitgood, Shen and Mcmahon, 1996; Clark, Garland and Russell, 2000; Pierucci-Alves, Clark and Russell, 2001). Defects in Leydig cell development were reported in $Dhh^{-/-}$ mice (Clark, Garland and Russell, 2000), which have been mimicked by use of broad hedgehog signalling inhibitors in *ex vivo* organ culture (Yao and Capel, 2002). In these models, Leydig cell development was greatly reduced compared to controls. In gonads lacking *Pdgfrα*, cell proliferation, migration and Leydig cell development are negatively affected (Brennan *et al.*, 2003). Similar to the phenotype observed in *Dhh*$^{-/-}$ mice, *Pdgfrα*$^{-/-}$ gonads exhibit reduced numbers of Leydig cells compared to wild-type or heterozygous states. This reduction in Leydig cell number is thought to be independent of the cell proliferation phenotype, thus implicating PDGF signalling in the specification of Leydig cells (Brennan *et al.*, 2003).

2.7 Germ cells to oocytes and sperm

As discussed, the gonadal primordium is unique in that it has the potential to form two completely different organs. Genetic cues direct differentiation as an ovary or a testis that then produces molecular cues to direct the fate of the germ cells. The gametes differentiate into oocytes or spermatozoa as directed by their somatic environment, regardless of their genetic sex (XX or XY). That is, XX germ cells have been observed differentiating into pro-spermatogonia when in a testis, and XY germ cells will develop as oocytes when in an ovary (Ford *et al.*, 1975; Palmer and Burgoyne, 1991b). The germ cells possess this bipotentiality until 12.5 dpc, when their developmental fate becomes fixed (McLaren and Southee, 1997). In this way, germ cells are not only completely dependent on the somatic cell environment for growth and survival, but also for their differentiation into functional gametes.

The first apparent signs that a germ cell has begun differentiation down the female or male pathway are changes in its cell cycle status. In female gonads, germ cell entry into meiosis prophase I at 13.5 dpc signifies commitment to the female pathway (Adams and McLaren, 2002). In the testis, male germ cells begin entry into mitotic arrest, denoting commitment to the male pathway, which is coordinated with their enclosure in the testis cords by 12.5 dpc (Hilscher *et al.*, 1974). As discussed below, these decisions are the starting points for the two different cascades of differentiation that male and female germ cells will undertake. Due to this divergence, the majority of male and female germ cell development will be dealt with separately, however as both apoptosis and meiosis are inevitable for male and female germ cells alike, the timing and mechanics of these processes are discussed below.

2.7.1 Apoptosis: maintaining the integrity of the germline

Timing of apoptosis

Extensive germ cell apoptosis is an event that takes place in both the ovary and the testis. During the transition into meiosis and eventual follicle development, up to 70% of the germ cells are lost due to apoptosis (Pepling and Spradling, 2001; McClellan, Gosden and Taketo, 2003). This cell death occurs both prenatally and postnatally in the ovary. During gonadogenesis, a population of mitotic oogonia and early meiotic oocytes can be observed undergoing programmed cell death at 13.5 dpc and 15.5–17.5 dpc, respectively (Coucouvanis *et al.*, 1993). The second and larger round of apoptosis is then observed during follicle formation by the third week of life (Rajah, Glaser and Hirshfield, 1992; Mazaud *et al.*, 2002). Similarly, in the male, spermatogonia undergo apoptosis in the foetal gonad between 13.5 and 17.5 dpc in addition to a second wave of apoptosis that occurs around the time of birth in the postnatal testis. As a consequence of this second round of apoptosis, only 25% of the expected numbers of preleptotene spermatocytes are produced from the spermatogonial stem cell.

Reasons for apoptosis

Despite the consequences, little is understood about this programmed cell death, although several theories have been proposed. Initially this process was thought to result randomly from nutritional and environmental factors (Pepling, 2006). However, most widely accepted now is the notion that any defect in the nuclear or mitochondrial genomes will target a germ cell for elimination (Morita *et al.*, 1999; Baker, 1972), to ensure high genomic integrity of all remaining germ cells that will potentially give rise to offspring (Bristol-Gould *et al.*, 2006). Additionally, germ cell loss could contribute to ensuring the appropriate ratio of germ cells to supporting cells required for functional oocytes and spermatozoa in the postnatal ovary (Mazaud *et al.*, 2005; Ohno and Smith, 1964) and testis (Sharpe, Millar and Mckinnell, 1993). Furthermore, a role for dying oocytes in transferring mitochondria and endoplasmic reticulum to living oocytes via intercellular bridges has been proposed (Pepling and Spradling, 2001).

Bcl2 family and germ cell apoptosis

Although the reasons behind germ cell apoptosis remain largely speculative, the cellular mechanisms driving this programmed cell death are now being identified. To date, the B-cell lymphoma/leukaemia-2 (Bcl2) family has been implicated in this process (Rucker *et al.*, 2000). Disruption of antiapoptotic Bcl2-like 1 (*Bcl-x*) was shown to result in complete male sterility and reduced oocyte numbers by 15.5 dpc, which could be rescued with simultaneous deletion of the proapoptotic gene Bcl2-associated X protein (*Bax*), suggesting that gonocyte survival is controlled by a balance of these two Bcl2 family members (Rucker *et al.*, 2000). Deletion of *bax* alone leads to increased oocyte numbers by adulthood, although only a small number of these follicles can be fertilized due to other defects in reproductive requirements not identified (Perez *et al.*, 1999). Another Bcl2 family member, the antiapoptotic gene *Bcl2,* has been observed to result in fewer oocytes when deleted (Ratts *et al.*, 1995), and overexpression was seen to increase oocyte numbers by 8 dpp (days *post partum*), although oocyte populations returned to control numbers by adulthood (Flaws *et al.*, 2001).

Caspase 2 (CASP2), a protease involved in *bax* activation in other cell types (Cao, Bennett and May, 2008), has also been implicated in germ cell apoptosis. Deletion of this gene resulted in increased numbers of primordial follicles, a phenotype comparable to those of *bax* deletion (Bergeron *et al.*, 1998; Perez *et al.*, 1999) and *Bcl2* overexpression (Flaws *et al.*, 2001). Inhibitors to caspase action have also recently been shown to slow down oocyte death in culture (De Felici, Lobascio and Klinger, 2008).

p53 Family and Germ Cell Apoptosis

Tumour protein p63 (p63), a member of the p53 family of tumour suppressors, has also been implicated in regulation of apoptosis in XX and XY foetal germ cells. Six isoforms of p63 exist (TA-alpha/beta/gamma and DeltaN-alpha/beta/gamma) which signal through the tumour protein p53 (p53)-mediated apoptotic pathway. Expression of TAp63 has been detected in germ cells of the adult ovary and testis (Kurita *et al.*, 2005), where it is believed to monitor DNA integrity during the prolonged period of meiotic arrest (Suh *et al.*, 2006). This is consistent with experiments showing p63-mediated apoptosis induced by ionizing radiation (Livera *et al.*, 2008; Suh *et al.*, 2006). *p63*-null females are fertile, and primordial follicles develop normally (Kurita *et al.*, 2005), however male mutants display increased germ cell numbers in the postnatal testis (Petre-Lazar *et al.*, 2007).

These studies have succeeded in identifying several factors required for normal gonocyte survival mechanisms using genetic manipulation in animal models. It is important to note, however, that in none of these cases was germ cell depletion completely penetrant, but rather a small population of oocytes/spermatozoa always persists. Importantly, and not surprisingly, this survival suggests that the ovary and testis utilize numerous levels of cell cycle control such that the absence or overabundance of one particular factor will not affect the entire population of gametes.

2.7.2 Meiosis – the fate of a germ cell

The mechanics of meiosis

Put simply, meiosis is the process where one diploid germ cell undergoes one round of DNA duplication followed by two cell divisions to create four haploid cells. This process is separated into the two phases of cell division termed meiosis I and meiosis II, the first with, and the second without, DNA duplication. Each phase can be broken into further stages (prophase, metaphase, anaphase, and telophase) that share many similarities with mitotic cell division.

Meiotic prophase I is divided further into subphases: sister chromatids condense whilst joined tightly to one another (leptotene). Condensed sister chromatid pairs align with homologous chromatid pairs at the synaptonemal complex (zygotene). This allows for ‘chiasmata’, in which homologous chromosomes crossover to exchange analogous fragments of DNA and facilitate genetic diversity (pachytene). The synaptonemal complex degrades such that sister chromatids separate slightly from each other and allow some transcription of DNA (diplotene). Lastly, the nuclear envelope disperses and the mitotic spindle is formed (diakinesis). During metaphase I, the homologous chromosome pairs align on the metaphase plate attached to microtubules that join them to the centromeres at opposite sides of the cell. Anaphase I sees homologous sister chromatid pairs move toward opposite poles. At each pole the microtubules disintegrate and a new nuclear envelope encompasses the chromosomes (telophase I) before cytokinesis (division of the cytoplasm) occurs to yield two daughter cells. Meiosis II is the final round of cell division required to achieve haploid gametes, which follows similarly to mitotic division, with the significant absence of DNA duplication.

As mentioned, there are both timing and biological differences between male and female germ cell meiosis (see Figure 2.6). In males this is a continual process occurring as described above, with four haploid spermatozoa produced from each gonocyte. Conversely, in the ovary, one germ cell gives rise to only one oocyte following meiosis (Peters, 1969). This is achieved as the second nucleus of each meiotic division is lost as a polar body before cytokinesis occurs. Consequently, following fertilization, one oocyte is present with two polar bodies.

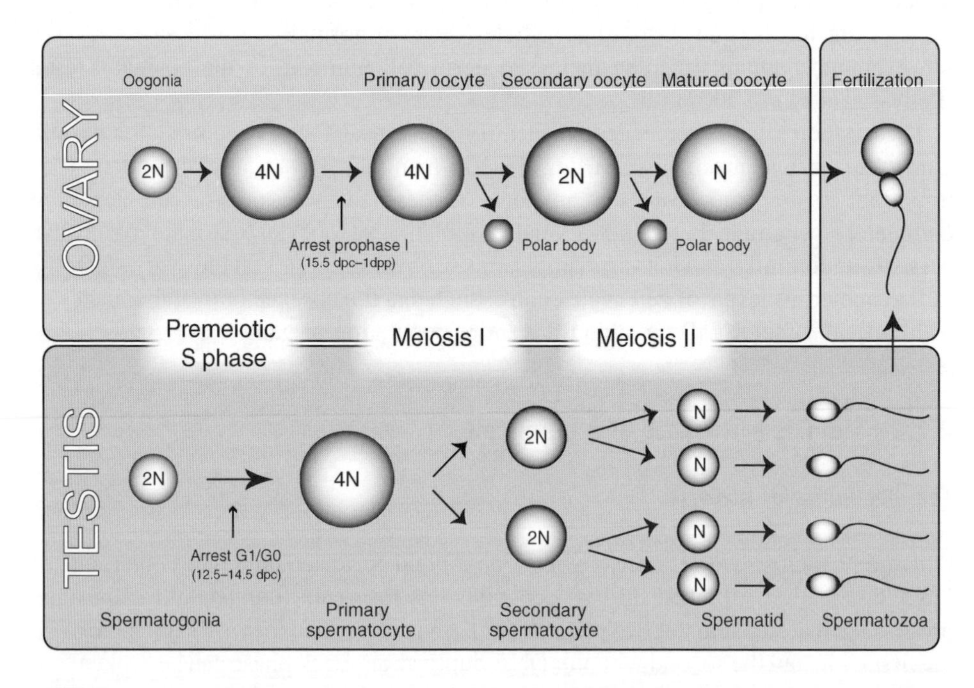

Figure 2.6 Schematic of meiosis. In the ovary, oogonia enter the first stages of meiosis I and begin to arrest in diplotene of prophase I by 17.5 dpc. Following follicle growth, meiosis I is completed with the exclusion of a polar body, and meiosis II is undertaken before arresting in metaphase II. The final stages of meiosis are not completed until fertilization, where the second polar body will be formed. In the testis, spermatogonia proliferate mitotically until 12.5 dpc, when they begin entry into G1/G0 arrest. This is maintained until several days after birth; mitosis is resumed at approximately 5–10 dpp, when they migrate to the basement membrane and become self-renewing spermatogonial stem cells. Following puberty, another round of mitosis yields primary spermatocytes that progress completely through meiosis I and II to produce four haploid spermatids. These cells must then undergo further maturational changes as they progress through to ejaculation and eventual fertilization. A full colour version of this figure appears in the colour plate section.

Markers of meiosis

Due to the unique nature of meiosis, there are several gene and protein markers useful for identifying various stages of this process. Stimulated by retinoic acid, gene 8 (STRA8) is required for premeiotic DNA replication and subsequent entry into meiosis prophase I (Oulad-Abdelghani *et al.*, 1996; Menke, Koubova and Page, 2003; Baltus *et al.*, 2006). SYCP3 is a structural protein involved in axial core formation during leptotene, the first phase of meiosis (Dobson *et al.*, 1994; Klink, Lee and Cooke, 1997; Heyting *et al.*, 1988). Dosage suppressor of MCK1 homologue (DMC1/DMC1H) is believed to participate in chromosomal recombination and synapsis during zygotene (Chuma and Nakatsuji, 2001; Sato *et al.*, 1995; Yoshida *et al.*, 1998; Pittman *et al.*, 1998). Whilst STRA8 can be used as an indicator of a cell preparing for entry into meiosis, both SYCP3 and DMC1 are believed to be true indicators of a cell undergoing meiosis. The robustness of these markers at the gene-expression level, however, is questionable (Novak *et al.*, 2006).

The timing of meiosis in the ovary

The time taken to complete meiosis for female oogonia extends over many months/years in mice, and decades in humans. Following gonadal sex differentiation, oogonia express *Stra8* in preparation for entry into meiosis at 12.5 dpc, (Menke, Koubova and Page, 2003). They are now referred to as oocytes (McLaren, 2000) and are clustered within ovigerous cords (Konishi *et al.*, 1986). Most oogonia will have entered meiosis by 15.5 dpc (Borum, 1961), although a small oogonia population has been observed undertaking this process postnatally (Hirshfield, 1992; Bristol-Gould *et al.*, 2006). Consistent with these reports of nonsynchronous meiosis entry, several studies have shown meiosis to proceed in an anterior to posterior wave (Bullejos and Koopman, 2004; Yao, DiNapoli and Capel, 2003; Menke, Koubova and Page, 2003) in response to signal(s) emanating in the same direction (to be discussed later). Oocytes progress through meiosis I until they enter the diplotene stage that occurs from 17.5 dpc to 5 dpp (Speed, 1982; Borum, 1961). Oocytes then remain arrested in diplotene while the ovigerous cords break down, so that flattened granulosa cells, also delineated by a basement membrane, enclose each meiotic oocyte as a primordial follicle. As the follicle begins to grow, the surrounding granulosa cells proliferate, and the follicle passes through primary, preantral and antral stages before resuming meiosis, to arrest again in metaphase II as a preovulatory follicle (Gougeon, 1996; Pedersen and Peters, 1968). Following ovulation, the theca and granulosa cells differentiate into luteal cells, and the final steps of meiosis are completed at fertilization.

The timing of meiosis in the testis

Spermatogenesis is undertaken in the postpubertal testis, with the noticeable absence of the long time delays during meiosis in the ovary. Following sex determination, male germ cells begin to arrest in G1/G0 of the mitotic cell cycle from 12.5 dpc onwards. This arrest is completed by approximately 14.5 dpc (Western *et al.*, 2008), and is maintained until 5–10 dpp, when male germ cells re-enter the cell cycle and undergo further rounds of mitosis. During this time, the germ cells, now referred to as pro-spermatogonia, migrate to the basement membrane of the seminiferous tubule and differentiate into spermatogonial stem cells (Setchell and Main, 1978). Following puberty, they again divide mitotically to produce two diploid cells, one of which remains as a stem cell to generate further spermatocytes, and the other a daughter cell that differentiates into spermatozoa. To achieve this, the daughter cell undergoes one further mitotic division to give rise to two primary spermatocytes. Meiosis I is then initiated to yield secondary spermatocytes, which undergo meiosis II to produce four early spermatids (see Figure 2.6). The final phase of spermatogenesis, termed spermiogenesis, involves extensive morphological modifications. These include the condensation of the nuclear material, and extensive cytoplasmic remodelling in which the round spermatid becomes elongated, comprising a tail/flagellum (for forward movement), midpiece (to house mitochondria) and a head (comprising the acrosome, nucleus, cytoskeletal structures and cytoplasm) (Eddy and O'Brien, 1994). As these elongated cells near the lumen, the supporting Sertoli cells strip them of excess cytoplasm to produce highly differentiated cells known as spermatozoa. This entire process from spermatogonial stem cell proliferation to spermatozoa takes approximately 35 days in the mouse (Cooke and Saunders, 2002) and 64 days in the human (Heller and Clermont, 1963).

2.7.3 Signals for germ cell sex

As two different cell states (meiosis or G1/G0 arrest) are viewed as the first indicators for germ cell sex differentiation, for almost 30 years researchers have tried to identify factor (s) (somatic or intrinsic to the germ cells) that are required to initiate these states in the female and male germ cells respectively. Two theories have dominated this field, one proposing that meiosis is cell-autonomously regulated, and the other proposing somatic cell induction of this event. Recently a factor originating from the mesonephros has been implicated in meiosis induction, and will be discussed with regard to the two long-standing theories.

Cell-autonomous theory of meiosis induction and G1/G0 arrest

In 1981, Anne McLaren proposed that both XX and XY germ cells are preprogrammed to enter meiosis at 13.5 dpc, as observed in the female gonad. This response was thought to require no external factor, and entry into mitotic arrest in the testis would be the result of a diffusible factor originating from the soma to inhibit this 'default' pathway (McLaren, 1981). This theory places significant emphasis on the gonadal environment in determining germ cell fate, and is supported by several studies.

In 1983, Zamboni and Upadhyay discovered that germ cells migrating erroneously to ectopic locations such as the adrenal gland proceeded to enter meiosis in parallel with female germ cells in the XX ovary, regardless of chromosomal sex. Here they are seen to form growing oocytes rather than arresting at the diplotene stage (Zamboni and Upadhyay, 1983; Upadhyay and Zamboni, 1982). Additionally, of those germ cells that migrated to the intervening mesonephric region in male gonads (as opposed to the gonad or adrenal gland), some entered meiosis and some mitotic arrest (McLaren, 1984). This indicated that a mitosis-arresting factor must be secreted from the testis to prevent nearby germ cells from entering meiosis (McLaren and Buehr, 1990). In addition to ectopic locations, both XX and XY germ cells will enter meiosis in various cultured environments, such as reaggregated lung cells (McLaren and Southee, 1997). McLaren and many others have interpreted these studies as indicating that meiosis is the default, cell-autonomous behaviour for male and female germ cells alike.

Several groups have demonstrated that a mitosis-arresting factor is required within the XY gonad for entry into mitotic arrest. Male germ cells in the testis appear to prepare for meiosis by entering the premeiotic stage, exhibiting an upregulation of meiotic genes *Sycp3* and *Dmc1* weakly (Chuma and Nakatsuji, 2001; Nakatsuji and Chuma, 2001). By 12.5 dpc, however, presumably in response to a male-specific gonadal factor, the meiotic genes are downregulated and the germ cells arrest in G1/G0. McLaren and Southee (1997) demonstrated that germ cells could be rescued from this signal if removed from the genital ridges at 11.5 dpc, and would subsequently develop as oocytes in cultured lung aggregates. In contrast, germ cells isolated from the XY gonad at 12.5 dpc are irreversibly committed to the male differentiation pathway (McLaren and Southee, 1997). These studies provide convincing evidence for the presence of a mitosis-arresting factor within the male gonad, functioning at the precise time to drive germ cells down a male differentiation pathway. Several candidates have been proposed, including transmembrane protein 184A (Tmem184a/Sdmg1) (Best *et al.*, 2008), prostaglandin D2 (Adams and McLaren, 2002), testis-specific β-defensin-like gene (*Tdl*) (Yamamoto and Matsui, 2002) and AMH (Vigier *et al.*, 1987); however these have not been convincingly shown to be involved in this process.

Somatic cell theory of meiosis induction and G1/G0 arrest

In 1985, Anne Grete Byskov proposed that germ cell entry into meiosis is induced by a diffusible factor secreted from the somatic cells that is present in both sexes, rather than being the default pathway for XX and XY germ cells. This putative factor has been termed a meiosis-activating substance. In the male gonad, this factor would be opposed by a meiosis-inhibiting factor to retain the cells in mitotic arrest until after birth. Alternatively, the meiosis-activating factor may be specific to the ovary during embryonic development, and only present in the testis after birth when entry into meiosis is triggered (Byskov, 1985).

Byskov and colleagues presented several lines of evidence for the presence of such a factor. Firstly, primitive ovaries were shown to be capable of inducing meiosis in undifferentiated male germ cells. Both whole ovary/testis cocultures and culture medium from ovaries were used, revealing that the induction of meiosis in male germ cells was dependent on both distance and dosage of ovarian cells (Byskov *et al.*, 1993; Byskov, 1978; Byskov and Saxen, 1976). In addition, a study on $Y^{DOM/POS}$ sex-reversed mice containing ovotestes revealed meiotic germ cells within the testis cords bordering the ovarian regions (Nagamine *et al.*, 1987). Similarly, in XX sex-reversed mice, meiotic germ cells were observed in the cranial portion of the gonads (McLaren, 1981). Byskov and colleagues have interpreted these observations as suggesting that the meiosis-activating substance is an ovarian-specific, diffusible factor.

More recent studies have also highlighted the involvement of the rete system and mesonephros in the timing of meiosis induction. Byskov and Hoyer (1994) initially identified the population of germ cells closest to the entry point of the rete ovarii into the gonad as the first to undergo meiosis. These oocytes are also the first to arrest in the diplotene stage and become enclosed in follicles (Byskov and Hoyer, 1994). This phenomenon has now been fully characterized as the 'rostrocaudal wave' of meiosis entry (Bullejos and Koopman, 2004; Yao, DiNapoli and Capel, 2003; Menke, Koubova and Page, 2003). Using both pluripotency and meiotic markers, meiosis entry was seen to begin at the cranial pole at 12.5 dpc and proceed through to the caudal pole of the ovary by 14.5–15.5 dpc. Expression of the pluripotency marker *Oct4* was seen to become downregulated concomitantly with the upregulation of meiosis markers *Stra8*; SYN/COR; H2A histone family, member X (H2AX); *Dmc1* and *Sycp3* in the rostrocaudal wave. This distinct pattern of meiotic entry further supports the existence of a diffusible substance originating from the somatic cells of the rete ovarii, inducing meiosis as it invades the length of the gonad (Menke, Koubova and Page, 2003; Bullejos and Koopman, 2004). Recently, retinoic acid (RA) has been identified as originating from both male and female mesonephroi in this way, but is degraded in the testis.

Retinoid signalling and meiosis

The RA-specific enzyme cytochrome P450, family 26, subfamily B, polypeptide 1 (CYP26B1) was first implicated in sex determination through several expression screens (Menke and Page, 2002; Bowles, Bullejos and Koopman, 2000). The *Cyp26b1* expression pattern was further characterized as displaying specific expression within the Sertoli cells from 12.5 dpc, with maximum levels reached by 13.5 dpc (Menke and Page, 2002; Bowles, Bullejos and Koopman, 2000). This finding suggested that RA might play a role in gonad development.

RA metabolism is a fundamental process involved in many aspects of embryo development (Reijntjes *et al.*, 2005). Synthesized by retinaldehyde dehydrogenases such as ALDH1A1, ALDH1A2 and ALDH1A3, and degraded by the enzymes CYP26B1, CYP26B2 and CYP26B3 (Niederreither *et al.*, 2002), RA levels are finely controlled in such environments (Reijntjes *et al.*, 2005; McCaffery *et al.*, 1999). RA signals through two families of nuclear receptors: retinoic acid receptors (RARs) and retinoid X receptors (RXRs) to modulate target gene transcription through RA response elements (RAREs).

The presence of RA in both male and female gonads was examined using a transgenic retinoic acid response element (RARE)-LacZ reporter line that revealed the mesonephroi of both sexes as rich sources of RA, concurrent with the expression of the gene encoding the major RA-synthesizing enzyme, *Aldh1a2* (Bowles, Bullejos and Koopman, 2000). It was subsequently proposed that RA is required in the ovary for germ cell entry into meiosis and is degraded in the testis by CYP26B1. To investigate this hypothesis, male genital ridges were cultured with exogenous RA, upon which upregulation of the meiosis-related genes *Sycp3*, *Dmc1* and *Stra8,* and downregulation of the pluripotency marker *Oct4* was evident (Bowles, Bullejos and Koopman, 2000; Koubova *et al.*, 2006). Conversely, on treatment of female genital ridges in culture with the RA receptor agonist AGN193109, the meiotic-specific genes became downregulated in accordance with the sustained expression of *Oct4* (Bowles, Bullejos and Koopman, 2000; Koubova *et al.*, 2006).

CYP26B1 was also investigated for its role in preventing meiosis in male germ cells. Culture experiments designed to antagonize CYP26B1 expression using both broad and specific cytochrome P450 inhibitors also showed upregulation of *Sycp3, Dmc1* and *Stra8* (Bowles, Bullejos and Koopman, 2000; Koubova *et al.*, 2006). Coculture with the CYP26B1 inhibitors and RAR panantagonists revealed no entry into meiosis, suggesting that CYP26B1 functions to degrade RA that would normally signal through RARs (Koubova *et al.*, 2006). These observations were supported by the analysis of the *Cyp26b1*$^{-/-}$ animal model, which revealed an increase of RA expression that was concurrent with entry of male germ cells into meiosis by 13.5 dpc, as detected by expression of SYCP3 and *Stra8* (McLean, Girvan and Munro, 2007; Bowles, Bullejos and Koopman, 2000). By 16.5 dpc, XY germ cells had progressed through to pachytene and this change was accompanied by a severe increase in apoptosis from 13.5 dpc onwards, such that neonates were essentially sterile, with no effect on somatic cell development of the testis or ovary. Additionally, meiosis was seen to progress earlier in the XX gonad, suggesting that CYP26B1 also functions in the ovary to prevent premature meiosis entry (Bowles, Bullejos and Koopman, 2000). The effects observed in the *Cyp26b1*$^{-/-}$ mutant were shown to be a result of RA overproduction rather than lack of CYP26B1-generated metabolites of RA, as a synthetic form of RA was also shown to induce meiosis (McLean, Girvan and Munro, 2007).

Collectively, these results provide evidence for the somatic cell induction of meiosis by RA in female germ cells. Now that this function has been recognized, the well-established theory of autonomous meiosis entry can be viewed in a different light. Indeed, the extensive production of RA throughout the developing embryo explains earlier observations of germ cells entering meiosis in extragonadal environments (McLaren and Southee, 1997; Zamboni and Upadhyay, 1983) and provides further support for RA-induced meiosis. Disaggregation experiments in which male germ cells are seen to enter meiosis (McLaren and Southee, 1997) were repeated in the presence of citral, an RA synthesis inhibitor, and lesser expression of meiotic markers was observed in the cultured germ cells (Bowles, Bullejos and Koopman, 2000).

From these studies, retinoic acid is now proposed as the meiosis-inducing substance postulated by Byskov and colleagues (Byskov, 1985) over two decades ago (Figure 2.7).

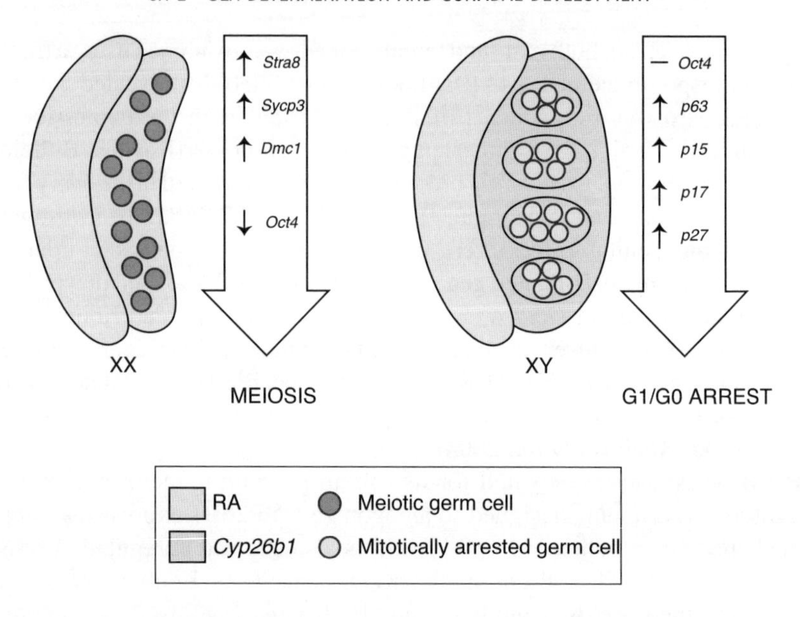

Figure 2.7 Retinoid signalling and meiosis induction. The mesonephroi of both male and female gonads are rich sources of RA. In the female, this diffuses into the gonad proper from the anterior pole to induce meiosis in the germ cells. This is concomitant with an upregulation of various meiotic markers and the downregulation of pluripotency marker *Oct4*. In the testis, Sertoli cells produce the retinoid-degrading enzyme gene *Cyp26b1* to degrade RA as it invades the gonad thereby preventing male germ cell entry into meiosis. Male germ cells enter G1/G0 arrest concomitant with the upregulation of several cell-cycle suppression genes. A full colour version of this figure appears in the colour plate section.

Furthermore, CYP26B1 also appears to fulfil the role of the meiosis-inhibiting substance, contrary to common assumption that this factor would be diffusible and capable of preventing meiosis whilst possibly also inducing mitotic arrest. As CYP26B1 is unlikely to fulfil this latter role, the search is ongoing to uncover what initiates mitotic arrest, the chief indicator of male sex differentiation during gonad development.

Some remaining questions and the way forward

Despite the apparent appropriateness for RA orchestrating germ cell meiosis entry, there are inevitably some inconsistencies and questions that still remain. Most notably, discrepancies between concentrations of RA used in the above-mentioned culture systems have been drawn to attention (Best *et al.*, 2008). In addition, the apparent lack of female germ cells that enter G1/G0 arrest and develop as pro-spermatogonia in the absence of RA poses another question as to the regulation of male germ cell progression. Lastly, the RA induction of meiosis in culture appears to contradict previous and long-standing reports that from 12.5 dpc onwards male germ cells are incapable of responding to meiosis-inducing substances (Adams and McLaren, 2002; McLaren and Southee, 1997).

In order to reconcile these findings, further animal models harbouring deletions within the RA system will be required. In particular, a mesonephros-specific deletion for the RA-synthesizing enzyme RALDH1A2 will shed light onto the ultimate *in vivo* role for RA in this system, within the constraints of biologically relevant concentrations. Additionally, loss of *Cyp26b1*/gain of RA across this developmental timeframe (10.5–13.5 dpc) in a temporally controlled, *in vivo* situation using genetic models is needed. This would eliminate possible artefacts of the *in vitro* systems from which many of these inferences have been taken.

The implication of RA modulating germ cell meiosis represents an enormous milestone in germ cell biology. These findings have only recently come to light and there is much work to be done before these interactions are fully elucidated. The next steps in this direction will see further characterization of the RA pathway that is active in the male and female gonads. In particular, the interacting/intermediate factors between the RARs and their downstream targets that eventually lead to meiosis modulation will be identified. Furthermore, additional factors responsible for controlling the degradation of RA (in addition to CYP26B1) should be sought. Given the complex nature of gene regulation utilized by the testis and ovary, it is unlikely that this event is reliant on one factor alone. The mitotic arrest-inducing factor has been proposed to directly regulate/inactivate the cell cycle machinery of the germ cells (Matsui, 1998) or, alternatively, to modulate a downregulation of germ cell growth receptors (Manova and Bachvarova, 1991). As with the case of the meiosis-inhibiting substance, it is likely that the answers to this question lie in a combination of factors with these properties.

2.7.4 Somatic cell regulation of mitosis in male germ cells

As meiosis is seen as the earliest discriminatory marker for progression down the female pathway, it follows that mitotic arrest should represent a definitive marker for the male pathway. Although years of research have failed to uncover factor(s) capable of inducing mitotic arrest in male germ cells, popular belief maintains that this substance is secreted from the Sertoli cells. At the time of mitotic arrest, the testis comprises Leydig cells undergoing differentiation, in addition to fully differentiated Sertoli cells that help form cord structures, implicating the Sertoli cells in the process of cell cycle arrest (reviewed by McLaren, 2003). To date, little is known about signalling from the gonadal somatic cells to germ cells, and even less about such factors that may initiate mitotic arrest. The next section discusses the requirement for male germ cells to enter mitotic arrest rather than meiosis as is observed in the ovary.

Markers for mitotic arrest

There are several markers useful for identifying the various stages of the mitotic cell cycle. Phosphohistone H3 is a nuclear histone that becomes phosphorylated during chromosome condensation during mitosis and is therefore utilized as a marker for cells in M phase (Hendzel *et al.*, 1997). Ki67 is a nuclear protein that marks cells that are actively cycling. It is detectable during the active phases of the cell cycle (G1, S, G2, and M), however it is absent from resting cells (G1/G0), making its absence a useful marker for mitotic arrest (Scholzen *et al.*, 2002). Bromodeoxyuridine (BrdU) incorporation is another marker used for identifying cells during the S phase of mitosis. This is achieved by the incorporation of this synthetic thymidine analogue into the DNA during synthesis (Hakala, 1959). Caspase 3 (CASP3) is a marker used to identify cells undergoing apoptosis. It is a member of the cysteine-aspartic acid protease family, and its activation leads to cleavage of critical cellular substrates that result in apoptosis (Cohen, 1995).

These markers were useful in identifying the changes in cell cycle state that male germ cells undergo from 12.5 dpc onwards as they begin entry into G1/G0 arrest. Whilst little is currently known about what induces this arrest, several other cell cycle modulators have recently been implicated in this process. Specifically, activation of the cyclin-dependent kinase 4 inhibitors p15(INK4b), p16(INK4a) and cyclin-dependent kinase inhibitor 1B (p27/Kip1), and dephosphorylation of the retinoblastoma protein occur during male germ cell arrest (Western *et al.*, 2008). Further information has been gleaned from knockout models that resulted in aberrant cell cycle states. For example, p63, a member of the p53 family that contains 6 isoforms, has been implicated in male germ cell apoptosis. *P63gamma* mRNA is upregulated as germ cells enter G1/G0 arrest, and the null mutation for all isoforms results in a reduced ability of germ cells to undergo apoptosis (Petre-Lazar *et al.*, 2007). PIN1 has been implicated in many aspects of the cell cycle including progression, DNA replication and checkpoint control by phosphorylation (Winkler *et al.*, 2000; Lu, Hanes and Hunter, 1996). Male germ cells in *Pin1*-null mutants displayed a prolonged cell cycle rate and an inability to enter G1/G0 arrest (Atchison, Capel and Means, 2003).

Signals for male germ cell arrest?

As mentioned previously, germ cell entry into mitotic arrest is not a cell-autonomous event, but is instead induced by signals originating from the surrounding somatic cells. Whilst we now know that RA is responsible for directing the female germ cells into meiosis, little is known about the factor(s) that direct male germ cell differentiation. As this search has progressed, a small number of signalling molecules between the soma and germ cells have been identified; however most have unknown functions.

FGFs are secreted by somatic cells, signalling through their receptors (FGFR-1 and -2) that have been identified on the surface of PGCs. The consequence of FGF signalling is modulation of gene expression, via the rat sarcoma viral oncogene (RAS) and mitogen-activated protein kinase (MAPK) signalling molecules (Resnick *et al.*, 1998). Expression of FGF-4 and -8 has been confirmed in somatic cells during the period of germ cell migration, suggesting some involvement in this process. Interestingly, however, this expression ceases during proliferation between 11.5 and 13.5 dpc, and is upregulated within the germ cells (Kawase, Hashimoto and

Pedersen, 2004). In addition to the role for FGF9 in somatic cell sex determination, discussed previously, FGF9 signalling has also been shown to promote male germ cell survival (DiNapoli, Batchvarov and Capel, 2006).

In addition to FGFs, the interleukin 6 (IL6) family, comprised of IL6, IL11, LIF, oncostatin M (OSM), cardiotrophin 1 (CT1) and cilliary neurotrophic factor (CNTF), has been shown to originate from somatic cells and signal to germ cells (Taga and Kishimoto, 1997). A certain level of redundancy has been detected within this family, and consequently mouse null mutants exhibit mild phenotypes. Each ligand has a specific receptor, and, upon binding, signals through the RAS/MAPK and JAK–STAT (janus kinase and signal transducer and activator of transcription) pathways. Of particular interest is the increased expression of the LIF receptor (LIFR) on male germ cells at 12.5 dpc, and LIF in the whole male gonad between 11.5 and 13.5 dpc. Also, the common receptor gp130 is similarly expressed in male PGCs at 10.5 and 12.5 dpc (Molyneaux *et al.*, 2003; Chuma and Nakatsuji, 2001; Hara *et al.*, 1998). The effects of this signalling pathway may represent a significant link between the IL6 family and mitotic arrest, given the close association between timing of expression and onset of arrest.

The Wnt pathway is another signalling mechanism functioning in the male primitive gonad. Briefly, it is strongly correlated with cell cycle control and differentiation in many cell types. The *Wnt4* mouse null mutant demonstrated the necessity for this signalling pathway in the development of the female, and suppression of the male, reproductive tracts, in addition to postmeiotic maintenance of oocytes (Jeays-Ward, Dandonneau and Swain, 2004; Vainio *et al.*, 1999). To date, the Wnt receptor Frizzled has not been identified on the germ cells of either sex, and expression appears specific to the female somatic cells.

Most recently, a Sertoli cell-specific gene encoding a novel transmembrane protein, Tmem184a/Sdmg1, was postulated to be the mitotic arrest-inducing factor (Best *et al.*, 2008); however, loss-of-function studies need to be carried out to establish this function.

Is there a biological significance for XY germ cell G1/G0 arrest?

Following germ cell proliferation at the time of gonadal colonization, both XY and XX germ cells have three different cell cycle paths available to them: (i) continue to divide mitotically; (ii) enter meiosis; or alternatively (iii) enter mitotic arrest. As discussed above, female germ cells immediately progress from mitotic divisions into meiosis once in the genital ridge. So why then would male germ cells remain in a quiescent state until puberty? Past and present literature is discussed below in order to understand why XY germ cells enter mitotic arrest.

Firstly, various studies have highlighted both the redundant role of germ cells in the developing testis and the negative effects of meiotic germ cells in this environment. In contrast to the female gonad, mitotically arrested germ cells are not required for either Sertoli cell differentiation or testis cord assembly. In their absence, predominantly normal testis morphology is achieved, with a slight delay in cord formation (Kurohmaru, Kanai and Hayashi, 1992; McLaren, 1988). Conversely, meiotic germ cells have been shown to have a detrimental effect on the testis environment. Transplantation studies performed by Yao, DiNapoli and Capel (2003) have shown that when meiotic germ cells are introduced into the testis, cord formation is disrupted to render the testis infertile (Yao, DiNapoli and Capel, 2003). In the natural testis environment, McLaren and others have observed a small number of germ cells that fail to become encapsulated within the testis cords and which enter meiosis and are subsequently apoptosed (Nakatsuji and Chuma, 2001; Coucouvanis *et al.*, 1993; McLaren, 1984). In the developing testis this apoptosis appears to provide a defence mechanism for the somatic cells, ensuring correct development.

It has also been demonstrated that mitotic arrest is not required for functional germ cells. Brinster and Avarbock (1994), using both genetic and chemotherapeutic means, rendered adult mouse testes sterile, prior to transplantation of 12.5 dpc germ cells containing LacZ into the recipient testis. Up to 80% of the progeny were sired by the transplanted cells, demonstrating that once committed to the male pathway at 12.5 dpc, germ cells are capable of responding to proliferation signals from an adult testis and subsequently producing live-born offspring (Brinster and Zimmermann, 1994). Therefore, the prolonged period in mitotic arrest (until approximately 5–10 dpp in mice (Bellve *et al.*, 1977)) is not required for germ cell development. Together, these studies have demonstrated that the germ cell entry into mitotic arrest is not a requirement for either a functional testis or spermatozoa, although meiotic germ cells are detrimental to both. Consequently, two cell cycle options remain for the male germ cells: continue to proliferate mitotically, or enter mitotic arrest.

Several lines of evidence suggest direction down the latter pathway is a consequence of the primitive testis being unable to both control and contain rapidly dividing cells. The period of time from mitotic arrest until the initiation of meiosis varies greatly between species and is most prolonged in higher primates and humans. During this time the somatic cells are undergoing a series of developmental changes that include a vast increase in testis volume through the growth of seminiferous cords and proliferation of Sertoli cells (Chemes, 2001). This is a critical step, as spermatozoa formation is dependent on the correct ratio of Sertoli cells to pro-spermatozoa/spermatids throughout all stages of testis development (Bendsen *et al.*, 2003; Orth, Gunsalus and Lamperti, 1988). This extensive remodelling must occur correctly to achieve the appropriate environment to support germ cell meiosis at the onset of puberty.

The fact that germ cells are extremely fast-dividing cells means that proliferation must be controlled precisely to avoid tumours. There are numerous cases in which ectopic male germ cells that fail to enter mitotic arrest or apoptosis proliferate to become paediatric germ cell tumours (Schneider *et al.*, 2001). If the germ cells were to continue dividing mitotically during testis development, tight regulation of the cell cycle would be required to prevent tumours, in addition to an extremely slow cell cycle rate, in order to maintain a manageable population and correct germ cell/somatic cell ratios. Mitotic arrest however, provides an efficient mechanism that minimizes the need for complicated cell cycle control, while still allowing the somatic cells to undergo their important developmental changes. These studies suggest that mitotic arrest, rather than a prerequisite for germ cell development, is a requirement for the somatic cells in order to achieve correct testis formation.

2.8 Summary

Early embryo development is the coordination of countless cellular interactions finely controlled by genetic cues to direct rapid proliferation and differentiation of specialized cell types. It is within the extraembryonic ectoderm of this rapidly changing environment that the primordial germ cells become specified in response to BMP signalling from the surrounding soma. The germ cell lineage is then reinforced by suppression of somatic cell markers along with activation of germ cell-specific genes. Over the course of several days the germ cells migrate through this environment to enter the embryo proper and finally colonize the newly formed genital ridges.

Coincident with the arrival of germ cells at the genital ridge, the sex determination programme is activated. In an XX gonad, Wnt signalling and *Foxl2* direct ovarian development through the granulosa cell lineage. In the XY gonad, testis development is initiated from the Y chromosome by expression of *Sry*. Once testis or ovarian fate is decided, the differentiation programme reinforces itself while antagonizing the other to ensure complete penetrance of either sexual phenotype. The outcome of the molecular struggle between the two opposing fates results in formation of organs with vast structural and molecular differences. Gonadal hormones then direct the development of sex-specific reproductive tracts and external genitalia to result in completion of the male and female phenotypes.

Germ cell fate is directed by changes in the somatic environment that occur during sex determination. In the ovary, the presence of RA initiates germ cell entry into meiosis. In the testis, RA-degrading enzymes protect germ cells from exposure to this signal and they undergo mitotic arrest. Once the germ cells have responded appropriately to these signals, the cascade of events comprising oogenesis or spermatogenesis can begin.

Sex determination, formation of gonads and the corresponding germ lines have profound implications for human development. These processes control the fundamental paradigm of gender as well as enabling the capacity to reproduce. The importance of understanding the functional genetics of sex determination and gonadal development becomes apparent when considering the high frequency of disorders of sexual development that are unexplained on the molecular level. Furthermore, understanding the molecular regulation of germ cell development may provide insight into the crisis of decreasing fertility around the world. While there is always more to understand, key genes and mechanisms regulating sex determination and germ cell development have been identified and provide a solid base for progression of future work.

Acknowledgements

We thank Terje Svingen, Josephine Bowles, and Dagmar Wilhelm for critically reading this chapter. This work was supported by research grants from the Australian Research Council (ARC) and National Health and Medical Research Council of Australia. Alex Combes and Cassy Spiller are supported by University of Queensland Postgraduate Research Scholarships, and Peter Koopman is a Federation Research Fellow of the ARC.

References

Adams, I.R. and McLaren, A. (2002) Sexually dimorphic development of mouse primordial germ cells: switching from oogenesis to spermatogenesis. *Development*, **129**, 1155–1164.

Albrecht, K.H. and Eicher, E.M. (2001) Evidence that Sry is expressed in pre-Sertoli cells and Sertoli and granulosa cells have a common precursor. *Dev. Biol.*, **240**, 92–107.

Albrecht, K.H., Young, M., Washburn, L.L., and Eicher, E.M. (2003) Sry expression level and protein isoform differences play a role in abnormal testis development in C57BL/6J mice carrying certain Sry alleles. *Genetics*, **164**, 277–288.

Allegrucci, C., Thurston, A., Lucas, E., and Young, L. (2005) Epigenetics and the germline. *Reproduction*, **129**, 137–149.

Ancelin, K., Lange, U.C., Hajkova, P. *et al.* (2006) Blimp1 associates with Prmt5 and directs histone arginine methylation in mouse germ cells. *Nat. Cell. Biol.*, **8**, 623–630.

Anderson, R., Copeland, T.K., Scholer, H. *et al.* (2000) The onset of germ cell migration in the mouse embryo. *Mech. Dev.*, **91**, 61–68.

Anderson, R., Fassler, R., Georges-Labouesse, E. *et al.* (1999) Mouse primordial germ cells lacking beta1 integrins enter the germline but fail to migrate normally to the gonads. *Development*, **126**, 1655–1664.

Arnold, S.J., Maretto, S., Islam, A. *et al.* (2006) Dose-dependent Smad1, Smad5 and Smad8 signaling in the early mouse embryo. *Dev. Biol.*, **296**, 104–118.

Atchison, F.W., Capel, B. and Means, A.R. (2003) Pin1 regulates the timing of mammalian primordial germ cell proliferation. *Development*, **130**, 3579–3586.

Baker, T.G. (1972) Gametogenesis. *Acta. Endocrinol. Suppl. (Copenh)*, **166**, 18–41.

Baltus, A.E., Menke, D.B., Hu, Y.C. *et al.* (2006) In germ cells of mouse embryonic ovaries, the decision to enter meiosis precedes premeiotic DNA replication. *Nat. Genet.*, **38**, 1430–1434.

Bardoni, B., Zanaria, E., Guioli, S. *et al.* (1994) A dosage sensitive locus at chromosome Xp21 is involved in male to female sex reversal. *Nat. Genet.*, **7**, 497–501.

Barrionuevo, F., Bagheri-Fam, S., Klattig, J. *et al.* (2006) Homozygous inactivation of Sox9 causes complete XY sex reversal in mice. *Biol. Reprod.*, **74**, 195–201.

Behringer, R.R., Cate, R.L., Froelick, G.J. *et al.* (1990) Abnormal sexual development in transgenic mice chronically expressing mullerian inhibiting substance. *Nature*, **345**, 167–170.

Bellve, A.R., Cavicchia, J.C., Millette, C.F. *et al.* (1977) Spermatogenic cells of the prepuberal mouse. Isolation and morphological characterization. *J. Cell. Biol.*, **74**, 68–85.

Bendel-Stenzel, M.R., Gomperts, M., Anderson, R. *et al.* (2000) The role of cadherins during primordial germ cell migration and early gonad formation in the mouse. *Mech. Dev.*, **91**, 143–152.

Bendsen, E., Byskov, A.G., Laursen, S.B. *et al.* (2003) Number of germ cells and somatic cells in human fetal testes during the first weeks after sex differentiation. *Hum. Reprod.*, **18**, 13–18.

Bergeron, L., Perez, G.I., Macdonald, G. *et al.* (1998) Defects in regulation of apoptosis in caspase-2-deficient mice. *Genes. Dev.*, **12**, 1304–1314.

Best, D., Sahlender, D.A., Walther, N. *et al.* (2008) Sdmg1 is a conserved transmembrane protein associated with germ cell sex determination and germline-soma interactions in mice. *Development*, **135**, 1415–1425.

Beverdam, A. and Koopman, P. (2006) Expression profiling of purified mouse gonadal somatic cells during the critical time window of sex determination reveals novel candidate genes for human sexual dysgenesis syndromes. *Hum. Mol. Genet.*, **15**, 417–431.

Bitgood, M.J., Shen, L. and Mcmahon, A.P. (1996) Sertoli cell signaling by Desert hedgehog regulates the male germline. *Curr. Biol.*, **6**, 298–304.

Boldajipour, B. and Raz, E. (2007) What is left behind—quality control in germ cell migration. *Sci. STKE*, **2007**, pe16.

Borum, K. (1961) Oogenesis in the mouse. A study of the meiotic prophase. *Exp. Cell. Res.*, **24**, 495–507.

Bourc'his, D. and Bestor, T.H. (2004) Meiotic catastrophe and retrotransposon reactivation in male germ cells lacking Dnmt3L. *Nature*, **431**, 96–99.

Bourc'his, D., Xu, G.L., Lin, C.S. *et al.* (2001) Dnmt3L and the establishment of maternal genomic imprints. *Science*, **294**, 2536–2539.

Bowles, J., Bullejos, M. and Koopman, P. (2000) A subtractive gene expression screen suggests a role for vanin-1 in testis development in mice. *Genesis*, **27**, 124–135.

Bowles, J., Knight, D., Smith, C. *et al.* (2006) Retinoid signaling determines germ cell fate in mice. *Science*, **312**, 596–600.

Brennan, J. and Capel, B. (2004) One tissue, two fates: molecular genetic events that underlie testis versus ovary development. *Nat. Rev. Genet.*, **5**, 509–521.

Brennan, J., Karl, J. and Capel, B. (2002) Divergent vascular mechanisms downstream of Sry establish the arterial system in the XY gonad. *Dev. Biol.*, **244**, 418–428.

Brennan, J., Tilmann, C. and Capel, B. (2003) Pdgfr-alpha mediates testis cord organization and fetal Leydig cell development in the XY gonad. *Genes. Dev.*, **17**, 800–810.

Brinster, R.L. and Avarbock, M.R. (1994) Germline transmission of donor haplotype following spermatogonial transplantation. *Proc. Natl. Acad. Sci. USA*, **91**, 11303–11307.

Brinster, R.L. and Zimmermann, J.W. (1994) Spermatogenesis following male germ-cell transplantation. *Proc. Natl. Acad. Sci. USA*, **91**, 11298–11302.

Bristol-Gould, S.K., Kreeger, P.K., Selkirk, C.G. *et al.* (2006) Fate of the initial follicle pool: empirical and mathematical evidence supporting its sufficiency for adult fertility. *Dev. Biol.*, **298**, 149–154.

Buehr, M., Gu, S. and McLaren, A. (1993a) Mesonephric contribution to testis differentiation in the fetal mouse. *Development*, **117**, 273–281.

Buehr, M., McLaren, A., Bartley, A. and Darling, S. (1993b) Proliferation and migration of primordial germ cells in We/We mouse embryos. *Dev. Dyn.*, **198**, 182–189.

Bullejos, M., Bowles, J. and Koopman, P. (2002) Extensive vascularization of developing mouse ovaries revealed by caveolin-1 expression. *Dev. Dyn.*, **225**, 95–99.

Bullejos, M. and Koopman, P. (2001) Spatially dynamic expression of Sry in mouse genital ridges. *Dev. Dyn.*, **221**, 201–205.

Bullejos, M. and Koopman, P. (2004) Germ cells enter meiosis in a rostro-caudal wave during development of the mouse ovary. *Mol. Reprod. Dev.*, **68**, 422–428.

Bullejos, M. and Koopman, P. (2005) Delayed Sry and Sox9 expression in developing mouse gonads underlies B6-Y(DOM) sex reversal. *Dev. Biol.*, **278**, 473–481.

Byskov, A.G. (1978) The anatomy and ultrastructure of the rete system in the fetal mouse ovary. *Biol. Reprod.*, **19**, 720–735.

Byskov, A.G. (1985) Control of meiosis. A summary. *Arch. Anat. Microsc. Morphol. Exp*, **74**, 17–18.

Byskov, A.G. (1986) Differentiation of mammalian embryonic gonad. *Physiol. Rev.*, **66**, 71–117.

Byskov, A.G., Fenger, M., Westergaard, L. and Andersen, C.Y. (1993) Forskolin and the meiosis inducing substance synergistically initiate meiosis in fetal male germ cells. *Mol. Reprod. Dev.*, **34**, 47–52.

Byskov, A.G. and Lintern-Moore, S. (1973) Follicle formation in the immature mouse ovary: the role of the rete ovarii. *J. Anat.*, **116**, 207–217.

Byskov, A.G. and Saxen, L. (1976) Induction of meiosis in fetal mouse testis in vitro. *Dev. Biol.*, **52**, 193–200.

Byskov, A.G. and Hoyer, P.E. (1994) *Embryology of Mammalian Gonads and Ducts: The Physiology of Reproduction*, Raven Press, New York.

Cao, X., Bennett, R.L. and May, W.S. (2008) c-Myc and caspase-2 are involved in activating Bax during cytotoxic drug-induced apoptosis. *J. Biol. Chem.*, **283**, 14490–14496.

Cederroth, C.R., Pitetti, J.L., Papaioannou, M.D. and Nef, S. (2007) Genetic programs that regulate testicular and ovarian development. *Mol. Cell. Endocrinol.*, **265–266**, 3–9.

Chaboissier, M.C., Kobayashi, A., Vidal, V.I. *et al.* (2004) Functional analysis of Sox8 and Sox9 during sex determination in the mouse. *Development*, **131**, 1891–1901.

Chambers, I., Silva, J., Colby, D. *et al.* (2007) Nanog safeguards pluripotency and mediates germline development. *Nature*, **450**, 1230–1234.

Chang, C., Holtzman, D.A., Chau, S. *et al.* (2001a) Twisted gastrulation can function as a BMP antagonist. *Nature*, **410**, 483–487.

Chang, H., Lau, A.L. and Matzuk, M.M. (2001b) Studying TGF-beta superfamily signaling by knockouts and knockins. *Mol. Cell. Endocrinol.*, **180**, 39–46.

Charpentier, G. and Magre, S. (1990) Masculinizing effect of testes on developing rat ovaries in organ culture. *Development*, **110**, 839–849.

Chassot, A.A., Ranc, F., Gregoire, E.P. *et al.* (2008) Activation of beta-catenin signaling by Rspo1 controls differentiation of the mammalian ovary. *Hum. Mol. Genet.*, **17**, 1264–1277.

Chemes, H.E. (2001) Infancy is not a quiescent period of testicular development. *Int. J. Androl.*, **24**, 2–7.

Chiquoine, A.D. (1954) The identification, origin, and migration of the primordial germ cells in the mouse embryo. *Anat. Rec.*, **118**, 135–146.

Chuma, S. and Nakatsuji, N. (2001) Autonomous transition into meiosis of mouse fetal germ cells in vitro and its inhibition by gp130-mediated signaling. *Dev. Biol.*, **229**, 468–479.

Clark, A.M., Garland, K.K. and Russell, L.D. (2000) Desert hedgehog (Dhh) gene is required in the mouse testis for formation of adult-type Leydig cells and normal development of peritubular cells and seminiferous tubules. *Biol. Reprod.*, **63**, 1825–1838.

Clark, J.M. and Eddy, E.M. (1975) Fine structural observations on the origin and associations of primordial germ cells of the mouse. *Dev. Biol.*, **47**, 136–155.

Cohen, J.J. (1995) Exponential growth in apoptosis. *Immunol. Today*, **16**, 346–348.

Collier, B., Gorgoni, B., Loveridge, C. *et al.* (2005) The DAZL family proteins are PABP-binding proteins that regulate translation in germ cells. *EMBO J.*, **24**, 2656–2666.

Colvin, J.S., Green, R.P., Schmahl, J. *et al.* (2001) Male-to-female sex reversal in mice lacking fibroblast growth factor 9. *Cell*, **104**, 875–889.

Combes, A.N., Wilhelm, D., Davidson, T. *et al.* (2009) Endothelial cell migration directs testis cord formation. *Dev. Biol.*, **326**, 112–120.

Cooke, H.J. and Saunders, P.T. (2002) Mouse models of male infertility. *Nat. Rev. Genet.*, **3**, 790–801.

Cooke, J.E., Heasman, J. and Wylie, C.C. (1996) The role of interleukin-4 in the regulation of mouse primordial germ cell numbers. *Dev. Biol.*, **174**, 14–21.

Cool, J., Carmona, F.D., Szucsik, J.C. and Capel, B. (2008) Peritubular myoid cells are not the migrating population required for testis cord formation in the XY gonad. *Sex. Dev.*, **2**, 128–133.

Cory, A.T., Boyer, A., Pilon, N. *et al.* (2007) Presumptive pre-Sertoli cells express genes involved in cell proliferation and cell signalling during a critical window in early testis differentiation. *Mol. Reprod. Dev.*, **74**, 1491–1504.

Coucouvanis, E.C., Sherwood, S.W., Carswell-Crumpton, C. *et al.* (1993) Evidence that the mechanism of prenatal germ cell death in the mouse is apoptosis. *Exp. Cell. Res.*, **209**, 238–247.

Coveney, D., Cool, J., Oliver, T. and Capel, B. (2008) Four-dimensional analysis of vascularization during primary development of an organ, the gonad. *Proc. Natl. Acad. Sci. USA*, **105**, 7212–7217.

Crisponi, L., Deiana, M., Loi, A. *et al.* (2001) The putative forkhead transcription factor FOXL2 is mutated in blepharophimosis/ptosis/epicanthus inversus syndrome. *Nat. Genet.*, **27**, 159–166.

Davis, T.L., Yang, G.J., Mccarrey, J.R. and Bartolomei, M.S. (2000) The H19 methylation imprint is erased and re-established differentially on the parental alleles during male germ cell development. *Hum. Mol. Genet.*, **9**, 2885–2894.

De Baere, E., Dixon, M.J., Small, K.W. *et al.* (2001) Spectrum of FOXL2 gene mutations in blepharophimosis-ptosis-epicanthus inversus (BPES) families demonstrates a genotype–phenotype correlation. *Hum. Mol. Genet.*, **10**, 1591–1600.

De Felici, M. (2000) Regulation of primordial germ cell development in the mouse. *Int. J. Dev. Biol.*, **44**, 575–580.

De Felici, M., Dolci, S. and Pesce, M. (1992) Cellular and molecular aspects of mouse primordial germ cell migration and proliferation in culture. *Int. J. Dev. Biol.*, **36**, 205–213.

De Felici, M., Lobascio, A.M. and Klinger, F.G. (2008) Cell death in fetal oocytes: many players for multiple pathways. *Autophagy*, **4**, 240–242.

Di Carlo, A. and De Felici, M. (2000) A role for E-cadherin in mouse primordial germ cell development. *Dev. Biol.*, **226**, 209–219.

DiNapoli, L., Batchvarov, J. and Capel, B. (2006) FGF9 promotes survival of germ cells in the fetal testis. *Development*, **133**, 1519–1527.

Dobson, M.J., Pearlman, R.E., Karaiskakis, A. *et al.* (1994) Synaptonemal complex proteins: occurrence, epitope mapping and chromosome disjunction. *J. Cell. Sci.*, **107** (Pt 10), 2749–2760.

Dong, J., Albertini, D.F., Nishimori, K. *et al.* (1996) Growth differentiation factor-9 is required during early ovarian folliculogenesis. *Nature*, **383**, 531–535.

Donovan, P.J., Stott, D., Cairns, L.A. *et al.* (1986) Migratory and postmigratory mouse primordial germ cells behave differently in culture. *Cell*, **44**, 831–838.

Eddy, E.M., O'Brien, D.A., Knobil, E. and Neill, J.D. (eds) (1994) *The Physiology of Reproduction*, Raven Press Ltd, New York.

Elvin, J.A., Clark, A.T., Wang, P. *et al.* (1999) Paracrine actions of growth differentiation factor-9 in the mammalian ovary. *Mol. Endocrinol.*, **13**, 1035–1048.

Enders, G.C. and May, J.J. 2nd (1994) Developmentally regulated expression of a mouse germ cell nuclear antigen examined from embryonic day 11 to adult in male and female mice. *Dev. Biol.*, **163**, 331–340.

Eppig, J.J. (2001) Oocyte control of ovarian follicular development and function in mammals. *Reproduction*, **122**, 829–838.

Erickson, G.F., Magoffin, D.A., Dyer, C.A. and Hofeditz, C. (1985) The ovarian androgen producing cells: a review of structure/function relationships. *Endocr. Rev.*, **6**, 371–399.

Flaws, J.A., Hirshfield, A.N., Hewitt, J.A. *et al.* (2001) Effect of bcl-2 on the primordial follicle endowment in the mouse ovary. *Biol. Reprod.*, **64**, 1153–1159.

Ford, C.E., Evans, E.P., Burtenshaw, M.D. *et al.* (1975) A functional 'sex-reversed' oocyte in the mouse. *Proc. R. Soc. Lond. B Biol. Sci.*, **190**, 187–197.

Foster, J.W., Dominguez-Steglich, M.A., Guioli, S. *et al.* (1994) Campomelic dysplasia and autosomal sex reversal caused by mutations in an SRY-related gene. *Nature*, **372**, 525–530.

Fox, N., Damjanov, I., Martinez-Hernandez, A. *et al.* (1981) Immunohistochemical localization of the early embryonic antigen (SSEA-1) in postimplantation mouse embryos and fetal and adult tissues. *Dev. Biol.*, **83**, 391–398.

Fraser, H.M. (2006) Regulation of the ovarian follicular vasculature. *Reprod. Biol. Endocrinol.*, **4**, 18.

Frojdman, K., Paranko, J., Kuopio, T. and Pelliniemi, L.J. (1989) Structural proteins in sexual differentiation of embryonic gonads. *Int. J. Dev. Biol.*, **33**, 99–103.

Frojdman, K., Paranko, J., Virtanen, I. and Pelliniemi, L.J. (1992) Intermediate filaments and epithelial differentiation of male rat embryonic gonad. *Differentiation*, **50**, 113–123.

Galloway, S.M., Mcnatty, K.P., Cambridge, L.M. *et al.* (2000) Mutations in an oocyte-derived growth factor gene (BMP15) cause increased ovulation rate and infertility in a dosage-sensitive manner. *Nat. Genet.*, **25**, 279–283.

Garcia-Castro, M.I., Anderson, R., Heasman, J. and Wylie, C. (1997) Interactions between germ cells and extracellular matrix glycoproteins during migration and gonad assembly in the mouse embryo. *J. Cell Biol.*, **138**, 471–480.

Ginsburg, M., Snow, M.H. and McLaren, A. (1990) Primordial germ cells in the mouse embryo during gastrulation. *Development*, **110**, 521–528.

Gobel, U., Schneider, D.T., Calaminus, G. *et al.* (2000) Germ-cell tumors in childhood and adolescence. GPOH MAKEI and the MAHO study groups. *Ann. Oncol.*, **11**, 263–271.

Godin, I., Wylie, C. and Heasman, J. (1990) Genital ridges exert long-range effects on mouse primordial germ cell numbers and direction of migration in culture. *Development*, **108**, 357–363.

Godin, I. and Wylie, C.C. (1991) TGF beta 1 inhibits proliferation and has a chemotropic effect on mouse primordial germ cells in culture. *Development*, **113**, 1451–1457.

Gomperts, M., Garcia-Castro, M., Wylie, C. and Heasman, J. (1994) Interactions between primordial germ cells play a role in their migration in mouse embryos. *Development*, **120**, 135–141.

Gougeon, A. (1996) Regulation of ovarian follicular development in primates: facts and hypotheses. *Endocr. Rev.*, **17**, 121–155.

Guigon, C.J., Coudouel, N., Mazaud-Guittot, S. *et al.* (2005) Follicular cells acquire sertoli cell characteristics after oocyte loss. *Endocrinology*, **146**, 2992–3004.

Guigon, C.J. and Magre, S. (2006) Contribution of germ cells to the differentiation and maturation of the ovary: insights from models of germ cell depletion. *Biol. Reprod.*, **74**, 450–458.

Hadley, M.A., Byers, S.W., Suarez-Quian, C.A. *et al.* (1985) Extracellular matrix regulates Sertoli cell differentiation, testicular cord formation, and germ cell development in vitro. *J. Cell Biol.*, **101**, 1511–1522.

Hajkova, P., Erhardt, S., Lane, N. *et al.* (2002) Epigenetic reprogramming in mouse primordial germ cells. *Mech. Dev.*, **117**, 15–23.

Hakala, M.T. (1959) Mode of action of 5-bromodeoxyuridine on mammalian cells in culture. *J. Biol. Chem.*, **234**, 3072–3076.

Hanley, N.A., Hagan, D.M., Clement-Jones, M. *et al.* (2000) SRY, SOX9, and DAX1 expression patterns during human sex determination and gonadal development. *Mech. Dev.*, **91**, 403–407.

Hara, T., Tamura, K., de Miguel, M.P. *et al.* (1998) Distinct roles of oncostatin M and leukemia inhibitory factor in the development of primordial germ cells and sertoli cells in mice. *Dev. Biol.*, **201**, 144–153.

Harley, V.R., Jackson, D.I., Hextall, P.J. *et al.* (1992) DNA binding activity of recombinant SRY from normal males and XY females. *Science*, **255**, 453–456.

Hata, K., Okano, M., Lei, H. and Li, E. (2002) Dnmt3L cooperates with the Dnmt3 family of de novo DNA methyltransferases to establish maternal imprints in mice. *Development*, **129**, 1983–1993.

Hayashi, K., de Sousa Lopes, S.M. and Surani, M.A. (2007) Germ cell specification in mice. *Science*, **316**, 394–396.

Hayashi, K., Kobayashi, T., Umino, T. *et al.* (2002) SMAD1 signaling is critical for initial commitment of germ cell lineage from mouse epiblast. *Mech. Dev.*, **118**, 99–109.

Heller, C.G. and Clermont, Y. (1963) Spermatogenesis in man: an estimate of its duration. *Science*, **140**, 184–186.

Hendzel, M.J., Wei, Y., Mancini, M.A. *et al.* (1997) Mitosis-specific phosphorylation of histone H3 initiates primarily within pericentromeric heterochromatin during G2 and spreads in an ordered fashion coincident with mitotic chromosome condensation. *Chromosoma*, **106**, 348–360.

Heyting, C., Dettmers, R.J., Dietrich, A.J. *et al.* (1988) Two major components of synaptonemal complexes are specific for meiotic prophase nuclei. *Chromosoma*, **96**, 325–332.

Hilscher, B., Hilscher, W., Bulthoff-Ohnolz, B. *et al.* (1974) Kinetics of gametogenesis. I. Comparative histological and autoradiographic studies of oocytes and transitional prospermatogonia during oogenesis and prespermatogenesis. *Cell Tissue Res.*, **154**, 443–470.

Hirshfield, A.N. (1991) Development of follicles in the mammalian ovary. *Int. Rev. Cytol.*, **124**, 43–101.

Hirshfield, A.N. (1992) Heterogeneity of cell populations that contribute to the formation of primordial follicles in rats. *Biol. Reprod.*, **47**, 466–472.

Hsieh, M., Johnson, M.A., Greenberg, N.M. and Richards, J.S. (2002) Regulated expression of Wnts and Frizzleds at specific stages of follicular development in the rodent ovary. *Endocrinology*, **143**, 898–908.

Hunt, S.E. and Mittwoch, U. (1987) Y-chromosomal and other factors in the development of testis size in mice. *Genet. Res.*, **50**, 205–211.

Ikeda, Y., Swain, A., Weber, T.J. *et al.* (1996) Steroidogenic factor 1 and Dax-1 colocalize in multiple cell lineages: potential links in endocrine development. *Mol. Endocrinol.*, **10**, 1261–1272.

Ikeda, Y., Takeda, Y., Shikayama, T. *et al.* (2001) Comparative localization of Dax-1 and Ad4BP/SF-1 during development of the hypothalamic-pituitary-gonadal axis suggests their closely related and distinct functions. *Dev. Dyn.*, **220**, 363–376.

Jager, R.J., Harley, V.R., Pfeiffer, R.A. *et al.* (1992) A familial mutation in the testis-determining gene SRY shared by both sexes. *Hum. Genet.*, **90**, 350–355.

Jamieson, R.V., Zhou, S.X., Wheatley, S.C. *et al.* (1998) Sertoli cell differentiation and Y-chromosome activity: a developmental study of X-linked transgene activity in sex-reversed X/XSxra mouse embryos. *Dev. Biol.*, **199**, 235–244.

Jaruzelska, J., Kotecki, M., Kusz, K. *et al.* (2003) Conservation of a Pumilio-Nanos complex from Drosophila germ plasm to human germ cells. *Dev. Genes. Evol.*, **213**, 120–126.

Jeays-Ward, K., Dandonneau, M. and Swain, A. (2004) Wnt4 is required for proper male as well as female sexual development. *Dev. Biol.*, **276**, 431–440.

Jeays-Ward, K., Hoyle, C., Brennan, J. *et al.* (2003) Endothelial and steroidogenic cell migration are regulated by WNT4 in the developing mammalian gonad. *Development*, **130**, 3663–3670.

Jeske, Y.W., Bowles, J., Greenfield, A. and Koopman, P. (1995) Expression of a linear Sry transcript in the mouse genital ridge. *Nat. Genet.*, **10**, 480–482.

Jho, E.H., Zhang, T., Domon, C. *et al.* (2002) Wnt/beta-catenin/Tcf signaling induces the transcription of Axin2, a negative regulator of the signaling pathway. *Mol. Cell. Biol.*, **22**, 1172–1183.

Jue, K., Bestor, T.H. and Trasler, J.M. (1995) Regulated synthesis and localization of DNA methyltransferase during spermatogenesis. *Biol. Reprod.*, **53**, 561–569.

Kanatsu-Shinohara, M. and Shinohara, T. (2006) The germ of pluripotency. *Nat. Biotechnol.*, **24**, 663–664.

Kaneda, M., Okano, M., Hata, K. *et al.* (2004) Essential role for de novo DNA methyltransferase Dnmt3a in paternal and maternal imprinting. *Nature*, **429**, 900–903.

Karl, J. and Capel, B. (1998) Sertoli cells of the mouse testis originate from the coelomic epithelium. *Dev. Biol.*, **203**, 323–333.

Kawase, E., Hashimoto, K. and Pedersen, R.A. (2004) Autocrine and paracrine mechanisms regulating primordial germ cell proliferation. *Mol. Reprod. Dev.*, **68**, 5–16.

Kent, J., Wheatley, S.C., Andrews, J.E. *et al.* (1996) A male-specific role for SOX9 in vertebrate sex determination. *Development*, **122**, 2813–2822.

Kezele, P.R., Nilsson, E.E. and Skinner, M.K. (2002) Insulin but not insulin-like growth factor-1 promotes the primordial to primary follicle transition. *Mol. Cell. Endocrinol.*, **192**, 37–43.

Kim, K.A., Zhao, J., Andarmani, S. *et al.* (2006a) R-Spondin proteins: a novel link to beta-catenin activation. *Cell Cycle*, **5**, 23–26.

Kim, Y., Kobayashi, A., Sekido, R. *et al.* (2006b) Fgf9 and Wnt4 act as antagonistic signals to regulate mammalian sex determination. *PLoS Biol.*, **4**, e187.

Kimura, T., Yomogida, K., Iwai, N. *et al.* (1999) Molecular cloning and genomic organization of mouse homologue of Drosophila germ cell-less and its expression in germ lineage cells. *Biochem. Biophys. Res. Commun.*, **262**, 223–230.

Klink, A., Lee, M. and Cooke, H.J. (1997) The mouse synaptosomal complex protein gene Sycp3 maps to band C of chromosome 10. *Mamm. Genome.*, **8**, 376–377.

Konishi, I., Fujii, S., Okamura, H. *et al.* (1986) Development of interstitial cells and ovigerous cords in the human fetal ovary: an ultrastructural study. *J. Anat.*, **148**, 121–135.

Koopman, P., Gubbay, J., Vivian, N. *et al.* (1991) Male development of chromosomally female mice transgenic for Sry. *Nature*, **351**, 117–121.

Koubova, J., Menke, D.B., Zhou, Q. *et al.* (2006) Retinoic acid regulates sex-specific timing of meiotic initiation in mice. *Proc. Natl. Acad. Sci. USA*, **103**, 2474–2479.

Kurita, T., Cunha, G.R., Robboy, S.J. *et al.* (2005) Differential expression of p63 isoforms in female reproductive organs. *Mech. Dev.*, **122**, 1043–1055.

Kurohmaru, M., Kanai, Y. and Hayashi, Y. (1992) A cytological and cytoskeletal comparison of Sertoli cells without germ cell and those with germ cells using the W/WV mutant mouse. *Tissue. Cell*, **24**, 895–903.

Lawson, K.A. and Hage, W.J. (1994) Clonal analysis of the origin of primordial germ cells in the mouse. *Ciba. Found. Symp.*, **182**, 68–84. Discussion 84–91.

Li, J.Y., Lees-Murdock, D.J., Xu, G.L. and Walsh, C.P. (2004) Timing of establishment of paternal methylation imprints in the mouse. *Genomics*, **84**, 952–960.

Little, M.H., Brennan, J., Georgas, K. *et al.* (2007) A high-resolution anatomical ontology of the developing murine genitourinary tract. *Gene. Expr. Patterns.*, **7**, 680–699.

Liu, C.F., Bingham, N., Parker, K. and Yao, H.H. (2009) Sex-specific roles of beta-catenin in mouse gonadal development. *Hum. Mol. Genet.*, **18**, 405–417.

Livera, G., Petre-Lazar, B., Guerquin, M.J. *et al.* (2008) p63 null mutation protects mouse oocytes from radio-induced apoptosis. *Reproduction*, **135**, 3–12.

Loffler, K.A., Zarkower, D. and Koopman, P. (2003) Etiology of ovarian failure in blepharophimosis ptosis epicanthus inversus syndrome: FOXL2 is a conserved, early-acting gene in vertebrate ovarian development. *Endocrinology*, **144**, 3237–3243.

Lu, K.P., Hanes, S.D. and Hunter, T. (1996) A human peptidyl-prolyl isomerase essential for regulation of mitosis. *Nature*, **380**, 544–547.

Ludbrook, L.M. and Harley, V.R. (2004) Sex determination: a 'window' of DAX1 activity. *Trends. Endocrinol. Metab.*, **15**, 116–121.

Lyet, L., Louis, F., Forest, M.G. *et al.* (1995) Ontogeny of reproductive abnormalities induced by deregulation of anti-mullerian hormone expression in transgenic mice. *Biol. Reprod.*, **52**, 444–454.

Maatouk, D.M., DiNapoli, L., Alvers, A. *et al.* (2008) Stabilization of beta-catenin in XY gonads causes male-to-female sex-reversal. *Hum. Mol. Genet.*, **17**, 2949–2955.

Maatouk, D.M., Kellam, L.D., Mann, M.R. *et al.* (2006) DNA methylation is a primary mechanism for silencing postmigratory primordial germ cell genes in both germ cell and somatic cell lineages. *Development*, **133**, 3411–3418.

MacGregor, G.R., Zambrowicz, B.P. and Soriano, P. (1995) Tissue non-specific alkaline phosphatase is expressed in both embryonic and extraembryonic lineages during mouse embryogenesis but is not required for migration of primordial germ cells. *Development*, **121**, 1487–1496.

Magoffin, D.A. (2005) Ovarian theca cell. *Int. J. Biochem. Cell. Biol*, **37**, 1344–1349.

Manova, K. and Bachvarova, R.F. (1991) Expression of c-kit encoded at the W locus of mice in developing embryonic germ cells and presumptive melanoblasts. *Dev. Biol.*, **146**, 312–324.

Martineau, J., Nordqvist, K., Tilmann, C. *et al.* (1997) Male-specific cell migration into the developing gonad. *Curr. Biol.*, **7**, 958–968.

Masuhara, M., Nagao, K., Nishikawa, M. *et al.* (2003) Enhanced degradation of MDM2 by a nuclear envelope component, mouse germ cell-less. *Biochem. Biophys. Res. Commun.*, **308**, 927–932.

Matsui, Y. (1992) [Regulation of murine primordial germ cell (PGC) growth by peptide growth factors]. *Tanpakushitsu Kakusan. Koso.*, **37**, 2935–2946.

Matsui, Y. (1998) Developmental fates of the mouse germ cell line. *Int. J. Dev. Biol.*, **42**, 1037–1042.

Matsui, Y., Toksoz, D., Nishikawa, S. *et al.* (1991) Effect of Steel factor and leukaemia inhibitory factor on murine primordial germ cells in culture. *Nature*, **353**, 750–752.

Mazaud, S., Guigon, C.J., Lozach, A. *et al.* (2002) Establishment of the reproductive function and transient fertility of female rats lacking primordial follicle stock after fetal gamma-irradiation. *Endocrinology*, **143**, 4775–4787.

Mazaud, S., Guyot, R., Guigon, C.J. *et al.* (2005) Basal membrane remodeling during follicle histogenesis in the rat ovary: contribution of proteinases of the MMP and PA families. *Dev. Biol.*, **277**, 403–416.

McCaffery, P., Wagner, E., O'Neil, J. *et al.* (1999) Dorsal and ventral rentinoic territories defined by retinoic acid synthesis, break-down and nuclear receptor expression. *Mech. Dev.*, **85**, 203–214.

McClellan, K.A., Gosden, R. and Taketo, T. (2003) Continuous loss of oocytes throughout meiotic prophase in the normal mouse ovary. *Dev. Biol.*, **258**, 334–348.

McElreavey, K., Vilain, E., Abbas, N. *et al.* (1993) A regulatory cascade hypothesis for mammalian sex determination: SRY represses a negative regulator of male development. *Proc. Natl. Acad. Sci. USA*, **90**, 3368–3372.

McGee, E.A. and Hsueh, A.J. (2000) Initial and cyclic recruitment of ovarian follicles. *Endocr. Rev.*, **21**, 200–214.

McLaren, A. (1981) The fate of germ cells in the testis of fetal Sex-reversed mice. *J. Reprod. Fertil.*, **61**, 461–467.

McLaren, A. (1983a) Primordial germ cells in mice. *Bibl. Anat.*, **24**, 59–66.

McLaren, A. (1983b) Studies on mouse germ cells inside and outside the gonad. *J. Exp. Zool.*, **228**, 167–171.

McLaren, A. (1984) Meiosis and differentiation of mouse germ cells. *Symp. Soc. Exp. Biol.*, **38**, 7–23.

McLaren, A. (1988) Somatic and germ-cell sex in mammals. *Philos. Trans. R. Soc. Lond. B Biol. Sci.*, **322**, 3–9.

McLaren, A. (2000) Germ and somatic cell lineages in the developing gonad. *Mol. Cell Endocrinol.*, **163**, 3–9.

McLaren, A. (2003) Primordial germ cells in the mouse. *Dev. Biol.*, **262**, 1–15.

McLaren, A. and Buehr, M. (1990) Development of mouse germ cells in cultures of fetal gonads. *Cell Differ. Dev.*, **31**, 185–195.

McLaren, A. and Monk, M. (1981) X-chromosome activity in the germ cells of sex-reversed mouse embryos. *J. Reprod Fertil.*, **63**, 533–537.

McLaren, A. and Southee, D. (1997) Entry of mouse embryonic germ cells into meiosis. *Dev. Biol.*, **187**, 107–113.

McLean, K.J., Girvan, H.M. and Munro, A.W. (2007) Cytochrome P450/redox partner fusion enzymes: biotechnological and toxicological prospects. *Expert. Opin. Drug. Metab. Toxicol*, **3**, 847–863.

Menke, D.B., Koubova, J. and Page, D.C. (2003) Sexual differentiation of germ cells in XX mouse gonads occurs in an anterior-to-posterior wave. *Dev. Biol.*, **262**, 303–312.

Menke, D.B. and Page, D.C. (2002) Sexually dimorphic gene expression in the developing mouse gonad. *Gene. Expr. Patterns*, **2**, 359–367.

Merchant, H. (1975) Rat gonadal and ovarian organogenesis with and without germ cells. An ultrastructural study. *Dev. Biol.*, **44**, 1–21.

Merchant-Larios, H. and Centeno, B. (1981) Morphogenesis of the ovary from the sterile W/Wv mouse. *Prog. Clin. Biol. Res.*, **59B**, 383–392.

Merchant-Larios, H., Moreno-Mendoza, N. and Buehr, M. (1993) The role of the mesonephros in cell differentiation and morphogenesis of the mouse fetal testis. *Int. J. Dev. Biol.*, **37**, 407–415.

Mintz, B. and Russell, E.S. (1957) Gene-induced embryological modifications of primordial germ cells in the mouse. *J. Exp. Zool.*, **134**, 207–237.

Mitchell, C.L. and Harley, V.R. (2002) Biochemical defects in eight SRY missense mutations causing XY gonadal dysgenesis. *Mol. Genet. Metab.*, **77**, 217–225.

Mittwoch, U., Delhanty, J.D. and Beck, F. (1969) Growth of differentiating testes and ovaries. *Nature*, **224**, 1323–1325.

Mittwoch, U. and Mahadevaiah, S. (1980) Additional growth – a link between mammalian testes, avian ovaries, gonadal asymmetry in hermaphrodites and the expression of H-Y antigen. *Growth*, **44**, 287–300.

Molyneaux, K.A., Stallock, J., Schaible, K. and Wylie, C. (2001) Time-lapse analysis of living mouse germ cell migration. *Dev. Biol.*, **240**, 488–498.

Molyneaux, K.A., Zinszner, H., Kunwar, P.S. *et al.* (2003) The chemokine SDF1/CXCL12 and its receptor CXCR4 regulate mouse germ cell migration and survival. *Development*, **130**, 4279–4286.

Monk, M. and McLaren, A. (1981) X-chromosome activity in foetal germ cells of the mouse. *J. Embryol. Exp. Morphol.*, **63**, 75–84.

Morais da Silva, S., Hacker, A., Harley, V. *et al.* (1996) Sox9 expression during gonadal development implies a conserved role for the gene in testis differentiation in mammals and birds. *Nat. Genet.*, **14**, 62–68.

Morita, Y., Perez, G.I., Maravei, D.V. *et al.* (1999) Targeted expression of Bcl-2 in mouse oocytes inhibits ovarian follicle atresia and prevents spontaneous and chemotherapy-induced oocyte apoptosis *in vitro*. *Mol. Endocrinol.*, **13**, 841–850.

Nagamine, C.M. and Carlisle, C. (1996) The dominant white spotting oncogene allele Kit(W-42J) exacerbates XY(DOM) sex reversal. *Development*, **122**, 3597–3605.

Nagamine, C.M., Michot, J.L., Roberts, C. *et al.* (1987) Linkage of the murine steroid sulfatase locus, Sts, to sex reversed, Sxr: a genetic and molecular analysis. *Nucleic. Acids. Res.*, **15**, 9227–9238.

Nakatsuji, N. and Chuma, S. (2001) Differentiation of mouse primordial germ cells into female or male germ cells. *Int. J. Dev. Biol.*, **45**, 541–548.

Natoli, T.A., Alberta, J.A., Bortvin, A. *et al.* (2004) Wt1 functions in the development of germ cells in addition to somatic cell lineages of the testis. *Dev. Biol.*, **268**, 429–440.

Nef, S., Schaad, O., Stallings, N.R. *et al.* (2005) Gene expression during sex determination reveals a robust female genetic program at the onset of ovarian development. *Dev. Biol.*, **287**, 361–377.

Nef, S., Verma-Kurvari, S., Merenmies, J. *et al.* (2003) Testis determination requires insulin receptor family function in mice. *Nature*, **426**, 291–295.

Nekola, M.V. and Nalbandov, A.V. (1971) Morphological changes of rat follicular cells as influenced by oocytes. *Biol. Reprod.*, **4**, 154–160.

Niederreither, K., Fraulob, V., Garnier, J.M. *et al.* (2002) Differential expression of retinoic acid-synthesizing (RALDH) enzymes during fetal development and organ differentiation in the mouse. *Mech. Dev.*, **110**, 165–171.

Nishino, K., Yamanouchi, K., Naito, K. and Tojo, H. (2001) Characterization of mesonephric cells that migrate into the XY gonad during testis differentiation. *Exp. Cell. Res.*, **267**, 225–232.

Noce, T., Okamoto-Ito, S. and Tsunekawa, N. (2001) Vasa homolog genes in mammalian germ cell development. *Cell. Struct. Funct.*, **26**, 131–136.

Novak, I., Lightfoot, D.A., Wang, H. *et al.* (2006) Mouse embryonic stem cells form follicle-like ovarian structures but do not progress through meiosis. *Stem. Cells*, **24**, 1931–1936.

Odor, D.L. and Blandau, R.J. (1969) Ultrastructural studies on fetal and early postnatal mouse ovaries. I. Histogenesis and organogenesis. *Am. J. Anat.*, **124**, 163–186.

Ohinata, Y., Payer, B., O'carroll, D. *et al.* (2005) Blimp1 is a critical determinant of the germ cell lineage in mice. *Nature*, **436**, 207–213.

Ohno, S. and Smith, J.B. (1964) Role of fetal follicular cells in meiosis of mammalian oocytes. *Cytogenetics*, **13**, 324–333.

Orth, J.M., Gunsalus, G.L. and Lamperti, A.A. (1988) Evidence from Sertoli cell-depleted rats indicates that spermatid number in adults depends on numbers of Sertoli cells produced during perinatal development. *Endocrinology*, **122**, 787–794.

Ostrer, H., Huang, H.Y., Masch, R.J. and Shapiro, E. (2007) A cellular study of human testis development. *Sex Dev.*, **1**, 286–292.

Ottolenghi, C., Omari, S., Garcia-Ortiz, J.E. *et al.* (2005) Foxl2 is required for commitment to ovary differentiation. *Hum. Mol. Genet.*, **14**, 2053–2062.

Ottolenghi, C., Pelosi, E., Tran, J. *et al.* (2007) Loss of Wnt4 and Foxl2 leads to female-to-male sex reversal extending to germ cells. *Hum. Mol. Genet.*, **16**, 2795–2804.

Oulad-Abdelghani, M., Bouillet, P., Decimo, D. *et al.* (1996) Characterization of a premeiotic germ cell-specific cytoplasmic protein encoded by Stra8, a novel retinoic acid-responsive gene. *J. Cell Biol.*, **135**, 469–477.

Pailhoux, E., Vigier, B., Chaffaux, S. *et al.* (2001) A 11.7-kb deletion triggers intersexuality and polledness in goats. *Nat. Genet.*, **29**, 453–458.

Palmer, S.J. and Burgoyne, P.S. (1991a) In situ analysis of fetal, prepuberal and adult XX—XY chimaeric mouse testes: Sertoli cells are predominantly, but not exclusively, XY. *Development*, **112**, 265–268.

Palmer, S.J. and Burgoyne, P.S. (1991b) XY follicle cells in the ovaries of XO/XY and XO/XY/XYY mosaic mice. *Development*, **111**, 1017–1019.

Parameswaran, M. and Tam, P.P. (1995) Regionalisation of cell fate and morphogenetic movement of the mesoderm during mouse gastrulation. *Dev. Genet.*, **17**, 16–28.

Paranko, J. (1987) Expression of type I and III collagen during morphogenesis of fetal rat testis and ovary. *Anat. Rec.*, **219**, 91–101.

Parma, P., Radi, O., Vidal, V. *et al.* (2006) R-spondin1 is essential in sex determination, skin differentiation and malignancy. *Nat. Genet.*, **38**, 1304–1309.

Pedersen, T. and Peters, H. (1968) Proposal for a classification of oocytes and follicles in the mouse ovary. *J. Reprod. Fertil.*, **17**, 555–557.

Pelliniemi, L.J. (1975) Ultrastructure of the early ovary and testis in pig embryos. *Am. J. Anat.*, **144**, 89–111.

Pelliniemi, L.J., Paranko, J., Grund, S.K. *et al.* (1984) Extracellular matrix in testicular differentiation. *Ann. N. Y. Acad. Sci.*, **438**, 405–416.

Pepling, M.E. (2006) From primordial germ cell to primordial follicle: mammalian female germ cell development. *Genesis*, **44**, 622–632.

Pepling, M.E. and Spradling, A.C. (1998) Female mouse germ cells form synchronously dividing cysts. *Development*, **125**, 3323–3328.

Pepling, M.E. and Spradling, A.C. (2001) Mouse ovarian germ cell cysts undergo programmed breakdown to form primordial follicles. *Dev. Biol.*, **234**, 339–351.

Perez, G.I., Robles, R., Knudson, C.M. *et al.* (1999) Prolongation of ovarian lifespan into advanced chronological age by Bax-deficiency. *Nat. Genet.*, **21**, 200–203.

Peters, H. (1969) The development of the mouse ovary from birth to maturity. *Acta. Endocrinol. (Copenh)*, **62**, 98–116.

Petre-Lazar, B., Livera, G., Moreno, S.G. *et al.* (2007) The role of p63 in germ cell apoptosis in the developing testis. *J. Cell. Physiol.*, **210**, 87–98.

Phelan, J.K. and McCabe, E.R. (2001) Mutations in NR0B1 (DAX1) and NR5A1 (SF1) responsible for adrenal hypoplasia congenita. *Hum. Mutat.*, **18**, 472–487.

Pierucci-Alves, F., Clark, A.M. and Russell, L.D. (2001) A developmental study of the Desert hedgehog-null mouse testis. *Biol. Reprod.*, **65**, 1392–1402.

Pittman, D.L., Cobb, J., Schimenti, K.J. *et al.* (1998) Meiotic prophase arrest with failure of chromosome synapsis in mice deficient for Dmc1, a germline-specific RecA homolog. *Mol. Cell*, **1**, 697–705.

Pontiggia, A., Rimini, R., Harley, V.R. *et al.* (1994) Sex-reversing mutations affect the architecture of SRY-DNA complexes. *EMBO J.*, **13**, 6115–6124.

Rajah, R., Glaser, E.M. and Hirshfield, A.N. (1992) The changing architecture of the neonatal rat ovary during histogenesis. *Dev. Dyn.*, **194**, 177–192.

Rajkovic, A., Pangas, S.A., Ballow, D. *et al.* (2004) NOBOX deficiency disrupts early folliculogenesis and oocyte-specific gene expression. *Science*, **305**, 1157–1159.

Ratts, V.S., Flaws, J.A., Kolp, R. *et al.* (1995) Ablation of bcl-2 gene expression decreases the numbers of oocytes and primordial follicles established in the post-natal female mouse gonad. *Endocrinology*, **136**, 3665–3668.

Reijntjes, S., Blentic, A., Gale, E. and Maden, M. (2005) The control of morphogen signalling: regulation of the synthesis and catabolism of retinoic acid in the developing embryo. *Dev. Biol.*, **285**, 224–237.

Resnick, J.L., Bixler, L.S., Cheng, L. and Donovan, P.J. (1992) Long-term proliferation of mouse primordial germ cells in culture. *Nature*, **359**, 550–551.

Resnick, J.L., Ortiz, M., Keller, J.R. and Donovan, P.J. (1998) Role of fibroblast growth factors and their receptors in mouse primordial germ cell growth. *Biol. Reprod.*, **59**, 1224–1229.

Rosselot, C., Kierszenbaum, A.L., Rivkin, E. and Tres, L.L. (2003) Chronological gene expression of ADAMs during testicular development: prespermatogonia (gonocytes) express fertilin beta (ADAM2). *Dev. Dyn.*, **227**, 458–467.

Rucker, E.B. 3rd, Dierisseau, P., Wagner, K.U. *et al.* (2000) Bcl-x and Bax regulate mouse primordial germ cell survival and apoptosis during embryogenesis. *Mol. Endocrinol.*, **14**, 1038–1052.

Saitou, M., Barton, S.C. and Surani, M.A. (2002) A molecular programme for the specification of germ cell fate in mice. *Nature*, **418**, 293–300.

Sasaki, H. and Matsui, Y. (2008) Epigenetic events in mammalian germ-cell development: reprogramming and beyond. *Nat. Rev. Genet.*, **9**, 129–140.

Sato, M., Kimura, T., Kurokawa, K. *et al.* (2002) Identification of PGC7, a new gene expressed specifically in preimplantation embryos and germ cells. *Mech. Dev.*, **113**, 91–94.

Sato, S., Kobayashi, T., Hotta, Y. and Tabata, S. (1995) Characterization of a mouse recA-like gene specifically expressed in testis. *DNA. Res.*, **2**, 147–150.

Saunders, P.T., Turner, J.M., Ruggiu, M. *et al.* (2003) Absence of mDazl produces a final block on germ cell development at meiosis. *Reproduction*, **126**, 589–597.

Schmahl, J., Eicher, E.M., Washburn, L.L. and Capel, B. (2000) Sry induces cell proliferation in the mouse gonad. *Development*, **127**, 65–73.

Schmahl, J., Yao, H.H., Pierucci-Alves, F. and Capel, B. (2003) Colocalization of WT1 and cell proliferation reveals conserved mechanisms in temperature-dependent sex determination. *Genesis*, **35**, 193–201.

Schmidt, D., Ovitt, C.E., Anlag, K. *et al.* (2004) The murine winged-helix transcription factor Foxl2 is required for granulosa cell differentiation and ovary maintenance. *Development*, **131**, 933–942.

Schmitt-Ney, M., Thiele, H., Kaltwasser, P. *et al.* (1995) Two novel SRY missense mutations reducing DNA binding identified in XY females and their mosaic fathers. *Am. J. Hum. Genet.*, **56**, 862–869.

Schneider, D.T., Schuster, A.E., Fritsch, M.K. *et al.* (2001) Genetic analysis of childhood germ cell tumors with comparative genomic hybridization. *Klin. Padiatr.*, **213**, 204–211.

Scholer, H.R., Ruppert, S., Suzuki, N. *et al.* (1990) New type of POU domain in germ line-specific protein Oct-4. *Nature*, **344**, 435–439.

Scholzen, T., Endl, E., Wohlenberg, C. *et al.* (2002) The Ki-67 protein interacts with members of the heterochromatin protein 1 (HP1) family: a potential role in the regulation of higher-order chromatin structure. *J. Pathol.*, **196**, 135–144.

Sekido, R., Bar, I., Narvaez, V. *et al.* (2004) SOX9 is up-regulated by the transient expression of SRY specifically in Sertoli cell precursors. *Dev. Biol.*, **274**, 271–279.

Sekido, R. and Lovell-Badge, R. (2008) Sex determination involves synergistic action of SRY and SF1 on a specific Sox9 enhancer. *Nature*, **453**, 930–934.

Setchell, B.P. and Main, S.J. (1978) Drugs and the blood-testis barrier. *Environ. Health. Perspect.*, **24**, 61–64.

Seydoux, G. and Braun, R.E. (2006) Pathway to totipotency: lessons from germ cells. *Cell*, **127**, 891–904.

Sharpe, R.M., Millar, M. and Mckinnell, C. (1993) Relative roles of testosterone and the germ cell complement in determining stage-dependent changes in protein secretion by isolated rat seminiferous tubules. *Int. J. Androl.*, **16**, 71–81.

Sinclair, A.H., Berta, P., Palmer, M.S. *et al.* (1990) A gene from the human sex-determining region encodes a protein with homology to a conserved DNA-binding motif. *Nature*, **346**, 240–244.

Skinner, M.K., Tung, P.S. and Fritz, I.B. (1985) Cooperativity between Sertoli cells and testicular peritubular cells in the production and deposition of extracellular matrix components. *J. Cell. Biol.*, **100**, 1941–1947.

Soyal, S.M., Amleh, A. and Dean, J. (2000) FIGalpha, a germ cell-specific transcription factor required for ovarian follicle formation. *Development*, **127**, 4645–4654.

Speed, R.M. (1982) Meiosis in the foetal mouse ovary. I. An analysis at the light microscope level using surface-spreading. *Chromosoma*, **85**, 427–437.

Stebler, J., Spieler, D., Slanchev, K. *et al.* (2004) Primordial germ cell migration in the chick and mouse embryo: the role of the chemokine SDF-1/CXCL12. *Dev. Biol.*, **272**, 351–361.

Suh, E.K., Yang, A., Kettenbach, A. *et al.* (2006) p63 protects the female germ line during meiotic arrest. *Nature*, **444**, 624–628.

Swain, A., Narvaez, V., Burgoyne, P. *et al.* (1998) Dax1 antagonizes Sry action in mammalian sex determination. *Nature*, **391**, 761–767.

Szabo, P.E. and Mann, J.R. (1995) Biallelic expression of imprinted genes in the mouse germ line: implications for erasure, establishment, and mechanisms of genomic imprinting. *Genes. Dev.*, **9**, 1857–1868.

Taga, T. and Kishimoto, T. (1997) Gp130 and the interleukin-6 family of cytokines. *Annu. Rev. Immunol.*, **15**, 797–819.

Taketo, T., Saeed, J., Manganaro, T. *et al.* (1993) Mullerian inhibiting substance production associated with loss of oocytes and testicular differentiation in the transplanted mouse XX gonadal primordium. *Biol. Reprod.*, **49**, 13–23.

Taketo-Hosotani, T., Merchant-Larios, H., Thau, R.B. and Koide, S.S. (1985) Testicular cell differentiation in fetal mouse ovaries following transplantation into adult male mice. *J. Exp. Zool.*, **236**, 229–237.

Tam, P.P., Kanai-Azuma, M. and Kanai, Y. (2003) Early endoderm development in vertebrates: lineage differentiation and morphogenetic function. *Curr. Opin. Genet. Dev.*, **13**, 393–400.

Tam, P.P. and Loebel, D.A. (2007) Gene function in mouse embryogenesis: get set for gastrulation. *Nat. Rev. Genet.*, **8**, 368–381.

Tam, P.P. and Snow, M.H. (1981) Proliferation and migration of primordial germ cells during compensatory growth in mouse embryos. *J. Embryol. Exp. Morphol.*, **64**, 133–147.

Tam, P.P. and Zhou, S.X. (1996) The allocation of epiblast cells to ectodermal and germ-line lineages is influenced by the position of the cells in the gastrulating mouse embryo. *Dev. Biol.*, **178**, 124–132.

Tam, P.P., Zhou, S.X. and Tan, S.S. (1994) X-chromosome activity of the mouse primordial germ cells revealed by the expression of an X-linked lacZ transgene. *Development*, **120**, 2925–2932.

Tanaka, S.S. and Matsui, Y. (2002) Developmentally regulated expression of mil-1 and mil-2, mouse interferon-induced transmembrane protein like genes, during formation and differentiation of primordial germ cells. *Gene. Expr. Patterns.*, **2**, 297–303.

Tang, H., Brennan, J., Karl, J. *et al.* (2008) Notch signaling maintains Leydig progenitor cells in the mouse testis. *Development*, **135**, 3745–3753.

Tilmann, C. and Capel, B. (1999) Mesonephric cell migration induces testis cord formation and Sertoli cell differentiation in the mammalian gonad. *Development*, **126**, 2883–2890.

Tomaselli, S., Megiorni, F., de Bernardo, C. *et al.* (2008) Syndromic true hermaphroditism due to an R-spondin1 (RSPO1) homozygous mutation. *Hum. Mutat.*, **29**, 220–226.

Tomizuka, K., Horikoshi, K., Kitada, R. *et al.* (2008) R-spondin1 plays an essential role in ovarian development through positively regulating Wnt-4 signaling. *Hum. Mol. Genet.*, **17**, 1278–1291.

Toyooka, Y., Tsunekawa, N., Takahashi, Y. *et al.* (2000) Expression and intracellular localization of mouse Vasa-homologue protein during germ cell development. *Mech. Dev.*, **93**, 139–149.

Tremblay, K.D., Dunn, N.R. and Robertson, E.J. (2001) Mouse embryos lacking Smad1 signals display defects in extra-embryonic tissues and germ cell formation. *Development*, **128**, 3609–3621.

Tsang, T.E., Khoo, P.L., Jamieson, R.V. *et al.* (2001) The allocation and differentiation of mouse primordial germ cells. *Int. J. Dev. Biol.*, **45**, 549–555.

Tsuda, M., Sasaoka, Y., Kiso, M. *et al.* (2003) Conserved role of nanos proteins in germ cell development. *Science*, **301**, 1239–1241.

Uda, M., Ottolenghi, C., Crisponi, L. *et al.* (2004) Foxl2 disruption causes mouse ovarian failure by pervasive blockage of follicle development. *Hum. Mol. Genet.*, **13**, 1171–1181.

Ueda, T., Abe, K., Miura, A. *et al.* (2000) The paternal methylation imprint of the mouse H19 locus is acquired in the gonocyte stage during foetal testis development. *Genes. Cells*, **5**, 649–659.

Upadhyay, S. and Zamboni, L. (1982) Ectopic germ cells: natural model for the study of germ cell sexual differentiation. *Proc. Natl. Acad. Sci. USA*, **79**, 6584–6588.

Vainio, S., Heikkila, M., Kispert, A. *et al.* (1999) Female development in mammals is regulated by Wnt-4 signalling. *Nature*, **397**, 405–409.

Vanderhyden, B.C., Telfer, E.E. and Eppig, J.J. (1992) Mouse oocytes promote proliferation of granulosa cells from preantral and antral follicles in vitro. *Biol. Reprod.*, **46**, 1196–1204.

Vidal, V.P., Chaboissier, M.C., de Rooij, D.G. and Schedl, A. (2001) Sox9 induces testis development in XX transgenic mice. *Nat. Genet.*, **28**, 216–217.

Vigier, B., Picard, J.Y., Tran, D. *et al.* (1984) Production of anti-Mullerian hormone: another homology between Sertoli and granulosa cells. *Endocrinology*, **114**, 1315–1320.

Vigier, B., Watrin, F., Magre, S. *et al.* (1988) Anti-mullerian hormone and freemartinism: inhibition of germ cell development and induction of seminiferous cord-like structures in rat fetal ovaries exposed in vitro to purified bovine AMH. *Reprod. Nutr. Dev.*, **28**, 1113–1128.

Vigier, B., Watrin, F., Magre, S. *et al.* (1987) Purified bovine AMH induces a characteristic freemartin effect in fetal rat prospective ovaries exposed to it in vitro. *Development*, **100**, 43–55.

Vincent, S.D., Dunn, N.R., Sciammas, R. *et al.* (2005) The zinc finger transcriptional repressor Blimp1/Prdm1 is dispensable for early axis formation but is required for specification of primordial germ cells in the mouse. *Development*, **132**, 1315–1325.

Wagner, T., Wirth, J., Meyer, J. *et al.* (1994) Autosomal sex reversal and campomelic dysplasia are caused by mutations in and around the SRY-related gene SOX9. *Cell*, **79**, 1111–1120.

Webster, K.E., O'bryan, M.K., Fletcher, S. *et al.* (2005) Meiotic and epigenetic defects in Dnmt3L-knockout mouse spermatogenesis. *Proc. Natl. Acad. Sci. USA*, **102**, 4068–4073.

Western, P.S., Miles, D.C., van den Bergen, J.A. *et al.* (2008) Dynamic regulation of mitotic arrest in fetal male germ cells. *Stem. Cells*, **26**, 339–347.

Whitworth, D.J., Shaw, G. and Renfree, M.B. (1996) Gonadal sex reversal of the developing marsupial ovary *in vivo* and *in vitro*. *Development*, **122**, 4057–4063.

Wilhelm, D., Hiramatsu, R., Mizusaki, H. *et al.* (2007) SOX9 regulates prostaglandin D synthase gene transcription in vivo to ensure testis development. *J. Biol. Chem.*, **282**, 10553–10560.

Wilhelm, D., Martinson, F., Bradford, S. *et al.* (2005) Sertoli cell differentiation is induced both cell-autonomously and through prostaglandin signaling during mammalian sex determination. *Dev. Biol.*, **287**, 111–124.

Winkler, K.E., Swenson, K.I., Kornbluth, S. and Means, A.R. (2000) Requirement of the prolyl isomerase Pin1 for the replication checkpoint. *Science*, **287**, 1644–1647.

Wylie, C. (1999) Germ cells. *Cell*, **96**, 165–174.

Wylie, C.C. (1993) The biology of primordial germ cells. *Eur. Urol.*, **23**, 62–66. Discussion 67.

Yabuta, Y., Kurimoto, K., Ohinata, Y. *et al.* (2006) Gene expression dynamics during germline specification in mice identified by quantitative single-cell gene expression profiling. *Biol. Reprod.*, **75**, 705–716.

Yamaguchi, S., Kimura, H., Tada, M. *et al.* (2005) Nanog expression in mouse germ cell development. *Gene. Expr. Patterns*, **5**, 639–646.

Yamamoto, M. and Matsui, Y. (2002) Testis-specific expression of a novel mouse defensin-like gene. *Tdl. Mech. Dev.*, **116**, 217–221.

Yao, H.H. and Capel, B. (2002) Disruption of testis cords by cyclopamine or forskolin reveals independent cellular pathways in testis organogenesis. *Dev. Biol.*, **246**, 356–365.

Yao, H.H., DiNapoli, L. and Capel, B. (2003) Meiotic germ cells antagonize mesonephric cell migration and testis cord formation in mouse gonads. *Development*, **130**, 5895–5902.

Yao, H.H., Matzuk, M.M., Jorgez, C.J. *et al.* (2004) Follistatin operates downstream of Wnt4 in mammalian ovary organogenesis. *Dev. Dyn.*, **230**, 210–215.

Yoshida, K., Kondoh, G., Matsuda, Y. *et al.* (1998) The mouse RecA-like gene Dmc1 is required for homologous chromosome synapsis during meiosis. *Mol. Cell*, **1**, 707–718.

Zamboni, L. and Upadhyay, S. (1983) Germ cell differentiation in mouse adrenal glands. *J. Exp. Zool.*, **228**, 173–193.

Zernicka-Goetz, M. (2002) Patterning of the embryo: the first spatial decisions in the life of a mouse. *Development*, **129**, 815–829.

Zhang, H. and Bradley, A. (1996) Mice deficient for BMP2 are nonviable and have defects in amnion/chorion and cardiac development. *Development*, **122**, 2977–2986.

3

Clytia hemisphaerica: A Cnidarian model for studying oogenesis

Aldine Amiel, Patrick Chang, Tsuyoshi Momose and Evelyn Houliston

UMR 7009, UPMC-CNRS, Developmental Biology Unit, Observatoire Océanologique, 06230 Villefranche sur mer, France

3.1 Introduction

This book demonstrates the success in using 'model' organisms to dissect the regulatory mechanisms responsible for the coordination of growth, meiosis and postfertilization events in animal oocytes. Elegant analyses at the cellular, molecular and biochemical levels in *Xenopus* and mouse, as well as starfish, ascidian and nematode, have greatly advanced our understanding of how these processes operate. It has transpired that many of the findings from these studies are 'universal' or at least widely applicable between species, such as cell cycle arrest in first meiotic prophase during oocyte growth, activation of the Cdk1/cyclin B complex (= MPF for maturation-promoting factor) at the onset of meiotic maturation, and the implication of Mos/MAP kinase in cytostatic arrest of the unfertilized egg. They have, however, also revealed many differences between models, such as in the signals that trigger meiotic maturation and initiate MPF activation, and the cell cycle stage at which cytostatic arrest occurs, as well as the molecules which mediate this arrest (see Section 3.5). Contributions from 'minor' models representing other branches of the animal kingdom can be of great value, both to assess which regulatory mechanisms are core components of oogenesis and which are species-specific specializations, and to gain insight into otherwise inaccessible or overlooked events.

We have recently started to develop a hydrozoan jellyfish, *Clytia hemisphaerica* as an experimental model for studying oogenesis and developmental mechanisms. It has long

Oogenesis: The Universal Process Marie-Hélène Verlhac and Anne Villeneuve

been recognized that hydrozoans provide attractive material for studying germ cell development, due to their simplicity of organization and accessibility to manipulation, as well as their transparency, with *Clytia* (= *Phialidium*) species proving a popular choice (Roosen-Runge, 1962; Roosen-Runge and Szollosi, 1965; Bodo and Bouillon, 1968; Honegger *et al.*, 1980; Freeman, 1987; Freeman and Ridgway, 1988; Carré and Carré, 2000; Freeman and Ridgway, 1993). In this chapter we will describe the main features of the *Clytia* system and our initial studies to characterize oogenesis, and summarize recent studies concerning maternal mRNA (messenger RNA) localization during the development of oocyte polarity and the role of the Mos/MAP kinase pathway in oocyte maturation, as illustrations of the experimental possibilities offered by the model.

The hydrozoans are a large group of aquatic animals showing a wide variety of morphologies and life cycles. They typically show alternation of generations between a free-swimming medusa phase and a fixed polyp stage (see Figure 3.1b), although species exist in which one or other phase has been abbreviated or eliminated (Boero, Bouillon and Piraino, 1992). The Hydrozoa is one of the subdivisions of the phylum Cnidaria. Together with other jellyfish groups, such as the Scyphozoa (true jellyfish), it forms the Medusozoa branch. The second cnidarian branch is the Anthozoa (corals, sea anemones etc.), which have polyp forms but no medusa phase (Ball *et al.*, 2004). Although the precise branching order of animal phyla at the base of the metazoan tree has been difficult to resolve, it now appears to be established that the Cnidaria, perhaps as part of a larger 'coelenterate' group including the ctenophores, form a sister group to the Bilateria (i.e. all the deuterostomes including vertebrates and echinoderms, and the protostomes including *C. elegans* and *Drosophila*) (Dunn *et al.*, 2008; Philippe *et al.*, 2009). Despite their overt simplicity, cnidarians possess many 'advanced' animal features including well-developed nervous system and musculature. Furthermore, it is becoming clear from the recent burst of interest in cnidarian genes and genomes, that their repertoire of developmental regulatory molecules is extremely similar to that of bilaterian species (Miller, Ball and Technau, 2005; Chevalier *et al.*, 2006; Jager *et al.*, 2006; Technau *et al.*, 2005; Putnam *et al.*, 2007; Miller and Ball, 2008). In the context of this chapter, it is also worth noting that despite frequent claims that cnidarians only have tissue-level organization, they have well-organized reproductive organs (Roosen-Runge and Szollosi, 1965), and thus offer a valuable perspective on the biology of gamete production and function in the animal kingdom.

3.2 *Clytia* as an experimental model

Clytia hemisphaerica has a typical three-phase hydrozoan life cycle (Figure 3.1). The free-swimming medusa is the sexual form. Fertilization is external and follows simultaneous release of gametes from separate male and female medusae into the seawater. The fertilized egg develops into a simple two-layered 'planula' larva, which swims directionally by means of ectodermal cilia (Bodo and Bouillon, 1968; Freeman, 1980) and shows morphological polarity along an axis termed oral–aboral (because the oral end gives rise to the mouth end of the primary polyp after metamorphosis

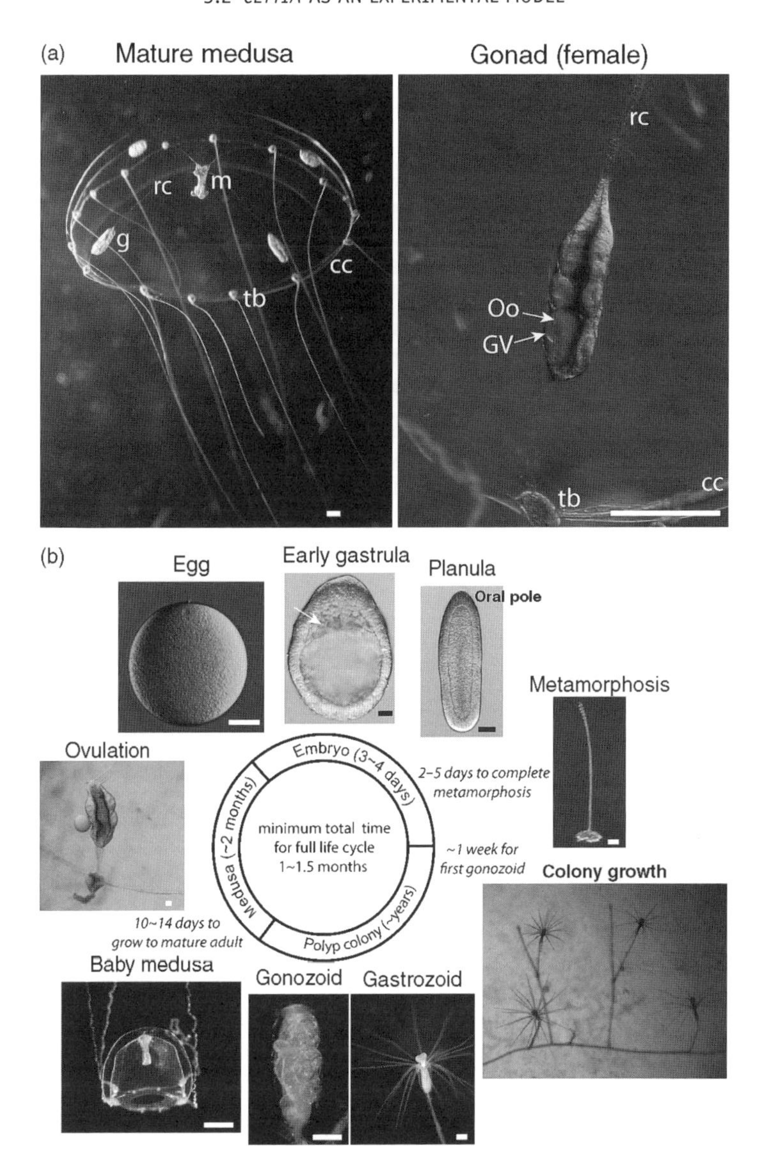

Figure 3.1 *Clytia hemisphaerica*. (a) Photo of an adult female medusa and details of the gonad. m = manubrium; rc = radial canal; g = gonad; cc = circular canal; tb = tentacle bulb; Oo = oocyte. (b) Other phases of the *C. hemisphaerica* life cycle, with the length of each phase indicated. The animal pole of the egg (top), marked by the position of the female pronucleus, gives rise to the site of cell ingression at gastrulation (arrow), the oral (= posterior) pole of the planula larva and, after metamorphosis, to the hydranth (feeding part) of the primary polyp (Freeman, 1980; Freeman, 2005). Connected polyp colonies form by vegetative stolon extension from the primary polyp, and contain two types of polyps: gastrozoids specialized for feeding, and gonozoids from which the clonal baby medusae bud. Scale bar = 0.5 mM in a, 50 μm in b

(Spindler and Müller, 1972; Schwoerer-Böhning *et al.*, 1990; Freeman, 2005)). After three to four days the planula settles onto a fixed substrate and metamorphoses into a feeding polyp ('gastrozoid'), resembling the well-known polyps of the related hydrozoan, *Hydra*. This primary polyp forms the basis of a connected colony of polyps which propagates vegetatively over the sea bed by stolon extension, generating new gastrozoids at regular intervals as well as interspersed 'gonozoids,' a second type of polyp specialized for the production of new medusae by budding. The colony is remarkable in that it has no finite lifespan, but can continue to produce genetically identical medusae for many years.

A key advantage of *Clytia* as a laboratory model is that all the steps of the life cycle, including spawning, fertilization, metamorphosis and medusa budding can be reproduced conveniently under laboratory conditions (Roosen-Runge, 1970; Kubota, 1978; Carré and Carré, 2000). All adult stages can be fed on *Artemia* larvae. The vegetative colonies are a particularly easy stage to maintain, requiring but a water change every two to three weeks. Gene function analysis is facilitated by the identical genetic composition of the clonally produced medusae from a single colony. Furthermore, the strains we use are self-crossed over several generations, providing high genetic homogeneity, which reduces problems due to polymorphism between alleles in wild populations. Self-crossing is made possible by the temperature dependence of sex determination, at least when the colony is young, such that lower temperatures (15 °C) favour the production of males, and higher temperatures (21–24 °C) females (Carré and Carré, 2000).

Clytia eggs and embryos are relatively large (around 200 μm in diameter), transparent and very well suited for experimental manipulation. Under laboratory conditions each medusa produces eggs daily, spawning being precisely controlled by the light–dark cycle, such that unfertilized eggs can be reliably collected 2 h after the beginning of a light period following at least 1 h of darkness. Depending on the feeding regime 4–20 eggs are spawned per medusa per day, so that a beaker of 30 females can produce 120–600 eggs. There are no protective egg envelopes, and the egg remains fertilizable for 60–90 minutes following spawning, providing ample time for microinjection or other manipulations prior to gamete mixing and analysis of developmental events (Momose and Houliston, 2007; Momose, Derelle and Houliston, 2008). Another experimental advantage of *Clytia* is that the medusae, embryos and larvae are very robust, and can easily accommodate the loss of cells or body parts (Maas, 1905; Teissier, 1933; Schmid and Tardent, 1971; Schmid *et al.*, 1976; Freeman, 1981b).

A final very remarkable particularity of the *Clytia* in the context of studies of oogenesis is the ability of the gonad to function autonomously. *Clytia* gonads can be isolated from the adult by simple dissection and cultured in filtered seawater. They undergo successive cycles of oocyte growth and ovulation for several days, responding normally to the light cues that induce spawning and maturation of competent oocytes (Honegger *et al.*, 1980; Freeman and Ridgway, 1988). This remarkable autonomy is a property shared by the medusa tentacle bulb (Denker *et al.*, 2008), which continues to support tentacle growth for many days when cultured in isolation. Living oocytes at all stages of oogenesis and meiotic maturation are accessible to observation and to manipulation, with growing oocytes injectable through the epithelial wall of the gonad (see Section 3.5). A similar analysis system involving 'umbrella-free medusae' has been used to study mechanisms of oocyte maturation in another hydrozoan, *Cytaeis uchidae* (Takeda, Kyozuka and Deguchi, 2006).

3.3 Characteristics of oogenesis in *Clytia*

The sexual stage of the *Clytia* life cycle is the medusa, which forms by budding from specialized polyps (gonozoids) of the vegetative colony (Figure 3.1). When the baby medusa is first released no gonads are visible, but as it grows, swellings appear on each of the four endodermal radial canals which connect the manubrium (mouth) to the circular canal running around the periphery of the bell (Figure 3.1a). As is typical in hydrozoans, the gonad consists of an organized collection of germline precursors, meiotic cells and vitellogenic oocyte stages, sandwiched between a layer of columnar endodermal cells, and a thin overlying ectoderm layer (Hertwig and Hertwig, 1895; Faulkner, 1929; Honegger *et al.*, 1980; Freeman, 1987; Carré and Carré, 2000). The germ cell precursors appear to derive from a population of stem cells or 'i-cells' (interstitial cells) that migrate into the medusa bud as it develops within the gonozoid (Weiler-Stolt, 1960). i-cells have been well characterized in *Hydra*, a hydrozoan which has lost the medusa phase, and provide not only germ cells but assorted somatic cell types including secretory cells, nerve cells and stinging cells (nematocytes) (Steele, 2002). Little is known about the cues that regulate proliferation and developmental choice of fate of i-cells and their descendants in *Clytia.* As in *Hydra* it is likely that local signals determine their behaviour and fate (Khalturin *et al.*, 2007), for instance directing i-cells positioned in the gonad region to produce only germ cells, and those at the base of the tentacle to produce nematocytes (Denker *et al.*, 2008). Changes in these signals during evolution could underlie life-cycle modifications: in *Clytia hemisphaerica,* the only putative i-cells identified in the female gonad contain nuage material typical of germline cells (see Figure 3.2b), while in *Clytia mccradyi*, in which the life cycle is truncated by formation of polyps in place of the gonads in adult medusae, i-cells with distinct morphologies are detectable in equivalent positions (Carré *et al.*, 1995).

Under laboratory feeding conditions, baby medusae complete growth and start spawning after 10–14 days. As the medusa grows, the female gonad takes on a characteristic organization, with putative i-cells and early differentiating oocytes positioned close to the radial canals, and vitellogenic stages of oocyte growth occupying more distal positions (Figure 3.2a/a′). Cohorts of small Stage I oocytes embark on their final growth phase each day following spawning (Amiel and Houliston, 2009), the number presumably depending on nutrient availability. During vitellogenesis the oocytes accumulate massive reserves of glycogen and lipid, yolk, ribosomal protein and mRNAs to support the early development of the embryo, likely by a combination of direct synthesis and uptake of nutrients supplied by digestive cells on the endodermal side. The large nucleus (or GV, for germinal vesicle) loses its central position and becomes positioned progressively closer to the future animal pole (see below). The nucleolus fragments and chromosomes partially decondense (Faulkner, 1929; Honegger *et al.*, 1980). 'Nuage' material and clustered

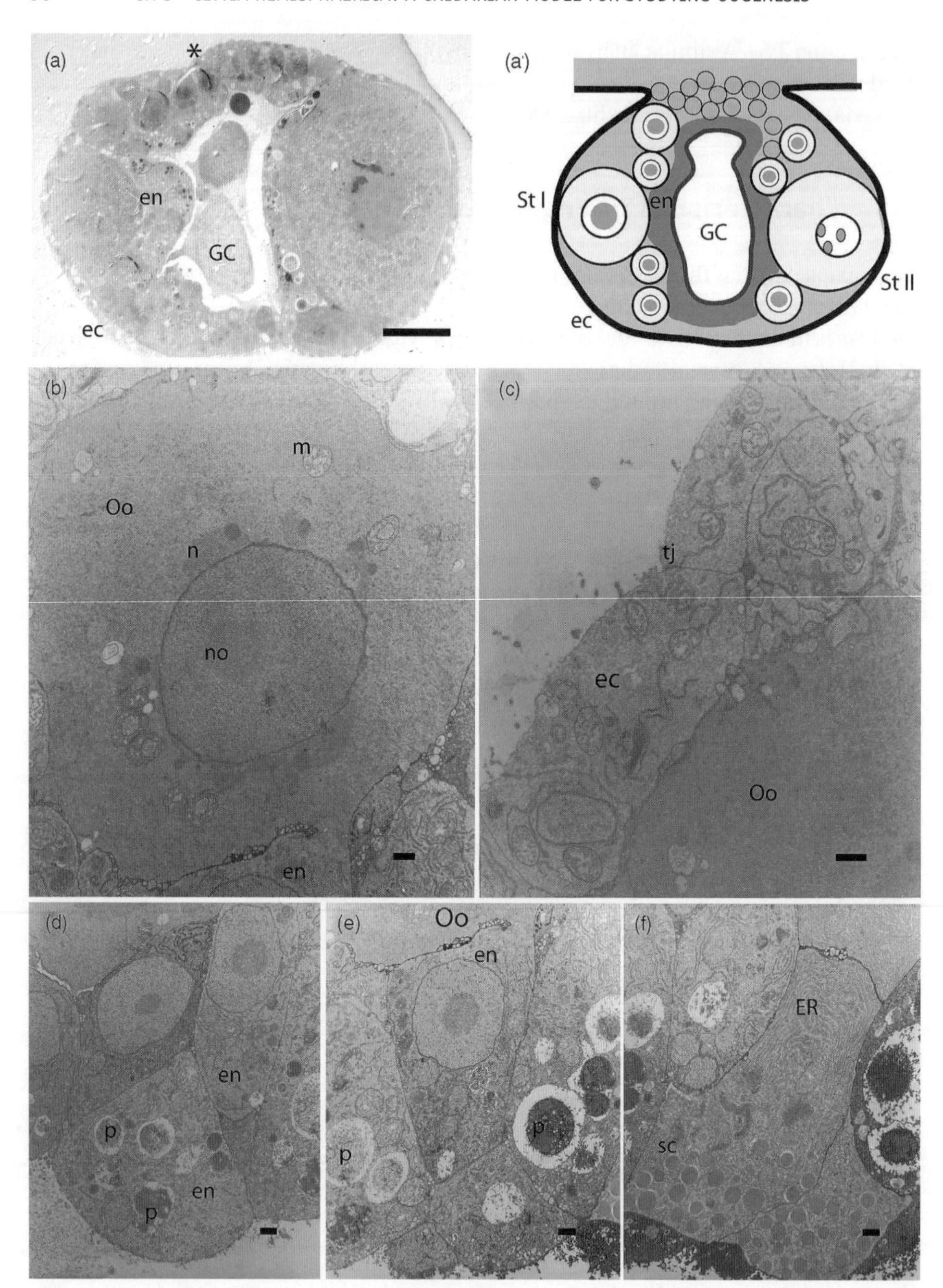

Figure 3.2 Ultrastructural features of the *Clytia* female gonad. Sections of an isolated gonad fixed using a protocol modified from Eisenman and Alfert (1981) and embedded in Spurr resin. (a) Overview: 0.5 μm thick section stained with methylene blue. GC = gastric cavity; ec = ectoderm; en = endoderm. Asterisk marks the region from which adjoining 80 nm thin sections were taken, shown in images b–f. (a′) Schematic diagram of gonad cross-section attached to the underside of the medusa bell. Putative i-cells (grey) and early stages of oogenesis are positioned proximally between the

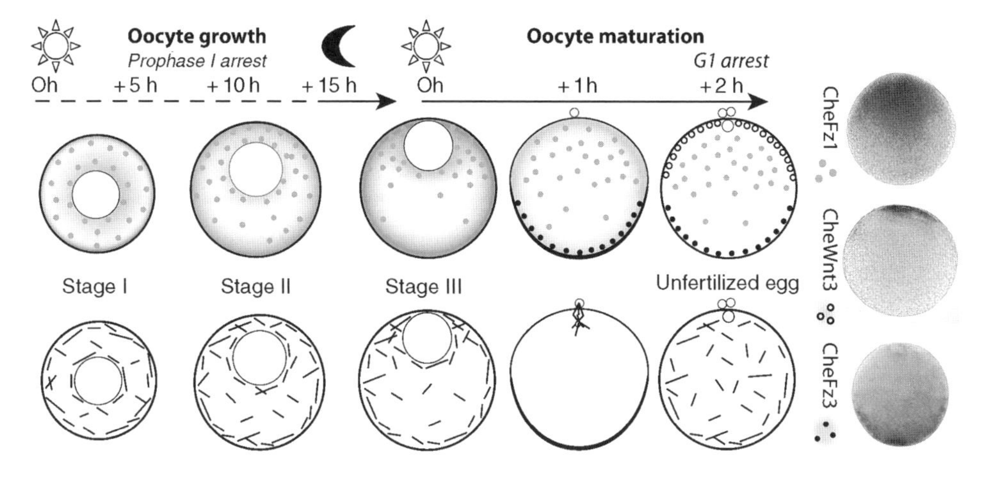

Figure 3.3 Development of polarity during oocyte growth and maturation. Oocyte polarity in *Clytia hemisphaerica* develops in two phases (Amiel and Houliston, 2009). The first covers stages II and III of vitellogenesis on the day preceding spawning, and involves microtubule-dependent repositioning of the GV to the animal cortex, and the parallel redistribution of CheFz1 RNA (grey circles) to form an animal–vegetal cytoplasmic gradient. The microtubule network, schematized in the bottom row of oocytes, shows a slight enhancement between the GV and the animal cortex at this time. The second polarization phase accompanies oocyte maturation, induced by a light signal after >2 h darkness. CheFz3 RNA adopts its final location in the vegetal cortex first polar body emission (around 50 minutes after the light signal), and CheWnt3 RNA its animal cortex location before second polar body emission (around 80 minutes after the light signal). CheFz3 but not CheWnt3 localization is microtubule dependent, and requires contacts and/or diffusible signals from the gonad tissue. *In situ* hybridization images on the right show the final localization patterns of the three RNAs in the unfertilized egg

mitochondria typical of germ cells (Eddy, 1975) can be detected around the oocyte nucleus from very early stages of oogenesis (Figure 3.2b; Carré *et al.*, 1995). Growth is completed after approximately 13–18 hours (Figure 3.3). About 2 hours after their first appearance, the prophase-arrested fully grown Stage III oocytes become competent to undergo meiotic maturation and complete meiotic division upon light stimulation (see Section 3.5). Similar timing for the development of maturation competence has been defined in *Cyteis* (Takeda, Kyozuka and Deguchi, 2006).

endoderm of the gastric cavity and the overlying ectoderm (Freeman, 1987), and vitellogenic oocytes more distally. During stage II of vitellogenesis the nucleolus fragments, and the oocyte nucleus loses its central position, such that by stage III (not shown) it is found at the oocyte periphery directly beneath the ectoderm, marking the oocyte animal pole. (b) Early stage I oocyte. Oo = oocyte; m = mitochondria; n = nuage; no = nucleolus; en = digestive cells of the endodermal layer (see e). (c) Ectodermal cells (ec) overlying the oocyte shown in b. tj = tight junction. (d–f) Three adjoining endodermal regions (regions e and f border the oocyte shown in b). (d), (e) Endodermal digestive cells (en) containing phagosome-like vesicles (p). (f) Putative secretory cell (sc) rich in ER. Scale bar = 50 μm in a, 1 μm in b–f

The mechanism by which nutrients are supplied to growing oocytes in *Clytia* has not been fully established, but likely involves direct or indirect transfer from the digestive endodermal cells that line the gastroendodermal cavity. Cells in this thickened endodermal layer, characterized by intense phagocytic nutrient uptake, are closely apposed to the oocyte vegetal surface (Roosen-Runge, 1962; Figure 3.2d–f). Furthermore, they frequently remain attached to oocytes following mechanical isolation (see Figure 3.4b, far right panel). The close relationship between endoderm cells and oocytes in *Clytia* is thus somewhat reminiscent of that described in various anthozoan and scyphozoan species, where a specialized structure called the trophonema forms from endodermal cells in contact with the young oocyte (Wedi and Dunn, 1983; Eckelbarger and Larson, 1992). During vitellogenesis, the tubular trophonema connects the developing oocyte to the gastroendodermal cavity through the mesoglea and the endoderm. A different situation has been described in the derived hydrozoan *Hydra*, where there is no well-defined gonad structure, and oocytes arise within patches of germ cells derived from the i-cell population. Large cytoplasmic connections have been demonstrated between the single oocyte and surrounding i-cell-derived 'nurse cells' (Miller *et al.*, 2000; Alexandrova *et al.*, 2005). The nurse cells have an unusual fate: they decrease in size as the oocyte grows and finally enter into apoptosis to become phagocytosed by the growing oocyte (Technau *et al.*, 2003; Alexandrova *et al.*, 2005). Similar phenomena have been described in some hydrozoan medusae (Kawaguti and Ogasawara, 1967; Meurer and Hündgen, 1978), however we have not detected any obvious specialized nurse cells, cytoplasmic bridges with neighbouring cells or evidence for phagocytosis of nurse cells by electron microscopy in *Clytia* (also Danièle Carré, personal communication). It is possible, however, that autodigestion of somatic and germ cells could contribute to recycling of cellular material in the gonad, since active circulation of visible digestive products in the gastroendodermal cavity continues for several days during culture of isolated gonads.

3.4 Development of oocyte polarity in *Clytia*

The fully grown *Clytia* oocyte shows a clear animal–vegetal (AV) polarity, with the GV positioned eccentrically close to the cortex at the animal pole. This manifest AV polarity is related to the position of the oocyte with respect to the somatic cell layers of the gonad, the GV always adopting a position opposite its contact with endodermal cells (Amiel and Houliston, 2009). This situation is common in hydrozoans (Teissier, 1931; Freeman, 1987; Rodimov, 2005), but different to that reported in some anthozoans and scyphozoans, where the GV is positioned close to the site of attachment of the endodermal trophonema. The transparent oocytes of *Clytia* show no other visible signs of AV polarity; however, other hydrozoan species show polarized distributions of pigment and other intracellular inclusions (Teissier, 1931; Hirose, Kinzie and Hidaka, 2000).

In species from the Bilateria, a common mechanism to establish polarity along one or more axes in the developing embryo is to prelocalize maternal 'determinant' factors with respect to the primary axis of the oocyte (Micklem, 1995; Bashirullah, Cooperstock and Lipshitz, 1998). In cnidarians, it has long been known that the animal pole is fated

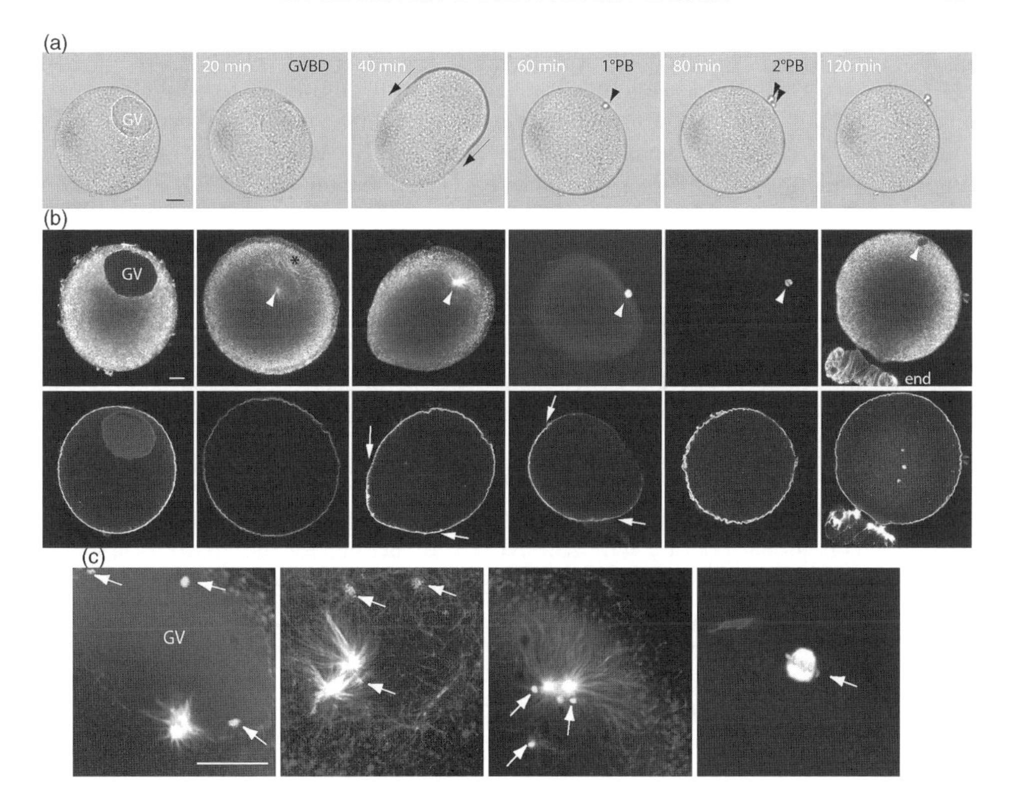

Figure 3.4 Meiotic maturation in *Clytia*. (a) Selected images from a time-lapse recording of oocyte maturation (available at http://biodev.obs-vlfr.fr/recherche/houliston/Clytia/ClytiaPhotosFilms.html) in an isolated oocyte triggered to mature using Br-cAMP. GVBD occurs 15 minutes after the start of maturation and is followed by an exaggerated contraction wave (40 minutes), which crosses the egg (arrows) prior to first polar body emission at 60 minutes and second polar body emission at 80 minutes (arrowheads indicate polar bodies). After 120 minutes maturation is complete and the cell cycle arrests in G1. (b) Confocal images of oocytes fixed at different times following Br-cAMP, corresponding approximately to the stages shown in a, labelled by antitubulin immunofluorescence (top row) and by rhodamine phalloidin for polymerized actin (bottom row). The dense microtubule network in fully-grown oocytes depolymerizes rapidly after the maturation signal. During GVBD, a cytoplasmic microtubule aster forms on the vegetal side of the GV, collects the chromosomes and migrates to the animal cortex where it reorganizes into the first meiotic spindle. A more disorganized microtubule structure forms transiently on the animal side of the GV (*). White arrowheads mark the position of the developing meiotic spindles, and, in the last panel, of the pronucleus, which lies opposite residual endodermal cells (end) attached to the egg vegetal pole. Cytoplasmic microtubules are sparse or absent during the meiotic period, but a dense network is restored by the end of maturation. Nuclear actin disperses during GVBD, while the actin-rich cortex shows transient local thickening (arrows) in parallel with the contractions that accompany first and second polar body formation. (c) Confocal images showing details of first meiotic spindle formation by combined antitubulin immunofluorescence and TOPRO-3 labelling of DNA (chromosomes arrowed) of oocytes fixed between 30 to 60 minutes following Br-cAMP treatment. All scale bars 20 μm

to give rise to the oral pole of the planula larva (Teissier, 1931), but the relationship between oocyte and embryo polarity appears unstable in many species (Rodimov, 2005) and has only recently been clarified. Indeed, the idea that oocyte animal–vegetal polarity might provide the basis for embryo polarity was largely abandoned as a result of an impressive and influential series of studies by Gary Freeman using *Clytia gregarium* and other hydrozoan species (Freeman, 1979; Freeman, 1980; Freeman, 1981b; Freeman, 1981a), showing that the site of first cleavage (dictated by zygote nucleus position) was a more reliable indicator of embryonic axis than the egg animal pole. Two key observations were (i) that experimental displacement of the zygote nucleus from the animal pole by low-speed centrifugation of fertilized eggs caused a corresponding respecification of the embryonic axis, and (ii) that experimental duplication of the zygote nucleus could lead to the formation of 'double-axis' larvae with duplicated posterior poles (Freeman, 1980; Freeman, 1981a). It thus came to be widely considered that cnidarian eggs were essentially unpolarized, and that a 'global' embryo and larval polarity was set up during the early cleavage stages in relation to the orientation of cell division. This global polarity was also evoked to account for the ability of embryo fragments cut at almost any stage of development to regulate and form normally proportioned larvae, retaining the polarity of the embryo from which they came (Teissier, 1931; Freeman, 1981b).

Over the last few years, the view of egg polarity in cnidarians has been brought sharply back into line with the bilaterian axiom of embryonic patterning by maternal determinants, with the identification of localized activators of the Wnt/ß-catenin signalling pathway within the embryo (Wikramanayake *et al.*, 2003; Momose and Houliston, 2007; Lee *et al.*, 2007; Plickert *et al.*, 2006; Momose, Derelle and Houliston, 2008). In *Clytia*, two localized RNAs acting upstream of this pathway have been shown experimentally to act as maternal axis determinants. These RNAs code for Wnt ligand receptors of thc Frizzled family, show opposite localizations and activities, and cooperate to direct the development of the embryonic oral–aboral axis (Momose and Houliston, 2007). CheFz1 is a classic Frizzled, and mediates activation of the canonical Wnt pathway. Its RNA is relatively concentrated in the animal half cytoplasm (see Figure 3.3 left panel), and can direct the development of oral fate when expressed ectopically. Since CheFz1 RNA is not tightly anchored in the fertilized egg it can be displaced by low-speed centrifugation (Amiel and Houliston, 2009), thus providing a possible explanation for embryonic axis respecification under these experimental conditions (Freeman, 1981a). CheFz3 RNA is tightly localized to the vegetal cortex of the egg, and codes for a divergent Frizzled which acts negatively to downregulate the canonical Wnt pathway in the future aboral territory. $CheF_33$ can also redirect axis development when expressed ectopically.

In addition to the Frizzled RNAs, mRNAs coding for Wnt3 family ligands have also been shown to be maternally localized in both *Hydractinia* and *Clytia*, exhibiting a distinct localization pattern at the animal cortex (Plickert *et al.*, 2006; Momose, Derelle and Houliston, 2008). We have shown that CheWnt3 has an essential role in embryonic polarity development but, in early stages at least, it is the two Frizzled RNAs rather than Wnt3 that provides the dominant spatial cues to direct axis orientation (Momose, Derelle and Houliston, 2008). In other cnidarians, localized Wnt pathway activation may be directed by alternative or additional determinants, for instance RNA for the downstream transcription factor TCF in *Hydractinia* (Plickert *et al.*, 2006), or protein for the cytoplasmic regulator Dishevelled in the anthozoan (sea anenome) *Nematostella* (Lee *et al.*, 2007). Other types of maternal localized molecules with potential determinant roles in early development are also being discovered in a variety of hydrozoan species, for example animal pole concentrations of mRNAs for the transcription factors Brachyury and Cnox4 in *Podocoryne* (Yanze *et al.*, 2001; Spring *et al.*, 2002) and of Vasa protein in *Hydractinia* egg (Rebscher *et al.*, 2008). Thus the molecular complexity of egg polarity in cnidarians is much richer than anticipated. The localization of oral fate determinants at the animal pole of cnidarian egg explains why vegetal fragments produced by early embryo bisection in both *Podocoryne* (Momose and Schmid, 2006) and *Nematostella* (Fritzenwanker *et al.*, 2007; Lee *et al.*, 2007) fail to develop embryonic polarity. *Clytia* embryos appear to have superior regenerative capacity precluding experimental demonstration of this localization (Freeman, 1981b), likely mediated by Wnt3 dependent reciprocal downregulation between the two Frizzled RNAs (Momose and Houliston, 2007; Momose, Derelle and Houliston, 2008).

It is remarkable that unfertilized *Clytia* eggs, despite their lack of visible polarity, contain maternal mRNAs with at least three distinct distributions along the animal–vegetal axis: CheFz1 exhibiting a declining animal–vegetal gradient in the cytoplasm, CheWnt3 mRNA localized at the animal cortex, and CheFz3 at the vegetal cortex (Figure 3.3). We have recently completed an analysis of the cellular basis of RNA localization during oogenesis in *Clytia,* focusing on the origin of the distinct localization patterns of these three mRNAs (Amiel and Houliston, 2009). This analysis revealed that CheFz1 RNA acquires its polarized cytoplasmic distribution in parallel with the repositioning of the GV to the animal pole during the latter phase of vitellogenesis. The repositioning both of the GV and of CheFz1 RNA away from contacts with the endoderm and towards the ectoderm requires an intact microtubule network, and these events may well be linked directly or indirectly.

The microtubule-dependent cell polarization during oocyte growth does not directly generate all the final asymmetry of the unfertilized egg, since CheFz3 and CheWnt3 RNAs in stage III fully grown oocytes remain distributed in a patchy but nonpolarized manner around the oocyte periphery. These two RNAs adopt their cortical polarized locations only during the process of meiotic maturation, during which massive polarized contraction waves cross the oocyte (see below, Figure 3.4). It had previously been shown using oocytes from other hydrozoan species that localized specializations of the surface at the animal pole, relating to sperm chemotaxis and/or localized sperm–egg fusion, also develop during the maturation process (Carré and Sardet, 1981; Freeman and Miller, 1982; Freeman, 1987; Freeman, 2005), this surface polarization being directed by the initial position of the GV. CheWnt3 RNA localization to the animal cortex, like the overlying surface glycoprotein localization (Carré and Sardet, 1981; Freeman and Miller, 1982; Freeman, 1987; Freeman, 2005), is a cell autonomous process that can occur in isolated oocytes (Amiel and Houliston, 2009). In contrast, CheFz3 RNA localization to the vegetal cortex does not occur in oocytes induced to mature following isolation, suggesting that cell contacts are required. Furthermore CheFz3 RNA localizes to the vegetal cortex by a mechanism which, like CheFz1, requires microtubules, while CheWnt3 RNA localization to the animal cortex cannot be prevented by either microtubule or microfilament disruption. Thus the localization of these two RNAs during oocyte maturation clearly involves distinct localization mechanisms.

To summarize, oocyte polarity in *Clytia* is acquired in successive and mechanistically separable steps (Figure 3.3), as is the case in the classically studied models for maternal RNA localization, *Drosophila* and *Xenopus* (St Johnston, 1995; King, Messitt and Mowry, 2005). Much remains to be learnt about the underlying cellular processes. *In vivo* analyses of these filament systems in conjunction with fluorescent-tagged RNAs should enable a detailed analysis of the underlying mechanisms. Cryptic or subtle polarity of the microtubule network in growing oocytes may contribute to GV relocalization and/or CheFz1 RNA localization, as it does for oscar RNA localization in *Drosophila* oocytes (Zimyanin *et al.*, 2008). Both the microtubule network and actin cortex show transitory asymmetries during oocyte maturation, which may contribute to the localization of CheFz3 and CheWnt3 RNAs (Figure 3.4b). Another interesting hypothesis to test is that differential RNA degradation is involved (Bashirullah *et al.*, 1999), since experimental treatments that prevent CheFz1 RNA localization during growth or CheFz3 localization during maturation appear to result in high, uniform RNA levels across the egg.

3.5 Regulation of oocyte maturation

As in other animals, the meiotic division cycle in hydrozoans is arrested in prophase of first meiosis during oocyte growth. Meiosis resumes at the time of spawning, as part of the maturation process by which oocytes acquire the ability to be fertilized. After completion of meiosis and emission of two polar bodies, the cell cycle arrests again in G1 until fertilization (Freeman and Ridgway, 1993; Kondoh, Tachibana and Deguchi, 2006). At the end of the maturation period, oocytes are released through rupturing of the overlying epithelium. Maturation and spawning are generally triggered in relation to the day–night cycle, either by a light cue after a dark period and/or by darkness after light (Ballard, 1942; Roosen-Runge, 1962; Honegger *et al.*, 1980; Takeda, Kyozuka and Deguchi, 2006). The light/dark stimulus causes the tissues of the gonad to release a diffusible factor, probably a peptide, which acts rapidly on the oocyte (Ikegami, Honji and Yoshida, 1978; Freeman, 1987; Takeda, Kyozuka and Deguchi, 2006). The exact source of this signal, its molecular identity and its manner of reception by the oocyte are unknown, but the immediate intracellular consequence is a rapid rise in cAMP concentrations (Takeda, Kyozuka and Deguchi, 2006). Elevated cAMP in turn leads to germinal vesicle breakdown (GVDB), due to activation of the universal M phase kinase Cdk1–cyclin B (MPF). The positive role for elevated cAMP in maturation in hydrozoans is shared with many invertebrate species, but contrasts with the inhibitory role in vertebrates and some echinoderms (Stricker and Smythe, 2001; Karaiskou *et al.*, 2001; Meijer *et al.*, 1989). The rapidity of GVBD, typically occurring 15–20 minutes after the light signal, suggests that MPF activation in hydrozoans, like that in starfish, may be regulated mainly by post-translational mechanisms. The dynamics of first meiotic spindle formation during GVBD in which chromosomes are gathered on centrosome nucleated asters before migrating to the egg cortex (Figure 3.4c), also show similarities with the starfish

(Lenart *et al.*, 2005), although the precise roles of actin and microtubules in this process require verification in *Clytia*. In *Clytia*, spawning occurs 110–120 minutes after the light signal, which can be as little as a few seconds following at least 1 hour of darkness. The first polar body forms after about 50–60 minutes and the second after 80–90 minutes (Honegger *et al.*, 1980; Freeman and Ridgway, 1988; Amiel and Houliston, 2009). This same sequence of maturation can conveniently be triggered experimentally by treatment of either intact gonads or manually isolated fully grown oocytes with the cell-permeable cAMP analogue, Br-cAMP (Freeman and Ridgway, 1988; Amiel and Houliston, 2009; Amiel *et al.*, 2009).

The 'cytostatic' arrest of the mature, unfertilized hydrozoan eggs in G1 has been shown to depend on MAP kinase activity (Kondoh, Tachibana and Deguchi, 2006), suggesting that this kinase may be universally involved in animal oocyte cytostatic arrest despite species-specific differences in its cell cycle stage (Sagata, 1998; Masui, 2000). In vertebrate and starfish oocytes, MAP kinase is activated as a consequence of the synthesis during oocyte maturation of Mos, a cytoplasmic kinase that phosphorylates and activates the MAP kinase kinase MEK. In vertebrates, Mos-activated MAP kinase contributes to cytostatic arrest in MII (Colledge *et al.*, 1994; Sagata *et al.*, 1989), operating in conjunction with Emi2, an APC/cyclosome inhibitor that prevents degradation of cyclin B (Inoue *et al.*, 2007; Liu *et al.*, 2006; Madgwick and Jones, 2007). In starfish, the Mos/MAPK cascade, including p90rsk, has been shown to mediate G1 cytostatic arrest (Mori *et al.*, 2006), and MAP kinase has also been implicated in MI cytostatic arrest in the sawfly (Yamamoto *et al.*, 2008), although this function appears to have been lost at least in part in *Drosophila* (Ivanovska *et al.*, 2004). Despite the generalized function of MAP kinase and perhaps of Mos in cytostatic arrest, there are a number of apparent differences between species, even when the cell cycle stage of cytostatic arrest is the same. Thus the MAP kinase substrate p90rsk is important for cytostatic arrest in *Xenopus*, but not in mouse, (Gross *et al.*, 1999; Dumont *et al.*, 2005). Furthermore, an additional role for Mos synthesis in the maturing oocyte has been revealed in *Xenopus*, with the resulting MAP kinase activity stimulating MPF activation and GVBD (Karaiskou *et al.*, 2001; Abrieu, Doree and Fisher, 2001). Mos synthesis is not essential for *Xenopus* oocyte maturation though, since cyclin B synthesis is able to assure MPF activation in its absence (Haccard and Jessus, 2006).

We have recently completed a first study of Mos function in *Clytia*, aimed at shedding light on the differences in results between other species (Amiel *et al.*, 2009). Curiously, we identified two distinct *Mos* genes from our EST collection, an unexpected finding since no animal had previously been found to possess more than one. It transpires that multiple *Mos* genes are not unusual in cnidarians; indeed the fully sequenced genome of *Nematostella* contains four. It is premature to speculate on how this situation arose; however, both *Clytia* Mos kinases had cytostatic activity when tested in *Xenopus* or *Clytia* embryos (Figure 3.5b), and their expression was detected exclusively in germ cells, suggesting that cnidarian *Mos* gene diversification was not related to acquisition of new functions in other tissues. Mos may ancestrally have had a general role in gametogenesis since both *Clytia* and mouse *Mos* genes are expressed in spermatids in males as well as in oocytes in females, although any function in males has apparently become nonessential in mice (Goldman *et al.*, 1987; Colledge *et al.*, 1994; Inselman and Handel, 2004).

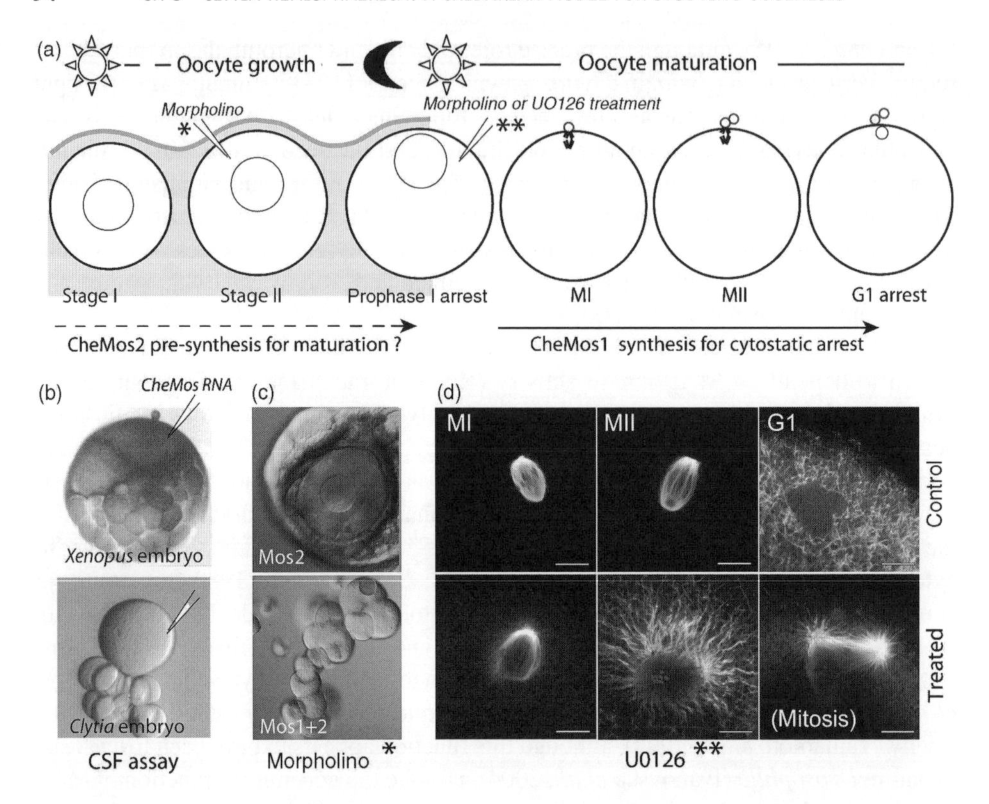

Figure 3.5 Mos in *Clytia* oocyte maturation. (a) Summary of *Clytia* Mos morpholino injection experiments. Injection of CheMos morpholino but not CheMos2 morpholino into growing stage II vitellogenic oocytes (*) through the epithelium of the gonad (grey) prevented subsequent spawning and GVBD in some cases, suggesting a possible role for CheMos2 synthesis upstream of maturation. Injection of CheMos1 morpholino but not CheMos2 morpholino into isolated immature oocytes (**) blocked the majority of MAP kinase activation during maturation and prevented polar body formation and cytostatic arrest in G1. Coinjection of CheMos2 morpholino enhanced this phenotype, with complete abolition of MAP kinase activation, phenocopying treatment with the MEK inhibitor U0126. (b) Demonstration of the cytostatic activity assay of *Clytia* Mos kinases. RNA from either gene injected into single blastomeres of *Xenopus* or *Clytia* can induce cell cycle arrest on the injected side. An interphase nucleus is visible in the arrested injected *Clytia* blastomere. (c) Demonstration that Mos2 morpholino can prevent spawning and maturation when injected into growing oocytes (* in a). The low incidence of this phenotype may be due in part to Mos2 synthesis starting at an earlier stage of oocyte growth. Oocytes injected with combined Mos1 and Mos2 morpholinos at this stage can enter into parthenogenetic mitotic cycles following maturation. (d) Demonstration that MAPK inhibition during oocyte maturation in isolated oocytes stimulated with Br-cAMP disrupts the morphology and positioning of both meiosis I (MI) and meiosis II (MII) spindles, explaining the failure of polar body emission. The treated oocytes do not arrest in G1 but attempt to enter into first mitosis with a multipolar aster. Equivalent effects were obtained by injection of CheMos1 or CheMos1 + CheMos2 morpholinos prior to Br-cAMP treatment (**). Scale bars = 10 μm

We showed that synthesis of *Clytia* Mos during oocyte maturation was responsible for MAP kinase activation during maturation, by coinjection of specific morpholino antisense oligonucleotides targeted to the two RNAs, into isolated oocytes prior to Br-cAMP treatment (Figure 3.5a). Following the end of the maturation period, the double morpholino-injected oocytes failed to arrest in G1 but passed spontaneously into a mitotic cycle, as seen in oocytes from Mos$^{-/-}$ mice (Hashimoto *et al.*, 1994; Colledge *et al.*, 1994) and Mos antisense-oligo-injected starfish oocytes (Tachibana *et al.*, 2000). A second striking phenotype, also obtained by prevention of MAP kinase activation using the MEK inhibitor U0126, was an absence of polar body formation, reflecting the failure of the first meiotic spindle to position correctly at the oocyte cortex and the second spindle to adopt a correct bipolar morphology (Figure 3.5d). We propose that spindle positioning at the cortex along with cytostatic arrest are ancestral and conserved roles for the Mos/MAP kinase cascade, similar phenotypes having been observed when the pathway is inhibited in mouse, frog and starfish (Verlhac *et al.*, 1996; Verlhac *et al.*, 2000; Bodart *et al.*, 2005; Tachibana *et al.*, 2000). It will be of great interest to use similar morpholino approaches to determine to what extent the downstream MAP kinase substrates mediating cytostatic arrest and spindle positioning are shared between *Clytia* and other species.

In *Clytia*, the proposed ancestral roles for Mos in cytostatic arrest and meiotic spindle dynamics are mostly accounted for by translation of one of the two genes, *CheMos1*. Injection of CheMos1 morpholino alone substantially reduced MAP kinase activity and was sufficient to cause spontaneous activation and polar body failure. The *CheMos2* gene may rather have adopted, during evolution, an earlier role in oogenesis. Preliminary observations suggest that CheMos RNA may undergo translation at an earlier stage of oogenesis, important for an unknown but essential preparatory step for oocyte maturation. Thus, injection of CheMos2 morpholino into stage II growing oocytes within isolated gonads through the ectodermal wall caused failure of spawning and of maturation the following day (see Figure 3.5a; Amiel *et al.*, 2009). Presynthesized protein could, for instance, be required for the acquisition of maturation competence, and/or provide a pool of inactive kinase to be activated post-translationally following reception of the maturation signal. This possible participation of CheMos2 in meiosis initiation in *Clytia* is reminiscent of that of *Xenopus* Mos in MPF activation at the beginning of maturation (Karaiskou *et al.*, 2001; Abrieu, Doree and Fisher, 2001). A role for Mos in maturation initiation is unlikely to be ancestral since mouse, starfish and *Drosophila* oocytes appear to enter meiosis normally in its the absence (Verlhac *et al.*, 1994; Tachibana *et al.*, 2000; Ivanovska *et al.*, 2004), but it is possible that *Xenopus* Mos and *Clytia* Mos2 kinases have been secondarily recruited during evolution to assist in this process.

For the moment, the evidence for CheMos2 translation during oocyte growth remains weak. The incidence of morpholino phenotypes following injection into growing ovarian oocytes (Figure 3.5c) was relatively low, perhaps reflecting dilution of the morpholinos during oocyte growth and/or prior protein synthesis in oocytes too small to be accessible to microinjection. We hope to further explore this question by using RNAi (RNA interference) approaches for gene knockdown (Chera *et al.*, 2006; Galliot *et al.*, 2007) and by monitoring of endogenous Mos protein levels following generation of specific antibodies. It should also be feasible to analyse the possibility of differential translational regulation of the Mos RNAs by experimental modification of UTR (untranslated region) motifs implicated in temporal control of translation during oocyte maturation in *Xenopus* (Belloc, Pique and Mendez, 2008). In this context it is interesting to note that CheMos2 RNA translation during oocyte growth may be mediated by the

5′TOP sequence detected at the extreme 5′ terminus, which in other systems including immature *Xenopus* oocytes has been shown to stimulate translation of growth-related mRNAs when the TOR pathway is active (Hamilton *et al.*, 2006; Schwab *et al.*, 1999).

3.6 Perspectives

Many interesting questions are now open for study in the simple, transparent and autonomous gonad of the female *Clytia* medusa: What signalling pathways control the selection of stage I oocytes for daily growth in response to nutrient availability? How does the dark–light signal trigger peptide release from the gonad, and how does this act on the oocyte to cause the cytoplasmic cAMP rise at maturation? What cis and trans factors assure the precise regulation of translation of different classes of maternal RNAs at each successive step of oocyte growth and maturation? Such questions have the potential both to inform us on the fascinating diversity of animal reproductive strategies, and to identify the fundamental features of mechanisms described in existing bilaterian models.

We have provided here an idea of the current experimental possibilities available for analyses of the molecular basis of oogenesis and oocyte maturation in *Clytia*. For molecular studies, many potentially interesting regulatory genes can be identified from existing EST and cDNA sequence collections, currently covering about 8000 different expressed transcripts. A full genome sequencing project is underway. It is possible to interfere with function of individual genes in mid-stage and full grown oocytes by injection of exogenous wild-type and mutated forms of RNAs as well as by morpholino antisense oligonucleotides to block RNA translation (Figure 3.5). Genes functioning during early stages of oogenesis are presently inaccessible because of the limits of microinjection. To circumvent this we are currently working to adapt the RNAi and transgenic techniques being developed in *Hydra* (Galliot *et al.*, 2007; Khalturin *et al.*, 2007) to *Clytia* adults. Another exciting direction will be the development of live imaging techniques to allow dynamic studies of regulatory protein and localized RNA within growing and maturing oocytes. We hope that this chapter will stimulate others to exploit the promising *Clytia* system, which can add a fresh perspective on the regulation of oogenesis and its evolutionary history.

Acknowledgements

We are indebted to all the members of our group past and present involved in developing the *Clytia* model, especially the other pioneers including Sandra Chevalier, Manon Quiquand, Emilie Peco, Lucie Robert and Cécile Fourrage. Crucial roles were also played by Michael Manuel and his group in Paris (especially, for the oocyte studies, Lucas Leclère), who joined forces with us in the development of tools and knowledge, and who showed us the usefulness of an evolutionary perspective for understanding biological processes. Finally, a special mention of Dany Carré, on whose recommendation we started using *Clytia*, and who has generously shared with us her great experience of hydrozoan biology.

Funding for the original research was gratefully accepted from the CNRS (Centre National de la Recherche Scientifique), ARC (Association pour la Recherche sur le Cancer) and ANR (Agence Nationale de la Recherche). EST sequences were generated by the Consortium National de Recherche en Genomique at the Génoscope (Evry, France).

References

Abrieu, A., Doree, M. and Fisher, D. (2001) The interplay between cyclin-B-Cdc2 kinase (MPF) and MAP kinase during maturation of oocytes. *J. Cell Sci.*, **114**, 257–267.

Alexandrova, O., Schade, M., Bottger, A. and David, C.N. (2005) Oogenesis in Hydra: nurse cells transfer cytoplasm directly to the growing oocyte. *Dev. Biol.*, **281**, 91–101.

Amiel, A. and Houliston, E. (2009) Three distinct RNA localization mechanisms contribute to oocyte polarity establishment in the cnidarian *Clytia hemisphaerica*. *Dev. Biol.*, **327**, 191–203.

Amiel, A., Leclère, L., Robert, L. *et al.* (2009) Conserved functions for Mos in eumetazoan oocyte maturation revealed by studies in a cnidarian. *Curr. Biol.*, **19**, 305–311.

Ball, E.E., Hayward, D.C., Saint, R. and Miller, D.J. (2004) A simple plan – cnidarians and the origins of developmental mechanisms. *Nat. Rev. Genet.*, **5**, 567–577.

Ballard, W.W. (1942) The mechanism for synchronous spawning in Hydractinia and Pennaria. *Biol. Bull.*, **82**, 329–339.

Bashirullah, A., Cooperstock, R.L. and Lipshitz, H.D. (1998) RNA localization in development. *Annu. Rev. Biochem.*, **67**, 335–394.

Bashirullah, A., Halsell, S.R., Cooperstock, R.L. *et al.* (1999) Joint action of two RNA degradation pathways controls the timing of maternal transcript elimination at the midblastula transition in Drosophila melanogaster. *EMBO J.*, **18**, 2610–2620.

Belloc, E., Pique, M. and Mendez, R. (2008) Sequential waves of polyadenylation and deadenylation define a translation circuit that drives meiotic progression. *Biochem. Soc. Trans.*, **36**, 665–670.

Bodart, J.F., Baert, F.Y., Sellier, C. *et al.* (2005) Differential roles of p39Mos-Xp42Mpk1 cascade proteins on Raf1 phosphorylation and spindle morphogenesis in *Xenopus* oocytes. *Dev. Biol.*, **283**, 373–383.

Bodo, F. and Bouillon, J. (1968) Etude histologique du développement embryonnaire de quelques Hydroméduses de Roscoff. *Cah. Biol. Mar.*, **IX**, 69–79.

Boero, F., Bouillon, J. and Piraino, S. (1992) On the origins and evolution of hydromedusan life cycles (*Cnidaria, Hydrozoa*). In *Sex Origin and Evolution* (ed. R. Dallai), Mucchi, Modena, pp. 59–68.

Carré, D. and Carré, C. (2000) Origin of germ cells, sex determination, and sex inversion in medusae of the genus Clytia (Hydrozoa, Leptomedusae): the influence of temperature. *J. Exp. Zool.*, **287**, 233–242.

Carré, D., Carré, C., Pagès, F. and Gili, J.-M. (1995) Asexual reproduction of the pelagic phase of Clytia mccradyii (Hydrozoa, Leptomedusae). *Sci. Mar.*, **59**, 193–202.

Carré, D. and Sardet, C. (1981) Sperm chemotaxis in siphonophores. *Biol. Cell*, **40**, 119–128.

Chera, S., de Rosa, R., Miljkovic-Licina, M. *et al.* (2006) Silencing of the hydra serine protease inhibitor Kazal1 gene mimics the human SPINK1 pancreatic phenotype. *J. Cell Sci.*, **119**, 846–857.

Chevalier, S., Martin, A., Leclère, L. *et al.* (2006) Polarised expression of FoxB and FoxQ2 genes during development of the hydrozoan *Clytia hemisphaerica*. *Dev. Genes. Evol.*, **216**, 709–720.

Colledge, W.H., Carlton, M.B., Udy, G.B. and Evans, M.J. (1994) Disruption of c-mos causes parthenogenetic development of unfertilized mouse eggs. *Nature*, **370**, 65–68.

Denker, E., Manuel, M., Leclere, L. *et al.* (2008) Ordered progression of nematogenesis from stem cells through differentiation stages in the tentacle bulb of *Clytia hemisphaerica* (Hydrozoa, Cnidaria). *Dev. Biol.*, **315**, 99–113.

Dumont, J., Umbhauer, M., Rassinier, P. *et al.* (2005) p90Rsk is not involved in cytostatic factor arrest in mouse oocytes. *J. Cell Biol.*, **169**, 227–231.

Dunn, C.W., Hejnol, A., Matus, D.Q. *et al.* (2008) Broad phylogenomic sampling improves resolution of the animal tree of life. *Nature*, **452**, 745–749.

Eckelbarger, K.J. and Larson, R.L. (1992) Ultrastructure of the ovary and oogenesis in the jellyfish *Linuche unguiculata* and *Stomolophus meleagris*, with a review of ovarian structure in the Scyphozoa. *Mar. Biol.*, **114**, 633–643.

Eddy, E.M. (1975) Germ plasm and the differentiation of the germ cell line. *Int. Rev. Cytol.*, **43**, 229–280.

Eisenman, E. and Alfert, M. (1981) A new fixation procedure for preserving the ultrastructure of marine invertebrate tissues. *J. Micros.*, **125**, 117–120.

Faulkner, G.H. (1929) The early prophases of the first oocyte division as seen in life, in Obelia Geniculata. *Q. J. Microsc. Sci.*, **s2-73**, 225–242.

Freeman, G. (1979) The multiple roles which division can play in the localization of developmental potential. In *Determinants of Spatial Organisation* (eds S. Subtelny and C. Bradt), Academic Press, pp. 53–76.

Freeman, G. (1980) The role of cleavage in the establishment of the anterior-posterior axis of the hydrozoan embryo. In *Developmental and Cellular Biology of Coelenterates* (eds P. Tardent and R. Tardent), Elsevier/North Holland Biomedical Press, Amsterdam, pp. 97–108.

Freeman, G. (1981a) The cleavage initiation site establishes the posterior pole of the hydrozoan embryo. *Wilhelm Roux's Arch.*, **190**, 123–125.

Freeman, G. (1981b) The role of polarity in the development of the hydrozoan planula larva. *Wilhelm Roux's Arch.*, **190**, 168–184.

Freeman, G. (1987) The role of oocyte maturation in the ontogeny of the fertilisation site in the hydrozoan *Hydractinia echinata. Roux's Arch. Dev. Biol.*, **196**, 83–92.

Freeman, G. (2005) The effect of larval age on developmental changes in the polyp prepattern of a hydrozoan planula. *Zoology*, **108**, 55–73.

Freeman, G. and Miller, R.L. (1982) Hydrozoan eggs can only be fertilized at the site of polar body formation. *Dev. Biol.*, **94**, 142–152.

Freeman, G. and Ridgway, E.B. (1988) The role of cAMP in oocyte maturation and the role of the germinal vesicle contents in mediating maturation and subsequent developmental events in hydrozoans. *Roux's Arch. Dev. Biol.*, **197**, 197–211.

Freeman, G. and Ridgway, E.B. (1993) The role of intracellular calcium and pH during fertilization and egg activation in the hydrozoan *Phialidium. Dev. Biol.*, **156**, 176–190.

Fritzenwanker, J.H., Genikhovich, G., Kraus, Y. and Technau, U. (2007) Early development and axis specification in the sea anemone *Nematostella vectensis. Dev. Biol.*, **310**, 264–279.

Galliot, B., Miljkovic-Licina, M., Ghila, L. and Chera, S. (2007) RNAi gene silencing affects cell and developmental plasticity in hydra. *C. R. Biol.*, **330**, 491–497.

Goldman, D.S., Kiessling, A.A., Millette, C.F. and Cooper, G.M. (1987) Expression of c-mos RNA in germ cells of male and female mice. *Proc. Natl. Acad. Sci. USA*, **84**, 4509–4513.

Gross, S.D., Schwab, M.S., Lewellyn, A.L. and Maller, J.L. (1999) Induction of metaphase arrest in cleaving Xenopus embryos by the protein kinase p90Rsk. *Science*, **286**, 1365–1367.

Haccard, O. and Jessus, C. (2006) Oocyte maturation, Mos and cyclins—a matter of synthesis: two functionally redundant ways to induce meiotic maturation. *Cell Cycle*, **5**, 1152–1159.

Hamilton, T.L., Stoneley, M., Spriggs, K.A. and Bushell, M. (2006) TOPs and their regulation. *Biochem. Soc. Trans.*, **34**, 12–16.

Hashimoto, N., Watanabe, N., Furuta, Y. *et al.* (1994) Parthenogenetic activation of oocytes in c-mos-deficient mice. *Nature*, **370**, 68–71.

Hertwig, O. and Hertwig, R. (1895) *Der Organismus der Medusen und seine Stellung zur Keimblättertheorie*, Jena.

Hirose, M., Kinzie, R.A. 3rd and Hidaka, M. (2000) Early development of zooxanthella-containing eggs of the corals Pocillopora verrucosa and P. eydouxi with special reference to the distribution of zooxanthellae. *Biol. Bull.*, **199**, 68–75.

Honegger, T.G., Achermann, J., Littlefield, R.J. *et al.* (1980) Light-controlled spawning in *Phialidium hemisphaericum* (Leptomedusae). In *Developmental and Cellular Biology of Coelenterates* (eds P. Tardent and R. Tardent), Elsevier/North Holland Biomedical Press, Amsterdam, pp. 83–88.

Ikegami, S., Honji, N. and Yoshida, M. (1978) Light-controlled production of spawning-inducing substance in jellyfish ovary. *Nature*, **272**, 611–612.

Inoue, D., Ohe, M., Kanemori, Y. *et al.* (2007) A direct link of the Mos-MAPK pathway to Erp1/Emi2 in meiotic arrest of Xenopus laevis eggs. *Nature*, **446**, 1100–1104.

Inselman, A. and Handel, M.A. (2004) Mitogen-activated protein kinase dynamics during the meiotic G2/MI transition of mouse spermatocytes. *Biol. Reprod.*, **71**, 570–578.

Ivanovska, I., Lee, E., Kwan, K.M. *et al.* (2004) The Drosophila MOS ortholog is not essential for meiosis. *Curr. Biol.*, **14**, 75–80.

Jager, M., Queinnec, E., Houliston, E. and Manuel, M. (2006) Expansion of the SOX gene family predated the emergence of the Bilateria. *Mol. Phylogenet. Evol.*, **39**, 468–477.

Karaiskou, A., Dupre, A., Haccard, O. and Jessus, C. (2001) From progesterone to active Cdc2 in Xenopus oocytes: a puzzling signalling pathway. *Biol. Cell*, **93**, 35–46.

Kawaguti, S. and Ogasawara, Y. (1967) Electron microscopy on the ovary of an anthomedusa, *Spirocodon saltatrix*. *Biol. J. Okayama. Univ.*, **13**, 115–129.

Khalturin, K., Anton-Erxleben, F., Milde, S. *et al.* (2007) Transgenic stem cells in Hydra reveal an early evolutionary origin for key elements controlling self-renewal and differentiation. *Dev. Biol.*, **309**, 32–44.

King, M.L., Messitt, T.J. and Mowry, K.L. (2005) Putting RNAs in the right place at the right time: RNA localization in the frog oocyte. *Biol. Cell*, **97**, 19–33.

Kondoh, E., Tachibana, K. and Deguchi, R. (2006) Intracellular Ca2 + increase induces post-fertilization events via MAP kinase dephosphorylation in eggs of the hydrozoan jellyfish *Cladonema pacificum*. *Dev. Biol.*, **293**, 228–241.

Kubota, S. (1978) The life-history of *Clytia edwardsi* (Hydrozoa; Campanulariidae) in Hokkaido, Japan. *Jour. Fac. Sci. Hokkaido Univ. Ser. VI, Zool.*, **21**, 317–354.

Lee, P.N., Kumburegama, S., Marlow, H.Q. *et al.* (2007) Asymmetric developmental potential along the animal-vegetal axis in the anthozoan cnidarian, *Nematostella vectensis*, is mediated by Dishevelled. *Dev. Biol.*, **310**, 169–186.

Lenart, P., Bacher, C.P., Daigle, N. *et al.* (2005) A contractile nuclear actin network drives chromosome congression in oocytes. *Nature*, **436**, 812–818.

Liu, J., Grimison, B., Lewellyn, A.L. and Maller, J.L. (2006) The anaphase-promoting complex/cyclosome inhibitor Emi2 is essential for meiotic but not mitotic cell cycles. *J. Biol. Chem.*, **281**, 34736–34741.

Maas, O. (1905) Experimentelle Beiträge zur Entwicklungsgeschichte der Medusen. *Zeitschr. f. Wiss. Zool.*, **82**, 601–610.

Madgwick, S. and Jones, K.T. (2007) How eggs arrest at metaphase II: MPF stabilisation plus APC/C inhibition equals Cytostatic Factor. *Cell Div.*, **2**, 4.

Masui, Y. (2000) The elusive cytostatic factor in the animal egg. *Nat. Rev. Mol. Cell Biol.*, **1**, 228–232.

Meijer, L., Dostmann, W., Genieser, H.G. *et al.* (1989) Starfish oocyte maturation: evidence for a cyclic AMP-dependent inhibitory pathway. *Dev. Biol.*, **133**, 58–66.

Meurer, M. and Hündgen, M. (1978) Licht- und elektronenmikroskopischer Bau der Süßwassermeduse *Craspedacusta sowerbii* (Hydrozoa, Limnohydrina). *Zool. Jb. Anat.*, **100**, 485–508.

Micklem, D.R. (1995) mRNA localisation during development. *Dev. Biol.*, **172**, 377–395.

Miller, D.J. and Ball, E.E. (2008) Cryptic complexity captured: the Nematostella genome reveals its secrets. *Trends Genet.*, **24**, 1–4.

Miller, D.J., Ball, E.E. and Technau, U. (2005) Cnidarians and ancestral genetic complexity in the animal kingdom. *Trends Genet.*, **21**, 536–539.

Miller, M.A., Technau, U., Smith, K.M. and Steele, R.E. (2000) Oocyte development in *Hydra* involves selection from competent precursor cells. *Dev. Biol.*, **224**, 326–338.

Momose, T., Derelle, R. and Houliston, E. (2008) A maternally localised Wnt ligand required for axial patterning in the cnidarian *Clytia hemisphaerica*. *Development*, **135**, 2105–2113.

Momose, T. and Houliston, E. (2007) Two oppositely localised Frizzled RNAs as axis determinants in a Cnidarian embryo. *PLoS Biology*, **5**, E70.

Momose, T. and Schmid, V. (2006) Animal pole determinants define oral-aboral axis polarity and endodermal cell-fate in hydrozoan jellyfish *Podocoryne carnea*. *Dev. Biol.*, **292**, 371–380.

Mori, M., Hara, M., Tachibana, K. and Kishimoto, T. (2006) p90Rsk is required for G1 phase arrest in unfertilized starfish eggs. *Development*, **133**, 1823–1830.

Philippe, H., Derelle, R., Lopez, P. *et al.* (2009) Phylogenomics revives traditional views on deep animal relationships. *Curr. Biol.*, **19**, 706–712.

Plickert, G., Jacoby, V., Frank, U. *et al.* (2006) Wnt signaling in hydroid development: formation of the primary body axis in embryogenesis and its subsequent patterning. *Dev. Biol.*, **298**, 368–378.

Putnam, N.H., Srivastava, M., Hellsten, U. *et al.* (2007) Sea anemone genome reveals ancestral eumetazoan gene repertoire and genomic organization. *Science*, **317**, 86–94.

Rebscher, N., Volk, C., Teo, R. and Plickert, G. (2008) The germ plasm component Vasa allows tracing of the interstitial stem cells in the cnidarian Hydractinia echinata. *Dev. Dyn.*, **237**, 1736–1745.

Rodimov, A.A. (2005) Development of morphological polarity in embryogenesis of Cnidaria. *Russ J. Dev. Biol.*, **36**, 298–303.

Roosen-Runge, E.C. (1962) On the biology of sexual reproduction of the Hydromedusae, Genus Phialidium Leuckhart. *Pac. Sci.*, **XVI**, 15–24.

Roosen-Runge, E.C. (1970) Life cycle of the hydromedusa Philidium Gregarium (A Agassiz 1862) in the laboratory. *Biol. Bull.*, **139**, 203–221.

Roosen-Runge, E.C. and Szollosi, D. (1965) On biology and structure of the testis of Philidium Leuckhart (Leptomedusae). *Z. Zellforsch. Mikrosk. Anat.*, **68**, 597–610.

Sagata, N. (1998) Introduction: meiotic maturation and arrest in animal oocytes. *Semin. Cell Dev. Biol.*, **9**, 535–537.

Sagata, N., Watanabe, N., Vande Woude, G.F. and Ikawa, Y. (1989) The c-mos proto-oncogene product is a cytostatic factor responsible for meiotic arrest in vertebrate eggs. *Nature*, **342**, 512–518.

Schmid, V., Schmid, B., Schneider, B. *et al.* (1976) Factors effecting manubrium regeneration in hydromedusae (Coelenterata). *Wilhelm Roux's Arch.*, **179**, 41–56.

Schmid, V. and Tardent, P. (1971) The reconstitutional performances of the Leptomedusa *Campanularia johnstoni*. *Mar. Biol.*, **8**, 99–104.

Schwab, M.S., Kim, S.H., Terada, N. *et al.* (1999) p70(S6K) controls selective mRNA translation during oocyte maturation and early embryogenesis in Xenopus laevis. *Mol. Cell Biol.*, **19**, 2485–2494.

Schwoerer-Böhning, B., Kroiher, M. and Müller, W. (1990) Signal transmission and covert pre-pattern in the metamorphosis of Hydractinia echinata. *Roux's Arch. Dev. Biol.*, **203**, 422–428.

Spindler, K. and Müller, W. (1972) Induction of metamorphosis by bacteria and by a lithium pulse in the larvae of Hydractinia ecinata (Hydrozoa). *Roux's Arch. Dev. Biol.*, **169**, 271–280.

Spring, J., Yanze, N., Josch, C. *et al.* (2002) Conservation of Brachyury, Mef2, and Snail in the myogenic lineage of jellyfish: a connection to the mesoderm of bilateria. *Dev. Biol.*, **244**, 372–384.

St Johnston, D. (1995) The intracellular localisation of messenger RNAs. *Cell*, **81**, 161–170.

Steele, R.E. (2002) Developmental signaling in Hydra: what does it take to build a "simple" animal? *Dev. Biol.*, **248**, 199–219.

Stricker, S.A. and Smythe, T.L. (2001) 5-HT causes an increase in cAMP that stimulates, rather than inhibits, oocyte maturation in marine nemertean worms. *Development*, **128**, 1415–1427.

Tachibana, K., Tanaka, D., Isobe, T. and Kishimoto, T. (2000) c-Mos forces the mitotic cell cycle to undergo meiosis II to produce haploid gametes. *Proc. Natl. Acad. Sci. USA*, **97**, 14301–14306.

Takeda, N., Kyozuka, K. and Deguchi, R. (2006) Increase in intracellular cAMP is a prerequisite signal for initiation of physiological oocyte meiotic maturation in the hydrozoan *Cytaeis uchidae*. *Dev. Biol.*, **298**, 248–258.

Technau, U., Miller, M.A., Bridge, D. and Steele, R.E. (2003) Arrested apoptosis of nurse cells during *Hydra* oogenesis and embryogenesis. *Dev. Biol.*, **260**, 191–206.

Technau, U., Rudd, S., Maxwell, P. *et al.* (2005) Maintenance of ancestral complexity and non-metazoan genes in two basal cnidarians. *Trends Genet.*, **21**, 633–639.

Teissier, G. (1931) Etude expérimentale du développement de quelques hydraires. *Ann. Sci. Nat. Ser. X*, **14**, 5–60.

Teissier, G. (1933) Recherches sur les potentialités de l'œuf des hydraires. Polarité des larves complexes produites par greffe embryonnaire. *C. R. Soc. Biol.*, **113**, 26–27.

Verlhac, M.H., Kubiak, J.Z., Clarke, H.J. and Maro, B. (1994) Microtubule and chromatin behavior follow MAP kinase activity but not MPF activity during meiosis in mouse oocytes. *Development*, **120**, 1017–1025.

Verlhac, M.H., Kubiak, J.Z., Weber, M. *et al.* (1996) Mos is required for MAP kinase activation and is involved in microtubule organization during meiotic maturation in the mouse. *Development*, **122**, 815–822.

Verlhac, M.H., Lefebvre, C., Guillaud, P. *et al.* (2000) Asymmetric division in mouse oocytes: with or without Mos. *Curr. Biol.*, **10**, 1303–1306.

Wedi, S.E. and Dunn, D.F. (1983) Gametogenesis and reproductive periodicity of the subtidal sea anemone *Urticina lofotensis* (Coelenterata: Actiniaria) in California. *Biol. Bull.*, **165**, 458–472.

Weiler-Stolt, B. (1960) Uber die Bedeutung der interstitiellen Zellen fur die Entwicklung und Fortpflanzung mariner Hydroiden. *Wilhelm Roux Arch. EntwMech.*, **152**, 389–454.

Wikramanayake, A.H., Hong, M., Lee, P.N. *et al.* (2003) An ancient role for nuclear beta-catenin in the evolution of axial polarity and germ layer segregation. *Nature*, **426**, 446–450.

Yamamoto, D.S., Tachibana, K., Sumitani, M. *et al.* (2008) Involvement of Mos-MEK-MAPK pathway in cytostatic factor (CSF) arrest in eggs of the parthenogenetic insect, Athalia rosae. *Mech. Dev.*, **125**, 996–1008.

Yanze, N., Spring, J., Schmidli, C. and Schmid, V. (2001) Conservation of Hox/ParaHox-related genes in the early development of a cnidarian. *Dev. Biol.*, **236**, 89–98.

Zimyanin, V.L., Belaya, K., Pecreaux, J. *et al.* (2008) In vivo imaging of oskar mRNA transport reveals the mechanism of posterior localization. *Cell*, **134**, 843–853.

Section II

Oocyte growth

4

Soma–germline interactions in the ovary: an evolutionary perspective

David Albertini[1,2,3] **and John Bromfield**[1]

[1]Department of Molecular and Integrative Physiology, University of Kansas Cancer Centre, Kansas City, KS 66160, USA
[2]Department of Anatomy and Cell Biology, University of Kansas Cancer Centre, Kansas City, KS 66160, USA
[3]Marine Biological Laboratory, Woods Hole, MA 02543, USA

4.1 Introduction

A hallmark of stable speciation in animals is the ability to improve reproductive fitness in a changing environment. At a basic level, organisms must adapt to environmental change by ensuring the production of viable and reproductively competent offspring. Organisms that reproduce sexually are committed to developmental design principles that guarantee the union of distinct gametes and thus propagate future generations. To achieve this, the soma engages in a direct dialogue with the gonads. And, this interaction is typically reciprocated by signals of gonadal origin that regulate the functionality of various somatic organ systems. By assuring mating opportunities at times that are optimal for mature gamete production and fertilization, organisms successfully reproduce.

Neither oogenesis nor spermatogenesis is autonomous of the soma. Organisms support gametogenesis by providing an intragonadal somatic cell niche where germ cells are formed and stored as a finite reserve, as in the case of eutherian mammals (Gilchrist, Ritter and Armstrong, 2004; Rodrigues *et al.*, 2008), or where they are derived from a renewable population of stem cells, as is most commonly the case in lower vertebrates and invertebrates (Kiger, White-Cooper and Fuller, 2000). Using this framework, we consider below the structure and function of communication systems

Oogenesis: The Universal Process Marie-Hélène Verlhac and Anne Villeneuve

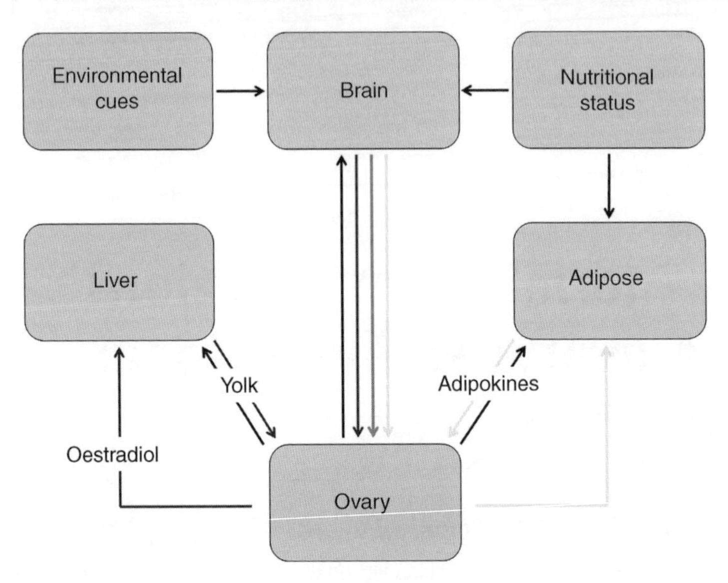

Figure 4.1 Diagram illustrating the basic network of interactions that are known to occur between the ovary and somatic compartments. Feedback loops between the brain and ovary are common to all organisms and integrate environmental cues with egg production and availability for fertilization. Oogenesis is dependent upon nutritional status for energy balance (adipose tissue equivalent) to support yolk synthesis in the liver, in response to ovarian oestrogens. Feedback and feed-forward pathways mediate these long-distance forms of communication between the soma and germline

that have evolved between thc ovarian follicle cell and the oocyte. It is a major tenet of this chapter that the process of oogenesis is intrinsically linked to folliculogenesis by a series of interactions involving paracrine feedback as well as junctional interactions at the germ cell–soma interface (Albertini and Barrett, 2003; Plancha *et al.*, 2005). In this way, a local communication system serves to coordinate the development and respective functions of the follicle and the oocyte. In the case of the oocyte, functionality is manifest by ovulation of full-grown and developmentally competent ova (Rodrigues *et al.*, 2008). In the case of the follicle, functionality is evidenced by the synchronization of endocrine output from the ovary aimed at supporting successful implantation and gestation in the event of fertilization (Rothchild, 2003). The follicle, directly or indirectly, is the fundamental unit that mediates long-distance communication between the ovary and multiple somatic targets (Figure 4.1).

4.2 Basic strategies for oogenesis: a phyllogenetic perspective

During the process of oogenesis, there are a number of phyllogenetic differences in the forms of communication established between the ovary and soma that appear to follow two basic strategies. In most organisms, oogenesis occurs seasonally and requires the investment of large energy resources to support the process of vitellogenesis in which yolk is synthesized and secreted by the liver or its equivalent (hepatopancreas) most typically in response to an oestrogenic stimulus received by the ovarian follicle

(Wallace and Selman, 1981, 1990; Rothchild, 2003; Webb *et al.*, 2002). From the blood, or after synthesis in follicle cells in some organisms (Marina *et al.*, 2004), yolk precursors or vitellogenins are retrieved into the growing oocyte by the process of receptor-mediated endocytosis (Anderson, 1972). Thus, the growth phase of oogenesis and the conversion from a previtellogenic to vitellogenic state is coupled to hormonal stimulation of yolk production. This form of long-distance communication between the germline and soma assures rapid oocyte hypertrophy through endocytosis and accumulation of yolk precursors within the ooplasm (Figure 4.1).

Many variations exist between species as to the mechanisms used to provide and store yolk in oocytes, but it is generally the case that yolky oocytes accommodate a pronounced expansion of the oolemma that must maintain active endocytosis until a postvitellogenic state is achieved (Wallace and Selman, 1990). This mechanism contrasts sharply with the oogenesis pathway for yolkless oocytes exhibited by organisms like eutherian mammals, which is the main focus of this chapter (Anderson, 1972). Rather than drawing upon a stem cell precursor (Kiger, White-Cooper and Fuller, 2000), mammals have adopted a strategy in which a finite oocyte pool is stored in the ovary within primordial follicles that are assembled prior to, at, or shortly after birth (Rodrigues *et al.*, 2008; Hertig and Barton, 1973). Oocytes of this kind tend to be relatively yolkless but likely require progressive changes in follicle cell structures that mediate the progressive needs of the developing oocyte within the ovarian follicle (Menkhorst *et al.*, 2009; Tanghe *et al.*, 2002; Su *et al.*, 2008). It is this form of soma–oocyte interaction that will be emphasized in this chapter and takes as its point of departure variations in the organization of the ovarian follicle.

4.3 Structural variations in interactions between oocytes and follicle cells

Cell contact is a widely used strategy for regulating homotypic or heterotypic cell activities. A major role played by cell contact, in its simplest form, is to provide an on/off switch through the engagement of specific surface signalling molecules. For example, neurons and T cells, in particular, elaborate highly differentiated membrane domains referred to as 'synapses'. These specialized domains convey information amongst and between neighbouring cells with which they make direct physical contact. Processes that occur within such domains include receptor aggregation (clusters), localized endocytosis, localized exocytosis and vectorial vesicle trafficking that guide delivery to, and retrieval from, sites of membrane apposition without disrupting the functional integrity of the synapse. This form of contact appears to be the most common amongst mammalian oocytes and uses a specialized extension of the granulosa cell known as transzonal projections (TZPs) (Albertini and Rider, 1994; Albertini and Barrett, 2003; Allworth and Albertini, 1993).

From an evolutionary perspective, it is instructive to consider the range of interactions seen at the oocyte soma interface to gain insight into changes in the nature of the dialogue between oocytes and their enveloping follicle cells. As shown in Figure 4.2, distinct organisms have adopted different strategies for establishing and

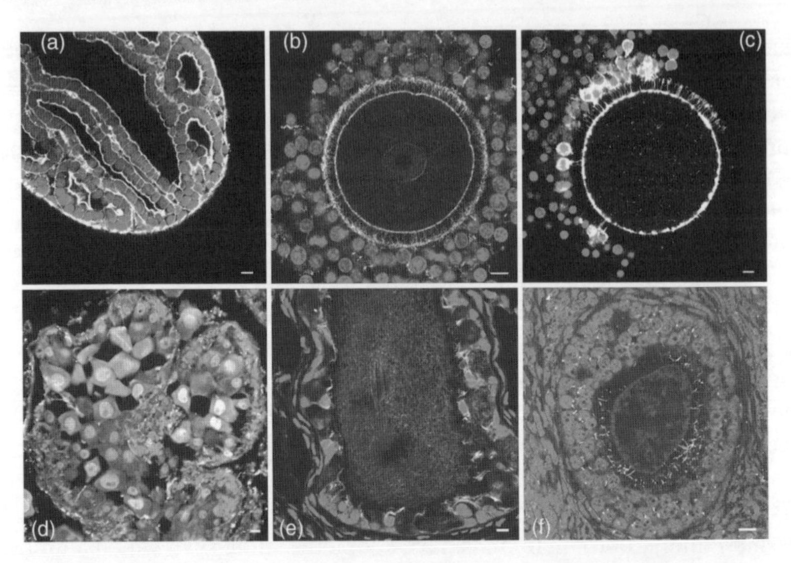

Figure 4.2 Overview of patterns of interaction between oocytes and follicle cells in diverse organisms. Molluscs (squid, a), amplify surface interactions within the follicle by extensive folding in the follicular epithelium which invaginates the oocyte; amplification in mammals (b, gerbil, c, bovine) involves formation of numerous TZPs that are attached to the actin-rich oocyte cortex. Panels a, b, and c are labelled with nuclear marker (red) and F-actin (white, phalloidin). The remaining panels illustrate acetylated tubulin labelling (white) and nuclei (red) in surf clam (d), dogfish (e), and baboon (f) follicles. Stable microtubule-rich TZPs link somatic cells to the oocytes in each of these species providing channels for direct communication. Scale bar = 10 μm, with the exception of d, where bar = 20 μm. A full colour version of this figure appears in the colour plate section.

maintaining contact. Molluscs, such as the squid, deploy a highly convoluted oolemma to accommodate maximal contact with a simple epithelium of follicle cells (Figure 4.2a). In other marine invertebrates such as the surf clam *Spissula sollidissima*, oocytes are released from the coelomic epithelium at the end of the growth stage of oogenesis and acquire a 'tear drop' shape as a result of the abscission of a projection that connects the oocyte to the follicular epithelium (Figure 4.2d). Organisms that formally use a follicle to contain oocytes that undergo vitellogenic growth display an orientated simple epithelium that is stabilized by acetylated microtubules at the apical/oolemmal surface (Figure 4.2e). Notably, mammalian oocytes amplify contact by increasing follicle number and the number of TZPs that interact with the oolemma (Figure 4.2b, c, e, and f).

Figure 4.3 summarizes three basic types of follicle cell–oocyte interactions that are seen in different organisms. Note that in all cases, these organisms contain oocytes within a follicle which is lined by a basement membrane that denotes the basal aspect of the follicle cell. It is also apparent that most oocytes assemble an extracellular matrix investment through which follicle cells must penetrate. In open forms of communication, there is direct cytoplasmic continuity between follicle cells and oocytes such that no filtration of somatic cell products would take place. Examples of this are manifold amongst invertebrates and lower vertebrates (Anderson, 1969; Anderson and Huebner, 1968; Neaves, 1971; Andreuccetti *et al.*, 1999; Grandi and

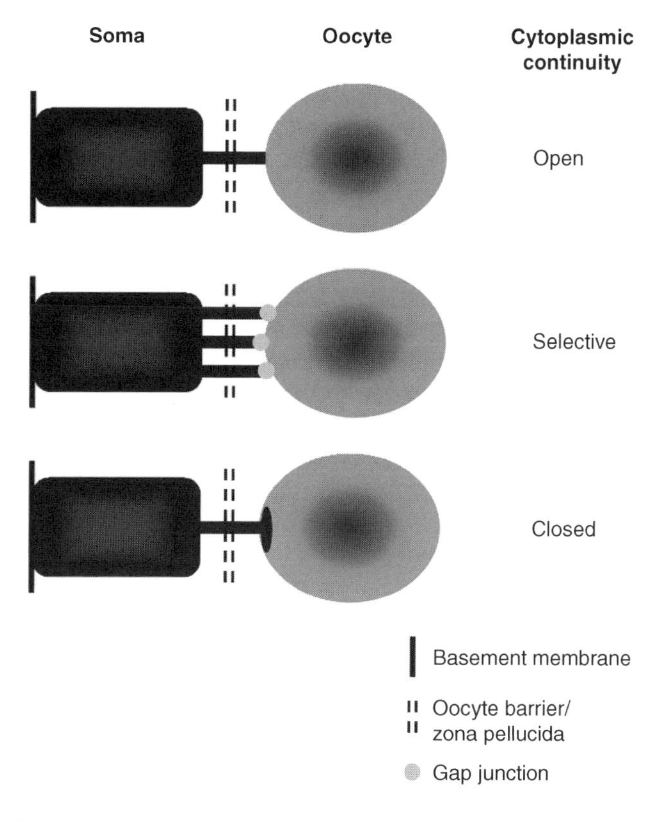

Figure 4.3 Schematic summarizing various forms of germ cell–somatic cell interactions. The schematic describes three orders of somatic–oocyte interactions in organisms of varying degrees of complexity. Cytoplasmic contiguity in lower organisms is complete with unrestricted passage (open) of organelles between the two cell types through cytoplasmic bridges. Many organisms regulate the passage of metabolites through gap junctions at cell contact points (filtered). In higher-order vertebrates, including humans, the soma–oocyte interactions are closed, but paracrine signalling is mediated through 'synaptic-like junctions'

Colombo, 1997; Andreuccetti, Taddei and Filosa, 1978; Gomez and Ramirez-Pinilla, 2004; Marina *et al.*, 2004). Filtered communication typifies those situations where connexin-based gap junctions have been defined that would permit selective passage of molecules with a molecular mass of 1000 kDa or less (Gilchrist, Ritter and Armstrong, 2004; Hertig and Barton, 1973; Murray *et al.*, 2008; Anderson and Albertini, 1976; Carabatsos *et al.*, 2000). A third class is present in mammals that consist of a solitary TZP that forms broad adhesive contacts at the oolemma (Albertini and Rider, 1994; Albertini and Barrett, 2003; Plancha *et al.*, 2005). While no cytoplasmic continuity is believed to occur at this interface, there is reason to believe that these contact domains are sites of active signalling and exchange of paracrine factors derived from the oocyte or follicle cell (Knight and Glister, 2006). Thus, amongst eutherian mammals, the common theme of TZPs appears. TZPs are seen as multiple radiating structures that are reinforced by both microtubule and microfilament components that can vary widely in density and form (Figures 4.2 and 4.4).

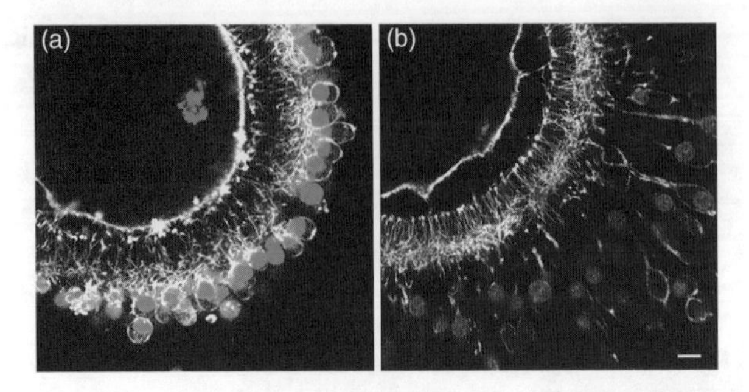

Figure 4.4 Representative confocal images demonstrating remodelling of TZPs during LH-induced meiotic maturation in horse follicles. (a) Organization of TZPs in an immature GV (germinal vesicle) stage oocyte; (b) TZP organization after LH exposure. Note the retraction of actin-rich TZPs but maintenance of contact between larger TZPs and oolemma. Nuclei are labelled in red and phalloidin-actin in white. Scale bar = 10 μm. A full colour version of this figure appears in the colour plate section.

Oocyte–granulosa interactions in mammals are not static structures but undergo changes in organization and function at different stages of oogenesis. In some cases, a direct role for hormones such as FSH (follicle stimulating hormone) has been implicated (Combelles *et al.*, 2004). Moreover, at ovulation, the role of the periovulatory surge of LH (luteinizing hormone) is to effect a dramatic remodelling of TZPs that varies widely between different mammals (Albertini, 2004). For example, in rodents there is a gradual retraction of TZPs during ovulation (Gilchrist, Ritter and Armstrong, 2004; Rodrigues *et al.*, 2008), whereas in bovine oocytes, actin TZPs are retracted and tubulin-based TZPs grow towards the oolemma establishing broad areas of membrane contact (Allworth and Albertini, 1993). Our recent studies in the horse suggest another variations in TZP remodelling during ovulation (Figure 4.4). Here, the retraction of TZPs appears to be selective, as most of these are withdrawn during cumulus expansion when the morphology of granulosa cells changes to a highly polarized state. Thus, variations exist over the course of oogenesis, both as a function of developmental stage and in relation to the species being studied. A schematic summarizing these events based on studies in the mouse is shown in Figure 4.5. The available data suggest that TZPs dominate between oocytes and granulosa cells in preantral follicles and that these are modified in response to FSH at the transition to an antral follicle state, and finally that widespread remodelling takes place during ovulation or *in vitro* maturation. This is an area of investigation that remains understudied and yet is central to understanding the determinants of oocyte quality which underscore the successful completion of oogenesis.

4.4 Conclusions

An extraordinary range of interactions are evident phyllogenetically at the interface of oocytes with somatic cells. Although the structural manifestations of these interactions imply the existence of developmental plasticity within a given species, it is difficult to ascertain the importance of such variations between organisms that have adopted widely divergent reproductive strategies. Two conserved physiological functions are subserved

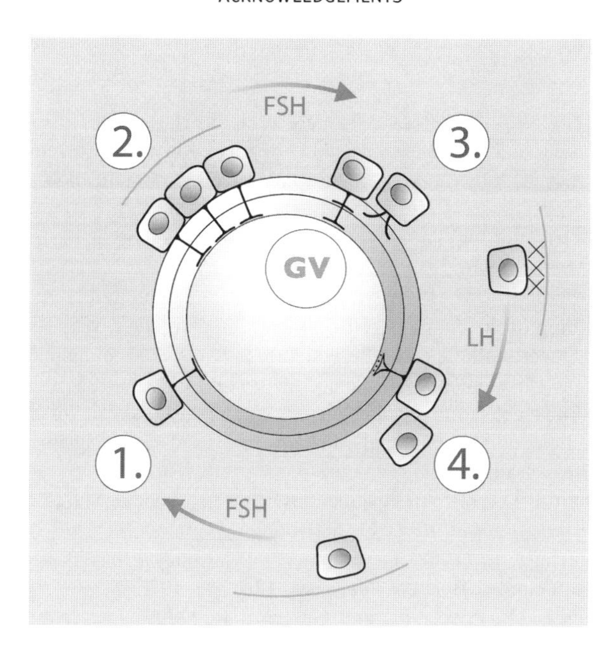

Figure 4.5 Schematic summarizing impact of gonadotrophins on TZP integrity in mammals. FSH initially induces remodelling of TZPs at early stages of follicle development and this is followed by an overall reduction in TZP density upon LH-induced oocyte maturation. This pattern reflects high metabolic support requirements during the growth phase of oogenesis, prior to FSH responsiveness, and the retention of contacts that serve to maintain cell cycle arrest until ovulation

by the follicle cell–oocyte interaction: to assure uptake, storage, and metabolic cooperation during the growth phase of oogenesis, and to link reproductive status and ovulation to the cell cycle state of the oocyte. There is immediate clinical relevance to these functions since the metabolic resources and maternal inheritance laid down during the growth phase of oogenesis are directly linked to the capacity of the ovum to sustain preimplantation development in eutherian mammals like the human. Moreover, the genetic stability of the oocyte is largely determined by the events of meiotic cell cycle progression coincident with ovulation, a time when dramatic alterations in the oocyte–granulosa cell communication are taking place. Sorting out the details of somatic cell–germline interactions during oogenesis poses a formidable challenge given species variation, but must be taken into account if improvements in animal and human reproductive fitness are to be obtained.

Acknowledgements

Past and present funding from the NIH, ESHE Fund, and the Hall Family Foundation have supported this work and provided the opportunity to study comparative aspects of oogenesis presented here. We recognize the input and encouragement of past and present members of the Albertini laboratory and especially thank Professor Everett Anderson for motivating and guiding the course of this work. We thank Stan Fernald for producing Figure 4.5.

References

Albertini, D.F. (2004) Oocyte-granulosa cell interactions, in *Essential IVF: Basic Research and Clinical Applications* (eds J. Van Blerkom and L. Gregory), Kluwer Academic Publishers, Boston.

Albertini, D.F. and Barrett, S.L. (2003) Oocyte-somatic cell communication. *Reprod. Suppl.*, **61**, 49–54.

Albertini, D.F. and Rider, V. (1994) Patterns of intercellular connectivity in the mammalian cumulus-oocyte complex. *Microsc. Res. Tech.*, **27**, 125–133.

Allworth, A.E. and Albertini, D.F. (1993) Meiotic maturation in cultured bovine oocytes is accompanied by remodelling of the cumulus cell cytoskeleton. *Dev. Biol.*, **158**, 101–112.

Anderson, E. (1969) Oocyte-follicle cell differentiation in two species of amphineurans (Mollusca), Mopalia mucosa and Chaetopleura apiculata. *J. Morphol.*, **129**, 89–125.

Anderson, E. (1972) The localisation of acid phosphatase and the uptake of horseradish peroxidase in the oocyte and follicle cells of mammals, in *Oogenesis* (eds J.D. Biggers and A.W. Schultz), University Park Press, Baltimore.

Anderson, E. and Albertini, D.F. (1976) Gap junctions between the oocyte and companion follicle cells in the mammalian ovary. *J. Cell Biol.*, **71**, 680–686.

Anderson, E. and Huebner, E. (1968) Development of the oocyte and its accessory cells of the polychaete, Diopatra cuprea (Bosc). *J. Morphol.*, **126**, 163–197.

Andreuccetti, P., Iodice, M., Prisco, M. and Gualtieri, R. (1999) Intercellular bridges between granulosa cells and the oocyte in the elasmobranch Raya asterias. *Anat. Rec.*, **255**, 180–187.

Andreuccetti, P., Taddei, C. and Filosa, S. (1978) Intercellular bridges between follicle cells and oocyte during the differentiation of follicular epithelium in Lacerta sicula Raf. *J. Cell Sci.*, **33**, 341–350.

Carabatsos, M.J., Sellitto, C., Goodenough, D.A. and Albertini, D.F. (2000) Oocyte-granulosa cell heterologous gap junctions are required for the coordination of nuclear and cytoplasmic meiotic competence. *Dev. Biol.*, **226**, 167.

Combelles, C.M., Carabatsos, M.J., Kumar, T.R. *et al.* (2004) Hormonal control of somatic cell oocyte interactions during ovarian follicle development. *Mol. Reprod. Dev.*, **69**, 347–355.

Gilchrist, R.B., Ritter, L.J. and Armstrong, D.T. (2004) Oocyte-somatic cell interactions during follicle development in mammals. *Anim. Reprod. Sci.*, **82–83**, 431–446.

Gomez, D. and Ramirez-Pinilla, M.P. (2004) Ovarian histology of the placentotrophic Mabuya mabouya (Squamata, Scincidae). *J. Morphol.*, **259**, 90–105.

Grandi, G. and Colombo, G. (1997) Development and early differentiation of gonad in the European eel (Anguilla anguilla [L.], Anguilliformes, Teleostei): A cytological and ultrastructural study. *J. Morphol.*, **231**, 195–216.

Hertig, A.T. and Barton, B.B. (1973) Fine structure of mammalian oocytes and ova, in *Handbook of Physiology: Endocrinology. Female Reproductive System, Part I* (eds O.R. Greep and E.B. Astwood), Williams & Wilkins, Baltimore.

Kiger, A.A., White-Cooper, H. and Fuller, M.T. (2000) Somatic support cells restrict germline stem cell self-renewal and promote differentiation. *Nature*, **407**, 750–754.

Knight, P.G. and Glister, C. (2006) TGF-{beta} superfamily members and ovarian follicle development. *Reproduction*, **132**, 191–206.

Marina, P., Salvatore, V., Maurizio, R. *et al.* (2004) Ovarian follicle cells in torpedo marmorata synthesize vitellogenin. *Mol. Reprod. Dev.*, **67**, 424–429.

Menkhorst, E., Nation, A., Cui, S. and Selwood, L. (2009) Evolution of the shell coat and yolk in amniotes: a marsupial perspective. *J. Exp. Zool. B Mol. Dev. Evol.*, **312**(6),625–638.

Murray, A.A., Swales, A.K.E., Smith, R.E. *et al.* (2008) Follicular growth and oocyte competence in the in vitro cultured mouse follicle: effects of gonadotrophins and steroids. *Mol. Hum. Reprod.*, **14**, 75–83.

Neaves, W.B. (1971) Intercellular bridges between follicle cells and oocyte in the lizard, Anolis carolinensis. *Anat. Rec.*, **170**, 285–301.

Plancha, C.E., Sanfins, A., Rodrigues, P. and Albertini, D. (2005) Cell polarity during folliculogenesis and oogenesis. *Reprod. Biomed. Online*, **10**, 478–484.

Rodrigues, P., Limback, D., Mcginnis, L.K. *et al.* (2008) Oogenesis: prospects and challenges for the future. *J. Cell Physiol.*, **216**, 355–365.

Rothchild, I. (2003) The yolkless egg and the evolution of eutherian viviparity. *Biol. Reprod.*, **68**, 337–357.

Su, Y.Q., Sugiura, K., Wigglesworth, K. *et al.* (2008) Oocyte regulation of metabolic cooperativity between mouse cumulus cells and oocytes: BMP15 and GDF9 control cholesterol biosynthesis in cumulus cells. *Development*, **135**, 111–121.

Tanghe, S., Van Soom, A., Nauwynck, H. *et al.* (2002) Minireview: functions of the cumulus oophorus during oocyte maturation, ovulation, and fertilisation. *Mol. Reprod. Dev.*, **61**, 414–424.

Wallace, R.A. and Selman, K. (1981) Cellular and dynamic aspects of oocyte growth in Teleosts. *Amer. Zool.*, **21**, 325–343.

Wallace, R.A. and Selman, K. (1990) Ultrastructural aspects of oogenesis and oocyte growth in fish and amphibians. *J. Electron. Microsc. Tech.*, **16**, 175–201.

Webb, M.A., Feist, G.W., Trant, J.M. *et al.* (2002) Ovarian steroidogenesis in white sturgeon (Acipenser transmontanus) during oocyte maturation and induced ovulation. *Gen. Comp. Endocrinol.*, **129**, 27–38.

Section III

Homologous chromosome pairing and recombination

5

Homologous chromosome pairing and synapsis during oogenesis

Susanna Mlynarczyk-Evans and Anne Villeneuve

Department of Developmental Biology, Stanford University School of Medicine, Beckman Center, B300, 279 Campus Drive, Stanford CA 94305-5329, USA

During gametogenesis, sexually reproducing organisms face the challenge of reducing their diploid chromosome number to a haploid complement so that, at fertilization, each gamete contributes precisely one set of chromosomes to the zygote, and diploid chromosome number is restored in the subsequent generation. This critical twofold reduction in chromosome number is accomplished by the specialized cell division programme of meiosis. Errors in chromosome inheritance during meiosis in human females represent a leading cause of miscarriage and birth defects, highlighting the importance of mechanisms that ensure accurate partitioning of chromosomes during oogenesis (Hassold and Hunt, 2001).

In meiosis, reduction of chromosome number is achieved by two successive nuclear divisions, termed meiosis I and meiosis II, following a single round of DNA replication. While the meiosis II division is similar to mitosis, in which sister chromatids segregate to opposite spindle poles, meiosis I is unique among cell divisions in that sister chromatids remain attached and homologous chromosomes segregate from one another. It is this event, the segregation of homologues at meiosis I, that is essential for the reduction of chromosome number to a haploid state.

At a fundamental level, the ability of homologous chromosomes to segregate from one another depends on the formation of pairwise associations between correct partner chromosomes. In this remarkable process, each chromosome must locate and recognize its one homologue among the many incorrect partners in the nucleus. In most organisms, chromosome pairing culminates in the side-by-side alignment of homologues, bridged along their lengths by a meiosis-specific structure known as the synaptonemal complex. This paired and synapsed chromosome organization promotes the formation of crossover

Oogenesis: The Universal Process Marie-Hélène Verlhac and Anne Villeneuve

recombination events in most organisms, creating physical linkages between the homologues that help to constrain them for bi-orientation on the meiosis I spindle. While examples of organisms in which meiosis occurs without synapsis and/or recombination have been identified, meiosis of all organisms invariably includes homologue pairing, underscoring the centrality of pairwise interactions between homologous chromosomes to the success of the meiotic programme.

The formation of stable interhomologue associations is especially important during oocyte meiosis, where the events of pairing, synapsis and recombination are often temporally uncoupled from the meiotic divisions. A characteristic feature of the oocyte developmental programme is an arrest prior to the meiosis I division, with resumption of the meiotic divisions occurring only after ovulation and/or fertilization. Thus, depending on the animal, associations between homologues must be maintained for hours, days, years or even decades to ensure correct meiotic chromosome inheritance.

In this chapter, we will discuss the events of early meiotic prophase that bring about and maintain stable associations between homologous chromosomes, highlighting how these events occur in the context of oogenesis. Further, we will consider evidence that these events are monitored to ensure oocyte quality. Our discussion will integrate lessons learned through a combination of genetic and cytological analyses in *Caenorhabditis elegans*, *Drosophila* and mammals.

5.1 Structure, composition and assembly of the synaptonemal complex

The synaptonemal complex (SC), a prominent zipperlike structure at the interface of paired and aligned homologous chromosomes, takes centre stage during meiotic prophase. Following its discovery in electron microscopy (EM) studies by M.J. Moses (Moses, 1956), the SC was soon recognized as a hallmark feature of the meiotic prophase nucleus, conserved across eukaryotes and present in most sexually reproducing organisms. While there is some variability between organisms in its cytological appearance by EM, a canonical SC can be described (Figure 5.1). It consists of a pair of electron-dense lateral elements (LEs), separated by approximately 100–200 nm, along which the chromatin of each homologue is organized in loops. The LEs are connected by a central region comprising a highly ordered lattice of transverse filaments flanking a central element that is decidedly pronounced in some organisms. Three-dimensional reconstruction reveals that this central region lattice is several layers thick (Schmekel and Daneholt, 1995).

Molecular components of the SC have been identified by a variety of approaches in the major animal meiosis systems. Biochemical purification of rat and hamster SCs yielded several rodent LE and central region proteins (Dobson *et al.*, 1994; Offenberg *et al.*, 1998; Meuwissen *et al.*, 1992; Lammers *et al.*, 1994); genetic mapping of mutations causing defects in meiosis led to the identification of proteins localizing to the LEs and/or central region in worms, flies, and mice (Page and Hawley, 2001; Webber, Howard and Bickel, 2004; Manheim and McKim, 2003; Bannister *et al.*, 2004; Zetka *et al.*, 1999; Martinez-Perez and Villeneuve, 2005; Couteau and Zetka, 2005; MacQueen *et al.*, 2002; Colaiacovo *et al.*, 2003; Smolikov *et al.*, 2007); and localization

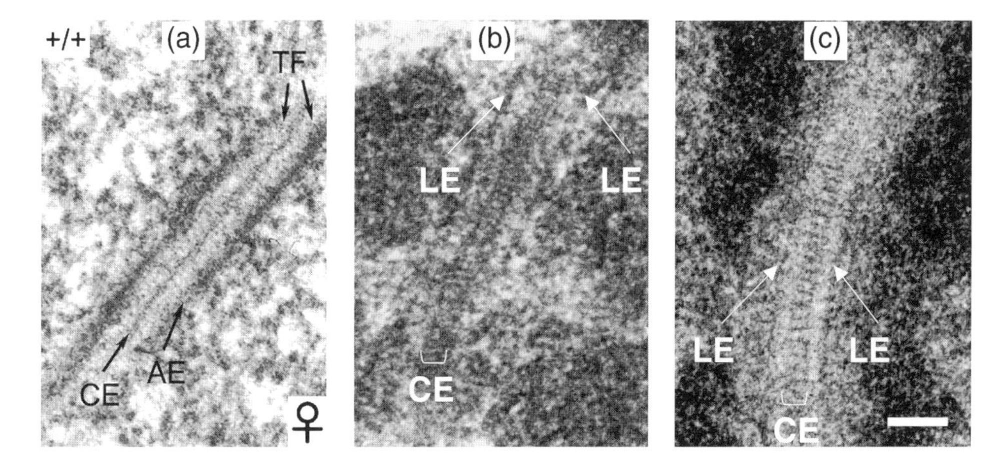

Figure 5.1 Morphology of synaptonemal complex (SC) in model organism oocytes. Transmission EM of sectioned mouse (a), *Drosophila* (b), and *C. elegans* (c) oocytes. Axial/lateral elements (AE/LE), central element (CE), and/or transverse filaments (TF) are indicated. Chromatin appears as dark staining to ether side of the SC. Scale bar = 100 nm in *C. elegans* image. Images adapted from (a) Hamer *et al.*, 2008, reproduced with permission from *The Journal of Cell Science*, doi: 10.1242/jcs.033233; (b) © Webber, Howard and Bickel, 2004, originally published in *The Journal of Cell Biology*, doi: 10.1083/jcb.200310077; and (c) Colaiacovo *et al.*, 2003, reproduced with permission from Elsevier, doi: 10.1016/S1534-5807(03)00232-6

studies of several germline-enriched proteins identified through functional genomics and biochemical approaches revealed additional SC components in mouse and worm (Chan *et al.*, 2003; Colaiacovo *et al.*, 2002; Prieto *et al.*, 2001; Revenkova *et al.*, 2004; Costa *et al.*, 2005; Hamer *et al.*, 2006; Pasierbek *et al.*, 2001; Pasierbek *et al.*, 2003; Goodyer *et al.*, 2008; Smolikov, Schild-Prufert and Colaiacovo, 2009). A theme emerging from this work is that the molecular components of the SC are quite poorly conserved between species, such that orthologues of many components have been difficult or impossible to identify based on sequence homology alone. Although the catalogue of SC structural components is probably not yet complete in any animal, several classes of proteins have been implicated (Table 5.1).

Proteins involved in sister chromatid cohesion represent one class of LE components. This class includes constituents of meiosis-specific cohesin complexes, the canonical mitotic form of which is comprised of a heterodimer of SMC1 and SMC3 plus the kleisin family member RAD21 and SCC3. At least one meiosis-specific subunit – usually the α-kleisin REC8 – is substituted in a meiotic version of the complex in most organisms, including worm and mouse (Pasierbek *et al.*, 2001; Bannister *et al.*, 2004). In the mouse, several additional meiosis-specific cohesin subunits have been identified (e.g. SMC1β in place of SMC1α (Revenkova *et al.*, 2004) or STAG3 in place of SCC3 (Prieto *et al.*, 2001)), and evidence suggests that several differentially composed meiotic cohesin complexes localize to distinct regions of the chromosomes (reviewed in Revenkova and Jessberger, 2006). Notably, the REC8 meiotic cohesin subunit is very poorly conserved, and functional studies have been required to support its identity in most organisms. In *Drosophila*, the C(2)M protein is an important LE

Table 5.1 Components of the synaptonemal complex in animal model systems

		Mouse	Worm	Fly
AE/LE	Cohesion proteins	SMC1β[a]	SMC-1[d]	SMC1
		SMC3	SMC-3	SMC3
		REC8[b]	REC-8[e]	?RAD21
		STAG3[c]	SCC-3	?
				ORD
	Non-cohesin components	SYCP2	HIM-3	C(2)M[h]
		SYCP3[f]	HTP-1	
		HORMAD1[g]	HTP-2	
		HORMAD2[g]	HTP-3	
Central region		SYCP1[i]	SYP-1	C(3)G
		SYCE1	SYP-2	
		SYCE2[j]	SYP-3	
		TEX12	SYP-4	

[a]Canonical SMC1α is present in some meiotic cohesin complexes.
[b]Mitotic cohesin RAD21 is present in some meiotic cohesin complexes.
[c]SA1/SA2 may be present in some meiotic cohesin complexes.
[d]Also known as HIM-1.
[e]Alternative kleisin subunits COH-3 and COH-4 are present in some meiotic cohesin complexes.
[f]Also known as COR1, SCP3.
[g]These components predominantly localize to unsynapsed AEs.
[h]Related to kleisin subunits of cohesin.
[i]Also known as SYN1, SCP1.
[j]Also known as CESC1.

component that displays some REC8 homology and has been shown to interact with SCC3, but plays only a minor role in cohesion (Manheim and McKim, 2003; Heidmann *et al.*, 2004). Instead, the non-cohesin ORD protein has assumed a major role in meiotic sister chromatid cohesion in this species (Miyazaki and Orr-Weaver, 1992; Bickel, Orr-Weaver and Balicky, 2002).

A second class of animal LE components belongs to the meiosis-enriched HORMA domain family, whose flagship member, Hop1, is a budding yeast LE component (Hollingsworth, Goetsch and Byers, 1990). The HORMA domain family is represented by different numbers of paralogues in different organisms and appears to be undergoing rapid evolution. Four paralogues – HIM-3, HTP-1, HTP-2, and HTP-3 – have been identified in *C. elegans*; in addition to their roles as LE structural components, studies have revealed numerous meiotic regulatory functions of these proteins (Zetka *et al.*, 1999; Nabeshima, Villeneuve and Hillers, 2004; Couteau *et al.*, 2004; Martinez-Perez and Villeneuve, 2005; Couteau and Zetka, 2005; Goodyer *et al.*, 2008). Two members of this family, *HORMAD1* and *HORMAD2*, are encoded in mammalian genomes, but this protein family appears to be absent in *Drosophila*. Interestingly, HORMAD1 and HORMAD2 are associated with unsynapsed axial elements (AEs, which are the precursors to the LEs of mature SC) in mouse oocytes both prior to synapsis and after desynapsis, but become depleted from synapsed regions of chromosomes upon installation of the SC central region (Fukuda *et al.*, 2009; Wojtasz *et al.*, 2009).

Many of the remaining SC components fall into a third class of proteins bearing prominent coiled-coil domains but otherwise displaying little homology to one another. This class includes two mouse LE components (SYCP2 (Yang *et al.*, 2006) and SYCP3 (Yuan *et al.*, 2000)), as well as the major central region proteins in all three organisms (mouse transverse filament protein SYCP1 (de Vries *et al.*, 2005) and central element proteins SYCE1 (Costa *et al.*, 2005), SYCE2 (Bolcun-Filas *et al.*, 2007) and TEX12 (Hamer *et al.*, 2006); worm central region proteins SYP-1 (MacQueen *et al.*, 2002), SYP-2 (Colaiacovo *et al.*, 2003), SYP-3 (Smolikov *et al.*, 2007) and SYP-4 (Smolikov, Schild-Prufert and Colaiacovo, 2009); and fly transverse filament protein C(3)G) (Page and Hawley, 2001). Given that the SC is a widespread and nearly universal feature of meiosis, the lack of conservation of its constituent proteins is likely to be significant. First, nonstructural roles of the SC components, such as meiotic regulatory functions, may be rapidly evolving. Second, the conservation of SC structural organization suggests that it is the overall architecture of the SC that is likely to be important for its function in meiosis.

Cytological studies employing immunofluorescence (IF) analysis to localize SC components have shown that assembly of the SC is coordinated with entry into and progression through meiotic prophase. While slight differences between species in the order and dependency of SC component localization have been documented, a general sequence of events can be described. Beginning in meiotic S phase, when homologous chromosomes are not yet associated in most organisms, cohesin complexes localize broadly to the chromatin (Khetani and Bickel, 2007; Chan *et al.*, 2003). Several non-cohesin LE components also localize diffusely to the chromatin at this stage (Hayashi, Chin and Villeneuve, 2007; Goodyer *et al.*, 2008; Khetani and Bickel, 2007). These early-loading SC components appear to be important for the subsequent morphogenesis of discrete AEs along the chromosome cores (Pasierbek *et al.*, 2001; Pasierbek *et al.*, 2003; Khetani and Bickel, 2007; Prieto *et al.*, 2004; Severson *et al.*, 2009). Distinct AEs begin to coalesce as nuclei enter the classical 'leptotene' stage of meiotic prophase, in which individual chromosomes become distinguishable and exhibit a convoluted, threadlike appearance. AEs first appear thin and patchy, but progressively thicken, shorten, and consolidate as additional AE/LE components load along each chromosome core in a process that continues in the subsequent 'zygotene' stage (Zetka *et al.*, 1999; Dobson *et al.*, 1994; Wojtasz *et al.*, 2009). The leptotene/zygotene transition is a noteworthy period of meiotic prophase marked in most organisms by the reorganization of chromosomes into polarized arrangements, and by the appearance of clear pairwise associations between the AEs of homologues (Zickler and Kleckner, 1998). Installation of the SC central region begins upon zygotene entry. The snapshots provided by analysis of fixed specimens suggest that SC central region protein loading initiates and spreads from a small number of nucleation sites per chromosome pair until full synapsis is achieved (MacQueen *et al.*, 2005; Tsubouchi, Macqueen and Roeder, 2008). Homologue pairs remain synapsed throughout the next 'pachytene' stage, in which the SC plays important roles in the completion of crossover recombination (see next chapter). During the late pachytene stage in *C. elegans* oocytes, a subset of LE proteins (HTP-1 and HTP-2) and the SC central region proteins become enriched on reciprocal chromosomal domains, in a crossover-dependent manner, as a prelude to desynapsis (Nabeshima, Villeneuve and Colaiacovo, 2005; Martinez-Perez *et al.*, 2008).

In the subsequent 'diplotene' stage, the SC disassembles, leaving behind bivalents consisting of homologous chromosomes that are now held together by crossovers in combination with flanking sister chromatid cohesion (Lee and Orr-Weaver, 2001). Bivalent associations are maintained during oocyte meiotic arrest, which occurs at the diplotene (dictyate) stage in mammals, at the subsequent diakinesis stage in *C. elegans*, and at metaphase I in *Drosophila*.

5.2 Role of the SC in homologous chromosome pairing

During the normal progression of meiosis, SC assembles between aligned homologous chromosomes, leading to early speculation that the SC might be involved in the process of establishing pairwise associations between homologues. We now know that installation of the SC central region is dispensable for homologue recognition and the initial establishment of pairing. Instead, synapsis functions in stabilizing and maintaining tight homologue association along the length of each chromosome pair during meiotic prophase.

The first evidence that synapsis is not required for establishing pairwise associations between homologues during oocyte meiosis came from studies in *C. elegans*. In this animal, the germline of each adult hermaphrodite contains hundreds of nuclei arranged in a spatiotemporal gradient of oocyte meiosis, an organization that allows detailed time-course analysis within a single specimen. Further, cytological studies can be performed in the context of intact nuclear architecture. In wild-type animals, fluorescence *in situ* hybridization (FISH) analysis reveals associations between homologous loci beginning at the leptotene/zygotene transition and persisting until the end of the pachytene stage (Dernburg *et al.*, 1998). In *syp-1* mutants, no SC is formed between chromosomes due to absence of this essential SC central region protein. However, FISH analysis revealed that homologous associations are nevertheless established at the leptotene/zygotene transition in *syp-1* mutants, indicating that synapsis is dispensable for homologue recognition and initial pairing (MacQueen *et al.*, 2002). Significantly, the colocalization of homologous loci became less frequent at later time points, demonstrating a role for SC in stabilizing intimate homologous associations as meiosis progresses. Interestingly, the degree of initial pairing detected depended upon the locus assayed, indicating that some chromosomal regions associate more tightly than others in asynaptic mutants. The most tightly associated loci were located near one end of each chromosome within genetically defined regions containing 'pairing centres' (PCs) (Herman, Kari and Hartman, 1982; Herman and Kari, 1989; McKim, Howell and Rose, 1988; McKim, Peters and Rose, 1993; Villeneuve, 1994). Observations in *syp-1* mutants indicated that PCs promote synapsis-independent stabilization of pairing in *C. elegans* (MacQueen *et al.*, 2002). Studies in mutants for additional SC central region components SYP-2, SYP-3, and SYP-4 yielded similar results, supporting the idea that synapsis is dispensable for initial homologue recognition and pairing, but is required to maintain associations between homologues along their lengths (Colaiacovo *et al.*, 2003; Smolikov *et al.*, 2007; Smolikov, Schild-Prufert and Colaiacovo, 2009).

Analysis of mouse meiocytes lacking SC central region components confirms that synapsis is also dispensable for homologue recognition in this animal. The mouse

Sycp1$^{-/-}$ mutant does not form any mature SC due to absence of this transverse filament protein. However, examination of spread preparations of meiotic nuclei by IF for AE components revealed aligned pairs of chromosomes of similar lengths (de Vries *et al.*, 2005). Furthermore, these unsynapsed chromosome pairs were connected by one or a few 'axial associations', visible by IF and in EM preparations. Rather than reflecting associations at predetermined chromosomal domains, as in *C. elegans*, these axial associations appear to correspond to sites of initiated recombination events (de Vries *et al.*, 2005). Mutants for central region proteins SYCE1, SYCE2 and TEX12, which fail to extend the SC beyond very limited, abnormal stretches as assayed by EM, also display associations between homologues (Bolcun-Filas *et al.*, 2007; Hamer *et al.*, 2008; Bolcun-Filas *et al.*, 2009). Thus, as in *C. elegans*, homologues associate in correct pairs and align in the absence of synapsis in the mouse; however, the SC central region is necessary for the intimate association of homologues beyond a limited number of sites per chromosome pair.

Work in *Drosophila* also supports the view that the SC central region is important for stabilizing interactions between homologous chromosomes during meiotic prophase. The association of specific loci has been probed by FISH or the lacI/lacO system in three-dimensionally preserved germaria, structures that contain premeiotic germ cells and several multinucleate oogenic cysts, each comprised of a developing oocyte surrounded by nurse cells, at progressive stages of meiosis. Compared to wild type, coincidence of homologous loci is observed less frequently in meiotic prophase nuclei of mutants that are defective for synapsis, such as *c(3)g* and *cona* (Sherizen *et al.*, 2005; Gong, McKim and Hawley, 2005; Page *et al.*, 2008). However, the ability to detect significant residual colocalization of homologous loci by these assays supports the idea that a pairing mechanism is still operational in the absence of synapsis. *Drosophila* is among a group of insects in which homologous chromosomes are paired in somatic nuclei (Hiraoka *et al.*, 1993; Fung *et al.*, 1998), and pairing is already established upon meiotic entry in this species (Vazquez, Belmont and Sedat, 2002; Sherizen *et al.*, 2005; Gong, McKim and Hawley, 2005). In contrast to wild type, *c(3)g* and *cona* mutants exhibit a drop in the frequencies of coincidence between homologous loci upon meiotic entry (Sherizen *et al.*, 2005; Page *et al.*, 2008). One interpretation of this result is that meiotic homologue pairing in *Drosophila* is unstable in the absence of synapsis, as in *C. elegans* (Page *et al.*, 2008). Insight into why synapsis is needed to stabilize pairing in *Drosophila* comes from analysis of *c(2)M* mutants. Work with mutants for this AE/LE component raises the possibility that homologue pairing can be destabilized by loading of AE proteins, and that installation of the SC central region normally counteracts this effect. *c(2)M* mutations prevent the formation of discrete AEs (Khetani and Bickel, 2007) and block synapsis by largely eliminating C(3)G localization to homologue pairs (Manheim and McKim, 2003). *c(3)G* is usually required to achieve high levels of meiotic recombination; however, this requirement is lifted in a *c(2)M* mutant background (Manheim and McKim, 2003). Together, these results suggest that, in the absence of synapsis, intimate associations between homologous chromosomes are better maintained when AE formation is prevented during *Drosophila* oogenesis.

Whereas mature SC and SC central region proteins are clearly dispensable for homologue recognition and initial establishment of pairing in all organisms studied, emerging evidence indicates that properly assembled AEs may play a role in these processes. In *C. elegans*, absence of certain AE/LE components causes failure to achieve pairing between all homologous sequences assayed, including pairing centres, in meiotic time-course analysis. These components include: SCC-3 cohesin, in the absence of which other known cohesin and non-cohesin AE/LE components also fail to load (Pasierbek *et al.*, 2003; Goodyer *et al.*, 2008; W. Zhang and A. Villeneuve, unpublished); HTP-3, which is also required for normal loading of all known cohesin and non-cohesin AE/LE proteins (Goodyer *et al.*, 2008; Severson *et al.*, 2009); HIM-3 (Couteau *et al.*, 2004); and HTP-1 in combination with close paralogue HTP-2 (Couteau and Zetka, 2005). Additionally, reduction of SMC-1 cohesin levels exacerbated the pairing defect in a genetic background in which pairing was partially compromised (Chan *et al.*, 2003). In most cases of AE/LE component deficiency, perturbations in nuclear reorganization at the leptotene/zygotene transition have also been noted. It is not yet understood whether the failure in pairing reflects a direct requirement for AE/LE proteins in homologue recognition *per se*, or an indirect effect through loss of nuclear reorganization, which may, in turn, promote homologue recognition (discussed in detail below).

Knockouts of mouse AE/LE components reported to date have not abolished homologous chromosome pairing. These mutants include: $Rec8^{-/-}$ and $Smc1\beta^{-/-}$, in which the other known cohesin and non-cohesin AE/LE components can still load (Bannister *et al.*, 2004; Xu *et al.*, 2005; Revenkova *et al.*, 2004); $Sycp2^{-/-}$, which also fails to load SYCP3 to the chromosome cores (Yang *et al.*, 2006); and $Sycp3^{-/-}$, which fails to load SYCP2 (Yuan *et al.*, 2000; Yuan *et al.*, 2002). Although some synapsis between chromosomes of similar length is achieved in $Smc1\beta^{-/-}$, $Sycp2^{-/-}$ and $Sycp3^{-/-}$ spermatocytes, a subset of chromosomes fails to exhibit SYCP1 transverse filament loading. Defects in the corresponding mutant oocytes have been more subtle, with chromosome pairs largely synapsed but SYCP1 stretches frequently exhibiting small gaps, and oocytes survive to meiosis II or beyond (Revenkova *et al.*, 2004; Yuan *et al.*, 2002; Yang *et al.*, 2006). Thus, synapsis may be more robust or less dependent on properly assembled AEs during oogenesis than spermatogenesis in the mouse. Analysis of $Sycp3^{-/-}$ spermatocytes indicated that although pairing is not abolished, it may be delayed in the absence of properly assembled AEs (Liebe *et al.*, 2004). Thus, pairing defects in mouse AE/LE component mutants are less severe than in *C. elegans*, perhaps reflecting species-specific differences in the involvement of AEs in pairing, or greater redundancy among mouse AE/LE components such that abrogation of AE function has not yet been achieved in any knockout in this species.

A clear role has emerged for mouse AE/LE components in defining meiotic chromosome axis length. In $Sycp3^{-/-}$ oocytes, the chromosome cores are approximately twice as long as normal, indicating that SYCP3 loading normally promotes meiotic chromosome compaction (Yuan *et al.*, 2002). Conversely, $Rec8^{-/-}$ and $Smc1\beta^{-/-}$ cohesin mutants exhibit shortened chromosome cores (Bannister *et al.*, 2004; Xu *et al.*, 2005; Revenkova *et al.*, 2004). SMC1β plays a role in the organization of meiotic chromatin into loops along each chromosome axis, illustrating a route by which cohesins could influence AE length (Revenkova *et al.*, 2004; Novak *et al.*, 2008). These observations reveal that the balance among the various AE/LE components determines the length of meiotic chromosome axes.

5.3 Mechanisms for coupling SC assembly to homologue identification

Installation of the SC central region needs to be carefully regulated to ensure that synapsis occurs only between homologues. Studies of meiocytes bearing altered karyotypes demonstrate that the SC itself is indifferent to homology. Synapsis can occur in haploid organisms where no homologues are present (Gillies, 1974; Loidl, Nairz and Klein, 1991). In heterozygotes for chromosomal translocations or inversions, polymerization of SC can bring nonhomologous chromosome segments into juxtaposition (MacQueen *et al.*, 2005; Loidl, 1990). Further, several observations suggest that SC central region loading occurs in a highly cooperative and processive manner. When transverse filament proteins are overexpressed, they spontaneously polymerize into polycomplexes that display structural features of the SC (Ollinger, Alsheimer and Benavente, 2005; Jeffress *et al.*, 2007). Conversely, under conditions where an SC component or synapsis-promoting factor is limiting, SC often assembles completely between a subset of chromosome pairs, rather than partially on all pairs (see e.g. Nabeshima, Villeneuve and Hillers, 2004; Couteau *et al.*, 2004; Yang *et al.*, 2006). In combination, these properties of SC central region assembly emphasize the need to regulate the nucleation step so that mature SC only assembles between correctly matched homologues. Sexually reproducing organisms appear to have solved this problem by evolving several distinct mechanisms for coupling initiation of synapsis to local homology verification.

In mouse, SC central region nucleation is mechanistically linked to the process of meiotic recombination. Prior to onset of synapsis in most organisms (Mahadevaiah *et al.*, 2001), meiotic recombination is initiated by the programmed introduction of double-strand breaks (DSBs) into the chromosomal DNA by a conserved nuclease, SPO11 (Keeney, 2001). Subsequently, these DSBs are repaired using the homologous chromosome as a template for recombinational repair, with a subset of recombination intermediates being repaired by a mechanism that yields crossover products. During gametogenesis in female and male mouse *Spo11* mutants, synapsis is severely defective: SYCP1 is absent from most chromosomes, and the few very short stretches of SYCP1 that form appear to occur between nonhomologues, as chromosomes of different lengths are involved and switches between synapsis partners occur (Romanienko and Camerini-Otero, 2000; Baudat *et al.*, 2000). These results indicate that initiation of recombination is important for normal assembly of SC between homologues in the mouse. Poor synapsis in *Spo11*$^{-/-}$ spermatocytes can be improved by the introduction of exogenous DSBs upon which the recombination machinery can act, suggesting that progression of recombination promotes synapsis, presumably in both sexes, in this organism (Romanienko and Camerini-Otero, 2000). Supporting this idea, a large class of mouse mutants affecting early or intermediate steps in the recombination process, including *Dmc1*$^{-/-}$, *Msh4*$^{-/-}$ and *Msh5*$^{-/-}$, shows a poor, nonhomologous synapsis phenotype similar to *Spo11*$^{-/-}$ (Yoshida *et al.*, 1998; Pittman *et al.*, 1998; Kneitz *et al.*, 2000; Edelmann *et al.*, 1999; de Vries *et al.*, 1999). However, maturation of recombination intermediates to yield crossover products is not required to promote synapsis, as mutants defective for late-acting crossover-promoting factors MLH1 and MLH3 exhibit full homologous synapsis despite a lack of crossovers for most chromosome pairs (Baker *et al.*, 1996; Edelmann *et al.*, 1996; Lipkin *et al.*, 2002). In the mouse, coupling SC formation with early steps in the recombination pathway apparently serves as a means of homology verification, ensuring that SC central region assembly is only nucleated between homologous chromosome pairs.

Although synapsis requires recombination in mammals, not all DSBs serve as sites for synapsis initiation. This point is illustrated by the fact that cytologically detectable DSB sites are in vast excess over synapsis initiation sites. Zygotene-stage nuclei exhibit partially synapsed chromosomes consistent with one or a few initiations per chromosome arm (see e.g. Baudat *et al.*, 2000), whereas factors that mark DSBs localize to numerous foci decorating the AEs/LEs of each homologue pair (Baudat and de Massy, 2007). The preferred locations of synapsis initiation appear to differ between the sexes, with most apparent initiations occurring near the telomeres during spermatogenesis and more internally during oogenesis (Scherthan *et al.*, 1996; Tankimanova, Hulten and Tease, 2004).

While coupling initiation of synapsis to the establishment of recombinational interactions is one way to ensure that SC is built between correctly paired homologues, flies and worms have found additional ways to solve this problem that do not require recombination. The existence of such mechanisms was made clear by the observation that loss of function of the *Drosophila* and *C. elegans* SPO11 homologues completely eliminates meiotic recombination but does not block the formation of morphologically normal SC between correctly aligned homologues (McKim *et al.*, 1998; McKim and Hayashi-Hagihara, 1998; Dernburg *et al.*, 1998). Thus, synapsis can proceed independently of recombination in flies and worms.

In worms, the PC located on each chromosome plays a prominent role in coupling SC assembly to pairing-partner choice. These chromosome domains not only function to stabilize pairing in the absence of synapsis, as discussed above; they also promote SC installation (MacQueen *et al.*, 2002; MacQueen *et al.*, 2005). Moreover, genetic analysis of reciprocal translocations in which pairing centres are exchanged between heterologous chromosomes indicates that PCs play a dominant role in partner choice (McKim, Howell and Rose, 1988; McKim, Peters and Rose, 1993), and cytological analysis of translocation heterozygotes suggests that synapsis initiates predominantly in the segment containing the PC and then proceeds to juxtapose heterologous segments (MacQueen *et al.*, 2005). Both the synapsis-independent stabilization of pairing and the synapsis-promoting functions of PCs require the HIM-8/ZIM-1/2/3 family of zinc finger proteins. One member of this four-protein family concentrates at the PC of each of the six chromosomes (ZIM-1 and -3 concentrate at two PCs each) (Phillips *et al.*, 2005; Phillips and Dernburg, 2006; Phillips *et al.*, 2009). Available data support a model in which PCs stabilize interactions between prospective pairing partners to permit local assessment of homology, and that synapsis proceeds if homology is verified (MacQueen *et al.*, 2005). Further, recent work suggests that the coupling between pairing and synapsis at PCs may operate in a manner analogous to the spindle assembly checkpoint, which delays anaphase in response to unattached kinetochores: PCs of chromosomes that have not yet identified a suitable pairing partner appear to impart a 'wait synapsis' signal that inhibits SC installation (Martinez-Perez and Villeneuve, 2005). Interestingly, this checkpoint-like mechanism requires HIM-3 and HTP-1, AE components that contain a HORMA domain, a feature that is shared with Mad2, a central component of the spindle assembly checkpoint (Gorbsky, Chen and Murray, 1998; Aravind and Koonin, 1998).

Although PCs play a predominant role in determining synapsis partner choice in *C. elegans*, several lines of evidence indicate that the information content for homologue recognition is not limited to PCs in this organism. Specifically, when one set of homologous chromosomes is heterozygous for a PC deletion, pairing and synapsis of these chromosomes are successful approximately half the time, indicating that interactions between two PCs are not strictly required for homologous synapsis (Villeneuve, 1994; MacQueen *et al.*, 2005). Furthermore, when PC function is compromised for two different chromosome pairs, the synapsis that does occur takes place between correctly matched chromosomes in the vast majority of cases. This observation indicates that even in the absence of PC-mediated synapsis-independent stabilization of pairing, homologues compete much more efficiently than nonhomologues to become synapsis partners (Phillips and Dernburg, 2006). Thus, although homologue recognition in *C. elegans* relies heavily on PCs, chromosomal regions outside the PC can also contribute to this process.

Drosophila provides a clear example of the use of multiple domains per chromosome to establish pairing between homologues via a recombination-independent mechanism. During oogenesis in flies heterozygous for translocations and complex chromosomal rearrangements, interactions between homologous chromosome segments are observed (Sherizen *et al.*, 2005; Gong, McKim and Hawley, 2005), demonstrating that information content used for homologue alignment is dispersed along the chromosome (McKee, 2009). Through genetic analysis, multiple sites have been identified per chromosome that appear to define synapsis intervals and stimulate SC formation (Hawley, 1980; Sherizen *et al.*, 2005). These observations suggest that recombination-independent homology verification and SC nucleation sites are both distributed at multiple sites along the *Drosophila* chromosomes.

The predominance of different mechanisms for coupling homology assessment to synapsis initiation in different organisms does not preclude the possibility that multiple mechanisms normally contribute to the establishment of homologous synapsis in a given system. Indeed, there is now strong evidence that this is the case in budding yeast. In this organism, as in mouse, SC assembly is dependent on initiation of recombination (Giroux, Dresser and Tiano, 1989; Alani, Padmore and Kleckner, 1990). However, recent analysis has revealed that a single domain on each chromosome also contributes to initiation of homologous synapsis in yeast meiosis. In particular, centromeres associate pairwise in early meiotic prophase nuclei; initially, these interactions are mostly nonhomologous, but partner switching takes place until full homologous centromere pairing is achieved in a recombination-dependent manner (Tsubouchi and Roeder, 2005). This centromere coupling process requires *zip1*, which encodes the SC transverse filament protein (Sym, Engebrecht and Roeder, 1993). Furthermore, centromeres are among the sites at which SC assembly is nucleated once a homologue is identified (Tsubouchi, Macqueen and Roeder, 2008). These data are consistent with a model in which homology assessment at the centromere is coupled to initiation of synapsis in budding yeast. Thus, rather than requiring the incredibly complex proposition that every sequence query the entire genome to identify its homologue, this mechanism may promote efficient homologue identification by focusing homology assessment on a limited pool of sequences. Once a match is found, initiation of SC polymerization could lock in partner selection and remove paired chromosomes from the pool still engaged in the homology search. This example illustrates how promoting interactions between specific chromosomal sites that also serve to nucleate SC assembly may be employed to simplify the homology search.

5.4 Roles for cytoskeleton-driven chromosome movements in meiotic prophase

It has long been clear that meiotic prophase must involve a substantial amount of chromosome motion. Classical cytological analysis in many species has revealed dramatic, large-scale changes in spatial organization of chromosomes within the nucleus beginning around the leptotene/zygotene transition. In most organisms, attachment of chromosome ends to the nuclear envelope (NE) coincides with these organizational changes, mediating markedly polarized nuclear organizations (Figure 5.2). In mammals, the telomeres cluster on the NE adjacent to the centrosome, while the chromosome arms

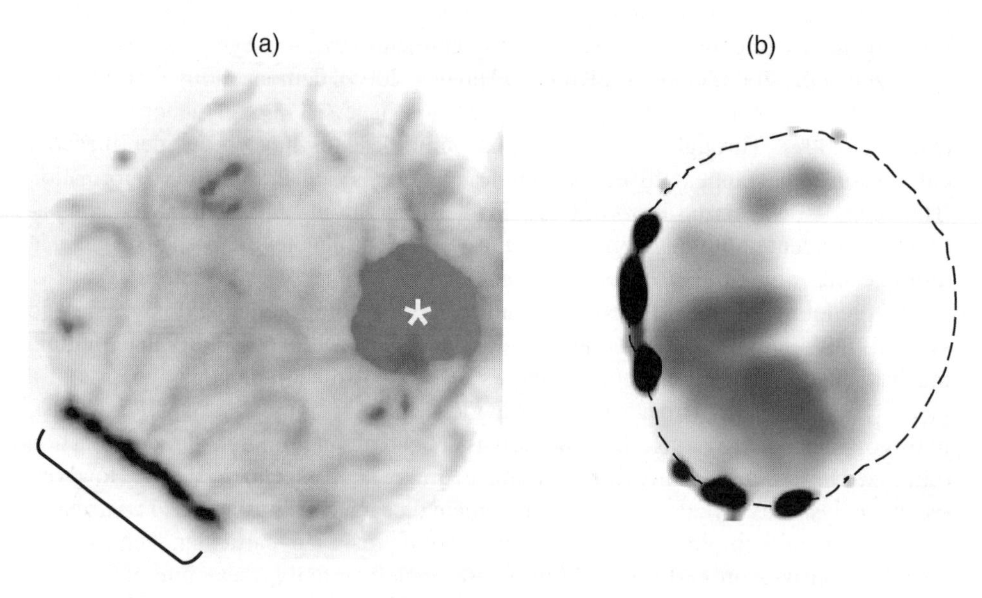

Figure 5.2 Polarized nuclear organizations mediated by tethering of chromosomes to the nuclear envelope in early meiotic prophase oocytes. Composite fluorescence microscope images showing chromosome organization in mammalian and nematode leptotene/zygotene oocytes. (a) Bovine oocyte nucleus displaying the chromosomal bouquet. Telomeres, identified by FISH (bracket, black), cluster tightly at the nuclear periphery, anchoring the AEs, marked by SYCP3 IF (grey ribbon-like staining), which loop into the nuclear interior. One large SYCP3 aggregate (*) is a characteristic marker of this stage. Image courtesy of H. Scherthan. (b) *C. elegans* oocyte nucleus displaying chromosome clustering. Chromosomes, stained with DAPI (grey), cluster in one hemisphere of the nucleus via anchorage to the nuclear envelope mediated by associations with the NE protein ZYG-12, detected by IF (black patches). Dashed line delineates the nuclear periphery. Image courtesy of A. Sato and A.F. Dernburg

loop towards the opposite side of the nucleus in the classical 'bouquet' configuration (Scherthan, 2001). In *C. elegans*, each chromosome attaches to the NE near only one of its two ends, and the chromosomes become clustered in one hemisphere of the nucleus (Goldstein and Slaton, 1982; MacQueen and Villeneuve, 2001). Observations such as these suggested a link between NE attachment and chromosome movement. Recent work has now provided compelling evidence that conserved mechanisms mobilize meiotic chromosomes by connecting them through the NE to the cytoskeleton.

Live imaging in the fission yeast *Schizosaccharomyces pombe* led the way in our emerging understanding of the nature and mechanism of chromosome movement during meiotic prophase. In a groundbreaking study, Hiraoka and colleagues demonstrated that the entire fission yeast nucleus undergoes dramatic movement during meiosis, oscillating between the cell poles. Further, the clustered telomeres and spindle pole body (yeast equivalent of the centrosome) are found at the leading edge of this movement (Chikashige *et al.*, 1994). Subsequent work has implicated inner and outer NE proteins containing SUN and KASH domains, respectively, in linking chromosomes via their telomeres to the cytoplasmic microtubule cytoskeleton (reviewed in Chikashige, Haraguchi and Hiraoka, 2007). In recent years, this SUN/KASH domain protein-mediated mechanism for moving meiotic chromosomes has been found to be conserved from yeasts to animals. In *C. elegans*, the SUN-1 protein has been shown to interact with a KASH domain protein, ZYG-12, which interfaces with the microtubule cytoskeleton (Malone *et al.*, 2003). Available data support a model in which the chromosomes' PCs, bound by HIM-8 or ZIM-1, -2, or -3, localize to NE patches containing SUN-1/ZYG-12, connecting the chromosomes to the microtubule cytoskeleton at the leptotene/zygotene transition (Penkner *et al.*, 2007). Similarly, at the leptotene/zygotene transition in mouse, telomeres localize to NE patches containing the SUN1 protein; as in other systems, it is thought that SUN1 probably interacts with an outer NE protein that interfaces with components of the cytoskeleton that have yet to be identified (Ding *et al.*, 2007). In budding yeast, the cytoskeletal network to which the NE protein complex connects the chromosomes is actin based (Trelles-Sticken *et al.*, 2005), demonstrating use of an alternative to the microtubule cytoskeleton in some systems. Recent analysis of mouse and worm mutants supports a conserved role for the SUN domain protein family in mediating chromosome reorganization in meiotic prophase, and has provided insight into the roles of NE attachment and chromosome movement in homologue paring and synapsis.

Upon deletion of mouse *Sun1*, telomeres fail to localize to the NE at the leptotene/zygotene transition, and the bouquet configuration is not displayed, consistent with loss of cytoskeleton-driven chromosome movements in meiotic prophase. *Sun1*$^{-/-}$ oocytes display very little synapsis (Ding *et al.*, 2007), recalling observations in the mouse *Smc1β* cohesin mutant, in which a few defective telomere attachments per nucleus correlated with failure of a few chromosomes to synapse (Revenkova *et al.*, 2004). The small amount of SC observed in the *Sun1* mutant appears to be installed between homologues (Ding *et al.*, 2007), indicating that the fundamental homologue recognition mechanism is still intact. Therefore, these results suggest that attaching telomeres to the NE and promoting movement of chromosomes may improve the efficiency of pairing and synapsis. This idea is consistent with work in budding yeast, where disrupting the system of chromosome linkage to the cytoskeletal apparatus results in a delay in achieving full homologous synapsis (Trelles-Sticken, Dresser and Scherthan, 2000). Several ways can be envisioned in which this system might promote efficient pairing and synapsis. Tethering chromosomes to the NE may facilitate homologue identification through reducing the homology search from a three-dimensional problem, where homologous sequences could be located anywhere in the nuclear volume, to a two-dimensional one, where homologous sequences are located on or at a fixed distance from a surface. Mobilizing chromosomes may improve the efficiency of this search through increasing the kinetics of interchromosomal interactions (Harper, Golubovskaya and Cande, 2004).

Analysis in worm meiosis extends these ideas further. In worms carrying a hypomorphic *sun-1* mutation, the clustered chromosome configuration is essentially eliminated, consistent with loss of chromosome movements. Very little coincidence of homologous sequences is detected by FISH, but extensive synapsis occurs between nonhomologous chromosomes. Experimental prevention of SC installation appears to restore a low level of homologous pairing (Penkner *et al.*, 2007), suggesting that the fundamental homologue recognition mechanism remains intact, but functions inefficiently in the absence of chromosome mobilization. The worm *htp-1* AE/LE component mutant, which shows a dramatic reduction in the frequency or duration of chromosome clustering, shows a similar nonhomologous synapsis phenotype (Martinez-Perez and Villeneuve, 2005). Together, these studies suggest that, in worms, cytoskeleton-driven chromosome movement may prevent SC central region protein loading until homologous pairing has been achieved. Such a mechanism is likely to be important in this species because pairing centres are capable of nucleating SC assembly between apposed sequences even when a homologous PC is not available (MacQueen *et al.*, 2005). Chromosome mobilization may prevent inappropriate synapsis by taking apart nonhomologous interactions more quickly than SC assembly can be nucleated, whereas the increased stability of interactions between matching pairing centres may allow enough time for nucleation of SC assembly between homologues (MacQueen *et al.*, 2005).

Taken together, the phenotypes of mouse *Sun1* and worm *sun-1* mutants provide evidence for two roles of early meiotic prophase chromosome movements. Incomplete synapsis in the mouse *Sun1* mutant emphasizes a role in promoting interactions between chromosomes to provide opportunities for homologue recognition, whereas nonhomologous synapsis in the worm *sun-1* mutant reveals a second role in disrupting incorrect associations between nonhomologues. In view of these ideas, the apparent lack of telomere attachment to the NE or chromosomal bouquet in *Drosophila* (Zickler and Kleckner, 1998) – and presumably the system for mobilizing meiotic chromosomes for which these phenomena are proxy – may be the exception that proves the rule: active chromosome motion may not be necessary in meiotic prophase of this species where alignment of homologues is already established at meiotic entry.

Live imaging studies of meiotic prophase in fission and budding yeasts have shown that chromosome movements driven by cytoskeletal machinery can be quite dramatic. A striking finding is that chromosome movement continues long after the establishment of pairing (Ding *et al.*, 2004) and (the bulk of) synapsis (Koszul *et al.*, 2008; Conrad *et al.*, 2008). Recent work has suggested that this mid-prophase movement may be important for completion of synapsis by helping to resolve entanglements between nonhomologous chromosomes that frequently arise during meiosis (Koszul *et al.*, 2008;

Conrad *et al.*, 2008). In addition, studies have implicated these movements in the progression of meiotic recombination (Kosaka, Shinohara and Shinohara, 2008; Wanat *et al.*, 2008; Conrad *et al.*, 2008). Live imaging of meiotic prophase chromosome dynamics in an animal oogenesis system (e.g. *C. elegans*), where large chromosomes provide opportunities for detailed cytological analysis, will surely contribute to our understanding of the mechanics and roles of chromosome movement in homologue pairing, synapsis, and other meiotic events.

5.5 Checkpoints for monitoring synapsis during oocyte development

The chromosomal events of meiosis are carried out within the larger context of the oocyte developmental programme. Homologue pairing and synapsis take place at the beginning of meiotic prophase, placing them very early in oogenesis, prior to overt gamete differentiation. A common theme in animals is that many more meioctyes attempt pairing and synapsis than ultimately mature into oocytes. In mammalian oogenesis, pairing and synapsis take place mid-gestation within one relatively synchronous cohort of meiocytes, and these events are completed before birth (Cohen, Pollack and Pollard, 2006). During their prolonged dictyate arrest until puberty, when follicle formation and ovulation begin, oocytes are reduced in number by an order of magnitude through apoptosis (Ghafari, Gutierrez and Hartshorne, 2007). In flies and worms, pairing and synapsis are early events in the ongoing oogenesis programmes of the adults. In the fly, synapsis initiates in at least 4 of the 16 cells within each cyst in the germarium before a single cell is specified to become the oocyte (Page and Hawley, 2001). In the worm, pairing and synapsis precede an extended pachytene stage, at the end of which approximately half of all meiocytes are eliminated by a wave of apoptosis. The survivors undergo cell growth and take up yolk proteins plus gene expression products contributed by the eliminated meiocytes as part of the oocyte differentiation process (Grant and Hirsh, 1999; Wolke, Jezuit and Priess, 2007). It is likely that the majority of prospective oocytes that are eliminated in each organism do not display defects in pairing and synapsis, and indeed there is evidence to support this idea in *C. elegans* (Gumienny *et al.*, 1999). However, the early timing of homologue pairing and synapsis within the oogenesis programme, prior to culling events that limit the oocyte pool, suggests that an opportunity may exist to assess the success of pairing and synapsis and influence the progression of a particular meiocyte in the oogenesis programme. Indeed, meiotic defects have been found to stall progression and/or trigger apoptosis during oogenesis in all three organisms. In addition to a checkpoint documented in mice, flies, and worms that responds to unrepaired DNA damage (Di Giacomo *et al.*, 2005; Staeva-Vieira, Yoo and Lehmann, 2003; Gartner *et al.*, 2000), evidence in worms and mice suggests that two checkpoints, operating at distinct points in meiotic prophase, monitor the synapsis status of chromosomes within an oocyte.

A growing body of evidence supports the existence of a checkpoint-like mechanism that delays exit from polarized chromosome organizations at the leptotene/zygotene transition until synapsis is complete. This idea developed out of a series of experimental observations in *C. elegans*. In wild-type meiosis, completion of synapsis is correlated temporally with the release of chromosomes from the clustered arrangement discussed above (MacQueen *et al.*, 2002). Furthermore, blocking SC assembly through *syp-1*, *-2*, *-3*, or *-4* mutations leads to persistence of chromosome clustering (MacQueen *et al.*, 2002; Colaiacovo *et al.*, 2003; Smolikov *et al.*, 2007; Smolikov, Schild-Prufert and Colaiacovo, 2009). Together, these observations raised the possibility that completion of synapsis might be coupled to release of chromosome clustering. In principle, synapsis could play a direct mechanical role in dispersing chromosomes; alternatively, synapsis status could be monitored by a checkpoint-like mechanism that, in turn, controls dispersal (MacQueen *et al.*, 2002). The observation that SC assembly is not required for release of chromosome clustering in the *htp-1* mutant background supports the latter hypothesis; further, it implicates the HTP-1 AE/LE component in a signal that blocks chromosome dispersal from the clustered arrangement when synapsis has not progressed on all chromosome pairs (Martinez-Perez and Villeneuve, 2005).

Coupling exit from the polarized chromosome organization to completion of synapsis is likely to be a general feature of animal meiosis (Scherthan *et al.*, 1996). Although extensive analysis of the bouquet stage in mouse oogenesis has not been reported, a number of mouse mutants that fail to achieve complete synapsis exhibit bouquet-stage enrichment during spermatogenesis, implying that exit from polarized chromosome organization is delayed when synapsis is incomplete (Liebe *et al.*, 2004; Liebe *et al.*, 2006; Mark *et al.*, 2008). Furthermore, the highest enrichment for bouquet stage so far reported is found in mice mutant for the ATM kinase (Liebe *et al.*, 2006), implicating a factor that has been proposed to play roles in monitoring meiotic as well as mitotic cell cycle progression (Barlow *et al.*, 1998) in the duration of polarized chromosome organization. These parallels between worm and mouse support the existence of a checkpoint-like mechanism that delays release from the polarized chromosome organization of the leptotene/zygotene transition – likely representing a period of chromosome mobilization during which the homology search is active – until all chromosomes have recognized and begun to form stable associations with their partners. Interestingly, the duration of the bouquet stage has been inferred to be substantially longer during oogenesis than spermatogenesis in several mammals, suggesting potential differences in regulation of this important period of chromosome mobilization between the sexes (Pfeifer, Scherthan and Thomsen, 2003; Roig *et al.*, 2004).

A second type of checkpoint monitors synapsis status at later stages of meiotic prophase, resulting in eventual apoptosis of oocytes containing chromosomes that display synapsis defects. The first evidence for the operation of such a checkpoint in animal meiosis came from investigations in the mouse. In this animal, meiotic mutants defective in pairing, synapsis, and/or recombination often cause complete elimination of germ cells after the zygotene stage of meiotic prophase, resulting in sterility (Barchi *et al.*, 2005; Di Giacomo *et al.*, 2005). Because synapsis is coupled to recombination in the mouse, it was initially unclear whether synapsis defects are monitored independently of the DNA damage (in the form of persistent DSBs) that also characterizes most asynaptic mutants. The existence of a distinct synapsis checkpoint was suggested by the discovery that defects in synapsis between the sex chromosomes during spermatogenesis trigger apoptosis by a mechanism that is molecularly distinct from the mechanism that triggers apoptosis in response to unrepaired DNA damage (Odorisio *et al.*, 1998).

DNA damage-independent apoptosis has more recently been identified during oogenesis of a number of mouse meiotic mutants, suggesting that a checkpoint also monitors synapsis status in females. Oogenesis often progresses further than spermatogenesis in a given meiotic mutant, perhaps due to differences in checkpoint control between the sexes, and two distinct points of oocyte loss during the meiotic programme have been identified. When asynapsis is accompanied by unrepaired DNA damage, apoptosis occurs earlier (at or before dictyate arrest), whereas when DSBs are experimentally eliminated, apoptosis occurs later (after dictyate arrest, but at or before follicle formation) (Di Giacomo *et al.*, 2005). Therefore, the mouse oogenesis programme appears capable of detecting asynapsis *per se* and responding by inducing apoptosis.

The operation of a second checkpoint for monitoring synapsis status has been most directly demonstrated in *C. elegans*, where synapsis does not depend on recombination. In genotypes in which synapsis of one or more chromosome pairs was experimentally blocked, elevated apoptosis was found to persist even when DSBs were eliminated using a *spo-11* mutant background to prevent activation of the DNA damage checkpoint. Further, this synapsis checkpoint has significant functional consequences for oocyte quality control, as checkpoint elimination substantially raises the frequency of chromosome segregation defects (Bhalla and Dernburg, 2005). It appears that by using two checkpoints to monitor and respond to synapsis status, animals are able not only to provide ample opportunity for homologous chromosomes to synapse, but also to eliminate oocytes that fail in this critical process.

In *C. elegans*, this synapsis checkpoint requires the function of the conserved PCH-2 protein (Bhalla and Dernburg, 2005), but it is not yet clear whether PCH2 orthologues carry out the same function during oogenesis in other systems. *Drosophila* PCH2 is required for a DNA damage-independent checkpoint that affects meiotic prophase progression in this organism, but asynapsis does not appear to serve as a trigger for this checkpoint (Joyce and McKim, 2009). Studies employing a hypomorphic allele of the mouse Pch2 homologue, *Trip13*, have implicated this gene both in normal timing or efficiency of meiotic recombination (Li and Schimenti, 2007), and in promoting removal of HORMAD1 and HORMAD2 from LEs in response to synapsis (Wojtasz *et al.*, 2009). It is clear that further analysis will be required both to clarify the precise meiotic functions of PCH2 and to elucidate the mechanisms by which oocytes detect unsynapsed chromosomes and respond by inducing apoptosis.

5.6 Concluding remarks

It has long been clear that pairwise alignment between homologous chromosomes is essential for successful chromosome inheritance during gametogenesis. However, until recently, the mechanisms underlying the homologue pairing process have remained largely mysterious. In this chapter, we have highlighted investigations in animal meiosis model systems that have contributed mechanistic insights into this process. Together, both similarities and differences among the systems have helped to illuminate the fundamental principles that govern homologue pairing, thus allowing the following framework to emerge: (i) Chromosomes assemble meiosis-specific structures – the AEs and mature SC – that enable establishment, and then stabilization of homologue pairing.

(ii) Multiple mechanisms have evolved (and likely operate in parallel in many organisms) to couple homology verification at a limited number of chromosomal sites to the nucleation of synapsis, thereby solidifying pairwise associations. (iii) Chromosome mobility driven by tethering of chromosome sites through the NE to the cytoskeletal motility apparatus contributes to the success of this process, both by facilitating interactions between chromosomes and by taking apart inappropriate interactions. (iv) Quality control mechanisms operate both to prevent synapsis errors and to eliminate defective meiocytes if errors do occur, thereby channelling reproductive resources towards oocytes in which pairing and synapsis were successful. Future work will clarify the mechanisms of these important facets of the homologue pairing programme and will reveal the interrelationships between them. Perhaps through the course of this work we will come closer to understanding the most mysterious aspect of meiotic pairing and synapsis; that is, the fundamental nature of homologue recognition.

References

Alani, E., Padmore, R. and Kleckner, N. (1990) Analysis of wild-type and rad50 mutants of yeast suggests an intimate relationship between meiotic chromosome synapsis and recombination. *Cell*, **61**, 419–436.

Aravind, L. and Koonin, E.V. (1998) The HORMA domain: a common structural denominator in mitotic checkpoints, chromosome synapsis and DNA repair. *Trends Biochem. Sci.*, **23**, 284–286.

Baker, S.M., Plug, A.W., Prolla, T.A. *et al.* (1996) Involvement of mouse Mlh1 in DNA mismatch repair and meiotic crossing over. *Nat. Genet.*, **13**, 336–342.

Bannister, L.A., Reinholdt, L.G., Munroe, R.J. and Schimenti, J.C. (2004) Positional cloning and characterization of mouse mei8, a disrupted allele of the meiotic cohesin Rec8. *Genesis*, **40**, 184–194.

Barchi, M., Mahadevaiah, S., Di Giacomo, M. *et al.* (2005) Surveillance of different recombination defects in mouse spermatocytes yields distinct responses despite elimination at an identical developmental stage. *Mol. Cell Biol.*, **25**, 7203–7215.

Barlow, C., Liyanage, M., Moens, P.B. *et al.* (1998) Atm deficiency results in severe meiotic disruption as early as leptonema of prophase I. *Development*, **125**, 4007–4017.

Baudat, F. and de Massy, B. (2007) Regulating double-stranded DNA break repair towards crossover or non-crossover during mammalian meiosis. *Chromosome Res.*, **15**, 565–577.

Baudat, F., Manova, K., Yuen, J.P. *et al.* (2000) Chromosome synapsis defects and sexually dimorphic meiotic progression in mice lacking Spo11. *Mol. Cell*, **6**, 989–998.

Bhalla, N. and Dernburg, A.F. (2005) A conserved checkpoint monitors meiotic chromosome synapsis in Caenorhabditis elegans. *Science*, **310**, 1683–1686.

Bickel, S.E., Orr-Weaver, T.L. and Balicky, E.M. (2002) The sister-chromatid cohesion protein ORD is required for chiasma maintenance in Drosophila oocytes. *Curr. Biol.*, **12**, 925–929.

Bolcun-Filas, E., Costa, Y., Speed, R. *et al.* (2007) SYCE2 is required for synaptonemal complex assembly, double strand break repair, and homologous recombination. *J. Cell Biol.*, **176**, 741–747.

Bolcun-Filas, E., Speed, R., Taggart, M. *et al.* (2009) Mutation of the mouse Syce1 gene disrupts synapsis and suggests a link between synaptonemal complex structural components and DNA repair. *PLoS Genet.*, **5**, e1000393.

Chan, R.C., Chan, A., Jeon, M. *et al.* (2003) Chromosome cohesion is regulated by a clock gene paralogue TIM-1. *Nature*, **423**, 1002–1009.

Chikashige, Y., Ding, D.Q., Funabiki, H. *et al.* (1994) Telomere-led premeiotic chromosome movement in fission yeast. *Science*, **264**, 270–273.

Chikashige, Y., Haraguchi, T. and Hiraoka, Y. (2007) Another way to move chromosomes. *Chromosoma*, **116**, 497–505.

Cohen, P.E., Pollack, S.E. and Pollard, J.W. (2006) Genetic analysis of chromosome pairing, recombination, and cell cycle control during first meiotic prophase in mammals. *Endocr. Rev.*, **27**, 398–426.

Colaiacovo, M.P., Stanfield, G.M., Reddy, K.C. *et al.* (2002) A targeted RNAi screen for genes involved in chromosome morphogenesis and nuclear organization in the Caenorhabditis elegans germline. *Genetics*, **162**, 113–128.

Colaiacovo, M.P., MacQueen, A.J., Martinez-Perez, E. *et al.* (2003) Synaptonemal complex assembly in C. elegans is dispensable for loading strand-exchange proteins but critical for proper completion of recombination. *Dev. Cell*, **5**, 463–474.

Conrad, M.N., Lee, C.Y., Chao, G. *et al.* (2008) Rapid telomere movement in meiotic prophase is promoted by NDJ1, MPS3, and CSM4 and is modulated by recombination. *Cell*, **133**, 1175–1187.

Costa, Y., Speed, R., Ollinger, R. *et al.* (2005) Two novel proteins recruited by synaptonemal complex protein 1 (SYCP1) are at the centre of meiosis. *J. Cell Sci.*, **118**, 2755–2762.

Couteau, F., Nabeshima, K., Villeneuve, A. and Zetka, M. (2004) A component of C. elegans meiotic chromosome axes at the interface of homolog alignment, synapsis, nuclear reorganization, and recombination. *Curr. Biol.*, **14**, 585–592.

Couteau, F. and Zetka, M. (2005) HTP-1 coordinates synaptonemal complex assembly with homolog alignment during meiosis in C. elegans. *Genes Dev.*, **19**, 2744–2756.

Dernburg, A.F., McDonald, K., Moulder, G. *et al.* (1998) Meiotic recombination in C. elegans initiates by a conserved mechanism and is dispensable for homologous chromosome synapsis. *Cell*, **94**, 387–398.

Di Giacomo, M., Barchi, M., Baudat, F. *et al.* (2005) Distinct DNA-damage-dependent and -independent responses drive the loss of oocytes in recombination-defective mouse mutants. *Proc. Natl. Acad. Sci. USA*, **102**, 737–742.

Ding, D.Q., Yamamoto, A., Haraguchi, T. and Hiraoka, Y. (2004) Dynamics of homologous chromosome pairing during meiotic prophase in fission yeast. *Dev. Cell*, **6**, 329–341.

Ding, X., Xu, R., Yu, J. *et al.* (2007) SUN1 is required for telomere attachment to nuclear envelope and gametogenesis in mice. *Dev. Cell*, **12**, 863–872.

Dobson, M.J., Pearlman, R.E., Karaiskakis, A. *et al.* (1994) Synaptonemal complex proteins: occurrence, epitope mapping and chromosome disjunction. *J. Cell Sci.*, **107**(10),2749–2760.

Edelmann, W., Cohen, P.E., Kane, M. *et al.* (1996) Meiotic pachytene arrest in MLH1-deficient mice. *Cell*, **85**, 1125–1134.

Edelmann, W., Cohen, P.E., Kneitz, B. *et al.* (1999) Mammalian MutS homologue 5 is required for chromosome pairing in meiosis. *Nat. Genet.*, **21**, 123–127.

Fukuda, T., Daniel, K., Wojtasz, L. *et al.* (2009) A novel mammalian HORMA domain-containing protein, HORMAD1, preferentially associates with unsynapsed meiotic chromosomes. *Exp. Cell Res.*, in press.

Fung, J.C., Marshall, W.F., Dernburg, A. *et al.* (1998) Homologous chromosome pairing in Drosophila melanogaster proceeds through multiple independent initiations. *J. Cell Biol.*, **141**, 5–20.

Gartner, A., Milstein, S., Ahmed, S. *et al.* (2000) A conserved checkpoint pathway mediates DNA damage-induced apoptosis and cell cycle arrest in C. elegans. *Mol. Cell*, **5**, 435–443.

Ghafari, F., Gutierrez, C.G. and Hartshorne, G.M. (2007) Apoptosis in mouse fetal and neonatal oocytes during meiotic prophase one. *BMC Dev. Biol.*, **7**, 87.

Gillies, C.B. (1974) The nature and extent of synaptonemal complex formation in haploid barley. *Chromosoma*, **48**, 441–453.

Giroux, C.N., Dresser, M.E. and Tiano, H.F. (1989) Genetic control of chromosome synapsis in yeast meiosis. *Genome*, **31**, 88–94.

Goldstein, P. and Slaton, D.E. (1982) The synaptonemal complexes of Caenorhabditis elegans: comparison of wild-type and mutant strains and pachytene karyotype analysis of wild-type. *Chromosoma*, **84**, 585–597.

Gong, W.J., McKim, K.S. and Hawley, R.S. (2005) All paired up with no place to go: pairing, synapsis, and DSB formation in a balancer heterozygote. *PLoS Genet.*, **1**, e67.

Goodyer, W., Kaitna, S., Couteau, F. *et al.* (2008) HTP-3 links DSB formation with homolog pairing and crossing over during C. elegans meiosis. *Dev. Cell*, **14**, 263–274.

Gorbsky, G.J., Chen, R.H. and Murray, A.W. (1998) Microinjection of antibody to Mad2 protein into mammalian cells in mitosis induces premature anaphase. *J. Cell Biol.*, **141**, 1193–1205.

Grant, B. and Hirsh, D. (1999) Receptor-mediated endocytosis in the Caenorhabditis elegans oocyte. *Mol. Biol. Cell*, **10**, 4311–4326.

Gumienny, T.L., Lambie, E., Hartwieg, E. *et al.* (1999) Genetic control of programmed cell death in the Caenorhabditis elegans hermaphrodite germline. *Development*, **126**, 1011–1022.

Hamer, G., Gell, K., Kouznetsova, A. *et al.* (2006) Characterization of a novel meiosis-specific protein within the central element of the synaptonemal complex. *J. Cell Sci.*, **119**, 4025–4032.

Hamer, G., Wang, H., Bolcun-Filas, E. *et al.* (2008) Progression of meiotic recombination requires structural maturation of the central element of the synaptonemal complex. *J. Cell Sci.*, **121**, 2445–2451.

Harper, L., Golubovskaya, I. and Cande, W.Z. (2004) A bouquet of chromosomes. *J. Cell Sci.*, **117**, 4025–4032.

Hassold, T. and Hunt, P. (2001) To err (meiotically) is human: the genesis of human aneuploidy. *Nat. Rev. Genet.*, **2**, 280–291.

Hawley, R.S. (1980) Chromosomal sites necessary for normal levels of meiotic recombination in Drosophila melanogaster. *Genetics*, **94**, 625–646.

Hayashi, M., Chin, G.M. and Villeneuve, A.M. (2007) C. elegans germ cells switch between distinct modes of double-strand break repair during meiotic prophase progression. *PLoS Genet.*, **3**, e191.

Heidmann, D., Horn, S., Heidmann, S. *et al.* (2004) The Drosophila meiotic kleisin C(2)M functions before the meiotic divisions. *Chromosoma*, **113**, 177–187.

Herman, R.K. and Kari, C.K. (1989) Recombination between small X chromosome duplications and the X chromosome in Caenorhabditis elegans. *Genetics*, **121**, 723–737.

Herman, R.K., Kari, C.K. and Hartman, P.S. (1982) Dominant X-chromosome nondisjunction mutants of Caenorhabditis elegans. *Genetics*, **102**, 379–400.

Hiraoka, Y., Dernburg, A.F., Parmelee, S.J. *et al.* (1993) The onset of homologous chromosome pairing during Drosophila melanogaster embryogenesis. *J. Cell Biol.*, **120**, 591–600.

Hollingsworth, N.M., Goetsch, L. and Byers, B. (1990) The HOP1 gene encodes a meiosis-specific component of yeast chromosomes. *Cell*, **61**, 73–84.

Jeffress, J.K., Page, S.L., Royer, S.K. *et al.* (2007) The formation of the central element of the synaptonemal complex may occur by multiple mechanisms: the roles of the N- and C-terminal domains of the Drosophila C(3)G protein in mediating synapsis and recombination. *Genetics*, **177**, 2445–2456.

Joyce, E.F. and McKim, K.S. (2009) Drosophila PCH2 is required for a pachytene checkpoint that monitors double-strand-break-independent events leading to meiotic crossover formation. *Genetics*, **181**, 39–51.

Keeney, S. (2001) Mechanism and control of meiotic recombination initiation. *Curr. Top. Dev. Biol.*, **52**, 1–53.

Khetani, R.S. and Bickel, S.E. (2007) Regulation of meiotic cohesion and chromosome core morphogenesis during pachytene in Drosophila oocytes. *J. Cell Sci.*, **120**, 3123–3137.

Kneitz, B., Cohen, P.E., Avdievich, E. *et al.* (2000) MutS homolog 4 localization to meiotic chromosomes is required for chromosome pairing during meiosis in male and female mice. *Genes. Dev.*, **14**, 1085–1097.

Kosaka, H., Shinohara, M. and Shinohara, A. (2008) Csm4-dependent telomere movement on nuclear envelope promotes meiotic recombination. *PLoS Genet.*, **4**, e1000196.

Koszul, R., Kim, K.P., Prentiss, M. *et al.* (2008) Meiotic chromosomes move by linkage to dynamic actin cables with transduction of force through the nuclear envelope. *Cell*, **133**, 1188–1201.

Lammers, J.H., Offenberg, H.H., van Aalderen, M. *et al.* (1994) The gene encoding a major component of the lateral elements of synaptonemal complexes of the rat is related to X-linked lymphocyte-regulated genes. *Mol. Cell Biol.*, **14**, 1137–1146.

Lee, J.Y. and Orr-Weaver, T.L. (2001) The molecular basis of sister-chromatid cohesion. *Annu. Rev. Cell Dev. Biol.*, **17**, 753–777.

Li, X.C. and Schimenti, J.C. (2007) Mouse pachytene checkpoint 2 (trip13) is required for completing meiotic recombination but not synapsis. *PLoS Genet.*, **3**, e130.

Liebe, B., Alsheimer, M., Hoog, C. *et al.* (2004) Telomere attachment, meiotic chromosome condensation, pairing, and bouquet stage duration are modified in spermatocytes lacking axial elements. *Mol. Biol. Cell*, **15**, 827–837.

Liebe, B., Petukhova, G., Barchi, M. *et al.* (2006) Mutations that affect meiosis in male mice influence the dynamics of the mid-preleptotene and bouquet stages. *Exp. Cell Res.*, **312**, 3768–3781.

Lipkin, S.M., Moens, P.B., Wang, V. *et al.* (2002) Meiotic arrest and aneuploidy in MLH3-deficient mice. *Nat. Genet.*, **31**, 385–390.

Loidl, J. (1990) The initiation of meiotic chromosome pairing: the cytological view. *Genome*, **33**, 759–778.

Loidl, J., Nairz, K. and Klein, F. (1991) Meiotic chromosome synapsis in a haploid yeast. *Chromosoma*, **100**, 221–228.

MacQueen, A.J., Colaiacovo, M.P., McDonald, K. and Villeneuve, A.M. (2002) Synapsis-dependent and -independent mechanisms stabilize homolog pairing during meiotic prophase in C. elegans. *Genes. Dev.*, **16**, 2428–2442.

MacQueen, A.J., Phillips, C.M., Bhalla, N. *et al.* (2005) Chromosome sites play dual roles to establish homologous synapsis during meiosis in C. elegans. *Cell*, **123**, 1037–1050.

MacQueen, A.J. and Villeneuve, A.M. (2001) Nuclear reorganization and homologous chromosome pairing during meiotic prophase require C. elegans chk-2. *Genes. Dev.*, **15**, 1674–1687.

Mahadevaiah, S.K., Turner, J.M., Baudat, F. *et al.* (2001) Recombinational DNA double-strand breaks in mice precede synapsis. *Nat. Genet.*, **27**, 271–276.

Malone, C.J., Misner, L., Le Bot, N. *et al.* (2003) The C. elegans hook protein, ZYG-12, mediates the essential attachment between the centrosome and nucleus. *Cell*, **115**, 825–836.

Manheim, E.A. and McKim, K.S. (2003) The synaptonemal complex component C(2)M regulates meiotic crossing over in Drosophila. *Curr. Biol.*, **13**, 276–285.

Mark, M., Jacobs, H., Oulad-Abdelghani, M. *et al.* (2008) STRA8-deficient spermatocytes initiate, but fail to complete, meiosis and undergo premature chromosome condensation. *J. Cell Sci.*, **121**, 3233–3242.

Martinez-Perez, E., Schvarzstein, M., Barroso, C. *et al.* (2008) Crossovers trigger a remodeling of meiotic chromosome axis composition that is linked to two-step loss of sister chromatid cohesion. *Genes Dev.*, **22**, 2886–2901.

Martinez-Perez, E. and Villeneuve, A.M. (2005) HTP-1-dependent constraints coordinate homolog pairing and synapsis and promote chiasma formation during C. elegans meiosis. *Genes. Dev.*, **19**, 2727–2743.

McKee, B.D. (2009) Homolog pairing and segregation in Drosophila meiosis. *Genome Dyn.*, **5**, 56–68.

McKim, K.S., Green-Marroquin, B.L., Sekelsky, J.J. *et al.* (1998) Meiotic synapsis in the absence of recombination. *Science*, **279**, 876–878.

McKim, K.S. and Hayashi-Hagihara, A. (1998) mei-W68 in Drosophila melanogaster encodes a Spo11 homolog: evidence that the mechanism for initiating meiotic recombination is conserved. *Genes. Dev.*, **12**, 2932–2942.

McKim, K.S., Howell, A.M. and Rose, A.M. (1988) The effects of translocations on recombination frequency in Caenorhabditis elegans. *Genetics*, **120**, 987–1001.

McKim, K.S., Peters, K. and Rose, A.M. (1993) Two types of sites required for meiotic chromosome pairing in Caenorhabditis elegans. *Genetics*, **134**, 749–768.

Meuwissen, R.L., Offenberg, H.H., Dietrich, A.J. *et al.* (1992) A coiled-coil related protein specific for synapsed regions of meiotic prophase chromosomes. *EMBO J.*, **11**, 5091–5100.

Miyazaki, W.Y. and Orr-Weaver, T.L. (1992) Sister-chromatid misbehavior in Drosophila ord mutants. *Genetics*, **132**, 1047–1061.

Moses, M.J. (1956) Chromosomal structures in crayfish spermatocytes. *J. Biophys. Biochem. Cytol.*, **2**, 215–218.

Nabeshima, K., Villeneuve, A.M. and Colaiacovo, M.P. (2005) Crossing over is coupled to late meiotic prophase bivalent differentiation through asymmetric disassembly of the SC. *J. Cell Biol.*, **168**, 683–689.

Nabeshima, K., Villeneuve, A.M. and Hillers, K.J. (2004) Chromosome-wide regulation of meiotic crossover formation in Caenorhabditis elegans requires properly assembled chromosome axes. *Genetics*, **168**, 1275–1292.

Novak, I., Wang, H., Revenkova, E. *et al.* (2008) Cohesin Smc1beta determines meiotic chromatin axis loop organization. *J. Cell Biol.*, **180**, 83–90.

Odorisio, T., Rodriguez, T.A., Evans, E.P. *et al.* (1998) The meiotic checkpoint monitoring synapsis eliminates spermatocytes via p53-independent apoptosis. *Nat. Genet.*, **18**, 257–261.

Offenberg, H.H., Schalk, J.A., Meuwissen, R.L. *et al.* (1998) SCP2: a major protein component of the axial elements of synaptonemal complexes of the rat. *Nucleic Acids Res.*, **26**, 2572–2579.

Ollinger, R., Alsheimer, M. and Benavente, R. (2005) Mammalian protein SCP1 forms synaptonemal complex-like structures in the absence of meiotic chromosomes. *Mol. Biol. Cell*, **16**, 212–217.

Page, S.L. and Hawley, R.S. (2001) c(3)G encodes a Drosophila synaptonemal complex protein. *Genes. Dev.*, **15**, 3130–3143.

Page, S.L., Khetani, R.S., Lake, C.M. *et al.* (2008) Corona is required for higher-order assembly of transverse filaments into full-length synaptonemal complex in Drosophila oocytes. *PLoS Genet.*, **4**, e1000194.

Pasierbek, P., Fodermayr, M., Jantsch, V. *et al.* (2003) The Caenorhabditis elegans SCC-3 homologue is required for meiotic synapsis and for proper chromosome disjunction in mitosis and meiosis. *Exp. Cell Res.*, **289**, 245–255.

Pasierbek, P., Jantsch, M., Melcher, M. *et al.* (2001) A Caenorhabditis elegans cohesion protein with functions in meiotic chromosome pairing and disjunction. *Genes. Dev.*, **15**, 1349–1360.

Penkner, A., Tang, L., Novatchkova, M. *et al.* (2007) The nuclear envelope protein Matefin/SUN-1 is required for homologous pairing in C. elegans meiosis. *Dev. Cell*, **12**, 873–885.

Pfeifer, C., Scherthan, H. and Thomsen, P.D. (2003) Sex-specific telomere redistribution and synapsis initiation in cattle oogenesis. *Dev. Biol.*, **255**, 206–215.

Phillips, C.M. and Dernburg, A.F. (2006) A family of zinc-finger proteins is required for chromosome-specific pairing and synapsis during meiosis in C. elegans. *Dev. Cell*, **11**, 817–829.

Phillips, C.M., Meng, X., Zhang, L. *et al.* (2009) Identification of chromosome sequence motifs that mediate meiotic pairing and synapsis in C. elegans. *Nat. Cell Biol.*, **11**, 934–942.

Phillips, C.M., Wong, C., Bhalla, N. *et al.* (2005) HIM-8 binds to the X chromosome pairing center and mediates chromosome-specific meiotic synapsis. *Cell*, **123**, 1051–1063.

Pittman, D.L., Cobb, J., Schimenti, K.J. *et al.* (1998) Meiotic prophase arrest with failure of chromosome synapsis in mice deficient for Dmc1, a germline-specific RecA homolog. *Mol. Cell*, **1**, 697–705.

Prieto, I., Suja, J.A., Pezzi, N. *et al.* (2001) Mammalian STAG3 is a cohesin specific to sister chromatid arms in meiosis I. *Nat. Cell Biol.*, **3**, 761–766.

Prieto, I., Tease, C., Pezzi, N. *et al.* (2004) Cohesin component dynamics during meiotic prophase I in mammalian oocytes. *Chromosome Res.*, **12**, 197–213.

Revenkova, E., Eijpe, M., Heyting, C. *et al.* (2004) Cohesin SMC1 beta is required for meiotic chromosome dynamics, sister chromatid cohesion and DNA recombination. *Nat. Cell Biol.*, **6**, 555–562.

Revenkova, E. and Jessberger, R. (2006) Shaping meiotic prophase chromosomes: cohesins and synaptonemal complex proteins. *Chromosoma*, **115**, 235–240.

Roig, I., Liebe, B., Egozcue, J. *et al.* (2004) Female-specific features of recombinational double-stranded DNA repair in relation to synapsis and telomere dynamics in human oocytes. *Chromosoma*, **113**, 22–33.

Romanienko, P.J. and Camerini-Otero, R.D. (2000) The mouse Spo11 gene is required for meiotic chromosome synapsis. *Mol. Cell*, **6**, 975–987.

Scherthan, H. (2001) A bouquet makes ends meet. *Nat. Rev. Mol. Cell Biol.*, **2**, 621–627.

Scherthan, H., Weich, S., Schwegler, H. *et al.* (1996) Centromere and telomere movements during early meiotic prophase of mouse and man are associated with the onset of chromosome pairing. *J. Cell Biol.*, **134**, 1109–1125.

Schmekel, K. and Daneholt, B. (1995) The central region of the synaptonemal complex revealed in three dimensions. *Trends Cell Biol.*, **5**, 239–242.

Severson, A.F., Ling, L., van Zuylen, V. and Meyer, B.J. (2009) The axial element protein HTP-3 promotes cohesin loading and meiotic axis assembly in C. elegans to implement the meiotic program of chromosome segregation. *Genes Dev.*, **23**, 1763–1778.

Sherizen, D., Jang, J.K., Bhagat, R. *et al.* (2005) Meiotic recombination in Drosophila females depends on chromosome continuity between genetically defined boundaries. *Genetics*, **169**, 767–781.

Smolikov, S., Eizinger, A., Schild-Prufert, K. *et al.* (2007) SYP-3 restricts synaptonemal complex assembly to bridge paired chromosome axes during meiosis in Caenorhabditis elegans. *Genetics*, **176**, 2015–2025.

Smolikov, S., Schild-Prufert, K. and Colaiacovo, M.P. (2009) A yeast two-hybrid screen for SYP-3 interactors identifies SYP-4, a component required for synaptonemal complex assembly and chiasma formation in Caenorhabditis elegans meiosis. *PLoS Genet.*, **5**, e1000669.

Staeva-Vieira, E., Yoo, S. and Lehmann, R. (2003) An essential role of DmRad51/SpnA in DNA repair and meiotic checkpoint control. *EMBO J.*, **22**, 5863–5874.

Sym, M., Engebrecht, J.A. and Roeder, G.S. (1993) ZIP1 is a synaptonemal complex protein required for meiotic chromosome synapsis. *Cell*, **72**, 365–378.

Tankimanova, M., Hulten, M.A. and Tease, C. (2004) The initiation of homologous chromosome synapsis in mouse fetal oocytes is not directly driven by centromere and telomere clustering in the bouquet. *Cytogenet. Genome Res.*, **105**, 172–181.

Trelles-Sticken, E., Adelfalk, C., Loidl, J. and Scherthan, H. (2005) Meiotic telomere clustering requires actin for its formation and cohesin for its resolution. *J. Cell Biol.*, **170**, 213–223.

Trelles-Sticken, E., Dresser, M.E. and Scherthan, H. (2000) Meiotic telomere protein Ndj1p is required for meiosis-specific telomere distribution, bouquet formation and efficient homologue pairing. *J. Cell Biol.*, **151**, 95–106.

Tsubouchi, T., Macqueen, A.J. and Roeder, G.S. (2008) Initiation of meiotic chromosome synapsis at centromeres in budding yeast. *Genes. Dev.*, **22**, 3217–3226.

Tsubouchi, T. and Roeder, G.S. (2005) A synaptonemal complex protein promotes homology-independent centromere coupling. *Science*, **308**, 870–873.

Vazquez, J., Belmont, A.S. and Scdat, J.W. (2002) The dynamics of homologous chromosome pairing during male Drosophila meiosis. *Curr. Biol.*, **12**, 1473–1483.

Villeneuve, A.M. (1994) A cis-acting locus that promotes crossing over between X chromosomes in Caenorhabditis elegans. *Genetics*, **136**, 887–902.

de Vries, F.A., de Boer, E., van den Bosch, M. *et al.* (2005) Mouse Sycp1 functions in synaptonemal complex assembly, meiotic recombination, and XY body formation. *Genes Dev.*, **19**, 1376–1389.

de Vries, S.S., Baart, E.B., Dekker, M. *et al.* (1999) Mouse MutS-like protein Msh5 is required for proper chromosome synapsis in male and female meiosis. *Genes Dev.*, **13**, 523–531.

Wanat, J.J., Kim, K.P., Koszul, R. *et al.* (2008) Csm4, in collaboration with Ndj1, mediates telomere-led chromosome dynamics and recombination during yeast meiosis. *PLoS Genet.*, **4**, e1000188.

Webber, H.A., Howard, L. and Bickel, S.E. (2004) The cohesion protein ORD is required for homologue bias during meiotic recombination. *In. J. Cell Biol.*, **164**, 819–829.

Wojtasz, L., Daniel, K., Roig, I. *et al.* (2009) Mouse HORMAD1 and HORMAD2, two conserved meiotic chromosomal proteins, are depleted from synapsed chromosome axes with the help of TRIP13 AAA-ATPase. *PLoS Genet.*, **5**, e1000702.

Wolke, U., Jezuit, E.A. and Priess, J.R. (2007) Actin-dependent cytoplasmic streaming in C. elegans oogenesis. *Development*, **134**, 2227–2236.

Xu, H., Beasley, M.D., Warren, W.D. *et al.* (2005) Absence of mouse REC8 cohesin promotes synapsis of sister chromatids in meiosis. *Dev. Cell*, **8**, 949–961.

Yang, F., De La Fuente, R., Leu, N.A. *et al.* (2006) Mouse SYCP2 is required for synaptonemal complex assembly and chromosomal synapsis during male meiosis. *J. Cell Biol.*, **173**, 497–507.

Yoshida, K., Kondoh, G., Matsuda, Y. *et al.* (1998) The mouse RecA-like gene Dmc1 is required for homologous chromosome synapsis during meiosis. *Mol. Cell*, **1**, 707–718.

Yuan, L., Liu, J.G., Hoja, M.R. *et al.* (2002) Female germ cell aneuploidy and embryo death in mice lacking the meiosis-specific protein SCP3. *Science*, **296**, 1115–1118.

Yuan, L., Liu, J.G., Zhao, J. *et al.* (2000) The murine SCP3 gene is required for synaptonemal complex assembly, chromosome synapsis, and male fertility. *Mol. Cell*, **5**, 73–83.

Zetka, M.C., Kawasaki, I., Strome, S. and Muller, F. (1999) Synapsis and chiasma formation in Caenorhabditis elegans require HIM-3, a meiotic chromosome core component that functions in chromosome segregation. *Genes. Dev.*, **13**, 2258–2270.

Zickler, D. and Kleckner, N. (1998) The leptotene-zygotene transition of meiosis. *Annu. Rev. Genet.*, **32**, 619–697.

6

Meiotic recombination in mammals

Sabine Santucci-Darmanin[1] and Frédéric Baudat[2]

[1]*FRE 3086, CNRS, Faculté de Médecine, Université de Nice Sophia-Antipolis, Nice CEDEX 2, France*
[2]*CNRS UPR 1142, Institut de Génétique Humaine, Montpellier CEDEX 5, France*

6.1 Introduction

In sexually reproducing organisms, meiosis is the process that converts a diploid cell into genetically distinct haploid gametes. To achieve this end, a single round of DNA replication is followed by two successive divisions: a reductional (meiosis I) and an equational (meiosis II) division. A specificity of oogenesis is that both the first and the second divisions are asymmetrical, giving rise to only one gamete (the egg) and two abortive products, the first and second polar bodies. During the reductional division, both homologous chromosomes (homologues) of each pair segregate. Their proper segregation depends on the physical connections between them, provided by the chiasmata, which are essential for their bipolar attachment to the meiosis I spindle. The chiasmata result from reciprocal exchanges of large fragments of genetic material, or crossovers (COs) between homologues. Therefore, COs play a crucial mechanical role during meiosis, and defects in their formation can result in aneuploidy due to the missegregation of homologues at the first division. Beyond this mechanical role, meiotic COs are also important to promote genetic diversity by producing new combinations of alleles in offspring.

Much of our knowledge on the molecular mechanism of meiotic recombination comes from studies in the yeast *Saccharomyces cerevisiae*. However, efforts have been made in the last 15 years to improve our understanding of recombination mechanisms in mammals. Indeed, many mouse genes involved in meiosis have been characterized, and the generation and analysis of mutant animals has given insights into their role in

Oogenesis: The Universal Process Marie-Hélène Verlhac and Anne Villeneuve

recombination. Studies in yeasts, mammals, and also in other eukaryotes, have revealed that meiotic recombination is a highly complicated molecular process, proceeding through several steps and separate pathways, most of them being conserved amongst species. In Section 6.2, we will give an overview of the current knowledge on the mechanism of meiotic recombination in mammals, and the proteins involved, with the help of the framework provided by the detailed data on molecular mechanisms coming from studies in yeast.

Given the importance of COs for the accurate segregation of chromosomes, it is crucial to ensure the formation of COs on every chromosome pair, which implies that molecular events (the formation of recombination products) are controlled in relation to big objects in the nucleus (the chromosomes). Indeed, it has long been known that not only the number of COs, but also their distribution along chromosomes, is tightly controlled. A defect in this control is associated with an increase of abnormal segregation of homologous chromosomes, the prevalence of which is particularly high during human female meiosis. Mechanisms governing crossover control remain poorly understood. However, recent studies have provided new information on the fine scale distribution of COs in mouse and human genomes. In Section 6.3, we summarize the various genetic and cytological approaches that enable the study of the frequency and the distribution of COs, and we review recent advances in understanding the factors involved in the control of CO distribution in mammals, with some emphasis placed on those that may explain the differences between sexes in CO distribution.

Finally, the last section of this chapter focuses on recent findings related to the relationship between meiotic recombination and meiotic prophase progression in mammals.

6.2 Meiotic DNA recombination events and proteins involved

6.2.1 Overview of the process

DNA recombination events in Saccharomyces Cerevisiae

Molecular events of meiotic recombination are not yet fully elucidated in mammals, but have been extensively characterized in *Saccharomyces cerevisiae*, which led to a consensus model (Figure 6.1). Meiotic recombination is initiated by the formation of DNA double-strand breaks (DSBs) catalyzed by the conserved Spo11 protein. After Spo11 is removed from DNA ends, one or more exonucleases process DSBs to generate 3′ single-stranded overhangs (Keeney, 2008; Neale, Pan and Keeney, 2005). Then, the two recombinases Rad51 and Dmc1 (some organisms, such as *Drosophila melanogaster*, *Caenorhabditis elegans* and *Neurospora crassa* lack a Dmc1 orthologue) bind the single-stranded tails, promote interaction with homologous duplex sequences and catalyze strand exchange. The majority of processed DSBs interact with a chromatid from the homologous chromosome, rather than with the sister chromatid (Zickler and Kleckner, 1999), a bias that

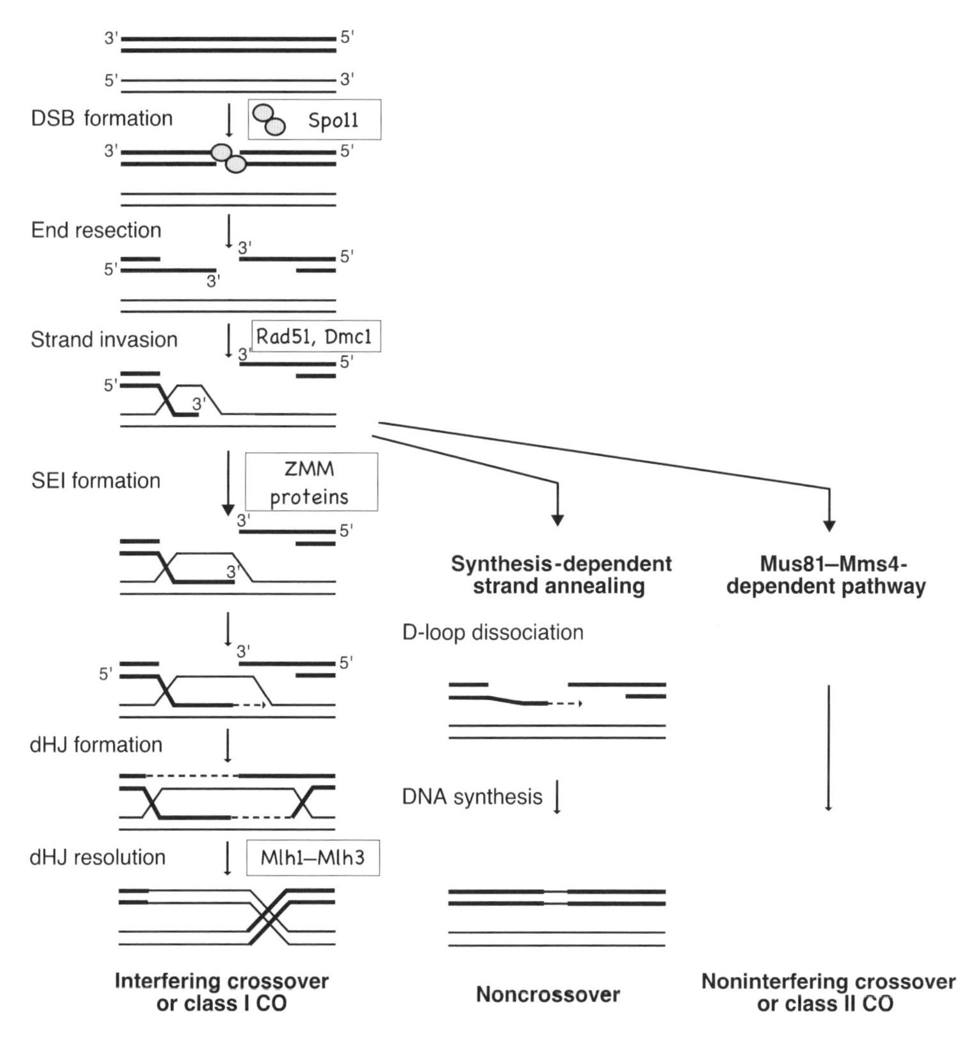

Figure 6.1 Model of pathways involved in meiotic CO and NCO formation, based on studies in *S. cerevisiae*. dHJ = double Holliday junction; SEI = single-end invasion. For details, see the text

contrasts with mitotic recombination in which DNA exchange occurs preferentially between sister chromatids. Further processing of the strand-exchange intermediates yields two kinds of recombination products: a subset of DNA recombination intermediates are designated to become COs, while remaining interactions are processed to yield NCO (noncrossover) products. COs resulting from this pathway (class I COs) are not distributed randomly along chromosomes. Notably, the presence of one CO decreases the probability of getting another CO nearby, a phenomenon called positive CO interference. A second pathway for CO formation appears to involve Mus81 and Mms4 and is not subject to interference (Cromie and Smith, 2007).

Physical, temporal and functional relationship between DNA events and chromosome organization

Meiotic recombination occurs during the prophase of the first meiotic division. The meiotic prophase I is temporally divided into substages defined by changes in chromosome organization. During leptonema, proteinaceous axial elements (AE) begin to form along each pair of sister chromatids. At zygonema, AEs of homologous chromosomes start to align, and a protein structure known as the central element (CE) forms between homologues and tethers them together, a process referred to as synapsis. The two AEs, now termed lateral elements (LEs), together with the central element constitute the synaptonemal complex (SC). The pairs of homologous autosomes remain fully synapsed throughout pachynema. During diplonema, the central element of the SC disassembles and homologous chromosomes remain held together at chiasmata, the cytological manifestation of COs. Prophase concludes with diakinesis, at which time much of the SC structure is lost.

Biochemical recombination complexes are physically associated with chromosome axes and the SC at many stages. Moreover, DNA recombination events are temporally coordinated with changes in chromosome organization and juxtaposition (Zickler and Kleckner, 1999; Blat *et al.*, 2002). The temporal relationship between DNA events and chromosome structure changes has been elucidated by direct analysis of DNA recombination intermediates in yeast (Padmore, Cao and Kleckner, 1991; Hunter and Kleckner, 2001) and appears to be conserved in many other organisms as judged by molecular studies of DNA recombination in mouse (Guillon *et al.*, 2005) and immunolocalization of recombination complexes along meiotic chromosomes (Moens *et al.*, 2002; Kolas *et al.*, 2005a). DSBs occur at leptonema and are followed by formation of bridges between the axes of homologous chromosomes. These bridges include recombination complexes and mark the sites of nascent DNA exchange between the homologues (e.g. Zickler and Kleckner, 1999; Tarsounas *et al.*, 1999). The CO/NCO decision occurs at the transition between leptonema and zygonema, concomitant to the initiation of SC formation. COs are formed by the end of pachynema (Guillon *et al.*, 2005; Allers and Lichten, 2001; Borner, Kleckner and Hunter, 2004; Terasawa *et al.*, 2007).

DNA recombination events and changes in meiotic chromosome structure are not only physically and temporally correlated, but are also functionally connected. In many organisms (but not in *D. melanogaster* and *C. elegans*) SC formation is dependent upon recombination initiation and processing of early recombination intermediates (e.g. Alani, Padmore and Kleckner, 1990; Pittman, Weinberg and Schimenti, 1998a; Baudat *et al.*, 2000; Romanienko and Camerini-Otero, 2000; Grelon *et al.*, 2001). Reciprocally, genetic analyses have revealed that the structure of the meiotic chromosomes is important for the formation of COs (Kleckner, 2006). For example, several mouse mutants defective for cohesin subunits (required for keeping sister chromatids together until their segregation (Suja and Barbero, 2009)) or SC components are partially or totally defective for the repair of meiotic recombination intermediates (e.g. Bolcun-Filas *et al.*, 2007; Bolcun-Filas *et al.*, 2009; de Vries *et al.*, 2005; Hamer *et al.*, 2008; Wang and Hoog, 2006; Revenkova *et al.*, 2004).

6.2.2 Initiation of meiotic recombination: Spo11-dependent double-strand break formation

A universal mechanism for the initiation of meiotic recombination

A large body of evidence shows that meiotic recombination in *S. cerevisiae* is initiated by the formation of DSBs, catalyzed by the Spo11 protein (reviewed in Keeney, (2001)). Thereafter, evidence based on studies in several other organisms (fission yeast, multicellular fungi, flies, worms, plants, mammals) supports the conclusion that Spo11-dependent programmed DSBs are a universal mechanism for the initiation of meiotic recombination. First, Spo11 orthologues have been identified in all species tested, and in every case *spo11*-null mutation abolishes meiotic recombination. Second, the phosphorylated form of histone H2AX, known to accumulate at sites of DSBs, forms Spo11-dependent transient foci on meiotic chromatin from leptonema to early pachynema (Mahadevaiah *et al.*, 2001; Jang *et al.*, 2003). Finally, some evidence for DNA breaks and Spo11-dependent DNA ends with 3′ overhangs have been obtained in mouse testicular germ cells by PCR (polymerase chain reaction) and *in situ* DNA labelling assays, respectively (Zenvirth *et al.*, 2003; Qin *et al.*, 2004).

The Spo11 protein

S. cerevisiae spo11 mutants make no DSBs and generate aneuploid inviable spores (Cao *et al.*, 1990; Klapholz, Waddell and Esposito, 1985). The Spo11 protein is covalently attached to the 5′ strand termini on either side of the DSBs in mutants accumulating unresected meiotic DSBs (Keeney, Giroux and Kleckner, 1997). Spo11 shows sequence similarity with the catalytic subunit (TopVIA) of the archeal type II topoisomerase VI (Bergerat *et al.*, 1997). These findings strongly suggest that Spo11 catalyzes the formation of meiotic DSBs through a topoisomerase-like transesterification reaction. However, it should be noted that the DNA-cleaving activity of Spo11 has not been demonstrated *in vitro* yet.

Mutational analyses of *S. cerevisiae* Spo11 led to the identification of residues necessary for the formation of meiotic DSBs, and indicate that Spo11 is involved not only in cleavage, but also in selection of DSBs sites (Bergerat *et al.*, 1997; Diaz *et al.*, 2002; Arora *et al.*, 2004; Nag *et al.*, 2006). Moreover, these analyses also suggest that Spo11 forms dimeric structures *in vivo*, a prediction that is supported by recent biochemical data (Sasanuma *et al.*, 2007).

A model for Spo11-induced DSB formation has been proposed in which a Spo11 homodimer creates two single-strand DNA breaks, resulting in a DSB with each 5′ DNA strand terminus covalently linked to a Spo11 monomer (Keeney, 2008). Thereafter, Spo11 is removed by endonucleolytic cleavage a few bases away from the break site (Neale, Pan and Keeney, 2005).

In *S. cerevisiae,* nine other proteins of poorly known function are required for the formation of meiotic DSBs in addition to Spo11 (Keeney, 2001). Most are not conserved across kingdoms. The Mre11 complex (Mre11, Rad50 and Xrs2/NBS1), as well as Ski8, have been identified in several organisms, including mammals. However, their function in the formation of meiotic DNA breaks might not be conserved in higher eukaryotes (Borde, 2007; Jolivet *et al.*, 2006).

Mouse models with impaired initiation of meiotic recombination

Disruption of the mouse *Spo11* gene causes male and female infertility (Baudat *et al.*, 2000; Romanienko and Camerini-Otero, 2000). Evidence from several studies suggests that meiotic DSBs are not formed in $Spo11^{-/-}$ meiocytes (Mahadevaiah *et al.*, 2001), and SC formation is profoundly impaired, supporting the view that, as in *S. cerevisiae*, initiation of recombination precedes synapsis and is required for SC formation. $Spo11^{-/-}$ spermatocytes and oocytes are eliminated by apoptosis, but at different stages of meiotic prophase, which highlights a sexual dimorphism also observed in several other meiotic recombination mouse mutants (see Section 6.4).

The mouse meiotic mutant $Mei1^{m1Jcs}$ (meiosis defective 1) has been isolated in a screen for infertile mice following a chemical mutagenesis of ES (embryonic stem) cells (Munroe *et al.*, 2000). The phenotype of $Mei1^{m1Jcs/m1Jcs}$ mice indicates that MEI1 is required for meiotic DSB formation (Libby *et al.*, 2002; Libby *et al.*, 2003; Reinholdt and Schimenti, 2005). The human MEI1 protein exhibits similarity with AtPRD1, a protein required for DSB formation in *Arabidopsis thaliana*. The conserved N-terminal region of AtPRD1 interacts with AtSPO11-1 in a yeast two-hybrid assay (De Muyt *et al.*, 2007), which supports the possibility that MEI1 interacts with SPO11 and promotes meiotic DSB formation in mammals. No yeast homologue of MEI1 has been found to date, and MEI1 does not contain any recognizable functional domains. Thus, biochemical approaches will be necessary to investigate the molecular function of MEI1 in mammalian meiosis.

6.2.3 Dna strand-exchange proteins

RecA strand-exchange reaction

In recombination reactions, single-strand DNA is used to initiate genetic exchange with a homologous duplex. The RecA protein from *Escherichia coli* is the first identified recombinase (McEntee *et al.*, 1976). RecA is the prototype for a ubiquitous family of proteins that function in recombination by assembling into a helical protein filament on overhanging 3′ single-stranded DNA (ssDNA) tails resulting from 5′ nucleolytic resection at DSBs (reviewed in Wang, Chen and Wang (2008)). The resulting nucleoprotein filament, referred to as the presynaptic filament, captures a duplex DNA molecule, forming a three-stranded complex (also called the synaptic complex). It is within this ternary complex that homology is thought to be probed. Once homology is detected, a stable DNA joint is formed. The joint is then extended by DNA strand exchange, forming what is known as a D-loop structure. Subsequent steps involve DNA synthesis, the capture of the second 3′ ssDNA end, the migration of branched DNA structures followed by their resolution and ligation, leading to the formation of mature recombinant products.

Two RecA homologues: Rad51 and Dmc1

In most eukaryotes, two RecA homologues are present: the Rad51 recombinase needed for both mitotic and meiotic homologous recombination, and the meiosis-specific recombinase Dmc1. The discovery of Dmc1 raised several questions: why are two recombinases needed for meiotic recombination? What is the specific role of each recombinase? How are their functions coordinated?

Rad51 and Dmc1 biochemical activities Human RAD51 and DMC1 share 45% amino acid identity (Masson and West, 2001). Several biochemical analyses have established that, overall, the intrinsic activities of the purified RAD51 and DMC1 proteins are similar (Sung and Robberson, 1995; Baumann, Benson and West, 1996; Li *et al.*, 1997; Hong, Shinohara and Bishop, 2001; Sehorn *et al.*, 2004; Sauvageau *et al.*, 2005). RAD51 and DMC1 helical filaments are identical as regards several structural parameters (Sheridan *et al.*, 2008), and both of them promote ATP-dependent homologous DNA pairing and strand exchange. Thus, analyzing the biochemical properties of RAD51 and DMC1 does not provide clues about the specificity and differences of their *in vivo* functions.

Cooperation of the two recombinases suggested by genetic, physical and cytological analyses In *S. cerevisiae*, both *rad51* and *dmc1* single mutants accumulate processed DSBs to levels higher than normal, and exhibit delayed and inefficient chromosome synapsis and decreased spore viability (Shinohara, Ogawa and Ogawa, 1992; Bishop *et al.*, 1992). Physical analyses of recombination intermediates in various mutants have shown that Dmc1 specifically promotes exchange between homologous non-sister chromatids and also that Rad51 is needed for this strong homologue bias (Schwacha and Kleckner, 1997). Therefore, it appears that Rad51 and Dmc1 may play distinct roles and cooperate to promote an interhomologue recombination pathway. On the other hand, recombination defects observed in *dmc1* yeast mutants can be partially suppressed by overexpression of either Rad51 or Rad54 (a protein that stimulates Rad51 activity), suggesting a functional overlap between the two recombinases (Bishop *et al.*, 1999; Tsubouchi and Roeder, 2003). Taken together, these observations led to the proposal that two distinct meiotic recombination pathways may operate in *S. cerevisiae*, one being dependent on Rad51 alone and the other on both Rad51 and Dmc1 (Tsubouchi and Roeder, 2003). It remains to be determined whether the Rad51-only pathway functions during meiosis in wild-type yeast cells.

Cytological analyses also support the view that Rad51 and Dmc1 cooperate in the repair of DSBs. In both mouse and yeast, Rad51 and Dmc1 assemble as Spo11-dependent cytologically visible complexes (foci) at the same sites on meiotic chromosomes (e.g. Tarsounas *et al.*, 1999; Baudat *et al.*, 2000; Bishop, 1994). In yeast, Rad51 is required for the normal assembly of Dmc1 complexes, while Rad51 foci are formed independently of Dmc1, suggesting a temporal control in the loading of the two recombinases (Bishop, 1994; Shinohara *et al.*, 1997). In mouse and human meiocytes, RAD51/DMC1 foci localize to AE and SC from leptonema to early pachynema (Moens *et al.*, 2002; Kolas *et al.*, 2005a; Tarsounas *et al.*, 1999; Plug *et al.*, 1996; Barlow *et al.*, 1997; Lenzi *et al.*, 2005; Oliver-Bonet *et al.*, 2005). DMC1 is not required for localization of RAD51 onto chromosomal axes, as suggested by the presence of RAD51 foci in DMC1-deficient spermatocytes (Pittman *et al.*, 1998b; Yoshida *et al.*, 1998). Unfortunately, the role of RAD51 in DMC1 recruitment to chromatin in mammals is not known because of the early embryonic lethality of the *Rad51*-null mutation in mice (Lim and Hasty, 1996; Tsuzuki *et al.*, 1996).

Dmc1 and infertility in mammals

Mice bearing a homozygous null mutation in *Dmc1* are sterile and exhibit a severe meiotic disruption in early prophase I (Pittman *et al.*, 1998b; Yoshida *et al.*, 1998). $Dmc1^{-/-}$ spermatocytes exhibit features characteristic of the persistence of unrepaired DSBs (Barchi *et al.*, 2005) and a strong synapsis defect, although axial elements are formed and appear mostly normal. $Dmc1^{-/-}$ meiocytes are eliminated by apoptosis.

Bannister *et al.* (2007) have analyzed a point mutation of *Dmc1* ($Dmc1^{Mei11}$), which confers a male-specific dominant sterility phenotype, similar to that of $Dmc1^{-/-}$ males. In contrast, $Dmc1^{Mei11}/+$ females are fertile, although the oocytes display moderate defects in SC formation and progression of recombination, resulting in a partial depletion of the pool of oocytes in adults. Interestingly, in each sex, the phenotype is slightly more severe in one genetic background (C57BL/6J) than in another (C3H), giving evidence of the role of the genetic environment, even for a process as conserved as this key step in meiotic recombination. *In vitro* experiments suggest that the $\mathrm{DMC1^{Mei11}}$ protein is still able to self-interact, but has a reduced affinity for DNA and is unable to perform a strand invasion reaction. The reason for this sexual dimorphism is unknown, but could be compared to several mutations affecting meiotic recombination and SC formation in mice (see Section 6.4).

Sequencing of candidate genes from a set of infertile patients has identified an infertile woman with premature ovarian failure, homozygous for the *Dmc1-M200V* polymorphism (Mandon-Pepin *et al.*, 2008). Structural biochemical and genetic analyses have provided evidence that this polymorphism impairs the function of DMC1, supporting the view that this single-nucleotide polymorphism (SNP) can be a cause of human infertility (Hikiba *et al.*, 2008).

Rad51 and Dmc1 accessory proteins

Several homologous recombination factors that stimulate RAD51 and/or DMC1-dependent strand-exchange reaction have been identified. These factors can be divided into two classes, those that act to favour the formation of RAD51 and/or DMC1 nucleoprotein filaments, termed recombination mediators, and those that act downstream by facilitating the formation of the synaptic complex and/or directly facilitating the strand-exchange reaction. Amongst recombination mediators, some mediate specifically the assembly of the RAD51 nucleofilament (RAD52 and the RAD55–RAD57 heterodimer), while others, such as the budding yeast Mei5–Sae3 complex, specifically promote the formation of the Dmc1 presynaptic filament. Interestingly, growing evidence suggests that in mammals, BRCA2 may serve to nucleate both RAD51 and

DMC1 presynaptic filament assemblies. BRCA2 interacts with RAD51 (Sharan *et al.*, 1997; Wong *et al.*, 1997), and a large body of results has provided evidence that BRCA2 acts as a recombination mediator by helping the assembly of RAD51 into active nucleoprotein filaments (San Filippo, Sung and Klein, 2008). BRCA2 also interacts with DMC1, suggesting a role for this protein in meiotic recombination (Thorslund and West, 2007). Consistent with these biochemical data, BRCA2 has been found to localize along meiotic chromosomes (Chen *et al.*, 1998), and viable mice with impaired BRCA2 expression are infertile and exhibit a similar defect in RAD51 and DMC1 focus formation along meiotic chromosomes (Sharan *et al.*, 2004). Taken together, these data support the possibility that BRCA2 also serves to nucleate DMC1 presynaptic filament assembly. The questions arising from these findings are whether, and how, BRCA2 plays a role in coordinating the activities of the two recombinases.

The Hop2–Mnd1 heterodimeric complex acts downstream of the recombination mediators both in *S. cerevisiae* and mammals. Male and female *Hop2* knockout mice are sterile. Mutant spermatocytes arrest prior to pachynema, display a strong defect in chromosome synapsis and exhibit features characteristic of the persistence of unrepaired DSBs (Petukhova, Romanienko and Camerini-Otero, 2003). Two recent studies have elucidated the action mechanism of HOP2–MND1 in mammals, by showing that HOP2–MND1 stabilizes both RAD51 and DMC1 presynaptic filaments, and stimulates the ability of the nucleofilaments to capture duplex DNA (Pezza *et al.*, 2007; Chi *et al.*, 2007).

Rad54 and its paralogues are members of the Swi2/Snf2 family. Members of this family are ATPases that promote chromatin remodelling, DNA topology alterations and displacement of proteins from DNA. In *S. cerevisiae*, Tid1/Rdh54, a Rad54 paralogue, promotes dissociation of Dmc1 from nonrecombinogenic sites on meiotic chromatin, and is required for Rad51 and Dmc1 colocalization *in vivo* (Shinohara *et al.*, 2000; Holzen *et al.*, 2006). RAD54 and its paralogue RAD54B are present in mammals. Interestingly, in *Rad54*$^{-/-}$ spermatocytes (but not in *Rad54B*$^{-/-}$), RAD51 forms aberrant foci persisting until diplonema on meiotic chromosomes (Wesoly *et al.*, 2006). On the other hand, RAD54B has been found to enhance the DNA strand-exchange activity of DMC1 by stabilizing the DMC1–ssDNA complex (Sarai *et al.*, 2006). However, the significance of these findings is unclear since deficiency of RAD54B, RAD54 or both does not induce meiotic recombination defects in mouse (Wesoly *et al.*, 2006).

Other recombinase accessory factors have been identified and there has been recent progress on elucidating their mechanisms of action, extensively discussed in an excellent review by San Filippo, Sung and Klein (2008).

6.2.4 Processing of the strand-exchange intermediates: crossover and noncrossover pathways

Several pathways coexist for the processing of DNA strand-exchange intermediates

CO and NCO are processed via separate pathways In the DSB repair model for recombination of Szostak (Szostak *et al.*, 1983), a single pathway of DNA intermediates generates both CO and NCO products, depending on the strands cleaved during the resolution of Holliday junctions (HJs). However, studies in *S. cerevisiae* have shown that early steps in CO and NCO formation proceed along the same pathway, but that, soon after nascent DNA–DNA interactions between homologues, the pathway branches to generate either COs or NCOs through different DNA intermediates (Hunter and Kleckner, 2001; Allers and Lichten, 2001; Borner, Kleckner and Hunter, 2004; Terasawa *et al.*, 2007). Two recombination intermediates have been identified that appear to be specific to the crossover pathway (Figure 6.1): single-end invasions (SEIs) that are asymmetric strand-exchange intermediates involving one DSB end and its homologue, and double HJs (dHJs). To date, DNA intermediates specific to the NCO pathway have not been reported. Nevertheless, it has been suggested that a major fraction of NCO products are produced by synthesis-dependent strand annealing (Allers and Lichten, 2001; Terasawa *et al.*, 2007; McMahill, Sham and Bishop, 2007). In mammals as well, both CO and NCO products have been detected and several lines of evidence indirectly suggest that NCO products are in large excess relative to the number of COs. Moreover, some data support the view that in mammals, as in *S. cerevisiae,* NCOs and COs arise from different pathways (reviewed in Baudat and de Massy (2007a)).

In budding yeast, most COs (class I COs) are subjected to interference and depend on a group of proteins referred to as ZMM proteins (for Zip1, Zip2, Zip3, Zip 4, Mer3, Msh4, Msh5) (Borner, Kleckner and Hunter, 2004). Mlh1 and Mlh3 proteins are also required for the formation of class I CO. In mammals, orthologues of several ZMM proteins and MLH1–MLH3 have been identified and are involved in CO formation.

There are several CO pathways In *S. cerevisiae*, *zmm* mutants exhibit residual COs, suggesting that one or more additional pathways contribute to the wild-type level of CO. Genetic analyses suggest that most non-class I COs do not exhibit interference and are dependent on the structure-specific endonuclease Mus81 and its heterodimeric partner Mms4 (de los Santos *et al.*, 2003; Hollingsworth and Brill, 2004). However, the Mus81/Mms4-dependent pathway (also called class II CO pathway) is still poorly defined.

In other organisms, various situations have been reported relative to the presence of these CO pathways. Some of them appear to utilize both ZMM- and Mus81-dependent CO pathways, while others exhibit only one of these two pathways. In mammals, the vast majority of COs (>90%) appear to be dependent upon the ZMM pathway.

Proteins involved in the interference-dependent crossover pathway

The ZMM proteins The ZMM group comprises different classes of proteins, and presumed ZMM orthologues have also been identified in plants and mammals. Mer3, Msh4 and Msh5 are highly conserved proteins. Mer3 is an ATP-dependent DNA helicase, which is thought to stabilize the first-strand invasion intermediate (Nakagawa and Kolodner, 2002; Mazina *et al.*, 2004). Msh4 and Msh5 are two homologues of the bacterial MutS protein that functions as a heterodimeric complex. The purified human MSH4–MSH5 heterodimer binds to three-armed progenitor HJs and to HJs, and forms a sliding clamp that embraces homologous chromosomes (Snowden *et al.*, 2004). It has been postulated that repeated loading of MSH4–MSH5 heterodimers stabilizes the DNA structure associated with strand invasion, and thereby promotes the formation of dHJ intermediates. Zip1 is a major component of the synaptonemal complex. Zip2, Zip3 and Zip4 (Spo22) are thought to be implicated in ubiquitinylation and SUMOylation (Perry, Kleckner and Borner, 2005). A recent work suggests that Zip3 is a SUMO E3 ligase, which activity might be required for early assembly of the SC in budding yeast (Cheng *et al.*, 2006). However, a direct functional link between Zip3-mediated SUMO modifications and DNA recombination has not yet been established.

Detailed analysis of various *S. cerevisiae zmm* mutants (*mer3*, *msh5*, *zip1*, *zip2*, *zip3*) has shown that the corresponding proteins are required for the processing of DSBs toward stable SEI intermediates (Borner, Kleckner and Hunter, 2004). Accumulating data suggest that ZMM proteins function together during the leptotene to zygotene transition at sites of future COs and SC nucleation (Agarwal and Roeder, 2000; Fung *et al.*, 2004; Henderson and Keeney, 2004). Interestingly, a recent study suggests that the ZMM proteins promote the formation of COs, in part by protecting the nascent CO-designated recombination intermediates from dissolution by the RecQ-helicase Sgs1 (discussed below) (Jessop *et al.*, 2006). ZMM proteins also play an important role in the assembly of the synaptonemal complex. Zip1 is an integral component of the SC, but the molecular functions of the other ZMM proteins in synapsis remain unclear and have been recently discussed (Lynn, Soucek and Borner, 2007).

The mammalian SYCP1 protein is a key component of the SC central element. For this function at least, it is the homologue of the budding yeast Zip1 protein. *Sycp1*$^{-/-}$ mice show defects in prophase progression, SC formation and DSB repair. Only a few spermatocytes reach metaphase I and most chromosomes form univalents, suggesting a CO defect (de Vries *et al.*, 2005). ZIP4H (Zip4 orthologue) deficiency in mice results in delayed repair of DSBs and in decreased CO formation (Adelman and Petrini, 2008; Yang *et al.*, 2008). However, unlike in yeast, ZIP4H is not required for normal synapsis, supporting the view that the role of Zip proteins in synapsis is not universal (Jantsch *et al.*, 2004; Chelysheva *et al.*, 2007). MSH4- and MSH5-deficient mice exhibit a strong synapsis defect, and apoptosis of spermatocytes and oocytes in early prophase and before the dictyate stage, respectively (de Vries *et al.*, 1999; Edelmann *et al.*, 1999; Kneitz *et al.*, 2000). Interestingly, the depletion of oocytes in *Msh5*$^{-/-}$ mice can be partially suppressed by deletion of *Spo11*, suggesting that oocyte loss is driven by a failure in the repair of DSBs (Di Giacomo *et al.*, 2005).

It has been proposed that in budding yeast ZMM proteins mark the sites of future COs. This is not the case in mammals, since the number of MSH4 foci (and presumably MSH5) along mouse meiotic chromosomes greatly exceeds the number of COs. Indeed, the number of MSH4 foci decreases from approximately 150 at zygonema (at this stage MSH4 colocalizes and most probably interacts with RAD51/DMC1 proteins), to 50 at mid-pachynema where MSH4 colocalizes with the MLH1 protein that marks the sites of COs (Kneitz *et al.*, 2000; Santucci-Darmanin *et al.*, 2000; Neyton *et al.*, 2004). The MSH4 foci are more evenly spaced than expected if they were randomly distributed, indicating that they display a low level of positive interference (de Boer *et al.*, 2006). Thus, in mammals the role of MSH4 and MSH5 is not restricted to the formation of class I COs, a possibility being that MSH4–MSH5 also participates in the formation of NCO products. Based on the spatiotemporal distribution of MSH4, it has been speculated that, in mammals, the selection of CO sites operates in two successive steps at different stages of DSB repair (de Boer *et al.*, 2006).

In *S. cerevisiae*, sites designated to give rise to COs are also sites of SC nucleation (Henderson and Keeney, 2005). Whether this is the case in mammals remains to be determined.

Mlh1 and Mlh3 Mlh1 and Mlh3 are two homologues of the bacterial MutL protein that function as a heterodimeric complex and are required for the formation of class I COs in the budding yeast. They function downstream of the ZMM proteins, most probably in the processing of dHJs (Hunter and Borts, 1997; Wang, Kleckner and Hunter, 1999; Argueso *et al.*, 2004). Both *Mlh1*- and *Mlh3*- knockout mice exhibit a strong defect in the formation of chiasmata in both male and female meioses (Baker *et al.*, 1996; Edelmann *et al.*, 1996; Woods *et al.*, 1999; Lipkin *et al.*, 2002; Kan *et al.*, 2008). Direct analyses at the DNA level in mouse have shown that MLH1 and MLH3 are necessary for the formation of around 90% of COs but not for NCO formation (Guillon *et al.*, 2005; Svetlanov *et al.*, 2008). Consistent with these findings, these proteins have been shown to colocalize at sites of chiasmata at the mid-pachytene stage (Marcon and Moens, 2003; Kolas *et al.*, 2005b). Several studies suggest that MLH3 and MLH1 are recruited sequentially to a subset of MSH4–MSH5 foci through direct protein–protein interactions (Santucci-Darmanin *et al.*, 2000; Lipkin *et al.*, 2002; Kolas *et al.*, 2005b; Santucci-Darmanin *et al.*, 2002).

Concerning the role of Mlh1 and Mlh3 in CO formation, various hypotheses can be formulated. One of them is that Mlh1–Mlh3 might act on Msh4–Msh5 sliding clamp structures to impose a dHJ conformation that ensures CO formation. Alternatively, Mlh1–Mlh3 might be directly involved in the resolution of dHJs through its endonuclease activity (Nishant, Plys and Alani, 2008). Finally, Mlh1–Mlh3 might recruit and/or activate a downstream factor that resolves intermediates into COs. In this regard, an interesting candidate is the GEN1/Yen1 resolvase newly identified in human and *S. cerevisiae*, which promotes HJ resolution in a manner analogous to that shown by the bacterial resolvase RuvC (Ip *et al.*, 2008).

The role of Mus81 and Mms4 (Eme1) in meiotic recombination

Mus81 is an evolutionarily conserved endonuclease, which forms a complex with a second protein, Mms4/Eme1 that is required for nuclease activity. Extensive analysis of the substrate specificity of Mus81–Mms4/Eme1 from both budding and fission yeasts, as well as from humans, has shown that this enzyme has a cleavage preference for structures such as nicked HJs, D-loops and 3′ flaps (for review see Hollingsworth and Brill (2004)). Nevertheless, recent studies suggest that this enzyme can also cleave intact HJs *in vitro* (Gaskell *et al.*, 2007; Taylor and McGowan, 2008). In *Schizosaccharomyces pombe*, the major pathway to form COs depends on the Mus81 complex. Cromie *et al.* (2006) have shown that most of the recombination intermediates detected in *S. pombe* are single HJs, and have provided evidence that Mus81–Eme1 promotes CO formation by resolving single HJs. These findings led to the proposal that, in *S. cerevisiae*, the major ZMM protein-dependent CO pathway that involves double HJs coexists with a minor pathway that involves rare single HJs and Mus81–Mms4. However, recent studies suggest that the primary function of Mus81–Mms4 in budding yeast meiosis is rather to resolve aberrant recombination intermediates that escaped disassembly by the Sgs1 helicase (see below) (Oh *et al.*, 2008; Jessop and Lichten, 2008). Whether or not Mus81–Mms4 resolves these aberrant joint molecules (JMs) directly to CO is unclear. Thus, it remains to clarify whether, in budding yeast, Mus81–Mms4 promotes CO formation by resolving single HJs or aberrant JMs, or either by promoting the formation of or stabilizing a subset of interhomologue JMs, as suggested by Oh *et al.* (2007).

Although *Mus81* deficiency does not affect mouse fertility (McPherson *et al.*, 2004; Dendouga *et al.*, 2005), a recent genetic study suggests that in mammals, MUS81 participates in generating a small subset of COs by an MLH3-independent pathway, and that a regulatory cross-talk operates between the MUS81- and the MLH3-dependant CO pathways (Holloway *et al.*, 2008).

RecQ helicase involvement in the processing of recombination intermediates

A large body of evidence suggests that the budding yeast Sgs1 RecQ-like helicase and its human homologue, BLM, have an anti-CO activity. BLM is capable of disrupting D-loop DNA structures *in vitro*, and both Sgs1 and BLM promote branch migration of HJs (e.g. Bennett *et al.*, 1999; Bachrati, Borts and Hickson, 2006). Moreover, both Sgs1 and BLM, in conjunction with topoisomerase III and RMI1/BLAP75, can disassemble synthetic dHJs to produce NCO products (reviewed in Mankouri and Hickson (2007)).

Recent studies have focused on the role of Sgs1 in meiotic recombination. Genetic data and physical analyses of meiotic recombination intermediates have shown that: (i) Sgs1 is not required for the formation of NCO products; (ii) Sgs1 has an anti-CO activity that is antagonized by the ZMM CO-promoting proteins at sites where DNA–DNA interactions are designated to mature into COs; (iii) Sgs1 limits the accumulation of aberrant recombination intermediates structure, such as intersister JMs or multichromatid JMs (Jessop *et al.*, 2006; Oh *et al.*, 2008; Jessop and Lichten, 2008; Oh *et al.*, 2007). Whether Sgs1 prevents JM accumulation by unwinding early strand-exchange intermediates before stable JM formation, by disassembling stable JMs after they form, or by doing a combination of both remains to be clarified. Taken together, these data suggest that Sgs1 is needed for accurate metabolism of recombination intermediates during meiosis. To date, it is unknown whether BLM exerts a similar function in mammalian meiosis. Luo *et al.* (2000) have reported that viable BLM-deficient mice exhibit a normal level of COs. Nevertheless, immunocytological analyses and the reduced fertility of Bloom syndrome patients suggest that BLM participates in meiotic recombination mechanisms (e.g. Walpita *et al.*, 1999).

6.3 Frequency and distribution of meiotic recombination events

6.3.1 Detection and mapping of recombination events

The repair of the Spo11-dependent DSBs through the pathways described above generates two types of recombination products (COs and NCOs), which differ from each other in several aspects, as discussed above. The distribution of COs results from the combination of two factors: the first is the distribution of initiating DSBs; the second is the proportion of precursors directed toward producing a CO, which varies over the genome. Thus, it is necessary to describe the frequencies and distributions of both COs and NCOs in order to understand the control of CO distribution.

The frequency and distribution of COs can be determined with good accuracy by several methods, each of them having its own advantages and limits (discussed in Arnheim, Calabrese and Tiemann-Boege, 2007; Lynn, Ashley and Hassold, 2004; Buard and de Massy, 2007; Kauppi, Jeffreys and Keeney, 2004). In addition, several methods providing some insight into the rate and distribution of NCOs are also mentioned below.

Cytological approaches

The cytological methods allow for a genome-wide estimate of the number of CO events and their distribution along chromosomes. The two main advantages of this approach are the ability to perform analyses within homozygous (inbred lines) or sterile individuals (providing that the meiocytes reach the appropriate stage of meiosis), and to detect CO events without the selection biases that could affect the transmission of CO products to the progeny. However, the resolution is limited and the cells can be difficult, sometimes almost impossible (human oocytes at the diakinesis/metaphase I stage), to obtain in sufficient number. Two types of markers have been used. The chiasmata are detected on diakinesis/metaphase I chromosome preparations, but their mapping is rough. More recently, the proteins MLH1 and MLH3 have been shown to form foci on the SC at positions corresponding to sites of chiasmata (Marcon and Moens, 2003; Anderson *et al.*, 1999; Froenicke *et al.*, 2002; Sun *et al.*, 2004). It should be kept in mind, however, that the overall distribution of COs might differ slightly from the one of MLH1 foci, because of the formation of a small proportion of COs ($<10\%$) independent from this protein (Baker *et al.*, 1996; Woods *et al.*, 1999; Lipkin *et al.*, 2002; Kan *et al.*, 2008; Holloway *et al.*, 2008).

Pedigree analysis

In pedigree analysis, the transmission of a series of markers from parents to offspring is studied to build whole genome sex-averaged or sex-specific genetic maps. In theory, the resolution is determined by the density of heterozygous markers. However, recent efforts towards identifying polymorphisms (especially SNPs) in the mouse and human genomes massively increased the amount of available markers, so that the major limitation for the accuracy and the operational resolution of genetic maps comes from the number of meioses subjected to analysis. In human for example, the average recombination rate is about 1.3cM/Mb, corresponding to an average of little more than a single event per 100 kb interval if 1000 meioses are analyzed (Kong *et al.*, 2002; Matise *et al.*, 2007). These analyses might suffer from selection biases due to the fact that they take into account only the COs that are transmitted to live offspring. For example, there could be selection on the number and distribution of COs required for the formation of euploid gametes.

Analyzing the patterns of genetic variation in populations

The distribution of historical COs in a population can be inferred from the analysis of the variation in genetic diversity along the genome in that population. These approaches take advantage of the fact that the recombination events that occurred during the history of a population shape the fine-scale pattern of genetic diversity in the present population (Stumpf and McVean, 2003). An obvious limitation of population-based approaches is that only sex-averaged recombination can be analyzed (with the exception of the non-pseudoautosomal region of the X chromosome). Other limitations include, notably, the possibility of bias introduced by selection or genetic drift (Stumpf and McVean, 2003; Coop and Przeworski, 2007). The pattern of linkage disequilibrium (LD) displays relatively long segments of high association (called haplotype blocks) between markers, interrupted by short regions of LD breakdown, which have been shown to correspond to CO hotspots (defined as short intervals experiencing a CO rate significantly higher than the adjacent regions) in many cases. The analysis of LD pattern has therefore been used for identifying potential recombination hotspots; some of them are characterized by directly measuring the exchange rate afterward (Kauppi, Jeffreys and Keeney, 2004; Jeffreys, Kauppi and Neumann, 2001). Recently, new methods based on coalescence have been introduced, allowing the detection and, importantly, the quantification of fine-scale variation in the LD-based rate of recombination in a population (Stumpf and McVean, 2003; Li and Stephens, 2003; Fearnhead *et al.*, 2004). These methods have been especially efficient in revealing the fine-scale variations of the recombination rate in the human genome, thanks to the enormous amount of information on SNPs that has been generated in the past few years (International HapMap Consortium, 2007; McVean *et al.*, 2004; Myers *et al.*, 2005). They have also been used for comparing fine-scale recombination rates between humans and chimpanzees (Ptak *et al.*, 2005; Winckler *et al.*, 2005).

Physical analysis of recombination hotspots by 'sperm' typing

Methods based on allele-specific PCR have been developed for detecting COs in very short intervals (<15 kb) in pooled sperms. Millions of sperms from a single male can be screened to detect and to characterize hundreds of CO products at fine scale (Jeffreys *et al.*, 2004). This approach has been developed initially for the analysis of CO hotspots in sperm, but has recently been adapted for analyzing recombination in oocytes as well (Guillon *et al.*, 2005; Ng *et al.*, 2008; Baudat and de Massy, 2007b).

Detection of NCOs

The tools allowing detection of NCO events are much scarcer than those for COs. There is no proven cytological marker specific for NCOs. However, several proteins involved in early steps of the recombination mechanism form foci, which are thought to localize either at sites of both ongoing COs and NCOs or at NCO sites only. The markers that have been used for estimating the number of recombination events in mammals are the phosphorylated form of H2AX (γH2AX), RAD51, DMC1, RPA, BLM, MSH4 and MSH5 (Moens *et al.*, 2002; Kolas *et al.*, 2005a; Tarsounas *et al.*, 1999; Lenzi *et al.*, 2005; de Boer *et al.*, 2006). Foci formed by γH2AX, RPA and MSH4 have been mapped along chromosomes (de Boer *et al.*, 2006; Grey, Baudat and de Massy, 2009). However, it is not known whether these foci mark all or only a fraction of recombination events, or whether they mark all sites simultaneously in a given nucleus (synchrony). Some caution should also be taken from the fact that, besides interhomologue COs and NCOs, the DSBs can lead to additional outcomes, including interactions with the sister chromatid or intrachromatid interactions, in proportions that are unknown. Some of the proteins mentioned above might also mark some of these unexplored events (discussed in Baudat and de Massy (2007a)).

The method of sperm typing, developed first for characterizing CO hotspots (see below), was modified for detecting and characterizing NCO products (Guillon *et al.*, 2005; Ng *et al.*, 2008; Baudat and de Massy, 2007b; Guillon and de Massy, 2002). Nevertheless, this method is restricted to very short intervals (the size of a PCR amplicon), limiting its use to previously identified CO hotspots. Finally, the presence of NCO hotspots can be detected by analyses of LD pattern and genetic diversity in populations, with lower efficiency and accuracy than for COs (Frisse *et al.*, 2001; Gay, Myers and McVean, 2007).

6.3.2 Nonrandom distribution of recombination events

Several factors control the overall number of COs per chromosome, as well as their distribution along chromosomes. Each chromosome pair must experience at least one chiasma, resulting from 1 CO (the 'obligate CO'), in order to ensure the correct segregation of homologous chromosomes at meiosis I (reviewed in Petronczki, Siomos and Nasmyth (2003)). On the other hand, having too many COs on one chromosome, or having them placed inadequately along the chromosome, may also become deleterious. Thus, the number and the distribution of recombination events are regulated, but the underlying mechanisms are still unknown, for the large part.

Control of the number of COs

One 'obligate' CO per bivalent is required for ensuring the segregation of homologous chromosomes. In fact, there is a minimum of one chiasma per euchromatic chromosome arm in mammals, while the short arm of acrocentric chromosomes does not display any chiasma usually (Laurie and Hulten, 1985a, 1985b; Hassold *et al.*, 2004; Lawrie, Tease and Hulten, 1995; Tease, Hartshorne and Hulten, 2002; Codina-Pascual *et al.*, 2006). Indeed, the number of chromosome arms (rather than the number of chromosomes) correlates strongly with the length of genetic maps across mammals, suggesting that this constitutes a major constraint on the whole genome CO rate (Coop and Przeworski, 2007).

The relationship between the number of COs and the size of the chromosome arms in a given organism is biphasic: the smallest chromosome arms experience a single CO, independently of their size, whereas a linear relationship is observed for chromosome arms whose size exceeds a certain threshold (Froenicke *et al.*, 2002; Matise *et al.*, 2007; Laurie and Hulten, 1985b; Lawrie, Tease and Hulten, 1995; Kaback, 1996; Broman *et al.*, 2002; Sun *et al.*, 2006). This threshold corresponds to approximately 100 Mb in mouse and 30 Mb in human oocytes (50 Mb in human spermatocytes), reflecting the difference between the organisms in average CO density.

Strikingly, the correlation between the number of COs and the size of the chromosomes is stronger if the length of SC, rather than the chromosomal DNA content, is taken into account (Froenicke *et al.*, 2002; Sun *et al.*, 2006; Lynn *et al.*, 2002). This can be connected with other situations showing a coordinate variation in CO number and pachytene chromosome axis length, both genome-wide and for individual chromosomes (Kleckner, Storlazzi and Zickler, 2003). Indeed, co-variation in SC length and CO rate (measured with MLH1 foci) has been observed between individuals and even between meiocytes from a single individual. These results suggest that the CO density is controlled by a factor or a mechanism, not identified yet, linked to the length of meiotic prophase I chromosome axes. An additional example is provided by the fact that a coordinate intersex variation in SC length (longer in female) and CO rate (higher in female) has been observed in several species, including mouse and human (Tease, Hartshorne and Hulten, 2002; Lynn *et al.*, 2002; Tease and Hulten, 2004).

The CO interference refers to the fact that the presence of one CO somewhere along a chromosome reduces the probability of occurrence of a second CO nearby on the same bivalent (positive interference). This phenomenon has consequences for both the frequency and the distribution of COs, especially along chromosomes having small genetic length and high level of interference, as are most mouse and human chromosomes (Falque *et al.* (2007), and references therein). In mouse and human, the interference is weaker in female than in male when the physical distances are measured in terms of DNA content (Mb), but it becomes similar when the distances are measured in terms of SC length (μm) (Tease, Hartshorne and Hulten, 2002; Petkov *et al.*, 2007). This difference in interference level, dependent on chromosome axis length, was sufficient to explain the intersex difference in CO rate along mouse chromosome 1, leading Petkov *et al.* (2007) to propose that the sex difference in interference level is the main cause for the intersex difference in CO rate in mouse.

Recombination events lie in hotspots

In budding yeast, most meiotic recombination events, if not all, occur in small regions (about 2 kb), as the result of the clustering of Spo11-generated DSBs in 50–200 bp intervals (Petes, 2001). In *S. pombe* too, recombination is initiated by meiotic DSBs localized at preferential sites, though it is not clear yet whether the identified DSB sites are at the origin of all meiotic recombination (Cromie and Smith, 2007; Cervantes, Farah and Smith, 2000; Ludin *et al.*, 2008).

As in *S. cerevisiae*, the major part of meiotic recombination in mammals appears to take place in hotspots. Indeed, CO hotspots, defined as short intervals experiencing a CO rate higher than the adjacent regions, have been identified in both human and mouse. The detailed molecular characterization of a few of them brought the demonstration that mammalian CO hotspots are sites of recombination initiation (reviewed in Arnheim, Calabrese and Tiemann-Boege, 2007; Kauppi, Jeffreys and Keeney, 2004; Jeffreys *et al.*, 2004; de Massy, 2003). Indeed, NCOs are detected at the centre of several CO hotspots, demonstrating that these are sites of recombination initiation (Guillon *et al.*, 2005; Ng *et al.*, 2008; Baudat and de Massy, 2007b; Guillon and de Massy, 2002; Jeffreys and May, 2004; Jeffreys and Neumann, 2005). In accord with this, DNA breaks have been detected in the *Eα* mouse hotspot (Qin *et al.*, 2004). Noticeably, NCO events are more tightly clustered at the centre of hotspots than CO exchange points, which are spread over a ~1.5 kb interval with a higher density in the central region (Figure 6.2a). Thus, gene conversion tracts associated with NCOs are shorter (<300 bp) than those associated with COs (500 bp).

The major role of hotspots in shaping the pattern of recombination has been demonstrated along a few intervals in mouse and human (Figure 6.2b) (Jeffreys, Kauppi and Neumann, 2001; Jeffreys *et al.*, 2005; Kelmenson *et al.*, 2005; Paigen *et al.*, 2008; Tiemann-Boege *et al.*, 2006). For example, six CO hotspots were identified and analyzed by sperm typing in a 216 kb segment in the human MHC (major histocompatibility complex), with a total CO frequency equal to the one estimated by pedigree analysis for the same interval (Jeffreys, Kauppi and Neumann, 2001). Strikingly, the location of these hotspots correlates with regions of LD breakdown, indicating that LD patterns are profoundly affected by the clustering of COs in hotspots. In this interval, the fraction of COs occurring outside hotspots has been estimated to represent only ~1/20th of the COs lying in hotspots. Genome-wide surveys of genetic variation in human populations made a breakthrough by permitting the identification of tens of thousands of hotspots (McVean *et al.*, 2004; Myers *et al.*, 2005). The most recent genome-wide analysis of SNP variation, taking advantage of the characterization of 3.1 million SNPs in four human populations, identified 32 996 LD-based CO hotspots (23 307 of them mapped to within 5 kb) accounting for about 60% of COs and 6% of the sequence of the human genome (International HapMap Consortium, 2007; Myers *et al.*, 2008). Surprisingly, despite the extremely high density of SNPs analyzed in this survey, about 0.5–1% of the SNPs analyzed are still untaggable, which means that they do not display significant LD with other SNPs, not even with the closest ones. Ninety percent of these untaggable SNPs are located within 5 kb of the centre of an identified CO hotspot. This suggests that most of these untaggable SNPs can be included into gene conversion tracts generated at CO hotspots, and therefore that most hotspots giving rise to NCOs also generate COs (International HapMap Consortium, 2007).

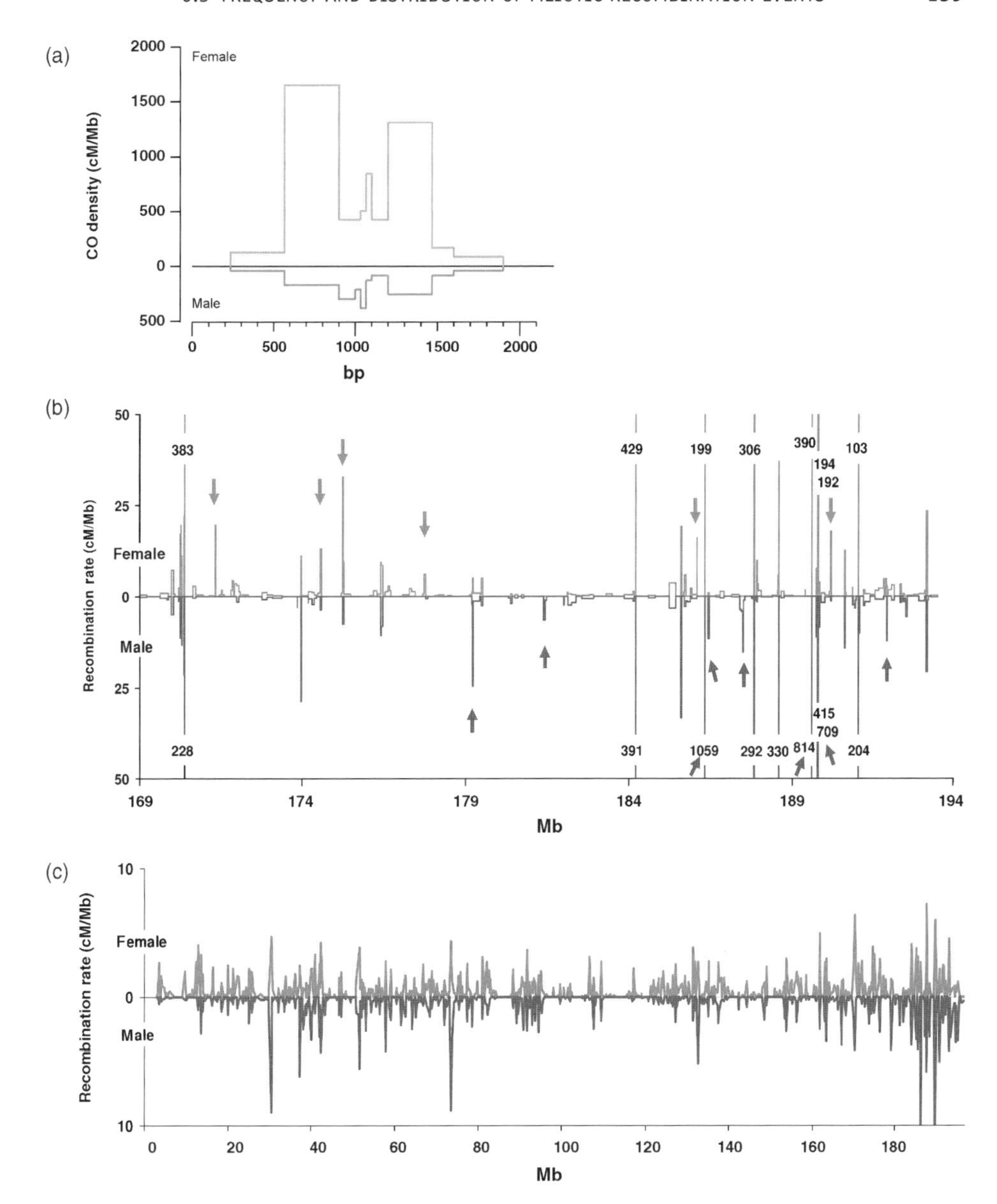

Figure 6.2 Sex-specific distribution of COs on mouse chromosomes. (a) High-resolution mapping of COs in the *Psmb9* hotspot on chromosome 17 (from Baudat and de Massy (2007b), with modifications). (b) Distribution of CO hotspots on a 25 Mb fragment on chromosome 1 (from Paigen *et al.*, 2008). (c) Chromosome-wide distribution of COs on mouse chromosome 1 (from Paigen *et al.*, 2008)

Variation in hotspot distribution

Meiotic DSB sites in *S. cerevisiae* are situated in their vast majority in intergenic intervals containing a transcription promoter. In turn, a large fraction of the promoter-containing intervals host an initiation site (reviewed in Petes (2001)). In *S. pombe*, DSB sites are particularly enriched in a few unusually wide intergenic regions (Cromie and Smith, 2007; Ludin *et al.*, 2008). Notwithstanding these discrepancies, both species share the fact that the structure of the chromatin (local accessibility, histone modifications) plays a major role in controlling the activity of recombination initiation sites (Petes, 2001; Ohta *et al.*, 1994; Wu and Lichten, 1994; Hirota *et al.*, 2008; Borde *et al.*, 2009).

In human, the combined analyses of COs detected by sperm typing over limited regions, and LD-based CO hotspots give an idea of the recombinational landscape, which appears to be composed of 1–2 kb-wide hotspots of variable intensity separated by an average of 50–100 kb with background CO density (reviewed in Buard and de Massy 2007). In mouse, CO hotspots have been detected at the position of haplotype block boundaries, suggesting that the situation is similar to that in human (Kauppi, Jasin and Keeney, 2007).

A significant correlation between CO rates and several parameters of sequence composition has been observed at various scales. However, these correlations were small, implying that these parameters are not responsible for a major part of the variation of CO rates (Kong *et al.*, 2002; McVean *et al.*, 2004; Myers *et al.*, 2005; Myers *et al.*, 2006; Shifman *et al.*, 2006; Spencer *et al.*, 2006). In both human and mouse, COs have a tendency to occur typically in Mb-long domains with a relatively low GC content, but rich in CpG dimers (though outside CpG islands). Human LD-based CO hotspots tend to be located within 50 kb of genes, but outside the transcribed units (Myers *et al.*, 2005). However, this is a trend, not a rule, and hotspots have been identified in introns (de Massy, 2003; Paigen *et al.*, 2008; Webb, Berg and Jeffreys, 2008).

Although in yeasts no clear sequence motif determining the presence of recombination hotspots has been identified, the analysis of LD-based CO hotspots in human revealed a few sequence motifs that were significantly associated with hotspots (Myers *et al.*, 2005). The authors have recently extended their analysis and found a degenerate13 bp motif (CCNCCNTNNCCNC), the presence of which explains at least 40% of human CO hotspots (Myers *et al.*, 2008). This finding suggests that the location of a significant fraction of human CO hotspots might be determined by a single DNA motif-binding protein. Additional factors must come into play, however, since the sole presence of this motif is not sufficient to determine the activity of a CO hotspot (Myers *et al.*, 2008; Webb, Berg and Jeffreys, 2008).

Several lines of evidence suggest that the hotspots are highly dynamic and have a relatively short lifespan in terms of evolutionary time (Buard and de Massy, 2007; Coop and Przeworski, 2007; Jeffreys *et al.*, 2005; Coop and Myers, 2007). First, the CO rate at many hotspots varies between individuals (in human) or lines (in mouse) (e.g. Webb, Berg and Jeffreys, 2008; Yauk, Bois and Jeffreys, 2003; Shiroishi *et al.*, 1991; Neumann and Jeffreys, 2006). The activity of hotspots appears to be regulated by several layers of controls, because the variation has been found to depend on various factors, including local polymorphisms acting in *cis*, and elements located elsewhere in the genome, which could be *trans*-acting factors (Baudat and de Massy, 2007b; Grey, Baudat and de Massy, 2009; Jeffreys and Neumann, 2005; Jeffreys and Neumann, 2002; Heine *et al.*, 1994; Parvanov *et al.*, 2009). Second, the correlation between LD-based hotspots and sperm-typing analyses is good but not perfect, suggesting that there are 'old' hotspots that are no longer active in the present time, as well as 'young' hotspots that are active but have not yet left their imprint on haplotype diversity (Jeffreys *et al.*, 2005; Kauppi, Stumpf and Jeffreys, 2005). Third, there has been little conservation of hotspots during evolution, even between closely related species as in human and chimpanzee, suggesting that fine-scale recombination rates evolve faster than the DNA sequence (Ptak *et al.*, 2005; Winckler *et al.*, 2005).

6.3.3 Sex-dependent control

Noticeably, both the number and the distribution of recombination events display striking sex-specific discrepancies in many organisms. In eutherian mammals, the overall CO rate is either similar in both sexes or higher in females. For example, the female genetic map is 1.6-fold longer than the male one in human, and 1.3-fold in mouse (Kong *et al.*, 2002; Matise *et al.*, 2007; Shifman *et al.*, 2006). In addition, the CO distribution differs between sexes.

Sex-specific control of hotspot activity

Most data on hotspot position and intensity are either male-specific (sperm-typing experiment) or sex-averaged (analysis of the genetic variation in populations), and little is known directly from the analysis of hotspots in females. Nevertheless, the available data suggest that the same hotspots are used in both sexes overall. First, amongst the few autosomal hotspots analyzed by sperm typing for which information on recombination is also available in female, some are more active in one sex or another, but to date none has proven to be fully sex-specific (Ng *et al.*, 2008; Baudat and de Massy, 2007b; Paigen *et al.*, 2008; Holloway, Lawson and Jeffreys, 2006). Second, the good correlation between the location and activity of hotspots determined by sperm typing and by population genetics is consistent with most hotspots being active in both sexes (Jeffreys, Kauppi and Neumann, 2001; McVean *et al.*, 2004; Jeffreys *et al.*, 2005). However, it is quite possible that a minority of hotspots are sex-specific.

The intersex differences in the distribution of recombination, observed at various scales (see below), can readily be explained by a sex-specific difference in the use of the same hotspots, as has been shown at several mouse hotspots (Figure 6.2a and b) (Ng *et al.*, 2008; Baudat and de Massy, 2007b; Paigen *et al.*, 2008). Interestingly, at the mouse *Psmb9* hotspot, CO and NCO rate variations were correlated, indicating that a regulatory process occurs at the initiation step.

Large-scale control of CO distribution

A consequence of the predominant localization of recombination events in hotspots is that the distribution of recombination events at any scale results mainly from the distribution and the intensity of hotspots along chromosomes. Nevertheless, the large-scale structure of the chromosomes is determinant in shaping the chromosome-wide pattern of CO distribution.

At the chromosome scale, CO formation is strongly repressed across centromeres, and their distribution shows a succession of Mb-long intervals with higher and lower density along the arms (Figure 6.2c) (Kong *et al.*, 2002; Paigen *et al.*, 2008; Shifman *et al.*, 2006; Laurent *et al.*, 2003; Choo, 1998). The global distribution of COs on chromosome arms differs between males and females in eutherian mammals. In males, the CO rate is usually low in intervals relatively close to the centromere, and particularly high in subtelomeric intervals. Conversely, COs are more uniformly distributed in females, beyond the repression in a short interval across the centromere. Both the frequency and the pattern of COs along chromosomes are identical in XY sex-reversed female mice and normal female mice, indicating that the oocyte-specific pattern results from the phenotypic sex, not from the genotype (Lynn *et al.*, 2005). Interestingly, the genetic map of the grey, short-tailed opossum (*Monodelphis domestica*) is longer in male than in female, with a seemingly more distal distribution of COs in female, suggesting that the converse situation might be true in marsupials (Samollow *et al.*, 2007).

The structure of chromosomes influences the distribution of COs, independently from the DNA sequence. Indeed, the arms of metacentric chromosomes experience less COs, with a more distal distribution, than similarly sized long arms of acrocentric chromosomes (e.g. Laurie and Hulten, 1985b; Sun *et al.*, 2006). Moreover, the fusion of two acrocentric chromosomes by their centromeres (Robertsonian translocation) results into a lower number of chiasmata per arm, with a more distal distribution, showing that this effect is sequence independent (Bidau *et al.*, 2001; Castiglia and Capanna, 2002; Dumas and Britton-Davidian, 2002).

de Boer *et al.* (2006) examined the distribution of MSH4 and MLH1 foci on mouse SCs in oocytes and spermatocytes. MSH4 foci are thought to mark sites of intermediate stages of recombination, leading to both COs and NCOs. The distribution of MSH4 foci on SCs was similar in both sexes: uniform, with a short interval at the centromeric end devoid of foci. In contrast, the density of MLH1 foci (which mark specifically the position of COs) was particularly high towards the telomere in spermatocytes, while in oocytes it was more-or-less uniform along the SC. This observation suggests that the overall distribution of recombination initiation events (leading to both COs and NCOs) is similar in oocytes and spermatocytes, and therefore that the sex-specific variation in the pattern of CO distribution might be controlled mainly at the step of the CO/NCO decision.

Changes in crossover frequency and distribution enhance the risk of producing aneuploid gametes

Oogenesis in human is characterized by the unusually high incidence of aneuploid gametes ($\geq$10% of the oocytes), while no more than 1–2% of sperm are aneuploid (Hassold and Hunt, 2001; Pellestor, Anahory and Hamamah, 2005). This results in a high frequency of monosomic or trisomic zygotes and embryos, and is therefore the leading cause of pregnancy loss, since most of them do not survive to term. In addition, the few aneuploidies compatible with live birth (trisomies 18, 21, XXY and XXX) represent the leading cause of congenital birth defects and mental retardation (reviewed in Hassold and Hunt, 2001; Hassold, Hall and Hunt, 2007). Although there is variation between chromosomes, the origin of trisomy is most often maternal (except for XXY trisomy), and amongst them most result from a meiosis I error – that is, both homologous chromosomes from one parent have been transmitted to the gamete. The frequency of aneuploidies of maternal origin increases dramatically with mother's age: for example, the birth-prevalence rate for trisomy 21 rises from 1/1400 births for women 20–24 years of age to 1/25 births for women $\geq$45 years (Sherman *et al.*, 2005; Yoon *et al.*, 1996). Several susceptible recombination patterns associated with maternal trisomy 21 have been identified (Lamb *et al.*, 1996; Oliver *et al.*, 2008 and references therein); trisomies 21 originating from meiosis I nondisjunctions are often associated with either the absence of any CO (no chiasma) or the presence of a single CO close to the telomere, which might not efficiently promote the formation of a fully functional chiasma. On the other hand, trisomy 21 originating from maternal meiosis II error – that is, having inherited two sister chromatids from the same maternal chromosome 21 – is often associated with the presence of $\geq$2 COs. The above configurations are at risk at all maternal ages. In addition, the presence of a single pericentromeric CO is associated specifically with the age-dependent increase in frequency of meiosis II trisomy 21. Altered recombination patterns (especially the lack of CO) appear to be frequently associated with other trisomies as well (Hassold, Hall and Hunt, 2007). Interestingly, some of these patterns of recombination have been found to be prone to nondisjunctions also in *D. melanogaster* and *S. cerevisiae* (Koehler *et al.*, 1996; Rockmill, Voelkel-Meiman and Roeder, 2006).

The origin of the particularly high rate of aneuploidy during human oogenesis, which is not observed in mouse, is still unclear (Hassold, Hall and Hunt, 2007; Hunt and Hassold, 2002; Morelli and Cohen, 2005). Lenzi *et al.* (2005) have found that the numbers of MLH1 and MLH3 foci are extremely heterogeneous among human oocytes (10–107 foci per oocyte, to be compared to the 23 pairs of homologues that must segregate at meiosis I). This level of heterogeneity has not been observed among human spermatocytes or mouse oocytes (Lenzi *et al.*, 2005; Tease, Hartshorne and Hulten, 2002; Topping *et al.*, 2006). With the assumption that the MLH1 foci represent the vast majority of COs in all oocytes, this observation suggests that oocytes with one or more achiasmatic chromosomes, prone to aneuploidy, are particularly frequent during human oogenesis. In addition, several observations suggest that the oocytes giving rise to offspring are selected for having experienced more COs than average: first, the female pedigree-based genetic map is $\sim$1.4-fold longer than expected from the count of MLH1 foci, whereas both counts are similar in male (Coop and Przeworski, 2007). Second, Kong *et al.* (2004) have detected the age-dependent lengthening of the pedigree-based female genetic map, which they suggest is due to increasing selection for a highest global CO rate. Thus, a significant fraction of human oocytes might be unable to achieve reproduction successfully, because of a defect in recombination activity.

As mentioned above, a rate of aneuploidy as high as in human oogenesis is never observed during normal mouse oogenesis. However, the mechanisms that lead to aneuploidy during human oogenesis might also apply to mouse oogenesis, at lower frequency. Indeed, age-dependent increase of the rate of aneuploidy has been observed cytologically in mouse oocytes, associated with a decrease of the mean number of chiasmata, and their more distal positioning (Henderson and Edwards, 1968; Jagiello and Fang, 1979; Speed, 1977; Cukurcam *et al.*, 2007). In addition, experimental conditions that alter the pattern of meiotic recombination and dramatically increase the rate of aneuploidy in mouse oocytes have been obtained. These include interspecies breeding and the environmental exposure of pregnant mothers to a chemical (bisphenol A) (Susiarjo *et al.*, 2007; Hunt *et al.*, 2003; Koehler *et al.*, 2006). Finally, oocytes from mice lacking the meiosis-specific SMC1β cohesin (involved in maintaining the cohesion between sister chromatids) display a nearly normal distribution of MLH1 foci at pachynema, but a profoundly altered distribution of chiasmata, which are shifted towards the centromere-distal end of the chromosomes. This is associated with an elevated rate of univalents and isolated chromatids in metaphase I oocytes, predictive of aneuploidies (this rate becomes even higher with ageing). This study demonstrates the importance of sister-chromatid cohesion for maintaining the physical link provided by the chiasmata and ensuring accurate segregation of chromosomes at both meiotic divisions. In addition, it suggests that defective sister-chromatid cohesion might be one of the underlying causes of human age-related aneuploidy (Hodges *et al.*, 2005).

6.4 Meiotic prophase progression and checkpoint

6.4.1 The elimination of spermatocytes displaying a synapsis defect

The interpretation of the phenotype of several mutants defective for meiotic recombination has been complicated by the fact that the phenotype is often more severe for spermatocytes than for oocytes, even if the primary defect is thought to be the same in both sexes. Specifically, it has been noted that conditions resulting in a defect in the formation of the synaptonemal complex and/or the formation of the sex body lead to a stronger phenotype for the male germ line than for the female germ line, both in mice and humans (Kolas *et al.*, 2005a; Hunt and Hassold, 2002; Morelli and Cohen, 2005). Such conditions include interspecies breeding, heterozygous chromosome rearrangements, and mutations in genes encoding either components of the synaptonemal complex or proteins participating in meiotic recombination.

A noticeable male-specific feature of mammalian meiosis is the sex body, which encompasses the transcriptionally silenced chromatin of the unsynapsed regions of the sex chromosomes during the pachytene stage (MSCI, for meiotic sex chromosome inactivation) (Turner, 2007). A series of recent studies in mouse, discussed extensively in Burgoyne, Mahadevaiah and Turner (2009), have demonstrated that the MSCI is a particular example of a more general phenomenon occurring in both male and female pachytene meiocytes, called meiotic silencing of unsynapsed chromosomes (MSUC) (Baarends *et al.*, 2005; Turner *et al.*, 2004; Turner *et al.*, 2005; Turner *et al.*, 2006; Mahadevaiah *et al.*, 2008; Homolka *et al.*, 2007). The MSUC consists of the transcriptional silencing of any region of the genome that is not synapsed during the pachytene stage, but can apply only to a limited portion of the genome. Thus, there is a failure to silence the sex chromosomes in conditions where a much larger portion of the chromosomal content remains unsynapsed at pachynema. This failure to silence the X and Y chromosomes is sufficient to induce spermatocyte apoptosis at the mid-pachytene stage, explaining why this early elimination is specific to spermatocytes (Mahadevaiah *et al.*, 2008). The mechanism of MSUC appears to apply similarly to human meiocytes (Ferguson, Chow and Ma, 2008; Garcia-Cruz *et al.*, 2009).

Many studies of meiotic mutants have focused on male meiosis for practical reasons. However, the early elimination of spermatocytes results often from a secondary defect affecting chromosome synapsis, thereby contributing to masking of some of the direct effects of these mutations. This pinpoints the importance of examining in detail mutant phenotypes in both male and female germlines.

6.4.2 Different meiotic recombination mutants display different arrest points

The overall conservation of the meiotic recombination mechanism (and proteins) has been of great help for identifying mammalian genes involved in the process, and analyzing the intricate phenotype of the corresponding mutants (for instance *Spo11*, *Dmc1*, *Mlh1*). The phenotype of these mutants has provided a framework for the analysis of mutants suffering from a less predictable defect. To simplify, the meiotic recombination mutants described so far can be divided into three main classes, based on the step in the recombination process that is impaired. They each lead to a distinct phenotype, even if subtle differences still exist between mutants of the same class.

The first class includes genes required for recombination initiation (*Spo11*, *Mei1*). These mutants do not form recombination intermediates and fail to make homologous synapsis. The spermatocytes are eliminated by apoptosis at a stage corresponding to middle pachynema, while the oocytes progress toward diplonema. Many are lost shortly after birth at the time of follicle formation, but a significant proportion progress to the dictyate arrest, and a few reach metaphase I, which fails due to the lack of chiasmata (Baudat *et al.*, 2000; Romanienko and Camerini-Otero, 2000; Libby *et al.*, 2003; Reinholdt and Schimenti, 2005; Barchi *et al.*, 2005; Di Giacomo *et al.*, 2005).

The second class contains several genes encoding proteins required for the repair of the *Spo11*-dependent DNA breaks (*Dmc1*, *Hop2*, *Msh4*, *Msh5*, *Atm*). Class 2 mutant meiocytes accumulate unrepaired recombination intermediates and are defective for the formation of homologous synapsis. Like class 1 mutants, these spermatocytes are eliminated at the mid-pachytene stage. There is a slight worsening of their phenotype, however, dependent upon *Spo11*, suggesting that the presence of unrepaired *Spo11*-dependent DNA damages *per se* might trigger a response in pachytene spermatocytes (Barchi *et al.*, 2005; Mahadevaiah *et al.*, 2008). Differently from class 1 mutants, oocytes are lost at an early stage, before or at the time of follicle formation, as a result of a DNA damage-dependent response (Pittman *et al.*, 1998b; Yoshida *et al.*, 1998; Petukhova, Romanienko and Camerini-Otero, 2003; de Vries *et al.*, 1999; Edelmann *et al.*, 1999; Kneitz *et al.*, 2000; Barlow *et al.*, 1998).

The third category contains genes not required for the repair of recombination intermediates, but necessary for their processing through the major mammalian CO-generating pathway (*Mlh1*, *Mlh3*, *Exo1*). These mutants produce recombination intermediates that are repaired, form normal SC, but fail to form chiasmata, because of a profound defect in generating CO. Both spermatocytes and oocytes progress in normal numbers to metaphase I, which fails because of the absence of chiasmata, though a very small proportion of oocytes is able to progress further and to perform both meiotic divisions (Baker *et al.*, 1996; Edelmann *et al.*, 1996; Woods *et al.*, 1999; Lipkin *et al.*, 2002; Wei *et al.*, 2003).

6.4.3 A link between recombination and the control of the progression of the meiotic cell cycle

Two CDK proteins have been shown to form foci on mouse meiotic chromosome axes, which colocalize with recombination intermediates. CDK4 foci colocalize with RPA on late zygonema–early pachynema synapsed regions of chromosomes. CDK2 forms three classes of foci during meiotic prophase, including one which colocalizes with MLH1 foci on mid–late pachynema SC (Ashley, Walpita and de Rooij, 2001). Interestingly, a mutation in *Hei10*, which encodes a putative B-type cyclin E3 ubiquitin ligase, displays a phenotype quite similar to that of MLH1 and MLH3 mutants, and an absence of MLH1, MLH3 and interstitial CDK2 foci (Ward *et al.*, 2007). The authors of this study proposed that HEI10 might regulate the interaction of CDK2 with the recombination machinery by promoting the degradation of an associated B-type cyclin, which might be the cyclin B3. If this hypothesis is true, CDK2 has a role in promoting the maturation of recombination intermediates into CO, and therefore the formation of chiasmata. *Cdk2* should have one or more additional function(s) necessary for the meiotic progression, because its inactivation leads to a synapsis defect and an early meiotic elimination in both sexes, similar to the phenotype of the class 2 mutants described above (Ortega *et al.*, 2003).

Acknowledgements

We thank Christine Mézard, Bernard de Massy and Rajeev Kumar for critical reading of the manuscript. This work was supported by funding from the Centre National de la Recherche Scientifique.

References

Adelman, C.A. and Petrini, J.H. (2008) ZIP4H (TEX11) deficiency in the mouse impairs meiotic double strand break repair and the regulation of crossing over. *PLoS Genet.*, **4**, e1000042.

Agarwal, S. and Roeder, G.S. (2000) Zip3 provides a link between recombination enzymes and synaptonemal complex proteins. *Cell*, **102**, 245–255.

Alani, E., Padmore, R. and Kleckner, N. (1990) Analysis of wild-type and *rad50* mutants of yeast suggests an intimate relationship between meiotic chromosome synapsis and recombination. *Cell*, **61**, 419–436.

Allers, T. and Lichten, M. (2001) Differential timing and control of noncrossover and crossover recombination during meiosis. *Cell*, **106**, 47–57.

Anderson, L.K., Reeves, A., Webb, L.M. *et al.* (1999) Distribution of crossing over on mouse synaptonemal complexes using immunofluorescent localization of MLH1 protein. *Genetics*, **151**, 1569–1579.

Argueso, J.L., Wanat, J., Gemici, Z. *et al.* (2004) Competing crossover pathways act during meiosis in Saccharomyces cerevisiae. *Genetics*, **168**, 1805–1816.

Arnheim, N., Calabrese, P. and Tiemann-Boege, I. (2007) Mammalian meiotic recombination hot spots. *Annu. Rev. Genet.*, **41**, 369–399.

Arora, C., Kee, K., Maleki, S. *et al.* (2004) Antiviral protein Ski8 is a direct partner of Spo11 in meiotic DNA break formation, independent of its cytoplasmic role in RNA metabolism. *Mol. Cell*, **13**, 549–559.

Ashley, T., Walpita, D. and de Rooij, D.G. (2001) Localization of two mammalian cyclin dependent kinases during mammalian meiosis. *J. Cell Sci.*, **114**, 685–693.

Baarends, W.M., Wassenaar, E., van der Laan, R. *et al.* (2005) Silencing of unpaired chromatin and histone H2A ubiquitination in mammalian meiosis. *Mol. Cell Biol.*, **25**, 1041–1053.

Bachrati, C.Z., Borts, R.H. and Hickson, I.D. (2006) Mobile D-loops are a preferred substrate for the Bloom's syndrome helicase. *Nucleic Acids Res.*, **34**, 2269–2279.

Baker, S.M., Plug, A.W., Prolla, T.A. *et al.* (1996) Involvement of mouse Mlh1 in DNA mismatch repair and meiotic crossing over. *Nat. Genet.*, **13**, 336–342.

Bannister, L.A., Pezza, R.J., Donaldson, J.R. *et al.* (2007) A dominant, recombination-defective allele of Dmc1 causing male-specific sterility. *PLoS Biol.*, **5**, e105.

Barchi, M., Mahadevaiah, S., Di Giacomo, M. *et al.* (2005) Surveillance of different recombination defects in mouse spermatocytes yields distinct responses despite elimination at an identical developmental stage. *Mol. Cell Biol.*, **25**, 7203–7215.

Barlow, A.L., Benson, F.E., West, S.C. *et al.* (1997) Distribution of the Rad51 recombinase in human and mouse spermatocytes. *EMBO J.*, **16**, 5207–5215.

Barlow, C., Liyanage, M., Moens, P.B. *et al.* (1998) Atm deficiency results in severe meiotic disruption as early as leptonema of prophase I. *Development*, **125**, 4007–4017.

Baudat, F. and de Massy, B. (2007a) Regulating double-stranded DNA break repair towards crossover or non-crossover during mammalian meiosis. *Chromosome Res.*, **15**, 565–577.

Baudat, F. and de Massy, B. (2007b) Cis- and trans-acting elements regulate the mouse Psmb9 meiotic recombination hotspot. *PLoS Genet.*, **3**, e100.

Baudat, F., Manova, K., Yuen, J.P. *et al.* (2000) Chromosome synapsis defects and sexually dimorphic meiotic progression in mice lacking spo11. *Mol. Cell*, **6**, 989–998.

Baumann, P., Benson, F.E. and West, S.C. (1996) Human Rad51 protein promotes ATP-dependent homologous pairing and strand transfer reactions in vitro. *Cell*, **87**, 757–766.

Bennett, R.J., Keck, J.L. and Wang, J.C. (1999) Binding specificity determines polarity of DNA unwinding by the Sgs1 protein of S. cerevisiae. *J. Mol. Biol.*, **289**, 235–248.

Bergerat, A., de Massy, B., Gadelle, D. *et al.* (1997) An atypical topoisomerase II from Archaea with implications for meiotic recombination. *Nature*, **386**, 414–417.

Bidau, C.J., Gimenez, M.D., Palmer, C.L. *et al.* (2001) The effects of Robertsonian fusions on chiasma frequency and distribution in the house mouse (Mus musculus domesticus) from a hybrid zone in northern Scotland. *Heredity*, **87**, 305–313.

Bishop, D.K. (1994) RecA homologs Dmc1 and Rad51 interact to form multiple nuclear complexes prior to meiotic chromosome synapsis. *Cell*, **79**, 1081–1092.

Bishop, D.K., Park, D., Xu, L. *et al.* (1992) DMC1: a meiosis-specific yeast homolog of*E. coli* recA required for recombination, synaptonemal complex formation, and cell cycle progression. *Cell*, **69**, 439–456.

Bishop, D.K., Nikolski, Y., Oshiro, J. *et al.* (1999) High copy number suppression of the meiotic arrest caused by a dmc1 mutation: REC114 imposes an early recombination block and RAD54 promotes a DMC1-independent DSB repair pathway. *Genes Cells*, **4**, 425–444.

Blat, Y., Protacio, R.U., Hunter, N. *et al.* (2002) Physical and functional interactions among basic chromosome organizational features govern early steps of meiotic chiasma formation. *Cell*, **111**, 791–802.

de Boer, E., Stam, P., Dietrich, A.J. *et al.* (2006) Two levels of interference in mouse meiotic recombination. *Proc. Natl. Acad. Sci. USA*, **103**, 9607–9612.

Bolcun-Filas, E., Costa, Y., Speed, R. *et al.* (2007) SYCE2 is required for synaptonemal complex assembly, double strand break repair, and homologous recombination. *J. Cell Biol.*, **176**, 741–747.

Bolcun-Filas, E., Speed, R., Taggart, M. *et al.* (2009) Mutation of the mouse Syce1 gene disrupts synapsis and suggests a link between synaptonemal complex structural components and DNA repair. *PLoS Genet.*, **5**, e1000393.

Borde, V. (2007) The multiple roles of the Mre11 complex for meiotic recombination. *Chromosome Res.*, **15**, 551–563.

Borde, V., Robine, N., Lin, W. *et al.* (2009) Histone H3 lysine 4 trimethylation marks meiotic recombination initiation sites. *EMBO J.*, **28**, 99–111.

Borner, G.V., Kleckner, N. and Hunter, N. (2004) Crossover/noncrossover differentiation, synaptonemal complex formation, and regulatory surveillance at the leptotene/zygotene transition of meiosis. *Cell*, **117**, 29–45.

Broman, K.W., Rowe, L.B., Churchill, G.A. *et al.* (2002) Crossover interference in the mouse. *Genetics*, **160**, 1123–1131.

Buard, J. and de Massy, B. (2007) Playing hide and seek with mammalian meiotic crossover hotspots. *Trends Genet.*, **23**, 301–309.

Burgoyne, P.S., Mahadevaiah, S.K. and Turner, J.M. (2009) The consequences of asynapsis for mammalian meiosis. *Nat. Rev. Genet.*, **10**, 207–216.

Cao, L., Alani, E. and Kleckner, N. (1990) A pathway for generation and processing of double-strand breaks during meiotic recombination in S. cerevisiae. *Cell*, **61**, 1089–1101.

Castiglia, R. and Capanna, E. (2002) Chiasma repatterning across a chromosomal hybrid zone between chromosomal races of Mus musculus domesticus. *Genetica*, **114**, 35–40.

Cervantes, M.D., Farah, J.A. and Smith, G.R. (2000) Meiotic DNA breaks associated with recombination in S. pombe. *Mol. Cell*, **5**, 883–888.

Chelysheva, L., Gendrot, G., Vezon, D. *et al.* (2007) Zip4/Spo22 is required for class I CO formation but not for synapsis completion in Arabidopsis thaliana. *PLoS Genet.*, **3**, e83.

Chen, J., Silver, D.P., Walpita, D. *et al.* (1998) Stable interaction between the products of the BRCA1 and BRCA2 tumor suppressor genes in mitotic and meiotic cells. *Mol. Cell*, **2**, 317–328.

Cheng, C.H., Lo, Y.H., Liang, S.S. *et al.* (2006) SUMO modifications control assembly of synaptonemal complex and polycomplex in meiosis of Saccharomyces cerevisiae. *Genes. Dev.*, **20**, 2067–2081.

Chi, P., San Filippo, J., Sehorn, M.G. *et al.* (2007) Bipartite stimulatory action of the Hop2-Mnd1 complex on the Rad51 recombinase. *Genes. Dev.*, **21**, 1747–1757.

Choo, K.H. (1998) Why is the centromere so cold? *Genome. Res.*, **8**, 81–82.

Codina-Pascual, M., Campillo, M., Kraus, J. *et al.* (2006) Crossover frequency and synaptonemal complex length: their variability and effects on human male meiosis. *Mol. Hum. Reprod.*, **12**, 123–133.

Coop, G. and Myers, S.R. (2007) Live hot, die young: transmission distortion in recombination hotspots. *PLoS Genet.*, **3**, e35.

Coop, G. and Przeworski, M. (2007) An evolutionary view of human recombination. *Nat. Rev. Genet.*, **8**, 23–34.

Cromie, G.A. and Smith, G.R. (2007) Branching out: meiotic recombination and its regulation. *Trends Cell Biol.*, **17**, 448–455.

Cromie, G.A., Hyppa, R.W., Taylor, A.F. *et al.* (2006) Single Holliday junctions are intermediates of meiotic recombination. *Cell*, **127**, 1167–1178.

Cukurcam, S., Betzendahl, I., Michel, G. *et al.* (2007) Influence of follicular fluid meiosis-activating sterol on aneuploidy rate and precocious chromatid segregation in aged mouse oocytes. *Hum. Reprod.*, **22**, 815–828.

De Muyt, A., Vezon, D., Gendrot, G. *et al.* (2007) AtPRD1 is required for meiotic double strand break formation in Arabidopsis thaliana. *EMBO J.*, **26**, 4126–4137.

Dendouga, N., Gao, H., Moechars, D. *et al.* (2005) Disruption of murine Mus81 increases genomic instability and DNA damage sensitivity but does not promote tumorigenesis. *Mol. Cell Biol.*, **25**, 7569–7579.

Di Giacomo, M., Barchi, M., Baudat, F. *et al.* (2005) Distinct DNA-damage-dependent and -independent responses drive the loss of oocytes in recombination-defective mouse mutants. *Proc. Natl. Acad. Sci. USA*, **102**, 737–742.

Diaz, R.L., Alcid, A.D., Berger, J.M. *et al.* (2002) Identification of residues in yeast Spo11p critical for meiotic DNA double-strand break formation. *Mol. Cell Biol.*, **22**, 1106–1115.

Dumas, D. and Britton-Davidian, J. (2002) Chromosomal rearrangements and evolution of recombination: comparison of chiasma distribution patterns in standard and robertsonian populations of the house mouse. *Genetics*, **162**, 1355–1366.

Edelmann, W., Cohen, P.E., Kane, M. *et al.* (1996) Meiotic pachytene arrest in MLH1-deficient mice. *Cell*, **85**, 1125–1134.

Edelmann, W., Cohen, P.E., Kneitz, B. *et al.* (1999) Mammalian MutS homologue 5 is required for chromosome pairing in meiosis. *Nat. Genet.*, **21**, 123–127.

Falque, M., Mercier, R., Mezard, C. *et al.* (2007) Patterns of recombination and MLH1 foci density along mouse chromosomes: modeling effects of interference and obligate chiasma. *Genetics*, **176**, 1453–1467.

Fearnhead, P., Harding, R.M., Schneider, J.A. *et al.* (2004) Application of coalescent methods to reveal fine-scale rate variation and recombination hotspots. *Genetics*, **167**, 2067–2081.

Ferguson, K.A., Chow, V. and Ma, S. (2008) Silencing of unpaired meiotic chromosomes and altered recombination patterns in an azoospermic carrier of a t(8;13) reciprocal translocation. *Hum. Reprod.*, **23**, 988–995.

Frisse, L., Hudson, R.R., Bartoszewicz, A. *et al.* (2001) Gene conversion and different population histories may explain the contrast between polymorphism and linkage disequilibrium levels. *Am. J. Hum. Genet.*, **69**, 831–843.

Froenicke, L., Anderson, L.K., Wienberg, J. *et al.* (2002) Malc mousc rccombination maps for cach autosome identified by chromosome painting. *Am. J. Hum. Genet.*, **71**, 1353–1368.

Fung, J.C., Rockmill, B., Odell, M. *et al.* (2004) Imposition of crossover interference through the nonrandom distribution of synapsis initiation complexes. *Cell*, **116**, 795–802.

Garcia-Cruz, R., Roig, I., Robles, P. *et al.* (2009) ATR, BRCA1 and gammaH2AX localize to unsynapsed chromosomes at the pachytene stage in human oocytes. *Reprod. Biomed. Online*, **18**, 37–44.

Gaskell, L.J., Osman, F., Gilbert, R.J. *et al.* (2007) Mus81 cleavage of Holliday junctions: a failsafe for processing meiotic recombination intermediates? *EMBO J.*, **26**, 1891–1901.

Gay, J., Myers, S. and McVean, G. (2007) Estimating meiotic gene conversion rates from population genetic data. *Genetics*, **177**, 881–894.

Grelon, M., Vezon, D., Gendrot, G. *et al.* (2001) AtSPO11-1 is necessary for efficient meiotic recombination in plants. *EMBO J.*, **20**, 589–600.

Grey, C., Baudat, F. and de Massy, B. (2009) Genome-wide control of the distribution of meiotic recombination. *PLoS Biol.*, **7**, e35.

Guillon, H. and de Massy, B. (2002) An initiation site for meiotic crossing-over and gene conversion in the mouse. *Nat. Genet.*, **32**, 296–299.

Guillon, H., Baudat, F., Grey, C. *et al.* (2005) Crossover and noncrossover pathways in mouse meiosis. *Mol. Cell*, **20**, 563–573.

Hamer, G., Wang, H., Bolcun-Filas, E. *et al.* (2008) Progression of meiotic recombination requires structural maturation of the central element of the synaptonemal complex. *J. Cell Sci.*, **121**, 2445–2451.

Hassold, T. and Hunt, P. (2001) To err (meiotically) is human: the genesis of human aneuploidy. *Nat. Rev. Genet.*, **2**, 280–291.

Hassold, T., Judis, L., Chan, E.R. *et al.* (2004) Cytological studies of meiotic recombination in human males. *Cytogenet. Genome. Res.*, **107**, 249–255.

Hassold, T., Hall, H. and Hunt, P. (2007) The origin of human aneuploidy: where we have been, where we are going. *Hum. Mol. Genet.*, **16**(2), R203–R208.

Heine, D., Khambata, S., Wydner, K.S. *et al.* (1994) Analysis of recombinational hot spots associated with the p haplotype of the mouse MHC. *Genomics*, **23**, 168–177.

Henderson, S.A. and Edwards, R.G. (1968) Chiasma frequency and maternal age in mammals. *Nature*, **218**, 22–28.

Henderson, K.A. and Keeney, S. (2004) Tying synaptonemal complex initiation to the formation and programmed repair of DNA double-strand breaks. *Proc. Natl. Acad. Sci. USA*, **101**, 4519–4524.

Henderson, K.A. and Keeney, S. (2005) Synaptonemal complex formation: where does it start? *Bioessays*, **27**, 995–998.

Hikiba, J., Hirota, K., Kagawa, W. *et al.* (2008) Structural and functional analyses of the DMC1-M200V polymorphism found in the human population. *Nucleic Acids Res.*, **36**, 4181–4190.

Hirota, K., Mizuno, K., Shibata, T. *et al.* (2008) Distinct chromatin modulators regulate the formation of accessible and repressive chromatin at the fission yeast recombination hotspot ade6-M26. *Mol. Biol. Cell*, **19**, 1162–1173.

Hodges, C.A., Revenkova, E., Jessberger, R. *et al.* (2005) SMC1beta-deficient female mice provide evidence that cohesins are a missing link in age-related nondisjunction. *Nat. Genet.*, **37**, 1351–1355.

Hollingsworth, N.M. and Brill, S.J. (2004) The Mus81 solution to resolution: generating meiotic crossovers without Holliday junctions. *Genes. Dev.*, **18**, 117–125.

Holloway, K., Lawson, V.E. and Jeffreys, A.J. (2006) Allelic recombination and de novo deletions in sperm in the human beta-globin gene region. *Hum. Mol. Genet.*, **15**, 1099–1111.

Holloway, J.K., Booth, J., Edelmann, W. *et al.* (2008) MUS81 generates a subset of MLH1-MLH3-independent crossovers in mammalian meiosis. *PLoS Genet.*, **4**, e1000186.

Holzen, T.M., Shah, P.P., Olivares, H.A. *et al.* (2006) Tid1/Rdh54 promotes dissociation of Dmc1 from nonrecombinogenic sites on meiotic chromatin. *Genes. Dev.*, **20**, 2593–2604.

Homolka, D., Ivanek, R., Capkova, J. *et al.* (2007) Chromosomal rearrangement interferes with meiotic X chromosome inactivation. *Genome Res.*, **17**, 1431–1437.

Hong, E.L., Shinohara, A. and Bishop, D.K. (2001) Saccharomyces cerevisiae Dmc1 protein promotes renaturation of single-strand DNA (ssDNA) and assimilation of ssDNA into homologous super-coiled duplex DNA. *J. Biol. Chem.*, **276**, 41906–41912.

Hunt, P.A. and Hassold, T.J. (2002) Sex matters in meiosis. *Science*, **296**, 2181–2183.

Hunt, P.A., Koehler, K.E., Susiarjo, M. *et al.* (2003) Bisphenol A exposure causes meiotic aneuploidy in the female mouse. *Curr. Biol.*, **13**, 546–553.

Hunter, N. and Borts, R.H. (1997) Mlh1 is unique among mismatch repair proteins in its ability to promote crossing-over during meiosis. *Genes Dev.*, **11**, 1573–1582.

Hunter, N. and Kleckner, N. (2001) The single-end invasion: an asymmetric intermediate at the double-strand break to double-Holliday junction transition of meiotic recombination. *Cell*, **106**, 59–70.

International HapMap Consortium (2007) A second generation human haplotype map of over 3.1 million SNPs. *Nature*, **449**, 851–861.

Ip, S.C., Rass, U., Blanco, M.G. *et al.* (2008) Identification of Holliday junction resolvases from humans and yeast. *Nature*, **456**, 357–361.

Jagiello, G. and Fang, J.S. (1979) Analyses of diplotene chiasma frequencies in mouse oocytes and spermatocytes in relation to ageing and sexual dimorphism. *Cytogenet. Cell Genet.*, **23**, 53–60.

Jang, J.K., Sherizen, D.E., Bhagat, R. *et al.* (2003) Relationship of DNA double-strand breaks to synapsis in Drosophila. *J. Cell Sci.*, **116**, 3069–3077.

Jantsch, V., Pasierbek, P., Mueller, M.M. *et al.* (2004) Targeted gene knockout reveals a role in meiotic recombination for ZHP-3, a Zip3-related protein in Caenorhabditis elegans. *Mol. Cell Biol.*, **24**, 7998–8006.

Jeffreys, A.J. and May, C.A. (2004) Intense and highly localized gene conversion activity in human meiotic crossover hot spots. *Nat. Genet.*, **36**, 151–156.

Jeffreys, A.J. and Neumann, R. (2002) Reciprocal crossover asymmetry and meiotic drive in a human recombination hot spot. *Nat. Genet.*, **31**, 267–271.

Jeffreys, A.J. and Neumann, R. (2005) Factors influencing recombination frequency and distribution in a human meiotic crossover hotspot. *Hum. Mol. Genet.*, **14**, 2277–2287.

Jeffreys, A.J., Kauppi, L. and Neumann, R. (2001) Intensely punctate meiotic recombination in the class II region of the major histocompatibility complex. *Nat. Genet.*, **29**, 217–222.

Jeffreys, A.J., Holloway, J.K., Kauppi, L. *et al.* (2004) Meiotic recombination hot spots and human DNA diversity. *Philos. Trans. R. Soc. Lond. B. Biol. Sci.*, **359**, 141–152.

Jeffreys, A.J., Neumann, R., Panayi, M. *et al.* (2005) Human recombination hot spots hidden in regions of strong marker association. *Nat. Genet.*, **37**, 601–606.

Jessop, L. and Lichten, M. (2008) Mus81/Mms4 endonuclease and Sgs1 helicase collaborate to ensure proper recombination intermediate metabolism during meiosis. *Mol. Cell*, **31**, 313–323.

Jessop, L., Rockmill, B., Roeder, G.S. *et al.* (2006) Meiotic chromosome synapsis-promoting proteins antagonize the anti-crossover activity of sgs1. *PLoS Genet.*, **2**, e155.

Jolivet, S., Vezon, D., Froger, N. *et al.* (2006) Non conservation of the meiotic function of the Ski8/Rec103 homolog in Arabidopsis. *Genes Cells*, **11**, 615–622.

Kaback, D.B. (1996) Chromosome-size dependent control of meiotic recombination in humans. *Nat. Genet.*, **13**, 20–21.

Kan, R., Sun, X., Kolas, N.K. *et al.* (2008) Comparative analysis of meiotic progression in female mice bearing mutations in genes of the DNA mismatch repair pathway. *Biol. Reprod.*, **78**, 462–471.

Kauppi, L., Jeffreys, A.J. and Keeney, S. (2004) Where the crossovers are: recombination distributions in mammals. *Nat. Rev. Genet.*, **5**, 413–424.

Kauppi, L., Stumpf, M.P. and Jeffreys, A.J. (2005) Localized breakdown in linkage disequilibrium does not always predict sperm crossover hot spots in the human MHC class II region. *Genomics*, **86**, 13–24.

Kauppi, L., Jasin, M. and Keeney, S. (2007) Meiotic crossover hotspots contained in haplotype block boundaries of the mouse genome. *Proc. Natl. Acad. Sci. USA*, **104**, 13396–13401.

Keeney, S. (2001) Mechanism and control of meiotic recombination initiation. *Curr. Top. Dev. Biol.*, **52**, 1–53.

Keeney, S. (2008) Spo11 and the formation of DNA double-strand breaks in meiosis, in *Recombination and Meiosis* (eds R. Egel and D.-H. Lankenau), Springer, Berlin, pp. 81–123.

Keeney, S., Giroux, C.N. and Kleckner, N. (1997) Meiosis-specific DNA double-strand breaks are catalyzed by Spo11, a member of a widely conserved protein family. *Cell*, **88**, 375–384.

Kelmenson, P.M., Petkov, P., Wang, X. *et al.* (2005) A torrid zone on mouse chromosome 1 containing a cluster of recombinational hotspots. *Genetics*, **169**, 833–841.

Klapholz, S., Waddell, C.S. and Esposito, R.E. (1985) The role of the SPO11 gene in meiotic recombination in yeast. *Genetics*, **110**, 187–216.

Kleckner, N. (2006) Chiasma formation: chromatin/axis interplay and the role(s) of the synaptonemal complex. *Chromosoma*, **115**, 175–194.

Kleckner, N., Storlazzi, A. and Zickler, D. (2003) Coordinate variation in meiotic pachytene SC length and total crossover/chiasma frequency under conditions of constant DNA length. *Trends Genet.*, **19**, 623–628.

Kneitz, B., Cohen, P.E., Avdievich, E. *et al.* (2000) MutS homolog 4 localization to meiotic chromosomes is required for chromosome pairing during meiosis in male and female mice. *Genes. Dev.*, **14**, 1085–1097.

Koehler, K.E., Boulton, C.L., Collins, H.E. *et al.* (1996) Spontaneous X chromosome MI and MII nondisjunction events in Drosophila melanogaster oocytes have different recombinational histories. *Nat. Genet.*, **14**, 406–414.

Koehler, K.E., Schrump, S.E., Cherry, J.P. *et al.* (2006) Near-human aneuploidy levels in female mice with homeologous chromosomes. *Curr. Biol.*, **16**, R579–R580.

Kolas, N.K., Marcon, E., Crackower, M.A. *et al.* (2005a) Mutant meiotic chromosome core components in mice can cause apparent sexual dimorphic endpoints at prophase or X-Y defective male-specific sterility. *Chromosoma*, **114**, 92–102.

Kolas, N.K., Svetlanov, A., Lenzi, M.L. *et al.* (2005b) Localization of MMR proteins on meiotic chromosomes in mice indicates distinct functions during prophase I. *J. Cell Biol.*, **171**, 447–458.

Kong, A., Gudbjartsson, D.F., Sainz, J. *et al.* (2002) A high-resolution recombination map of the human genome. *Nat. Genet.*, **31**, 241–247.

Kong, A., Barnard, J., Gudbjartsson, D.F. *et al.* (2004) Recombination rate and reproductive success in humans. *Nat. Genet.*, **36**, 1203–1206.

Lamb, N.E., Freeman, S.B., Savage-Austin, A. *et al.* (1996) Susceptible chiasmate configurations of chromosome 21 predispose to non-disjunction in both maternal meiosis I and meiosis II. *Nature Genet.*, **14**, 400–405.

Laurent, A.M., Li, M., Sherman, S. *et al.* (2003) Recombination across the centromere of disjoined and non-disjoined chromosome 21. *Hum. Mol. Genet.*, **12**, 2229–2239.

Laurie, D.A. and Hulten, M.A. (1985a) Further studies on chiasma distribution and interference in the human male. *Ann. Hum. Genet.*, **49**, 203–214.

Laurie, D.A. and Hulten, M.A. (1985b) Further studies on bivalent chiasma frequency in human males with normal karyotypes. *Ann. Hum. Genet.*, **49**, 189–201.

Lawrie, N.M., Tease, C. and Hulten, M.A. (1995) Chiasma frequency, distribution and interference maps of mouse autosomes. *Chromosoma*, **104**, 308–314.

Lenzi, M.L., Smith, J., Snowden, T. *et al.* (2005) Extreme heterogeneity in the molecular events leading to the establishment of chiasmata during meiosis I in human oocytes. *Am. J. Hum. Genet.*, **76**, 112–127.

Li, N. and Stephens, M. (2003) Modeling linkage disequilibrium and identifying recombination hotspots using single-nucleotide polymorphism data. *Genetics*, **165**, 2213–2233.

Li, Z., Golub, E.I., Gupta, R. *et al.* (1997) Recombination activities of HsDmc1 protein, the meiotic human homolog of RecA protein. *Proc. Natl. Acad. Sci. USA*, **94**, 11221–11226.

Libby, B.J., De La Fuente, R., O'Brien, M.J. *et al.* (2002) The mouse meiotic mutation mei1 disrupts chromosome synapsis with sexually dimorphic consequences for meiotic progression. *Dev. Biol.*, **242**, 174–187.

Libby, B.J., Reinholdt, L.G. and Schimenti, J.C. (2003) Positional cloning and characterization of Mei1, a vertebrate-specific gene required for normal meiotic chromosome synapsis in mice. *Proc. Natl. Acad. Sci. USA*, **100**, 15706–15711.

Lim, D.S. and Hasty, P. (1996) A mutation in mouse rad51 results in an early embryonic lethal that is suppressed by a mutation in p53. *Mol. Cell Biol.*, **16**, 7133–7143.

Lipkin, S.M., Moens, P.B., Wang, V. *et al.* (2002) Meiotic arrest and aneuploidy in MLH3-deficient mice. *Nat. Genet.*, **31**, 385–390.

Ludin, K., Mata, J., Watt, S. *et al.* (2008) Sites of strong Rec12/Spo11 binding in the fission yeast genome are associated with meiotic recombination and with centromeres. *Chromosoma*, **117**, 431–444.

Luo, G., Santoro, I.M., McDaniel, L.D. *et al.* (2000) Cancer predisposition caused by elevated mitotic recombination in Bloom mice. *Nat. Genet.*, **26**, 424–429.

Lynn, A., Koehler, K.E., Judis, L. *et al.* (2002) Covariation of synaptonemal complex length and mammalian meiotic exchange rates. *Science*, **296**, 2222–2225.

Lynn, A., Ashley, T. and Hassold, T. (2004) Variation in human meiotic recombination. *Annu. Rev. Genomics. Hum. Genet.*, **5**, 317–349.

Lynn, A., Schrump, S., Cherry, J. *et al.* (2005) Sex, not genotype, determines recombination levels in mice. *Am. J. Hum. Genet.*, **77**, 670–675.

Lynn, A., Soucek, R. and Borner, G.V. (2007) ZMM proteins during meiosis: crossover artists at work. *Chromosome Res.*, **15**, 591–605.

Mahadevaiah, S.K., Turner, J.M., Baudat, F. *et al.* (2001) Recombinational DNA double-strand breaks in mice precede synapsis. *Nat. Genet.*, **27**, 271–276.

Mahadevaiah, S.K., Bourc'his, D., de Rooij, D.G. *et al.* (2008) Extensive meiotic asynapsis in mice antagonises meiotic silencing of unsynapsed chromatin and consequently disrupts meiotic sex chromosome inactivation. *J. Cell Biol.*, **182**, 263–276.

Mandon-Pepin, B., Touraine, P., Kuttenn, F. *et al.* (2008) Genetic investigation of four meiotic genes in women with premature ovarian failure. *Eur. J. Endocrinol.*, **158**, 107–115.

Mankouri, H.W. and Hickson, I.D. (2007) The RecQ helicase-topoisomerase III-Rmi1 complex: a DNA structure-specific 'dissolvasome'? *Trends Biochem. Sci.*, **32**, 538–546.

Marcon, E. and Moens, P. (2003) MLH1p and MLH3p localize to precociously induced chiasmata of okadaic-acid-treated mouse spermatocytes. *Genetics*, **165**, 2283–2287.

Masson, J.Y. and West, S.C. (2001) The Rad51 and Dmc1 recombinases: a non-identical twin relationship. *Trends Biochem. Sci.*, **26**, 131–136.

de Massy, B. (2003) Distribution of meiotic recombination sites. *Trends Genet.*, **19**, 514–522.

Matise, T.C., Chen, F., Chen, W. *et al.* (2007) A second-generation combined linkage physical map of the human genome. *Genome. Res.*, **17**, 1783–1786.

Mazina, O.M., Mazin, A.V., Nakagawa, T. *et al.* (2004) Saccharomyces cerevisiae Mer3 helicase stimulates 3′-5′ heteroduplex extension by Rad51; implications for crossover control in meiotic recombination. *Cell*, **117**, 47–56.

McEntee, K., Hesse, J.E. and Epstein, W. (1976) Identification and radiochemical purification of the recA protein of *Escherichia coli* K-12. *Proc. Natl. Acad. Sci. USA*, **73**, 3979–3983.

McMahill, M.S., Sham, C.W. and Bishop, D.K. (2007) Synthesis-dependent strand annealing in meiosis. *PLoS Biol.*, **5**, e299.

McPherson, J.P., Lemmers, B., Chahwan, R. *et al.* (2004) Involvement of mammalian Mus81 in genome integrity and tumor suppression. *Science*, **304**, 1822–1826.

McVean, G.A., Myers, S.R., Hunt, S. *et al.* (2004) The fine-scale structure of recombination rate variation in the human genome. *Science*, **304**, 581–584.

Moens, P.B., Kolas, N.K., Tarsounas, M. *et al.* (2002) The time course and chromosomal localization of recombination-related proteins at meiosis in the mouse are compatible with models that can resolve the early DNA-DNA interactions without reciprocal recombination. *J. Cell Sci.*, **115**, 1611–1622.

Morelli, M.A. and Cohen, P.E. (2005) Not all germ cells are created equal: aspects of sexual dimorphism in mammalian meiosis. *Reproduction*, **130**, 761–781.

Munroe, R.J., Bergstrom, R.A., Zheng, Q.Y. *et al.* (2000) Mouse mutants from chemically mutagenized embryonic stem cells. *Nat. Genet.*, **24**, 318–321.

Myers, S., Bottolo, L., Freeman, C. *et al.* (2005) A fine-scale map of recombination rates and hotspots across the human genome. *Science*, **310**, 321–324.

Myers, S., Spencer, C.C., Auton, A. *et al.* (2006) The distribution and causes of meiotic recombination in the human genome. *Biochem. Soc. Trans.*, **34**, 526–530.

Myers, S., Freeman, C., Auton, A. *et al.* (2008) A common sequence motif associated with recombination hot spots and genome instability in humans. *Nat. Genet.*, **40**, 1124–1129.

Nag, D.K., Pata, J.D., Sironi, M. *et al.* (2006) Both conserved and non-conserved regions of Spo11 are essential for meiotic recombination initiation in yeast. *Mol. Genet. Genomics*, **276**, 313–321.

Nakagawa, T. and Kolodner, R.D. (2002) The MER3 DNA helicase catalyzes the unwinding of Holliday junctions. *J. Biol. Chem.*, **277**, 28019–28024.

Neale, M.J., Pan, J. and Keeney, S. (2005) Endonucleolytic processing of covalent protein-linked DNA double-strand breaks. *Nature*, **436**, 1053–1057.

Neumann, R. and Jeffreys, A.J. (2006) Polymorphism in the activity of human crossover hotspots independent of local DNA sequence variation. *Hum. Mol. Genet.*, **15**, 1401–1411.

Neyton, S., Lespinasse, F., Moens, P.B. *et al.* (2004) Association between MSH4 (MutS homologue 4) and the DNA strand-exchange RAD51 and DMC1 proteins during mammalian meiosis. *Mol. Hum. Reprod.*, **10**, 917–924.

Ng, S.H., Parvanov, E., Petkov, P.M. *et al.* (2008) A quantitative assay for crossover and noncrossover molecular events at individual recombination hotspots in both male and female gametes. *Genomics*, **92**, 204–209.

Nishant, K.T., Plys, A.J. and Alani, E. (2008) A mutation in the putative MLH3 endonuclease domain confers a defect in both mismatch repair and meiosis in Saccharomyces cerevisiae. *Genetics*, **179**, 747–755.

Oh, S.D., Lao, J.P., Hwang, P.Y. *et al.* (2007) BLM ortholog, Sgs1, prevents aberrant crossing-over by suppressing formation of multichromatid joint molecules. *Cell*, **130**, 259–272.

Oh, S.D., Lao, J.P., Taylor, A.F. *et al.* (2008) RecQ helicase, Sgs1, and XPF family endonuclease, Mus81-Mms4, resolve aberrant joint molecules during meiotic recombination. *Mol. Cell*, **31**, 324–336.

Ohta, K., Shibata, T. and Nicolas, A. (1994) Changes in chromatin structure at recombination initiation sites during yeast meiosis. *EMBO J.*, **13**, 5754–5763.

Oliver, T.R., Feingold, E., Yu, K. *et al.* (2008) New insights into human nondisjunction of chromosome 21 in oocytes. *PLoS Genet.*, **4**, e1000033.

Oliver-Bonet, M., Turek, P.J., Sun, F. *et al.* (2005) Temporal progression of recombination in human males. *Mol. Hum. Reprod.*, **11**, 517–522.

Ortega, S., Prieto, I., Odajima, J. *et al.* (2003) Cyclin-dependent kinase 2 is essential for meiosis but not for mitotic cell division in mice. *Nat. Genet.*, **35**, 25–31.

Padmore, R., Cao, L. and Kleckner, N. (1991) Temporal comparison of recombination and synaptonemal complex formation during meiosis in S. cerevisiae. *Cell*, **66**, 1239–1256.

Paigen, K., Szatkiewicz, J.P., Sawyer, K. *et al.* (2008) The recombinational anatomy of a mouse chromosome. *PLoS Genet.*, **4**, e1000119.

Parvanov, E.D., Ng, S.H., Petkov, P.M. *et al.* (2009) Trans-regulation of mouse meiotic recombination hotspots by Rcr1. *PLoS Biol.*, **7**, e36.

Pellestor, F., Anahory, T. and Hamamah, S. (2005) Effect of maternal age on the frequency of cytogenetic abnormalities in human oocytes. *Cytogenet Genome. Res.*, **111**, 206–212.

Perry, J., Kleckner, N. and Borner, G.V. (2005) Bioinformatic analyses implicate the collaborating meiotic crossover/chiasma proteins Zip2, Zip3, and Spo22/Zip4 in ubiquitin labeling. *Proc. Natl. Acad. Sci. USA*, **102**, 17594–17599.

Petes, T.D. (2001) Meiotic recombination hot spots and cold spots. *Nat. Rev. Genet.*, **2**, 360–369.

Petkov, P.M., Broman, K.W., Szatkiewicz, J.P. *et al.* (2007) Crossover interference underlies sex differences in recombination rates. *Trends Genet.*, **23**, 539–542.

Petronczki, M., Siomos, M.F. and Nasmyth, K. (2003) Un ménage à quatre: the molecular biology of chromosome segregation in meiosis. *Cell*, **112**, 423–440.

Petukhova, G.V., Romanienko, P.J. and Camerini-Otero, R.D. (2003) The Hop2 protein has a direct role in promoting interhomolog interactions during mouse meiosis. *Dev. Cell*, **5**, 927–936.

Pezza, R.J., Voloshin, O.N., Vanevski, F. *et al.* (2007) Hop2/Mnd1 acts on two critical steps in Dmc1-promoted homologous pairing. *Genes. Dev.*, **21**, 1758–1766.

Pittman, D.L., Weinberg, L.R. and Schimenti, J.C. (1998a) Identification, characterization, and genetic mapping of Rad51d, a new mouse and human RAD51/RecA-related gene. *Genomics*, **49**, 103–111.

Pittman, D.L., Cobb, J., Schimenti, K.J. *et al.* (1998b) Meiotic prophase arrest with failure of chromosome synapsis in mice deficient for Dmc1, a germline-specific RecA homolog. *Mol. Cell*, **1**, 697–705.

Plug, A.W., Xu, J., Reddy, G. *et al.* (1996) Presynaptic association of Rad51 protein with selected sites in meiotic chromatin. *Proc. Natl. Acad. Sci. USA*, **93**, 5920–5924.

Ptak, S.E., Hinds, D.A., Koehler, K. *et al.* (2005) Fine-scale recombination patterns differ between chimpanzees and humans. *Nat. Genet.*, **37**, 429–434.

Qin, J., Richardson, L.L., Jasin, M. *et al.* (2004) Mouse strains with an active H2-Ea meiotic recombination hot spot exhibit increased levels of H2-Ea-specific DNA breaks in testicular germ cells. *Mol. Cell Biol.*, **24**, 1655–1666.

Reinholdt, L.G. and Schimenti, J.C. (2005) Mei1 is epistatic to Dmc1 during mouse meiosis. *Chromosoma*, **114**, 127–134.

Revenkova, E., Eijpe, M., Heyting, C. *et al.* (2004) Cohesin SMC1 beta is required for meiotic chromosome dynamics, sister chromatid cohesion and DNA recombination. *Nat. Cell Biol.*, **6**, 555–562.

Rockmill, B., Voelkel-Meiman, K. and Roeder, G.S. (2006) Centromere-proximal crossovers are associated with precocious separation of sister chromatids during meiosis in Saccharomyces cerevisiae. *Genetics*, **174**, 1745–1754.

Romanienko, P.J. and Camerini-Otero, R.D. (2000) The mouse spo11 gene is required for meiotic chromosome synapsis. *Mol. Cell*, **6**, 975–987.

Samollow, P.B., Gouin, N., Miethke, P. *et al.* (2007) A microsatellite-based, physically anchored linkage map for the gray, short-tailed opossum (Monodelphis domestica). *Chromosome Res.*, **15**, 269–281.

San Filippo, J., Sung, P. and Klein, H. (2008) Mechanism of eukaryotic homologous recombination. *Annu. Rev. Biochem.*, **77**, 229–257.

de los Santos, T., Hunter, N., Lee, C. *et al.* (2003) The Mus81/Mms4 endonuclease acts independently of double-Holliday junction resolution to promote a distinct subset of crossovers during meiosis in budding yeast. *Genetics*, **164**, 81–94.

Santucci-Darmanin, S., Walpita, D., Lespinasse, F. *et al.* (2000) MSH4 acts in conjunction with MLH1 during mammalian meiosis. *FASEB J.*, **14**, 1539–1547.

Santucci-Darmanin, S., Neyton, S., Lespinasse, F. *et al.* (2002) The DNA mismatch-repair MLH3 protein interacts with MSH4 in meiotic cells, supporting a role for this MutL homolog in mammalian meiotic recombination. *Hum. Mol. Genet.*, **11**, 1697–1706.

Sarai, N., Kagawa, W., Kinebuchi, T. *et al.* (2006) Stimulation of Dmc1-mediated DNA strand exchange by the human Rad54B protein. *Nucleic Acids Res.*, **34**, 4429–4437.

Sasanuma, H., Murakami, H., Fukuda, T. *et al.* (2007) Meiotic association between Spo11 regulated by Rec102, Rec104 and Rec114. *Nucleic Acids Res.*, **35**, 1119–1133.

Sauvageau, S., Stasiak, A.Z., Banville, I. *et al.* (2005) Fission yeast rad51 and dmc1, two efficient DNA recombinases forming helical nucleoprotein filaments. *Mol. Cell Biol.*, **25**, 4377–4387.

Schwacha, A. and Kleckner, N. (1997) Interhomolog bias during meiotic recombination: meiotic functions promote a highly differentiated interhomolog-only pathway. *Cell*, **90**, 1123–1135.

Sehorn, M.G., Sigurdsson, S., Bussen, W. *et al.* (2004) Human meiotic recombinase Dmc1 promotes ATP-dependent homologous DNA strand exchange. *Nature*, **429**, 433–437.

Sharan, S.K., Morimatsu, M., Albrecht, U. *et al.* (1997) Embryonic lethality and radiation hypersensitivity mediated by Rad51 in mice lacking Brca2. *Nature*, **386**, 804–810.

Sharan, S.K., Pyle, A., Coppola, V. *et al.* (2004) BRCA2 deficiency in mice leads to meiotic impairment and infertility. *Development*, **131**, 131–142.

Sheridan, S.D., Yu, X., Roth, R. *et al.* (2008) A comparative analysis of Dmc1 and Rad51 nucleoprotein filaments. *Nucleic Acids Res.*, **36**, 4057–4066.

Sherman, S.L., Freeman, S.B., Allen, E.G. *et al.* (2005) Risk factors for nondisjunction of trisomy 21. *Cytogenet. Genome Res.*, **111**, 273–280.

Shifman, S., Bell, J.T., Copley, R.R. *et al.* (2006) A high-resolution single nucleotide polymorphism genetic map of the mouse genome. *PLoS Biol.*, **4**, e395.

Shinohara, A., Ogawa, H. and Ogawa, T. (1992) Rad51 protein involved in repair and recombination in S. cerevisiae is a RecA-like protein. *Cell*, **69**, 457–470.

Shinohara, A., Gasior, S., Ogawa, T. *et al.* (1997) Saccharomyces cerevisiae recA homologues RAD51 and DMC1 have both distinct and overlapping roles in meiotic recombination. *Genes Cells*, **2**, 615–629.

Shinohara, M., Gasior, S.L., Bishop, D.K. *et al.* (2000) Tid1/Rdh54 promotes colocalization of rad51 and dmc1 during meiotic recombination. *Proc. Natl. Acad. Sci. USA*, **97**, 10814–10819.

Shiroishi, T., Sagai, T., Hanzawa, N. *et al.* (1991) Genetic control of sex-dependent meiotic recombination in the major histocompatibility complex of the mouse. *EMBO J.*, **10**, 681–686.

Snowden, T., Acharya, S., Butz, C. *et al.* (2004) hMSH4-hMSH5 recognizes Holliday Junctions and forms a meiosis-specific sliding clamp that embraces homologous chromosomes. *Mol. Cell*, **15**, 437–451.

Speed, R.M. (1977) The effects of ageing on the meiotic chromosomes of male and female mice. *Chromosoma*, **64**, 241–254.

Spencer, C.C., Deloukas, P., Hunt, S. *et al.* (2006) The influence of recombination on human genetic diversity. *PLoS Genet.*, **2**, e148.

Stumpf, M.P. and McVean, G.A. (2003) Estimating recombination rates from population-genetic data. *Nat. Rev. Genet.*, **4**, 959–968.

Suja, J.A. and Barbero, J.L. (2009) Cohesin complexes and sister chromatid cohesion in mammalian meiosis. *Genome. Dyn.*, **5**, 94–116.

Sun, F., Oliver-Bonet, M., Liehr, T. *et al.* (2004) Human male recombination maps for individual chromosomes. *Am. J. Hum. Genet.*, **74**, 521–531.

Sun, F., Oliver-Bonet, M., Liehr, T. *et al.* (2006) Variation in MLH1 distribution in recombination maps for individual chromosomes from human males. *Hum. Mol. Genet.*, **15**, 2376–2391.

Sung, P. and Robberson, D.L. (1995) DNA strand exchange mediated by a RAD51-ssDNA nucleoprotein filament with polarity opposite to that of RecA. *Cell*, **82**, 453–461.

Susiarjo, M., Hassold, T.J., Freeman, E. *et al.* (2007) Bisphenol A exposure in utero disrupts early oogenesis in the mouse. *PLoS Genet.*, **3**, e5.

Svetlanov, A., Baudat, F., Cohen, P.E. *et al.* (2008) Distinct functions of MLH3 at recombination hot spots in the mouse. *Genetics*, **178**, 1937–1945.

Szostak, J.W., Orr-Weaver, T.L., Rothstein, R.J. *et al.* (1983) The double-strand-break repair model for recombination. *Cell*, **33**, 25–35.

Tarsounas, M., Morita, T., Pearlman, R.E. *et al.* (1999) RAD51 and DMC1 form mixed complexes associated with mouse meiotic chromosome cores and synaptonemal complexes. *J. Cell Biol.*, **147**, 207–220.

Taylor, E.R. and McGowan, C.H. (2008) Cleavage mechanism of human Mus81-Eme1 acting on Holliday-junction structures. *Proc. Natl. Acad. Sci. USA*, **105**, 3757–3762.

Tease, C. and Hulten, M.A. (2004) Inter-sex variation in synaptonemal complex lengths largely determine the different recombination rates in male and female germ cells. *Cytogenet. Genome Res.*, **107**, 208–215.

Tease, C., Hartshorne, G.M. and Hulten, M.A. (2002) Patterns of meiotic recombination in human fetal oocytes. *Am. J. Hum. Genet.*, **70**, 1469–1479.

Terasawa, M., Ogawa, H., Tsukamoto, Y. *et al.* (2007) Meiotic recombination-related DNA synthesis and its implications for cross-over and non-cross-over recombinant formation. *Proc. Natl. Acad. Sci. USA*, **104**, 5965–5970.

Thorslund, T. and West, S.C. (2007) BRCA2: a universal recombinase regulator. *Oncogene*, **26**, 7720–7730.

Tiemann-Boege, I., Calabrese, P., Cochran, D.M. *et al.* (2006) High-resolution recombination patterns in a region of human chromosome 21 measured by sperm typing. *PLoS Genet.*, **2**, e70.

Topping, D., Brown, P., Judis, L. *et al.* (2006) Synaptic defects at meiosis I and non-obstructive azoospermia. *Hum. Reprod.*, **21**, 3171–3177.

Tsubouchi, H. and Roeder, G.S. (2003) The importance of genetic recombination for fidelity of chromosome pairing in meiosis. *Dev. Cell*, **5**, 915–925.

Tsuzuki, T., Fujii, Y., Sakumi, K. *et al.* (1996) Targeted disruption of the Rad51 gene leads to lethality in embryonic mice. *Proc. Natl. Acad. Sci. USA*, **93**, 6236–6240.

Turner, J.M. (2007) Meiotic sex chromosome inactivation. *Development*, **134**, 1823–1831.

Turner, J.M., Aprelikova, O., Xu, X. *et al.* (2004) BRCA1, histone H2AX phosphorylation, and male meiotic sex chromosome inactivation. *Curr. Biol.*, **14**, 2135–2142.

Turner, J.M., Mahadevaiah, S.K., Fernandez-Capetillo, O. *et al.* (2005) Silencing of unsynapsed meiotic chromosomes in the mouse. *Nat. Genet.*, **37**, 41–47.

Turner, J.M., Mahadevaiah, S.K., Ellis, P.J. *et al.* (2006) Pachytene asynapsis drives meiotic sex chromosome inactivation and leads to substantial postmeiotic repression in spermatids. *Dev. Cell*, **10**, 521–529.

de Vries, S.S., Baart, E.B., Dekker, M. *et al.* (1999) Mouse MutS-like protein Msh5 is required for proper chromosome synapsis in male and female meiosis. *Genes Dev.*, **13**, 523–531.

de Vries, F.A., de Boer, E., van den Bosch, M. *et al.* (2005) Mouse Sycp1 functions in synaptonemal complex assembly, meiotic recombination, and XY body formation. *Genes Dev.*, **19**, 1376–1389.

Walpita, D., Plug, A.W., Neff, N.F. *et al.* (1999) Bloom's syndrome protein, BLM, colocalizes with replication protein A in meiotic prophase nuclei of mammalian spermatocytes. *Proc. Natl. Acad. Sci. USA*, **96**, 5622–5627.

Wang, H. and Hoog, C. (2006) Structural damage to meiotic chromosomes impairs DNA recombination and checkpoint control in mammalian oocytes. *J. Cell Biol.*, **173**, 485–495.

Wang, T.F., Kleckner, N. and Hunter, N. (1999) Functional specificity of MutL homologs in yeast: evidence for three Mlh1-based heterocomplexes with distinct roles during meiosis in recombination and mismatch correction. *Proc. Natl. Acad. Sci. USA*, **96**, 13914–13919.

Wang, T.F., Chen, L.T. and Wang, A.H. (2008) Right or left turn? RecA family protein filaments promote homologous recombination through clockwise axial rotation. *Bioessays*, **30**, 48–56.

Ward, J.O., Reinholdt, L.G., Motley, W.W. *et al.* (2007) Mutation in mouse hei10, an e3 ubiquitin ligase, disrupts meiotic crossing over. *PLoS Genet.*, **3**, e139.

Webb, A.J., Berg, I.L. and Jeffreys, A. (2008) Sperm cross-over activity in regions of the human genome showing extreme breakdown of marker association. *Proc. Natl. Acad. Sci. USA*, **105**, 10471–10476.

Wei, K., Clark, A.B., Wong, E. *et al.* (2003) Inactivation of Exonuclease 1 in mice results in DNA mismatch repair defects, increased cancer susceptibility, and male and female sterility. *Genes. Dev.*, **17**, 603–614.

Wesoly, J., Agarwal, S., Sigurdsson, S. *et al.* (2006) Differential contributions of mammalian Rad54 paralogs to recombination, DNA damage repair, and meiosis. *Mol. Cell Biol.*, **26**, 976–989.

Winckler, W., Myers, S.R., Richter, D.J. *et al.* (2005) Comparison of fine-scale recombination rates in humans and chimpanzees. *Science*, **308**, 107–111.

Wong, A.K., Pero, R., Ormonde, P.A. *et al.* (1997) RAD51 interacts with the evolutionarily conserved BRC motifs in the human breast cancer susceptibility gene brca2. *J. Biol. Chem.*, **272**, 31941–31944.

Woods, L.M., Hodges, C.A., Baart, E. *et al.* (1999) Chromosomal influence on meiotic spindle assembly: abnormal meiosis I in female Mlh1 mutant mice. *J. Cell Biol.*, **145**, 1395–1406.

Wu, T.C. and Lichten, M. (1994) Meiosis-induced double-strand break sites determined by yeast chromatin structure. *Science*, **263**, 515–518.

Yang, F., Gell, K., van der Heijden, G.W. *et al.* (2008) Meiotic failure in male mice lacking an X-linked factor. *Genes Dev.*, **22**, 682–691.

Yauk, C.L., Bois, P.R. and Jeffreys, A.J. (2003) High-resolution sperm typing of meiotic recombination in the mouse MHC Ebeta gene. *EMBO J.*, **22**, 1389–1397.

Yoon, P.W., Freeman, S.B., Sherman, S.L. *et al.* (1996) Advanced maternal age and the risk of Down syndrome characterized by the meiotic stage of chromosomal error: a population-based study. *Am. J. Hum. Genet.*, **58**, 628–633.

Yoshida, K., Kondoh, G., Matsuda, Y. *et al.* (1998) The mouse RecA-like gene Dmc1 is required for homologous chromosome synapsis during meiosis. *Mol. Cell*, **1**, 707–718.

Zenvirth, D., Richler, C., Bardhan, A. *et al.* (2003) Mammalian meiosis involves DNA double-strand breaks with 3′ overhangs. *Chromosoma*, **111**, 369–376.

Zickler, D. and Kleckner, N. (1999) Meiotic chromosomes: integrating structure and function. *Annu. Rev. Genet.*, **33**, 603–754.

Section IV

Meiosis resumption

7

Initiation of the meiotic prophase-to-metaphase transition in mammalian oocytes

Laurinda A. Jaffe and Rachael P. Norris

Department of Cell Biology, University of Connecticut Health Center, Farmington, CT 06032, USA

7.1 Introduction

Mammalian oocytes, like those of other vertebrates, begin meiosis but then become arrested in prophase until luteinizing hormone (LH) from the pituitary causes the meiotic cell cycle to resume, in preparation for fertilization. This prophase arrest begins during foetal development, and lasts for months in mice, and years in women. Periodically, one or more oocytes, and the somatic cells of the follicles surrounding them, complete their growth and synthesize proteins required for meiotic resumption (see Eppig *et al.*, 2004). In response to the LH signal, the oocyte proceeds from prophase to second metaphase, where it remains arrested until fertilization causes the completion of meiosis. This chapter concerns only the first of these meiotic transitions, at the prophase-to-metaphase border.

The prophase-to-metaphase transition is characterized by the breakdown of the nuclear envelope (NEBD, also known as germinal vesicle breakdown or GVBD), and by preceding changes in chromatin configuration and formation of microtubule organizing centres (see Schuh and Ellenberg, 2007). The initiation of NEBD is mediated by the activation of cyclin dependent kinase 1 (CDK1, or CDC2A) (see Eppig *et al.*, 2004). This chapter focuses on the LH signalling events that lead to the activation of CDK1, and emphasizes mice and rats, the mammalian species that have been studied most extensively.

Oogenesis: The Universal Process Marie-Hélène Verlhac and Anne Villeneuve

7.2 Maintenance of meiotic arrest in fully grown mammalian oocytes by somatic cell signals and cyclic AMP

Mammalian oocytes are surrounded by a complex of somatic cells, making up a spherical follicle (Figure 7.1), and the presence of these somatic cells is essential for maintaining prophase arrest in the fully grown oocyte. The outer layers of somatic cells (mural granulosa), rather than the inner cumulus cells that directly surround the oocyte, are the primary source of the meiosis-inhibitory signal. The cumulus–oocyte complex is connected on one side to the mural granulosa cells, and surrounded elsewhere by a fluid-filled antrum. Oocytes or cumulus–oocyte complexes that are removed from the follicle resume meiosis spontaneously (Pincus and Enzmann, 1935; Edwards, 1965). If the cumulus–oocyte complex is dislodged from the mural cells, using a fine needle, meiosis resumes in the oocyte, even though the follicle is otherwise intact (Racowsky and Baldwin, 1989). Thus physical continuity of the cellular layers connecting the mural granulosa cells and oocyte is required to convey the inhibitory signal. If gap junction communication between the oocyte and somatic cells of the follicle is inhibited, the oocyte also resumes meiosis (Piontkewitz and Dekel, 1993; Sela-Abramovich *et al.*, 2006; Norris *et al.*, 2008). These findings indicate that the inhibitory signal passes through gap junctions from the mural granulosa cells to the oocyte.

As will be discussed below, the binding of LH to receptors on the somatic cells (Amsterdam *et al.*, 1975) causes meiosis to resume, at least largely by reversing the inhibitory signal that is provided to the oocyte by the somatic cells. Therefore, understanding of how LH causes meiosis to resume relies on understanding of how the somatic cells act on the fully grown oocyte to maintain prophase arrest. Before considering the mechanisms by which LH triggers meiotic resumption, we will first review some additional studies pertaining to how the oocyte is kept in prophase arrest until the follicle is acted on by the hormone. We will restrict this discussion to fully grown oocytes within follicles that have developed the ability to respond to LH; the somewhat different mechanisms that operate during the period of oocyte and follicle growth are discussed elsewhere (Chesnel and Eppig, 1995; Kovo *et al.*, 2006).

In addition to the presence of the somatic cells, maintenance of prophase arrest in the fully grown mammalian oocyte requires that cyclic AMP (cAMP) in the oocyte be maintained at a high level. This conclusion has been supported by many studies, going back to early findings that the presence of a membrane-permeant cAMP analogue (Cho, Stern and Biggers, 1974), or an inhibitor of cAMP phosphodiesterases (Magnusson and Hillensjo, 1977), prevents meiotic resumption in isolated oocytes. The cAMP is generated in the oocyte by way of a constitutively active G_s-linked receptor, GPR3 or GPR12, which acts to stimulate adenylyl cyclase (Mehlmann, Jones and Jaffe, 2002; Mehlmann *et al.*, 2004; Horner *et al.*, 2003; Kalinowski *et al.*, 2004; Freudzon *et al.*, 2005; Hinckley *et al.*, 2005; Ledent *et al.*, 2005; Mehlmann, 2005; DiLuigi *et al.*, 2008). Interfering with the function of the receptor, G_s, or adenylyl cyclase, causes cAMP to decrease, and correspondingly causes the resumption of meiosis. Likewise, injection of follicle-enclosed oocytes with a cAMP phosphodiesterase, which lowers cAMP in the oocyte, causes meiosis to resume (Norris *et al.*, 2009). Injection of a cAMP phosphodiesterase can also overcome the prophase arrest in isolated mouse oocytes in

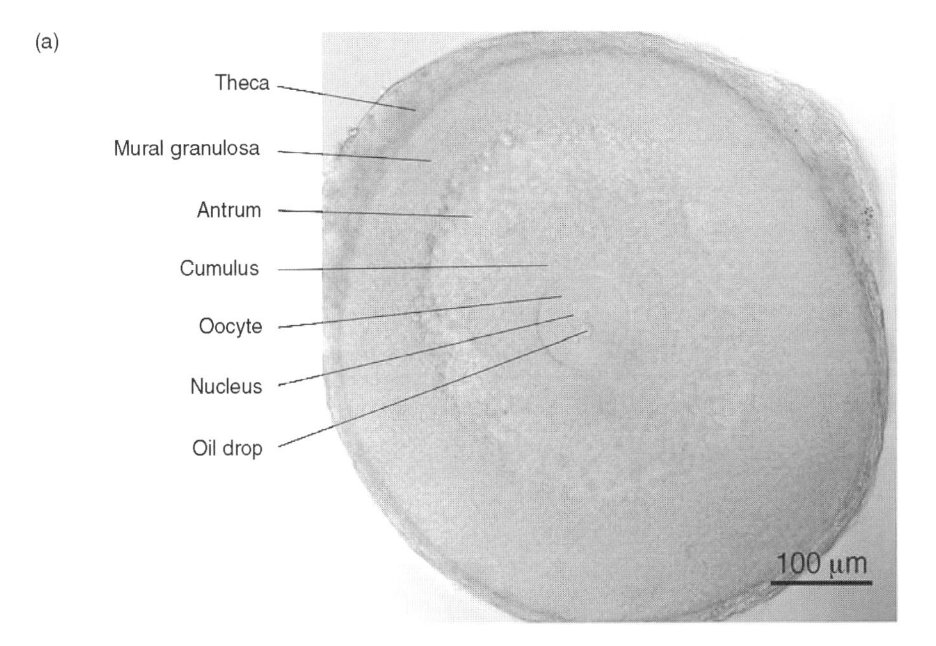

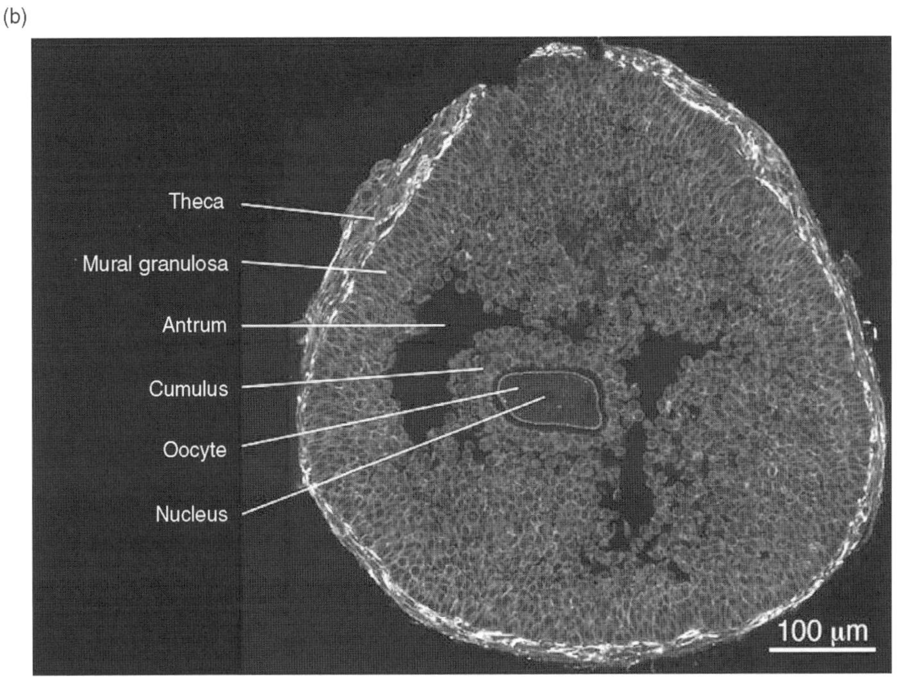

Figure 7.1 Antral follicles isolated from mouse ovaries. (a) A live follicle compressed between coverslips, as described by Norris *et al.* (2008). The oil drop is present as a consequence of microinjection. (b) A frozen section of a follicle, stained with an antibody against the G_s G-protein, as described by Norris *et al.* (2007). G_s is present in the plasma membranes of both the somatic cells and the oocyte, as well as in the theca cells that surround the follicle. G_s is coupled to the LH receptor in the mural granulosa cells and to the constitutively active GPR3 receptor in the oocyte

the presence of a phosphodiesterase inhibitor (Bornslaeger, Mattei and Schultz, 1986) or in which the oocyte's primary cAMP phosphodiesterase, PDE3A, is genetically deleted (Han *et al.*, 2006).

Recent measurements of the cAMP concentration in intact follicle-enclosed mouse oocytes, made using an optical sensor, indicate that the basal cAMP concentration is ~700 nM (Norris *et al.*, 2009). This cAMP concentration is sufficient to activate both forms of the cAMP-dependent kinase (PKA), PKAI and PKAII; half-maximal activation of these enzymes requires ~70–500 nM cAMP (Dostmann and Taylor, 1991; Viste *et al.*, 2005). Both PKAI and PKAII are present in the oocyte (Newhall *et al.*, 2006). Cyclic AMP maintains meiotic prophase arrest by way of PKA-mediated phosphorylation of proteins that regulate CDK1 activity (Maller and Krebs, 1977; Bornslaeger, Mattei and Schultz, 1986; Choi *et al.*, 1991; Kovo *et al.*, 2006). When CDK1 is phosphorylated, it is inactive, and the oocyte remains in prophase. PKA is not thought to directly phosphorylate CDK1, but rather to phosphorylate a phosphatase, CDC25 (CDC25B in mouse), causing CDC25 to be sequestered in the cytoplasm in a complex with a 14-3-3 protein such that CDC25 cannot dephosphorylate CDK1 in the nucleus (Duckworth, Weaver and Ruderman, 2002; Lincoln *et al.*, 2002; Zhang *et al.*, 2008; Pirino, Wescott and Donovan, 2009; Oh, Han, and Conti, 2010). In some species (such as pigs, but not rodents), protein synthesis in the oocyte is required in the sequence of events leading to NEBD (Wassarman, Josefowicz and Letourneau, 1976; Anger *et al.*, 2004), and in these species, PKA may also regulate proteins that control translation (see Haccard and Jessus, 2006).

Removal of the oocyte from its follicle (Vivarelli *et al.*, 1983; Törnell, Billig and Hillensjö, 1990a), or inhibition of the gap junction permeability within the follicle (Sela-Abramovich *et al.*, 2006), causes cAMP to decrease in the oocyte. These findings support the conclusion that the meiosis inhibitory signal that passes from the somatic cells to the oocyte through gap junctions acts by maintaining cAMP in the oocyte at a high level. One possibility is that this inhibitory signal from the somatic cells is additional cAMP (Anderson and Albertini, 1976). Consistent with this possibility, data from immunoassays, considered together with estimated cellular volumes of mouse and rat follicles, indicate that the concentration of cAMP in the somatic cells, averaged for the whole follicle, is ~10 μM (Tsafriri *et al.*, 1972; Schultz, Montgomery and Belanoff, 1983; Hashimoto, Kishimoto and Nagahama, 1985; Dekel, Galiani and Beers, 1988; Hsieh *et al.*, 2007). The spatial distribution of cAMP within the follicle is unknown.

This high cAMP concentration is surprising, because it is also known that PKA activity in the somatic cells prior to LH exposure is low (Tsafriri *et al.*, 1972; Hunzicker-Dunn, 1981). A possible explanation of this apparent discrepancy is that PKA in the somatic cells is sequestered by A kinase anchoring proteins (AKAPs), which could also bind the cAMP phosphodiesterase PDE4D, such that the microdomains where PKA is present have low cAMP concentrations. PDE4D, which is the primary cAMP phosphodiesterase in the somatic cells (Tsafriri *et al.*, 1996; Jin *et al.*, 1999), has been shown to bind AKAPs in other cells (Dodge-Kafka and Kapiloff, 2006).

Assuming that the cAMP concentration in the somatic cell cytosol outside of the AKAP domains is as high, as indicated by immunoassays, cAMP from the somatic cells could potentially diffuse into the oocyte. However, since the total volume of the somatic cells is ~100-fold greater than the oocyte volume, it is somewhat difficult to reconcile this idea with evidence that cAMP from the somatic cells is insufficient to maintain cAMP in the oocyte at an inhibitory level when cAMP generation in the oocyte is prevented (Mehlmann, Jones and Jaffe, 2002; Mehlmann, 2005). It could be that despite the large cAMP gradient, the permeability of the Cx37 junctions at the oocyte surface is insufficient to allow the somatic cells to control cAMP in the oocyte.

An alternative mechanism by which the somatic cells could maintain oocyte cAMP at the ~700 nM level is by providing a signal that inhibits the oocyte's cAMP phosphodiesterase, PDE3A. This idea is supported by evidence that if PDE3A is pharmacologically inhibited (Tsafriri *et al.*, 1996) or genetically deleted (Masciarelli *et al.*, 2004), the oocyte remains arrested in prophase even when removed from the follicle. As will be reviewed in the next section, current evidence indicates that cyclic GMP (cGMP) is an essential inhibitory molecule that diffuses from the somatic cells to the oocyte, and that cGMP keeps cAMP elevated in the oocyte by reducing hydrolysis of cAMP by PDE3A, which is known to be competitively inhibited by cGMP (Hambleton *et al.*, 2005). Based on studies of other gap junctions, there is precedent for thinking that Cx37 gap junctions could be more permeable to cGMP than to cAMP (Bevans *et al.*, 1998; Locke *et al.*, 2004).

7.3 Cyclic GMP diffusion from the somatic cells to the oocyte maintains oocyte cyclic AMP at a level that suppresses meiotic resumption

The concept of cGMP as an inhibitor of meiotic progression was originally proposed by Törnell, Billig and Hillensjö (1991), based on evidence that cGMP in the oocyte decreases with time after it is removed from the follicle, and that cGMP injection into an isolated oocyte delays spontaneous meiotic resumption (Törnell, Billig and Hillensjö, 1990a). It was also known at that time that cGMP in the oocyte and in the whole follicle decreases in response to LH (Hubbard, 1986), that gap junctions are present between the cells of the follicle (Larsen, Wert and Brunner, 1987), and that cGMP can inhibit cAMP phosphodiesterase activity in oocyte lysates (Bornslaeger, Wilde and Schultz, 1984). In addition, it was known that NEBD occurs if the follicle is exposed to inhibitors of inosine monophosphate dehydrogenase, an enzyme required for formation of cGMP (Downs and Eppig, 1987), and that activators of both soluble and transmembrane guanylyl cyclases partially inhibit spontaneous NEBD in cumulus–oocyte complexes (Törnell, Carlsson and Billig, 1990b). All of these observations supported the proposed model, but at the time, means were not available to test it definitively.

This hypothesis has gained further experimental support from several recent studies: (i) Pharmacological inhibition of the soluble guanylyl cyclase GUCY1, using the compound 1H-[1,2,4]oxadiazolo[4,3-a]quinoxalin-1-one (ODQ), causes NEBD in follicle-enclosed oocytes, and this effect is reversed by 8-Br-cGMP (a membrane-permeant and hydrolysis-resistant cGMP analogue), indicating that production of cGMP by GUCY1 is required to maintain meiotic arrest (Sela-Abramovich *et al.*,

2008). (ii) Inhibition of gap junction permeability in the follicle, using carbenoxolone, lowers cGMP in the follicle-enclosed oocyte from ~900 nM to ~90 nM, supporting the idea that cGMP enters the oocyte from the somatic cells (Norris *et al.*, 2009). Based on the enzymatic properties of PDE3A, this decrease in cGMP would increase the activity of oocyte PDE3A by approximately fourfold (Norris *et al.*, 2009). Consistently, the measured cAMP phosphodiesterase activity of an oocyte lysate also increases over this cGMP concentration range (Vaccari *et al.*, 2009). (iii) Injection of the follicle-enclosed oocyte with the cGMP-specific phosphodiesterase PDE9A decreases cGMP, and as a consequence also decreases cAMP to a level that would lower PKA activity. Correspondingly, injection of PDE9A, or another cGMP-specific phosphodiesterase PDE5A, causes meiosis to resume, but not if PDE3A in the oocyte is inhibited or genetically deleted (Norris *et al.*, 2009; Vaccari *et al.*, 2009). Thus, although cGMP diffusion into the oocyte is required to maintain prophase arrest, the cGMP acts indirectly, through inhibition of the hydrolysis of cAMP by PDE3A.

In summary, accumulating evidence supports the conclusion that prophase arrest is maintained by the coordinated function of cyclic nucleotide regulatory systems in both the somatic cells and oocyte, compartments that are connected by gap junctions. Cyclic AMP is generated in the oocyte by a constitutively active G_s-linked receptor that activates adenylyl cyclase. The cAMP is prevented from degradation because cGMP from the somatic cells enters the oocyte through gap junctions, and inhibits the hydrolysis of cAMP by PDE3A. Thus cAMP is maintained at a level that activates the cAMP-dependent kinase, PKA, and PKA-mediated phosphorylation of cell cycle regulatory proteins maintains prophase arrest.

7.4 Unanswered questions about maintenance of prophase arrest

Much remains unknown about how prophase arrest is maintained. In particular, it is incompletely understood which guanylyl cyclases in the mural granulosa cells, and possibly in the oocyte as well, are responsible for the production of cGMP, and how they are regulated. Results mentioned above, using the inhibitor ODQ, point to a role for the soluble guanylyl cyclase, GUCY1 (Sela-Abramovich *et al.*, 2008), but transmembrane guanylyl cyclases are also expressed in somatic cells (Törnell, Carlsson and Billig, 1990b; Noubani, Farookhi and Gutkowska, 2000), and may play a role as well. ODQ is quite specific for GUCY1 (Garthwaite *et al.*, 1995), but the possibility of inhibition of other guanylyl cyclases under the conditions used has not been completely eliminated. There are also unanswered questions about the spatial distribution of guanylyl cyclases, cGMP and cAMP within the somatic cells, and about the relative cGMP/cAMP permeability of the Cx37 junctions between the cumulus cells and oocyte.

At the level of the oocyte membrane, it is unknown if the GPR3 receptor that activates G_s is truly constitutively active. The presence of the somatic cells is not required to maintain GPR3 activity (Freudzon *et al.*, 2005), but there could be an agonist in the oocyte itself that is essential to maintain this activity. Additionally, more studies are needed of the phosphatases, kinases and other proteins that are regulated by PKA in the oocyte to maintain CDK1 in its phosphorylated and inactive state, particularly in light of evidence that in oocytes of some other organisms, high cAMP stimulates rather than inhibits meiotic resumption (see Takeda, Kyozuka and Deguchi, 2006). Finally, much remains to be determined about which of the mechanisms that control prophase arrest in rodents do or do not pertain to larger species including humans. In the human oocyte, cAMP, G_s activity and PDE3 inhibition are essential for maintaining meiotic arrest, and RNA that encodes the GPR3 receptor is present (Nogueira *et al.*, 2006; DiLuigi *et al.*, 2008), but other aspects of the regulatory pathway described above remain to be tested.

7.5 LH causes the prophase-to-metaphase transition in mammalian oocytes by activating a G-protein-coupled receptor in the somatic cells, which then causes cyclic AMP to decrease in the oocyte

Luteinizing hormone, which is released from the pituitary, is the primary stimulus for meiotic resumption in vertebrate oocytes. In mice and rats, NEBD occurs approximately two to four hours after exposure of isolated follicles to LH (Tsafriri, 1985; Park *et al.*, 2004), or after injection of the animal with the LH receptor agonist, human chorionic gonadotrophin (hCG) (Schultz, Montgomery and Belanoff, 1983; Larsen, Wert and Brunner, 1986). LH receptors are not found on the oocyte itself, but rather on the somatic cells surrounding the oocyte; both the outer layers of the mural granulosa cells of the follicle, and the inner layer of theca cells that surround the follicle have these receptors (Amsterdam *et al.*, 1975). While most studies of LH stimulation of meiotic resumption have focused on the action of LH on the mural granulosa cells, actions at the level of theca cells cannot be excluded.

The LH receptor is a well-characterized G-protein-coupled receptor that activates G_s, as well as G_i and $G_{q/11}$ (Herrlich *et al.*, 1996; Rajagopalan-Gupta *et al.*, 1998; Lee *et al.*, 2002). G_s activation causes the production of cAMP in the granulosa cells (Tsafriri *et al.*, 1972; Schultz, Montgomery and Belanoff, 1983; Hashimoto, Kishimoto and Nagahama, 1985; Dekel, Galiani and Beers, 1988; Hsieh *et al.*, 2007), stimulating their PKA activity (Tsafriri *et al.*, 1972; Hunzicker-Dunn, 1981). LH receptor activation of G_i and $G_{q/11}$ causes the activation of phospholipase C, leading to calcium release (Herrlich *et al.*, 1996; Rajagopalan-Gupta *et al.*, 1998; Lee *et al.*, 2002). Increasing cAMP in the follicle, using forskolin (Dekel and Sherizly, 1983; Hashimoto, Kishimoto and Nagahama, 1985) or a PDE4 inhibitor (Tsafriri *et al.*, 1996), is sufficient to cause NEBD in the oocyte. Likewise, increasing calcium in the follicle, using ionophores, is also sufficient to cause NEBD in the oocyte (Tsafriri and Bar-Ami, 1978; Goren, Oron and Dekel, 1990).

LH signalling could be imagined to counteract the meiosis inhibitory pathway described above either by lowering oocyte cAMP, or by bypassing its inhibitory effect. Radioimmunoassays have indicated that LH causes a decrease in cAMP in mouse and rat oocytes, prior to NEBD (Schultz, Montgomery and Belanoff, 1983; Sela-Abramovich *et al.*, 2006), but questions have remained about whether the magnitude of the cAMP decrease was sufficient to cause a significant decrease in PKA activity. The effect of LH on oocyte cAMP was recently re-examined by use of an optical sensor for cAMP in live follicle-enclosed oocytes (Norris *et al.*, 2009). At 1–1.4 hours after LH exposure, cAMP has decreased from ~700 nM to ~140 nM, a change that would decrease PKA activity. These findings, together with the evidence reviewed above that decreasing oocyte cAMP is sufficient to cause meiotic resumption, support the conclusion that the primary means by which LH causes meiotic resumption is by decreasing oocyte cAMP.

How then does LH signalling in the mural granulosa cells lead to a fall in cAMP in the oocyte? Although this question has not been completely answered, there has been considerable progress recently, which we review in the next two sections.

7.6 LH decreases cyclic AMP in the oocyte by decreasing cyclic GMP, thus relieving the inhibition of cyclic AMP phosphodiesterase activity in the oocyte

The LH-stimulated decrease in the concentration of cAMP in the oocyte could be caused either by a decrease in cAMP synthesis by adenylyl cyclase (as occurs in frog oocytes; see references cited in Gallo *et al.*, 1995), or by an increase in cAMP degradation by a phosphodiesterase. Cyclic AMP synthesis could be regulated at the level of the GPR3 receptor, if LH action caused a decrease in a GPR3 agonist or an increase in a GPR3 antagonist. However, recent evidence indicates that G_s activity in the oocyte does not change in response to LH, arguing against regulation of GPR3–G_s signalling (Norris *et al.*, 2007). Cyclic AMP synthesis could also be regulated by a separate receptor in the oocyte that acts through G_i or a decrease in intracellular Ca^{2+} to inhibit adenylyl cyclase, but available evidence based on the use of pertussis toxin and Ca^{2+} chelators argues against these possibilities as well (Mehlmann *et al.*, 2006).

Instead, current evidence indicates that LH signalling acts to stimulate cAMP phosphodiesterase activity in the oocyte. The primary cAMP phosphodiesterase in the mouse oocyte is PDE3A (Richard, Tsafriri and Conti, 2001; Masciarelli *et al.*, 2004), an enzyme that can be regulated in at least two ways; it is competitively inhibited by cGMP (Hambleton *et al.*, 2005; Vaccari *et al.*, 2009), and it can be stimulated by phosphorylation (Han *et al.*, 2006). So far, there is no direct evidence that LH causes an increase in phosphorylation of PDE3A. However, it has been thought that this may occur, based on measurements of a ~2.5-fold increase in PDE3 activity in lysates of cumulus–oocyte complexes from hCG-stimulated mice (Richard, Tsafriri and Conti, 2001). These measurements were made by determining the difference between total cAMP PDE activity of the complexes and the PDE activity in the presence of the PDE3-specific inhibitor, cilostamide. While this result remains to be confirmed by direct measurements of PDE activity in oocytes, it cannot be explained by a decrease in cGMP, since cGMP would be diluted in the phosphodiesterase assay mixture. An increase in the activity of the kinase AKT/PKB is seen in mouse oocytes after LH receptor stimulation (Han *et al.*, 2006;

Kalous *et al.*, 2006), and AKT can phosphorylate and activate PDE3A (Han *et al.*, 2006). Antibodies specific for phosphorylated PDE3A could potentially be used to determine whether PDE3A is phosphorylated in response to LH.

With regard to the possibility of cGMP regulation of PDE3A in response to LH, recent evidence from use of an optical sensor for cGMP in live follicle-enclosed oocytes indicates that cGMP in the oocyte decreases in response to LH (Norris *et al.*, 2009). Within one hour, cGMP falls from ~900 nM to ~40 nM. Immunoassays of isolated oocytes have also shown a decrease in cGMP in response to LH receptor stimulation *in vivo* (Hubbard, 1986; Vaccari *et al.*, 2009). Based on the cGMP measurements from the follicle-enclosed oocytes, and on the enzymatic properties of PDE3A, this cGMP decrease would increase the rate of hydrolysis of cAMP by approximately fivefold (Norris *et al.*, 2009). Consistently, cAMP phosphodiesterase activity, as measured in an oocyte lysate, increases over this cGMP concentration range (Vaccari *et al.*, 2009). As described above, a comparable decrease in oocyte cGMP in response to injection of PDE9A is sufficient to cause meiotic resumption, supporting the conclusion that the LH-induced cGMP decrease reverses the inhibition of PDE3A, leading to meiotic resumption.

The evidence discussed above, that cGMP enters the oocyte from the somatic cells by way of gap junctions, suggests two possibilities as to how LH could decrease cGMP in the oocyte: (i) by decreasing the concentration of cGMP in the mural granulosa cells; and (ii) by decreasing the permeability of the gap junctions. Both of these events were proposed by Törnell, Billig and Hillensjö (1991), and, as discussed below, both have been found to occur.

A decrease in cGMP in the somatic cells in response to LH was reported by Hubbard (1986), and recently confirmed (Norris *et al.*, 2009; Vaccari *et al.*, 2009). The cGMP concentration falls from ~2 μM to ~80 nM at one hour after LH application (Norris *et al.*, 2009). Thus, one mechanism by which LH decreases cGMP in the oocyte appears to be by way of decreasing cGMP in the interconnected somatic compartment. If cGMP can diffuse freely out of the oocyte into the somatic cells through gap junctions, and if the oocyte does not generate a significant amount of cGMP by itself, a fall in cGMP in the somatic cells would cause a fall in cGMP in the oocyte. The LH regulation of somatic cell cGMP could involve inhibition of a guanylyl cyclase and/or stimulation of a cGMP phosphodiesterase; some evidence supports regulation at the level of the guanylyl cyclase rather than the phosphodiesterase (Patwardhan and Lanthier, 1984). As discussed above, the somatic cells of the follicle express both soluble and transmembrane guanylyl cyclases. Amongst several phosphodiesterases that can hydrolyze cGMP, PDE5A accounts for much but not all of the basal activity (Vaccari *et al.*, 2009). Which of these or other cGMP-regulatory enzymes are regulated by LH, and how this occurs, is unknown.

A second way by which LH causes cGMP to decrease in the oocyte is by closing gap junctions in the follicle. Within 30 minutes of applying LH to the follicle, the Cx43 channels between the somatic cells close, as was detected by monitoring the diffusion of a fluorescent tracer injected into follicle-enclosed oocytes (Norris *et al.*, 2008 (Figure 7.2). Although the Cx37 channels between the somatic cells and the oocyte remain open, the closure of the Cx43 channels between the somatic cells prevents cGMP from the mural granulosa cells from diffusing into the oocyte, as indicated by

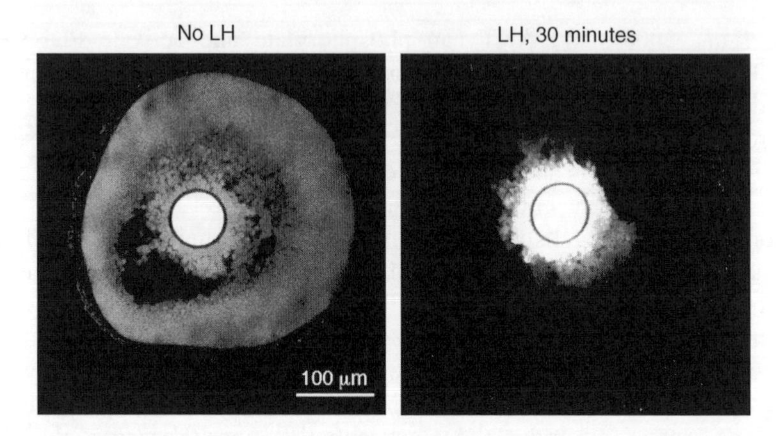

Figure 7.2 Antral follicles with or without a 30 minute LH treatment, illustrating the closure of gap junctions between the somatic cells in response to LH. Twenty minutes before imaging of the live follicles by two-photon microscopy, the oocyte was injected with the small fluorescent tracer Alexa-350 (see Norris *et al.*, 2008). Without LH exposure, all of the cells of the follicle were coupled by gap junctions, allowing the tracer to spread throughout the follicle. After LH exposure, the Cx37 junctions at the oocyte surface remained open, but the Cx43 junctions between the somatic cells were closed

the decrease in cGMP in the oocyte when gap junction communication is inhibited (Norris *et al.*, 2009). The closure of the Cx43 channels results from phosphorylation of several Cx43 serines by MAP kinase (Sela-Abramovich *et al.*, 2005; Norris *et al.*, 2008). Importantly, while gap junction closure is sufficient to cause NEBD, it is not required. If gap junction closure is prevented by use of an inhibitor of MAP kinase activation, LH still causes meiotic resumption (Norris *et al.*, 2008), and cGMP still decreases in the oocyte (Norris *et al.*, 2009), most likely due to the decrease in somatic cell cGMP.

These findings support the conclusion that LH acts by reversing the signalling system that maintains meiotic arrest prior to LH exposure. However, other studies have argued that LH does more than reverse an inhibitory signal from the somatic cells, and instead (or in addition) provides a positive stimulus (Downs, Daniel and Eppig, 1988; Downs and Chen, 2008). The basis for this concept has been that application of follicle stimulating hormone (FSH) or epidermal growth factor (EGF) to cumulus–oocyte complexes partially overcomes the inhibition of NEBD by a membrane-permeant cAMP analogue or cAMP phosphodiesterase inhibitor, resulting in a greater fraction of the oocytes undergoing NEBD than what is seen with cumulus-free oocytes in these same media. A possible explanation is that FSH and EGF could act on the cumulus cells to reduce their level of cGMP, such that cGMP would diffuse out of the oocyte. Thus the 'positive' stimulus seen under these circumstances could involve a relief of inhibition. Supporting this interpretation, EGF receptor agonists cause cGMP to decrease when applied to follicles (Vaccari *et al.*, 2009).

In summary, LH acts to decrease cGMP in the oocyte, both by decreasing cGMP in the somatic cells, and by closing gap junctions between the somatic cells. In the following section, we will discuss the intermediate events between LH receptor activation and these somatic cell responses, and how LH receptor activation in the outer layers of the follicle is coupled to these events in the follicle interior.

7.7 The LH signal is spread throughout the somatic compartment of the follicle by the release of EGF receptor ligands

In frogs and fish, LH stimulation of meiotic resumption is mediated by the synthesis of progesterone and its derivatives (Masui, 1967; Fortune, 1983; Nagahama and Yamashita, 2008), suggesting that a similar mechanism could function in mammals as well. However, while LH causes mammalian follicles to produce progesterone, which is essential for ovulation and subsequent implantation of the embryo, progesterone and its derivatives do not have a significant function in stimulating the prophase-to-metaphase transition in mammalian oocytes (reviewed by Tsafriri and Motola, 2007). In contrast, another steroid, follicular fluid meiosis activating sterol (FF-MAS), which is also elevated in response to LH prior to NEBD, can stimulate meiotic resumption in isolated oocytes, overcoming the inhibitory effect of a cAMP phosphodiesterase inhibitor (Hegele-Hartung *et al.*, 1999; Baltsen, 2001). However, NEBD in response to FF-MAS occurs slowly, with a halftime of ~14 hours in mouse oocytes (Hegele-Hartung *et al.*, 1999), compared to ~2–4 hours for LH. Additional evidence that FF-MAS is unlikely to be a primary mechanism by which LH causes the prophase-to-metaphase transition is reviewed by Tsafriri *et al.* (2005). FF-MAS does have a significant function in promoting the normal progression of meiosis to metaphase II and improving preimplantation development (Marin Bivens *et al.*, 2004).

Thus, another process must account for how the LH signal is conveyed from the outer layers of the mural granulosa cells to the follicle interior and oocyte. LH-induced signalling processes leading to meiotic resumption, such as MAP kinase activation, Cx43 phosphorylation, and gap junction closure, occur throughout all of the somatic cells of the follicle, not just in the mural granulosa cells where the LH receptors are located (Panigone *et al.*, 2008; Norris *et al.*, 2008). As will be discussed below, current evidence indicates that an essential mechanism for signal transduction from the follicle surface inwards is the synthesis and release of EGF receptor ligands, leading to EGF receptor activation throughout the somatic cells of the follicle. The release of ligands for other receptors may also contribute (Kawamura *et al.*, 2009).

Within 30 minutes after exposure of the follicle to LH, EGF receptors are activated, as indicated by their phosphorylation, and most of this phosphorylation depends on the activation of PKA (Panigone *et al.*, 2008). EGF receptors are present on both the mural granulosa cells and cumulus cells, but not on the oocyte (Park *et al.*, 2004; Panigone *et al.*, 2008). Application of EGF receptor ligands to antral follicles causes meiotic resumption (Dekel and Sherizly, 1985; Park *et al.*, 2004), and a requirement for EGF receptor activation in the LH signalling pathway leading to meiotic resumption is supported by both pharmacological and genetic studies. Most importantly, studies of a mutant mouse with reduced EGF receptor kinase activity, and mice with genetic deletions of EGF receptor ligands, have shown that reducing EGF receptor activation largely, but not completely, inhibits meiotic resumption in response to LH (Hsieh *et al.*, 2007).

Of the six known EGF receptor ligands, the two that appear to be important for LH signalling are epiregulin and amphiregulin, since their transcription increases by 30 minutes after LH application (Park *et al.*, 2004; Panigone *et al.*, 2008). Other ligands – EGF, HB-EGF, TGFα and betacellulin – do not show such rapid changes in expression (Park *et al.*, 2004; Hsieh, Zamah and Conti, 2009). At least in primary cultured granulosa cells, epiregulin appears to be the main EGF receptor ligand that mediates LH stimulation of meiotic resumption, based on RNA expression levels and the use of a neutralizing antibody (Andric and Ascoli, 2008). Granulosa cells from women as well as nonhuman primates also synthesize EGF receptor ligands in response to LH (reviewed by Hsieh, Zamah and Conti, 2009).

Epiregulin and amphiregulin are $\sim$5 kDa polypeptides that are released into the extracellular space from transmembrane precursors, by the action of a transmembrane metalloendoprotease, ADAM-17/TACE (Blobel, Carpenter and Freeman, 2009). Correspondingly, the metalloendoprotease inhibitor galardin prevents meiotic resumption in response to LH (Ashkenazi *et al.*, 2005). In addition to stimulating their transcription, LH signalling stimulates the release of EGF receptor ligands from their precursors (Andric and Ascoli, 2008).

One consequence of EGF receptor activation is the activation of MAP kinase, although MAP kinase may also be activated by an EGF receptor-independent pathway (Panigone *et al.*, 2008). MAP kinase activation acts in a positive feedback loop to cause additional release of epiregulin (Andric and Ascoli, 2008). Through MAP kinase, EGF receptor activation is likely to be an important component of how LH causes the Cx43 phosphorylation and gap junction closure that contribute to reinitiating meiosis. Activation of the EGF receptor can also cause cGMP to decrease in the somatic cells, and pharmacological evidence indicates that EGF receptor activity is required for the LH-induced cGMP decrease (Vaccari *et al.*, 2009).

7.8 Conclusions

How the somatic cells of the mammalian ovarian follicle act to maintain prophase arrest in the oocyte, and how luteinizing hormone causes meiosis to resume, have been long-standing questions. Current evidence indicates that the somatic cells of the follicle maintain prophase arrest by supplying the oocyte with cGMP, by way of gap junctions. Cyclic GMP inhibits PDE3A in the oocyte, thus preventing the degradation of cAMP that is generated in the oocyte by the activity of a constitutively active G_s-linked receptor, (GPR3 in mouse oocytes, GPR12 in rat oocytes). These processes keep cAMP in the oocyte at an elevated level that suppresses meiotic progression. LH signalling reverses the arrest by lowering cGMP in the somatic cells, and by causing MAP kinase-dependent phosphorylation and closure of the Cx43 gap junctions between the somatic cells. As a consequence, cGMP in the oocyte decreases, PDE3A activity increases, cAMP decreases, and meiosis resumes. The LH signal is initiated by binding of LH to a G-protein-coupled receptor on the outer mural granulosa cells, which elevates cAMP and calcium in these cells. The signal is then conveyed to the inner regions of the follicle by the production and release of EGF-receptor ligands. Much remains to be determined about the intermediate steps linking these many events to each other and to the eventual breakdown of the nuclear envelope that marks the transition to metaphase.

Acknowledgements

The images shown in Figures 7.1a and 7.2 were generated in studies done together with Marina Freudzon (Norris *et al.*, 2008). We thank Marco Conti, John Eppig, Marina Freudzon, Keith Jones, Lisa Mehlmann and William Ratzan for helpful discussions and comments on the manuscript. Work in the authors' laboratory is supported by grants from the National Institutes of Health (NIH) to Laurinda Jaffe (HD014939, DK073499).

References

Amsterdam, A., Koch, Y., Lieberman, M.E. and Lindner, H.R. (1975) Distribution of binding sites for human chorionic gonadotropin in the preovulatory follicle of the rat. *J. Cell Biol.*, **67**, 894–900.

Anderson, E. and Albertini, D.F. (1976) Gap junctions between the oocyte and companion follicle cells in the mammalian ovary. *J. Cell Biol.*, **71**, 680–686.

Andric, N. and Ascoli, M. (2008) The luteinizing hormone receptor-activated extracellularly regulated kinase-1/2 cascade stimulates epiregulin release from granulosa cells. *Endocrinology*, **149**, 5549–5556.

Anger, M., Klima, J., Kubelka, M. *et al.* (2004) Timing of Plk1 and MPF activation during porcine oocyte maturation. *Mol. Reprod. Dev.*, **69**, 11–16.

Ashkenazi, H., Cao, X., Motola, S. *et al.* (2005) Epidermal growth factor family members: endogenous mediators of the ovulatory response. *Endocrinology*, **146**, 77–84.

Baltsen, M. (2001) Gonadotropin-induced accumulation of 4,4-dimethylsterols in mouse ovaries and its temporal relation to meiosis. *Biol. Reprod.*, **65**, 1743–1750.

Bevans, C.G., Kordel, M., Rhee, S.K. and Harris, A.L. (1998) Isoform composition of connexin channels determines selectivity among second messengers and uncharged molecules. *J. Biol. Chem.*, **273**, 2808–2816.

Blobel, C.P., Carpenter, G. and Freeman, M. (2009) The role of protease activity in ErbB biology. *Exp. Cell Res.*, **315**, 671–682.

Bornslaeger, E.A., Wilde, M.W. and Schultz, R.M. (1984) Regulation of mouse oocyte maturation: involvement of cyclic AMP phosphodiesterase and calmodulin. *Dev. Biol.*, **105**, 488–499.

Bornslaeger, E.A., Mattei, P. and Schultz, R.M. (1986) Involvement of cAMP-dependent protein kinase and protein phosphorylation in regulation of mouse oocyte maturation. *Dev. Biol.*, **114**, 453–462.

Chesnel, F. and Eppig, J.J. (1995) Synthesis and accumulation of $p34^{cdc2}$ and cyclin B in mouse oocytes during acquisition of competence to resume meiosis. *Mol. Reprod. Dev.*, **40**, 503–508.

Cho, W.K., Stern, S. and Biggers, J.D. (1974) Inhibitory effect of dibutyryl cAMP on mouse oocyte maturation *in vitro*. *J. Exp. Zool.*, **187**, 383–386.

Choi, T., Aoki, F., Mori, M. *et al.* (1991) Activation of $p34^{cdc2}$ protein kinase activity in meiotic and mitotic cell cycles in mouse oocytes and embryos. *Development*, **113**, 789–795.

Dekel, N. and Sherizly, I. (1983) Induction of maturation in rat follicle-enclosed oocyte by forskolin. *FEBS Lett.*, **151**, 153–155.

Dekel, N. and Sherizly, I. (1985) Epidermal growth factor induces maturation of rat follicle-enclosed oocytes. *Endocrinology*, **116**, 406–409.

Dekel, N., Galiani, D. and Beers, W.H. (1988) Induction of maturation in follicle-enclosed oocytes: the response to gonadotropins at different stages of follicular development. *Biol. Reprod.*, **38**, 517–521.

DiLuigi, A., Weitzman, V.N., Pace, M.C. *et al.* (2008) Meiotic arrest in human oocytes is maintained by a G_s signaling pathway. *Biol. Reprod.*, **78**, 667–672.

Dodge-Kafka, K.L. and Kapiloff, M.S. (2006) The mAKAP signaling complex: integration of cAMP, calcium, and MAP kinase signaling pathways. *Eur. J. Cell Biol.*, **85**, 593–602.

Dostmann, W.R. and Taylor, S.S. (1991) Identifying the molecular switches that determine whether (R_p)-cAMPS functions as an antagonist or an agonist in the activation of cAMP-dependent protein kinase I. *Biochemistry*, **30**, 8710–8716.

Downs, S.M. and Chen, J. (2008) EGF-like peptides mediate FSH-induced maturation of cumulus cell-enclosed mouse oocytes. *Mol. Reprod. Dev.*, **75**, 105–114.

Downs, S.M. and Eppig, J.J. (1987) Induction of mouse oocyte maturation in vivo by perturbants of purine metabolism. *Biol. Reprod.*, **36**, 431–437.

Downs, S.M., Daniel, S.A. and Eppig, J.J. (1988) Induction of maturation in cumulus cell-enclosed mouse oocytes by follicle-stimulating hormone and epidermal growth factor: evidence for a positive stimulus of somatic cell origin. *J. Exp. Zool.*, **245**, 86–96.

Duckworth, B.C., Weaver, J.S. and Ruderman, J.V. (2002) G_2 arrest in Xenopus oocytes depends on phosphorylation of cdc25 by protein kinase A. *Proc. Natl. Acad. Sci. USA*, **99**, 16794–16799.

Edwards, R.G. (1965) Maturation in vitro of mouse, sheep, cow, pig, rhesus monkey and human ovarian oocytes. *Nature*, **208**, 349–351.

Eppig, J.J., Viveiros, M.M., Marin Bivens, C.L. and de la Fuente, R. (2004) Regulation of mammalian oocyte maturation, in *The Ovary* (eds P.C.K. Leung and E.Y. Adashi), Elsevier/Academic Press, San Diego, pp. 113–129.

Fortune, J.E. (1983) Steroid production by *Xenopus* ovarian follicles at different developmental stages. *Dev. Biol.*, **99**, 502–509.

Freudzon, L., Norris, R.P., Hand, A.R. *et al.* (2005) Regulation of meiotic prophase arrest in mouse oocytes by GPR3, a constitutive activator of the G_s G protein. *J. Cell Biol.*, **171**, 255–265.

Gallo, C.J., Hand, A.R., Jones, T.L.Z. and Jaffe, L.A. (1995) Stimulation of *Xenopus* oocyte maturation by inhibition of the G-protein α_S subunit, a component of the plasma membrane and yolk platelet membranes. *J. Cell Biol.*, **130**, 275–284.

Garthwaite, J., Southam, E., Boulton, C.L. *et al.* (1995) Potent and selective inhibition of nitric oxide-sensitive guanylyl cyclase by 1*H*-[1,2,4]oxadiazolo[4,3-a]quinoxalin-1-one. *Mol. Pharmacol.*, **48**, 184–188.

Goren, S., Oron, Y. and Dekel, N. (1990) Rat oocyte maturation: role of calcium in hormone action. *Mol. Cell Endocrinol.*, **72**, 131–138.

Haccard, O. and Jessus, C. (2006) Oocyte maturation, Mos and cyclins—a matter of synthesis: two functionally redundant ways to induce meiotic maturation. *Cell Cycle*, **5**, 1152–1159.

Hambleton, R., Krall, J., Tikishvili, E. *et al.* (2005) Isoforms of cyclic nucleotide phosphodiesterase PDE3 and their contribution to cAMP hydrolytic activity in subcellular fractions of human myocardium. *J. Biol. Chem.*, **280**, 39168–39174.

Han, S.J., Vaccari, S., Nedachi, T. *et al.* (2006) Protein kinase B/Akt phosphorylation of PDE3A and its role in mammalian oocyte maturation. *EMBO J.*, **25**, 5716–5725.

Hashimoto, N., Kishimoto, T. and Nagahama, Y. (1985) Induction and inhibition of meiotic maturation in follicle-enclosed mouse oocytes by forskolin. *Develop. Growth Differ.*, **27**, 709–716.

Hegele-Hartung, C., Kuhnke, J., Lessl, M. *et al.* (1999) Nuclear and cytoplasmic maturation of mouse oocytes after treatment with synthetic meiosis-activating sterol in vitro. *Biol. Reprod.*, **61**, 1362–1372.

Herrlich, A., Kuhn, B., Grosse, R. *et al.* (1996) Involvement of G_s and G_i proteins in dual coupling of the luteinizing hormone receptor to adenylyl cyclase and phospholipase C. *J. Biol. Chem.*, **271**, 16764–16772.

Hinckley, M., Vaccari, S., Horner, K. *et al.* (2005) The G-protein-coupled receptors GPR3 and GPR12 are involved in cAMP signaling and maintenance of meiotic arrest in rodent oocytes. *Dev. Biol.*, **287**, 249–261.

Horner, K., Livera, G., Hinckley, M. *et al.* (2003) Rodent oocytes express an active adenylyl cyclase required for meiotic arrest. *Dev. Biol.*, **258**, 385–396.

Hsieh, M., Lee, D., Panigone, S. *et al.* (2007) Luteinizing hormone-dependent activation of the epidermal growth factor network is essential for ovulation. *Mol. Cell Biol.*, **27**, 1914–1924.

Hsieh, M., Zamah, A.M. and Conti, M. (2009) Epidermal growth factor-like growth factors in the follicular fluid: role in oocyte development and maturation. *Semin. Reprod. Med.*, **27**, 52–61.

Hubbard, C.J. (1986) Cyclic AMP changes in the component cells of Graafian follicles: possible influences on maturation in the follicle-enclosed oocytes of hamsters. *Dev. Biol.*, **118**, 343–351.

Hunzicker-Dunn, M. (1981) Selective activation of rabbit ovarian protein kinase isozymes in rabbit ovarian follicles and corpora lutea. *J. Biol. Chem.*, **256**, 12185–12193.

Jin, S.L., Richard, F.J., Kuo, W.P. *et al.* (1999) Impaired growth and fertility of cAMP-specific phosphodiesterase PDE4D-deficient mice. *Proc. Natl. Acad. Sci. USA*, **96**, 11998–12003.

Kalinowski, R.R., Berlot, C.H., Jones, T.L.Z. *et al.* (2004) Maintenance of meiotic prophase arrest in vertebrate oocytes by a G_s protein-mediated pathway. *Dev. Biol.*, **267**, 1–13.

Kalous, J., Solc, P., Baran, V. *et al.* (2006) PKB/AKT is involved in resumption of meiosis in mouse oocytes. *Biol. Cell*, **98**, 111–123.

Kawamura, K., Ye, Y., Liang, C.G. *et al.* (2009) Paracrine regulation of the resumption of oocyte meiosis by endothelin-1. *Dev. Biol.*, **327**, 62–70.

Kovo, M., Kandli-Cohen, M., Ben-Haim, M. *et al.* (2006) An active protein kinase A (PKA) is involved in meiotic arrest of rat growing oocytes. *Reproduction*, **132**, 33–43.

Larsen, W.J., Wert, S.E. and Brunner, G.D. (1986) A dramatic loss of cumulus cell gap junctions is correlated with germinal vesicle breakdown in rat oocytes. *Dev. Biol.*, **113**, 517–521.

Larsen, W.J., Wert, S.E. and Brunner, G.D. (1987) Differential modulation of rat follicle cell gap junction populations at ovulation. *Dev. Biol.*, **122**, 61–71.

Ledent, C., Demeestere, I., Blum, D. *et al.* (2005) Premature ovarian aging in mice deficient for Gpr3. *Proc. Natl. Acad. Sci. USA*, **102**, 8922–8926.

Lee, P.S.N., Buchan, A.M.J., Hsueh, A.J.W. *et al.* (2002) Intracellular calcium mobilization in response to the activation of human wild-type and chimeric gonadotropin receptors. *Endocrinology*, **143**, 1732–1740.

Lincoln, A.J., Wickramasinghe, D., Stein, P. *et al.* (2002) Cdc25b phosphatase is required for resumption of meiosis during oocyte maturation. *Nat. Genet.*, **30**, 446–449.

Locke, D., Stein, T., Davies, C. *et al.* (2004) Altered permeability and modulatory character of connexin channels during mammary gland development. *Exp. Cell Res.*, **298**, 643–660.

Magnusson, C. and Hillensjo, T. (1977) Inhibition of maturation and metabolism in rat oocytes by cyclic AMP. *J. Exp. Zool.*, **201**, 139–147.

Maller, J.L. and Krebs, E.G. (1977) Progesterone-stimulated meiotic cell division in *Xenopus* oocytes. Induction by regulatory subunit and inhibition by catalytic subunit of adenosine 3′:5′-monophosphate-dependent protein kinase. *J. Biol. Chem.*, **252**, 1712–1718.

Marin Bivens, C.L., Grondahl, C., Murray, A. *et al.* (2004) Meiosis-activating sterol promotes the metaphase I to metaphase II transition and preimplantation developmental competence of mouse oocytes maturing *in vitro*. *Biol. Reprod.*, **70**, 1458–1464.

Masciarelli, S., Horner, K., Liu, C. *et al.* (2004) Cyclic nucleotide phosphodiesterase 3A-deficient mice as a model of female infertility. *J. Clin. Invest.*, **114**, 196–205.

Masui, Y. (1967) Relative roles of the pituitary, follicle cells, and progesterone in the induction of oocyte maturation in *Rana pipiens*. *J. Exp. Zool.*, **166**, 365–375.

Mehlmann, L.M. (2005) Oocyte-specific expression of *Gpr3* is required for the maintenance of meiotic arrest in mouse oocytes. *Dev. Biol.*, **288**, 397–404.

Mehlmann, L.M., Jones, T.L.Z. and Jaffe, L.A. (2002) Meiotic arrest in the mouse follicle maintained by a G_s protein in the oocyte. *Science*, **297**, 1343–1345.

Mehlmann, L.M., Saeki, Y., Tanaka, S. *et al.* (2004) The G_s-linked receptor GPR3 maintains meiotic arrest in mammalian oocytes. *Science*, **306**, 1947–1950.

Mehlmann, L.M., Kalinowski, R.R., Ross, L.F. *et al.* (2006) Meiotic resumption in response to luteinizing hormone is independent of a G_i family G protein or calcium in the mouse oocyte. *Dev. Biol.*, **299**, 345–355.

Nagahama, Y. and Yamashita, M. (2008) Regulation of oocyte maturation in fish. *Dev. Growth Differ.*, **50**(Suppl 1), S195–S219.

Newhall, K.J., Criniti, A.R., Cheah, C.S. *et al.* (2006) Dynamic anchoring of PKA is essential during oocyte maturation. *Curr. Biol.*, **16**, 321–327.

Nogueira, D., Ron-El, R., Friedler, S. *et al.* (2006) Meiotic arrest in vitro by phosphodiesterase 3-inhibitor enhances maturation capacity of human oocytes and allows subsequent embryonic development. *Biol. Reprod.*, **74**, 177–184.

Norris, R.P., Freudzon, L., Freudzon, M. *et al.* (2007) A G_s-linked receptor maintains meiotic arrest in mouse oocytes, but luteinizing hormone does not cause meiotic resumption by terminating receptor-G_s signaling. *Dev. Biol.*, **310**, 240–249.

Norris, R.P., Freudzon, M., Mehlmann, L.M. *et al.* (2008) Luteinizing hormone causes MAP kinase-dependent phosphorylation and closure of connexin 43 gap junctions in mouse ovarian follicles: one of two paths to meiotic resumption. *Development*, **135**, 3229–3238.

Norris, R.P., Ratzan, W.J., Freudzon, M. *et al.* (2009) Cyclic GMP from the surrounding somatic cells regulates cyclic AMP and meiosis in the mouse oocyte. *Development*, **136**, 1869–1878.

Noubani, A., Farookhi, R. and Gutkowska, J. (2000) B-type natriuretic peptide receptor expression and activity are hormonally regulated in rat ovarian cells. *Endocrinology*, **141**, 551–559.

Oh, J.S., Han, S.J. and Conti, M. (2010) Wee1B, Myt1, and Cdc25 function in distinct compartments of the mouse oocyte to control meiotic resumption. *J. Cell Biol.* (in press).

Panigone, S., Hsieh, M., Fu, M. *et al.* (2008) Luteinizing hormone signaling in preovulatory follicles involves early activation of the epidermal growth factor receptor pathway. *Mol. Endocrinol.*, **22**, 924–936.

Park, J.Y., Su, Y.Q., Ariga, M. *et al.* (2004) EGF-like growth factors as mediators of LH action in the ovulatory follicle. *Science*, **303**, 682–684.

Patwardhan, V.V. and Lanthier, A. (1984) Cyclic GMP phosphodiesterase and guanylate cyclase activities in rabbit ovaries and the effect of in-vivo stimulation with LH. *J. Endocrinol.*, **101**, 305–310.

Pincus, G. and Enzmann, E.V. (1935) The comparative behavior of mammalian eggs in vivo and in vitro. *J. Exp. Med.*, **62**, 665–675.

Piontkewitz, Y. and Dekel, N. (1993) Heptanol, an alkanol that blocks gap junctions, induces oocyte maturation. *Endocrine. J.*, **1**, 365–372.

Pirino, G., Wescott, M.P. and Donovan, P.J. (2009) Protein kinase A regulates resumption of meiosis by phosphorylation of Cdc25B in mammalian oocytes. *Cell Cycle*, **8**, 665–670.

Racowsky, C. and Baldwin, K.V. (1989) *In vitro* and *in vivo* studies reveal that hamster oocyte meiotic arrest is maintained only transiently by follicular fluid, but persistently by membrana/cumulus granulosa cell contact. *Dev. Biol.*, **134**, 297–306.

Rajagopalan-Gupta, R.M., Lamm, M.L.G., Mukherjee, S. *et al.* (1998) Luteinizing hormone/choriogonadotropin receptor-mediated activation of heterotrimeric guanine nucleotide binding proteins in ovarian follicular membranes. *Endocrinology*, **139**, 4547–4555.

Richard, F.J., Tsafriri, A. and Conti, M. (2001) Role of phosphodiesterase type 3A in rat oocyte maturation. *Biol. Reprod.*, **65**, 1444–1451.

Schuh, M. and Ellenberg, J. (2007) Self-organization of MTOCs replaces centrosome function during acentrosomal spindle assembly in live mouse oocytes. *Cell*, **130**, 484–498.

Schultz, R.M., Montgomery, R.R. and Belanoff, J.R. (1983) Regulation of mouse oocyte meiotic maturation: implication of a decrease in oocyte cAMP and protein dephosphorylation in commitment to resume meiosis. *Dev. Biol.*, **97**, 264–273.

Sela-Abramovich, S., Chorev, E., Galiani, D. and Dekel, N. (2005) Mitogen-activated protein kinase mediates luteinizing hormone-induced breakdown of communication and oocyte maturation in rat ovarian follicles. *Endocrinology*, **146**, 1236–1244.

Sela-Abramovich, S., Edry, I., Galiani, D. *et al.* (2006) Disruption of gap junctional communication within the ovarian follicle induces oocyte maturation. *Endocrinology*, **147**, 2280–2286.

Sela-Abramovich, S., Galiani, D., Nevo, N. and Dekel, N. (2008) Inhibition of rat oocyte maturation and ovulation by nitric oxide: mechanism of action. *Biol. Reprod.*, **78**, 1111–1118.

Takeda, N., Kyozuka, K. and Deguchi, R. (2006) Increase in intracellular cAMP is a prerequisite signal for initiation of physiological oocyte meiotic maturation in the hydrozoan *Cytaeis uchidae*. *Dev. Biol.*, **298**, 248–258.

Törnell, J., Billig, H. and Hillensjö, T. (1990a) Resumption of rat oocyte meiosis is paralleled by a decrease in guanosine 3′,5′-cyclic monophosphate (cGMP) and is inhibited by microinjection of cGMP. *Acta. Physiol. Scand.*, **139**, 511–517.

Törnell, J., Carlsson, B. and Billig, H. (1990b) Atrial natriuretic peptide inhibits spontaneous rat oocyte maturation. *Endocrinology*, **126**, 1504–1508.

Törnell, J., Billig, H. and Hillensjö, T. (1991) Regulation of oocyte maturation by changes in ovarian levels of cyclic nucleotides. *Hum. Reprod.*, **6**, 411–422.

Tsafriri, A. (1985) The control of meiotic maturation in mammals, in *Biology of Fertilization* (eds C. B. Metz and A. Monroy), Academic Press, Orlando, pp. 221–252.

Tsafriri, A. and Bar-Ami, S. (1978) Role of divalent cations in the resumption of meiosis in rat oocytes. *J. Exp. Zool.*, **205**, 293–300.

Tsafriri, A. and Motola, S. (2007) Are steroids dispensable for meiotic resumption in mammals? *Trends Endocrinol. Metab.*, **18**, 321–327.

Tsafriri, A., Lindner, H.R., Zor, U. and Lamprecht, S.A. (1972) *In-vitro* induction of meiotic division in follicle-enclosed rat oocytes by LH, cyclic AMP and prostaglandin E_2. *J. Reprod. Fertil.*, **31**, 39–50.

Tsafriri, A., Chun, S.Y., Zhang, R. *et al.* (1996) Oocyte maturation involves compartmentalization and opposing changes of cAMP levels in follicular somatic and germ cells: studies using selective phosphodiesterase inhibitors. *Dev. Biol.*, **178**, 393–402.

Tsafriri, A., Cao, X., Ashkenazi, H. *et al.* (2005) Resumption of oocyte meiosis in mammals: on models, meiosis activating sterols, steroids and EGF-like factors. *Mol. Cell Endocrinol.*, **234**, 37–45.

Vaccari, S., Weeks, J.L., Hsieh, M. *et al.* (2009) Cyclic GMP signaling is involved in the luteinizing hormone-dependent meiotic maturation of mouse oocytes. *Biol. Reprod.*, **81**, 595–604.

Viste, K., Kopperud, R.K., Christensen, A.E. and Doskeland, S.O. (2005) Substrate enhances the sensitivity of type I protein kinase A to cAMP. *J. Biol. Chem.*, **280**, 13279–13284.

Vivarelli, E., Conti, M., De Felici, M. and Siracusa, G. (1983) Meiotic resumption and intracellular cAMP levels in mouse oocytes treated with compounds which act on cAMP metabolism. *Cell Differ.*, **12**, 271–276.

Wassarman, P.M., Josefowicz, W.J. and Letourneau, G.E. (1976) Meiotic maturation of mouse oocytes in vitro: inhibition of maturation at specific stages of nuclear progression. *J. Cell Sci.*, **22**, 531–545.

Zhang, Y., Zhang, Z., Xu, X.-Y. *et al.* (2008) Protein kinase A modulates Cdc25B activity during meiotic resumption of mouse oocytes. *Dev. Dyn.*, **237**, 3777–3786.

8

Oocyte-specific translational control mechanisms

Isabel Novoa, Carolina Eliscovich, Eulàlia Belloc and Raúl Méndez

Gene Regulation Program, Centre for Genomic Regulation (CRG), C/Dr Aiguader, 88, 08003, Barcelona, Spain

Vertebrate immature oocytes are arrested at prophase of meiosis I (PI). During this growth period (oogenesis), the oocytes synthesize and store large quantities of dormant mRNAs (Lamarca, Smith and Strobel, 1973; Rodman and Bachvarova, 1976), which will later drive the oocyte's re-entry into meiosis (Mendez and Richter, 2001; Schmitt and Nebreda, 2002; De Moor, Meijer and Lissenden, 2005). The resumption of meiosis marks the onset of oocyte maturation and in *Xenopus* is stimulated by progesterone. Meiotic maturation is comprised of two consecutive M phases, metaphase I (MI) and metaphase II (MII), without an intervening S phase. Then, at MII, the oocytes become arrested for a second time and await fertilization before concluding the second meiotic division (Sagata, 1996). Remarkably, oocyte maturation occurs in the absence of transcription (Newport and Kirschner, 1982; Clegg and Piko, 1982), and is fully dependent on cascades of kinases/phosphatases and on the sequential translational activation of the maternal mRNAs accumulated during the PI arrest (reviewed in Schmitt and Nebreda (2002); Belloc, Pique and Mendez (2008)). Given that during this transcriptionally silent period the oocytes/embryos have to go through the meiotic and embryonic mitotic divisions and establish the main body patterns, it is not surprising that the oocytes store mRNAs of a large portion of the genome (45% in mice (Wang *et al.*, 2004), and 55% in *Drosophila* (Tadros, Westwood and Lipshitz, 2007)), and display massive translational reprogramming (Potireddy *et al.*, 2006). As oogenesis and oocyte maturation proceed, the basic embryonic axis formation and the establishment of germ cells are also defined by 'symmetry-breaking' events based on mRNA localization within the oocyte. For protein synthesis to be spatially restricted, translation of localized

Oogenesis: The Universal Process Marie-Hélène Verlhac and Anne Villeneuve

mRNAs must be repressed during their transport, to be later activated once they have reached their final destination and at the appropriate time.

The precise localization and translation regulation for each mRNA are dictated by the combination of cis-acting elements and trans-acting factors present in their untranslated regions (UTRs), usually at the 3′UTRs (Kuersten and Goodwin, 2003). Indeed, the 3′UTRs seem to be the primary source of gene expression regulation in the germline (Merritt *et al.*, 2008). The coordinated temporal and spatial regulation of large numbers of RNAs during oocyte maturation and early embryogenesis is ensured by the combinatorial regulation of groups of mRNAs implicated in a similar function, by translational control cascades and by positive and negative translational feedback loops.

For maternal mRNAs, it is a common feature that a particular mRNA will be regulated by multiple redundant mechanisms targeting different translation steps and mediated by different proteins (Vardy and Orr-Weaver, 2007). Conversely, many of the regulatory RNA-binding proteins (RNA-BPs) have multiple functions in assembling repression, localization and translational activation complexes. By this combination of mechanisms not only is the translation of key mRNAs ensured by failsafe mechanisms, but also very precise timing/strength/localization control can be accomplished. The other consequence of this combinatorial mechanism to control translation is that many different mRNAs will share at least one regulatory element assuring the coordinated regulation in multidimensional networks. Indeed, recent genome-wide analyses for mRNAs associated with specific RNA-BPs have illustrated this principle by showing that each RNA-BP is associated with hundreds of mRNAs encoding functionally related proteins, and that most mRNAs are potentially regulated by more than one RNA-BP, establishing networks to coordinate translational regulation (reviewed in Keene (2007)).

Another recurrent scenario during oocyte maturation and early development is the translational regulatory cascades or sequential waves of translational activation/repression. This hierarchical organization is used to control discrete temporal (i.e. meiotic phase transitions) or spatial (i.e. local translation during axis determination) complex protein expression patterns. The general principle consists of a cascade where the translation of the mRNAs encoding translation regulators is itself spatially or temporally controlled. These translational cascades are, in turn, reinforced by positive and negative translational feedback loops, ensuring coordinated and unidirectional regulation of gene expression. The feedback loops allow for full activation or repression of groups of genes, ensuring discrete and irreversible switch-like phase transitions and defined spatial expression patterns (Ferrell, 2002; Xiong and Ferrell, 2003).

In *Xenopus* oocytes, a brief exposure to the hormone progesterone triggers an irreversible switch, where the oocyte irrevocably commits to maturation by establishing a bistable signalling system that converts a transient stimulus into a reliable, self-sustaining pattern of protein activities (Ferrell, 2002). At the core of the transition from PI arrest to MI prevail the numerous positive feedbacks established in the Mos/MAPK/Cdc2/cyclin B network. While many of these feedbacks involve rapid phosphorylation/dephosphorylation events, translational regulation of maternal mRNAs is required to show an all-or-none, bistable response to progesterone.

The molecular mechanisms sustaining these networks of temporal and spatial translational control in the oocytes have been mostly studied in the *Drosophila* and

Xenopus systems. Although the picture is as yet far from complete, some of the best-characterized examples are described below.

8.1 Combinatorial mechanisms of maternal mRNA translational control

Translation of an mRNA is divided into three steps: initiation, elongation and termination, the first one being the most common target for protein synthesis regulation (reviewed in Hershey and Merrick (2000)). Translation initiation requires the association of the eukaryotic translation initiation factor-4F (eIF4F) complex with the cap structure at the mRNA 5′ end. eIF4F consists of the cap-binding factor eIF4E, the RNA helicase eIF4A and the scaffolding protein eIF4G. The recruitment of eIF4F is stabilized by the binding of eIF4G to the 3′ poly(A) tail-binding protein (PABP) and the resulting circularization of mRNA molecules (Wells *et al.*, 1998). Then, the 43S preinitiation complex (which includes the 40S ribosomal subunit) is recruited through the interaction between eIF3 and eIF4G. After scanning along the 5′UTR for an appropriate AUG start codon, the preinitiation complex is dissolved and the 60S ribosomal subunit joins the 40S subunit to form a translationally competent 80S ribosome, and translation of the ORF (open reading frame) starts to produce the encoded polypeptide. Both the cap and the poly(A) tail of mRNAs act synergically to facilitate translation initiation, through the stabilization of the closed loop formed by the interaction of factors bound to both ends of the transcript.

8.1.1 Translational repression

Blocking the cap and/or shortening the poly(A) tail are the two most common mechanisms for silencing/repressing maternal mRNAs. In many cases, this mRNA-specific translational repression is accompanied by sequestration of the silenced mRNAs in large ribonucleoprotein complexes (mRNPs) inaccessible to the ribosomes. The assembly of the repressed mRNPs starts in the nucleus, and the nuclear life of many mRNAs dictates their subsequent cytoplasmic fate. Indeed, several components of the exon junction complex and the nuclear export machinery are shuttling proteins that bind hnRNA (heterogenous nuclear RNA) in the nucleus and participate in the repression and localization of the mature mRNAs in the cytoplasm (Palacios *et al.*, 2004; Hachet and Ephrussi, 2001; Kalifa, Armenti and Gavis, 2009; Huynh *et al.*, 2004). Other factors associated to maternal mRNAs include general RNA-BPs present in cytoplasmic granules implicated in mRNA silencing and turnover (i.e. P bodies or stress granules) (Noble *et al.*, 2008; Boag *et al.*, 2008). This is the case for the Y-box proteins, such as the *Xenopus* FRGY2 (Bouvet and Wolffe, 1994). FRGY2 is added co-transcriptionally to mRNAs, and together with RAP55, a member of the Scd6 or Lsm14 family, and Xp54, localizes to P-body-like granules and represses translation (Tanaka *et al.*, 2006). Other members of this Y-box protein family are Yps (*Drosophila*), and MSY2 and MSY4 (mouse). All members of this family contain a cold shock nucleic-acid binding domain and are highly expressed in germ cells, where they are required for translational repression/localization of maternal mRNAs (Mansfield, Wilhelm and Hazelrigg, 2002). The DDX6-like RNA helicases (Xp54 in *Xenopus*, Me31B in *Drosophila*, RCK/p54 in mammals, and CGH-1 in *Caenorhabditis elegans*) are also integral components of silenced mRNPs and are found in the mitochondrial cloud, sponge bodies, polar granules, P bodies and stress granules (Weston and Sommerville, 2006). These proteins are expressed in germ cells and early stages of embryogenesis, and their depletion results in widespread derepression of protein synthesis in oocytes. In *Drosophila*, Me31B, together with Yps, plays an essential role in translational silencing of oocyte-localizing mRNAs during their transport to the oocyte (Mansfield, Wilhelm and Hazelrigg, 2002; Nakamura *et al.*, 2001).

But, even if these core components of repressed/transported mRNPs are essential, the specificity of the translational regulation is conferred by RNA-BPs that recognize specific cis-acting elements. Only then, the repressed mRNPs may be assembled in larger heterogeneous particles containing differentially regulated mRNAs and common core components (Gao *et al.*, 2008). Below we describe in detail some of the best-characterized mRNA-binding complexes that inhibit translation of specific mRNAs during oogenesis and oocyte maturation.

8.1.2 Cytoplasmic polyadenylation element binding protein (CPEB1)-mediated repressed complex

CPEB1, the founding member of the CPEB family of proteins, is a zinc finger and RRM-type RNA-binding protein (Hake, Mendez and Richter, 1998) that recognizes the cytoplasmic polyadenylation element (CPE), with a consensus sequence of UUUUAU or UUUUAAU, present in the 3′UTR of the targeted mRNAs (Richter, 2007). CPEB1 has multiple functions in the regulation of translation. First, in prophase I (PI)-arrested oocytes, CPEB1 mediates translational silencing, by shortening the poly(A) tail, and/or active repression, by blocking the access of the 40S ribosomal subunit to the cap. Second, in response to progesterone-induced meiotic resumption, CPEB1 drives cytoplasmic polyadenylation and translational stimulation. Active repression requires a particular arrangement of CPEs in the target mRNA with at least two CPEs spaced by less than 50 nucleotides, which probably reflects the formation of a CPEB1 dimer, whereas, for silencing, a single CPE may be sufficient (Pique *et al.*, 2008). Translational silencing is the consequence of the cytoplasmic shortening of the long poly(A) tail, acquired during the nuclear cleavage and polyadenylation of the pre-mRNA, from 200–500 to 20–40 nucleotides. This deadenylation is the result of the direct recruitment of the deadenylase PARN by CPEB1 (Kim and Richter, 2006; Figure 8.1a). Active repression (masking) is accomplished by the recruitment of Maskin through direct interaction with CPEB1. In turn, Maskin binds to the cap-bound eIF4E, precluding the recruitment of the eIF4G and therefore of the 43S ribosomal complex (Stebbins-Boaz *et al.*, 1999; Figure 8.1b).

But the closed loop driven by CPEB1–Maskin–eIF4E is not the only repression complex assembled by CPEB1. In early oogenesis (where Maskin and PARN are not expressed), CPEB1 fractionates with very large mRNP complexes containing CPEB1 associated with eIF4E-T (an eIF4-E binding protein involved in nucleocytoplasmic

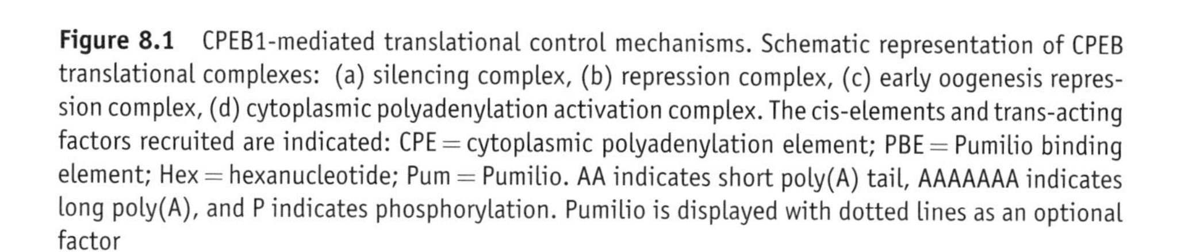

Figure 8.1 CPEB1-mediated translational control mechanisms. Schematic representation of CPEB translational complexes: (a) silencing complex, (b) repression complex, (c) early oogenesis repression complex, (d) cytoplasmic polyadenylation activation complex. The cis-elements and trans-acting factors recruited are indicated: CPE = cytoplasmic polyadenylation element; PBE = Pumilio binding element; Hex = hexanucleotide; Pum = Pumilio. AA indicates short poly(A) tail, AAAAAAA indicates long poly(A), and P indicates phosphorylation. Pumilio is displayed with dotted lines as an optional factor

transport and present in P bodies (Andrei *et al.*, 2005) and an ovary-specific eIF4E1b that binds the cap weakly (Minshall *et al.*, 2007). The identification of this complex, that does not contain Maskin, suggests an additional model for repression where the recruitment of eIF4E-T by CPEB1, and its association with eIF4E1b, would compete for eIF4G association, thus blocking translation initiation. This large mRNP also includes the RNA helicase RCK/Xp54, and the P-body components P100 (Pat1) and Rap55 (Figure 8.1c). Interestingly, Xp54 has been described itself as a CPEB1 and eIF4E interacting protein, providing an additional mechanism to repress translation (Minshall and Standart, 2004).

Another trans-acting factor recruited by repressed CPE-containing 3′UTRs is *Xenopus* Pumilio (Pum), an RNA-binding protein (see below) that interacts with CPEB1. Although Pum has a very weak effect on the translational repression on CPE-containing reporters (Pique *et al.*, 2008; Nakahata *et al.*, 2003), it may play a critical role in the silencing by deadenylation. Accordingly, Pum is present in CPEB1 complexes containing Maskin, but not in the ones containing the cytoplasmic poly(A) polymerase GLD2 (Rouhana *et al.*, 2005; Figure 8.1b).

8.1.3 Bruno-mediated mRNA repressed complex

Bruno is an RNA-BP of the RRM type (Webster *et al.*, 1997) that binds to sequences named Bruno response elements (BREs), found in the 3′UTR of oskar mRNA, and represses its translation (Kim-HA, Kerr and Macdonald, 1995). Bruno mediates translational repression of oskar mRNA by two different mechanisms. First, Bruno recruits an eIF4E-binding protein named Cup, an eIF4E-T homologue that prevents the eIF4E–eIF4G interaction and therefore inhibits the 43S ribosomal complex recruitment (Nakamura, Sato and Hanyu-Nakamura, 2004; Wilhelm *et al.*, 2003; Figure 8.2a). In addition, Bruno assembles, independently of Cup, a large mRNP complex formed by mRNA oligomers that renders oskar mRNA inaccessible to ribosomes and therefore translationally inert (Chekulaeva, Hentze and Ephrussi, 2006). Bruno also regulates the translation of cyclin A mRNA in *Drosophila* oocytes, although the mechanism is not known (Sugimura and Lilly, 2006).

Following the trend of combinatorial composition of mRNPs, Cup can also be recruited to other mRNAs, during oogenesis and early embryogenesis, by different RNA-BPs (Figure 8.2a). Thus, Cup is recruited by Bruno and Squid to repress the translation of gurken mRNA in the oocyte (Filardo and Ephrussi, 2003; Clouse,

(a) **Repression by blocking accesibility to the cap**

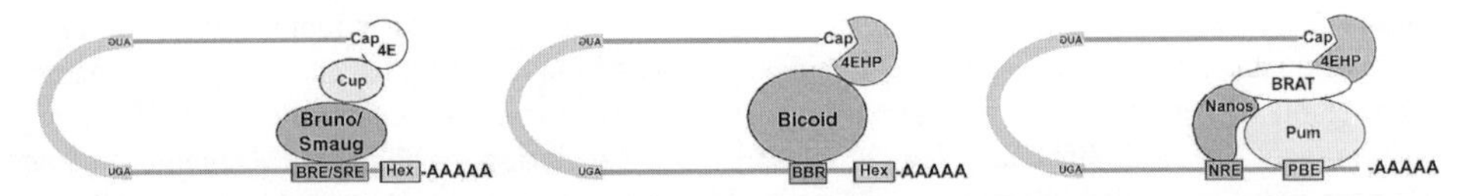

(b) **Silencing by deadenylation**

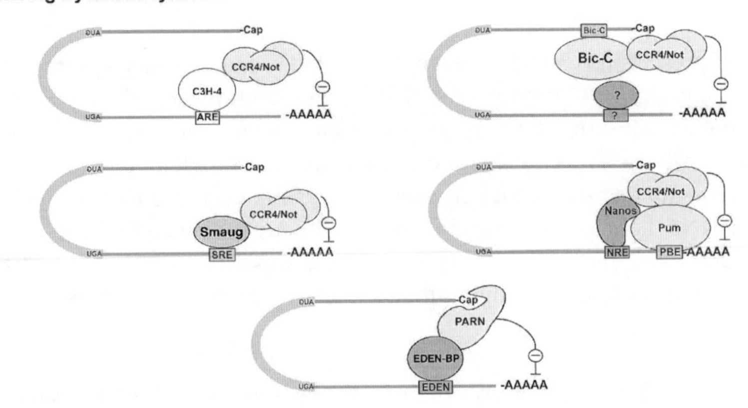

(c) **Poly(A) tail independent translational activation**

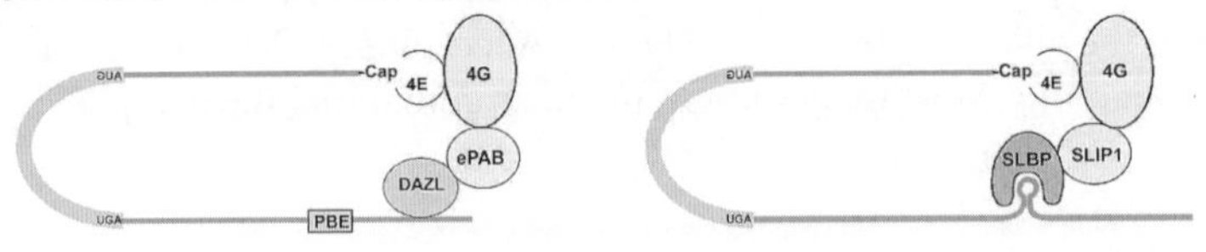

Figure 8.2 Maternal mRNAs translational control mechamisms. (a) Repression complexes that block the accessibility to the cap: Cup and 4EHP complexes. (b) Translational silencing of mRNAs by deadenylation mechanisms: C3H-4, Bic-C, Smaug and Nanos–Pumilio recruitment of CCR4/Not deadenylation complex, and EDEN-BP recruitment of PARN deadenylase. (c) Poly(A) tail independent translational activation mechanisms mediated by DAZL and SLBP. Cis-elements indicated: BRE = Bruno response element; SRE = Smaug recognition element; Hex = hexanucleotide; ARE = AU-rich element; PBE = Pumilio binding element; NRE = Nanos response element; EDEN = Embryonic deadenylation element

Ferguson and Schupbach, 2008) and later on, during early embryogenesis, Cup interacts with Smaug to repress the translation of nanos mRNA (Nelson, Leidal and Smibert, 2004). Also in embryogenesis, 4EHP, another weak cap-binding protein, is directly recruited, through the dual DNA/RNA-binding protein Bicoid, to the BBR (bcd-binding region) in caudal 3′UTR, blocking the translation of this mRNA at the anterior pole of the embryo by preventing eIF4E from binding to the cap (Cho *et al.*, 2005; Figure 8.2a). Similarly, 4EHP is also recruited to hunchback mRNA by the protein BRAT which interacts with the RNA-BP Nanos and Pumilio at the 3′UTR (Cho *et al.*, 2006; Figure 8.2a).

8.1.4 Translational silencing by deadenylation

In addition to the initial deadenylation of maternal mRNAs during oogenesis in the PI-arrested oocytes, shortening of the poly(A) tail of both housekeeping and maternal mRNAs is used during the resumption of meiosis to regulate gene expression in the MI to MII transition and to finish meiosis upon fertilization. Indeed, a recent study identified more than 500 mRNAs (out of 3000 analyzed) undergoing changes in their poly(A) tail length during oocyte maturation (Graindorge *et al.*, 2006). In metazoans, deadenylation is mediated by two major complexes, the Pan2/Pan3 and the CCR4 multisubunit complex. These two deadenylation complexes can be recruited to different mRNAs and at different times, by interacting with a large variety of RNA-BPs that provide specificity to the regulation of translation by deadenylation. Thus, a large number of multifunctional deadenylase complexes can be generated to provide mRNA-specific temporal control of translational silencing (Goldstrohm and Wickens, 2008).

In *Xenopus*, PARN mediates the deadenylation of housekeeping mRNAs, lacking CPE elements, during oocyte maturation (Copeland and Wormington, 2001). In addition, PARN is recruited by CPEB to CPE-silenced mRNAs during oogenesis (Kim and Richter, 2006). In mature (metaphase I) oocytes, the CCR4 complex is recruited by an RNA-BP named C3H-4 (see below) to mRNAs containing AU-rich elements (AREs) (Belloc and Mendez, 2008; Figure 8.2b). In *Drosophila*, CCR4 has been shown to interact with different RNA-BPs such as Pum, Nos (Kadyrova *et al.*, 2007), Bic-C (Chicoine *et al.*, 2007) and Smaug (Zaessinger, Busseau and Simonelig, 2006) (Figure 8.2b). Fertilization triggers the completion of meiosis and the concomitant deadenylation by the RNA-binding protein EDEN-BP (embryonic deadenylation element binding protein) also known as CUG-binding protein (CUG-BP) (reviewed in Osborne *et al.* (2005)) (Figure 8.2b). Although in *Xenopus* oocytes the deadenylase recruited by EDEN-BP is still unknown, in mammalian cells CUG-BP is bound by PARN (Moraes, Wilusz and Wilusz, 2006).

8.1.5 Pumilio-mediated repressed complex

The PUF (Pumilio–FBF) family of RNA-BPs promotes mRNA translational silencing by recruiting CCR4/Not deadenylation complexes (Kadyrova *et al.*, 2007), and in *C. elegans*, members of this family of proteins are at the core of the mitosis to meiosis transition and in the sperm/oocyte decision (reviewed in Kimble and Crittenden (2007)). Vertebrates have only two members of the family, Pum1 and Pum2. Pum1 binds to the classical NRE (Nanos response element, a bipartite sequence consisting of a 5′box A (GUUGU) and a 3′box B (UGUA)), while Pum2 binds the Pumilio binding element (PBE; UGUANAUA) (Zamore *et al.*, 1999). In addition to its function in deadenylation, Pum has been implicated in translational repression by poly(A) tail-independent mechanisms. In PI-arrested oocytes, Pum2 binds to two conserved PBEs within *Xenopus* ringo 3′UTR mRNA, assembling a translational repression complex together with the embryonic poly(A)-binding protein (ePAB) and Deleted for Azoospermia-like protein (DAZL) (Padmanabhan and Richter, 2006). In *Xenopus*, several CPE-containing mRNAs, such as the one encoding cyclin B1, also contain PBEs, and Pum has been described to interact with both the *Xenopus* Nanos homologue and also CPEB1 to repress translation in PI-arrested oocytes, by an undefined mechanism (Nakahata *et al.*, 2003; Nakahata *et al.*, 2001; Rouhana and Wickens, 2007).

8.1.6 Small noncoding RNA-mediated repression

Small noncoding mRNAs include three major families: small interfering RNAs (siRNAs), microRNAs (miRNAs) and RNAs associated with the Piwi-class Argonaute proteins (piRNAs). The siRNAs and miRNAs are 19–25 nucleotide-long RNAs generated from double-stranded long dsRNAs or hairpin precursors by the Dicer endonucleases, and function with Argonaute-family proteins to target transcript destruction or to silence translation, respectively. piRNAs consist of 24–30 nucleotide-long RNAs, produced by a Dicer-independent mechanism, which associate with the Piwi-class Argonaute proteins. Mouse ovaries express 122 microRNAs (miRNAs), and 79 piwi-interacting RNAs (piRNAs) (Ro *et al.*, 2007; Tang *et al.*, 2007; Lykke-Andersen *et al.*, 2008).

Although these RNAs are key regulators of gene expression during development, it is still unclear as to whether they function as translational regulators during oogenesis and oocyte maturation. The function of miRNAs and siRNAs in early development can be inferred from animals that lack Dicer1 or the drosha cofactor DGCR8, where defects in gastrulation and embryonic axis formation are found (reviewed in Stern (2006)). In maternal-zygotic dicer mutant zebrafish, which lack mature miRNAs, oogenesis proceeds normally but morphogenesis is affected. These results lead the authors to suggest that miRNAs facilitate the deadenylation and clearance of maternal mRNAs during early embryogenesis (Giraldez *et al.*, 2006) and mediate the translational control of nanos in primordial germ cells (PGCs) (Mishima *et al.*, 2006). In addition, *Drosophila* mutants in armitage, aubergine, maelstrom and spindle-E, which are required for the siRNA/miRNA pathways, display defects in oocyte polarization and oocyte oskar translational repression (Tomari *et al.*, 2004; Cook *et al.*, 2004). It remains to be demonstrated whether there is a direct effect on translation of maternal mRNAs or just an indirect effect mediated by the ATR/Chk2 DNA damage signal transduction pathway (Klattenhoff *et al.*, 2007). In *Drosophila*, only 4% of oocyte proteins increase in Dicer mutants (Nakahara *et al.*, 2005), suggesting that only a minority of mRNAs are regulated by miRNAs in oocytes. In mice, a subset of pseudogenes generates endogenous small interfering RNAs (endo-siRNAs) in oocytes, regulating both protein-coding transcripts and retrotransposons (Tam *et al.*, 2008; Watanabe *et al.*, 2008). Accordingly, tissue-restricted Dicer loss in mice oocytes results in meiosis I arrest with multiple disorganized spindles and severe chromosome congression defects (Murchison *et al.*, 2007).

piRNAs have been implicated in germline development, by maintaining germline DNA integrity. However, whether piRNAs control RNA stability or translation is as yet unclear (reviewed in Klattenhoff and Theurkauf (2008). In *Drosophila*, mutations in piRNA-pathway genes disrupt both stem cell maintenance and oocyte production, and the localization of morphogenetic RNAs in the oocyte during axis specification (Klattenhoff and Theurkauf, 2008). Although the molecular functions of Piwi and the piRNA pathway in germline development have not been defined, the primary function of this pathway, both in *Drosophila* and mice, seems to be maintaining germline DNA integrity (Klattenhoff and Theurkauf, 2008). However, some observations may point to a translational function for piRNAs (Cook *et al.*, 2004; Pane, Wehr and Schupbach, 2007). It is therefore possible that piRNA–Piwi-class Argonaute complexes also trigger the translational silencing of imperfectly matched targets (Klattenhoff and Theurkauf, 2008). Consistent with this speculation, a subset of piRNAs associates with polysomes in the mouse (Grivna, Pyhtila and Lin, 2006; Unhavaithaya *et al.*, 2009).

8.1.7 Translational activation

Upon meiotic resumption and the early stages of the embryonic development, all repressed maternal mRNAs have to be sequentially activated at the right time and in the right place. This means that the cap must be made again available to assemble the eIF4F complex and, if the poly(A) tail was shortened, the mRNA has to be cytoplasmically polyadenylated or at least the PABP recruitment enhanced. Thus, the closed loop eIF4E–eIF4G–PABP can be formed again to recruit the 43S ribosomal complex.

Rendering the cap available for the eIF4F means decreasing the affinity of the translational repressors for their target mRNAs, either by phosphorylation of the repressing factors or by competitive binding of factors that squelch the repressor and stabilize the eIF4F. In oocytes, probably the best-characterized example of this regulation is the dissociation of the repressor closed loop formed by CPEB1–Maskin–eIF4E. Upon hormone stimulation, masking is phosphorylated, reducing its affinity for the eIF4E (Barnard, Cao and Richter, 2005; Pascreau *et al.*, 2005). At the same time, the elongation of the poly(A) tail (see below) results in enhanced binding of the PABP, which helps in the recruitment of eIF4G and outcompeting the Maskin–eIF4E interaction (Cao, Kim and Richter, 2006). In addition, concomitantly with the polyadenylation, CPEB1 mediates the 2′-*O*-methylation of the cap (m^7GpppNN) to form cap1 (m^7GpppNmN) and cap2 (m^7GpppNmNm) to activate translation of mos mRNA (Kuge *et al.*, 1998; Kuge and Richter, 1995). However, not every CPEB-regulated mRNA undergoes this modification of the cap (Gillian-Daniel *et al.*, 1998).

At the opposite end of the mRNA, the translational silencing due to the short poly(A) tail of the maternal mRNAs can be overcome by either cytoplasmic polyadenylation or by the poly(A)-independent recruitment of PABP. The latter is illustrated by the DAZL family members, which do not modify the poly(A) tail but are able to directly recruit PABP (Figure 8.2c). DAZL family members encode for proteins that contain an RNA recognition motif (RRM) and a varying number of copies of a DAZ motif, which is believed to mediate protein–protein interactions (Tsui *et al.*, 2000). DAZL is required for male and female gametogenesis in *Drosophila*, *C. elegans*, *Xenopus* and mouse; a number of potential mRNA targets have been described: twine (cdc25C), Tpx-1, GRSF-1, TRF2 and SDAD1, but the binding sites within these mRNAs are not well defined. Its potential role as a translational regulator is derived from studies showing association of these proteins with polysomes. Recently it has been found that tethered DAZL stimulates translation of a reporter mRNA by recruiting PABP to the mRNA, increasing the formation of 80S (Collier *et al.*, 2005; Figure 8.2c). DAZL can also be recruited to mRNAs by Pum2 (Moore *et al.*, 2003). Pum2 and DAZL have been shown to bind the 3′UTR of ringo mRNA and inhibit its translation. Pum2 dissociates from this repressed complex upon progesterone treatment, and translation initiation of ringo mRNA is derepressed by the action of DAZL and the embryonic poly(A)-binding protein (ePAB), the predominant cytoplasmic PABP in *Xenopus* oocytes and early embryos (Padmanabhan and Richter, 2006).

An even more extreme case of poly(A) tail-independent reactivation of translation is the case of the histone mRNAs, where the need for PABP is eliminated altogether by the recruitment of the stem-loop binding protein (SLBP), which binds to a stem-loop structure present at the end of the histone 3′UTR. These mRNAs do not acquire a long poly(A) tail in the nucleus, but SLBP can overcome the need for PABP both during the export to the cytoplasm and for translational initiation activation (reviewed in Marzluff, 2007). *Xenopus* oocytes express two SLBP species (xSLBP1 and xSLBP2), and translational control is affected by a change in the type of SLBP bound to the histone mRNAs (Sanchez and Marzluff, 2002), xSLBP1 being the translationally active protein. SLBP stimulates translation by recruiting the SLBP-interacting protein, SLIP1, which in turn interacts with eIF4GI and eIF4GII, supporting translational activation (Cakmakci *et al.*, 2008; Figure 8.2c).

However, the most common mechanism to activate translation of stored maternal mRNAs is by cytoplasmic elongation of their poly(A) tails (Figure 8.1d). As the result of progesterone stimulation and meiotic resumption, the repressing CPEB1–mRNP is remodelled to drive cytoplasmic polyadenylation. This polyadenylation requires two elements in the 3′UTRs of responding mRNAs: the polyadenylation signal hexanucleotide AAUAAA (Hex) (Sheets *et al.*, 1994), which is bound by the cleavage and polyadenylation specificity factor (CPSF) (Dickson *et al.*, 1999); and the CPE, which recruits CPEB (Hake and Richter, 1994). The triggering of polyadenylation is the phosphorylation of CPEB1 at serine 174 by Aurora-A (Eg2) kinase (Mendez *et al.*, 2000a). CPEB1–S174 phosphorylation increases the affinity of CPEB1 for CPSF (Mendez *et al.*, 2000b), which in turn recognizes the Hex, and both together recruit the cytoplasmic poly(A) polymerase GLD2 to elongate the poly(A) tail of the CPE-containing mRNAs. This complex is stabilized by Symplekin, a core component of the nuclear and cytoplasmic polyadenylation complexes that contacts directly with CPEB1 and CPSF (Barnard *et al.*, 2004). In addition, CPEB1–S174 phosphorylation decreases the affinity of CPEB1 for the deadenylase PARN. PARN is preloaded in a

CPEB1–CPSF–GLD2–PARN complex, where the activities of GLD2 and PARN neutralize each other, keeping the poly(A) tail short. Thus, upon the expulsion of PARN from the complex, GLD2 would be free to elongate the poly(A) tail (Kim and Richter, 2006). It should be noted that this silencing complex is different from the repression complex formed with Maskin and Pum, which does not contain GLD2 (Rouhana *et al.*, 2005). This differential composition is in good agreement with the fact that the deadenylase activity of PARN requires the interaction with the cap (Gao *et al.*, 2000). ePAB is also present in the unstimulated CPEB1 complex. Upon progesterone stimulation, ePAB is released from CPEB1 and associates with the elongating poly(A) tail (Kim and Richter, 2007).

Other components of the CPEB1–mRNP complex, with less-defined functions include CstF77, xGEF and APLP (Rouget, Papin and Mandart, 2006; Reverte *et al.*, 2003; Cao *et al.*, 2005). Pum also interacts with CPEB1, enhancing its binding to weak CPEs. This stabilization does not affect the cytoplasmic polyadenylation, but is required for the translational activation (Pique *et al.*, 2008). In *Drosophila,* Orb (CPEB) interacts, in a sequential manner, with both the canonical PAP and GLD2 (Benoit *et al.*, 2008).

Besides the CPEs, two other elements have been described to mediate cytoplasmic polyadenylation of mRNAs during oocyte maturation: the polyadenylation response element (PRE), and the translational control sequence (TCS). The PRE was initially identified within mos 3′UTR as an element that mediates early polyadenylation of mos mRNA in PI (Charlesworth *et al.*, 2002) and later found in several CPE-containing mRNAs, such as the ones encoding Histone-like B4, D7, G10 FGF receptor 1 and Eg2 (Charlesworth, Cox and Macnicol, 2004). The PRE recruits Musashi, an RNA-BP previously identified as a translational repressor (Charlesworth *et al.*, 2006). Although the effect of the PRE seems to require the Hex, it is still unknown whether Musashi interacts with CPSF or any other component of the cytoplasmic polyadenylation machinery. The TCS is present in the 3′UTR of Wee1 and Pcm-1 mRNAs and confers translational repression in immature oocytes and early cytoplasmic polyadenylation in progesterone-stimulated oocytes (Wang *et al.*, 2008).

8.2 Temporal control of maternal mRNA translation: regulatory cascades and feedback loops

Meiotic progression from the PI arrest until the MII arrest is controlled by three key activities (Figure 8.3a). First, the maturation-promoting factor (MPF), a heterodimer of Cdc2 kinase and cyclin B (Gautier *et al.*, 1990; Gautier *et al.*, 1988) catalyzes entry into M phase of meiosis I and II. This heterodimer is initially formed as an inactive pre-MPF, with cyclins B1 and B5 (Hochegger *et al.*, 2001), and is activated as the result of new synthesis of RINGO and Mos (Schmitt and Nebreda, 2002). Then the anaphase-promoting complex/cyclosome (APC/C) is directly activated by MPF, and induces the ubiquitination and destruction of cyclins B (Peters, 2006). Activation of APC/C during interkinesis is combined with the increased synthesis of cyclins B1 and B4, resulting in only a partial inactivation of MPF at anaphase I, thus preventing entry into S phase (Iwabuchi *et al.*, 2000). Finally, the cytostatic factor (CSF) inhibits the

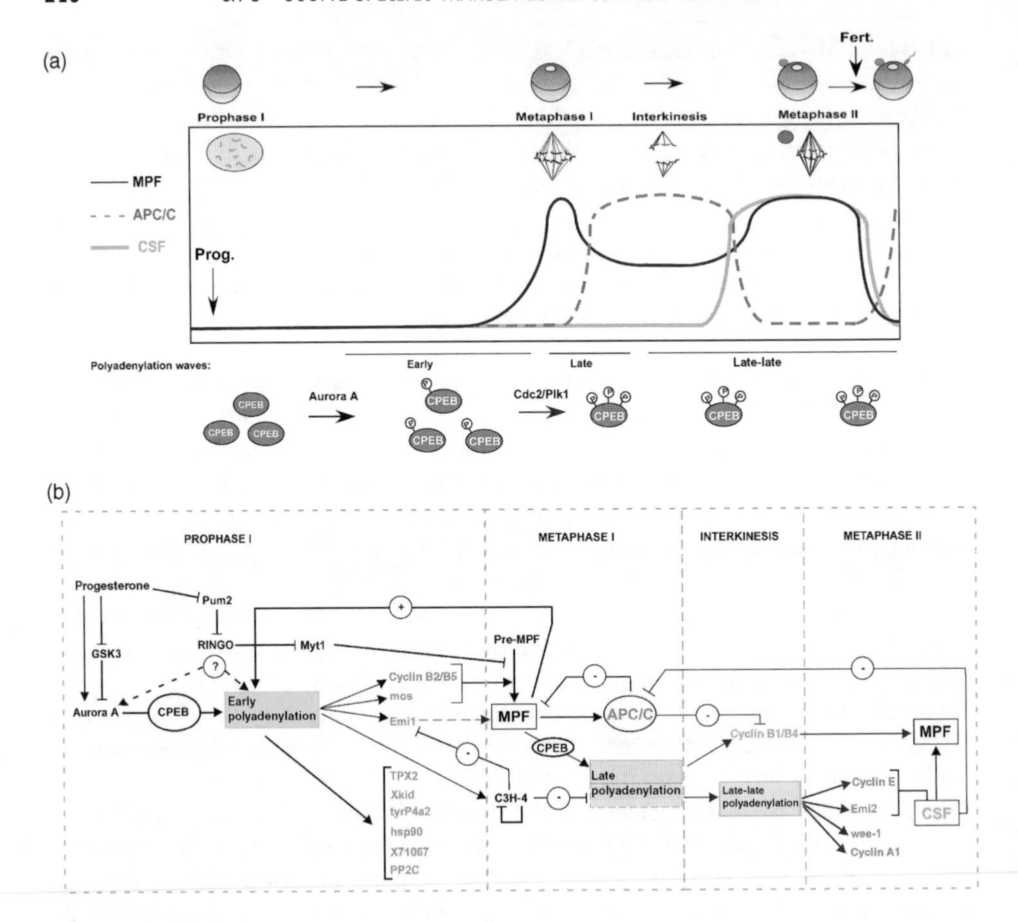

Figure 8.3 Schematic representation of meiotic progression from PI arrest to fertilization. (a) maturation-promoting factor (MPF), anaphase-promoting complex/cyclosome (APC/C) and cytostatic factor (CSF) activities are indicated. Oocyte morphology, chromosome dynamics, mitotic spindles and polar body are shown. Cytoplasmic polyadenylation element binding protein (CPEB protein) levels and phosphorylation regulation and the three waves of cytoplasmic polyadenylation (early, late and late-late) are also depicted. P indicates phosphorylation; Fert. indicates fertilization. (b) Circuit showing the sequential waves of polyadenylation and deadenylation driving meiotic progression. Positive and negative feedback loops are indicated

APC/C-stabilizing high MPF activity in MII. CSF requires the new synthesis of Emi2, cyclin E and high levels of Mos (Liu *et al.*, 2006).

The correct and sequential regulation of the MPF, APC/C and CSF requires protein synthesis at three different stages (Figure 8.3a). First, for the activation of MPF and the transition from the prophase I (PI) arrest to metaphase I (MI) (Schmitt and Nebreda, 2002). Then, at MI, translation is required for the transition to metaphase II (MII) (Hochegger *et al.*, 2001). Last, new protein synthesis at interkinesis is required for the CSF arrest (Belloc and Mendez, 2008; Liu *et al.*, 2006).

The first known mRNA to be translationally activated upon progesterone stimulation is the one encoding the Cdc2-activator ringo mRNA that is derepressed by the release of Pum2 from the mRNA; in turn RINGO activates CPEB (Padmanabhan and

Richter, 2006) and, by inactivating Myt1, promotes the activation of MPF (Liu *et al.*, 2006; Ruiz, Hunt and Nebreda, 2008) (Figure 8.3b). This initial polyadenylation-independent translational activation triggers the ordered activation of CPEB-regulated mRNAs in three sequential waves of cytoplasmic polyadenylation (Figure 8.3b). The sequence of activation of these maternal mRNAs is defined by the number and position of CPEs present in their 3′UTRs and by the levels and activity of CPEB, combined with deadenylation-driven negative feedback loops (Pique *et al.*, 2008; Belloc and Mendez, 2008). But meiotic progression also requires that the extent of translational activation will be finely regulated, resulting in differential rates of product accumulation that, combined with the control of protein degradation, establish phase-specific peaks of expression of the factors that drive meiotic progression.

The first (early) wave of cytoplasmic polyadenylation, which takes place during PI and is required to enter MI (i.e. to activate MPF), is induced by the activation of Aurora-A kinase, which phosphorylates and activates CPEB (Mendez *et al.*, 2000a). This early wave targets mRNAs with, at least, a single consensus CPE or a nonconsensus CPE together with a PBE. The CPE must be closer than 100 nucleotides from the Hex, but not overlapping, and the distance between the CPE and Hex elements determines the extent of polyadenylation and translational activation with an optimal distance of 25 nucleotides (Pique *et al.*, 2008) (Figure 8.4). The mRNAs activated during the early polyadenylation event include mos, cyclin B5, cyclin B2, emi1, c3h4, PP2C, tyrP4a2, hsp90, X71067, TPX2 and Xkid mRNAs (Pique *et al.*, 2008; Belloc and Mendez, 2008; Eliscovich *et al.*, 2008) (Figure 8.3b). Cyclin B5 and cyclin B2 are themselves part of MPF in the first meiotic division (Hochegger *et al.*, 2001), whereas Mos is required to activate MPF (Sagata *et al.*, 1989), and Hsp90 activates Mos (Fisher, Mandart and Doree, 2000). Emi1 is an APC/C inhibitor and therefore prevents MPF inhibition by blocking cyclin B destruction (Tung and Jackson, 2005). TPX2 and Xkid are required for proper spindle formation and chromosome segregation (Eliscovich *et al.*, 2008; Perez *et al.*, 2002). PP2C, tyrP4a2 and X71067 are also required for meiotic progression, although the mechanisms are unknown (Belloc and Mendez, 2008).

In MI, CPEB is phosphorylated by MPF (Cdc2/cyclin B) and Polo-like kinase 1 (Plk1), resulting in the degradation of up to 90% of this factor (Mendez, Barnard and Richter, 2002; Setoyama, Yamashita and Sagata, 2007; Reverte, Ahearn and Hake, 2001). Reduced levels of CPEB trigger the second (late) wave of cytoplasmic polyadenylation, which is required for the MI–MII transition. This late wave targets mRNAs containing at least two CPEs, with one of them overlapping the Hex. As for the early wave, the extent of polyadenylation is defined by the position of the CPE(s) not overlapping the Hex (Pique *et al.*, 2008; Figure 8.4). These late-activated transcripts include cyclin B1 and cyclin B4 mRNAs (Pique *et al.*, 2008), which are the main components of MPF in the second meiotic division (Hochegger *et al.*, 2001) (Figure 8.3b). Continuous synthesis of cyclins B is also required during interkinesis to maintain intermediate levels of active MPF and prevent DNA replication between the two meiotic divisions (Hochegger *et al.*, 2001; Iwabuchi *et al.*, 2000). To compensate for the APC/C-mediated cyclin degradation, cytoplasmic polyadenylation is reinforced during this meiotic phase by positive feedback loops from the MAPK/MPF pathway into the activation of CPEB, and the synthesis of components of the

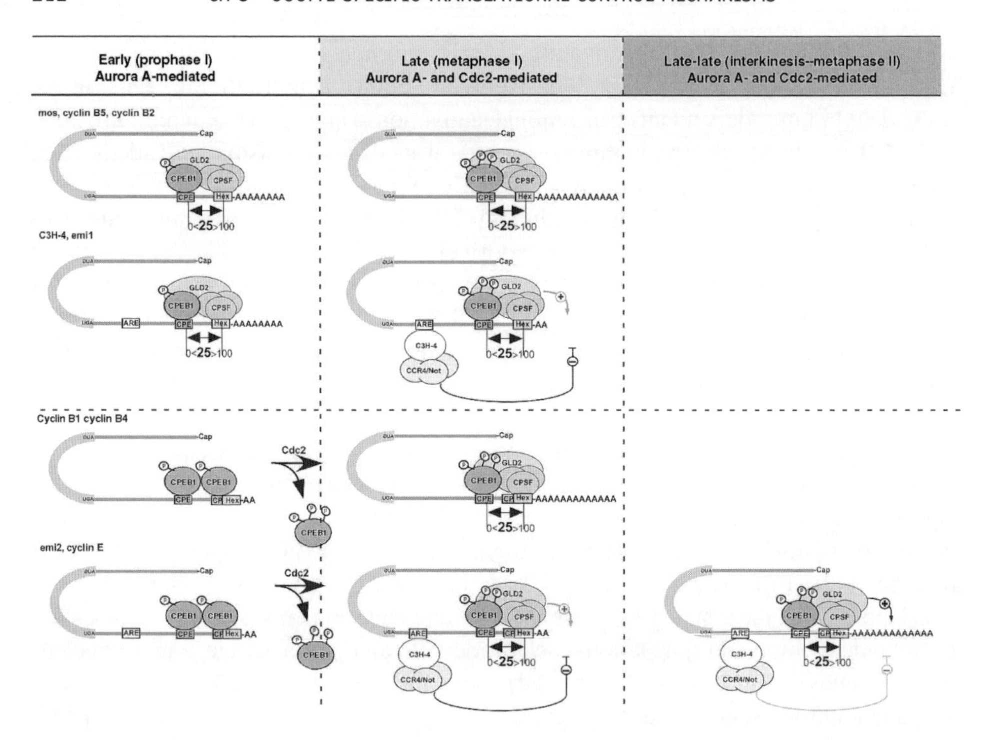

Figure 8.4 Model for CPE/ARE-mediated temporal translational control by CPEB1 and C3H-4 proteins. Schematic representation of the cis-elements and trans-acting factors recruited, with their covalent modifications. The distances required for translational repression and activation, as well as the time of activation, are indicated. CPE = cytoplasmic polyadenylation element; CPEB = CPE-binding protein; ARE = AU-rich element; PBE = Pumilio binding element. AA indicates short Poly(A), AAAAAAA indicates long poly(A), and P indicates phosphorylation

cytoplasmic polyadenylation machinery (such as Orb, GLD2 and Aurora A), which are themselves encoded by mRNAs activated by cytoplasmic polyadenylation (Pique *et al.*, 2008; Rouhana and Wickens, 2007; Howard *et al.*, 1999; Tan *et al.*, 2001) (Figure 8.3b).

At the same time, the early translational activation of the c3h-4 mRNA leads to the accumulation of this protein at MI, which in turn establishes a negative feedback loop to inactivate transcripts with AU-rich elements (AREs) in their 3′UTRs. The active form of C3H-4 recruits the CCR4/Not deadenylase complex to the ARE-containing mRNAs, triggering their deadenylation and allowing MI exit (Belloc and Mendez, 2008; Figure 8.4). Thus, for mRNAs containing AREs and early CPEs, C3H-4 overrides the CPEB-mediated polyadenylation after MI, inactivating the mRNA. This is the case of the mRNAs encoding the APC/C inhibitor, Emi1, which has to be degraded to allow the activation of the APC/C and the meiotic progression to anaphase, and C3H-4 itself, thus establishing a negative feedback of the negative feedback (Belloc and Mendez, 2008). This translational negative feedback loop cooperates with another one at the post-translational level, where active MPF brings about the activation of the APC/C, which in turn results in the *polyubiquitination* and proteolysis of Cyclin B

(King *et al.*, 1996), inactivating MPF and allowing the exit from MI into interkinesis (Figure 8.3b).

The third or late-late wave of cytoplasmic polyadenylation, required to enter the second meiotic division and to mediate the CSF arrest at MII, is generated by the combination of the late polyadenylation, driven by CPEs overlapping the Hex, and the ARE-mediated deadenylation (Figure 8.4). Both events are activated concomitantly at MI by the destruction of CPEB and the accumulation of C3H-4, neutralizing each other. Thus, for mRNAs containing both AREs and late-strong CPEs, polyadenylation is displaced from MI to interkinesis (Belloc and Mendez, 2008), when C3H-4 is probably inactivated. Examples of this third group of cytoplasmically polyadenylated transcripts include emi2, wee1, cyclin E1 and cyclin A1 mRNAs (Belloc and Mendez, 2008; Sheets *et al.*, 1994; Wang *et al.*, 2008; Charlesworth, Welk and Macnicol, 2000). Emi2 is the main component of the CSF activity (Liu, Grimison and Maller, 2007), whereas cyclin E contributes to the establishment of the CSF arrest (Tunquist *et al.*, 2002). Wee1 participates in the repression of Cdc2 by phosphorylating Tyr14 during interkinesis (Nakajo *et al.*, 2000). Cyclin A1 is not required for meiotic progression but for later events in development (Minshull *et al.*, 1991).

At least in mouse, there is an even earlier stage of polyadenylation, which takes place during the pachytene stage, during early oogenesis prior to the PI arrest. Two mRNAs, SCP-1 and SCP-3, which code for proteins of the synaptonemal complex required for chromosomal recombination, are polyadenylated by a transient activation of Aurora-A (Tay and Richter, 2001; Tay *et al.*, 2003).

8.3 mRNA localization and translational control during oogenesis

Asymmetric localization of silenced mRNAs within the oocytes defines the basis for the embryonic axis formation and the establishment of germ cells. Although translational repression and localization are coupled events, in most cases they are mediated by separated elements in the 3′UTRs that recruit specific RNA-BPs. The most common mechanisms for localizing repressed mRNAs are active transport along the polarized cytoskeleton, diffusion and local trapping, or local protection from degradation. At the appropriate time and place, either in the oocyte itself or later in the embryo, the localized mRNAs will be reactivated by disassembling the repression complex and/or by cytoplasmic polyadenylation. Similarly to the temporal translational cascades described above, spatial cascades can be defining by generating gradients of translational regulators encoded by localized mRNAs (St Johnston, 2005; King, Messitt and Mowry, 2005; Bashirullah, Cooperstock and Lipshitz, 1998). While this phenomena is probably widespread, with up to 71% of the mRNAs localized in the embryos (Lecuyer *et al.*, 2007), some of the best-characterized examples are the mRNAs encoding the *Drosophila* determinants of embryonic polarity, *bicoid, oskar, nanos* and *gurken* (Johnstone and Lasko, 2001; Huynh and St Johnston, 2004; Kloc and Etkin, 2005). These maternal mRNAs are made in the nurse cells, translationally repressed and transported into the adjacent oocyte, where they are localized and translationally regulated.

8.3.1 bicoid (bcd), an anteriorly localizing mRNA

During oogenesis bicoid (bcd) mRNA undergoes a multistep localization pathway in the oocyte. First localizing in the posterior cytoplasm, later in the anterior part and finally in the dorsal-anterior region of the oocyte. bcd mRNA localization depends on its 3′UTR, which forms a complex secondary structure including multiple stem-loops (five independent domains I–V) (Macdonald and Struhl, 1988). Deletion analysis identified the bcd localization element 1 (BLE1), which is sufficient to direct RNA accumulation into the oocyte and to its anterior cortex (Macdonald *et al.*, 1993). Genetic analyses have led to the identification of trans-acting factors required for bcd mRNA anterior localization, including Exuperantia (Exu), Exuperantia-like (Exl), Swallow (Swa) and, for the final stages of localization, Staufen (Stau) proteins (Johnstone and Lasko, 2001; Kloc and Etkin, 2005). bcd mRNA correct localization is microtubule-dependent, and both motors, Dynein and Kinesin, cooperate in the polar transport of bcd to its cortical domain (Johnstone and Lasko, 2001; Kloc and Etkin, 2005). However, it has been recently described that microtubule-dependent movement of bcd mRNA is replaced by a stable actin-based anchoring mechanism by the end of oogenesis (Weil, Forrest and Gavis, 2006; Weil *et al.*, 2008). bcd mRNA is initially repressed through Nanos response elements (NREs) (Wharton and Struhl, 1991) and later on, in the embryo, undergoes cytoplasmic polyadenylation and is translationally activated (Johnstone and Lasko, 2001).

8.3.2 oskar (osk), an early localizing posterior mRNA

oskar (osk) mRNA is localized in the oocyte in a multistep manner. In previtellogenic oocytes, osk mRNA is abundant throughout the cytoplasm, to be later transiently localized to the anterior pole from where it migrates to its final destination at the posterior cortex (Ephrussi, Dickinson and Lehmann, 1991; Kim-Ha, Smith and Macdonald, 1991). Although multiple subelements within its 3′UTR are required for its localization to the posterior (Johnstone and Lasko, 2001), osk 3′UTR is necessary but not sufficient for its transport, and splicing is also required for proper localization of the mRNA (Hachet and Ephrussi, 2004). Genetic analyses have revealed multiple RNA-BPs required for osk mRNA localization including Hrp48, the exon junction complex (EJC) proteins Barentsz, Mago nashi, Y14 (Tsunagi) and Stau. Many different mechanisms have been postulated to localize osk mRNA, including diffusion and trapping to the posterior (Glotzer *et al.*, 1997), active transport (Januschke *et al.*, 2002; Brendza *et al.*, 2002; Palacios and St Johnston, 2002), or exclusion from the anterior and lateral cortex (Cha *et al.*, 2002). However, Zimyanin *et al.* have recently demonstrated by following osk mRNA particles in living oocytes, that the mRNA is actively transported along microtubules in all directions, with a slight bias toward the posterior (Zimyanin *et al.*, 2008). Accordingly, transport of osk mRNA from the nurse cells into the oocyte is coupled to Dynein, which moves the mRNAs toward the minus ends of microtubules in the oocyte (Clark, Meignin and Davis, 2007). The localization of osk mRNA within the oocyte is also actin- and microtubule-dependent, and requires the recruitment of PAR-1 to the posterior cortex of the oocyte (Clark *et al.*, 1994;

Doerflinger *et al.*, 2006). Other cytoskeletal proteins involved are Tropomyosin II (TmII) and the Kinesin heavy chain (Khc) (Kloc and Etkin, 2005).

As described above, osk mRNA is initially repressed by Bruno and then translated as soon as it reaches the posterior pole (Gavis, 1997). Genetic analyses have identified several translational activators of osk mRNA, including Stau, Vasa (Vas) and Aubergine (Aub) (Johnstone and Lasko, 2001). Moreover, Orb has been shown to mediate localization and translational regulation by cytoplasmic polyadenylation of osk mRNA (Castagnetti and Ephrussi, 2003; Chang, Tan and Schedl, 1999). Interestingly, Osk protein is itself required for maintenance of its own mRNA localization by stabilizing and amplifying microtubule polarity within the oocyte, producing a positive feedback loop leading to more localization of osk mRNA (Zimyanin, Lowe and St Johnston, 2007).

8.3.3 nanos (nos), a late-localizing posterior mRNA

nanos (nos) mRNA accumulates at the pole cells in the posterior pole during late oogenesis. Posterior localization and translation activation of nos mRNA are controlled by cis-acting regulatory elements present in its 3′UTR. (Gavis, Curtis and Lehmann, 1996a; Bergsten and Gavis, 1999). A 500-nucleotide region within the 3′UTR consisting of partially redundant elements is required to direct all stages of nos localization throughout oogenesis (Gavis *et al.*, 1996b). This mRNA localization does not seem to be mediated by active localization but rather by specific anchoring (Bergsten and Gavis, 1999) and by degradation of the unlocalized mRNA (Smibert *et al.*, 1996; Dahanukar and Wharton, 1996). Translational repression of nos mRNA is mediated by two stem-loops that together comprise the Translational Control Element or TCE (Crucs, Chatterjee and Gavis, 2000), overlapping but distinct from the localization element (Crucs, Chatterjee and Gavis, 2000). Stem-loop II recruits Smaug (Smg) (Johnstone and Lasko, 2001), whereas stem-loop III binds to Glorund (Glo; Kalifa *et al.*, 2006). As mentioned earlier, Cup is also required to mediate translational repression of nos mRNA (Wilhelm *et al.*, 2003). Once nos mRNA reaches the posterior pole it is translationally derepressed. Cyclin B1 mRNA localizes to the posterior pole with kinetics similar to nos mRNA, and is also incorporated into pole cells where it is translationally repressed by Nos and Pum (Johnstone and Lasko, 2001).

8.3.4 gurken (grk), an mRNA associated with the oocyte nucleus

When the nucleus migrates to the dorsal anterior corner of the oocyte, gurken (grk) mRNA and protein become localized to the exterior surface of the dorsal anterior corner of the nucleus, establishing dorsal cell fates in lateral follicle cells (Johnstone and Lasko, 2001). Sequences within the 5′ and 3′UTRs, as well as within the coding region, have been implicated in directing grk localization throughout oogenesis (Johnstone and Lasko, 2001). The sequences recruit a number of factors and RNA-BPs, such as Squid (Sqd), Hrp48/Hrb27C, Out, Enc, K10 and Orb (Clouse, Ferguson and Schupbach, 2008; Johnstone and Lasko, 2001; Kloc and Etkin, 2005; Goodrich,

Clouse and Schupbach, 2004; Hawkins *et al.*, 1997). grk mRNA is not only actively transported along microtubules (Clark, Meignin and Davis, 2007; Delanoue *et al.*, 2007; Macdougall *et al.*, 2003), but also anchored by Dynein to large cytoplasmic structures called sponge bodies at the dorsal anterior corner (Delanoue *et al.*, 2007). Accordingly, genetic analyses have identified several factors involved in cytoskeletal assembly and integrity required for grk localization, such as Maelstrom (Mael), Cappuccino (Capu), Spire (actin nucleator) and Spindle-F (Spin-F; microtubule organizer) (Johnstone and Lasko, 2001; Abdu, Bar and Schupbach, 2006). Additionally, the molecular motors Kinesin and Dynein are required for correct grk mRNA localization (Kloc and Etkin, 2005; Clark, Meignin and Davis, 2007; Delanoue *et al.*, 2007).

grk mRNA is repressed by Bruno and its interacting protein, Cup, together with Sqd and Hrb27C/Hrp48 (Filardo and Ephrussi, 2003; Clouse, Ferguson and Schupbach, 2008; Saunders and Cohen, 1999). Once the mRNA reaches its final destination, PABP directs translational activation together with Enc (Hawkins *et al.*, 1997) and Vas and maybe Orb (Chang *et al.*, 2001; Christerson and Mckearin, 1994).

8.3.5 mRNA localization in Xenopus oocytes

Xenopus embryonic axis formation and establishment of germ cells is also defined by asymmetric mRNA localization within the oocyte. mRNA localization to the oocyte vegetal cortex follows two temporally defined pathways: the early pathway (also know as the messenger transport organizer (METRO)) and the late or Vg1 pathway. The METRO pathway functions during early oogenesis and transports germinal granules, several RNAs involved in germ-cell specification, and germline-specific mitochondria in a specialized structure called the mitochondrial cloud or Balbiani body (King, Messitt and Mowry, 2005; Kloc and Etkin, 2005). Cis-acting elements, such as the mitochondrial cloud localization elements (MCLEs) or the germinal granule localization element (GGLE), present in those RNAs, are responsible for their localization pattern by diffusion and selective entrapment (Kloc *et al.*, 2000; Claussen, Horvay and Pieler, 2004). mRNAs localized by the late pathway require an intact microtubule cytoskeleton (Kloc and Etkin, 1995; Yisraeli, Sokol and Melton, 1990), and Kinesins II (Betley *et al.*, 2004) and I (Yoon and Mowry, 2004; Messitt *et al.*, 2008). Actin and cytokeratin filaments (Kloc and Etkin, 1995; Alarcon and Elinson, 2001), and localized RNAs (Heasman *et al.*, 2001; Kloc and Etkin, 1994) help to anchor the mRNAs close to the plasma membrane in the vegetal cortex. Vg1 mRNA late localization is mediated by the Vg1 localization element (VLE) (Mowry and Melton, 1992), whereas the VTE (Vg1 translation element) controls its translation (Wilhelm, Vale and Hegde, 2000; Otero, Devaux and Standart, 2001). The RNA-BPs hnRNP I, Vg RBP/Vera and Stau are required for the Vg1 localization (King, Messitt and Mowry, 2005; Kloc and Etkin, 2005).

Other RNA-BPs, such as CPEB1 and C3H-4, localize preferentially to the animal pole of the oocyte (Groisman *et al.*, 2000; De *et al.*, 1999; Bally-Cuif, Schatz and Ho, 1998), together with CPE-containing mRNAs (Eliscovich *et al.*, 2008; Groisman *et al.*, 2000). This localization seems to be microtubule directed and, indeed, these mRNAs and their binding proteins later localize to the meiotic and mitotic spindles

(Eliscovich *et al.*, 2008; Groisman *et al.*, 2000; Blower *et al.*, 2007). The spindle-associated RNAs include structural components of the spindle (Alliegro, Alliegro and Palazzo, 2006; Blower *et al.*, 2005), repressed centrosome-localized mRNAs for asymmetrical inheritance in embryonic divisions (Alliegro, Alliegro and Palazzo, 2006; Lambert and Nagy, 2002) and CPE-regulated mRNAs encoding factors directly involved in chromosome segregation, spindle formation and meiotic progression (Eliscovich *et al.*, 2008).

Acknowledgements

This work was supported by grants from the MEC (Ministero de Educación y Ciencia). Raúl Méndez is a recipient of a contract from the 'Programa I3' (MEC). Isabel Novoa is a recipient of a contract from 'Programa Ramón y Cajal'. Carolina Eliscovich is recipient of a fellowship from the Departament d'Universitats Recerca i Societat de la Informació (DURSI, Generalitat de Catalunya) and from the European Social Fund.

References

Abdu, U., Bar, D. and Schupbach, T. (2006) spn-F encodes a novel protein that affects oocyte patterning and bristle morphology in Drosophila. *Development*, **133**(8), 1477–1484.

Alarcon, V.B. and Elinson, R.P. (2001) RNA anchoring in the vegetal cortex of the Xenopus oocyte. *J. Cell Sci.*, **114**(Pt 9), 1731–1741.

Alliegro, M.C., Alliegro, M.A. and Palazzo, R.E. (2006) Centrosome-associated RNA in surf clam oocytes. *Proc. Natl. Acad. Sci. USA*, **103**(24), 9034–9038.

Andrei, M.A., Ingelfinger, D., Heintzmann, R. *et al.* (2005) A role for eIF4E and eIF4E-transporter in targeting mRNPs to mammalian processing bodies. *RNA*, **11**(5), 717–727.

Bally-Cuif, L., Schatz, W.J. and Ho, R.K. (1998) Characterization of the zebrafish Orb/CPEB-related RNA binding protein and localization of maternal components in the zebrafish oocyte. *Mech. Dev.*, **77**(1), 31–47.

Barnard, D.C., Ryan, K., Manley, J.L. and Richter, J.D. (2004) Symplekin and xGLD-2 are required for CPEB-mediated cytoplasmic polyadenylation. *Cell*, **119**(5), 641–651.

Barnard, D.C., Cao, Q. and Richter, J.D. (2005) Differential phosphorylation controls Maskin association with eukaryotic translation initiation factor 4E and localization on the mitotic apparatus. *Mol. Cell Biol.*, **25**(17), 7605–7615.

Bashirullah, A., Cooperstock, R.L. and Lipshitz, H.D. (1998) RNA localization in development. *Annu. Rev. Biochem.*, **67**, 335–394.

Belloc, E. and Mendez, R. (2008) A deadenylation negative feedback mechanism governs meiotic metaphase arrest. *Nature*, **452**(7190), 1017–1021.

Belloc, E., Pique, M. and Mendez, R. (2008) Sequential waves of polyadenylation and deadenylation define a translation circuit that drives meiotic progression. *Biochem. Soc. Trans.*, **36**(Pt 4), 665–670.

Benoit, P., Papin, C., Kwak, J.E. *et al.* (2008) PAP- and GLD-2-type poly(A) polymerases are required sequentially in cytoplasmic polyadenylation and oogenesis in Drosophila. *Development*, **135**(11), 1969–1979.

Bergsten, S.E. and Gavis, E.R. (1999) Role for mRNA localization in translational activation but not spatial restriction of nanos RNA. *Development*, **126**(4), 659–669.

Betley, J.N., Heinrich, B., Vernos, I. *et al.* (2004) Kinesin II mediates Vg1 mRNA transport in Xenopus oocytes. *Curr. Biol.*, **14**(3), 219–224.

Blower, M.D., Nachury, M., Heald, R. and Weis, K. (2005) A Rae1-containing ribonucleoprotein complex is required for mitotic spindle assembly. *Cell*, **121**(2), 223–234.

Blower, M.D., Feric, E., Weis, K. and Heald, R. (2007) Genome-wide analysis demonstrates conserved localization of messenger RNAs to mitotic microtubules. *J. Cell Biol.*, **179**(7), 1365–1373.

Boag, P.R., Atalay, A., Robida, S. *et al.* (2008) Protection of specific maternal messenger RNAs by the P body protein CGH-1 (Dhh1/RCK) during Caenorhabditis elegans oogenesis. *J. Cell Biol.*, **182**(3), 543–557.

Bouvet, P. and Wolffe, A.P. (1994) A role for transcription and FRGY2 in masking maternal mRNA within Xenopus oocytes. *Cell*, **77**(6), 931–941.

Brendza, R.P., Serbus, L.R., Saxton, W.M. and Duffy, J.B. (2002) Posterior localization of dynein and dorsal-ventral axis formation depend on kinesin in Drosophila oocytes. *Curr. Biol.*, **12**(17), 1541–1545.

Cakmakci, N.G., Lerner, R.S., Wagner, E.J. *et al.* (2008) SLIP1, a factor required for activation of histone mRNA translation by the stem-loop binding protein. *Mol. Cell Biol.*, **28**(3), 1182–1194.

Cao, Q., Huang, Y.S., Kan, M.C. and Richter, J.D. (2005) Amyloid precursor proteins anchor CPEB to membranes and promote polyadenylation-induced translation. *Mol. Cell Biol.*, **25**(24), 10930–10939.

Cao, Q., Kim, J.H. and Richter, J.D. (2006) CDK1 and calcineurin regulate Maskin association with eIF4E and translational control of cell cycle progression. *Nat. Struct. Mol. Biol.*, **13**(12), 1128–1134.

Castagnetti, S. and Ephrussi, A. (2003) Orb and a long poly(A) tail are required for efficient oskar translation at the posterior pole of the Drosophila oocyte. *Development*, **130**(5), 835–843.

Cha, B.J., Serbus, L.R., Koppetsch, B.S. and Theurkauf, W.E. (2002) Kinesin I-dependent cortical exclusion restricts pole plasm to the oocyte posterior. *Nat. Cell Biol.*, **4**(8), 592–598.

Chang, J.S., Tan, L. and Schedl, P. (1999) The Drosophila CPEB homolog, orb, is required for oskar protein expression in oocytes. *Dev. Biol.*, **215**(1), 91–106.

Chang, J.S., Tan, L., Wolf, M.R. and Schedl, P. (2001) Functioning of the Drosophila orb gene in gurken mRNA localization and translation. *Development*, **128**(16), 3169–3177.

Charlesworth, A., Welk, J. and Macnicol, A.M. (2000) The temporal control of Wee1 mRNA translation during Xenopus oocyte maturation is regulated by cytoplasmic polyadenylation elements within the 3′-untranslated region. *Dev. Biol.*, **227**(2), 706–719.

Charlesworth, A., Ridge, J.A., King, L.A. *et al.* (2002) A novel regulatory element determines the timing of Mos mRNA translation during Xenopus oocyte maturation. *EMBO J.*, **21**(11), 2798–2806.

Charlesworth, A., Cox, L.L. and Macnicol, A.M. (2004) Cytoplasmic polyadenylation element (CPE)- and CPE-binding protein (CPEB)-independent mechanisms regulate early class maternal mRNA translational activation in Xenopus oocytes. *J. Biol. Chem.*, **279**(17), 17650–17659.

Charlesworth, A., Wilczynska, A., Thampi, P. *et al.* (2006) Musashi regulates the temporal order of mRNA translation during Xenopus oocyte maturation. *EMBO J.*, **25**(12), 2792–2801.

Chekulaeva, M., Hentze, M.W. and Ephrussi, A. (2006) Bruno acts as a dual repressor of oskar translation, promoting mRNA oligomerization and formation of silencing particles. *Cell*, **124**(3), 521–533.

Chicoine, J., Benoit, P., Gamberi, C. *et al.* (2007) Bicaudal-C recruits CCR4-NOT deadenylase to target mRNAs and regulates oogenesis, cytoskeletal organization, and its own expression. *Dev. Cell*, **13**(5), 691–704.

Cho, P.F., Poulin, F., Cho-Park, Y.A. *et al.* (2005) A new paradigm for translational control: inhibition via 5′-3′ mRNA tethering by Bicoid and the eIF4E cognate 4EHP. *Cell*, **121**(3), 411–423.

Cho, P.F., Gamberi, C., Cho-Park, Y.A. *et al.* (2006) Cap-dependent translational inhibition establishes two opposing morphogen gradients in Drosophila embryos. *Curr. Biol.*, **16**(20), 2035–2041.

Christerson, L.B. and Mckearin, D.M. (1994) orb is required for anteroposterior and dorsoventral patterning during Drosophila oogenesis. *Genes. Dev.*, **8**(5), 614–628.

Clark, I., Giniger, E., Ruohola-Baker, H. *et al.* (1994) Transient posterior localization of a kinesin fusion protein reflects anteroposterior polarity of the Drosophila oocyte. *Curr. Biol.*, **4** (4), 289–300.

Clark, A., Meignin, C. and Davis, I. (2007) A Dynein-dependent shortcut rapidly delivers axis determination transcripts into the Drosophila oocyte. *Development*, **134**(10), 1955–1965.

Claussen, M., Horvay, K. and Pieler, T. (2004) Evidence for overlapping, but not identical, protein machineries operating in vegetal RNA localization along early and late pathways in Xenopus oocytes. *Development*, **131**(17), 4263–4273.

Clegg, K.B. and Piko, L. (1982) RNA synthesis and cytoplasmic polyadenylation in the one-cell mouse embryo. *Nature*, **295**(5847), 343–344.

Clouse, K.N., Ferguson, S.B. and Schupbach, T. (2008) Squid, Cup, and PABP55B function together to regulate gurken translation in Drosophila. *Dev. Biol.*, **313**(2), 713–724.

Collier, B., Gorgoni, B., Loveridge, C. *et al.* (2005) The DAZL family proteins are PABP-binding proteins that regulate translation in germ cells. *EMBO J.*, **24**(14), 2656–2666.

Cook, H.A., Koppetsch, B.S., Wu, J. and Theurkauf, W.E. (2004) The Drosophila SDE3 homolog armitage is required for oskar mRNA silencing and embryonic axis specification. *Cell*, **116**(6), 817–829.

Copeland, P.R. and Wormington, M. (2001) The mechanism and regulation of deadenylation: identification and characterization of Xenopus PARN. *RNA*, **7**(6), 875–886.

Crucs, S., Chatterjee, S. and Gavis, E.R. (2000) Overlapping but distinct RNA elements control repression and activation of nanos translation. *Mol. Cell*, **5**(3), 457–467.

Dahanukar, A. and Wharton, R.P. (1996) The Nanos gradient in Drosophila embryos is generated by translational regulation. *Genes. Dev.*, **10**(20), 2610–2620.

De, J., Lai, W.S., Thorn, J.M. *et al.* (1999) Identification of four CCCH zinc finger proteins in Xenopus, including a novel vertebrate protein with four zinc fingers and severely restricted expression. *Gene*, **228**(1–2), 133–145.

De Moor, C.H., Meijer, H. and Lissenden, S. (2005) Mechanisms of translational control by the 3′ UTR in development and differentiation. *Semin. Cell Dev. Biol.*, **16**(1), 49–58.

Delanoue, R., Herpers, B., Soetaert, J. *et al.* (2007) Drosophila Squid/hnRNP helps Dynein switch from a gurken mRNA transport motor to an ultrastructural static anchor in sponge bodies. *Dev. Cell*, **13**(4), 523–538.

Dickson, K.S., Bilger, A., Ballantyne, S. and Wickens, M.P. (1999) The cleavage and polyadenylation specificity factor in Xenopus laevis oocytes is a cytoplasmic factor involved in regulated polyadenylation. *Mol. Cell Biol.*, **19**(8), 5707–5717.

Doerflinger, H., Benton, R., Torres, I.L. *et al.* (2006) Drosophila anterior-posterior polarity requires actin-dependent PAR-1 recruitment to the oocyte posterior. *Curr. Biol.*, **16**(11), 1090–1095.

Eliscovich, C., Peset, I., Vernos, I. and Mendez, R. (2008) Spindle-localized CPE-mediated translation controls meiotic chromosome segregation. *Nat. Cell Biol.*, **10**(7), 858–865.

Ephrussi, A., Dickinson, L.K. and Lehmann, R. (1991) Oskar organizes the germ plasm and directs localization of the posterior determinant nanos. *Cell*, **66**(1), 37–50.

Ferrell, J.E. Jr (2002) Self-perpetuating states in signal transduction: positive feedback, double-negative feedback and bistability. *Curr. Opin. Cell Biol.*, **14**(2), 140–148.

Filardo, P. and Ephrussi, A. (2003) Bruno regulates gurken during Drosophila oogenesis. *Mech. Dev.*, **120**(3), 289–297.

Fisher, D.L., Mandart, E. and Doree, M. (2000) Hsp90 is required for c-Mos activation and biphasic MAP kinase activation in Xenopus oocytes. *EMBO J.*, **19**(7), 1516–1524.

Gao, M., Fritz, D.T., Ford, L.P. and Wilusz, J. (2000) Interaction between a poly(A)-specific ribonuclease and the 5′ cap influences mRNA deadenylation rates *in vitro*. *Mol. Cell*, **5**(3), 479–488.

Gao, Y., Tatavarty, V., Korza, G. *et al.* (2008) Multiplexed dendritic targeting of alpha calcium calmodulin-dependent protein kinase II, neurogranin, and activity-regulated cytoskeleton-associated protein RNAs by the A2 pathway. *Mol. Biol. Cell*, **19**(5), 2311–2327.

Gautier, J., Norbury, C., Lohka, M. *et al.* (1988) Purified maturation-promoting factor contains the product of a Xenopus homolog of the fission yeast cell cycle control gene cdc2 +. *Cell*, **54**(3), 433–439.

Gautier, J., Minshull, J., Lohka, M. *et al.* (1990) Cyclin is a component of maturation-promoting factor from Xenopus. *Cell*, **60**(3), 487–494.

Gavis, E.R. (1997) Expeditions to the pole: RNA localization in Xenopus and Drosophila. *Trends Cell Biol.*, **7**(12), 485–492.

Gavis, E.R., Curtis, D. and Lehmann, R. (1996a) Identification of cis-acting sequences that control nanos RNA localization. *Dev. Biol.*, **176**(1), 36–50.

Gavis, E.R., Lunsford, L., Bergsten, S.E. and Lehmann, R. (1996b) A conserved 90 nucleotide element mediates translational repression of nanos RNA. *Development*, **122**(9), 2791–2800.

Gillian-Daniel, D.L., Gray, N.K., Astrom, J. *et al.* (1998) Modifications of the 5′ cap of mRNAs during Xenopus oocyte maturation: independence from changes in poly(A) length and impact on translation. *Mol. Cell Biol.*, **18**(10), 6152–6163.

Giraldez, A.J., Mishima, Y., Rihel, J. *et al.* (2006) Zebrafish MiR-430 promotes deadenylation and clearance of maternal mRNAs. *Science*, **312**(5770), 75–79.

Glotzer, J.B., Saffrich, R., Glotzer, M. and Ephrussi, A. (1997) Cytoplasmic flows localize injected oskar RNA in Drosophila oocytes. *Curr. Biol.*, **7**(5), 326–337.

Goldstrohm, A.C. and Wickens, M. (2008) Multifunctional deadenylase complexes diversify mRNA control. *Nat. Rev. Mol. Cell Biol.*, **9**(4), 337–344.

Goodrich, J.S., Clouse, K.N. and Schupbach, T. (2004) Hrb27C, Sqd and Otu cooperatively regulate gurken RNA localization and mediate nurse cell chromosome dispersion in Drosophila oogenesis. *Development*, **131**(9), 1949–1958.

Graindorge, A., Thuret, R., Pollet, N. *et al.* (2006) Identification of post-transcriptionally regulated Xenopus tropicalis maternal mRNAs by microarray. *Nucleic Acids Res.*, **34**(3), 986–995.

Grivna, S.T., Pyhtila, B. and Lin, H. (2006) MIWI associates with translational machinery and PIWI-interacting RNAs (piRNAs) in regulating spermatogenesis. *Proc. Natl. Acad. Sci. USA*, **103**(36), 13415–13420.

Groisman, I., Huang, Y.S., Mendez, R. *et al.* (2000) CPEB, maskin, and cyclin B1 mRNA at the mitotic apparatus: implications for local translational control of cell division. *Cell*, **103**(3), 435–447.

Hachet, O. and Ephrussi, A. (2001) Drosophila Y14 shuttles to the posterior of the oocyte and is required for oskar mRNA transport. *Curr. Biol.*, **11**(21), 1666–1674.

Hachet, O. and Ephrussi, A. (2004) Splicing of oskar RNA in the nucleus is coupled to its cytoplasmic localization. *Nature*, **428**(6986), 959–963.

Hake, L.E. and Richter, J.D. (1994) CPEB is a specificity factor that mediates cytoplasmic polyadenylation during Xenopus oocyte maturation. *Cell*, **79**(4), 617–627.

Hake, L.E., Mendez, R. and Richter, J.D. (1998) Specificity of RNA binding by CPEB: requirement for RNA recognition motifs and a novel zinc finger. *Mol. Cell Biol.*, **18**(2), 685–693.

Hawkins, N.C., Van Buskirk, C., Grossniklaus, U. and Schupbach, T. (1997) Post-transcriptional regulation of gurken by encore is required for axis determination in Drosophila. *Development*, **124**(23), 4801–4810.

Heasman, J., Wessely, O., Langland, R. *et al.* (2001) Vegetal localization of maternal mRNAs is disrupted by VegT depletion. *Dev. Biol.*, **240**(2), 377–386.

Hershey, J.W.B. and Merrick, W.C. (2000) Pathway and mechanism of initiation of protein synthesis, in *Translational Control of Gene Expression* (eds N. Sonenberg, J.W.B. Hershey, M.B. Mathews), Cold Spring Harbor Laboratory Press, Cold Spring Harbor, NY, pp. 33–88.

Hochegger, H., Klotzbucher, A., Kirk, J. *et al.* (2001) New B-type cyclin synthesis is required between meiosis I and II during Xenopus oocyte maturation. *Development*, **128**(19), 3795–3807.

Howard, E.L., Charlesworth, A., Welk, J. and Macnicol, A.M. (1999) The mitogen-activated protein kinase signaling pathway stimulates mos mRNA cytoplasmic polyadenylation during Xenopus oocyte maturation. *Mol. Cell Biol.*, **19**(3), 1990–1999.

Huynh, J.R. and St Johnston, D. (2004) The origin of asymmetry: early polarisation of the Drosophila germline cyst and oocyte. *Curr. Biol.*, **14**(11), R438–R449.

Huynh, J.R., Munro, T.P., Smith-Litiere, K. *et al.* (2004) The Drosophila hnRNPA/B homolog, Hrp48, is specifically required for a distinct step in osk mRNA localization. *Dev. Cell*, **6**(5), 625–635.

Iwabuchi, M., Ohsumi, K., Yamamoto, T.M. *et al.* (2000) Residual Cdc2 activity remaining at meiosis I exit is essential for meiotic M-M transition in Xenopus oocyte extracts. *EMBO J.*, **19**(17), 4513–4523.

Januschke, J., Gervais, L., Dass, S. *et al.* (2002) Polar transport in the Drosophila oocyte requires Dynein and Kinesin I cooperation. *Curr. Biol.*, **12**(23), 1971–1981.

Johnstone, O. and Lasko, P. (2001) Translational regulation and RNA localization in Drosophila oocytes and embryos. *Annu. Rev. Genet.*, **35**, 365–406.

Kadyrova, L.Y., Habara, Y., Lee, T.H. and Wharton, R.P. (2007) Translational control of maternal Cyclin B mRNA by Nanos in the Drosophila germline. *Development*, **134**(8), 1519–1527.

Kalifa, Y., Huang, T., Rosen, L.N. *et al.* (2006) Glorund, a Drosophila hnRNP F/H homolog, is an ovarian repressor of nanos translation. *Dev. Cell*, **10**(3), 291–301.

Kalifa, Y., Armenti, S.T. and Gavis, E.R. (2009) Glorund interactions in the regulation of gurken and oskar mRNAs. *Dev. Biol.*, **326**(1), 68–74.

Keene, J.D. (2007) RNA regulons: coordination of post-transcriptional events. *Nat. Rev.*, **8**(7), 533–543.

Kim, J.H. and Richter, J.D. (2006) Opposing polymerase-deadenylase activities regulate cytoplasmic polyadenylation. *Mol. Cell*, **24**(2), 173–183.

Kim, J.H. and Richter, J.D. (2007) RINGO/cdk1 and CPEB mediate poly(A) tail stabilization and translational regulation by ePAB. *Genes. Dev.*, **21**(20), 2571–2579.

Kimble, J. and Crittenden, S.L. (2007) Controls of germline stem cells, entry into meiosis, and the sperm/oocyte decision in Caenorhabditis elegans. *Annu. Rev. Cell Dev. Biol.*, **23**, 405–433.

Kim-Ha, J., Smith, J.L. and Macdonald, P.M. (1991) oskar mRNA is localized to the posterior pole of the Drosophila oocyte. *Cell*, **66**(1), 23–35.

Kim-Ha, J., Kerr, K. and Macdonald, P.M. (1995) Translational regulation of oskar mRNA by bruno, an ovarian RNA-binding protein, is essential. *Cell*, **81**(3), 403–412.

King, R.W., Deshaies, R.J., Peters, J.M. and Kirschner, M.W. (1996) How proteolysis drives the cell cycle. *Science*, **274**(5293), 1652–1659.

King, M.L., Messitt, T.J. and Mowry, K.L. (2005) Putting RNAs in the right place at the right time: RNA localization in the frog oocyte. *Biol. Cell*, **97**(1), 19–33.

Klattenhoff, C. and Theurkauf, W. (2008) Biogenesis and germline functions of piRNAs. *Development*, **135**(1), 3–9.

Klattenhoff, C., Bratu, D.P., Mcginnis-Schultz, N. *et al.* (2007) Drosophila rasiRNA pathway mutations disrupt embryonic axis specification through activation of an ATR/Chk2 DNA damage response. *Dev. Cell*, **12**(1), 45–55.

Kloc, M. and Etkin, L.D. (1994) Delocalization of Vg1 mRNA from the vegetal cortex in Xenopus oocytes after destruction of Xlsirt RNA. *Science*, **265**(5175), 1101–1103.

Kloc, M. and Etkin, L.D. (1995) Two distinct pathways for the localization of RNAs at the vegetal cortex in Xenopus oocytes. *Development*, **121**(2), 287–297.

Kloc, M. and Etkin, L.D. (2005) RNA localization mechanisms in oocytes. *J. Cell Sci.*, **118**(Pt 2), 269–282.

Kloc, M., Bilinski, S., Pui-Yee Chan, A. and Etkin, L.D. (2000) The targeting of Xcat2 mRNA to the germinal granules depends on a cis-acting germinal granule localization element within the 3′UTR. *Dev. Biol.*, **217**(2), 221–229.

Kuersten, S. and Goodwin, E.B. (2003) The power of the 3′ UTR: translational control and development. *Nat. Rev.*, **4**(8), 626–637.

Kuge, H. and Richter, J.D. (1995) Cytoplasmic 3′ poly(A) addition induces 5′ cap ribose methylation: implications for translational control of maternal mRNA. *EMBO J.*, **14**(24), 6301–6310.

Kuge, H., Brownlee, G.G., Gershon, P.D. and Richter, J.D. (1998) Cap ribose methylation of c-mos mRNA stimulates translation and oocyte maturation in Xenopus laevis. *Nucleic Acids Res.*, **26**(13), 3208–3214.

Lamarca, M.J., Smith, L.D. and Strobel, M.C. (1973) Quantitative and qualitative analysis of RNA synthesis in stage 6 and stage 4 oocytes of Xenopus laevis. *Dev. Biol*, **34**(1), 106–118.

Lambert, J.D. and Nagy, L.M. (2002) Asymmetric inheritance of centrosomally localized mRNAs during embryonic cleavages. *Nature*, **420**(6916), 682–686.

Lecuyer, E., Yoshida, H., Parthasarathy, N. *et al.* (2007) Global analysis of mRNA localization reveals a prominent role in organizing cellular architecture and function. *Cell*, **131**(1), 174–187.

Liu, J., Grimison, B., Lewellyn, A.L. and Maller, J.L. (2006) The anaphase-promoting complex/cyclosome inhibitor Emi2 is essential for meiotic but not mitotic cell cycles. *J. Biol. Chem.*, **281**(46), 34736–34741.

Liu, J., Grimison, B. and Maller, J.L. (2007) New insight into metaphase arrest by cytostatic factor: from establishment to release. *Oncogene*, **26**(9), 1286–1289.

Lykke-Andersen, K., Gilchrist, M.J., Grabarek, J.B. *et al.* (2008) Maternal Argonaute 2 is essential for early mouse development at the maternal-zygotic transition. *Mol. Biol. Cell*, **19**(10), 4383–4392.

Macdonald, P.M. and Struhl, G. (1988) cis-acting sequences responsible for anterior localization of bicoid mRNA in Drosophila embryos. *Nature*, **336**(6199), 595–598.

Macdonald, P.M., Kerr, K., Smith, J.L. and Leask, A. (1993) RNA regulatory element BLE1 directs the early steps of bicoid mRNA localization. *Development*, **118**(4), 1233–1243.

Macdougall, N., Clark, A., Macdougall, E. and Davis, I. (2003) Drosophila gurken (TGFalpha) mRNA localizes as particles that move within the oocyte in two dynein-dependent steps. *Dev. Cell*, **4**(3), 307–319.

Mansfield, J.H., Wilhelm, J.E. and Hazelrigg, T. (2002) Ypsilon Schachtel, a Drosophila Y-box protein, acts antagonistically to Orb in the oskar mRNA localization and translation pathway. *Development*, **129**(1), 197–209.

Marzluff, W.F. (2007) U2 snRNP: not just for poly(A) mRNAs. *Mol. Cell*, **28**(3), 353–354.

Mendez, R. and Richter, J.D. (2001) Translational control by CPEB: a means to the end. *Nat. Rev. Mol. Cell Biol.*, **2**(7), 521–529.

Mendez, R., Hake, L.E., Andresson, T. *et al.* (2000a) Phosphorylation of CPE binding factor by Eg2 regulates translation of c-mos mRNA. *Nature*, **404**(6775), 302–307.

Mendez, R., Murthy, K.G., Ryan, K. *et al.* (2000b) Phosphorylation of CPEB by Eg2 mediates the recruitment of CPSF into an active cytoplasmic polyadenylation complex. *Mol. Cell*, **6**(5), 1253–1259.

Mendez, R., Barnard, D. and Richter, J.D. (2002) Differential mRNA translation and meiotic progression require Cdc2-mediated CPEB destruction. *EMBO J.*, **21**(7), 1833–1844.

Merritt, C., Rasoloson, D., Ko, D. and Seydoux, G. (2008) 3′ UTRs are the primary regulators of gene expression in the C. elegans germline. *Curr. Biol.*, **18**(19), 1476–1482.

Messitt, T.J., Gagnon, J.A., Kreiling, J.A. *et al.* (2008) Multiple kinesin motors coordinate cytoplasmic RNA transport on a subpopulation of microtubules in Xenopus oocytes. *Dev. Cell*, **15**(3), 426–436.

Minshall, N. and Standart, N. (2004) The active form of Xp54 RNA helicase in translational repression is an RNA-mediated oligomer. *Nucleic Acids Res.*, **32**(4), 1325–1334.

Minshall, N., Reiter, M.H., Weil, D. and Standart, N. (2007) CPEB interacts with an ovary-specific eIF4E and 4E-T in early Xenopus oocytes. *J. Biol. Chem.*, **282**(52), 37389–37401.

Minshull, J., Murray, A., Colman, A. and Hunt, T. (1991) Xenopus oocyte maturation does not require new cyclin synthesis. *J. Cell Biol.*, **114**(4), 767–772.

Mishima, Y., Giraldez, A.J., Takeda, Y. *et al.* (2006) Differential regulation of germline mRNAs in soma and germ cells by zebrafish miR-430. *Curr. Biol.*, **16**(21), 2135–2142.

Moore, F.L., Jaruzelska, J., Fox, M.S. *et al.* (2003) Human Pumilio-2 is expressed in embryonic stem cells and germ cells and interacts with DAZ (Deleted in AZoospermia) and DAZ-like proteins. *Proc. Natl. Acad. Sci. USA*, **100**(2), 538–543.

Moraes, K.C., Wilusz, C.J. and Wilusz, J. (2006) CUG-BP binds to RNA substrates and recruits PARN deadenylase. *RNA*, **12**(6), 1084–1091.

Mowry, K.L. and Melton, D.A. (1992) Vegetal messenger RNA localization directed by a 340-nt RNA sequence element in Xenopus oocytes. *Science*, **255**(5047), 991–994.

Murchison, E.P., Stein, P., Xuan, Z. *et al.* (2007) Critical roles for Dicer in the female germline. *Genes. Dev.*, **21**(6), 682–693.

Nakahara, K., Kim, K., Sciulli, C. *et al.* (2005) Targets of microRNA regulation in the Drosophila oocyte proteome. *Proc. Natl. Acad. Sci. USA*, **102**(34), 12023–12028.

Nakahata, S., Katsu, Y., Mita, K. *et al.* (2001) Biochemical identification of Xenopus Pumilio as a sequence-specific cyclin B1 mRNA-binding protein that physically interacts with a Nanos homolog, Xcat-2, and a cytoplasmic polyadenylation element-binding protein. *J. Biol. Chem.*, **276**(24), 20945–20953.

Nakahata, S., Kotani, T., Mita, K. *et al.* (2003) Involvement of Xenopus Pumilio in the translational regulation that is specific to cyclin B1 mRNA during oocyte maturation. *Mech. Dev.*, **120**(8), 865–880.

Nakajo, N., Yoshitome, S., Iwashita, J. *et al.* (2000) Absence of Wee1 ensures the meiotic cell cycle in Xenopus oocytes. *Genes. Dev.*, **14**(3), 328–338.

Nakamura, A., Amikura, R., Hanyu, K. and Kobayashi, S. (2001) Me31B silences translation of oocyte-localizing RNAs through the formation of cytoplasmic RNP complex during Drosophila oogenesis. *Development*, **128**(17), 3233–3242.

Nakamura, A., Sato, K. and Hanyu-Nakamura, K. (2004) Drosophila cup is an eIF4E binding protein that associates with Bruno and regulates oskar mRNA translation in oogenesis. *Dev. Cell*, **6**(1), 69–78.

Nelson, M.R., Leidal, A.M. and Smibert, C.A. (2004) Drosophila Cup is an eIF4E-binding protein that functions in Smaug-mediated translational repression. *EMBO J.*, **23**(1), 150–159.

Newport, J. and Kirschner, M. (1982) A major developmental transition in early Xenopus embryos: II. Control of the onset of transcription. *Cell*, **30**(3), 687–696.

Noble, S.L., Allen, B.L., Goh, L.K. *et al.* (2008) Maternal mRNAs are regulated by diverse P body-related mRNP granules during early Caenorhabditis elegans development. *J. Cell Biol.*, **182**(3), 559–572.

Osborne, H.B., Gautier-Courteille, C., Graindorge, A. *et al.* (2005) Post-transcriptional regulation in Xenopus embryos: role and targets of EDEN-BP. *Biochem. Soc. Trans.*, **33**(Pt 6), 541–543.

Otero, L.J., Devaux, A. and Standart, N. (2001) A 250-nucleotide UA-rich element in the 3′ untranslated region of Xenopus laevis Vg1 mRNA represses translation both *in vivo* and *in vitro*. *RNA*, **7**(12), 1753–1767.

Padmanabhan, K. and Richter, J.D. (2006) Regulated Pumilio-2 binding controls RINGO/Spy mRNA translation and CPEB activation. *Genes. Dev.*, **20**(2), 199–209.

Palacios, I.M. and St Johnston, D. (2002) Kinesin light chain-independent function of the Kinesin heavy chain in cytoplasmic streaming and posterior localisation in the Drosophila oocyte. *Development*, **129**(23), 5473–5485.

Palacios, I.M., Gatfield, D., St Johnston, D. and Izaurralde, E. (2004) An eIF4AIII-containing complex required for mRNA localization and nonsense-mediated mRNA decay. *Nature*, **427**(6976), 753–757.

Pane, A., Wehr, K. and Schupbach, T. (2007) zucchini and squash encode two putative nucleases required for rasiRNA production in the Drosophila germline. *Dev. Cell*, **12**(6), 851–862.

Pascreau, G., Delcros, J.G., Cremet, J.Y. *et al.* (2005) Phosphorylation of maskin by Aurora-A participates in the control of sequential protein synthesis during Xenopus laevis oocyte maturation. *J. Biol. Chem.*, **280**(14), 13415–13423.

Perez, L.H., Antonio, C., Flament, S. *et al.* (2002) Xkid chromokinesin is required for the meiosis I to meiosis II transition in Xenopus laevis oocytes. *Nat. Cell Biol.*, **4**(10), 737–742.

Peters, J.M. (2006) The anaphase promoting complex/cyclosome: a machine designed to destroy. *Nat. Rev. Mol. Cell Biol.*, **7**(9), 644–656.

Pique, M., Lopez, J.M., Foissac, S. *et al.* (2008) A combinatorial code for CPE-mediated translational control. *Cell*, **132**(3), 434–448.

Potireddy, S., Vassena, R., Patel, B.G. and Latham, K.E. (2006) Analysis of polysomal mRNA populations of mouse oocytes and zygotes: dynamic changes in maternal mRNA utilization and function. *Dev. Biol.*, **298**(1), 155–166.

Reverte, C.G., Ahearn, M.D. and Hake, L.E. (2001) CPEB degradation during Xenopus oocyte maturation requires a PEST domain and the 26S proteasome. *Dev. Biol.*, **231**(2), 447–458.

Reverte, C.G., Yuan, L., Keady, B.T. *et al.* (2003) XGef is a CPEB-interacting protein involved in Xenopus oocyte maturation. *Dev. Biol.*, **255**(2), 383–398.

Richter, J.D. (2007) CPEB: a life in translation. *Trends Biochem. Sci.*, **32**(6), 279–285.

Ro, S., Song, R., Park, C. *et al.* (2007) Cloning and expression profiling of small RNAs expressed in the mouse ovary. *RNA*, **13**(12), 2366–2380.

Rodman, T.C. and Bachvarova, R. (1976) RNA synthesis in preovulatory mouse oocytes. *J. Cell Biol.*, **70**(1), 251–257.

Rouget, C., Papin, C. and Mandart, E. (2006) Cytoplasmic CstF-77 protein belongs to a masking complex with cytoplasmic polyadenylation element-binding protein in Xenopus oocytes. *J. Biol. Chem.*, **281**(39), 28687–28698.

Rouhana, L. and Wickens, M. (2007) Autoregulation of GLD-2 cytoplasmic poly(A) polymerase. *RNA*, **13**(2), 188–199.

Rouhana, L., Wang, L., Buter, N. *et al.* (2005) Vertebrate GLD2 poly(A) polymerases in the germline and the brain. *RNA*, **11**(7), 1117–1130.

Ruiz, E.J., Hunt, T. and Nebreda, A.R. (2008) Meiotic inactivation of Xenopus Myt1 by CDK/XRINGO, but not CDK/cyclin, via site-specific phosphorylation. *Mol. Cell*, **32**(2), 210–220.

Sagata, N. (1996) Meiotic metaphase arrest in animal oocytes: its mechanisms and biological significance. *Trends Cell Biol.*, **6**(1), 22–28.

Sagata, N., Daar, I., Oskarsson, M. *et al.* (1989) The product of the mos proto-oncogene as a candidate "initiator" for oocyte maturation. *Science*, **245**(4918), 643–646.

Sanchez, R. and Marzluff, W.F. (2002) The stem-loop binding protein is required for efficient translation of histone mRNA *in vivo* and *in vitro*. *Mol. Cell Biol.*, **22**(20), 7093–7104.

Saunders, C. and Cohen, R.S. (1999) The role of oocyte transcription, the 5′UTR, and translation repression and derepression in Drosophila gurken mRNA and protein localization. *Mol. Cell*, **3**(1), 43–54.

Schmitt, A. and Nebreda, A.R. (2002) Signalling pathways in oocyte meiotic maturation. *J. Cell Sci.*, **115**(Pt 12), 2457–2459.

Setoyama, D., Yamashita, M. and Sagata, N. (2007) Mechanism of degradation of CPEB during Xenopus oocyte maturation. *Proc. Natl. Acad. Sci. USA*, **104**(46), 18001–18006.

Sheets, M.D., Fox, C.A., Hunt, T. *et al.* (1994) The 3′-untranslated regions of c-mos and cyclin mRNAs stimulate translation by regulating cytoplasmic polyadenylation. *Genes. Dev.*, **8**(8), 926–938.

Smibert, C.A., Wilson, J.E., Kerr, K. and Macdonald, P.M. (1996) smaug protein represses translation of unlocalized nanos mRNA in the Drosophila embryo. *Genes. Dev.*, **10**(20), 2600–2609.

St Johnston, D. (2005) Moving messages: the intracellular localization of mRNAs. *Nat. Rev. Mol. Cell Biol.*, **6**(5), 363–375.

Stebbins-Boaz, B., Cao, Q., de Moor, C.H. *et al.* (1999) Maskin is a CPEB-associated factor that transiently interacts with elF-4E. *Mol. Cell*, **4**(6), 1017–1027.

Stern, C.D. (2006) Evolution of the mechanisms that establish the embryonic axes. *Curr. Opin. Genet. Dev.*, **16**(4), 413–418.

Sugimura, I. and Lilly, M.A. (2006) Bruno inhibits the expression of mitotic cyclins during the prophase I meiotic arrest of Drosophila oocytes. *Dev. Cell*, **10**(1), 127–135.

Tadros, W., Westwood, J.T. and Lipshitz, H.D. (2007) The mother-to-child transition. *Dev. Cell*, **12**(6), 847–849.

Tam, O.H., Aravin, A.A., Stein, P. *et al.* (2008) Pseudogene-derived small interfering RNAs regulate gene expression in mouse oocytes. *Nature*, **453**(7194), 534–538.

Tan, L., Chang, J.S., Costa, A. and Schedl, P. (2001) An autoregulatory feedback loop directs the localized expression of the Drosophila CPEB protein Orb in the developing oocyte. *Development*, **128**(7), 1159–1169.

Tanaka, K.J., Ogawa, K., Takagi, M. *et al.* (2006) RAP55, a cytoplasmic mRNP component, represses translation in Xenopus oocytes. *J. Biol Chem.*, **281**(52), 40096–40106.

Tang, F., Kaneda, M., O'carroll, D. *et al.* (2007) Maternal microRNAs are essential for mouse zygotic development. *Genes. Dev.*, **21**(6), 644–648.

Tay, J. and Richter, J.D. (2001) Germ cell differentiation and synaptonemal complex formation are disrupted in CPEB knockout mice. *Dev. Cell*, **1**(2), 201–213.

Tay, J., Hodgman, R., Sarkissian, M. and Richter, J.D. (2003) Regulated CPEB phosphorylation during meiotic progression suggests a mechanism for temporal control of maternal mRNA translation. *Genes. Dev.*, **17**(12), 1457–1462.

Tomari, Y., Du, T., Haley, B. *et al.* (2004) RISC assembly defects in the Drosophila RNAi mutant armitage. *Cell*, **116**(6), 831–841.

Tsui, S., Dai, T., Roettger, S. *et al.* (2000) Identification of two novel proteins that interact with germ-cell-specific RNA-binding proteins DAZ and DAZL1. *Genomics*, **65**(3), 266–273.

Tung, J.J. and Jackson, P.K. (2005) Emi1 class of proteins regulate entry into meiosis and the meiosis I to meiosis II transition in Xenopus oocytes. *Cell Cycle*, **4**(3), 478–482.

Tunquist, B.J., Schwab, M.S., Chen, L.G. and Maller, J.L. (2002) The spindle checkpoint kinase bub1 and cyclin e/cdk2 both contribute to the establishment of meiotic metaphase arrest by cytostatic factor. *Curr. Biol.*, **12**(12), 1027–1033.

Unhavaithaya, Y., Hao, Y., Beyret, E. *et al.* (2009) MILI, a PIWI-interacting RNA-binding protein, is required for germ line stem cell self-renewal and appears to positively regulate translation. *J. Biol. Chem.*, **284**(10), 6507–6519.

Vardy, L. and Orr-Weaver, T.L. (2007) Regulating translation of maternal messages: multiple repression mechanisms. *Trends Cell Biol.*, **17**(11), 547–554.

Wang, Q.T., Piotrowska, K., Ciemerych, M.A. *et al.* (2004) A genome-wide study of gene activity reveals developmental signaling pathways in the preimplantation mouse embryo. *Dev. Cell*, **6**(1), 133–144.

Wang, Y.Y., Charlesworth, A., Byrd, S.M. *et al.* (2008) A novel mRNA 3′ untranslated region translational control sequence regulates Xenopus Wee1 mRNA translation. *Dev. Biol.*, **317**(2), 454–466.

Watanabe, T., Totoki, Y., Toyoda, A. *et al.* (2008) Endogenous siRNAs from naturally formed dsRNAs regulate transcripts in mouse oocytes. *Nature*, **453**(7194), 539–543.

Webster, P.J., Liang, L., Berg, C.A. *et al.* (1997) Translational repressor bruno plays multiple roles in development and is widely conserved. *Genes Dev.*, **11**(19), 2510–2521.

Weil, T.T., Forrest, K.M. and Gavis, E.R. (2006) Localization of bicoid mRNA in late oocytes is maintained by continual active transport. *Dev. Cell*, **11**(2), 251–262.

Weil, T.T., Parton, R., Davis, I. and Gavis, E.R. (2008) Changes in bicoid mRNA anchoring highlight conserved mechanisms during the oocyte-to-embryo transition. *Curr. Biol.*, **18**(14), 1055–1061.

Wells, S.E., Hillner, P.E., Vale, R.D. and Sachs, A.B. (1998) Circularization of mRNA by eukaryotic translation initiation factors. *Mol. Cell*, **2**(1), 135–140.

Weston, A. and Sommerville, J. (2006) Xp54 and related (DDX6-like) RNA helicases: roles in messenger RNP assembly, translation regulation and RNA degradation. *Nucleic Acids Res.*, **34**(10), 3082–3094.

Wharton, R.P. and Struhl, G. (1991) RNA regulatory elements mediate control of Drosophila body pattern by the posterior morphogen nanos. *Cell*, **67**(5), 955–967.

Wilhelm, J.E., Vale, R.D. and Hegde, R.S. (2000) Coordinate control of translation and localization of Vg1 mRNA in Xenopus oocytes. *Proc. Natl. Acad. Sci. USA*, **97**(24), 13132–13137.

Wilhelm, J.E., Hilton, M., Amos, Q. and Henzel, W.J. (2003) Cup is an eIF4E binding protein required for both the translational repression of oskar and the recruitment of Barentsz. *J. Cell Biol.*, **163**(6), 1197–1204.

Xiong, W. and Ferrell, J.E. Jr (2003) A positive-feedback-based bistable 'memory module' that governs a cell fate decision. *Nature*, **426**(6965), 460–465.

Yisraeli, J.K., Sokol, S. and Melton, D.A. (1990) A two-step model for the localization of maternal mRNA in Xenopus oocytes: involvement of microtubules and microfilaments in the translocation and anchoring of Vg1 mRNA. *Development*, **108**(2), 289–298.

Yoon, Y.J. and Mowry, K.L. (2004) Xenopus Staufen is a component of a ribonucleoprotein complex containing Vg1 RNA and kinesin. *Development*, **131**(13), 3035–3045.

Zaessinger, S., Busseau, I. and Simonelig, M. (2006) Oskar allows nanos mRNA translation in Drosophila embryos by preventing its deadenylation by Smaug/CCR4. *Development*, **133**(22), 4573–4583.

Zamore, P.D., Bartel, D.P., Lehmann, R. and Williamson, J.R. (1999) The PUMILIO-RNA interaction: a single RNA-binding domain monomer recognizes a bipartite target sequence. *Biochemistry*, **38**(2), 596–604.

Zimyanin, V., Lowe, N. and St Johnston, D. (2007) An oskar-dependent positive feedback loop maintains the polarity of the Drosophila oocyte. *Curr. Biol.*, **17**(4), 353–359.

Zimyanin, V.L., Belaya, K., Pecreaux, J. *et al.* (2008) *In vivo* imaging of oskar mRNA transport reveals the mechanism of posterior localization. *Cell*, **134**(5), 843–853.

9

MPF and the control of meiotic divisions: old problems, new concepts

Catherine Jessus

Biologie du Développement – UMR 7622, UPMC-CNRS, Paris CEDEX 05, France

'This whole field, so interesting both from the point of view of embryology and of cellular physiology, remains to be explored' (Brachet, 1944). The whole field mentioned by Jean Brachet in 1944 corresponds to oocyte maturation, a cell system at the crossroad between embryology and cell biology that has long been used by the scientific community (Heilbrunn, Daugherty and Wilbur, 1939). Indeed, the physiological approach of cell division control was championed by researchers who favoured marine and amphibian oocytes. Half a century after Brachet's comment, the Nobel prize for Physiology or Medicine was awarded in 2001 to three scientists who made seminal discoveries concerning control of the cell cycle, among them Tim Hunt, who discovered cyclins in sea urchin eggs.

9.1 MPF and the autoamplification mechanism; the discovery

In the animal kingdom, oocytes growing in the ovaries are arrested at prophase of the first meiotic division. At the time of ovulation, they resume meiosis in response to external stimuli such as hormones, and become mature oocytes or fertilizable eggs whose cell cycle is halted and that await for fertilization. The frog oocyte has been widely used to study the biochemical regulation of cell division. In amphibians, maturing oocytes complete the first meiotic division in response to steroid hormones secreted by the follicle cells, such as progesterone, and arrest at metaphase of meiosis II. Fertilization relieves the metaphase arrest and initiates a series of embryonic cell

Oogenesis: The Universal Process Marie-Hélène Verlhac and Anne Villeneuve

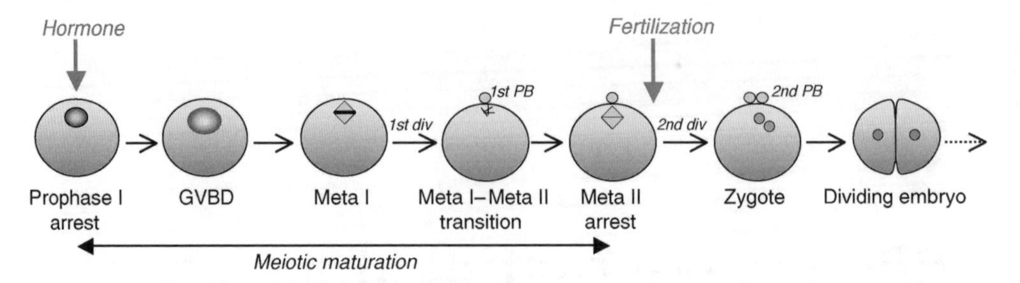

Figure 9.1 Oocyte meiotic maturation. In the ovaries, oocytes are arrested at prophase I. At the time of ovulation, a hormonal signal triggers meiotic maturation: germinal vesicle breakdown (GVBD); formation of the metaphase I spindle (meta I); first meiotic division (1st div) and extrusion of the first polar body (1st PB); formation of the metaphase II spindle (meta II). In vertebrates, the oocyte arrests at metaphase II. Fertilization releases this arrest: the egg completes the second meiotic division (2nd div) by extruding the second polar body (2nd PB) and initiates a series of embryonic divisions. A full colour version of this figure appears in the colour plate section.

divisions that proceed synchronously without any detectable G1 and G2 phases (Figure 9.1).

Experiments based on nuclear transplantation and cytoplasmic transfer in oocytes demonstrated the existence of a cytoplasmic factor responsible for the completion of the first meiotic division. In a pioneering paper, Yoshio Masui showed that injection of cytoplasm from metaphase II-arrested oocytes into prophase-arrested oocytes of the frog *Rana pipiens* induces meiotic maturation (Masui and Markert, 1971). He named 'maturation-promoting factor' or MPF as the cytoplasmic activity produced in response to the steroid signal and responsible for the meiotic M phase induction, and he highlighted three important properties of MPF in oocytes.

The first one is that MPF appearance is initiated by cytoplasmic events without involvement of the nucleus. This was proved by experiments with oocytes whose nucleus (called the germinal vesicle) had been removed prior to progesterone stimulation. When treated with progesterone, the cytoplasm of these enucleated oocytes produced MPF. Before the important work of Masui, many experiments had emphasized the importance of the contribution of the germinal vesicle to the oocyte maturation process (Dettlaff, 1966; Dettlaff, Nikitina and Stroeva, 1964; Hirai, Kubota and Kanatani, 1971). Masui demonstrated that the production of MPF can be attributed entirely to a cytoplasmic process independent of nuclear function, implying for the first time that nuclear activities involved with the initiation of oocyte maturation are under the control of cytoplasmic activity, but not vice versa as thought before.

The second property is the auto-catalytic reaction of the production of MPF in amphibian oocytes. Masui showed that there is no decrease in the level of MPF activity generated after serial transfer of cytoplasm (meaning that recipient oocytes become donor oocytes), despite the fact that the cytoplasm of the original progesterone-treated donor was extensively diluted through the serial transfers. These results imply that prophase-blocked oocytes contain an MPF precursor, called pre-MPF, that is autocatalytically amplified by the MPF activity received via cytoplasmic transfer.

In 1975, the laboratory of Masui made a third important discovery using *Xenopus* oocytes (Wasserman and Masui, 1975). They showed that protein synthesis inhibition could not stop the oocyte maturation induced by MPF injection, unlike that induced by progesterone. Then it was concluded that, in the *Xenopus* oocyte, the initiation of oocyte maturation by progesterone depends on the synthesis of new protein(s), required for the activation of pre-MPF. Once the first active molecules of MPF appear, they further activate pre-MPF in an autocatalytical process with no protein synthesis requirement.

A few years after the publication of the work of Masui, MPF was found in both meiotic and mitotic cell cycles. The presence of MPF and its autocatalytic amplification were reported in oocytes of a variety of distantly related species, including starfish and other echinoderms (Kishimoto and Kanatani, 1976; Schatt, Moreau and Guerrier, 1983), *Spisula* and other molluscs (Kishimoto *et al.*, 1982), fishes (Dettlaff and Ryabova, 1986; Yamashita *et al.*, 1992), several amphibian species (Masui and Markert, 1971; Wasserman and Masui, 1975; Dettlaff and Ryabova, 1986) and mouse (Sorensen, Cyert and Pedersen, 1985). Moreover, MPF is present in extracts of yeast (Weintraub *et al.*, 1982; Tachibana, Yanagishima and Kishimoto, 1987), in mammalian cells (Sunkara, Wright and Rao, 1979; Nelkin, Nichols and Vogelstein, 1980) and frog blastomeres (Wasserman and Smith, 1978; Gerhart, Wu and Kirschner, 1984) that are in mitosis. All these reports led to the conclusion that MPF activity is not only an inducer of the meiotic state but also the ubiquitous key regulator of M phase of the eukaryotic cell cycle. It was therefore renamed 'M phase-promoting factor'.

If MPF appeared as the universal activity promoting entry into the first meiotic division in oocytes whatever their species origin, it became rapidly known that the requirement of protein synthesis for the induction of MPF activation during this process is not a universal feature. The mechanisms of oocyte MPF activation during entry into the first meiotic division can be divided in three types (Figure 9.2).

- The first case is represented in many animals, including mouse and starfish, where MPF activation and its first direct consequence, breakdown of the nuclear envelope (known as GVBD for germinal vesicle breakdown), do not require protein synthesis at all (Kishimoto, 2003).

- The second case corresponds to the situation of the oocyte of *Xenopus* and nearly all mammals (except small rodents such as mouse), where new protein(s) must be synthesized to initiate MPF activation, but then the autocatalytic property of MPF allows it to activate a pre-MPF stockpile in the absence of protein synthesis (Wasserman and Masui, 1975; Kalous *et al.*, 1993; Hunter and Moor, 1987).

- In the third case that corresponds to a variety of fish and amphibian species, the existence of pre-MPF is questionable. MPF would have to be *de novo* synthesized following hormonal stimulation to initiate maturation (Yamashita, 1998).

Therefore, from the mid 1970s, it became clear that MPF activation leading to the first meiotic division of oocytes obeys different mechanisms depending on species (Figure 9.2). Active MPF is either *de novo* synthesized (type 3, Figure 9.2) or originates from a pre-MPF stock (types 1 and 2, Figure 9.2). If a pre-MPF stock exists, it allows the implementation of an autocatalytical process (called 'autoamplification') initially triggered by a starter amount of MPF activity. Firing the positive loop between pre-MPF

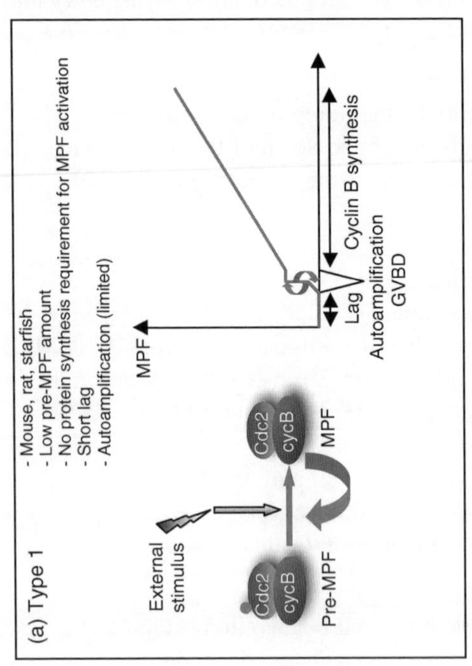
(a) Type 1
External stimulus
Pre-MPF
Cdc2
cycB
MPF
- Mouse, rat, starfish
- Low pre-MPF amount
- No protein synthesis requirement for MPF activation
- Short lag
- Autoamplification (limited)
MPF
Lag
Autoamplification
GVBD
Cyclin B synthesis

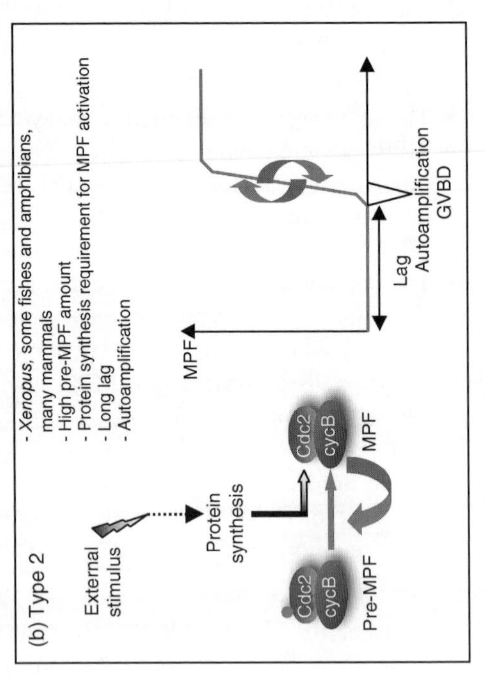
(b) Type 2
External stimulus
Protein synthesis
Pre-MPF
Cdc2
cycB
MPF
- Xenopus, some fishes and amphibians, many mammals
- High pre-MPF amount
- Protein synthesis requirement for MPF activation
- Long lag
- Autoamplification
MPF
Lag
Autoamplification
GVBD

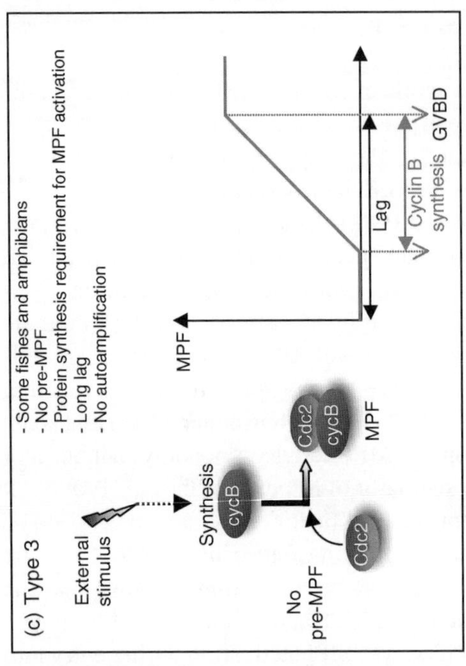
(c) Type 3
External stimulus
Synthesis
cycB
No pre-MPF
Cdc2
MPF
- Some fishes and amphibians
- No pre-MPF
- Protein synthesis requirement for MPF activation
- Long lag
- No autoamplification
MPF
Lag
Cyclin B synthesis
GVBD

and MPF relies (type 2, Figure 9.2), or does not rely (type 1, Figure 9.2), on protein synthesis.

This diversity is restricted to the initiation of the first meiotic division. After this step, MPF regulation follows universal rules. In all species, MPF activity declines at the end of the first meiotic cycle. It reappears rapidly before second metaphase, when maturation arrests in vertebrates. It was found that protein synthesis is required in *Xenopus* and starfish oocytes for reappearance of MPF activity in the second meiotic cell cycle (Gerhart, Wu and Kirschner, 1984; Picard *et al.*, 1985). Using starfish oocytes that arrest in G1 after the completion of the second meiotic division, Picard and co-workers showed that MPF activation at entry into the second meiotic M phase requires protein synthesis, and MPF disappearance at exit from this M phase requires proteolysis (Picard *et al.*, 1985). During the same period, it was demonstrated that protein synthesis is required at each cell cycle of sea urchin and *Xenopus* early embryos for MPF activation to occur (Gerhart, Wu and Kirschner, 1984; Wagenaar, 1983). In conclusion, MPF regulation operating during the second meiotic division of oocytes is similar to what happens in mitosis: its activation depends on protein translation, whereas its inactivation at the exit of meiosis II requires proteolysis.

9.2 From the biological activity to the molecular identification

Obviously, such features of MPF were reminiscent of cyclins, which had been identified in 1983 by Tim Hunt and his collaborators (Evans *et al.*, 1983). They had taken advantage of the rapid and synchronous divisions occurring after fertilization of the sea urchin and the clam embryos, and described proteins that are destroyed every time the cells divide. They proposed to call these proteins the cyclins and gave the definition in the title of their article: 'Cyclin, a protein specified by maternal mRNA in sea urchin eggs that is destroyed at each cleavage division'. The view that cyclins are in some way linked with MPF was strengthened when it was shown that expression of clam cyclin A induces meiotic M phase of *Xenopus* oocytes in the absence of hormonal stimulation (Swenson, Farrell and Ruderman, 1986). In 1989, Murray and Kirschner discovered that

Figure 9.2 MPF activation at entry in first meiotic division: principles and diversity among species. (a) First type (mouse, rat, starfish): MPF activation by the meiotic inducer does not require protein synthesis and occurs after a short lag period. MPF activity is generated from the pre-MPF stock and involves an autoamplification mechanism. The pre-MPF stockpile is low. Once it has been converted into MPF, synthesis of cyclin B (cycB) takes place and increases MPF activity. (b) Second type (*Xenopus*, some fishes and amphibians, many mammals): MPF activation by the meiotic inducer requires protein synthesis and occurs after a few hours' lag period. Newly synthesized proteins allow the generation of the first MPF molecules. This MPF trigger allows the conversion of the pre-MPF stock into MPF according to an autoamplification mechanism independent of protein synthesis. (c) Third type (some fishes and amphibians): the prophase oocyte does not possess any pre-MPF stockpile. MPF activation by the meiotic inducer requires synthesis of cyclin B that binds pre-existing Cdc2 molecules and directly generates active MPF after a lag period of several hours. MPF activity accumulates as a function of the cyclin B synthesis rate, without involving an autoamplification mechanism. A full colour version of this figure appears in the colour plate section.

cyclin plays a pivotal role in the control of mitosis. Using extracts of frog eggs that can perform multiple cell cycles *in vitro*, they showed that newly synthesized cyclin protein accumulates during each interphase and is necessary to enter into M phase (Murray and Kirschner, 1989a). Conversely, its degradation is required to exit from mitosis (Murray, Solomon and Kirschner, 1989). However, the connection between cyclins and MPF was far from clear.

Besides frog and marine oocyte studies, genetically tractable organisms such as the fungi had provided a different avenue for study of the mechanisms of cell cycle progression. Analysis of the G2/M transition had been primarily carried out with the fission yeast, *Schizosaccharomyces pombe*. In this species, a key component of the network of interacting genes that regulate the onset of mitosis is the *Cdc2* gene that encodes a protein kinase (Nurse and Bissett, 1981). It had been shown that the human homologue of Cdc2 was present in a complex and that this complex possessed elevated kinase activity during mitosis (Draetta and Beach, 1988). Then, Cdc2 and MPF both appeared to act as fundamental regulators of entry into mitosis and were found in a wide range of eukaryotic cells (Lee and Nurse, 1987). However, the relationship between the Cdc2 protein and MPF was unclear due to the lack of a biochemical approach.

Although MPF was first described in 1971 (Masui and Markert, 1971), attempts at purifying MPF met with considerable difficulties owing to its instability and to the lack of an easy biological assay. Thanks to the development of a cell-free system from amphibian eggs in which nuclei can be induced to undergo early mitotic events by the addition of crude or partially purified preparations of MPF, Lohka *et al.* succeeded in obtaining a purified fraction able to induce GVBD when injected into *Xenopus* oocytes in the presence of a protein synthesis inhibitor (Lohka, Hayes and Maller, 1988). This MPF fraction contained two major proteins, of mass 34 kDa and 45 kDa, and displayed protein kinase activity. These biochemical data provided interesting clues for the molecular identification of MPF.

By combining the respective technical advantages of *Xenopus* and yeast, it was rapidly shown that Cdc2 protein corresponds to the 34 kDa component of MPF that displays the kinase activity of the complex (Dunphy *et al.*, 1988; Gautier *et al.*, 1988). In these studies, the 45 kDa component was not yet identified. At that time, only a few protein kinases that contain regulatory subunits had been identified, the paradigm for such kinases being the cAMP-dependent protein kinase, PKA, whose regulatory subunit inhibits the activity of the catalytic subunit. Thus, the positive control of cyclin on MPF activity did not imply that it was necessarily a subunit of this kinase. The laboratory of Marcel Doree provided decisive evidence that cyclin B is a genuine subunit of Cdc2 kinase, by purifying MPF from starfish oocytes to homogeneity and showing that 'MPF from starfish oocytes at first meiotic metaphase is a heterodimer containing one molecule of Cdc2 and one molecule of cyclin B' as stated in the title of the article (Labbe *et al.*, 1989). In 1990, identification of Cdc2–cyclin B as MPF was extended to *Xenopus* oocytes (Gautier *et al.*, 1990). It was then shown that mammalian Cdc2 homologues can also bind to cyclins (Pines and Hunter, 1990), leading to a new convention for naming the kinases that are associated with cyclins: they were called 'cyclin-dependent kinases' or CDKs, Cdc2 being Cdk1.

A simple cell cycle model emerged from these findings (Murray and Kirschner, 1989b). During interphase, cyclins accumulate to a threshold at which MPF is activated.

Active MPF phosphorylates proteins involved in the structural changes characteristic of mitosis. One of the consequences of MPF activation is to promote cyclin degradation. Disappearance of cyclin causes MPF inactivation. Phosphatases reverse the phosphorylation of MPF substrates and result in the reestablishment of interphase structures. The loss of MPF activity would also turn off cyclin proteolysis, so that cyclin can accumulate for another round of the cell cycle. This model nicely explains the second meiotic cell cycle and the first synchronous cell division cycles that follow egg fertilization. All these cycles are characterized by MPF oscillations that depend on cyclin synthesis and cyclin destruction.

However, the model is not consistent with entry into the first meiotic division in oocytes. As mentioned previously, in many species, MPF activation does not require protein synthesis (as in mouse or starfish, Figure 9.2). Moreover, in the case where protein synthesis is needed, as in the *Rana pipiens* or *Xenopus* oocyte, the mechanism of MPF autoamplification operates in the absence of protein synthesis. Indeed, it was rapidly evidenced that a stockpile of inactive Cdc2–cyclin B complexes is present in G2-arrested oocytes (Kobayashi *et al.*, 1991) (Figure 9.2). Then the induction of the first meiotic division would not rely on cyclin synthesis, with the exception of a few fish and amphibian species devoid of a pre-MPF stockpile, where the *de novo* synthesis of cyclin B from the stored maternal mRNA would be induced by the hormonal trigger and required to lead to MPF production (Yamashita, 1998) (Figure 9.2). Besides these cases, the full grown prophase-arrested oocytes appear to contain inactive Cdc2–cyclin B complexes, a quite ubiquitous rule in the animal kingdom. What is the nature of the mechanism that blocks M phase entry under conditions where cyclins are present and associated with Cdc2?

9.3 Cdc2 regulators

The mechanism that restrains MPF activation is the phosphorylation of Cdc2 on two sites overlapping the ATP-binding site. In G2 phase, Cdc2 is phosphorylated on Tyr15 and Thr14, which maintains Cdc2 in a catalytically inactive state. Dephosphorylation either *in vivo* or *in vitro* allows Cdc2 activation in mitotic cells, from yeast to man (Draetta and Beach, 1988; Gould and Nurse, 1989; Norbury, Blow and Nurse, 1991; Ferrell *et al.*, 1991). In 1989, Gautier *et al.* showed that Cdc2 activation depends on its dephosphorylation on tyrosine in the *Xenopus* oocyte (Gautier *et al.*, 1989). We also know that there is a positive phosphorylation requirement for MPF activation: Cdc2 must be phosphorylated on Thr161, a phosphorylation event universally required for CDK activation and catalyzed by the Cdc2-activating kinase (CAK) (Solomon, Harper and Shuttleworth, 1993; Poon *et al.*, 1993; Fesquet *et al.*, 1993). Since CAK activity does not appear to be tightly regulated during the cell cycle, it is unlikely that the level of Thr161 phosphorylation of Cdc2 controls MPF activation. Indeed, Cdc2 is already phosphorylated on Thr161 in the inactive pre-MPF molecules of the G2-arrested oocytes (De Smedt *et al.*, 2002). Therefore, more emphasis has been placed on the enzymes that control the negative tyrosine phosphorylation that restrains MPF activity during the very long period of prophase arrest in oocytes.

A first kinase able to phosphorylate Tyr15 of Cdc2, called Wee1, was identified in fission yeast (Russell and Nurse, 1987). Later on, Wee1 was identified as the kinase that phosphorylates Tyr15 in various organisms but was not efficient as a Thr14-specific kinase. In 1995, Dunphy and collaborators cloned from *Xenopus* a new member of the Wee1 family, a kinase called Myt1 able to phosphorylate Cdc2 efficiently on both Thr14 and Tyr15 (Mueller *et al.*, 1995). Therefore, two kinases are able to restrain MPF activation: Wee1 and Myt1. In *Xenopus*, the full-grown prophase oocytes express only Myt1 at the protein level. The Wee1 mRNA is not translated at this stage. During the *Xenopus* meiotic cell cycles, Wee1 protein expression starts only at meiosis II (Murakami and VandeWoude, 1998). Then, Myt1 is the only player responsible for the inactivation of MPF in prophase-arrested full-grown oocytes. Myt1 becomes heavily phosphorylated during mitosis and *Xenopus* oocyte maturation, and these phosphorylations correlate with a decline of its catalytic activity (Mueller *et al.*, 1995; Palmer, Gavin and Nebreda, 1998).

Since tyrosine dephosphorylation of Cdc2 represents a crucial step in MPF activation, much work has focused on identification and regulation of the antagonistic enzyme of Wee1/Myt1, the tyrosine phosphatase of Cdc2. Once again, the combination of the yeast genetics and the amphibian oocyte biochemistry proved to be a powerful strategy. It was known that in the fission yeast, the dephosphorylation pathway which activates Cdc2 requires the *Cdc25* gene product (Gould *et al.*, 1990; Russell and Nurse, 1986), and that the regulation of Cdc2 by Cdc25 is widely conserved, as functional homologues of Cdc25 had been isolated in the budding yeast (Russell, Moreno and Reed, 1989), *Drosophila* (Edgar and O'Farrell, 1989) and humans (Sadhu *et al.*, 1990). Although the genetic evidence showed that Cdc25 activates Cdc2 at the onset of mitosis, the biochemistry of Cdc25 protein activity was not understood and it had been proposed that Cdc25 would act indirectly on Cdc2 by activating a cryptic phosphatase. However, the observation that a highly conserved region of Cdc25 protein shows discernable homology with a group of dual-specificity (Tyr/Ser) phosphatases supported the possibility that Cdc25 could act directly as a tyrosine phosphatase to activate Cdc2. In 1991, using a biochemical approach based on *Xenopus* oocyte extracts, it was indeed discovered that Cdc25 is the specific protein phosphatase that dephosphorylates Tyr15 and possibly Thr14 residues on Cdc2, and regulates MPF activation (Gautier *et al.*, 1991; Dunphy and Kumagai, 1991; Jessus and Beach, 1992). One year later, the activity of the Cdc25 phosphatase at the G2/M transition was shown to be directly regulated through changes in its phosphorylation state, phosphorylation of several Ser and Thr residues accompanying its activation, while a treatment with either phosphatase 1 or 2A decreases its ability to activate Cdc2 kinase (Izumi, Walker and Maller, 1992; Kumagai and Dunphy, 1992).

In most species (types 1 and 2, Figure 9.2), the full-grown prophase-blocked oocyte is then equipped with pre-MPF molecules formed of complexes between cyclin B and Cdc2, where Cdc2 is kept inactive by inhibitory Tyr15 and Thr14 phosphates. The two direct regulators of MPF are also expressed in the oocyte, Myt1 kinase being active and Cdc25 phosphatase being inactive, both of them under a non-phosphorylated state. The activation of MPF is a two-step mechanism: the first step requires the formation of a trigger that can result in the appearance of a few molecules of active MPF. In the second step, the small pool of newly formed active MPF is able to bring about Cdc25 activation, and inactivation of Myt1, and hence to establish the positive feedback loop of MPF activation known as MPF autoamplification: the conversion of inactive pre-MPF into MPF is strongly accelerated by MPF itself. The next paragraphs focus on this mechanism that allows entry into the first meiotic division.

9.4 MPF activation, act I: formation of a starter

Clearly, the mechanisms operating in oocytes where protein synthesis is required for MPF activation differ greatly from those involved in oocytes where translation is not needed.

9.4.1 The starfish and mouse paradigm: oocytes with pre-MPF, and no need of protein synthesis to activate MPF

The starfish oocyte is a model system where MPF is activated at meiosis I independently of protein synthesis. Under the stimulus of 1-methyladenine, the maturation-inducing hormone, MPF is activated within 10 minutes with no requirement for new protein synthesis (Kanatani *et al.*, 1969). Hence it is clear that the initiation of MPF activation is promoted by the reversion of the balance between the two opposing MPF regulators, Cdc25 and Myt1, without the need of new formed Cdc2–cyclin B complexes. The mechanisms involved in this regulation have been quite well clarified. Binding of 1-methyladenine to its putative surface receptor releases $G_{\beta\gamma}$ that activates PI3K and, in turn, PDK (Jaffe *et al.*, 1993; Hiraoka *et al.*, 2004; Sadler and Ruderman, 1998). Finally, the kinase Akt is phosphorylated and activated in a PDK-dependent manner (Hiraoka *et al.*, 2004; Okumura *et al.*, 2002). Akt functions as the trigger kinase, able to activate MPF for meiotic resumption. It phosphorylates both Myt1 and Cdc25, inactivating the first one and activating the second one (Kishimoto, 2003; Okumura *et al.*, 2002). Akt therefore switches the balance of both regulator activities, causing the initial activation of Cdc2–cyclin B (Figure 9.3).

As in starfish, Cdc2–cyclin B activation at meiotic resumption of the mouse oocyte does not require new protein synthesis. The mouse prophase oocyte contains a pool of inactive Cdc2–cyclin B complexes, that is, a pre-MPF stock. However, MPF activation during mouse oocyte meiosis I relies on an original two-step mechanism that resembles neither the starfish nor the *Xenopus* models. In a first phase, pre-MPF molecules are dephosphorylated and activated by Cdc25, rapidly leading to GVBD and chromatin condensation without requirement for the synthesis of cyclin B or any other protein (Ledan *et al.*, 2001; Polanski *et al.*, 1998). However, the MPF activity level generated by pre-MPF dephosphorylation is not sufficient to establish conditions of a full M phase, especially the formation of a functional meiosis I microtubular spindle. This first period is followed by the massive synthesis of cyclin B which gradually increases the level of active MPF (Winston, 1997; Hampl and Eppig, 1995; Hoffmann *et al.*, 2006). This second phase, which depends on cyclin B synthesis, produces the high MPF activity levels required to organize the first meiotic spindle. So, the first meiotic M phase of the mouse oocyte results from a biphasic activation of MPF: the initial phase relying on

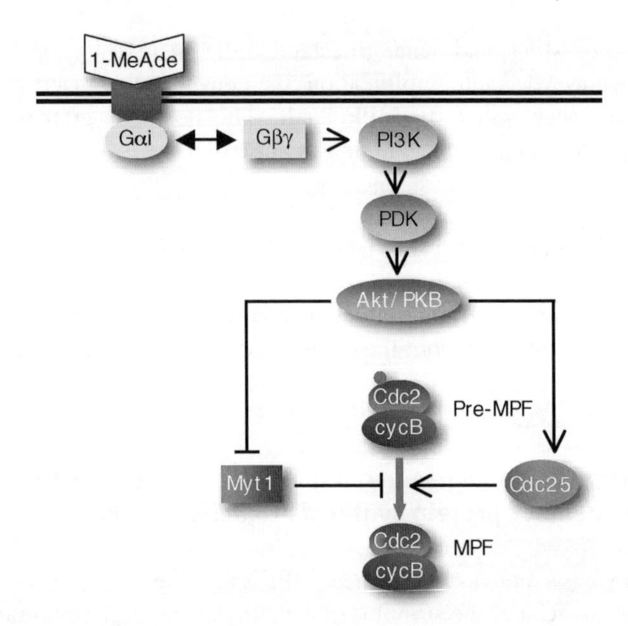

Figure 9.3 Starfish oocyte model. Binding of 1-methyladenine (1-MeAde) to its receptor releases $G\beta\gamma$ that activates PI3K and, in turn, PDK. Then PDK activates Akt/PKB that suppresses Myt1 activity and activates Cdc25. The first molecules of active MPF originate from the pre-MPF stockpile and initiate the autoamplification mechanism. cycB = cyclin B. A full colour version of this figure appears in the colour plate section.

Cdc25 and pre-MPF activation independently of protein synthesis, and correlated with GVBD and chromatin condensation; the second one depending on cyclin B translation and allowing the formation of the first meiotic spindle (Figure 9.2). The molecular pathway inducing Cdc25 activation in a translation-independent manner is not yet elucidated in the mouse oocyte, but does not rely on the PI3K/PDK/Akt cascade operating in starfish oocytes. Interestingly, although it is widely accepted that the Cdc25 activity acting to dephosphorylate Cdc2 during oocyte meiotic maturation corresponds to the Cdc25C member of the Cdc25 family, knockout experiments have revealed that Cdc25B is necessary and sufficient for meiotic resumption in mouse oocytes, even though Cdc25A and Cdc25C are present (Lincoln *et al.*, 2002), and that Cdc25C is clearly dispensable (Chen *et al.*, 2001).

9.4.2 The *Xenopus* paradigm: oocytes with pre-MPF, and need of protein synthesis to activate MPF

Let's consider now the case of *Xenopus* oocytes that contain a stock of pre-MPF but nevertheless do require newly synthesized proteins to generate the triggering activity leading to MPF activation. Three good candidates have been proposed to help account for the protein synthesis requirement during oocyte maturation.

The first one is the product of the *c-mos* proto-oncogene. Mos is a Ser/Thr protein kinase that is specifically expressed and functions during meiotic maturation of oocytes.

Oocytes arrested at prophase I lack detectable levels of Mos, which is synthesized from a pool of maternal mRNAs at the time of MPF activation (Sagata *et al.*, 1988). Mos activates a MAP kinase kinase, MEK, which in turn activates MAPK, resulting finally in the activation of the ribosomal S6 kinase, p90^{Rsk} (Posada *et al.*, 1993; Nebreda and Hunt, 1993; Hsiao *et al.*, 1994; Shibuya and Ruderman, 1993). Mos and its constitutively active downstream targets – MEK, MAPK, p90^{rsk} – are able to induce meiotic maturation in the absence of progesterone when microinjected into prophase oocyte (Sagata *et al.*, 1989; Gross, Lewellyn and Maller, 2001; Huang, Kessler and Erikson, 1995; Haccard *et al.*, 1995). MAPK is the mandatory link between Mos and Cdc2 activation, as injected Mos is not able to promote MPF activation in the presence of the MEK inhibitor, U0126, that prevents MAPK phosphorylation and activation (Gross *et al.*, 2000). Moreover, the destruction of Mos mRNA by antisense oligodeoxynucleotides was shown to prevent progesterone-induced meiotic maturation (Sagata *et al.*, 1988). Mos was thus proposed to be a prime candidate to control the entry into meiosis in amphibians. It was then shown that p90^{Rsk} phosphorylates the regulatory domain of Myt1, downregulates Myt1 kinase activity on Cdc2–cyclin B complexes and associates with Myt1 in mature oocytes (Palmer, Gavin and Nebreda, 1998). In 2002, another link between Mos and MPF was revealed in *Xenopus* oocyte. Mos was shown to bind Myt1 to trigger its phosphorylation *in vivo*, even in the absence of MAPK activation (Peter *et al.*, 2002). Altogether, these results led to a simple scenario: Mos is synthesized in response to progesterone and leads to p90^{Rsk} activation. The interaction between Mos and Myt1 would facilitate Myt1 inactivation by direct phosphorylation and by allowing the complete phosphorylation of Myt1 by other kinases, such as p90^{Rsk}. Inactivation of Myt1 would allow the formation of the first molecules of active dephosphorylated Cdc2 that can initiate the positive feedback loop. However, the biological relevance of this pathway was then strongly questioned. First, Mos synthesis is not sufficient for the MPF activation process. Indeed, in the absence of protein synthesis, injected Mos cannot induce MPF activation (Nebreda and Hunt, 1993; Yew, Mellini and Vande Woude, 1992; Daar, Yew and Vande Woude, 1993). Second, MAPK is dispensable for MPF activation because the MEK inhibitor U0126 does not block oocyte maturation induced by progesterone (Gross *et al.*, 2000; Dupre *et al.*, 2002). How to explain that Mos is required for MPF activation through MAPK recruitment, while the only Mos downstream target, MAPK, is dispensable for the same process? Third, inhibition of Mos synthesis by morpholino antisense oligonucleotides fails to block progesterone-stimulated GVBD (Dupre *et al.*, 2002; Baert *et al.*, 2003), in contrast with the reported effects of the destruction of Mos mRNA by conventional antisense oligonucleotides (Sagata *et al.*, 1988). Fourth, Mos protein accumulation and MAPK activation are not detectable in progesterone-stimulated oocytes where MPF activation is impaired by a direct Cdc2 inhibitor, showing that the Mos/MAPK/p90^{Rsk} pathway is under the control of MPF (Frank-Vaillant *et al.*, 1999). Fifth, it is questionable that the unique inhibition of Myt1 activity would be sufficient to drive MPF activation. The prophase-blocked oocyte contains a stock of pre-MPF where Cdc2 is already phosphorylated on Tyr15, cyclin B synthesis is almost inactive, so no new complexes are formed, and the Cdc25 phosphatase is inactive. Therefore, since the Myt1 substrate, Cdc2, is already stably phosphorylated, the inhibition of Myt1 should not modify its phosphorylation level, unless Cdc25 is activated. All these data, in conjunction with the observation that in mouse, starfish, and goldfish, neither Mos synthesis nor the downstream MAPK activity are required for Cdc2 activation or progression through the first meiosis (Tachibana *et al.*, 2000; Colledge *et al.*, 1994; Hashimoto *et al.*, 1994; Kajiura-Kobayashi *et al.*, 2000), led to the conclusion that the Mos/MAPK/p90^{Rsk} pathway is not necessary for MPF activation in *Xenopus* oocytes. However, one has to keep in mind that, although not necessary for GVBD, Mos remains a powerful inducer of meiotic maturation when microinjected.

The second conspicuous protein whose synthesis would account for MPF initiation is an activating partner of Cdc2. Based on a study showing that a mutant of Cdc2, able to bind cyclins but lacking kinase activity, blocks MPF activation, it has been proposed that newly synthesized Cdc2-binding activator could trigger the activation of MPF (Nebreda, Gannon and Hunt, 1995). The newly synthesized Cdc2 partner would bind and activate Cdc2, generating a small pool of active MPF able to bring about Cdc25 activation and inactivation of Myt1. There are only a limited number of known regulatory subunits of Cdc2 in the frog oocyte as candidates for this role, namely cyclins B, cyclins A and Ringo/Speedy. Cyclin A synthesis is unlikely to be responsible for triggering MPF activation, as the levels of the protein are very low at the time of GVBD and it was shown by antisense injection that cyclin A is not required for oocyte maturation (Kobayashi *et al.*, 1991; Minshull *et al.*, 1991). B-type cyclins are newly synthesized in response to progesterone and independently of MPF activity (Kobayashi *et al.*, 1991; Frank-Vaillant *et al.*, 1999), and are able to induce entry into meiosis even in the absence of protein synthesis (Roy *et al.*, 1991; Huchon *et al.*, 1993). However, it was shown by an antisense strategy that synthesis of all B-type cyclins expressed in the *Xenopus* oocyte (B1, B2, B4 and B5) is indeed not required for the activation of MPF (Hochegger *et al.*, 2001; Haccard and Jessus, 2006).

Besides cyclins, another Cdc2-binding protein named Ringo/Speedy has been identified. This protein accumulates in response to progesterone and its injection promotes GVBD and MPF activation in *Xenopus* oocytes (Lenormand *et al.*, 1999; Ferby *et al.*, 1999). Moreover, the Cdc2–Ringo complexes are less susceptible to regulation by Myt1 and CAK, and could be active under conditions where cyclin-bound Cdc2 is inhibited (Karaiskou *et al.*, 2001). Recently, it has been proposed that Myt1 would be downregulated by Cdc2–Ringo and not by Cdc2–cyclin (Ruiz, Hunt and Nebreda, 2008). Ringo is therefore a good candidate to help account for the generation of a small pool of active Cdc2 kinase able to fire the autoamplification loop. However, if its translation seems sufficient for entry into meiosis I, it appears unlikely to be required, since ablation of its mRNA by antisense injection delays GVBD but does not suppress MPF activation (Haccard and Jessus, 2006; Lenormand *et al.*, 1999).

Recent data contributed to unravelling the question of the identification of the newly synthesized proteins required for Cdc2 activation in *Xenopus* oocytes. The experiments consisted of inhibiting the Mos/MAPK cascade, cyclin B and Ringo/Speedy synthesis either separately or concomitantly, and monitoring the effects on meiotic maturation. Remarkably, it was observed that Cdc2 activation induced by progesterone was completely abolished when cyclin B synthesis and the Mos/MAPK pathway were simultaneously impaired (Haccard and Jessus, 2006). In light of these new findings, Cdc2 activation can result from either of the pathways triggered, cyclin B or Mos. According to this model, either pathway is sufficient to trigger Cdc2 activation if

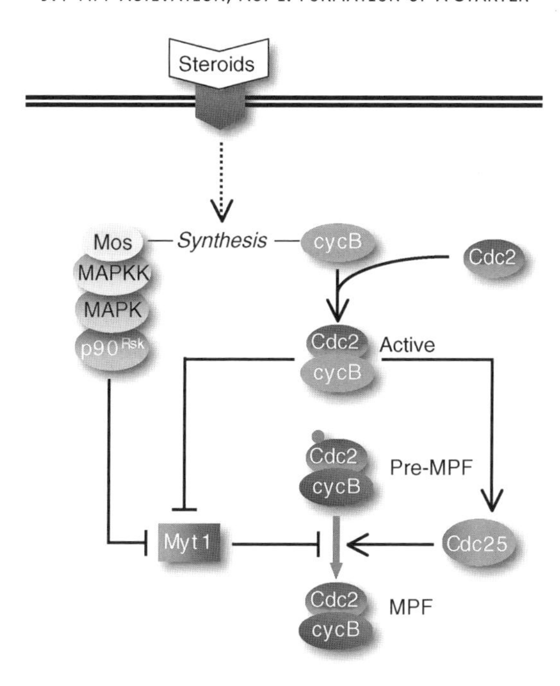

Figure 9.4 *Xenopus* oocyte model. Binding of progesterone to an unidentified membrane receptor leads to the synthesis of cyclin B (cycB) that binds and activates Cdc2. The newly formed active complexes bring about Cdc25 activation and Myt1 suppression. Then, the first molecules of active MPF that initiate the autoamplification mechanism do not originate from the pre-MPF stockpile but are formed *de novo* by synthesizing cyclin B. If cyclin B translation is impaired, the Mos/MAPK/p90^{Rsk} pathway, that is turned on by Mos synthesis, can lead to MPF activation by inhibiting Myt1. In this case, the first molecules of active MPF originate from the pre-MPF stockpile. A full colour version of this figure appears in the colour plate section.

the other fails. Under physiological conditions, the initial starter able to fire the autoamplification loop would be the synthesis of cyclin B, and the Mos/MAPK/p90^{Rsk} pathway would not be recruited or would exert an ancillary function. Cyclin B synthesis would lead to newly formed active Cdc2–cyclin B complexes (Figure 9.4). How these new complexes are prevented from inactivation by immediate phosphorylation by Myt1 is an important question. Progesterone treatment could downregulate Myt1 activity in parallel to the stimulation of cyclin B synthesis. Alternatively, inactivation or sequestration of Myt1 in the full-grown oocyte would allow Cdc2–cyclin B produced in response to progesterone to escape inhibitory phosphorylation (see Section 9.6). If cyclin B synthesis is impaired, the frog oocyte can use the activation of Mos/MAPK/p90^{Rsk} as a rescue mechanism to lead to MPF activation. Under these conditions, since no new Cdc2–cyclin B complexes are formed, the only way for the Mos pathway to generate a starter amount of MPF is to activate the pre-existing store of inactive Cdc2–cyclin B complexes. Activating the Cdc25 phosphatase is likely to be the key. It is therefore more than probable that the Mos/MAPK/p90^{Rsk} pathway, able to downregulate Myt1, also positively regulates Cdc25; a hypothesis that has yet to be explored. The functional redundancy of the distinct pathways, one depending on cyclin B synthesis and the other one on Mos synthesis, resulting in the formation of a trigger for MPF activation, probably ensures the physiological robustness of the process of MPF activation in the frog oocyte.

9.4.3 Fishes and some amphibians: oocytes without pre-MPF

MPF has been highly purified from eggs of several species of fish, such as the carp *Cyprinus carpio*, the catfish *Clarias batrachus* and the perch *Anabus testudineus*, where it exhibits the universal molecular structure of a complex of Cdc2 and cyclin B (Yamashita *et al.*, 1992; Basu *et al.*, 2004; Balamurugan and Haider, 1998). However, cyclin B is not expressed at the protein level in the prophase-blocked oocytes of many fish species, such as goldfish, carp, catfish, zebrafish and lamprey, and of several amphibians, such as the frog *Rana japonica*, the toad *Bufo japonicus* and the newt *Cynops pyrrhogaster* (Tanaka and Yamashita, 1995; Sakamoto *et al.*, 1998; Ihara *et al.*, 1998; Kondo *et al.*, 1997). All Cdc2 molecules exist as a monomer, and the protein content of Cdc2 is constant during oocyte maturation. In response to the maturation-inducing steroid, cyclin B protein is synthesized from its stored mRNA and binds to pre-existing Cdc2 (Hirai *et al.*, 1992). In fishes, induction of oocyte meiotic maturation requires the synthesis of proteins and is independent of transcription (Katsu, Yamashita and Nagahama, 1999). It has been clearly demonstrated that cyclin B is the only protein whose translation is necessary and sufficient to promote MPF activation in goldfish oocyte. Injection of cyclin B protein into immature goldfish oocytes promotes GVBD independently of protein synthesis (Katsu *et al.*, 1993). Conversely, MPF activation is blocked by injection of oligonucleotide antisense targeted against cyclin B in the oocyte of *Rana japonica* (Ihara *et al.*, 1998). Clearly, also, the translation of the Ser/Thr protein kinase Mos, that had been proposed to play an important role in MPF activation in *Xenopus* oocytes, is dispensable for MPF activation in oocytes in which pre-MPF is absent. In goldfish and *Rana japonica*, ectopic expression of Mos is unable to initiate oocyte maturation, whereas inhibition of Mos synthesis and MAPK activation does not prevent MPF activation induced by the maturation-inducing steroids (Kajiura-Kobayashi *et al.*, 2000; Yoshida, Mita and Yamashita, 2000).

During goldfish oocyte maturation, Cdc2 becomes phosphorylated on Thr161 by CAK upon cyclin binding, but the new complexes are not phosphorylated on Tyr15 or Thr14 of Cdc2 (Kondo *et al.*, 1997; Yamashita *et al.*, 1995). Consequently, Cdc2 is activated solely by cyclin binding and Thr161 phosphorylation, and escapes the critical regulation exerted by the Wee1/Myt1 and Cdc25 enzymes (Figure 9.5). Since CAK is active throughout the process of zebrafish oocyte meiotic maturation (Kondo *et al.*, 1997), the translational control of cyclin B appears to be the key event in MPF activation in the fish oocyte. The translational activation of cyclin B mRNA involves microfilament-dependent change in the mRNA distribution, from an aggregated form to a dispersed form (Kondo, Kotani and Yamashita, 2001), and the polyadenylation of the mRNA (Katsu, Yamashita and Nagahama, 1999; Katsu, Yamashita and Nagahama, 1997).

However, the absence of pre-MPF in prophase-arrested oocytes is not universal among fishes and amphibians. Cyclin B is present in immature oocytes of the freshwater

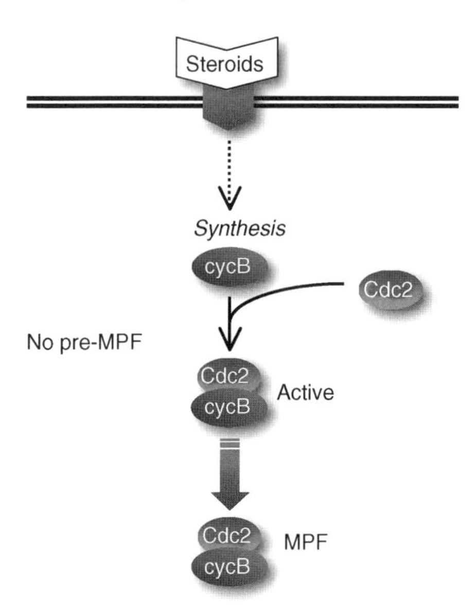

Figure 9.5 Fish oocyte model. Binding of steroids to an unidentified membrane receptor leads to the synthesis of cyclin B (cycB). Cdc2 is activated solely by cyclin binding, and escapes the regulation by Myt1 and Cdc25. Pre-MPF molecules are absent. MPF activity results from the accumulation of the newly formed complexes between synthesized cyclin B and pre-existing monomeric Cdc2, and consequently does not involve the autoamplification process. A full colour version of this figure appears in the colour plate section.

perch (Basu *et al.*, 2004), the rainbow trout (Qiu *et al.*, 2008), the urodele axolotl (Pelczar *et al.*, 2007; Vaur *et al.*, 2004) and at least two *Xenopus* species: *laevis* (Kobayashi *et al.*, 1991) and *tropicalis* (Stanford *et al.*, 2003; Bodart *et al.*, 2002). Therefore, the molecular mechanisms of MPF activation are different depending on fish and amphibian species, some of them devoid of pre-MPF and relying on the synthesis of cyclin B, while others require the synthesis of new proteins to activate the stockpiled pre-MPF. This could reflect a difference in the stage at which the prophase full-grown oocytes are arrested in these different species. Oocytes that have the germinal vesicle located deeply in the cytoplasm and migrating toward the animal pole in response to the steroid inducer, such as goldfish, could be arrested at a stage more distal to GVBD than oocytes where the germinal vesicle is near the animal pole, as in *Xenopus* (Yamashita *et al.*, 1995; Vaur *et al.*, 2004). The extreme models, the goldfish oocyte without pre-MPF and the *Xenopus* oocyte with its high amount of pre-MPF, would be separated by intermediate categories, represented by oocytes of axolotl, perch and rainbow trout, characterized by low levels of pre-MPF. MPF activation in the oocytes of these intermediate categories could require both dephosphorylation of Tyr15 of Cdc2 contained in the pre-MPF complexes, and cyclin B synthesis to elevate the MPF content. Then, the translational control of cyclin B is a crucial step in MPF activation, either to form, with Cdc2, a starter that fires the conversion of pre-MPF into MPF (Figure 9.4), or to form *de novo* MPF complexes in oocytes in which pre-MPF is absent (Figure 9.5).

9.5 MPF activation, act II: inside the loop

The positive feedback loop of MPF activation allows a small pool of newly formed active MPF to bring about the conversion of the inactive pre-MPF stock into MPF independently of protein synthesis. This process therefore relies on the existence of a significant store of pre-MPF in the prophase-blocked oocyte, able to amplify MPF activation and to sharpen the transition to meiotic M phase. A high pre-MPF concentration implies an efficient positive feedback loop and then a short lag between the formation of a trigger threshold level and the full activation of MPF. Consequently, MPF activation in prophase-arrested oocytes that do not contain a pre-MPF stockpile (some fishes and amphibians) does not involve any autoamplification mechanism. In these cases, MPF activation is a slow process depending mainly on cyclin B synthesis (Figure 9.2, type 3). It relies on a mechanism inhibiting tyrosine phosphorylation of Cdc2 molecules that associate with newly synthesized cyclin B molecules. Such a mechanism could be supported by active Cdc2 itself in a positive feedback control, but does not correspond to the all-or-nothing nature of MPF activation which is characteristic of the autoamplification concept.

The main features of MPF autoamplification have been established using the *Xenopus* oocyte, which is equipped with a high pre-MPF endowment. They rely on the ability of an active Cdc2 threshold level to strongly accelerate the conversion of the tyrosine-phosphorylated Cdc2 molecules of the pre-MPF stockpile into tyrosine-dephosphorylated active Cdc2 kinase. Cdc25 and Myt1 are clearly the major determinants of the positive feedback loop controlling MPF activation. What are the regulatory pathways that connect Cdc2 to its own regulators, Myt1 and Cdc25?

9.5.1 The Cdc25 side

Cdc2

Cdc25 is regulated by several complex pathways. As the oocyte enters the first meiotic M phase, Cdc25 undergoes an elevation in phosphatase activity that depends on an extensive phosphorylation on serine and threonine residues located in the N-terminal region of the protein (Izumi, Walker and Maller, 1992; Kumagai and Dunphy, 1992). Several of these residues are included in serine/threonine–proline sequences that are a feature of the consensus site for mitotically active kinases such as Cdc2 (Nigg, 1991). Indeed, several lines of evidence point out a direct control of Cdc25 by Cdc2. Cdc25 has high affinity for the Cdc2–cyclin B complex and *in vivo* forms a transient protein complex with MPF (Jessus and Beach, 1992). Interestingly, cyclin B contains a small region, known as P-box, that displays slight sequence similarity to certain phosphotyrosyl phosphatases and has been postulated to contribute in trans to the overall phosphatase activity of Cdc25 that lacks this region (Galaktionov and Beach, 1991; Zheng and Ruderman, 1993). P-box is required for a productive interaction between Cdc2–cyclin B and Cdc25, and Cdc25 activation (Zheng and Ruderman, 1993). Clearly, Cdc2–cyclin B activates and phosphorylates Cdc25 on several residues whose elimination decreases its activity (Izumi and Maller, 1993),

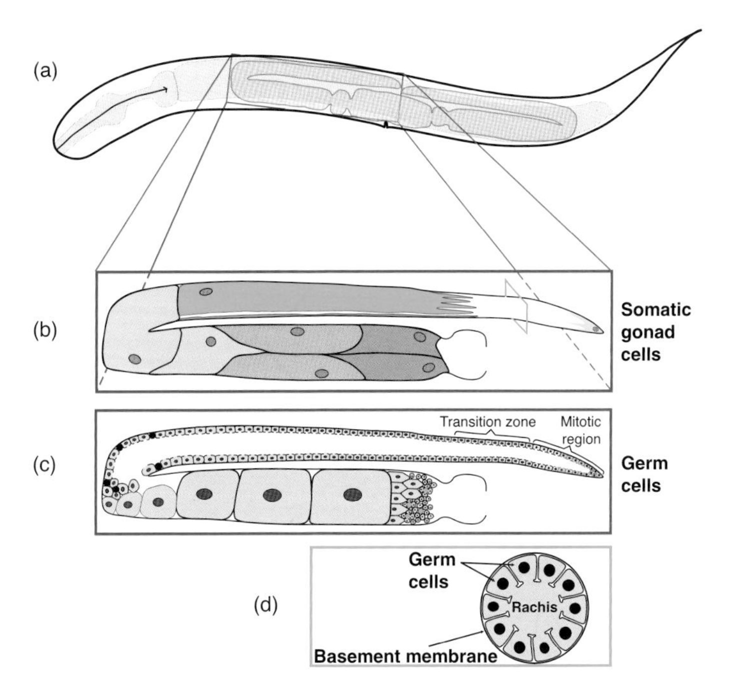

Figure 1.1 Structure of the hermaphrodite gonad. (a) Diagram of a young adult hermaphrodite, showing the digestive system in light green, and the gonad in grey. Anterior is to the left, and ventral is down. (b) Inset diagram of the anterior ovotestis, showing cells of the somatic gonad. The distal tip cell is yellow. Sheath cell 1 is dark blue, sheath cell 2 is light blue, and sheath cell 3 is tan. The second member of each pair is on the opposite side of the gonad, with only the edge of sheath cell 1 visible. sheath cell pair 4 is peach, and sheath cell pair 5 is orange. (c) Inset diagram of the anterior ovotestis, showing the germ cells. Cells expressing female transcripts and proteins are pink, and those expressing male transcripts are blue. Cell corpses are black circles, and residual bodies are blue circles. (d) Cross-section of the gonad

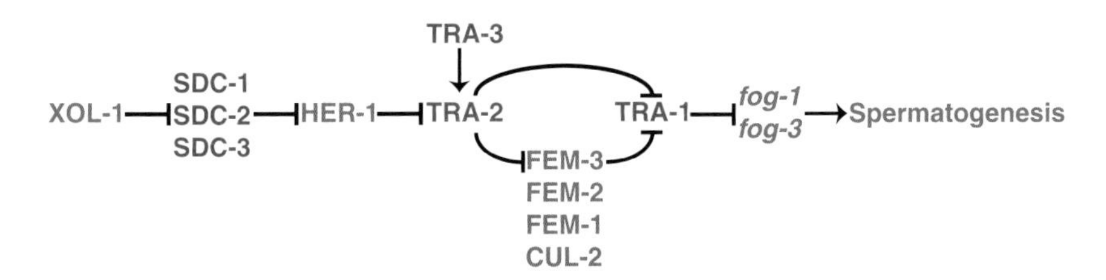

Figure 1.2 The core sex-determination pathway. Genes promoting male fates are blue, and those promoting female fates are pink. Arrows indicate positive interactions, and '—|' indicates negative interactions. Proteins are indicated by capital letters, and genes by lowercase italics

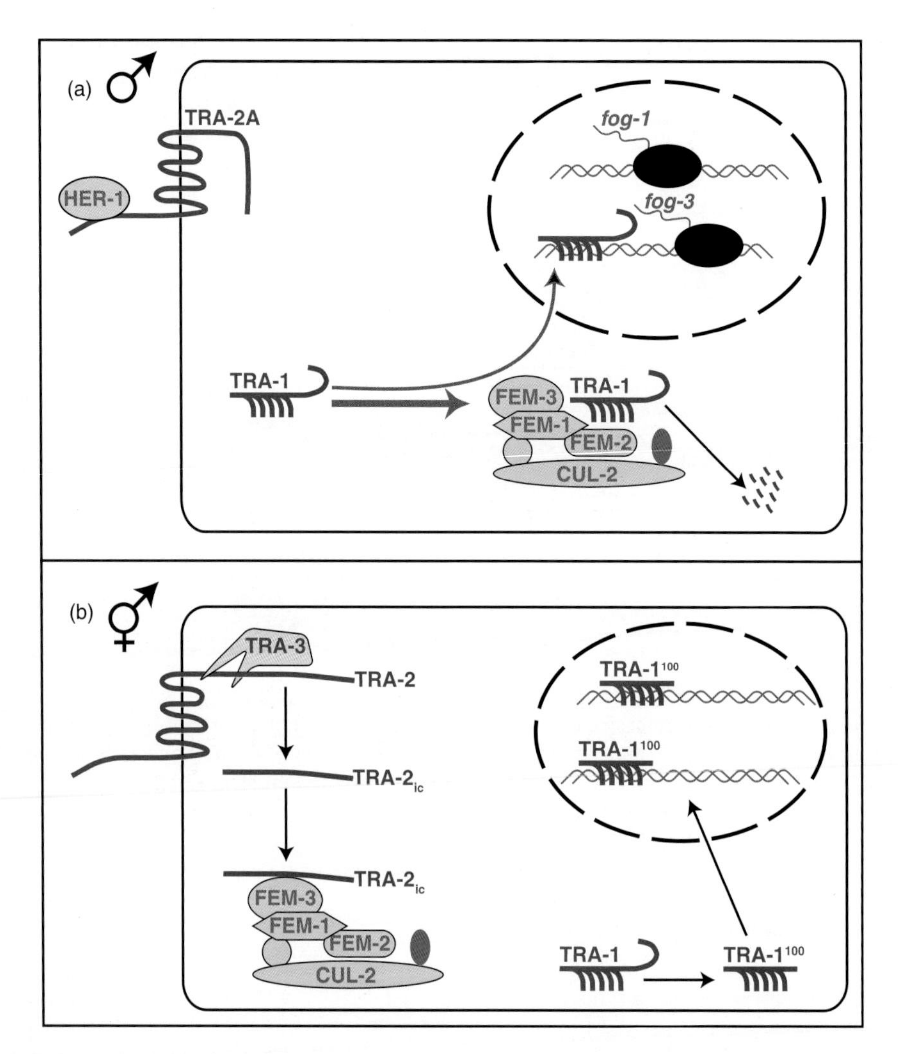

Figure 1.3 Model for the sperm/oocyte decision in adults. (a) In males, HER-1 binds to and represses the TRA-2A receptor; in this diagram, we do not depict cleavage of TRA-2A, but it has not yet been proven that HER-1 prevents this cleavage. The FEM/CUL-2 complex degrades full length TRA-1, which is needed to maintain spermatogenesis in older animals; thus, some TRA-1A is shown being degraded, and some entering the nucleus and regulating targets. The *fog-1* and *fog-3* genes are transcribed and promote spermatogenesis. In the figure, the black ellipses represent RNA polymerase, and the dark blue ellipsis represents ubiquitin. (b) In adult hermaphrodites, TRA-2 and TRA-3 are active, and prevent the FEM/CUL-2 complex from degrading TRA-1A. One possibility is that cleavage of TRA-2A by TRA-3 releases an intracellular fragment that inhibits the FEM complex by binding FEM-3. TRA-1 is cleaved to produce an aminoterminal fragment that represses transcription

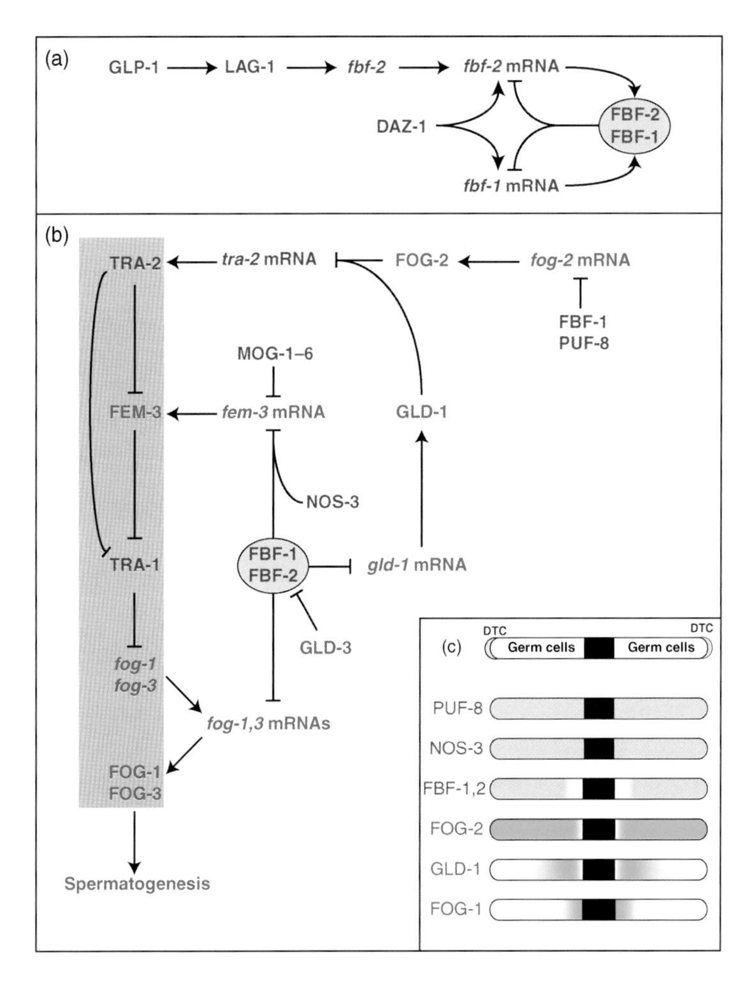

Figure 1.4 Translational regulation of germ cell fates. (a) The distal tip cell promotes FBF activity. In germ cells, the GLP-1 (Notch) receptor is activated by a signal from the distal tip cells (reviewed by Kimble and Crittenden, 2007). Working through the transcription factor LAG-1, it promotes transcription of *fbf-2*. The FBF proteins in turn promote mitotic proliferation or female germ cell fates. Through a feedback loop, they also inhibit their own translation; repression of *fbf-1* by FBF-2 and repression of *fbf-2* by FBF-1 have been demonstrated, and auto-repression is inferred. Proteins are shown in uppercase, and genes in lower case. Arrows indicate positive interactions, and '—|' indicates negative interactions. (b) Modulation of the core sex-determination pathway by translational regulators (highlighted in grey; see text). The FBF proteins act at several points in the sex-determination pathway to prevent the translation of messenger RNAs that promote spermatogenesis. Similarly, GLD-1 acts with FOG-2 to prevent translation of *tra-2* messages, which normally promote oogenesis. GLD-1 also binds *tra-1* messages. All molecules that promote male fates are blue, and those that promote female fates are pink. (c) Expression of translational regulators in L3 hermaphrodites. A schematic of the L3 gonad is shown at top, with the distal tip cells (DTC, yellow) at either end, and other somatic cells (black) in the centre. Rough sketches of the protein levels of key translational regulators are shown below; since none of these studies compared different proteins in the same animals, the regions shown are only approximate. The PUF-8 expression pattern is based on a PUF-8::GFP transgene (Ariz, Mainpal and Subramaniam, 2009). NOS-3 is based on antibody staining (Kraemer *et al.*, 1999), as are FBF (Zhang *et al.*, 1997), FOG-2 (Clifford *et al.*, 2000), GLD-1 (Jones, Francis and Schedl, 1996) and FOG-1 (Lamont and Kimble, 2007)

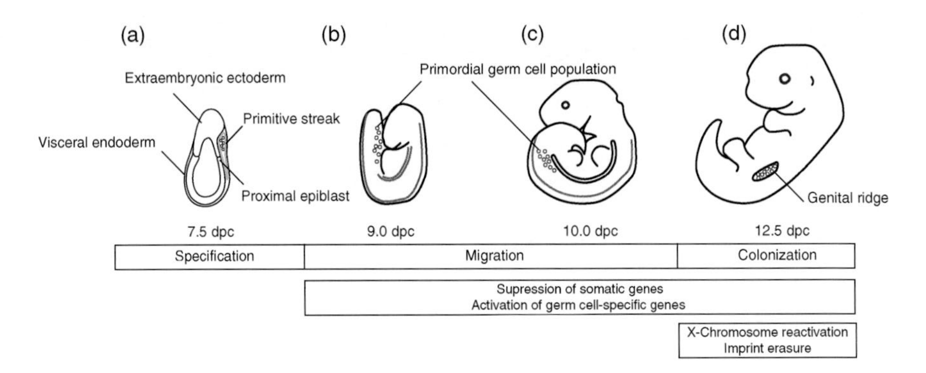

Figure 2.1 Germ cell specification and migration during early mouse development. The primordial germ cells are first identified at 7.25 dpc within the proximal epiblast (a). This population proliferates and migrates through the hindgut (b and c) to colonize the genital ridges by 11.0–12.5 dpc (d). Throughout this process, genetic regulation reinforces the germ cell lineage with suppression of somatic cell genes and upregulation of germ cell-specific genes. X-Chromosome reactivation occurs in female gonads prior to imprint erasure in both sexes. Cartoons for the mouse embryos were adapted from Sasaki and Matsui (2008) and Boldajipour and Raz (2007)

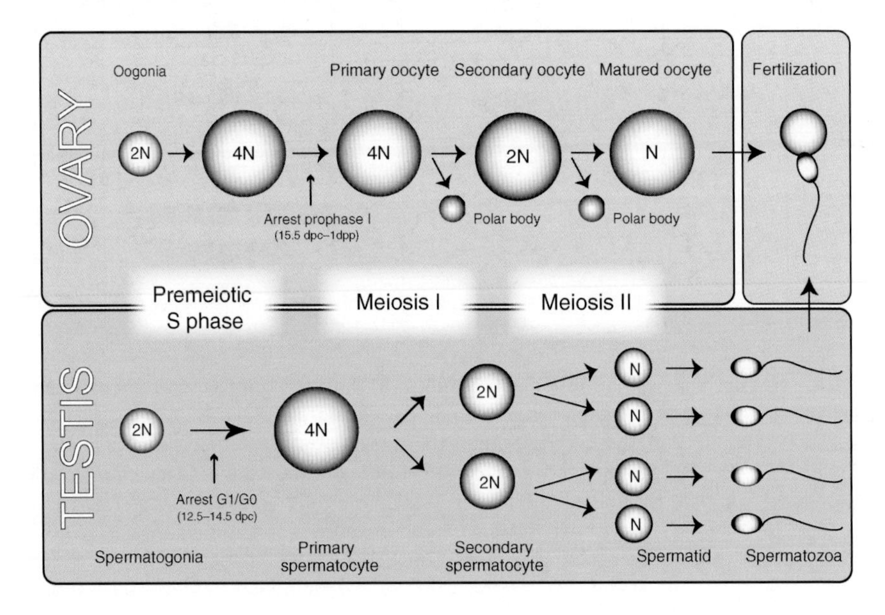

Figure 2.6 Schematic of meiosis. In the ovary, oogonia enter the first stages of meiosis I and begin to arrest in diplotene of prophase I by 17.5 dpc. Following follicle growth, meiosis I is completed with the exclusion of a polar body, and meiosis II is undertaken before arresting in metaphase II. The final stages of meiosis are not completed until fertilization, where the second polar body will be formed. In the testis, spermatogonia proliferate mitotically until 12.5 dpc, when they begin entry into G1/G0 arrest. This is maintained until several days after birth; mitosis is resumed at approximately 5–10 dpp, when they migrate to the basement membrane and become self-renewing spermatogonial stem cells. Following puberty, another round of mitosis yields primary spermatocytes that progress completely through meiosis I and II to produce four haploid spermatids. These cells must then undergo further maturational changes as they progress through to ejaculation and eventual fertilization

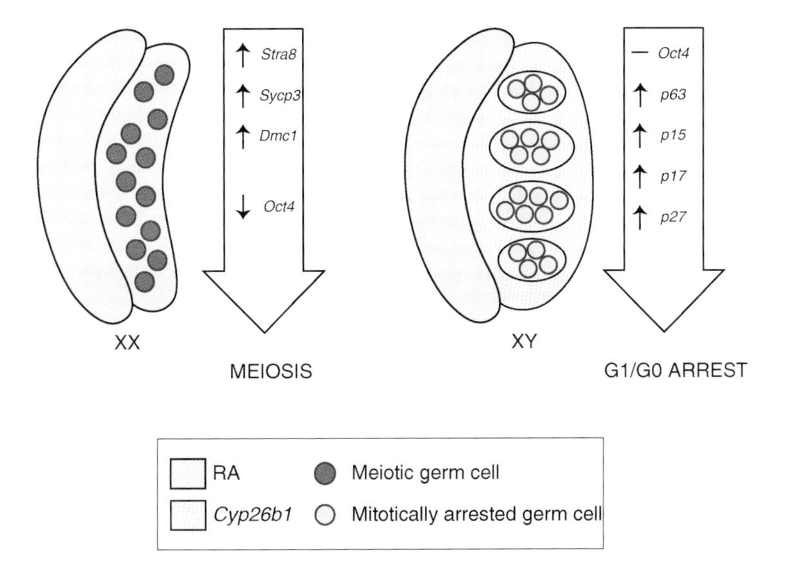

Figure 2.7 Retinoid signalling and meiosis induction. The mesonephroi of both male and female gonads are rich sources of RA. In the female, this diffuses into the gonad proper from the anterior pole to induce meiosis in the germ cells. This is concomitant with an upregulation of various meiotic markers and the downregulation of pluripotency marker *Oct4*. In the testis, Sertoli cells produce the retinoid-degrading enzyme gene *Cyp26b1* to degrade RA as it invades the gonad thereby preventing male germ cell entry into meiosis. Male germ cells enter G1/G0 arrest concomitant with the upregulation of several cell-cycle suppression genes

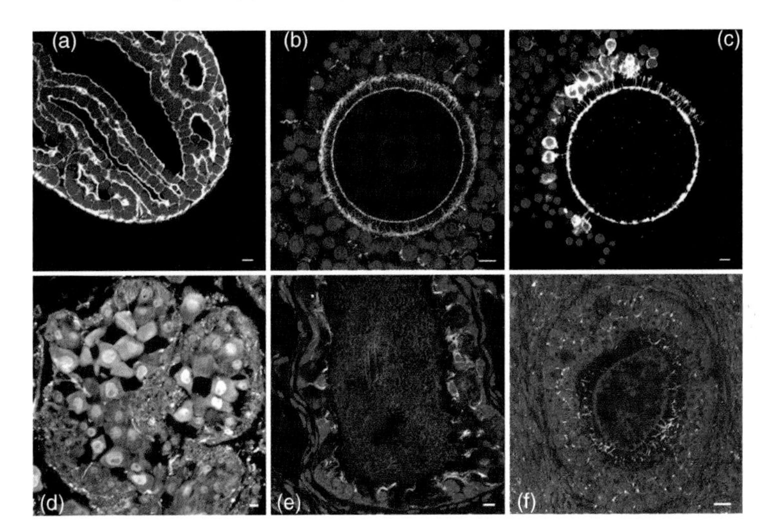

Figure 4.2 Overview of patterns of interaction between oocytes and follicle cells in diverse organisms. Molluscs (squid, a), amplify surface interactions within the follicle by extensive folding in the follicular epithelium which invaginates the oocyte; amplification in mammals (b, gerbil, c, bovine) involves formation of numerous TZPs that are attached to the actin-rich oocyte cortex. Panels a, b, and c are labelled with nuclear marker (red) and F-actin (white, phalloidin). The remaining panels illustrate acetylated tubulin labelling (white) and nuclei (red) in surf clam (d), dogfish (e), and baboon (f) follicles. Stable microtubule-rich TZPs link somatic cells to the oocytes in each of these species providing channels for direct communication. Scale bar = 10 μm, with the exception of d, where bar = 20 μm

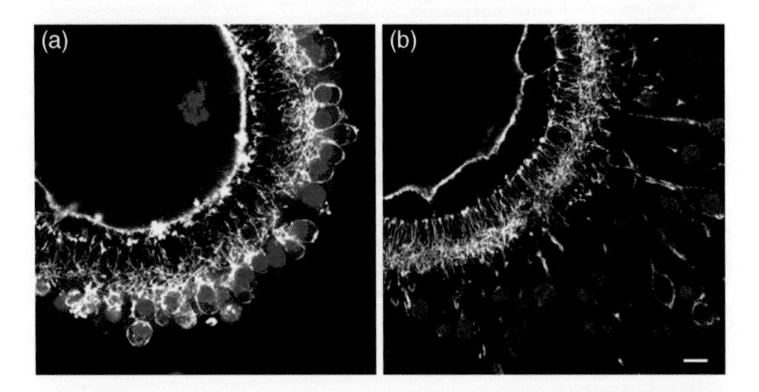

Figure 4.4 Representative confocal images demonstrating remodelling of TZPs during LH-induced meiotic maturation in horse follicles. (a) Organization of TZPs in an immature GV (germinal vesicle) stage oocyte; (b) TZP organization after LH exposure. Note the retraction of actin-rich TZPs but maintenance of contact between larger TZPs and oolemma. Nuclei are labelled in red and phalloidin-actin in white. Scale bar = 10 μm

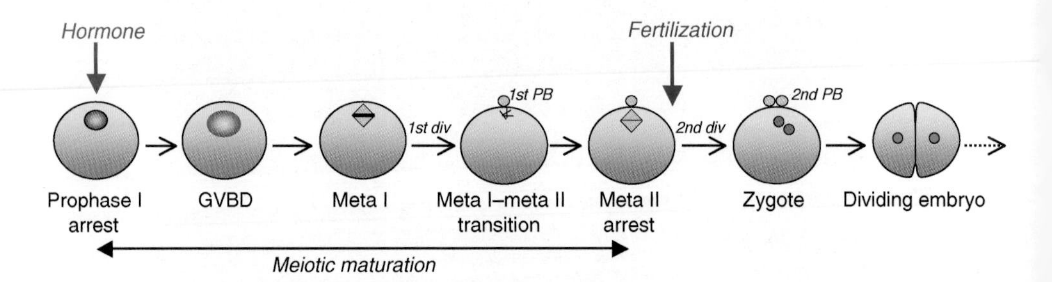

Figure 9.1 Oocyte meiotic maturation. In the ovaries, oocytes are arrested at prophase I. At the time of ovulation, a hormonal signal triggers meiotic maturation: germinal vesicle breakdown (GVBD); formation of the metaphase I spindle (meta I); first meiotic division (1st div) and extrusion of the first polar body (1st PB); formation of the metaphase II spindle (meta II). In vertebrates, the oocyte arrests at metaphase II. Fertilization releases this arrest: the egg completes the second meiotic division (2nd div) by extruding the second polar body (2nd PB) and initiates a series of embryonic divisions

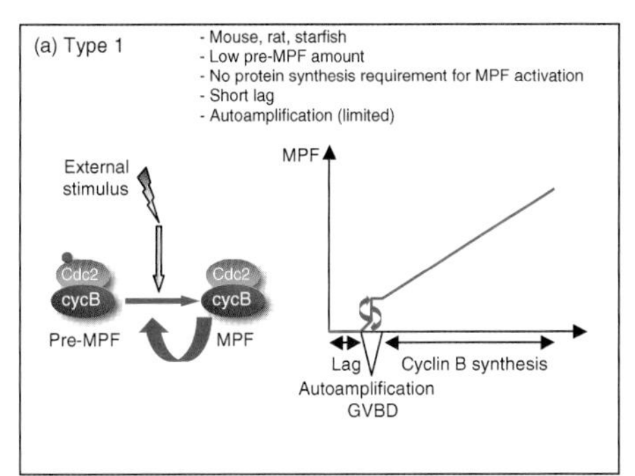

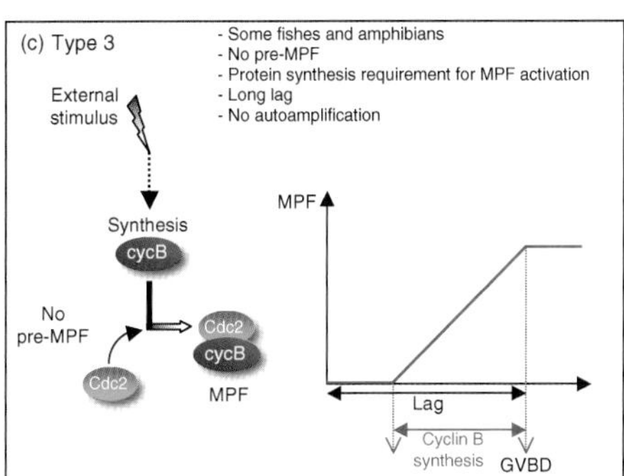

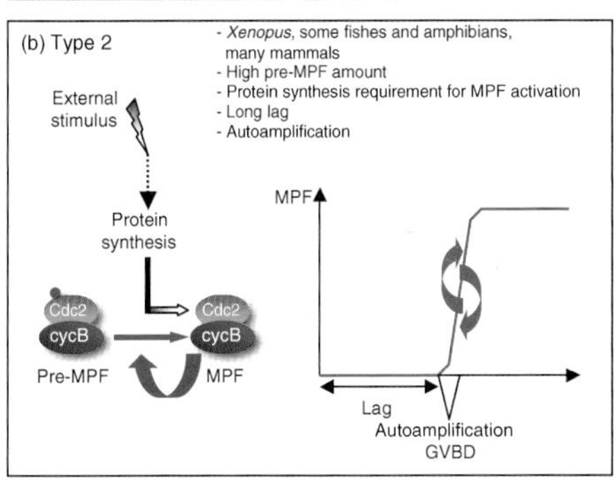

Figure 9.2 MPF activation at entry in first meiotic division: principles and diversity among species. (a) First type (mouse, rat, starfish): MPF activation by the meiotic inducer does not require protein synthesis and occurs after a short lag period. MPF activity is generated from the pre-MPF stock and involves an autoamplification mechanism. The pre-MPF stockpile is low. Once it has been converted into MPF, synthesis of cyclin B (cycB) takes place and increases MPF activity. (b) Second type (*Xenopus*, some fishes and amphibians, many mammals): MPF activation by the meiotic inducer requires protein synthesis and occurs after a few hours' lag period. Newly synthesized proteins allow the generation of the first MPF molecules. This MPF trigger allows the conversion of the pre-MPF stock into MPF according to an autoamplification mechanism independent of protein synthesis. (c) Third type (some fishes and amphibians): the prophase oocyte does not possess any pre-MPF stockpile. MPF activation by the meiotic inducer requires synthesis of cyclin B that binds pre-existing Cdc2 molecules and directly generates active MPF after a lag period of several hours. MPF activity accumulates as a function of the cyclin B synthesis rate, without involving an autoamplification mechanism

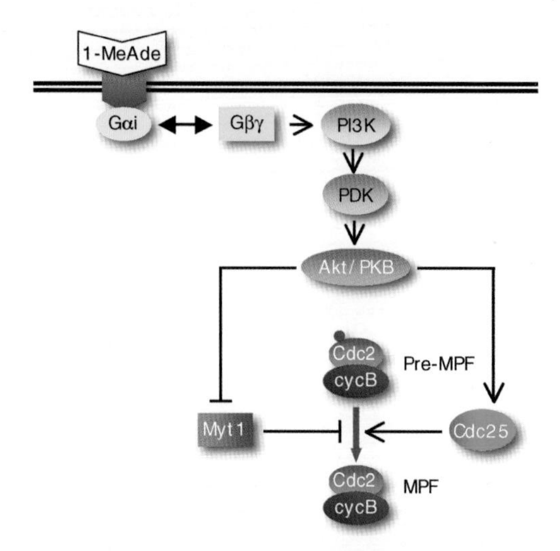

Figure 9.3 Starfish oocyte model. Binding of 1-methyladenine (1-MeAde) to its receptor releases $G\beta\gamma$ that activates PI3K and, in turn, PDK. Then PDK activates Akt/PKB that suppresses Myt1 activity and activates Cdc25. The first molecules of active MPF originate from the pre-MPF stockpile and initiate the autoamplification mechanism. cycB = cyclin B

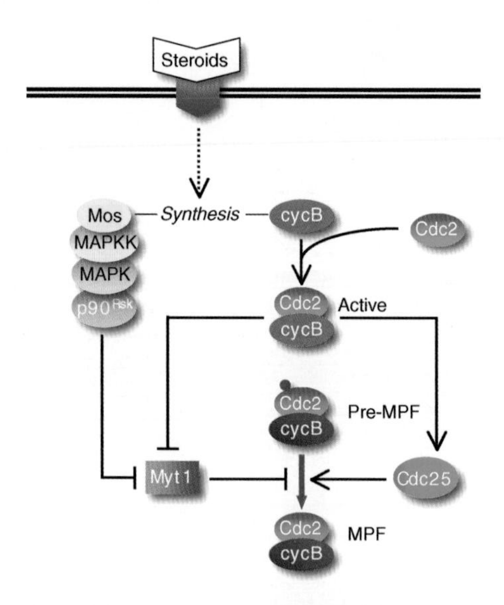

Figure 9.4 *Xenopus* oocyte model. Binding of progesterone to an unidentified membrane receptor leads to the synthesis of cyclin B (cycB) that binds and activates Cdc2. The newly formed active complexes bring about Cdc25 activation and Myt1 suppression. Then, the first molecules of active MPF that initiate the autoamplification mechanism do not originate from the pre-MPF stockpile but are formed *de novo* by synthesizing cyclin B. If cyclin B translation is impaired, the Mos/MAPK/$p90^{Rsk}$ pathway, that is turned on by Mos synthesis, can lead to MPF activation by inhibiting Myt1. In this case, the first molecules of active MPF originate from the pre-MPF stockpile

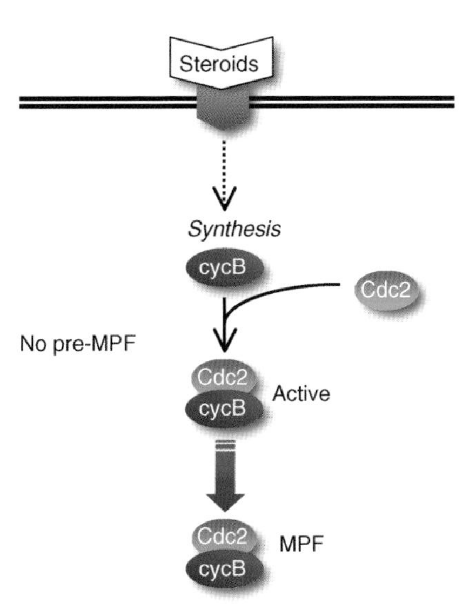

Figure 9.5 Fish oocyte model. Binding of steroids to an unidentified membrane receptor leads to the synthesis of cyclin B (cycB). Cdc2 is activated solely by cyclin binding, and escapes the regulation by Myt1 and Cdc25. Pre-MPF molecules are absent. MPF activity results from the accumulation of the newly formed complexes between synthesized cyclin B and pre-existing monomeric Cdc2, and consequently does not involve the autoamplification process

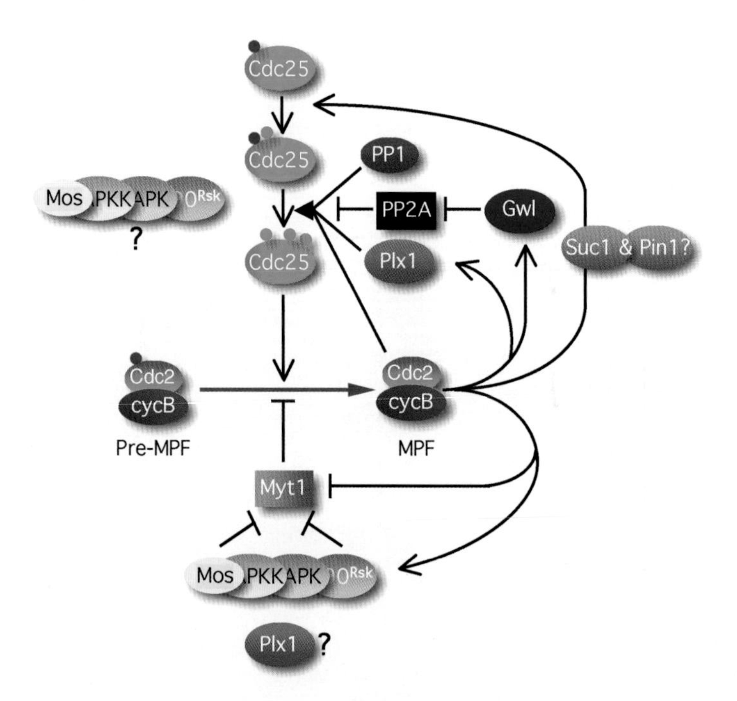

Figure 9.6 MPF autoamplification model. The autoamplification mechanism relies on the ability of active Cdc2–cyclin B complexes to activate their activator Cdc25, and inactivate their inactivator Myt1. Cdc2 partially phosphorylates Cdc25, enhancing the activity level of Cdc25, and Cdc25 dephosphorylation by PP1. Pin1 and Suc1 could regulate the interaction of Cdc2 with Cdc25 through conformational changes. Plx1 and Greatwall (Gwl) are activated under the control of Cdc2. The first one directly phosphorylates Cdc25 and counteracts the inhibitory effects of PP2A, whose activity is downregulated by Gwl. On the other side, Cdc2 plays a central role in inhibiting Myt1 both directly and through the activation of the Mos/MAPK/$p90^{Rsk}$ pathway. Whether the Mos/MAPK/$p90^{Rsk}$ pathway and Plx1 contribute respectively to Cdc25 upregulation and Myt1 downregulation is still a matter of debate. Red circle on Cdc25 = Ser287 inhibitory phosphate targeted by PP1; green circles on Cdc25 = activatory phosphates targeted by Cdc2, Plx1 and PP2A; cycB = cyclin B

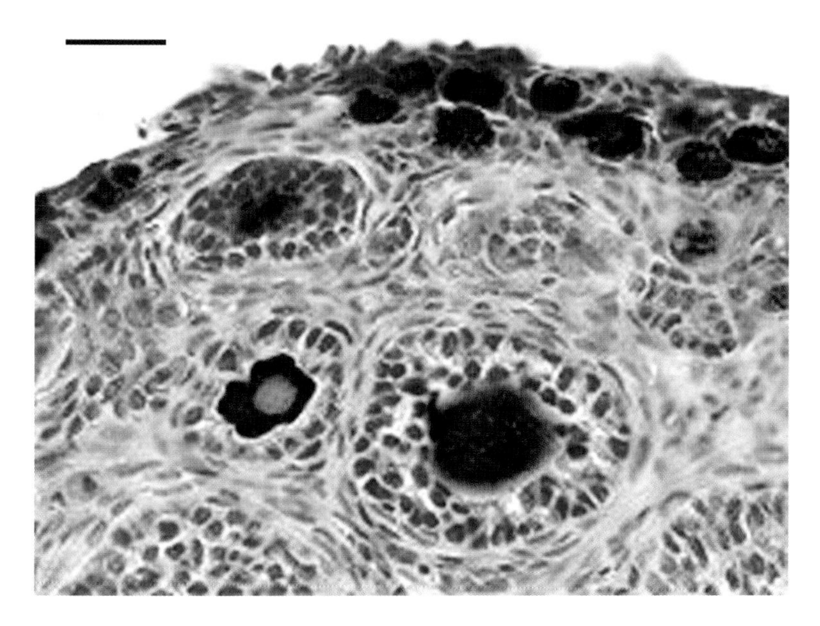

Figure 15.1 Juvenile mouse ovary showing abundant primordial follicles close to the surface epithelium, whereas the growing stages are deeper in the stroma. The oocyte cytoplasm is stained for Vasa and nuclei counterstained with methylene blue. Scale bar = 50 μm

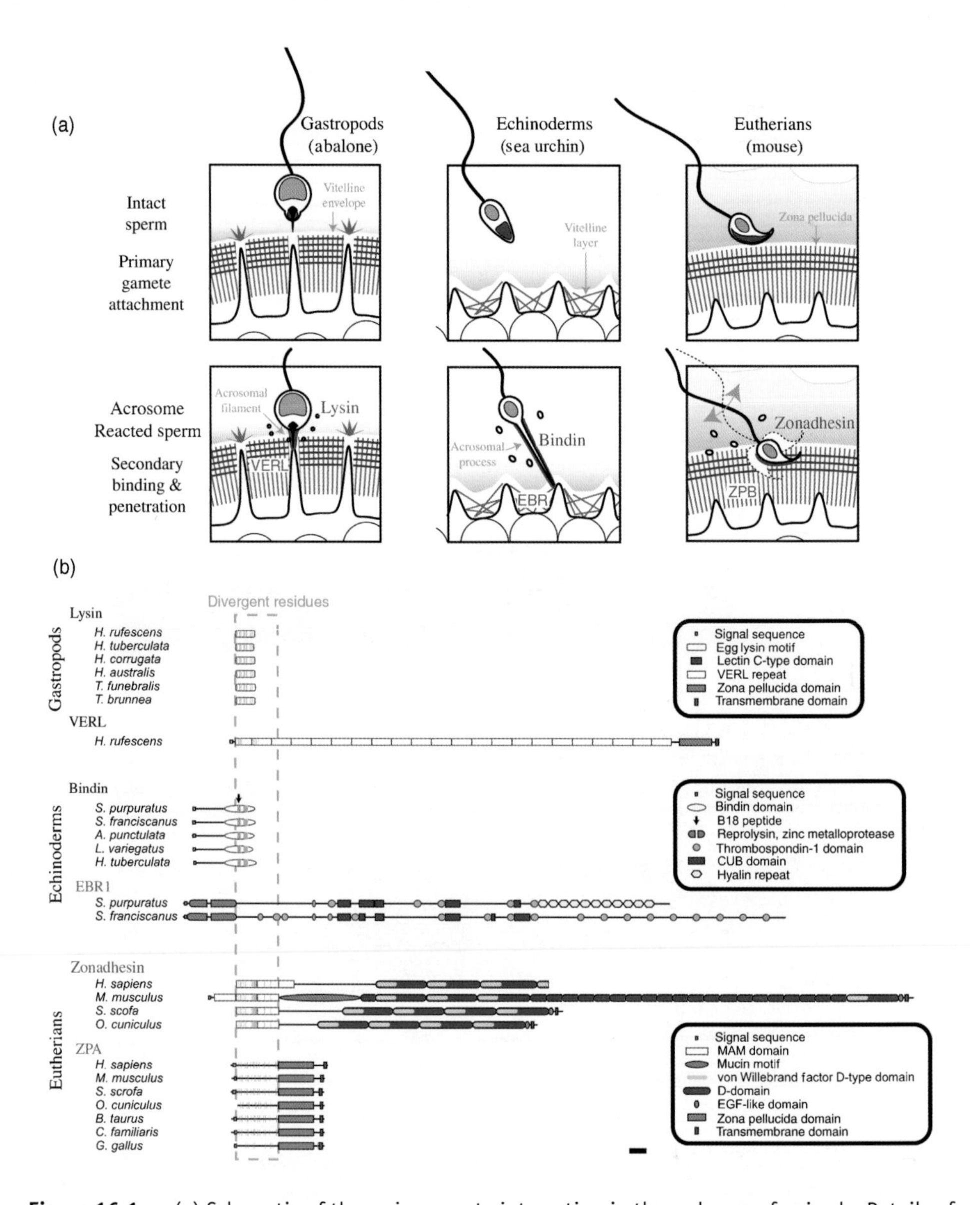

Figure 16.1 (a) Schematic of the major gamete interaction in three classes of animals. Details of egg cortex at the site of sperm binding are shown. Chemoattractant layer (yellow) covers the egg extracellular matrix (blue). The major sperm proteins (red) thought to contribute to the species-specific events are found first in the acrosome, but following exocytosis are relocated to the sperm surface. Basic images are modified from Wong and Wessel (2006). (b) Primary sequence maps of coevolving gamete-binding proteins from each class of animals. Domains specific to each orthologue are detailed in the legends. Most diverse residues (green) are clustered in select regions. Accession numbers include: [lysin] *Haliotis rufescens* (AAA29 196), *H. tuberculata* (AAB59 168), *H. corrugata* (P19 448), *H. australis* (AAA21 517), *Tegula funebralis* (AAD28 265), *T. brunnea* (AAD28 264); [VERL] *Haliotis rufescens* (AAL50 827); [bindin] *Strongylocentrotus purpuratus* (AAA30 038), *S. franciscanus* (AAA30 037), *Arbacia punctulata* (CAA38 094), *Lytechinus variegatus* (AAA29 997), *Heliocidaris tuberculata* (AAQ09 975); [EBR1] *S. purpuratus* (AAR03 494), *S. franciscanus* (AAP44 488); [zonadhesin] *Homo sapiens* (AAC78 790), *Mus musculus* (AAC26 680), *Sus scrofa* (Q28 983), *Oryctolagus cuniculus* (P57 999); [ZPA] *H. sapiens* (AAA61 335), *M. musculus* (P20 239), *S. scrofa* (P42 099), *O. cuniculus* (P48 829), *Bos taurus* (Q9BH10), *Canus familiaris* (P47 983), and *Gallus gallus* (NP_001 034 187). Bar represents 100 residues

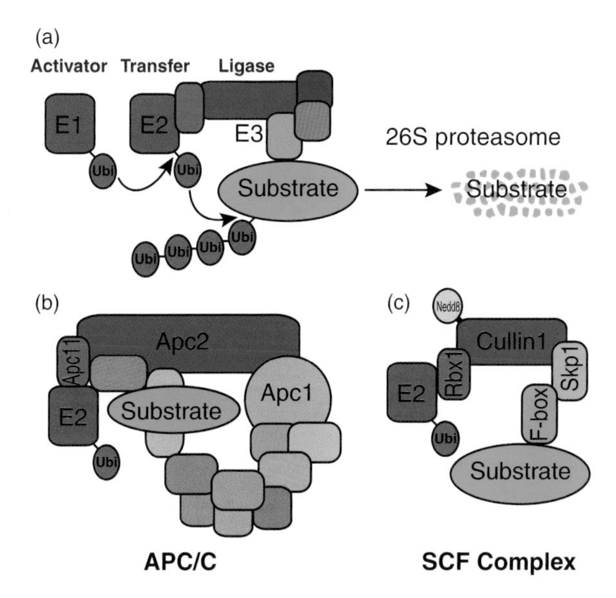

Figure 17.1 Ubiquitination and protein degradation. (a) Ubiquitin is sequentially conjugated onto the E1 activating enzyme, then onto E2 transfer enzyme, and finally E3 ligase brings the E2 close to the target protein, enabling its ubiquitination. Reiteration of these processes leads to the assembly of a polyubiquitin chain, and polyubiquitinated proteins are subsequently degraded by the 26S proteasome. (b) Basic structure of the SCF complex, the archetype of the CRL E3 type ligases. (c) A possible structure for the APC complex (adapted from Peters, 2006)

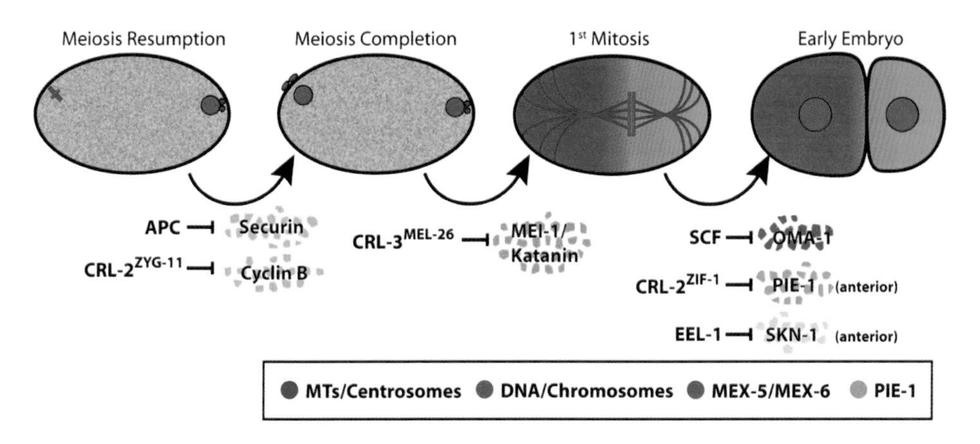

Figure 17.2 Main protein degradation events in the *C. elegans* oocyte-to-embryo transition. Triggering of degradation of specific proteins by E3 ligases is crucial for the main steps to take place. APC and CUL-$2^{ZYG\text{-}11}$ ubiquitinate securin and cyclin B, respectively, targeting them for degradation, thus enabling completion of meiosis. CUL-$3^{MEL\text{-}26}$ targets MEI-1/Katanin for degradation upon completion of meiosis, allowing the formation of a proper mitotic spindle. SCF targets OMA-1 for degradation, ensuring the start of zygotic transcription. In the anterior of the embryo, as part of cell fate patterning, CUL-$2^{ZIF\text{-}1}$ targets PIE-1 (as well as POS-1 and MEX-1, not shown on the figure) for degradation in a MEX-5/6-dependent manner, and EEL-1 targets SKN-1 for degradation

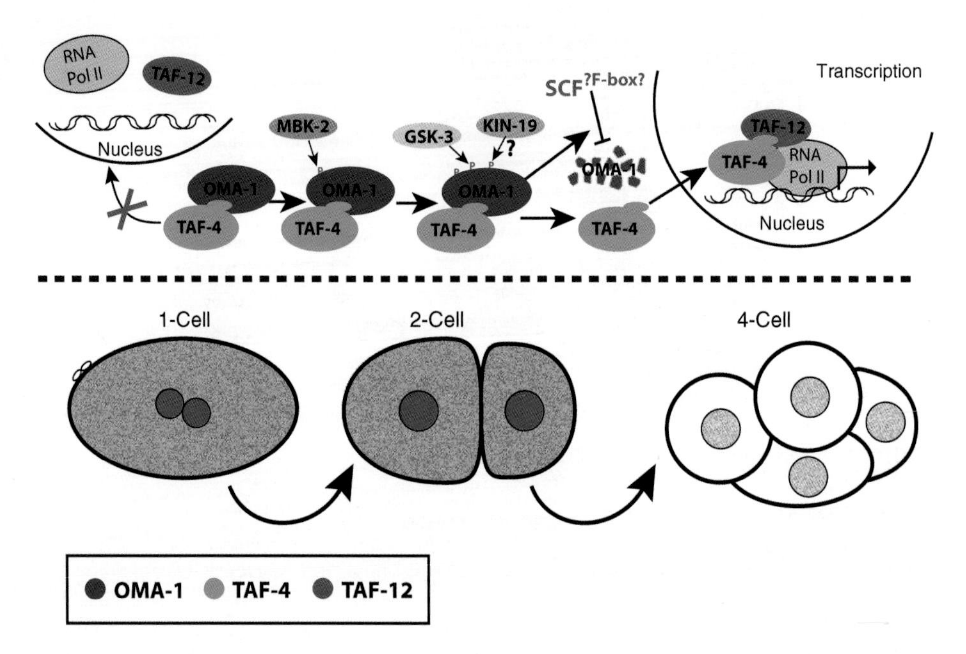

Figure 17.3 Activation of zygotic transcription. In one- and two-cell stage embryos, OMA-1 sequesters TAF-4 in the cytoplasm, precluding translocation to the nucleus, interaction with TAF-12 and RNA Pol II, thus repressing transcription. Both OMA-1 and TAF-12 bind TAF-4 via its histone fold, thus competing for binding through this domain. OMA-1 is phosphorylated by MBK-2, priming it for further phosphorylation by GSK-3, and probably KIN-19 as well. Phosphorylated OMA-1 is ubiquitinated by a SCF complex and degraded, releasing repression of TAF-4. TAF-4 is then free to be translocated to the nucleus, bind TAF-12 and RNA Pol II, leading to transcription activation in somatic cells. Note that although TAF-4/TAF-12 are present in the nucleus of germline precursors, transcription is repressed by PIE-1 (not represented in the figure)

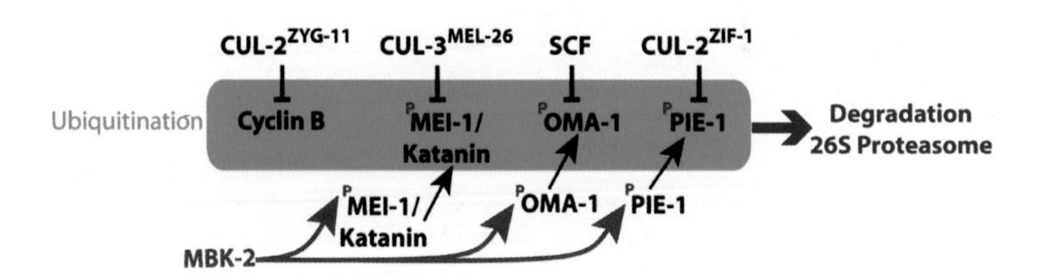

Figure 17.4 Cullin-based E3 ligases and phosphorylation. Four CRLs are involved in degradation of maternal products: CUL-2^{ZYG-11} ubiquitinates cyclin B, CUL-3^{MEL-26} ubiquitinates MEI-1/Katanin, SCF ubiquitinates OMA-1, and CUL-2^{ZIF-1} ubiquitinates PIE-1. MEI-1 and OMA-1 are phosphorylated by MBK-2, enabling recognition by the respective E3 ligase, and MEI-1/Katanin is phosphorylated by an unknown kinase in order to be recognized by the CRL

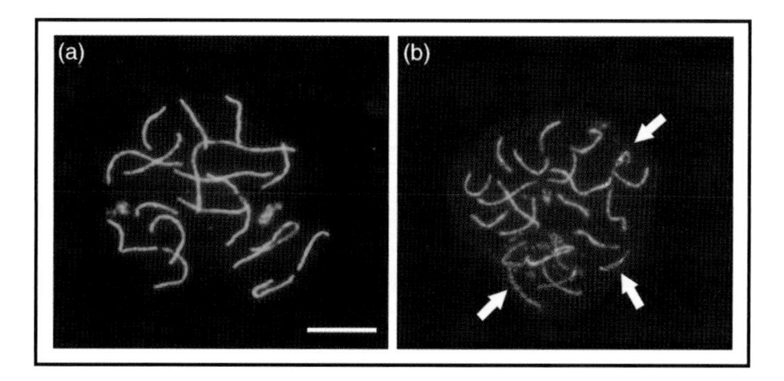

Figure 18.1 Incomplete homologous chromosome synapsis in *Lsh*$^{(-/-)}$ oocytes at the pachytene stage. (a) Control wild-type oocyte at the pachytene stage stained with SYCP3 antibody (green). SYCP3 is a component of the lateral elements of the synaptonemal complex. At this stage, control oocytes exhibit full synapsis of homologous chromosomes as indicated by the presence of 20 bivalents. (b) In contrast, following SYCP3 staining (green), *Lsh*-null oocytes exhibit incomplete homologous chromosome synapsis and persistence of double-strand DNA breaks as indicated by the colocalization of RAD51 foci (red) with asynapsed chromosomes (arrows). Note the absence of RAD51 foci in control wild-type oocytes. These results indicate that chromatin remodelling during prophase I of meiosis is required for proper chromosome synapsis in the female germline. Scale bar = 10 μM

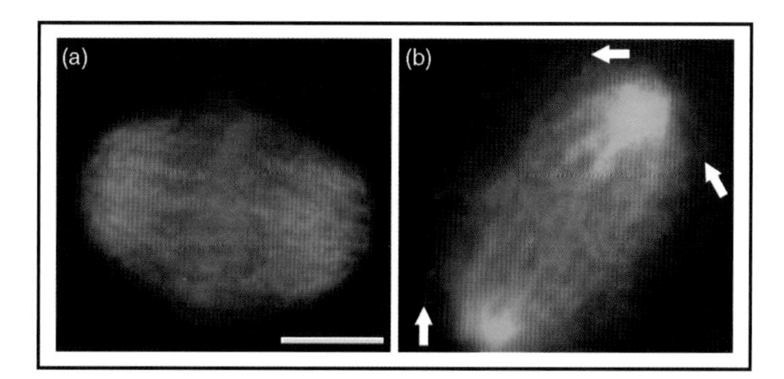

Figure 18.2 Inhibition of histone deacetylases (HDACs) disrupts meiotic progression and induces aberrant chromosome segregation. (a) Meiotic metaphase II spindle in control oocytes showing proper alignment of chromosomes (red) to the equatorial region. β-Tubulin staining (green) confirms the formation of a bipolar spindle. (b) Inhibition of HDACs with trichostatin A (TSA) results in the formation of abnormal meiotic spindles, elongated chromatids and a high incidence of chromosome lagging. Scale bar = 10 μM

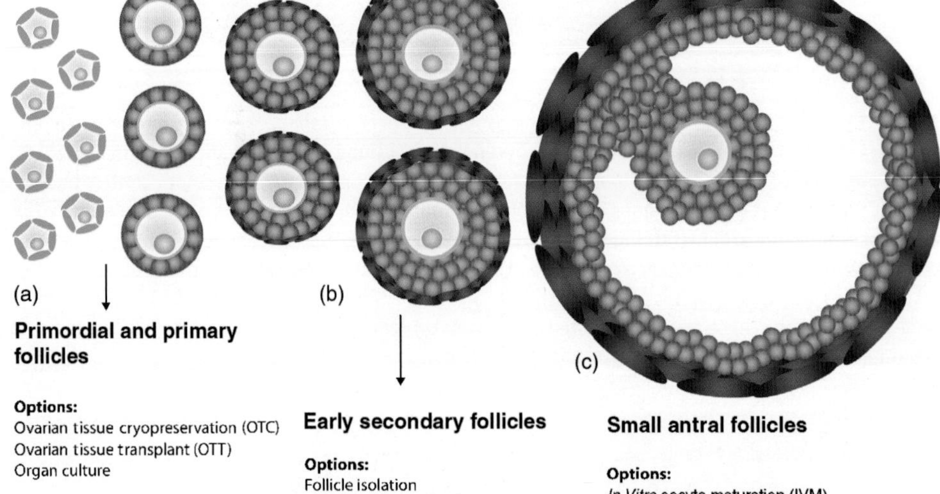

Figure 19.1 Fertility preservation options according to follicle stage. Key: granulosa cells (green); oocyte (tan); theca cells (purple)

thereby creating a feedback activation loop that contributes to the rapid initiation of M phase.

Plx1

Several results also suggested the existence of a Cdc25 triggering kinase distinct from Cdc2 (Kumagai and Dunphy, 1992; Izumi and Maller, 1993). In 1996, the *Xenopus* homologue of the *Drosophila* polo kinase, Plx1, was identified as the first kinase distinct from Cdc2 that phosphorylates Cdc25 and increases its phosphatase activity (Kumagai and Dunphy, 1996). To better understand the interactions between these players in the MPF autocatalytic activation, cell-free systems derived either from prophase or metaphase II-arrested *Xenopus* oocytes were developed. These approaches have led to the proposal of a two-step model for Cdc25 activation (Figure 9.6). During the first step, Cdc25 is partially phosphorylated by Cdc2 and acquires a basal catalytic activity. Plx1 is activated under the control of Cdc2, by a kinase called Plkk1 only identified in *Xenopus* (Erikson *et al.*, 2004). The full phosphorylation and activation of Cdc25 is achieved by Plx1 during a second step (Karaiskou *et al.*, 1999; Karaiskou *et al.*, 1998; Abrieu *et al.*, 1998).

PP2A

The identity of the protein phosphatases that catalyze the dephosphorylation of the activatory sites of Cdc25 and inactivate its phosphatase activity is still uncertain. Remarkably, injection of okadaic acid, a specific inhibitor of the Ser/Thr phosphatases PP1 and PP2A, in *Xenopus* oocytes induces MPF activation and GVBD independently of protein synthesis (Rime *et al.*, 1990; Goris *et al.*, 1989). In cell-free systems derived from oocytes or eggs, okadaic acid triggers MPF activation at concentrations not sufficient to inhibit PP1 (Karaiskou *et al.*, 1998; Felix, Cohen and Karsenti, 1990). Interestingly, a specific PP1 inhibitor has been reported to inhibit meiotic maturation (Huchon, Ozon and Demaille, 1981), indicating that the two phosphatases might control MPF activation in an opposite manner, negatively for PP2A and positively for PP1. Although Cdc25 can be dephosphorylated *in vitro* by either PP2A or PP1 phosphatases (Izumi, Walker and Maller, 1992), several findings strongly suggest that PP2A corresponds to the phosphatase that negatively controls the hyperphosphorylation required for Cdc25 activation. By using microcystin beads that allow the generation of *Xenopus* oocyte extracts depleted in PP2A and retaining PP1, it has been shown that PP2A depletion is sufficient to lead to Cdc2 activation, whereas the activity level of PP1 does not affect Cdc2 kinase activation promoted by PP2A removal (Maton *et al.*, 2005). In conclusion, it appears that PP2A could be the phosphatase responsible for the dephosphorylation of the Cdc25 activatory residues. More precisely, it would counteract Plx1 kinase activity that operates during the second step of the autoamplification process (Figure 9.6). A pending question is to understand how PP2A activity could be downregulated during this process.

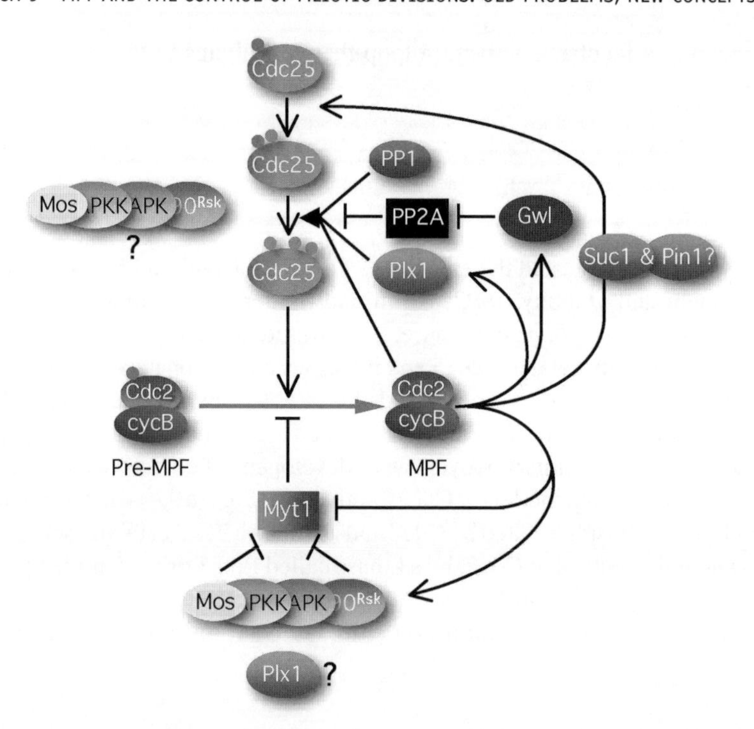

Figure 9.6 MPF autoamplification model. The autoamplification mechanism relies on the ability of active Cdc2–cyclin B complexes to activate their activator Cdc25, and inactivate their inactivator Myt1. Cdc2 partially phosphorylates Cdc25, enhancing the activity level of Cdc25, and Cdc25 dephosphorylation by PP1. Pin1 and Suc1 could regulate the interaction of Cdc2 with Cdc25 through conformational changes. Plx1 and Greatwall (Gwl) are activated under the control of Cdc2. The first one directly phosphorylates Cdc25 and counteracts the inhibitory effects of PP2A, whose activity is downregulated by Gwl. On the other side, Cdc2 plays a central role in inhibiting Myt1 both directly and through the activation of the Mos/MAPK/p90^{Rsk} pathway. Whether the Mos/MAPK/p90^{Rsk} pathway and Plx1 contribute respectively to Cdc25 upregulation and Myt1 downregulation is still a matter of debate. Red circle on Cdc25 = Ser287 inhibitory phosphate targeted by PP1; green circles on Cdc25 = activatory phosphates targeted by Cdc2, Plx1 and PP2A; cycB = cyclin B. A full colour version of this figure appears in the colour plate section.

Greatwall

A new player in the autoamplification loop has been recently discovered. The *Greatwall* gene initially described in *Drosophila* encodes an evolutionarily conserved protein kinase. Mutations in the *Drosophila Greatwall* gene cause improper chromosome condensation and delay cell cycle progression in larval neuroblasts. Cells take much longer to transit the period of chromosome condensation from late G2 through nuclear envelope breakdown (Yu *et al.*, 2004). These observations prompted the authors of this work to suggest that Greatwall helps activate cell cycle regulators that prepare interphase cells to enter mitosis, the Cdc2–cyclin B complex being the best candidate. To investigate Greatwall's mitotic function, they examined the behaviour of Greatwall in *Xenopus* egg extracts and during oocyte meiotic maturation. Their results strongly suggest that Greatwall is activated under the control of MPF and participates in the autoactivation loop that generates and maintains high levels of MPF activity by contributing to the activation of the Cdc25 phosphatase (Yu *et al.*, 2006). Interestingly, Greatwall can induce phosphorylations of Cdc25 in the absence of the activity of Cdc2 and Plx1. The effects of active Greatwall mimic, in many respects, those associated with addition of the phosphatase inhibitor, okadaic acid. These findings therefore support a model in which Greatwall negatively regulates the crucial phosphatase, a member of the PP2A family, that inhibits Cdc25 activation and M phase induction (Zhao *et al.*, 2008) (Figure 9.6). Besides this regulation of MPF activity, downregulation of PP2A by Greatwall would also contribute to the maintenance of the M phase state by preventing the dephosphorylation of the numerous mitotic Cdc2 substrates.

Pin1 and p9

Pin1 is a small and abundant protein that contains two domains: an N-terminal WW domain, which specifically binds the phosphorylated sequence P-Ser-Pro or P-Thr-Pro; and a C-terminal peptidyl-prolyl isomerisation domain able to interconvert the cis and trans isomeric forms of peptidyl-proline bonds (Lu, Hanes and Hunter, 1996; Lu *et al.*, 1999). Pin1 binds to the phosphorylated active form of *Xenopus* Cdc25 through its WW domain, and isomerizes the protein, indicating that Pin1 binding follows Cdc25 phosphorylation by Cdc2 and Plx1, and operates during the autoamplification process. Pin1 has been reported to either inhibit (Shen *et al.*, 1998) or induce a modest increase (Stukenberg and Kirschner, 2001) or to not affect (Crenshaw *et al.*, 1998) Cdc25 phosphatase activity. To summarize, three consequences for Cdc25 of the Pin1-catalysed isomerisation can be envisaged. Firstly, Pin1 could induce large changes in the structural conformation of Cdc25, modulating its catalytic activity (Stukenberg and Kirschner, 2001). Second, it could regulate the dephosphorylation of the activatory sites of Cdc25 by PP2A, which recognizes P-Ser-Pro and P-Thr-Pro specifically in the trans conformation (Zhou *et al.*, 2000). Third, Pin1 could antagonize the stimulatory effect of the Suc1/Cks protein, p9, on the phosphorylation of the main regulators of Cdc2.

Xenopus p9 is a small protein homologue of the Suc1/Cks yeast proteins that interacts with Cdc2 and is required for entry into mitosis. Mice lacking Cks2, one of two mammalian homologues of Suc1/Cks, are sterile in both sexes, due to failure of both male and female germ cells to progress past the first meiotic metaphase (Spruck *et al.*, 2003). These data suggest that, in mouse oocyte, Cks2 is not necessary for MPF activation but plays a critical role for homologous chromosome segregation at meiotic anaphase. The role of the other mammalian member of the family, Cks1, during oocyte meiosis has not been investigated yet. In *Xenopus* oocytes, p9 is required for the dephosphorylation of Cdc2 by Cdc25 (Patra and Dunphy, 1996), and conversely stimulates the ability of the Cdc2–cyclin B complex to phosphorylate Cdc25 (Patra *et al.*, 1999). It plays an indispensable role during the second step of the autocatalytic activation of MPF, by allowing the interaction between Cdc2 and the hyperphosphorylated form of Cdc25 (Karaiskou *et al.*, 1999). At a molecular level, Cdc2–cyclin B would stimulate the phosphorylation of Cdc25 through binding of its Suc1/Cks module to the phosphoepitope of the substrate. This process would be antagonized by the WW domain of Pin1, caused by competitive binding of both protein modules to the same phosphoepitope (Landrieu *et al.*, 2001). Nevertheless, further studies will be required to conclusively resolve the physiological role of Pin1 and p9; these regulators raising the question of the prevalence of conformational catalysis in MPF activation.

MAP kinases

As discussed previously, the activation of the Mos/MAPK/$p90^{Rsk}$ pathway is under the control of MPF. However, Mos is a powerful inducer of meiotic maturation when microinjected, and has the ability to contribute to MPF activation through the inhibition of the protein kinase Myt1 (Palmer, Gavin and Nebreda, 1998; Peter *et al.*, 2002). Besides the Myt1 control, it is tempting to speculate that the Mos/MAPK/$p90^{Rsk}$ pathway contributes to Cdc25 activation, since microinjected Mos is able to lead to tyrosine dephosphorylation of Cdc2 contained in the pre-existing stockpile of pre-MPF; an event relying on Cdc25 activity (Haccard and Jessus, 2006). Another MAP kinase family member, the p38γ/SAPK3 kinase, contributes to Cdc25 activation through the phosphorylation of the Ser205 residue of Cdc25 during *Xenopus* oocyte meiotic maturation (Perdiguero *et al.*, 2003). Hence these members of the MAP kinase family most probably participate in the loop between Cdc2 and Cdc25 (Figure 9.6).

PP1

Besides activatory phosphorylation sites, Cdc25 is also negatively controlled by phosphorylation on Ser287 (*Xenopus* numbering; Ser216 in humans). In somatic cells, Ser287 phosphorylation mediates the checkpoint suppression of Cdc25 that prevents mitosis in the presence of incompletely replicated or damaged DNA (Kumagai *et al.*, 1998a; Sanchez *et al.*, 1997). Ser287 phosphorylation is required for docking of the small acidic protein, 14-3-3, but the extent to which Cdc25 regulation by Ser287 phosphorylation strictly depends on 14-3-3 binding is not entirely clear (Sanchez *et al.*, 1997; Kumagai, Yakowec and Dunphy, 1998b). In somatic cells, Ser287 can be phosphorylated by several kinases, including both checkpoint kinases, Chk1 and Chk2/Cds1; Ca^{2+}/calmodulin-activated protein kinase CaMKII; and c-TAK1 kinase (Sanchez *et al.*, 1997; Matsuoka, Huang and Elledge, 1998; Peng *et al.*, 1998; Peng *et al.*, 1997; Hutchins, Dikovskaya and Clarke, 2003). In *Xenopus* prophase-arrested oocytes, Cdc25 is bound to 14-3-3 proteins and phosphorylated on Ser287 (Kumagai, Yakowec and Dunphy, 1998b; Yang *et al.*, 1999; Duckworth, Weaver and Ruderman, 2002). It is probable that this phosphorylation contributes to the downregulation of the Cdc25 phosphatase activity during the long-lasting period of the meiotic prophase arrest, independently of the DNA-responsive checkpoint. Different mechanisms of Cdc25 inhibition by 14-3-3 binding have been proposed, either related to the exclusion of Cdc25 from the germinal vesicle (Yang *et al.*, 1999; Oe *et al.*, 2001), a conflicting point that will be discussed further (see Section 9.6, below), or to the protection of Cdc25 from premature Ser287 dephosphorylation (Margolis *et al.*, 2003). Several kinases have been proposed to contribute to Ser287 phosphorylation of Cdc25 in the *Xenopus* oocyte, including Chk1, Chk2/Cds1 and PKA (Duckworth, Weaver and

Ruderman, 2002; Nakajo *et al.*, 1999; Gotoh *et al.*, 2001). Both the 14-3-3 binding and Ser287 phosphorylation of Cdc25 are abrogated at the time of MPF activation (Yang *et al.*, 1999; Duckworth, Weaver and Ruderman, 2002) and should contribute to Cdc25 activation. Upon entry into M phase, 14-3-3 removal from Cdc25 precedes Ser287 dephosphorylation, suggesting the existence of a phosphatase-independent pathway for 14-3-3 removal from Cdc25, which allows the unmasking of the phosphoserine and the subsequent access of the Ser287-specific phosphatase. PP1 is the primary Ser287-directed phosphatase in the oocyte, thereby derepressing Cdc25 at the time of MPF activation (Margolis *et al.*, 2003), explaining the finding that injection of *Xenopus* oocytes with the PP1 inhibitor-1 impedes Cdc2 activation and oocyte maturation (Huchon, Ozon and Demaille, 1981). Interestingly, Ser287 dephosphorylation by PP1 is indirectly dependent on Cdc2 activity. Indeed, the neighbouring Ser285 residue of *Xenopus* Cdc25 is phosphorylated by Cdc2–cyclin B, and this phosphorylation markedly enhances PP1-mediated Ser287 dephosphorylation by promoting the increased binding of Cdc25 to PP1 (Margolis *et al.*, 2006; Bulavin *et al.*, 2003a, 2003b) (Figure 9.6). These data illustrate the cooperation existing between the dual pathways leading to Cdc25 dephosphorylation on Ser287 and Cdc25 phosphorylation on activatory sites, and highlight Cdc2 as the master regulator coordinating all of these events.

Cdc2 and Cdc25 are therefore connected by multiple links inside the autoamplification feedback loop; mainly: (i) Direct phosphorylation of Cdc25 by Cdc2 on multiple sites, some of them increasing the catalytic activity of the phosphatase, one (Ser285) enabling the efficient reversal of Cdc25 suppression mediated by Ser287 phosphorylation; (ii) Phosphorylation of Cdc25 by the Cdc2-activated kinase Plx1 on additional sites; and (iii) Protection of these last phosphorylated sites from PP2A by Greatwall activation (Figure 9.6).

9.5.2 The Myt1 side

As with Cdc25, its opposing enzyme, the Myt1 kinase, is controlled by its own phosphorylation status. Depending on the organism and its stage of development, multiple kinases were shown to phosphorylate Myt1 and could contribute to its downregulation. Indeed, kinases as diverse as Plx1, Mos, $p90^{Rsk}$ or Cdc2–cyclin B have been reported as possible Myt1 regulators, and represent kinase candidates to play a role in the MPF amplification loop operating in the oocyte. Additionally, *Xenopus* Myt1-catalyzed autophosphorylation of residue Ser66 has been proposed to be a prerequisite for the further phosphorylation and inactivation of Myt1 during meiotic maturation, adding another layer of complexity to the phosphorylation-dependent mechanism of Myt1 regulation (Kristjansdottir *et al.*, 2006).

The current accepted 'working hypothesis' is that Myt1 is phosphorylated and inhibited at the onset of meiotic reinitiation by members of the MAPK pathway, Mos and $p90^{Rsk}$ (Palmer, Gavin and Nebreda, 1998; Peter *et al.*, 2002), and by Plx1 after fertilization (Inoue and Sagata, 2005). However, it has been shown that meiotic reinitiation can be triggered in the absence of the MAPK pathway (Gross *et al.*, 2000; Dupre *et al.*, 2002; Fisher *et al.*, 1999). Under these conditions, Myt1 undergoes a full phosphorylation shift and Cdc2 is totally dephosphorylated on tyrosine, indicating that the Mos/MAPK pathway is not necessary for Myt1 inactivation or is not the exclusive Myt1 regulator. Therefore, a kinase or a set of kinases, not belonging to the MAPK pathway, can account for full Myt1 phosphorylation in the oocyte. This alternative pathway could act in parallel with the MAPK pathway under physiological conditions, or could be a 'rescue' pathway recruited by the oocyte in case of failure of the activation of the MAPK cascade. The most probable candidates as regulators of Myt1 kinase, besides Mos and $p90^{Rsk}$, are Cdc2 and Plx1. Indeed, although $p90^{Rsk}$ associates with Myt1 in a complex excluding Plx1 during meiosis (Inoue and Sagata, 2005), this does not definitively rule out Plx1 from the role of a possible Myt1 regulator before fertilization. It is also established that human and *Xenopus* Myt1 can be directly phosphorylated by Cdc2 (Inoue and Sagata, 2005; Nakajima *et al.*, 2003). In parallel with Cdc25 regulation, it is tempting to propose that Cdc2 kinase could play a central role in inhibiting Myt1 and could cooperate either with the MAPK pathway or with Plx1 (Figure 9.6).

9.6 MPF activation, control by subcellular localization

Besides the fundamental mechanism by which Cdc2–cyclin B is regulated (through the antagonistic actions of Cdc25 and Myt1), several lines of evidence indicate that additional mechanisms are also involved, among them subcellular structures and MPF subcellular localization. In interphase cells, Cdc2–cyclin B accumulates in the cytoplasm. In late prophase, the complex moves rapidly into the nucleus, and it has been proposed that the intranuclear targeting induced by cyclin B phosphorylation would be required for MPF activation. Alternatively, several reports indicate that Cdc2–cyclin B is first activated at the centrosomes in the cytoplasm before entering the nucleus (Dutertre *et al.*, 2004; Jackman *et al.*, 2003). Therefore, two regions appear to be of particular significance for MPF activation at the onset of mitosis: the nucleus itself, and the perinuclear area where the centrosomes are located. In the fertilized egg, MPF activation initiates locally within the animal hemisphere (Perez-Mongiovi, Chang and Houliston, 1998; Rankin and Kirschner, 1997) and is stimulated by various subcellular structures including nuclei, centrosomes (provided by the sperm aster) and microtubule asters. However, little is known about the subcellular localization of MPF and its regulators in oocytes, and the importance of this localization on its activation at meiosis I. The question of the localization of MPF regulators in the amphibian oocyte was approached mainly by biochemical fractionation, and led to the conclusion that neither MPF itself nor its main regulators (Cdc25, Myt1 and Plx1) are differentially distributed between animal and vegetal regions. However, more subtle localizations of active subpopulations of these molecules may be involved in the localized initiation of MPF activation.

9.6.1 Which role for the intranuclear material?

In all species, prophase-arrested oocytes are equipped with a very large nucleus, called the germinal vesicle. The biological significance of the extraordinarily large volume of the oocyte nucleus has raised many questions. Experiments with amphibian enucleated oocytes proved that the appearance of MPF during maturation is a purely cytoplasmic event independent of the germinal vesicle. These results, obtained with *Rana* oocytes (Masui and Markert, 1971), and later confirmed using *Xenopus* oocytes (Reynhout and Smith, 1974; Schorderet-Slatkine and Drury, 1973), were based on cytoplasm transfer by microinjection. They showed that MPF activity can be recovered in enucleated oocytes stimulated by progesterone, but did not provide any information about the quantity and the rate of MPF production in the absence of a nucleus, and the efficiency of the autoamplification mechanism. This was done later by Fisher and collaborators who investigated at a molecular level the requirements for germinal vesicle material during the course of meiotic maturation in *Xenopus* oocytes (Fisher *et al.*, 1998). They showed that nuclear material is not required for the initial MPF activation induced by progesterone, nor for cyclin B degradation occurring at meiosis I, nor for activation of the newly assembled Cdc2–cyclin B complexes in meiosis II (Fisher *et al.*, 1998). However, other reports have challenged the general view that nuclear material is dispensable for the meiotic cell cycle in amphibian oocytes. Iwashita *et al.* reported that germinal vesicle material accelerates MPF activation and is essential for MPF reactivation at meiosis II by controlling the Tyr15 phosphorylation level of Cdc2 (Iwashita, Hayano and Sagata, 1998). In another amphibian species, the axolotl *Ambystoma mexicanum*, MPF production in oocytes after hormone treatment occurs later in enucleated oocytes than in nucleated controls, a delay attributed to the initiation of MPF activation rather than to MPF amplification (Gautier, 1987). Moreover, Li *et al.* showed that phosphorylation of Cyclin B1 in its cytoplasmic retention domain is required for its biological activity, by promoting the targeting of the Cdc2–cyclin B complex in the nucleus (Li, Meyer and Donoghue, 1997). This investigation supports the hypothesis that the control of subcellular localization of cyclins plays a key role in regulating the biological activity of Cdc2–cyclin B complexes. Then, in addition to its obvious role in facilitating access of MPF to its nuclear substrates (such as lamins, histones, condensins and nuclear pore proteins), the role of nuclear translocation may be to bring pre-MPF together with activators of MPF kinase activity. In agreement with this view, it has been proposed that in the prophase-arrested oocyte, Cdc25 shuttles in and out of the nucleus, but that its binding to 14-3-3 protein markedly reduces the nuclear import rate, allowing nuclear export to predominate. When 14-3-3 binding to Cdc25 is abolished in response to progesterone stimulation, nuclear export is inhibited and the coordinate nuclear accumulation of Cdc25 and Cdc2–cyclin B facilitates their mutual activation, thereby promoting oocyte maturation (Yang *et al.*, 1999). In contrast with this result, it has been shown that inhibition of Cdc25 function by Ser287 phosphorylation, the specific 14-3-3 binding residue, occurs in the cytoplasm of prophase-arrested oocytes, and that Cdc25 is activated exclusively in the cytoplasm of maturing oocytes (Oe *et al.*, 2001).

To summarize this conflicting literature, it appears that Cdc2 and its regulators are mainly localized in the cytoplasm (Izumi, Walker and Maller, 1992; Fisher *et al.*, 1998), and it can be clearly concluded that, in amphibian oocytes, the initial MPF activation triggered by the hormonal signal does not depend on the nucleus. Whether or not nuclear components contribute to MPF autoamplification and are necessary for cyclin B degradation and resynthesis during the metaphase I–metaphase II transition is still a matter of debate. It has to be noted that the interpretation of experiments with amphibian enucleated oocytes is complex: the large size and the yolky cytoplasm severely hamper microscopy techniques; the mechanical damage of the enucleation process can lead to some protein leak, and enucleated oocytes lack the usual external 'white spot' phenotype of Cdc2 kinase activation.

The participation of germinal vesicle material in the production of MPF was also investigated with oocytes of the starfish *Asterina pectinifera* (Kishimoto, Hirai and Kanatani, 1981). Cdc2–cyclin B is present exclusively in the cytoplasm in prophase oocytes, and accumulates in the nucleus just before GVBD (Ookata *et al.*, 1992; Picard and Peaucellier, 1998; Terasaki *et al.*, 2003). MPF is produced in enucleated oocytes treated with 1-methyladenine, showing that, similarly to the amphibian oocytes, the initial MPF activation does not require nuclear components. However, the amount of MPF produced is smaller than in the case of intact oocytes with germinal vesicles. Moreover, MPF injection in enucleated oocytes does not activate endogenous MPF molecules, suggesting that germinal vesicle material is required for MPF amplification in starfish oocyte (Kishimoto, Hirai and Kanatani, 1981). Strikingly, nuclear material is required for specific translation of cyclin B after the first meiotic cell cycle in starfish oocyte (Galas *et al.*, 1993). It has been proposed that the phosphatase PP1 suppresses cyclin B translation until breakdown of the nuclear envelope, which delivers to the cytoplasm a potent translational activator of cyclin B, most likely a PP1 inhibitor (Lapasset *et al.*, 2005).

Therefore, the oocyte nucleus does not critically regulate the initial activation of MPF in amphibian and starfish. Later on during the meiotic maturation process, it is probable that the nuclear envelope breakdown releases essential nuclear components for the completion of the second meiotic division.

9.6.2 Which role for the perinuclear area and the centrosome?

In cultured mammalian cells, it seems that the vicinity of the centrosomes is the decision-making region for entry into M phase (Jackman *et al.*, 2003). Similarly, in the fertilized *Xenopus* egg, the nuclear–centrosomal region localized in the animal part serves to accumulate MPF and/or its regulators to achieve locally a threshold level that seeds the initial activation of mitosis. After initiation of MPF activation in this specific region, MPF activity subsequently propagates across the egg (Perez-Mongiovi, Chang and Houliston, 1998; Rankin and Kirschner, 1997; Perez-Mongiovi *et al.*, 2000).

Obviously, the situation is strikingly different during oocyte meiosis as, in probably all vertebrates, centrioles disappear from the oocyte during early oogenesis, leading to acentriolar oocytes devoid of centrosomes. However, it is known that multiple structures known as microtubule organizing centres (MTOC) substitute for the centrosomes and form the spindle poles during meiosis of vertebrate oocytes. In mouse, multiple inactive cytoplasmic MTOCs are present in the cytoplasm of the oocyte (Maro, Howlett and Webb, 1985). *Xenopus* prophase oocytes contain an extensive network of cytoplasmic microtubules with no evidence of any functional MTOC (Jessus, Huchon and Ozon, 1986; Huchon and Ozon, 1985; Heidemann *et al.*, 1985). The analysis of the distribution of γ-tubulin reveals that inactive MTOCs surround the germinal vesicle and are also distributed in the cortex as a gradient along the animal–vegetal axis (Gard, 1994). There is no evidence that these inactive MTOCs contribute in MPF activation in *Xenopus* or mouse oocytes. Indeed, attempts to stimulate microtubule nucleation by microinjection of centrosomes or taxol, or conversely, treatments with microtubule depolymerizing drugs, such as colchicine or nocodazole, do not affect MPF activation and nuclear envelope breakdown (Huchon and Ozon, 1985; Heidemann and Kirschner, 1978; Heidemann and Gallas, 1980; Thibier *et al.*, 1997; Rime, Jessus and Ozon, 1987). If MTOCs do not significantly contribute to MPF activation, still the animal hemisphere, and more precisely the perinuclear area, appears important in the initiation of Cdc2 activation in the frog oocyte. MPF activity propagates through oocyte cytoplasm from the animal to the vegetal half, as it does during the first embryonic division in the fertilized egg. If the germinal vesicle is displaced by centrifugation to the vegetal half and the oocyte is then constricted by thin thread along the equatorial line and treated by progesterone, there is a significant delay in nuclear envelope breakdown, in proportion to the tightness of constriction (Masui, 1972). This indicates that the perinuclear area of the nucleus located in the animal hemisphere plays a role in the initiation of MPF activation in the amphibian oocyte, independently of the nuclear content and the presence of centrosomes or MTOCs. This is in agreement with the observation that cyclin B is concentrated in perinuclear rings in toad oocytes (Sakamoto *et al.*, 1998), as it is in *Drosophila* (Raff, Whitfield and Glover, 1990).

In contrast to vertebrates, oocytes of numerous invertebrate species, such as insects, nematodes and echinoderms, contain centrosomes. In starfish or sea urchin, the fully grown immature oocyte contains two centrosomes, each of them consisting of two centrioles, located between the animal pole and the germinal vesicle (Nakashima and Kato, 2001; Picard *et al.*, 1988; Miyazaki, Kamitsubo and Nemoto, 2000). In starfish oocyte as in frog, activation of Cdc2–cyclin B starts in the animal hemisphere and travels the cell to the vegetal half (Picard and Peaucellier, 1998). The first active complexes enter the nucleus presumably due to phosphorylation of cyclin B, which affects the balance of cyclin B import and export rates between the nucleus and cytoplasm. Active Cdc2–cyclin B complexes start to enter into the nucleus from the animal pole side, where the centrosomes are located, leading to the proposal that activation of MPF first occurs in the perinuclear area, and more precisely in the region of the centrosomes, as it is the case in mitotic cells (Ookata *et al.*, 1992; Picard and Peaucellier, 1998; Terasaki *et al.*, 2003). Centrosomes, MTOC and the perinuclear area then function as sites of integration for the regulators that trigger MPF activation, probably by increasing the local concentration of regulatory factors, such as Cdc25, Plx1 or Greatwall.

9.6.3 Which role for intracellular membranes?

Several reports indicate that inactive Cdc2–cyclin B is not free to diffuse in the oocyte cytoplasm, but instead is associated with intracellular structures and undergoes changes in associations when it becomes activated. Two types of localization have been proposed to play an important role in MPF activation during the first meiotic division. The first one is the localization of cyclin B in aggregates in the cytoplasm of immature oocytes of starfish and clam. In clam, fertilization triggers entry into meiotic divisions and results in the release of cyclin B to disperse in a soluble form, suggesting that fertilization-triggered unmasking of cyclin B drives cells into meiosis I (Westendorf, Swenson and Ruderman, 1989). In starfish, cyclin B is similarly present in aggregates in the cytoplasm of immature oocytes, and the aggregates disperse at the time of MPF activation, beginning from the region containing the centrosomes (Terasaki *et al.*, 2003). This led to a scenario based on a relatively simple mathematical formulation (Slepchenko and Terasaki, 2003). In the prophase-arrested oocyte, only inactive pre-MPF can be caged into aggregates, and the aggregates are in equilibrium with inactive molecules of pre-MPF in solution. During maturation, the hormone triggers inactivation of Myt1, leading to the depletion of the soluble inactive Cdc2–cyclin B complexes that drop below saturation level. Therefore, the aggregates would dissolve, thus increasing the total amount of MPF in the cytoplasm (Slepchenko and Terasaki, 2003). This model provides a robust bio-switch in agreement with the MPF positive feedback loop, thought to result in all-or-none activation that does not flicker on and off once it is turned on. The next issues that should be addressed experimentally concern the molecular content of the aggregates, the mechanisms that promote aggregation of inactive Cdc2–cyclin B and prevent aggregation of active MPF, and the existence of such spatial bistate equilibrium in oocytes of other species.

A second line of evidence has demonstrated that membranes are important for MPF activation in oocytes, even in cytoplasmic extracts. The negative regulator of Cdc2 kinase activation, the Myt1 kinase, possesses a putative membrane-targeting domain corresponding to a stretch of hydrophobic and uncharged amino acids located outside the catalytic domain, and is associated with membranes in oocytes as well as in mammalian cultured cell lines (Mueller *et al.*, 1995; Liu *et al.*, 1997; Nakajima *et al.*, 2008). Therefore, during the long-lasting period of prophase I arrest that supports the accumulation of pre-MPF in the growing oocyte, Cdc2–cyclin B molecules are formed by association between newly synthesized cyclin B and Cdc2, and have to be inactivated by constantly interacting with the membrane compartment where Myt1 is anchored. In contrast, progesterone treatment of the full-grown oocyte leads to the formation of new complexes that are prevented from Myt1 inactivation and fire the MPF autoamplification loop. The mechanism underlying this regulation is unknown. Progesterone could downregulate Myt1 catalytic activity and/or prevent any interaction between Myt1 and the new complexes by regulating the cellular localization of both of them. All these observations highlight the pivotal role that should be played by intracellular membranes and membrane traffic in the oocyte. A number of studies, showing that the inhibition of membrane trafficking in the oocyte provokes activation of the Cdc2 kinase, are consistent with this hypothesis (De Smedt *et al.*, 1995; Rime *et al.*, 1998; Mulner-Lorillon *et al.*, 1995). Moreover, among various reports, a binding inhibitor of Cdc2–cyclin B was found tightly associated with cell membranes in *Xenopus* extracts (Lee and Kirschner, 1996), cyclin B was observed in vesicle-like structures in starfish oocytes (Picard and Peaucellier, 1998), and inactive Cdc2 was recovered in association with the external layer of membrane vesicles isolated from *Xenopus* prophase oocytes (De Smedt, Crozet and Jessus, 1999). More recently, Beckhelling and collaborators demonstrated that pre-MPF is associated with annulate lamellae in *Xenopus* prophase oocytes (Beckhelling *et al.*, 2003). Annulate lamellae, most commonly observed in the cytoplasm of oocytes and rapidly dividing somatic cells, are flattened membrane cisternae arranged in stacks of parallel sheets, thought to form from excess nuclear membrane components, since they contain a high density of nuclear pores (Kessel, 1989). They disassemble in parallel with the nuclear envelope at meiosis (Terasaki, Runft and Hand, 2001). The observation that pre-MPF is associated with annulate lamellae, and released in a soluble fraction upon activation, uncovers the unexpected participation of nuclear membranes in activation of MPF. It provides explanation for the stimulation of MPF activation by nuclei, centrosomes and, more generally, the perinuclear area. It is also consistent with the insolubility of pre-MPF and its solubilization upon activation noted by several reports (Picard and Peaucellier, 1998; Terasaki *et al.*, 2003; Beckhelling *et al.*, 2003; Bailly *et al.*, 1992). Association of insoluble pre-MPF with a specific membrane compartment would keep it in close vicinity to its negative membrane-associated regulators, Myt1 and the Cdc2-binding inhibitor (Lee and Kirschner, 1996), hence maintaining its inactive state during the prophase arrest. Progesterone would enhance the synthesis of cyclin B and lead to the accumulation of new active Cdc2–cyclin B complexes in a soluble perinuclear compartment, away from the inhibitory action of the membrane stacks. These primary active molecules can then turn on the positive MPF feedback loop according to a bio-switch based on the dissolution of membrane-bound Cdc2–cyclin B molecules. A localization study of the various regulators of MPF may further illuminate the basis of localized MPF regulation in the perinuclear area of the oocyte.

9.7 The growing oocyte: an opportunity to uncover MPF regulation

Oogenesis depends upon two intricate events: meiosis, which results in the formation of a haploid genome; and accumulation of molecular factors necessary for early development. Oogenesis is initiated when the oogonia enter into the first meiotic prophase. When the oocyte reaches the diplotene stage, meiosis arrests and, during a long-lasting period of prophase arrest, the follicle-enclosed oocyte grows and accumulates a variety of molecules that are required for early embryonic development. In most of the species studied until now, the prophase-arrested oocyte is unable to activate MPF and re-enter meiosis until it has acquired its full size (mouse: Sorensen and Wassarman (1976); rat: Bar-Ami and Tsafriri (1981); pig: Motlik *et al.* (1984); sheep: Moor and Gandolfi (1987); cattle: Fuhrer *et al.* (1989); amphibian: Reynhout *et al.* (1975); Hanocq-Quertier, Baltus and Brachet (1976)). In some mammalian species such as goat, the acquisition of meiotic competence can be correlated with the follicular size rather than with the oocyte size, and develops progressively, as the oocytes acquire, first, the ability to break down the nuclear envelope and to condense chromosomes, then to reach metaphase I, and finally to reach metaphase II only when follicular growth is completed (De Smedt, Crozet and Gall, 1994). In *Xenopus*, steroid treatments are unable to release the prophase arrest of growing oocytes whose diameter is smaller than 1 mm. At the end of the growing period, the oocyte, still blocked in prophase I, becomes responsive to the steroids produced by the surrounding follicular cells by activating MPF. The inability of smaller oocytes to activate MPF, in response to the physiological meiosis triggers which function in full-grown oocytes, prevents premature meiotic maturation and fertilization; a mechanism essential to impede the generation of deficient embryos, and hence to the success of embryonic development. Smaller oocytes provide an interesting physiological model to study how cell growth is coupled to the cell cycle. However, most studies have been devoted to elucidation of the mechanisms of MPF activation in the full-grown oocyte, and only a few of them addressed the question of the inability of small growing oocytes to support MPF activation, despite the physiological importance of this process.

In the mammalian species analyzed so far, acquisition of meiotic competence at the end of oocyte/follicle growth is correlated with the synthesis and accumulation of Cdc2 and cyclin B proteins (Dedieu *et al.*, 1998; de Vantéry *et al.*, 1997; de Vantéry *et al.*, 1996; Chesnel and Eppig, 1995). The situation is very different in amphibians. The incompetence of small *Xenopus* oocytes to undergo meiotic maturation in response to steroids lies at the level of MPF activation. A progesterone receptor is already functional in small oocytes, since a decrease in cAMP can be induced by progesterone (Mulner, Belle and Ozon, 1983; Sadler and Maller, 1983). Moreover, pre-MPF as well as Cdc25 and Myt1 are present in incompetent growing oocytes (Rime, Jessus and Ozon, 1995; Furuno, Kawasaki and Sagata, 2003). The origins of the inability of small amphibian oocytes to support MPF activation can therefore depend on either the inability to generate a small amount of active MPF, triggering the autoamplification mechanism, and/or inability of the positive feedback loop to function.

The first limiting step for MPF activation in the growing oocyte is the inability of the transduction pathway induced by progesterone to connect to MPF regulators: cyclin B1 synthesis, that occurs in full-grown oocytes in response to progesterone, and independently of MPF (Frank-Vaillant *et al.*, 1999), is not inducible by progesterone in small growing oocytes (Karaiskou *et al.*, 2004). Similarly, Mos synthesis is not induced by progesterone in small oocytes (Karaiskou *et al.*, 2004). Therefore, the growing oocyte is unable to initiate the MPF autoamplification loop due to the lack of newly formed Cdc2–cyclin B complexes and/or of Mos/MAPK pathway activation.

Second, when cyclins are injected into small oocytes, they associate with endogenous free Cdc2, and the illegitimate complexes undergo phosphorylation on Tyr15, indicating a failure in small oocytes to prevent new complexes from Myt1 inactivation (Rime, Jessus and Ozon, 1995). Therefore, even if progesterone were able to generate a small amount of starter MPF by cyclin synthesis in growing oocytes (which it cannot do), the new complexes would be directly inactivated by Tyr15 phosphorylation, preventing the autoamplification loop from being initiated.

Third, the positive feedback loop operating between Cdc2 and Cdc25 is not functional in small growing oocytes. Although entry into M phase can be triggered in the growing oocytes by microinjection of cytoplasm taken from matured oocytes (Hanocq-Quertier, Baltus and Brachet, 1976; Sadler and Maller, 1983; Taylor and Smith, 1987), Tyr15 dephosphorylation of endogenous Cdc2 is not complete (Rime *et al.*, 1991). Moreover, inhibition of PP2A by okadaic acid, which triggers Cdc2 activation in fully grown oocytes, is inefficient in small oocytes (Rime, Jessus and Ozon, 1995). All these observations suggest that the autoamplification mechanism is not fully functional. More recently, a study revealed that Plx1, which is required for the autoamplification mechanism, is absent in incompetent oocytes, and represents a crucial limiting factor, accounting for the incompetence of small oocytes to re-enter meiosis in response to progesterone (Karaiskou *et al.*, 2004). The absence of Plx1 results in a double negative control on MPF activation: first, the formation of active complexes between Cdc2 and newly synthesized cyclins is prevented by a sustained activity of Myt1 that escapes downregulation by Plx1; second, Cdc25 activation, which is normally achieved through a feedback loop involving Plx1, is also prevented. Further investigation will be necessary to discover how Plx1 expression is controlled by cell size at the end of oogenesis. It is also probable that Plx1 is not the only limiting factor accounting for incompetence of small amphibian oocytes to resume meiosis, and much work is still needed to understand the molecular and cellular basis of this important physiological process.

Acknowledgements

I thank my colleagues of the team 'Biologie de l'ovocyte' for stimulating discussions. I am especially grateful to René Ozon for his invaluable input. This work was supported by funding from the Centre National de la Recherche Scientifique (CNRS), University Pierre et Marie Curie-Paris 6, Association pour la Recherche contre le Cancer (ARC, grant 3969) and the Agence Nationale de la Recherche (ANR, grant BLANC07-3 185404).

References

Abrieu, A., Brassac, T., Galas, S. *et al.* (1998) The Polo-like kinase Plx1 is a component of the MPF amplification loop at the G2/M-phase transition of the cell cycle in Xenopus eggs. *J. Cell Sci.*, **111**, 1751–1757.

Baert, F., Bodart, J.F., Bocquet-Muchembled, B. *et al.* (2003) Xp42(Mpk1) activation is not required for germinal vesicle breakdown but for Raf complete phosphorylation in insulin-stimulated Xenopus oocytes. *J. Biol. Chem.*, **278**, 49714–49720.

Bailly, E., Pines, J., Hunter, T. and Bornens, M. (1992) Cytoplasmic accumulation of cyclin-b1 in human cells – association with a detergent-resistant compartment and with the centrosome. *J. Cell Sci.*, **101**, 529–545.

Balamurugan, K. and Haider, S. (1998) Partial purification of maturation-promoting factor from catfish. Clarias batrachus: identification as the histone H1 kinase and its periodic activation. *Comp. Biochem. Physiol. C Pharmacol. Toxicol. Endocrinol.*, **120**, 329–342.

Bar-Ami, S. and Tsafriri, A. (1981) Acquisition of meiotic competence in the rat: role of gonadotropin and estrogen. *Gamete Res.*, **4**, 463–472.

Basu, D., Navneet, A.K., Dasgupta, S. and Bhattacharya, S. (2004) Cdc2-cyclin B-induced G2 to M transition in perch oocyte is dependent on Cdc25. *Biol. Reprod.*, **71**, 894–900.

Beckhelling, C., Chang, P., Chevalier, S. *et al.* (2003) Pre-M phase-promoting factor associates with annulate lamellae in Xenopus oocytes and egg extracts. *Mol. Biol. Cell*, **14**, 1125–1137.

Bodart, J.F., Gutierrez, D.V., Nebreda, A.R. *et al.* (2002) Characterization of MPF and MAPK activities during meiotic maturation of Xenopus tropicalis oocytes. *Dev. Biol.*, **245**, 348–361.

Brachet, J. (1944) *Embryologie Chimique*, Masson et Cie, Paris.

Bulavin, D.V., Demidenko, Z.N., Phillips, C. *et al.* (2003a) Phosphorylation of Xenopus Cdc25C at Ser285 interferes with ability to activate a DNA damage replication checkpoint in pre-midblastula embryos. *Cell Cycle*, **2**, 263–266.

Bulavin, D.V., Higashimoto, Y., Demidenko, Z.N. *et al.* (2003b) Dual phosphorylation controls Cdc25 phosphatases and mitotic entry. *Nat. Cell Biol.*, **5**, 545–551.

Chen, M.S., Hurov, J., White, L.S. *et al.* (2001) Absence of apparent phenotype in mice lacking Cdc25C protein phosphatase. *Mol. Cell Biol.*, **21**, 3853–3861.

Chesnel, F. and Eppig, J.J. (1995) Synthesis and accumulation of p34cdc2 and cyclin B in mouse oocytes during acquisition of competence to resume meiosis. *Mol. Reprod. Dev.*, **40**, 503–508.

Colledge, W.H., Carlton, M.B.L., Udy, G.B. and Evans, M.J. (1994) Disruption of c-mos causes parthenogenetic development of unfertilized mouse eggs. *Nature*, **370**, 65–68.

Crenshaw, D.G., Yang, J., Means, A.R. and Kornbluth, S. (1998) The mitotic peptidyl-prolyl isomerase, Pin1, interacts with Cdc25 and Plx1. *EMBO J.*, **17**, 1315–1327.

Daar, I., Yew, N. and Vande Woude, G.F. (1993) Inhibition of mos-induced oocyte maturation by protein kinase A. *J. Cell Biol.*, **120**, 1197–1202.

De Smedt, V., Crozet, N. and Gall, L. (1994) Morphological and functional changes accompanying the acquisition of meiotic competence in ovarian goat oocyte. *J. Exp. Zool.*, **269**, 128–139.

De Smedt, V., Rime, H., Jessus, C. and Ozon, R. (1995) Inhibition of glycosphingolipid synthesis induces p34cdc2 activation in Xenopus oocyte. *FEBS Lett.*, **375**, 249–253.

De Smedt, V., Crozet, N. and Jessus, C. (1999) *In vitro* binding of free cdc2 and raf kinase to membrane vesicles: a possible new regulatory mechanism for cdc2 kinase activation in Xenopus oocyte. *Microsc. Res. Tech.*, **45**, 13–30.

De Smedt, V., Poulhe, R., Cayla, X. *et al.* (2002) Thr-161 phosphorylation of monomeric Cdc2. Regulation by protein phosphatase 2C in Xenopus oocytes. *J. Biol. Chem.*, **277**, 28592–28600.

Dedieu, T., Gall, L., Hue, I. *et al.* (1998) p34cdc2 expression and meiotic competence in growing goat oocytes. *Mol. Reprod. Dev.*, **50**, 251–262.

Dettlaff, T.A. (1966) Action of actinomycin and puromycin upon frog oocyte maturation. *J. Embryol. Exp. Morphol.*, **16**, 183–195.

Dettlaff, T.A. and Ryabova, L.V. (1986) Maturation of Rana temporaria and Acipenser stellatus oocytes induced by the cytoplasm of embryos at different cell cycle phases and at different stages of cleavage and blastulation. *Cell Differ.*, **18**, 9–16.

Dettlaff, T.A., Nikitina, L.A. and Stroeva, O.G. (1964) The role of the germinal vesicle in oocyte maturation in anurans as revealed by the removal and transplantation of nuclei. *J. Embryol. Exp. Morphol.*, **12**, 851–873.

Draetta, G. and Beach, D. (1988) Activation of cdc2 protein kinase during mitosis in human cells: cell cycle-dependent phosphorylation and subunit rearrangement. *Cell*, **54**, 17–26.

Duckworth, B.C., Weaver, J.S. and Ruderman, J.V. (2002) G2 arrest in Xenopus oocytes depends on phosphorylation of cdc25 by protein kinase A. *Proc. Natl. Acad. Sci. USA*, **99**, 16794–16799.

Dunphy, W.G. and Kumagai, A. (1991) The cdc25 protein contains an intrinsic phosphatase activity. *Cell*, **67**, 189–196.

Dunphy, W.G., Brizuela, L., Beach, D. and Newport, J. (1988) The Xenopus cdc2 protein is a component of MPF, a cytoplasmic regulator of mitosis. *Cell*, **54**, 423–431.

Dupre, A., Jessus, C., Ozon, R. and Haccard, O. (2002) Mos is not required for the initiation of meiotic maturation in Xenopus oocytes. *EMBO J.*, **21**, 4026–4036.

Dutertre, S., Cazales, M., Quaranta, M. *et al.* (2004) Phosphorylation of CDC25B by Aurora-A at the centrosome contributes to the G2-M transition. *J. Cell Sci.*, **117**, 2523–2531.

Edgar, B.A. and O'Farrell, P.H. (1989) Genetic control of cell division patterns in the Drosophila embryo. *Cell*, **57**, 177–187.

Erikson, E., Haystead, T.A., Qian, Y.W. and Maller, J.L. (2004) A feedback loop in the polo-like kinase activation pathway. *J. Biol. Chem.*, **279**, 32219–32224.

Evans, T., Rosenthal, E.T., Youngblom, J. *et al.* (1983) Cyclin: a protein specified by maternal mRNA in sea urchin eggs that is destroyed at each cleavage division. *Cell*, **33**, 389–396.

Felix, M.A., Cohen, P. and Karsenti, E. (1990) Cdc2 H1 kinase is negatively regulated by a type 2A phosphatase in the Xenopus early embryonic cell cycle: evidence from the effects of okadaic acid. *EMBO J.*, **9**, 675–683.

Ferby, I., Blazquez, M., Palmer, A. *et al.* (1999) A novel p34(cdc2)-binding and activating protein that is necessary and sufficient to trigger G(2)/M progression in Xenopus oocytes. *Genes. Dev.*, **13**, 2177–2189.

Ferrell, J. Jr, Wu, M., Gerhart, J.C. and Martin, G.S. (1991) Cell cycle tyrosine phosphorylation of $p34^{cdc2}$ and a microtubule-associated protein kinase homolog in *Xenopus* oocytes and eggs. *Mol. Cell Biol.*, **11**, 1965–1971.

Fesquet, D., Labbe, J.C., Derancourt, J. *et al.* (1993) The MO15 gene encodes the catalytic subunit of a protein kinase that activates cdc2 and other cyclin-dependent kinases (CDKs) through phosphorylation of Thr161 and its homologues. *EMBO J.*, **12**, 3111–3121.

Fisher, D., Coux, O., Bompard-Marechal, G. and Doree, M. (1998) Germinal vesicle material is dispensable for oscillations in cdc2 and MAP kinase activities, cyclin B degradation and synthesis during meiosis in Xenopus oocytes. *Biol. Cell*, **90**, 497–508.

Fisher, D.L., Brassac, T., Galas, S. and Doree, M. (1999) Dissociation of MAP kinase activation and MPF activation in hormone-stimulated maturation of Xenopus oocytes. *Development*, **126**, 4537–4546.

Frank-Vaillant, M., Jessus, C., Ozon, R. *et al.* (1999) Two distinct mechanisms control the accumulation of cyclin B1 and Mos in Xenopus oocytes in response to progesterone. *Mol. Biol. Cell*, **10**, 3279–3288.

Fuhrer, F., Mayr, B., Schellander, K. *et al.* (1989) Maturation competence and chromatin behaviour in growing and fully grown cattle oocytes. *Zentralbl. Veterinarmed. A*, **36**, 285–291.

Furuno, N., Kawasaki, A. and Sagata, N. (2003) Expression of cell-cycle regulators during Xenopus oogenesis. *Gene Expr. Patterns*, **3**, 165–168.

Galaktionov, K. and Beach, D. (1991) Specific activation of cdc25 tyrosine phosphatases by B-type cyclins: evidence for multiple roles of mitotic cyclins. *Cell*, **67**, 1181–1194.

Galas, S., Barakat, H., Doree, M. and Picard, A. (1993) Nuclear factor required for specific translation of cyclin B may control the timing of first meiotic cleavage in starfish oocytes. *Mol. Biol. Cell*, **4**, 1295–1306.

Gard, D.L. (1994) Gamma-tubulin is asymmetrically distributed in the cortex of Xenopus oocytes. *Dev. Biol.*, **161**, 131–140.

Gautier, J. (1987) The role of the germinal vesicle for the appearance of maturation-promoting factor activity in the axolotl oocyte. *Dev. Biol.*, **123**, 483–486.

Gautier, J., Norbury, C., Lohka, M. *et al.* (1988) Purified maturation-promoting factor contains the product of a Xenopus homolog of the fission yeast cell cycle control gene cdc2 + . *Cell*, **54**, 433–439.

Gautier, J., Matsukawa, T., Nurse, P. and Maller, J. (1989) Dephosphorylation and activation of Xenopus p34cdc2 protein kinase during the cell cycle. *Nature*, **339**, 626–629.

Gautier, J., Minshull, J., Lohka, M. *et al.* (1990) Cyclin is a component of maturation-promoting factor from *Xenopus*. *Cell*, **60**, 487–494.

Gautier, J., Solomon, M.J., Booher, R.N. *et al.* (1991) cdc25 is a specific tyrosine phosphatase that directly activates p34cdc2. *Cell*, **67**, 197–211.

Gerhart, J., Wu, M. and Kirschner, M. (1984) Cell cycle dynamics of an M-phase-specific cytoplasmic factor in Xenopus laevis oocytes and eggs. *J. Cell Biol.*, **98**, 1247–1255.

Goris, J., Hermann, J., Hendrix, P. *et al.* (1989) Okadaic acid, a specific protein phosphatase inhibitor, induces maturation and MPF formation in Xenopus laevis oocytes. *FEBS Lett.*, **245**, 91–94.

Gotoh, T., Ohsumi, K., Matsui, T. *et al.* (2001) Inactivation of the checkpoint kinase Cds1 is dependent on cyclin B-Cdc2 kinase activation at the meiotic G(2)/M-phase transition in Xenopus oocytes. *J. Cell Sci.*, **114**, 3397–3406.

Gould, K.L. and Nurse, P. (1989) Tyrosine phosphorylation of the fission yeast cdc2 + protein kinase regulates entry into mitosis. *Nature*, **342**, 39–45.

Gould, K.L., Moreno, S., Tonks, N.K. and Nurse, P. (1990) Complementation of the mitotic activator, p80cdc25, by a human protein-tyrosine phosphatase. *Science*, **250**, 1573–1576.

Gross, S.D., Schwab, M.S., Taieb, F.E. *et al.* (2000) The critical role of the MAP kinase pathway in meiosis II in Xenopus oocytes is mediated by p90(Rsk). *Curr. Biol.*, **10**, 430–438.

Gross, S.D., Lewellyn, A.L. and Maller, J.L. (2001) A constitutively active form of the protein kinase p90Rsk1 is sufficient to trigger the G2/M transition in Xenopus oocytes. *J. Biol. Chem.*, **276**, 46099–46103.

Haccard, O. and Jessus, C. (2006) Redundant pathways for Cdc2 activation in Xenopus oocyte: either cyclin B or Mos synthesis. *EMBO Rep.*, **7**, 321–325.

Haccard, O., Lewellyn, A., Hartley, R.S. *et al.* (1995) Induction of Xenopus oocyte meiotic maturation by MAP kinase. *Dev. Biol.*, **168**, 677–682.

Hampl, A. and Eppig, J.J. (1995) Translational regulation of the gradual increase in histone H1 kinase activity in maturing mouse oocytes. *Mol. Reprod. Dev.*, **40**, 9–15.

Hanocq-Quertier, J., Baltus, E. and Brachet, J. (1976) Induction of maturation (meiosis) in small Xenopus laevis oocytes by injection of maturation promoting factor. *Proc. Natl. Acad. Sci. USA*, **73**, 2028–2032.

Hashimoto, N., Watanabe, N., Furuta, Y. *et al.* (1994) Parthenogenetic activation of oocytes in c-mos-deficient mice. *Nature*, **370**, 68–71.

Heidemann, S.R. and Gallas, P.T. (1980) The effect of taxol on living eggs of Xenopus laevis. *Dev. Biol.*, **80**, 489–494.

Heidemann, S.R. and Kirschner, M.W. (1978) Induced formation of asters and cleavage furrows in oocytes of Xenopus laevis during in vitro maturation. *J. Exp. Zool.*, **204**, 431–444.

Heidemann, S.R., Hamborg, M.A., Balasz, J.E. and Lindley, S. (1985) Microtubules in immature oocytes of Xenopus laevis. *J. Cell Sci.*, **77**, 129–141.

Heilbrunn, L.V., Daugherty, K. and Wilbur, K.M. (1939) Initiation of maturation in the frog egg. *Physiol. Zool.*, **12**, 97–100.

Hirai, S., Kubota, J. and Kanatani, H. (1971) Induction of cytoplasmic maturation by 1-methyladenine in starfish oocytes after removal of the germinal vesicle. *Exp. Cell Res.*, **68**, 137–143.

Hirai, T., Yamashita, M., Yoshikuni, M. *et al.* (1992) Cyclin-B in fish oocytes – its cDNA and amino acid sequences, appearance during maturation, and induction of $p34^{cdc2}$-activation. *Mol. Reprod. Dev.*, **33**, 131–140.

Hiraoka, D., Hori-Oshima, S., Fukuhara, T. *et al.* (2004) PDK1 is required for the hormonal signaling pathway leading to meiotic resumption in starfish oocytes. *Dev. Biol.*, **276**, 330–336.

Hochegger, H., Klotzbucher, A., Kirk, J. *et al.* (2001) New B-type cyclin synthesis is required between meiosis I and II during Xenopus oocyte maturation. *Development*, **128**, 3795–3807.

Hoffmann, S., Tsurumi, C., Kubiak, J.Z. and Polanski, Z. (2006) Germinal vesicle material drives meiotic cell cycle of mouse oocyte through the 3′UTR-dependent control of cyclin B1 synthesis. *Dev. Biol.*, **292**, 46–54.

Hsiao, K.M., Chou, S.Y., Shih, S.J. and Ferrell, J.E. (1994) Evidence that inactive p42 mitogen-activated protein kinase and inactive Rsk exist as a heterodimer *in vivo*. *Proc. Natl. Acad. Sci. USA*, **91**, 5480–5484.

Huang, W., Kessler, D. and Erikson, R. (1995) Biochemical and biological analysis of Mek1 phosphorylation site mutants. *Mol. Biol. Cell*, **6**, 237–245.

Huchon, D. and Ozon, R. (1985) Microtubules during germinal vesicle breakdown (GVBD) of Xenopus oocytes: effect of Ca2+ ionophore A-23187 and taxol. *Reprod. Nutr. Dev.*, **25**, 465–479.

Huchon, D., Ozon, R. and Demaille, J.G. (1981) Protein phosphatase-1 is involved in Xenopus oocyte maturation. *Nature*, **294**, 358–359.

Huchon, D., Rime, H., Jessus, C. and Ozon, R. (1993) Control of metaphase-I formation in Xenopus oocyte – effects of an indestructible cyclin-B and of protein synthesis. *Biol. Cell*, **77**, 133–141.

Hunter, A.G. and Moor, R.M. (1987) Stage-dependent effects of inhibiting ribonucleic acids and protein synthesis on meiotic maturation of bovine oocytes *in vitro*. *J. Dairy Sci.*, **70**, 1646–1651.

Hutchins, J.R., Dikovskaya, D. and Clarke, P.R. (2003) Regulation of Cdc2/cyclin B activation in *Xenopus* egg extracts via inhibitory phosphorylation of Cdc25C phosphatase by Ca^{2+}/calmodulin-dependent protein [corrected] kinase II. *Mol. Biol. Cell*, **14**, 4003–4014.

Ihara, J., Yoshida, N., Tanaka, T. *et al.* (1998) Either cyclin B1 or B2 is necessary and sufficient for inducing germinal vesicle breakdown during frog (Rana japonica) oocyte maturation. *Mol. Reprod. Dev.*, **50**, 499–509.

Inoue, D. and Sagata, N. (2005) The Polo-like kinase Plx1 interacts with and inhibits Myt1 after fertilization of Xenopus eggs. *EMBO J.*, **24**, 1057–1067.

Iwashita, J., Hayano, Y. and Sagata, N. (1998) Essential role of germinal vesicle material in the meiotic cell cycle of Xenopus oocytes. *Proc. Natl. Acad. Sci. USA*, **95**, 4392–4397.

Izumi, T. and Maller, J.L. (1993) Elimination of Cdc2 phosphorylation sites in the Cdc25 phosphatase blocks initiation of M-phase. *Mol. Biol. Cell*, **4**, 1337–1350.

Izumi, T., Walker, D.H. and Maller, J.L. (1992) Periodic changes in phosphorylation of the Xenopus cdc25 phosphatase regulate its activity. *Mol. Biol. Cell*, **3**, 927–939.

Jackman, M., Lindon, C., Nigg, E.A. and Pines, J. (2003) Active cyclin B1-Cdk1 first appears on centrosomes in prophase. *Nat. Cell Biol.*, **5**, 143–148.

Jaffe, L.A., Gallo, C.J., Lee, R.H. *et al.* (1993) Oocyte maturation in starfish is mediated by the beta gamma-subunit complex of a G-protein. *J. Cell Biol.*, **121**, 775–783.

Jessus, C. and Beach, D. (1992) Oscillation of MPF is accompanied by periodic association between cdc25 and cdc2-cyclin B. *Cell*, **68**, 323–332.

Jessus, C., Huchon, D. and Ozon, R. (1986) Distribution of microtubules during the breakdown of the nuclear envelope of the Xenopus oocyte: an immunocytochemical study. *Biol. Cell*, **56**, 113–120.

Kajiura-Kobayashi, H., Yoshida, N., Sagata, N. *et al.* (2000) The Mos/MAPK pathway is involved in metaphase II arrest as a cytostatic factor but is neither necessary nor sufficient for initiating oocyte maturation in goldfish. *Dev. Genes. Evol.*, **210**, 416–425.

Kalous, J., Kubelka, M., Rimkevicova, Z. *et al.* (1993) Okadaic acid accelerates germinal vesicle breakdown and overcomes cycloheximide- and 6-dimethylaminopurine block in cattle and pig oocytes. *Dev. Biol.*, **157**, 448–454.

Kanatani, H., Shirai, H., Nakanishi, K. and Kurokawa, T. (1969) Isolation and indentification on meiosis inducing substance in starfish Asterias amurensis. *Nature*, **221**, 273–274.

Karaiskou, A., Cayla, X., Haccard, O. *et al.* (1998) MPF amplification in Xenopus oocyte extracts depends on a two-step activation of cdc25 phosphatase. *Exp. Cell Res.*, **244**, 491–500.

Karaiskou, A., Jessus, C., Brassac, T. and Ozon, R. (1999) Phosphatase 2A and polo kinase, two antagonistic regulators of cdc25 activation and MPF auto-amplification. *J. Cell Sci.*, **112**, 3747–3756.

Karaiskou, A., Perez, L.H., Ferby, I. *et al.* (2001) Differential regulation of Cdc2 and Cdk2 by RINGO and cyclins. *J. Biol. Chem.*, **18**, 18

Karaiskou, A., Lepretre, A.C., Pahlavan, G. *et al.* (2004) Polo-like kinase confers MPF autoamplification competence to growing Xenopus oocytes. *Development*, **131**, 1543–1552.

Katsu, Y., Yamashita, M., Kajiura, H. and Nagahama, Y. (1993) Behavior of the components of maturation-promoting factor, cdc2 kinase and cyclin B, during oocyte maturation of goldfish. *Dev. Biol.*, **160**, 99–107.

Katsu, Y., Yamashita, M. and Nagahama, Y. (1997) Isolation and characterization of goldfish Y box protein, a germ-cell-specific RNA-binding protein. *Eur. J. Biochem.*, **249**, 854–861.

Katsu, Y., Yamashita, M. and Nagahama, Y. (1999) Translational regulation of cyclin B mRNA by 17alpha,20beta-dihydroxy-4-pregnen-3-one (maturation-inducing hormone) during oocyte maturation in a teleost fish, the goldfish (Carassius auratus). *Mol. Cell Endocrinol.*, **158**, 79–85.

Kessel, R.G. (1989) The annulate lamellae—from obscurity to spotlight. *Electron. Microsc. Rev.*, **2**, 257–348.

Kishimoto, T. (2003) Cell-cycle control during meiotic maturation. *Curr. Opin. Cell Biol.*, **15**, 654–663.

Kishimoto, T. and Kanatani, H. (1976) Cytoplasmic factor responsible for germinal vesicle breakdown and meiotic maturation in starfish oocyte. *Nature*, **260**, 321–322.

Kishimoto, T., Hirai, S. and Kanatani, H. (1981) Role of germinal vesicle material in producing maturation-promoting factor in starfish oocyte. *Dev. Biol.*, **81**, 177–181.

Kishimoto, T., Kuriyama, R., Kondo, H. and Kanatani, H. (1982) Generality of the action of various maturation-promoting factors. *Exp. Cell Res.*, **137**, 121–126.

Kobayashi, H., Minshull, J., Ford, C. *et al.* (1991) On the synthesis and destruction of A- and B-type cyclins during oogenesis and meiotic maturation in Xenopus laevis. *J. Cell Biol.*, **114**, 755–765.

Kondo, T., Yanagawa, T., Yoshida, N. and Yamashita, M. (1997) Introduction of cyclin B induces activation of the maturation-promoting factor and breakdown of germinal vesicle in growing zebrafish oocytes unresponsive to the maturation-inducing hormone. *Dev. Biol.*, **190**, 142–152.

Kondo, T., Kotani, T. and Yamashita, M. (2001) Dispersion of cyclin B mRNA aggregation is coupled with translational activation of the mRNA during zebrafish oocyte maturation. *Dev. Biol.*, **229**, 421–431.

Kristjansdottir, K., Safi, A., Shah, C. and Rudolph, J. (2006) Autophosphorylation of Ser66 on Xenopus Myt1 is a prerequisite for meiotic inactivation of Myt1. *Cell Cycle*, **5**, 421–427.

Kumagai, A. and Dunphy, W.G. (1992) Regulation of the cdc25 protein during the cell cycle in Xenopus extracts. *Cell*, **70**, 139–151.

Kumagai, A. and Dunphy, W.G. (1996) Purification and molecular cloning of Plx1, a Cdc25-regulatory kinase from Xenopus egg extracts. *Science*, **273**, 1377–1380.

Kumagai, A., Guo, Z., Emami, K.H. *et al.* (1998a) The Xenopus Chk1 protein kinase mediates a caffeine-sensitive pathway of checkpoint control in cell-free extracts. *J. Cell Biol.*, **142**, 1559–1569.

Kumagai, A., Yakowec, P.S. and Dunphy, W.G. (1998b) 14-3-3 proteins act as negative regulators of the inducer Cdc25 in Xenopus egg extracts. *Mol. Biol. Cell*, **9**, 345–354.

Labbe, J.C., Capony, J.P., Caput, D. *et al.* (1989) MPF from starfish oocytes at first meiotic metaphase is a heterodimer containing one molecule of cdc2 and one molecule of cyclin B. *EMBO J.*, **8**, 3053–3058.

Landrieu, I., Odaert, B., Wieruszeski, J.M. *et al.* (2001) p13(SUC1) and the WW domain of PIN1 bind to the same phosphothreonine-proline epitope. *J. Biol. Chem.*, **276**, 1434–1438.

Lapasset, L., Pradet-Balade, B., Lozano, J.C. *et al.* (2005) Nuclear envelope breakdown may deliver an inhibitor of protein phosphatase 1 which triggers cyclin B translation in starfish oocytes. *Dev. Biol.*, **285**, 200–210.

Ledan, E., Polanski, Z., Terret, M.E. and Maro, B. (2001) Meiotic maturation of the mouse oocyte requires an equilibrium between cyclin B synthesis and degradation. *Dev. Biol.*, **232**, 400–413.

Lee, T.H. and Kirschner, M.W. (1996) An inhibitor of p34cdc2/cyclin B that regulates the G2/M transition in Xenopus extracts. *Proc. Natl. Acad. Sci. USA*, **93**, 352–356.

Lee, M.G. and Nurse, P. (1987) Complementation used to clone a human homologue of the fission yeast cell cycle control gene cdc2. *Nature*, **327**, 31–35.

Lenormand, J.L., Dellinger, R.W., Knudsen, K.E. *et al.* (1999) Speedy: a novel cell cycle regulator of the G2/M transition. *EMBO J.*, **18**, 1869–1877.

Li, J., Meyer, A.N. and Donoghue, D.J. (1997) Nuclear localization of cyclin B1 mediates its biological activity and is regulated by phosphorylation. *Proc. Natl. Acad. Sci. USA*, **94**, 502–507.

Lincoln, A.J., Wickramasinghe, D., Stein, P. *et al.* (2002) Cdc25b phosphatase is required for resumption of meiosis during oocyte maturation. *Nat. Genet.*, **30**, 446–449.

Liu, F., Stanton, J.J., Wu, Z.Q. and PiwnicaWorms, H. (1997) The human Myt1 kinase preferentially phosphorylates Cdc2 on threonine 14 and localizes to the endoplasmic reticulum and Golgi complex. *Mol. Cell Biol.*, **17**, 571–583.

Lohka, M.J., Hayes, M.K. and Maller, J.L. (1988) Purification of maturation-promoting factor, an intracellular regulator of early mitotic events. *Proc. Natl. Acad. Sci. USA*, **85**, 3009–3013.

Lu, K.P., Hanes, S.D. and Hunter, T. (1996) A human peptidyl-prolyl isomerase essential for regulation of mitosis. *Nature*, **380**, 544–547.

Lu, P.J., Zhou, X.Z., Shen, M. and Lu, K.P. (1999) Function of WW domains as phosphoserine- or phosphothreonine-binding modules. *Science*, **283**, 1325–1328.

Margolis, S.S., Walsh, S., Weiser, D.C. *et al.* (2003) PP1 control of M phase entry exerted through 14-3-3-regulated Cdc25 dephosphorylation. *EMBO J.*, **22**, 5734–5745.

Margolis, S.S., Perry, J.A., Weitzel, D.H. *et al.* (2006) A role for PP1 in the Cdc2/Cyclin B-mediated positive feedback activation of Cdc25. *Mol. Biol. Cell*, **17**, 1779–1789.

Maro, B., Howlett, S.K. and Webb, M. (1985) Non-spindle microtubule organizing centers in metaphase II-arrested mouse oocytes. *J. Cell Biol.*, **101**, 1665–1672.

Masui, Y. (1972) Distribution of cytoplasmic activity inducing germinal vesicle breakdown in frog oocytes. *J. Exp. Zool.*, **179**, 365–377.

Masui, Y. and Markert, C.L. (1971) Cytoplasmic control of nuclear behavior during meiotic maturation of frog oocytes. *J. Exp. Zool.*, **177**, 129–145.

Maton, G., Lorca, T., Girault, J.A. *et al.* (2005) Differential regulation of Cdc2 and Aurora-A in Xenopus oocytes: a crucial role of phosphatase 2A. *J. Cell Sci.*, **118**, 2485–2494.

Matsuoka, S., Huang, M. and Elledge, S.J. (1998) Linkage of ATM to cell cycle regulation by the Chk2 protein kinase. *Science*, **282**, 1893–1897.

Minshull, J., Murray, A., Colman, A. and Hunt, T. (1991) Xenopus oocyte maturation does not require new cyclin synthesis. *J. Cell Biol.*, **114**, 767–772.

Miyazaki, A., Kamitsubo, E. and Nemoto, S.I. (2000) Premeiotic aster as a device to anchor the germinal vesicle to the cell surface of the presumptive animal pole in starfish oocytes. *Dev. Biol.*, **218**, 161–171.

Moor, R.M. and Gandolfi, F. (1987) Molecular and cellular changes associated with maturation and early development of sheep eggs. *J. Reprod. Fertil. Suppl.*, **34**, 55–69.

Motlik, J., Kopecny, V., Travnik, P. and Pivko, J. (1984) RNA synthesis in pig follicular oocytes. Autoradiographic and cytochemical study. *Biol. Cell*, **50**, 229–235.

Mueller, P.R., Coleman, T.R., Kumagai, A. and Dunphy, W.G. (1995) Myt1: a membrane-associated inhibitory kinase that phosphorylates Cdc2 on both threonine-14 and tyrosine-15. *Science*, **270**, 86–90.

Mulner, O., Belle, R. and Ozon, R. (1983) cAMP-dependent protein kinase regulates in ovo cAMP level of the Xenopus oocyte: evidence for an intracellular feedback mechanism. *Mol. Cell Endocrinol.*, **31**, 151–160.

Mulner-Lorillon, O., Belle, R., Cormier, P. *et al.* (1995) Brefeldin A provokes indirect activation of cdc2 kinase (MPF) in Xenopus oocytes, resulting in meiotic cell division. *Dev. Biol.*, **170**, 223–229.

Murakami, M.S. and VandeWoude, G.F. (1998) Analysis of the early embryonic cell cycles of Xenopus; regulation of cell cycle length by Xe-wee1 and Mos. *Development*, **125**, 237–248.

Murray, A.W. and Kirschner, M.W. (1989a) Cyclin synthesis drives the early embryonic cell cycle. *Nature*, **339**, 275–280.

Murray, A.W. and Kirschner, M.W. (1989b) Dominoes and clocks: the union of two views of the cell cycle. *Science*, **246**, 614–621.

Murray, A.W., Solomon, M.J. and Kirschner, M.W. (1989) The role of cyclin synthesis and degradation in the control of maturation promoting factor activity. *Nature*, **339**, 280–286.

Nakajima, H., Toyoshima-Morimoto, F., Taniguchi, E. and Nishida, E. (2003) Identification of a consensus motif for Plk (Polo-like kinase) phosphorylation reveals Myt1 as a Plk1 substrate. *J. Biol. Chem.*, **278**, 25277–25280.

Nakajima, H., Yonemura, S., Murata, M. *et al.* (2008) Myt1 protein kinase is essential for Golgi and ER assembly during mitotic exit. *J. Cell Biol.*, **181**, 89–103.

Nakajo, N., Oe, T., Uto, K. and Sagata, N. (1999) Involvement of Chk1 kinase in prophase I arrest of Xenopus oocytes. *Dev. Biol.*, **207**, 432–444.

Nakashima, S. and Kato, K.H. (2001) Centriole behavior during meiosis in oocytes of the sea urchin Hemicentrotus pulcherrimus. *Dev. Growth Differ.*, **43**, 437–445.

Nebreda, A.R. and Hunt, T. (1993) The c-mos proto-oncogene protein kinase turns on and maintains the activity of MAP kinase, but not MPF, in cell-free extracts of Xenopus oocytes and eggs. *EMBO J.*, **12**, 1979–1986.

Nebreda, A., Gannon, J. and Hunt, T. (1995) Newly synthesized protein(s) must associate with p34cdc2 to activate MAP kinase and MPF during progesterone-induced maturation of Xenopus oocytes. *EMBO J.*, **14**, 5597–5607.

Nelkin, B., Nichols, C. and Vogelstein, B. (1980) Protein factor(s) from mitotic CHO cells induce meiotic maturation in Xenopus laevis oocytes. *FEBS Lett.*, **109**, 233–238.

Nigg, E.A. (1991) The substrates of the cdc2 kinase. *Semin. Cell Biol.*, **2**, 261–270.

Norbury, C., Blow, J. and Nurse, P. (1991) Regulatory phosphorylation of the p34cdc2 protein kinase in vertebrates. *EMBO J.*, **10**, 3321–3329.

Nurse, P. and Bissett, Y. (1981) Gene required in G1 for commitment to cell cycle and in G2 for control of mitosis in fission yeast. *Nature*, **292**, 558–560.

Oe, T., Nakajo, N., Katsuragi, Y. *et al.* (2001) Cytoplasmic occurrence of the Chk1/Cdc25 pathway and regulation of Chk1 in Xenopus oocytes. *Dev. Biol.*, **229**, 250–261.

Okumura, E., Fukuhara, T., Yoshida, H. *et al.* (2002) Akt inhibits Myt1 in the signalling pathway that leads to meiotic G2/M-phase transition. *Nat. Cell Biol.*, **4**, 111–116.

Ookata, K., Hisanaga, S., Okano, T. *et al.* (1992) Relocation and distinct subcellular localization of $p34^{cdc2}$-cyclin-B complex at meiosis reinitiation in starfish oocytes. *EMBO J.*, **11**, 1763–1772.

Palmer, A., Gavin, A.C. and Nebreda, A.R. (1998) A link between MAP kinase and p(34cdc2) cyclin B during oocyte maturation: p90(rsk) phosphorylates and inactivates the p34(cdc2) inhibitory kinase Myt1. *EMBO J.*, **17**, 5037–5047.

Patra, D. and Dunphy, W.G. (1996) Xe-p9, a Xenopus Suc1/Cks homolog, has multiple essential roles in cell cycle control. *Genes. Dev.*, **10**, 1503–1515.

Patra, D., Wang, S.X., Kumagai, A. and Dunphy, W.G. (1999) The Xenopus Suc1/Cks protein promotes the phosphorylation of G(2)/M regulators. *J. Biol. Chem.*, **274**, 36839–36842.

Pelczar, H., Caulet, S., Thibier, C. *et al.* (2007) Characterization and expression of a maternal axolotl cyclin B1 during oogenesis and early development. *Dev. Growth Differ.*, **49**, 407–419.

Peng, C.Y., Graves, P.R., Thoma, R.S. *et al.* (1997) Mitotic and G(2) checkpoint control: Regulation of 14-3-3 protein binding by phosphorylation of Cdc25C on serine-216. *Science*, **277**, 1501–1505.

Peng, C.Y., Graves, P.R., Ogg, S. *et al.* (1998) C-TAK1 protein kinase phosphorylates human Cdc25C on serine 216 and promotes 14-3-3 protein binding. *Cell Growth Differ.*, **9**, 197–208.

Perdiguero, E., Pillaire, M.J., Bodart, J.F. *et al.* (2003) Xp38gamma/SAPK3 promotes meiotic G(2)/M transition in Xenopus oocytes and activates Cdc25C. *EMBO J.*, **22**, 5746–5756.

Perez-Mongiovi, D., Chang, P. and Houliston, E. (1998) A propagated wave of MPF activation accompanies surface contraction waves at first mitosis in Xenopus. *J. Cell Sci.*, **111** (3), 385–393.

Perez-Mongiovi, D., Beckhelling, C., Chang, P. *et al.* (2000) Nuclei and microtubule asters stimulate maturation/M phase promoting factor (MPF) activation in Xenopus eggs and egg cytoplasmic extracts. *J. Cell Biol.*, **150**, 963–974.

Peter, M., Labbe, J.C., Doree, M. and Mandart, E. (2002) A new role for Mos in Xenopus oocyte maturation: targeting Myt1 independently of MAPK. *Development*, **129**, 2129–2139.

Picard, A. and Peaucellier, G. (1998) Behavior of cyclin B and cyclin B-dependent kinase during starfish oocyte meiosis reinitiation: evidence for non-identity with MPF. *Biol. Cell*, **90**, 487–496.

Picard, A., Peaucellier, G., le Bouffant, F. *et al.* (1985) Role of protein synthesis and proteases in production and inactivation of maturation-promoting activity during meiotic maturation of starfish oocytes. *Dev. Biol.*, **109**, 311–320.

Picard, A., Harricane, M.C., Labbe, J.C. and Doree, M. (1988) Germinal vesicle components are not required for the cell-cycle oscillator of the early starfish embryo. *Dev. Biol.*, **128**, 121–128.

Pines, J. and Hunter, T. (1990) Human cyclin A is adenovirus E1A-associated protein p60 and behaves differently from cyclin B. *Nature*, **346**, 760–763.

Polanski, Z., Ledan, E., Brunet, S. *et al.* (1998) Cyclin synthesis controls the progression of meiotic maturation in mouse oocytes. *Development*, **125**, 4989–4997.

Poon, R.Y.C., Yamashita, K., Adamczewski, J.P. *et al.* (1993) The cdc2-related protein p40(mo15) is the catalytic subunit of a protein kinase that can activate p33(cdk2) and p34(cdc2). *EMBO J.*, **12**, 3123–3132.

Posada, J., Yew, N., Ahn, N.G. *et al.* (1993) Mos stimulates MAP kinase in Xenopus oocytes and activates a MAP kinase kinase *in vitro*. *Mol. Cell Biol.*, **13**, 2546–2553.

Qiu, G.F., Ramachandra, R.K., Rexroad, C.E. 3rd and Yao, J. (2008) Molecular characterization and expression profiles of cyclin B1, B2 and Cdc2 kinase during oogenesis and spermatogenesis in rainbow trout (Oncorhynchus mykiss). *Anim. Reprod. Sci.*, **105**, 209–225.

Raff, J.W., Whitfield, W.G. and Glover, D.M. (1990) Two distinct mechanisms localise cyclin B transcripts in syncytial Drosophila embryos. *Development*, **110**, 1249–1261.

Rankin, S. and Kirschner, M.W. (1997) The surface contraction waves of Xenopus eggs reflect the metachronous cell-cycle state of the cytoplasm. *Curr. Biol.*, **7**, 451–454.

Reynhout, J.K. and Smith, L.D. (1974) Studies on the appearance and nature of a maturation-inducing factor in the cytoplasm of amphibian oocytes exposed to progesterone. *Dev. Biol.*, **38**, 394–400.

Reynhout, J.K., Taddei, C., Smith, L.D. and LaMarca, M.J. (1975) Response of large oocytes of Xenopus laevis to progesterone in vitro in relation to oocyte size and time after previous HCG-induced ovulation. *Dev. Biol.*, **44**, 375–379.

Rime, H., Jessus, C. and Ozon, R. (1987) Distribution of microtubules during the first meiotic cell division in the mouse oocyte: effect of taxol. *Gamete Res.*, **17**, 1–13.

Rime, H., Huchon, D., Jessus, C. *et al.* (1990) Characterization of MPF activation by okadaic acid in Xenopus oocytes. *Cell Differ. Dev.*, **29**, 47–58.

Rime, H., Yang, J., Jessus, C. and Ozon, R. (1991) MPF is activated in growing immature Xenopus oocytes in absence of detectable tyrosine dephosphorylation of $p34^{cdc2.}$ *Exp. Cell Res.*, **196**, 241–245.

Rime, H., Jessus, C. and Ozon, R. (1995) Tyrosine phosphorylation of p34cdc2 is regulated by protein phosphatase 2A in growing immature Xenopus oocytes. *Exp. Cell Res.*, **219**, 29–38.

Rime, H., Talbi, N., Popoff, M.R. *et al.* (1998) Inhibition of small G proteins by Clostridium sordellii lethal toxin activates cdc2 and MAP kinase in Xenopus oocytes. *Dev. Biol.*, **204**, 592–602.

Roy, L.M., Swenson, K.I., Walker, D.H. *et al.* (1991) Activation of $p34^{cdc2}$ kinase by cyclin A. *J. Cell Biol.*, **113**, 507–514.

Ruiz, E.J., Hunt, T. and Nebreda, A.R. (2008) Meiotic inactivation of Xenopus Myt1 by CDK/XRINGO, but not CDK/cyclin, via site-specific phosphorylation. *Mol. Cell*, **32**, 210–220.

Russell, P. and Nurse, P. (1986) cdc25 + functions as an inducer in the mitotic control of fission yeast. *Cell*, **45**, 145–153.

Russell, P. and Nurse, P. (1987) Negative regulation of mitosis by wee1 +, a gene encoding a protein kinase homolog. *Cell*, **49**, 559–567.

Russell, P., Moreno, S. and Reed, S.I. (1989) Conservation of mitotic controls in fission and budding yeasts. *Cell*, **57**, 295–303.

Sadhu, K., Reed, S.I., Richardson, H. and Russell, P. (1990) Human homolog of fission yeast cdc25 mitotic inducer is predominantly expressed in G2. *Proc. Natl. Acad. Sci. USA*, **87**, 5139–5143.

Sadler, S.E. and Maller, J.L. (1983) The development of competence for meiotic maturation during oogenesis in Xenopus laevis. *Dev. Biol.*, **98**, 165–172.

Sadler, K.C. and Ruderman, J.V. (1998) Components of the signaling pathway linking the 1-methyladenine receptor to MPF activation and maturation in starfish oocytes. *Dev. Biol.*, **197**, 25–38.

Sagata, N., Oskarsson, M., Copeland, T. *et al.* (1988) Function of c-mos proto-oncogene product in meiotic maturation in Xenopus oocytes. *Nature*, **335**, 519–525.

Sagata, N., Daar, I., Oskarsson, M. *et al.* (1989) The product of the mos proto-oncogene as a candidate "initiator" for oocyte maturation. *Science*, **245**, 643–646.

Sakamoto, I., Takahara, K., Yamashita, M. and Iwao, Y. (1998) Changes in cyclin B during oocyte maturation and early embryonic cell cycle in the newt, Cynops pyrrhogaster: Requirement of germinal vesicle for MPF activation. *Dev. Biol.*, **195**, 60–69.

Sanchez, Y., Wong, C., Thoma, R.S. *et al.* (1997) Conservation of the Chk1 checkpoint pathway in mammals: Linkage of DNA damage to Cdk regulation through Cdc25. *Science*, **277**, 1497–1501.

Schatt, P., Moreau, M. and Guerrier, P. (1983) Variation cyclique de la phosphorylation des protéines and de l'activité MPF pendant la segmentation précoce de l'oeuf d'oursin. *CR Acad. Sci. Paris*, **296**, 551–554.

Schorderet-Slatkine, S. and Drury, K.C. (1973) Progesterone induced maturation in oocytes of Xenopus laevis. Appearance of a 'maturation promoting factor' in enucleated oocytes. *Cell Differ.*, **2**, 247–254.

Shen, M.H., Stukenberg, P.T., Kirschner, M.W. and Lu, K.P. (1998) The essential mitotic peptidyl-prolyl isomerase Pin1 binds and regulates mitosis-specific phosphoproteins. *Genes Dev.*, **12**, 706–720.

Shibuya, E.K. and Ruderman, J.V. (1993) Mos induces the in vitro activation of mitogen-activated protein kinases in lysates of frog oocytes and mammalian somatic cells. *Mol. Biol. Cell*, **4**, 781–790.

Slepchenko, B.M. and Terasaki, M. (2003) Cyclin aggregation and robustness of bio-switching. *Mol. Biol. Cell*, **14**, 4695–4706.

Solomon, M.J., Harper, J.W. and Shuttleworth, J. (1993) Cak, the p34(cdc2) activating kinase, contains a protein identical or closely related to p40(mo15). *EMBO J.*, **12**, 3133–3142.

Sorensen, R.A. and Wassarman, P.M. (1976) Relationship between growth and meiotic maturation of the mouse oocyte. *Dev. Biol.*, **50**, 531–536.

Sorensen, R.A., Cyert, M.S. and Pedersen, R.A. (1985) Active maturation-promoting factor is present in mature mouse oocytes. *J. Cell Biol.*, **100**, 1637–1640.

Spruck, C.H., de Miguel, M.P., Smith, A.P. *et al.* (2003) Requirement of Cks2 for the first metaphase/anaphase transition of mammalian meiosis. *Science*, **300**, 647–650.

Stanford, J.S., Lieberman, L.S., Wong, V.L. and Ruderman, J.V. (2003) Regulation of the G2/M transition in oocytes of Xenopus tropicalis. *Dev. Biol.*, **260**, 438–448.

Stukenberg, P.T. and Kirschner, M.W. (2001) Pin1 acts catalytically to promote a conformational change in Cdc25. *Mol. Cell*, **7**, 1071–1083.

Sunkara, P.S., Wright, D.A. and Rao, P.N. (1979) Mitotic factors from mammalian cells induce germinal vesicle breakdown and chromosome condensation in amphibian oocytes. *Proc. Natl. Acad. Sci. USA*, **76**, 2799–2802.

Swenson, K.I., Farrell, K.M. and Ruderman, J.V. (1986) The clam embryo protein cyclin A induces entry into M phase and the resumption of meiosis in Xenopus oocytes. *Cell*, **47**, 861–870.

Tachibana, K., Yanagishima, N. and Kishimoto, T. (1987) Preliminary characterization of maturation-promoting factor from yeast Saccharomyces cerevisiae. *J. Cell Sci.*, **88** (3), 273–281.

Tachibana, K., Tanaka, D., Isobe, T. and Kishimoto, T. (2000) c-Mos forces the mitotic cell cycle to undergo meiosis II to produce haploid gametes. *Proc. Natl. Acad. Sci. USA*, **97**, 14301–14306.

Tanaka, T. and Yamashita, M. (1995) Pre-MPF is absent in immature oocytes of fishes and amphibians except Xenopus. *Develop. Growth Differ.*, **37**, 387–393.

Taylor, M.A. and Smith, L.D. (1987) Induction of maturation in small Xenopus laevis oocytes. *Dev. Biol.*, **121**, 111–118.

Terasaki, M., Runft, L.L. and Hand, A.R. (2001) Changes in organization of the endoplasmic reticulum during Xenopus oocyte maturation and activation. *Mol. Biol. Cell*, **12**, 1103–1116.

Terasaki, M., Okumura, E., Hinkle, B. and Kishimoto, T. (2003) Localization and dynamics of Cdc2-cyclin B during meiotic reinitiation in starfish oocytes. *Mol. Biol. Cell*, **14**, 4685–4694.

Thibier, C., De Smedt, V., Poulhe, R. *et al.* (1997) In vivo regulation of cytostatic activity in Xenopus metaphase II-arrested oocytes. *Dev. Biol.*, **185**, 55–66.

de Vantéry, C., Gavin, A.C., Vassalli, J.D. and Schorderet-Slatkine, S. (1996) An accumulation of p34cdc2 at the end of mouse oocyte growth correlates with the acquisition of meiotic competence. *Dev. Biol.*, **174**, 335–344.

de Vantéry, C., Stutz, A., Vassalli, J.D. and Schorderet-Slatkine, S. (1997) Acquisition of meiotic competence in growing mouse oocytes is controlled at both translational and posttranslational levels. *Dev. Biol.*, **187**, 43–54.

Vaur, S., Poulhe, R., Maton, G. *et al.* (2004) Activation of Cdc2 kinase during meiotic maturation of axolotl oocyte. *Dev. Biol.*, **267**, 265–278.

Wagenaar, E.B. (1983) The timing of synthesis of proteins required for mitosis in the cell cycle of the sea urchin embryo. *Exp. Cell Res.*, **144**, 393–403.

Wasserman, W.J. and Masui, Y. (1975) Effects of cycloheximide on a cytoplasmic factor initiating meiotic naturation in Xenopus oocytes. *Exp. Cell Res.*, **91**, 381–388.

Wasserman, W.J. and Smith, L.D. (1978) The cyclic behavior of a cytoplasmic factor controlling nuclear membrane breakdown. *J. Cell Biol.*, **78**, R15–R22.

Weintraub, H., Buscaglia, M., Ferrez, M. *et al.* (1982) Demonstration of maturation promoting factor activity in Saccharomyces cerevisiae. *C R Seances Acad. Sci. III*, **295**, 787–790.

Westendorf, J., Swenson, K. and Ruderman, J. (1989) The role of cyclin B in meiosis I. *J. Cell Biol.*, **108**, 1431–1444.

Winston, N.J. (1997) Stability of cyclin B protein during meiotic maturation and the first mitotic cell division in mouse oocytes. *Biol. Cell*, **89**, 211–219.

Yamashita, M. (1998) Molecular mechanisms of meiotic maturation and arrest in fish and amphibian oocytes. *Semin. Cell Dev. Biol.*, **9**, 569–579.

Yamashita, M., Fukada, S., Yoshikuni, M. *et al.* (1992) Purification and characterization of maturation-promoting factor in fish. *Dev. Biol.*, **149**, 8–15.

Yamashita, M., Kajiura, H., Tanaka, T. *et al.* (1995) Molecular mechanisms of the activation of maturation-promoting factor during goldfish oocyte maturation. *Dev. Biol.*, **168**, 62–75.

Yang, J., Winkler, K., Yoshida, M. and Kornbluth, S. (1999) Maintenance of G2 arrest in the Xenopus oocyte: a role for 14-3-3-mediated inhibition of Cdc25 nuclear import. *EMBO J.*, **18**, 2174–2183.

Yew, N., Mellini, M.L. and Vande Woude, G.F. (1992) Meiotic initiation by the mos protein in Xenopus. *Nature*, **355**, 649–652.

Yoshida, N., Mita, K. and Yamashita, M. (2000) Function of the Mos/MAPK pathway during oocyte maturation in the Japanese brown frog Rana japonica. *Mol. Reprod. Dev.*, **57**, 88–98.

Yu, J., Fleming, S.L., Williams, B. *et al.* (2004) Greatwall kinase: a nuclear protein required for proper chromosome condensation and mitotic progression in Drosophila. *J. Cell Biol.*, **164**, 487–492.

Yu, J., Zhao, Y., Li, Z. *et al.* (2006) Greatwall kinase participates in the Cdc2 autoregulatory loop in Xenopus egg extracts. *Mol. Cell*, **22**, 83–91.

Zhao, Y., Haccard, O., Wang, R. *et al.* (2008) Roles of Greatwall kinase in the regulation of cdc25 phosphatase. *Mol. Biol. Cell*, **19**, 1317–1327.

Zheng, X.F. and Ruderman, J.V. (1993) Functional analysis of the P box, a domain in cyclin B required for the activation of Cdc25. *Cell*, **75**, 155–164.

Zhou, X.Z., Kops, O., Werner, A. *et al.* (2000) Pin1-dependent prolyl isomerization regulates dephosphorylation of Cdc25C and tau proteins. *Mol. Cell*, **6**, 873–883.

Section V

The cytological events of meiotic divisions

10
Meiotic spindle assembly and chromosome segregation in oocytes

Julien Dumont[1] and Stéphane Brunet[2]

[1]*Desai Laboratory, Department of Cellular and Molecular Medecine, Ludwig Institute for Cancer Research, University of California, 9500 Gilman Drive, La Jolla, CA 92093-0660, USA*
[2]*Biologie du Développement – UMR 7622, UPMC-CNRS, Paris CEDEX 05, France*

10.1 Introduction

Oogenesis is the process by which female haploid gametes or oocytes are formed. To achieve this haploidization of the genome, a single premeiotic phase of DNA replication is followed by two successive cell divisions in a process termed meiotic maturation (Figure 10.1a). The second division (meiosis II) resembles mitosis in that sister chromatids are segregated from each other. The first division (meiosis I), on the other hand, is unique in that homologous chromosomes are the cargo for segregation. Homologous chromosomes comprise a pair of sister chromatids and are connected together by a bond created by recombination (crossover) between nonsister chromatids. The oocyte is usually fertilized around meiotic maturation. In insects and vertebrates fertilization occurs at metaphase of the first and second meiotic division respectively. In echinoderms, fertilization occurs after the second division, while it takes place before the first division in worms. Meiotic maturation is thus crucial for the completion of oogenesis. Accuracy of the process ensures the formation of a competent egg capable, upon fertilization, of generating living euploid offspring. Errors during the haploidization process lead to gametes carrying an incorrect number of chromosomes and eventually to aneuploid embryos. Aneuploidy is a cause of birth defects and is a major obstacle in achieving reproductive success. Indeed, the vast majority of aneuploid embryos are nonviable and lead to spontaneous abortion, which is likely to be one cause of the recent decline in human fertility (Hassold, Hall and Hunt, 2007).

Oogenesis: The Universal Process Marie-Hélène Verlhac and Anne Villeneuve

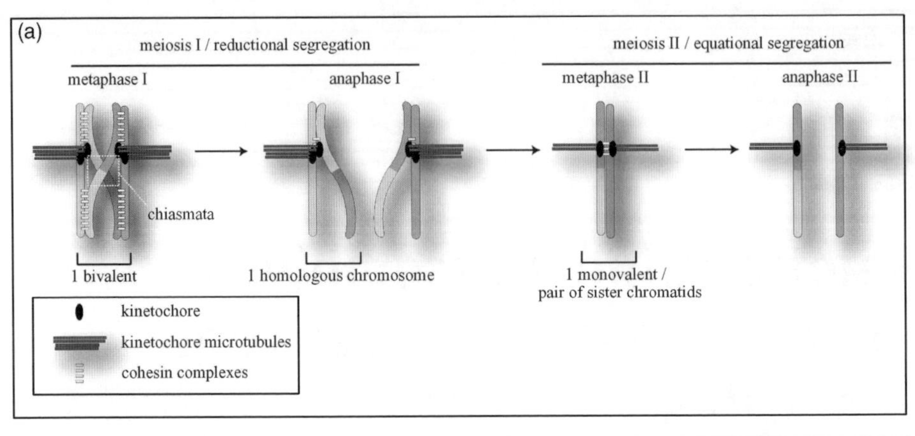

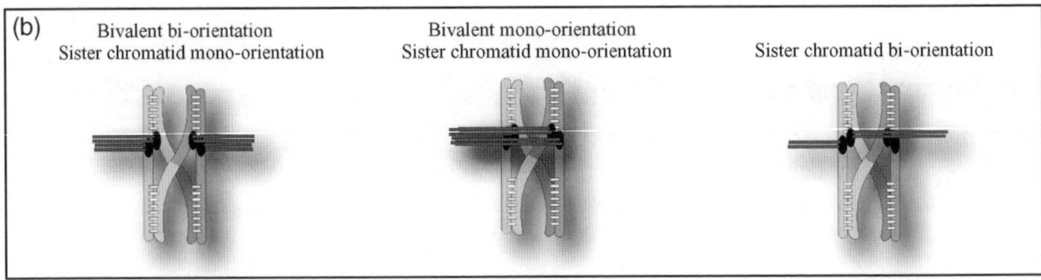

Figure 10.1 Chromosome attachment and segregation during meiosis. Recombined homologous chromosomes are depicted in light and dark grey. Recombination and exchange of genetic material occurred at the site of chiasmata (dotted white square in schematic a). Kinetochore fibres appear as thick dark grey lines connected to kinetochores (black ovals). Cohesin complexes are represented as white rectangles. (b) Different mode of homologous chromosome (or bivalent) attachment during meiosis I. Accurate chromosome segregation requires bivalent bi-orientation (left schematic). Bivalent mono-orientation or sister chromatid bi-orientation leads to meiotic chromosome missegregation and eventually to aneuploid oocytes (middle and right schematics). (a) Meiosis corresponds to two successive rounds of chromosome segregation. Meiosis I is a reductional division during which homologous chromosomes are segregated. Meiosis II is a classical equational division, like mitosis, which allows sister chromatid segregation. During meiosis I, cohesion is maintained at the centromere to prevent premature separation of sister chromatids

Meiotic maturation immediately follows the release of the oocyte from a universal arrest of the cell cycle in prophase I. During this arrest, the oocyte displays a nucleus containing partially decondensed chromosomes and an interphase-like cytoskeleton organization. Upon release of the prophase I arrest and entry into meiotic maturation, nuclear envelope breakdown (NEBD) and full chromosome condensation is achieved, accompanied by a dramatic reorganization of the microtubule cytoskeleton. Indeed, a highly sophisticated spindle-shaped structure, termed the meiotic spindle, is assembled around chromosomes. The spindle is a bipolar stable, yet dynamic, structure composed of labile microtubules and numerous associated proteins. Microtubules are filamentous polymers of α/β-tubulin dimers, and display a polarized organization with a slow polymerizing (−)-end and a fast polymerizing (+)-end. Typically, minus-ends are concentrated at the spindle poles, while plus-ends are oriented toward the chromosomes. The meiotic spindle is involved in various aspects of meiotic maturation including positioning of the cell division plane during cytokinesis (reviewed in Chapter 11) and ensuring accurate repartition of the genetic material during the two successive meiotic divisions. The first investigations on meiotic cell division were conducted in the early 1880s by E. Van Beneden and paralleled the pioneering observations of W. Flemming on mitosis. Yet, more than one century after, most of our current knowledge on spindle formation and function and more generally on cell division arises from studies performed on mitotic systems. A number of molecules and basic principles are obviously shared between mitosis and meiosis; however meiosis-specific adaptations, likely due to spatial and temporal constraints (the large size of the oocytes, their unusual cell cycle progression, etc.) have recently emerged from studies on female meiosis. In this chapter, we focus on advances in understanding meiotic spindle assembly mechanisms and we discuss recent findings on the meiotic chromosome–spindle interface.

10.2 From centrosome loss to microtubule self-organization

10.2.1 Microtubule assembly without canonical centrosome

In animal somatic cells or in spermatocytes the dominant microtubule-organizing centre is the centrosome, comprised of two orthogonal tubulin cylinders – the centrioles – embedded in a cloud of pericentriolar material (PCM). Microtubule nucleating and anchoring activities are displayed by the PCM (for a review see (Azimzadeh and Bornens, 2007). Centrosome duplication occurs only once per cell cycle during G1/S phase so that, prior to mitosis onset, the two centrosomes mature, split and move toward opposite sides of the nucleus under the effect of antiparallel microtubule sliding and cortical forces (Cytrynbaum, Scholey and Mogilner, 2003). As a result, the spindle poles and axis are already set up upon rupture of the nuclear envelope (Figure 10.2a).

However, in most animal species, oogenesis is associated with centriole loss. This process, termed 'centrosome reduction' remains poorly understood (Manandhar, Schatten and Sutovsky, 2005). In echinoderms, meiosis I spindle poles are organized from two regular centrosomes (i.e. containing two centrioles each). By contrast, meiosis II spindle poles display only one centriole each (Sluder *et al.*, 1989). Finally, the retained centriole degenerates in parthenogenetically activated eggs (Moy, Brandriff and Vacquier, 1977). In insect, worm and vertebrate oocytes, the loss of centrioles is achieved prior to meiotic divisions. *Drosophila* centrioles persist until mid oogenesis (around stage 9) but not later (Januschke *et al.*, 2006). In mammals, electron microscopy analyses (Szollosi, Calarco and Donahue, 1972; Hertig and Adams, 1967) demonstrated the absence of centrioles in fully-grown oocytes, confirming pioneer cytologists' observations (Kirkham and Burr, 1913). However, the exact cell cycle stage at which centrioles disappear in mammals is still unknown. Meiotic divisions thus proceed in the absence of canonical centrosomes. After fertilization, a reverse process called 'centrosome restoration' leads to the reassembly of centriole-containing centrosomes in the zygote or in the embryo. This process depends on the contribution of sperm-associated centriole in many species (Manandhar, Simerly and Schatten, 2000), with the noticeable exception of rodents, where the sperm centriole fully degenerates after

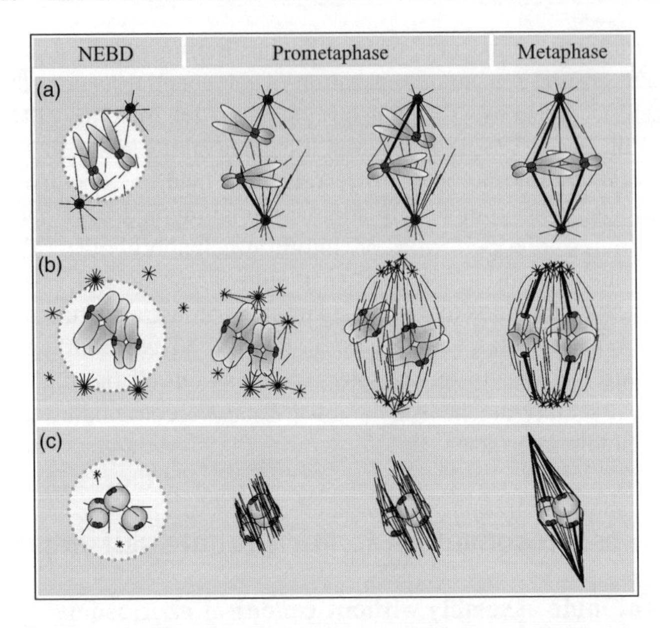

Figure 10.2 Comparative models for centriolar and acentriolar spindle assembly. The chromosomes are depicted in light grey and their associated kinetochores appear as dark grey discs. Microtubules appear as black lines and kinetochore fibres as thicker lines. Centrosomes in schematic a appear as grey discs at the centre of the radiating microtubules. The position of the nucleus prior to nuclear envelope breakdown (NEBD) is symbolized by a dotted circle. (a) Centriolar spindle assembly: case of somatic cells, spermatocytes and echinoderm oocytes. At NEBD, microtubules are nucleated in the vicinity of the chromosomes or emanate from the centrosomes and grow preferentially toward the chromosomes. Microtubule capture at the kinetochore leads to robust kinetochore fibre formation and triggers chromosome congression on the metaphase plate. (b) Acentriolar spindle assembly: case of the mouse oocyte. At NEBD, microtubule nucleation in the vicinity of chromosomes leads to the formation of multiple microtubule asters. Microtubule asters progressively organize into a bipolar and barrel-shaped spindle. (c) Acentriolar spindle assembly: case of the *Drosophila* oocyte. Microtubules nucleated around the chromosomes are first sorted into an antiparallel array called the 'central spindle'. Microtubule (−)-ends are then focused into sharp spindle poles

fertilization. In mouse, *de novo* centriole assembly begins in the 64-cells stage embryo and involves a completely unknown mechanism (for a review see Schatten, 1994).

In vertebrate oocytes, canonical centrosomes are replaced by multiple microtubule organizing centres (or MTOCs), which govern microtubule assembly. However, the exact nature, dynamics and function of these MTOCs remain elusive. In mouse oocytes, MTOCs contain classic constituents of PCM such as γ-tubulin and pericentrin (Gueth-Hallonet *et al.*, 1993; Carabatsos *et al.*, 2000) and are thus considered as aggregates of pericentriolar material displaying microtubule nucleation capacity. An original study combining γ-tubulin immunofluorescence in fixed oocytes and ultrastructural analysis concluded that MTOCs originate from the fragmentation of two large aggregates of PCM present in prophase I arrested oocytes (Calarco, 2000). This implies that the MTOC number and position at NEBD is determined by the initial fragmentation phase of the large aggregates. In this model, MTOCs act as centrosome-like structures; that is, they are the sites of initial microtubule nucleation. However, the initial PCM aggregates were not observed in a recent study using pericentrin immunofluorescence. Instead, MTOCs were proposed to form *de novo* from the interphase-like microtubule network (Schuh and Ellenberg, 2007). In this model, MTOCs only form as a consequence of nucleation and organization of microtubules into small aster-like structures that subsequently recruit high concentrations of PCM. This model is comparable to the *in vitro* situation of *Xenopus* egg extracts. In this system, aggregates of PCM are not initially observed (Buendia, Draetta and Karsenti, 1992). However, adding paclitaxel (taxol), a microtubule-stabilizing drug, to the extract leads to the production of multiple PCM-containing asters. Individual microtubules are assembled and sorted into an aster under the action of the (−)-end-directed microtubule motor Dynein (Verde *et al.*, 1991). This aster then concentrates PCM components. It is possible that similar processes are at play in the mouse oocyte. A third model would be a combination of these two mechanisms, with the pre-existence of some MTOCs that 'launch' the nucleation process, which in turn leads to formation of small asters that recruit high concentrations of PCM. However, in absence of analysis performed in live oocytes, the exact origin and function of MTOCs remain obscure. After NEBD, multiple asters assemble in the cytoplasm and around the chromosomes. All these asters are positive for PCM staining and are thus likely to arise from MTOCs. They first gather to form a single mass of microtubules around chromosomes and are then progressively organized into a bipolar spindle. During this process, MTOCs concentrate at the flattened meiotic spindle poles, conferring a specific 'barrel' shape to the meiotic spindle (Wassarman and Fujiwara, 1978) (Figure 10.2b). In fixed metaphase II-arrested oocytes, MTOCs are present in the cytoplasm at a distance from the spindle and are not associated with microtubules. When such oocytes are incubated in paclitaxel, multiple ectopic asters assemble in the cytoplasm, probably from these MTOCs, suggesting that they are in fact competent for microtubule nucleation. Altogether, these observations suggest that only a fraction of MTOCs actively contribute to spindle formation, probably because of their proximity to chromosomes (Maro, Howlett and Webb, 1985). In *Xenopus* oocytes, a lamellar structure, which likely represents a single MTOC, is present at the base of the nucleus at NEBD (Huchon *et al.*, 1981; Gard, 1992). Microtubules are assembled, from this structure, into a fibrillar network and progressively organized into a bipolar barrel-shaped spindle during meiosis I. However, the exact steps involved in this process are still poorly understood. Indeed, the large size of *Xenopus* oocytes, combined with their opaque yolk-filled cytoplasm, make live imaging microscopy challenging in this model.

By contrast, the use of time-lapse microscopy in live oocytes was critical to better understanding of microtubule assembly and organization in *Drosophila*. In this model, no MTOC-like structure has been described (Figure 10.2c). Meiotic spindles are elongated with sharp spindle poles devoid of γ-tubulin foci (Matthies *et al.*, 1996), while γ-tubulin itself is required for spindle assembly (Tavosanis *et al.*, 1997). Around NEBD, short and unorganized microtubules assemble around the mass of chromosomes. However it is still unclear whether or not tiny and transient microtubule asters are first assembled at distance from the chromosomes, which then coalesce around the mass of chromosomes (Matthies *et al.*, 1996; Skold, Komma and Endow, 2005; Colombie *et al.*, 2008). Similarly, in *Caenorhabditis elegans*, no MTOC-like structure is observed. However, before NEBD, γ-tubulin progressively concentrates on the nuclear envelope (McNally *et al.*, 2006). Upon NEBD, a few microtubule asters assemble in this area, which then connect to each other to form a round microtubule structure around the chromosomes (J. Dumont, unpublished observations). This structure progressively acquires a bipolar organization. During this process, γ-tubulin enters the nuclear region forming a diffuse cloud, but is never seen as concentrated foci on the initial asters or at the spindle poles. Interestingly, in *C. elegans*, γ-tubulin-mediated microtubule nucleation is not the only way of increasing microtubule density in the meiotic spindle. An additional mechanism relies on the activity of the ATP-dependent microtubule-severing enzyme, katanin. During spindle assembly, katanin converts long microtubule polymers into shorter microtubule fragments near meiotic chromosomes and thus increases the number of microtubule polymers and limits the size of the spindle (McNally *et al.*, 2006; Srayko *et al.*, 2006).

Taken together, these observations show that the basic mechanisms of meiotic spindle formation are similar between different species. The first step usually involves the assembly of tiny asters at distance from the chromosomes, which subsequently coalesce to form a single structure on or around the mass of chromosomes. This initial structure is then progressively 'remodelled' into a bipolar spindle. This succession of events occurs with or without the presence of visible MTOCs, but usually requires γ-tubulin.

10.2.2 Microtubule self-organization during meiosis

Based on the dissection of *in vitro* spindle assembly in *Xenopus* egg extracts and on *in vivo* studies in live oocytes, the mechanisms involved in the progressive bipolarization of the meiotic spindle are starting to be unravelled. *Xenopus* egg extracts are prepared from oocytes arrested in metaphase of the second meiotic division. Addition of chromatin to this 'meiotic' cytoplasm and 'cycling' of the reaction through interphase and back into a 'mitotic' state promotes formation of bipolar spindles. In this model system, spindle assembly does occur in the absence of canonical centrosome and is based on the central role of chromosomes as spatial organizers (see below) and on the self-organization of microtubules driven by microtubule-associated proteins (MAPs) and molecular motors (for a review see Karsenti and Vernos, 2001). Schematically, individual microtubules are nucleated with random orientation, at dispersed sites around the chromosomes. Microtubule (−)-ends are then sent away from the chromosomes under the activity of a tetrameric (+)-end-directed microtubule motor of the kinesin-5 family, leading to the formation of an antiparallel microtubule array that surrounds chromosomes. Finally, the (−)-end-directed microtubule motor Dynein focuses these (−)-ends allowing formation of spindle poles (Walczak *et al.*, 1998). In mouse oocyte, the kinesin-5 family member Kif 11 is similarly required for meiotic spindle bipolarization and Dynein is essential for spindle pole integrity (Mailhes, Mastromatteo and Fuseler, 2004; Zhang *et al.*, 2007). In *Drosophila* oocyte, microtubules nucleated around the chromosomes are first sorted into an antiparallel array called the 'central spindle' (distinct from the classical anaphase central spindle) (Figure 10.2c). Its formation and robustness depends on the activity of Subito, a member of the kinesin-6 family (Jang, Rahman and McKim, 2005) in cooperation with the chromosomal passenger protein Incenp (Colombie *et al.*, 2008). Subito is presumably a (+)-end-directed motor and Incenp was shown *in vitro* to display microtubule-bundling activity (Wheatley *et al.*, 2001). Interestingly, by combining *in vivo* experiments in yeast and *in silico* modelling, it was recently shown that activities of a (+)-end-directed microtubule motor and a microtubule-bundling protein are sufficient to generate stable bipolar bundles (Janson *et al.*, 2007). The oligomeric (−)-end-directed motor Ncd (for non-claret disjunctional) could also participate in the formation of this antiparallel array of microtubules (Skold *et al.*, 2005). Once the 'central spindle' is assembled, microtubules slide and become focused at the two extremities to give rise to the sharp spindle poles. This last step depends on the activity of the (−)-end-directed motor Ncd (Matthies *et al.*, 1996). Differences in initial conditions, that is, in the nucleation rates or in the balance between microtubule nucleation and 'sorting' activities, are likely to account for the variability in the initial events (presence or absence of visible MTOCs) and in the microtubule structures (barrel shape versus sharp spindle poles) that are observed between different species.

10.3 Chromosomes as spatial organizers of the meiotic spindle

All the processes described so far are spatially centred on chromosomes. Chromosomes are indeed central actors of spindle assembly. On one hand, they contribute to microtubule organization by an 'at distance' effect: they generate biochemical gradients that locally constrain the activity of molecular factors required for spindle assembly. On the other hand, they interact with labile spindle microtubules and stabilize them. This 'contact' effect contributes to spindle assembly and robustness.

10.3.1 The Ran pathway during meiosis

The main characterized 'at distance' effect is mediated by the small GTPase Ran, which is present in active GTP-bound (RanGTP) and inactive GDP-bound (RanGDP) forms in the cell and acts as a molecular switch. During M phase, the nucleotide exchange factor RCC1 concentrates on chromosomes where it loads Ran with GTP. RanGTP then diffuses away from chromosomes in the cytoplasm where the two Ran GTPase activating proteins, RanGAP1 and 2, catalyse the hydrolysis of GTP. This spatial partitioning of Ran's regulators leads to the formation of a RanGTP gradient centred on chromosomes. Downstream of RanGTP, specific factors required for spindle formation (SAFs or spindle-assembly factors) are activated locally around chromosomes and thus promote spindle assembly in this region (for a review, see Caudron *et al.*, 2005; Zheng, 2004). The list of Ran-regulated SAFs is constantly growing and contains microtubule-nucleating factors, such as TPX2, or kinesin-like proteins, such as XCTK2. Their common feature is the capacity to interact with the nucleocytoplasmic transport proteins, Importins, via a nuclear localization sequence (NLS). During interphase, this interaction regulates their shuttling between the nucleus and the cytoplasm and is tightly controlled by Ran. During M phase, the Importin-mediated sequestration inactivates these factors away from chromosomes and is released in the region of the chromosomes by RanGTP.

In mouse oocyte, a RanGTP gradient centred on chromosomes is observed throughout meiotic maturation. In this model as well as in frog oocytes, alterations of the RanGTP levels differently affect meiosis I and II spindle assembly. When RanGTP production is inhibited, a significant delay in meiosis I spindle bipolarization and defects in spindle structure are observed. However, these spindles are functional, as evidenced by accurate chromosome segregation. In contrast, during meiosis II, preventing RanGTP production leads to severe spindle defects with the assembly of ectopic cytoplasmic microtubule asters or monopolar spindle formation (Dumont *et al.*, 2007). Accordingly, one of the key RanGTP targets, the microtubule assembly factor TPX2, is absent in mouse oocyte during prophase I and slowly accumulates during meiotic maturation (Brunet *et al.*, 2008). Moreover, depleting TPX2 does not affect early steps of spindle assembly, but leads to spindle collapsing during late prometaphase I. This is consistent with a model in which early steps of spindle assembly are not strictly dependant on the Ran pathway. In addition, it had been previously shown that in mouse oocytes cytoplasts (halves lacking chromosomes generated by manually cutting the oocyte) that are unable to generate a RanGTP gradient, microtubules still assemble and organize into bipolar structures (Brunet *et al.*, 1998). Altogether, these observations suggest that meiotic spindle assembly involves parallel pathways that are partially redundant, the Ran pathway and at least another one, which probably involves MTOC activity.

10.3.2 Microtubule–chromosome arm interaction

In addition to their capacity to generate diffuse signals that influence microtubule dynamics and organization, chromosomes can interact directly with spindle microtubules and stabilize them. These interactions, which involve chromosome arms or a specialized region called the centromere, contribute to spindle architecture and control chromosome motility within the spindle, which is ultimately essential for their accurate segregation at anaphase. The microtubule–centromere interface is especially central to this process and will be discussed separately (see below).

Chromosome arm–microtubule interactions are mediated by a specialized type of molecular motors called ‘chromokinesins’; that is, kinesins that can bind both to microtubules and chromosomes (for a review see Mazumdar and Misteli, 2005).

In somatic cells, the activity of the chromokinesin Kid, a member of the kinesin-10 family, is required for proper spindle morphology (Levesque *et al.*, 2003) and spindle size (Tokai-Nishizumi *et al.*, 2005). In addition, Kid is necessary to generate ‘polar ejection forces’ (Funabiki and Murray, 2000; Antonio *et al.*, 2000). These forces, emanating from the poles, push on chromosome arms and contribute to their accurate alignment on the spindle equator (Rieder and Salmon, 1994). Surprisingly, Kid does not seem to play the same roles during meiosis. In mouse, Kid is dispensable during meiotic maturation (Tokai-Nishizumi *et al.*, 2005) and its *Xenopus* homologue is instead required for proper meiotic cell-cycle progression (Perez *et al.*, 2002). Other chromokinesins may thus substitute for Kid functions during meiosis. In *Drosophila*, Nod is a kinesin-10 family member (Carpenter, 1973) that is specifically required for accurate segregation of nonexchange chromosomes; that is, chromosomes that do not undergo meiotic exchange through crossing-over and remain connected only by heterochromatic pairings (Theurkauf and Hawley, 1992). Nod is a nonmotile kinesin (Matthies, Baskinand and Hawley, 2001). *In vitro* it binds preferentially to microtubule (+)-ends and stimulates microtubule growth (Cui *et al.*, 2005). Nod could thus play the role of Kid in *Drosophila* oocytes by generating polar ejection forces on chromosome arms.

The chromokinesin Xklp1 (*Xenopus* kinesin-like protein 1) of the kinesin-4 family is a (+)-end directed motor that can inhibit both microtubule growth and shrinkage *in vitro* (Bringmann *et al.*, 2004). In *Xenopus* egg extracts, depletion of Xklp1 leads to impaired spindle assembly, with a reduction in the bipolarization efficiency and an increase in microtubule density (Castoldi and Vernos, 2006). Xklp1 may ‘freeze’ the dynamics of spindle microtubules contacting chromosome arms and thus contribute to the overall spindle shaping and integrity. However, its exact function in a cellular context remains unknown.

In conclusion, while chromosome arm–microtubule interactions have been postulated to be critical for meiotic spindle assembly in the oocyte (Brunet *et al.*, 1999), specific factors involved in this process remain to be identified.

10.4 The kinetochore–microtubule interface

Each chromosome displays a specialized region called the centromere that provides a platform for the assembly of the microtubule-interaction structure known as the kinetochore.

In addition to their role in chromosomal attachment to spindle microtubules, kinetochores play an essential role in translating this attachment into coordinated chromosome movement. During prometaphase of mitosis, kinetochores are involved in chromosome congression on the equatorial plate, where they ensure chromosome bi-orientation by generating tension on paired sister chromatids. Eventually they are required for sister chromatid segregation to opposite spindle poles at anaphase. This succession of events requires the coordinated action of multiple microtubule-associated proteins at kinetochores to generate a core attachment site, couple kinetochore movement to disassembling microtubules, affect polymerization dynamics of kinetochore-bound microtubules and drive translocation along spindle microtubules (Cheeseman and Desai, 2008).

Meiosis adds another layer of complexity in this process. Indeed, during the reductional division of meiosis I, recombined homologous chromosomes, rather than sister chromatids, are the cargo for segregation. This places a special demand on kinetochores, as sister chromatids must connect and segregate to the same spindle pole. Thus, during meiosis, specific mechanisms have evolved to allow the oriented interaction between chromosomes and microtubules, as well as the timely coordination of the reductional meiosis I followed by a ‘classical’ equational division (Figure 10.1a).

10.4.1 Organizing chromosome structure for meiosis I

Following DNA replication, the genome is compacted into chromosomes comprised of two equal halves, termed sister chromatids. Before anaphase, a multisubunit complex called cohesin holds the two sisters together. During mitosis and meiosis II, the repartition of DNA into two equal genomic complements requires sister chromatids to be segregated to opposite spindle poles. Kinetochores assemble on the outer surface of each sister chromatid, promoting chromosome bi-orientation or amphitelic attachment (connection of sisters to microtubules emanating from opposite spindle poles; Figure 10.1b). At anaphase, removal of cohesin, coupled to forces exerted by spindle microtubules on sister kinetochores, allows segregation of sisters away from each other. The same mechanism, however, does not work during meiosis I, where segregation of homologous chromosomes, corresponding to a pair of sister chromatids, occurs. This involves sister chromatid cosegregation to the same spindle pole and thus sister kinetochore mono-orientation (also called co-orientation or syntelic attachment; Figure 10.1b). Micromanipulation experiments indicated that this specialized mechanism of chromosome segregation does not depend on the cell-cycle state or on a particular microtubule organization, but is an intrinsic property of the meiosis I chromosomes. A homologous pair taken from a meiosis I cell and introduced onto a meiosis II spindle segregates as in meiosis I (Paliulis and Nicklas, 2000). This implies that meiosis I kinetochores are modified or oriented specifically in order to allow sister chromatid mono-orientation.

Fundamental differences exist between meiosis I chromosomes and their mitotic counterpart. During prophase I, maternal and paternal chromosomes associate closely and exchange genetic material via crossover formation, in a process called meiotic recombination. This ensures that sister chromatid cohesion, established during the premeiotic DNA replication phase, holds not only sisters together, but also homologous chromosomes. The resulting structure is a bivalent comprised of two homologous chromosomes held together by cohesins around the site of recombination called the chiasm. These bivalents possess four kinetochores (one per sister chromatid) that are subjected to a specific regulation in order to allow bi-orientation of homologous chromosomes. Another major difference between mitosis or meiosis II and meiosis I is the regulation of cohesion. At the onset of mitotic anaphase or anaphase II, cohesin has to be removed from chromosome arms as well as from centromeres in order to allow sister chromatid segregation. During meiosis I, only chromosome arm cohesion has to be released at anaphase. Centromeric cohesion, which provides the link between sister chromatids, has to be protected until anaphase II (Hauf and Watanabe, 2004; Marston and Amon, 2004). This function is mediated by conserved proteins called MEI-S332/Shugoshins (Sgo, Japanese for 'guardian spirit') (Kitajima, Kawashima and Watanabe, 2004; Kerrebrock *et al.*, 1992). During meiosis I, in mouse oocytes, Sgo2 recruits protein phosphatase 2A (PP2A) at the centromere (Lee *et al.*, 2008). PP2A then protects centromeric cohesion, possibly by dephosphorylating a specific cohesin subunit called Rec8 (see below), which protects it from Separase-mediated degradation. During mitosis and meiosis II, Sgo and PP2A are also recruited at the centromere. However, the tension generated by sister chromatid bi-orientation relocates the Sgo/PP2A complex from the site of cohesion at the inner centromere to the kinetochore, allowing Separase action on centromeric cohesion (Lee *et al.*, 2008). In the holocentric nematode *C. elegans*, protection of cohesion during meiosis I is independent of Sgo, and instead relies on the activity of LAB-1 (long arms of the bivalent protein)(de Carvalho *et al.*, 2008). LAB-1 recruits protein phosphatase 1 (PP1), which counteracts AuroraB-mediated phosphorylation of Rec8 (Kaitna *et al.*, 2002; Rogers *et al.*, 2002). Thus protection of centromeric cohesion during meiosis I in monocentric and holocentric organisms involves different actors. Interestingly, both systems rely on the regulated interplay between kinases and phosphatases that control centromeric cohesion.

Electron microscopic analysis showed that in striking contrast to mitosis or meiosis II, sister kinetochores orient side by side during meiosis I (Goldstein, 1981; Lee *et al.*, 2000; Parra *et al.*, 2004). However, the molecular mechanisms for this specific behaviour of kinetochores are just beginning to be understood. In the budding yeast *S. cerevisiae*, mono-orientation of sister chromatids depends on the assembly of the monopolin complex comprised of the meiosis I-specific protein Mam1 (monopolar microtubule attachment during meiosis 1), the nucleolar proteins Lrs4 and Csm1, and the casein kinase Hrr25 (Toth *et al.*, 2000; Rabitsch *et al.*, 2003). In the absence of this complex, sister chromatids bi-orient during meiosis I, and homologous chromosome segregation completely fails leading mostly to unviable spores (Toth *et al.*, 2000; Rabitsch *et al.*, 2003; Petronczki *et al.*, 2006). Mam1 is a meiosis-specific protein localized at the kinetochore from pachytene to metaphase I, whereas Lrs4 and Csm1 are expressed during mitosis and meiosis. During meiosis, the Polo kinase Cdc5 releases them from the nucleolus in G2. After their release, they form a complex with Mam1 and Hrr25. This complex is then phosphorylated by the kinase Cdc7, which leads to its recruitment at kinetochores (Matos *et al.*, 2008). How the monopolin complex allows the side-by-side positioning of sister kinetochores is still unclear. However, the monopolin complex seems to physically join sister kinetochores in a cohesin-independent manner. This contrasts with the situation in the fission yeast *Schizosaccharomyces pombe* where cohesion is central to the process of homologous chromosome bi-orientation (Watanabe and Nurse, 1999). During meiosis, the cohesin complex consists of two SMC (structural maintenance of chromosome) family proteins, Smc1 and Smc3, an accessory subunit Scc3 and the meiosis-specific kleisin subunit Rec8 that replaces mitotic Scc1/Rad21 (Ahringer, 2003; Pasierbek *et al.*, 2003; Keefe *et al.*, 2003). In *S. pombe*, replacing Rec8 by its mitotic counterpart Scc1/Rad21 can sustain sister chromatid cohesion, however amphitelic attachment is established and sister chromatids segregate to opposite spindle poles at anaphase I (Yokobayashi, Yamamoto and Watanabe, 2003). This difference in the ability of Rec8 and Scc1/Rad21 to sustain sister chromatid mono-orientation, resides in their respective localizations. Rec8 localizes to chromosome arms and pericentromeric DNA as well as the central region of the centromere where the kinetochore assembles. By contrast, when ectopically expressed during meiosis, Scc1/Rad21 localizes to the chromosome arms and the pericentromeric DNA, but not to the central region of the centromere (Yokobayashi and Watanabe, 2005). This particular localization of Rec8 was recently shown to bring together sister kinetochores and force their side-by-side orientation (Sakuno, Tada and Watanabe, 2009). However, Rec8 is not sufficient to establish mono-orientation of sister chromatids. Ectopic expression of Rec8 during mitosis does not induce mono-orientation, even with Rec8 localized at the inner centromere. This suggests that additional factors must exist to promote mono-orientation in fission yeast. Genetic screening for mutations that confer equational segregation at meiosis I led to the identification of Moa1 (monopolar attachment 1), a meiosis I-specific protein localized to the inner centromere (Yokobayashi and Watanabe, 2005). Moa1 interacts with Rec8 and would facilitate establishment of inner centromere cohesion by promoting Rec8 proper localization. Thus in fission yeast Rec8, together with Moa1, and presumably additional factors that remain to be identified, generate a chromosomal architecture that forces side-by-side orientation of sister kinetochores and thus homologous chromosome bi-orientation.

Factors specifically involved in mono-orientation of sister chromatids have not been identified yet in higher eukaryotes. Neither the monopolin complex in budding yeast nor Moa1 in fission yeast have orthologues in metazoans. However, centromeric cohesion could be a common theme in mono-orientation, as absence of Rec8 in *Zea mays* and *Arabidopsis thaliana* leads to sister chromatid bi-orientation during meiosis I (Chelysheva *et al.*, 2005; Hamant *et al.*, 2005).

10.4.2 Specifying the site of chromosome–microtubule interaction

While centromere function is extraordinarily conserved among eukaryotes, their size and sequence composition vary considerably between different species. In the budding yeast *Saccharomyces cerevisiae*, the centromere is defined by a single 125 base pair (bp) sequence, while human centromeric DNA can extend to more than 4 megabases, and holocentric organisms display extended centromeres along the entire length of their chromosomes. The presence of sequence-dependent centromeres in budding yeast suggested an attractive model in which analogous sequence elements in other organisms would also define the assembly of kinetochores at distinct chromosomal sites. However, most organisms lack a specific centromeric DNA sequence that defines the site of kinetochore assembly (Sullivan, Blower and Karpen, 2001; Schueler and Sullivan, 2006). *Schizosaccharomyces pombe* centromeres display a central nonrepetitive region surrounded on either side by repeated elements termed innermost and outer repeats that are not conserved even between chromosomes. Human centromeres are enriched in repeated arrays of a 171 bp α-satellite DNA sequence (Ekwall, 2007). Interestingly, while the nature of centromeric DNA varies considerably between different species, most centromeric proteins are conserved (Talbert, Bryson and Henikoff, 2004). An explanation for this paradox is that the DNA sequence is not what specifies the centromere. Instead, the site of kinetochore assembly is thought to be primarily epigenetically determined, rather than sequence based (Karpen and Allshire, 1997). A primary candidate for this epigenetic mark is the CENP-A (for 'centromere protein A')-containing chromatin.

Kinetochores are built upon centromeric chromatin that universally features specialized nucleosomes containing the histone H3 variant, CENP-A. During mitosis, CENP-A-containing nucleosomes provide the scaffold for kinetochore assembly and are required for the proper targeting of all tested kinetochore components (Oegema *et al.*, 2001; Stoler *et al.*, 1995; Howman *et al.*, 2000; Blower and Karpen, 2001; Van Hooser *et al.*, 2001; Kallio, Eriksson and Gorbsky, 2000; Maddox *et al.*, 2004).

CENP-A, the closely associated centromeric protein, CENP-C, and outer kinetochore components form a linear assembly pathway (Desai *et al.*, 2003). CENP-A-containing chromatin directs the recruitment of CENP-C, which in turn interacts with and recruits a multisubunit complex corresponding to the microtubule-binding interface (Cheeseman *et al.*, 2004). In mouse, pig and *Drosophila melanogaster* oocytes, the localization of CENP-A is coincident with outer kinetochore components (Lee *et al.*, 2000; Kallio, Eriksson and Gorbsky, 2000; Brunet *et al.*, 2003; Wassmann, Niault and Maro, 2003; Gilliland *et al.*, 2007). This suggests that in monocentric organisms the site of meiotic kinetochore assembly is dictated by CENP-A localization. By striking contrast, during meiosis in the holocentric nematode *C. elegans*, CENP-A localization is coincident with chromosomal DNA, while outer kinetochore components are concentrated at the surface of the chromosomes where they form two cup-like structures that enclose the two halves of each bivalent (Moore, Morrison and Roth, 1999; Howe *et al.*, 2001; Monen *et al.*, 2005). CENP-C is absent from chromosomes in fertilized CENP-A-depleted oocytes, but, remarkably, outer kinetochore components localize normally to the cup-like structures on the surface of chromosomes during both meiotic divisions. Thus the linear assembly pathway formed by CENP-A, CENP-C and the outer kinetochore components during mitosis does not seem to be conserved during meiosis in *C. elegans*. Outer kinetochore components are targeted independently of CENP-A and CENP-C to the cup-like structures, which are likely to represent the holocentric meiotic kinetochores. Consistent with a mechanism of meiotic kinetochore specification independent of CENP-A and CENP-C, CENP-A depletion results in a severe chromosome missegregation phenotype (also called 'kinetochore null' phenotype) during the first mitosis but does not seem to prevent accurate meiotic chromosome segregation, nor polar bodies extrusion (Monen *et al.*, 2005). This targeting of outer kinetochore components uncoupled from CENP-A/C-containing chromatin during meiosis has not been described in other organisms.

However, this uncoupling mechanism is likely to be specific to holocentric species and probably helps prevent merotelic attachment, where a chromosome is connected simultaneously to both spindle poles (Figure 10.1b). Indeed, during the reductional division of meiosis I, recombined homologous chromosomes, rather than sister chromatids, are the cargo for segregation. This places a special demand on kinetochores, since sister chromatids must connect and segregate to the same spindle pole. In holocentric organisms kinetochore assembly along the length of each chromatid poses a topological problem for segregating recombined chromatids (Monen *et al.*, 2005). DNA flanking the recombination event that was originally from a single chromatid must now move in opposite directions. Thus it is likely that holocentric organisms have developed special mechanisms to prevent recombined chromatids from being simultaneously pulled toward both spindle poles. Initial observations had led to the assumption that meiosis is inverted in holocentric organisms with the equational division first and the reductional division occurring only at meiosis II (Nordenskiold, 1961; Nordenskiold, 1962). This would be a direct consequence of the equatorial orientation of the bivalents with each sister kinetochore being parallel to the equatorial plate and facing opposite spindle poles. However, recent examinations of this hypothesis have shown that inverted meiosis could be confined only to the behaviour of achiasmatic sex chromosomes of some holocentric insect species (Nokkala, Laukkanen and Nokkala, 2002). It is now clear that by late diakinesis, in every holocentric organism studied to date including *C. elegans*, paired homologous chromosomes have a characteristic end-to-end configuration where the long axis of each sister chromatid, and thus of the bivalent, is perpendicular to the equatorial plate. This orientation of the bivalents, with each homologous chromosome facing only one spindle pole, together with the particular localization of kinetochore components and the CENP-A uncoupling mechanism, are likely to favour bi-orientation by preventing microtubules from a spindle pole accessing both chromosomes of a pair. A future challenge will be to elucidate the nature of the CENP-A/C-independent mechanism of meiotic kinetochore specification in the holocentric nematode *C. elegans*.

10.4.3 Interaction between microtubules and the kinetochore

Kinetochore ultrastructure

The centromere is a specialized region of chromosomal DNA that was first described as the primary constriction of a condensed chromosome visible by light microscopy (Flemming, 1882). However, holocentric organisms, like the nematode *C. elegans*, have chromosomes devoid of this primary constriction; instead their centromeres extend along the entire length of the chromosomes (Maddox *et al.*, 2004). Thus, cytological and molecular analyses of this region over the past 15 years have led to a new definition of the centromere as the domain that directs the formation of the kinetochore.

Transmission electron microscopy coupled to chemical fixation techniques provided the first insight into the organization of mitotic and meiotic kinetochores. When longitudinally sectioned, kinetochores display a trilamelar structure closely apposed to a region of chromatin that is more electron dense than the rest of the chromosomes. Both the inner and outer layers of the trilamelar structure are ~20 nm thick and are separated by a ~30 nm thick low-electron-density region (Brunet *et al.*, 1999; Lee *et al.*, 2000; Ris and Witt, 1981; Comings and Okada, 1971; Rieder, 1982). In monocentric mammalian chromosomes, a fibrous corona, which radiates 100 nm or more from the poleward face of the trilamelar structure, is seen in the absence of microtubule attachment. Ultrastructural studies of meiotic holocentric chromosomes using chemical fixations have failed to identify a characteristic trilamelar structure (Buck, 1967; Comings and Okada, 1972; Goldstein, 1977; Goday, Ciofi-Luzzatto and Pimpinelli, 1985; Pimpinelli and Goday, 1989). Instead, microtubules projecting into the mass of the chromosome were shown, and the spindle was thought to attach directly to the chromatin (Albertson and Thomson, 1993). However, more recently, the use of high-pressure freezing followed by freezing substitution (HPF/FS), which preserves structures better than chemical fixation, reconciled monocentric and holocentric meiotic kinetochore structure (Howe *et al.*, 2001). With this technique in monocentric organisms, the kinetochore appears as a ~70 nm thick mat of light-staining fibrous material that is directly connected with the more electron-opaque surface of the centromeric chromatin. This line separates the chromatin from a clear zone of ~150 nm which excludes ribosomes and other cytoplasmic components. The fibrous mat corresponds to the outer plate defined by conventional electronic microscopy, while the clear zone correlates with the fibrous corona (McEwen *et al.*, 1998). Use of HPF/FS to study meiotic holocentric chromosomes in *C. elegans* spermatocytes has revealed the presence of a zone of ribosome exclusion surrounding the whole surface of homologous chromosomes, suggesting that meiotic holocentric chromosomes have a kinetochore structure similar to that of monocentric chromosomes (Howe *et al.*, 2001).

Kinetochore function

The kinetochore is a proteinaceous organelle essential for chromosomal attachment to spindle microtubules. As a consequence, the depletion of any core component that leads to the disruption of kinetochore function during mitosis gives rise to severe chromosome segregation defects in all systems that have been examined so far. The situation is not as clear during meiosis. Indeed, if the mitotic role of every kinetochore component has been extensively investigated, only very few have been studied for their role during meiosis.

In mouse oocyte, homologous chromosome congression does not involve the kinetochore (Brunet *et al.*, 1999). In contrast to mitosis, kinetochores are not competent for anchoring and/or stabilizing microtubules during the particularly long prometaphase, which can last up to 10 hours. Homologous chromosomes are nevertheless transported towards the equator of the spindle and oscillate in this region for several hours. This kinetochore-independent mechanism of chromosome congression probably involves the activity of chromokinesins; however the exact nature of the molecular mechanism involved remains unknown. After this extended prometaphase, the activation of kinetochores triggers their interaction with spindle microtubules, ending chromosome oscillation and leading to the formation of a tight metaphase plate. The nature of the late kinetochore-activating signal remains elusive. Meiotic kinetochores may be submitted to a very slow and unusual maturation involving the late recruitment of some crucial components. Interestingly, some components including members of the spindle-assembly checkpoint or the kinesin-7 family member, (+)-end-directed motor CENP-E, are already localized at the kinetochore in early prometaphase (Kallio, Eriksson and Gorbsky, 2000; Brunet *et al.*, 2003; Wassmann *et al.*, 2003; Gilliland *et al.*, 2007). However, the presence and/or the dynamics of the core microtubule attachment site, such as the Ndc80/HEC1 complex, or of proteins controlling microtubule dynamics at the kinetochore, such as CLASP, have never been investigated in this system (Hannak and Heald, 2006); for an extensive review of these proteins during mitosis see Cheeseman and Desai (2008). Alternatively, kinetochore activation could be controlled by post-translational modification of these components rather than by their recruitment. This late kinetochore activation has never been described in other organisms. However, it could be a general feature of female meiosis I that has not been observed in other model systems with faster meiosis I.

Once every chromosome is correctly attached to the spindle, the tension generated at the kinetochore satisfies the spindle-assembly checkpoint, allowing anaphase to occur. During mitotic anaphase, sister chromatids are pulled to opposite spindle poles via microtubules attached to their kinetochores. The exact function of kinetochore–

microtubule interaction during meiotic anaphase is not known. In *C. elegans*, only the role of HIM-10 (high incidence of male progeny-10 also called Nuf2 and part of the Ndc80/HEC1 complex) has been investigated during meiosis (Howe *et al.*, 2001). HIM-10 was first identified in a genetic screen for strains with a high incidence of male progeny (Hodgkin, Horvitz and Brenner, 1979). In *C. elegans*, hermaphrodites are genetically XX, while males are X0 and arise from rare chromosome nondisjunction events during meiosis. HIM-10 localizes to the cup-like structures during both meiotic divisions in spermatocytes and oocytes. Reduction of HIM-10 activity enhances the incidence of male progeny by promoting nondisjunction events. This suggests that HIM-10 does function in meiotic chromosome segregation. However, all the kinetochore components identified so far have been successfully depleted by RNAi (RNA interference), and their role has been studied during mitosis in *C. elegans* (Oegema *et al.*, 2001; Desai *et al.*, 2003; Cheeseman *et al.*, 2004; Moore, Morrison and Roth, 1999; Cheeseman *et al.*, 2005). In these one-cell mitotic embryos depleted of a kinetochore component, the normal appearance of the maternal pronucleus suggests that the preceding meiotic divisions took place without severe chromosome segregation defects. Compared to the severe 'kinetochore null' mitotic phenotype, this brings into question the real function of kinetochores during female meiosis and thus the potential existence of parallel mechanisms that would act together to achieve accurate meiotic chromosome segregation.

10.5 Concluding remarks

Understanding the mechanisms required for the formation of a functional oocyte competent for fertilization is a major goal for developmental and cell biologists. The assembly of meiotic spindles and the establishment of proper meiotic spindle–chromosome interactions is definitely one of the most critical of these processes.

After one century of investigations on spindles, mainly in mitotic systems, a plethora of molecules has been identified and we are only starting to unveil how they act together and are regulated in time and space in the dividing cell. Transposing molecules and principles at play during mitosis is essential but clearly not sufficient to fully understand the complex process of meiotic spindle assembly. Nevertheless, in the last few years, investigations on meiotic systems have expanded. Technical improvements, like the use of time-lapse microscopy in live oocytes, coupled to systematic RNAi screens, as well as the introduction of new mathematics and physics approaches (McGuinness *et al.*, 2009) are now contributing to the dissection of the meiotic mechanisms with temporal and spatial accuracy. In addition, large-scale analyses are now routinely performed on oocytes. Proteomic approaches have been developed using emerging models such as *Ciona* (Nomura, Nakajima and Inaba, 2009); transcriptomic analyses have been achieved using *Xenopus*, mouse (Evsikov *et al.*, 2006) and human oocytes (Jones *et al.*, 2008). Specific questions can now be raised and answered using such approaches. One crucial issue is to understand how homologous chromosome segregation is altered with increased maternal age in mammals. Expression profiling of transcripts in oocytes collected on 'young' versus 'old' mice recently revealed that some spindle assembly factors, like the kinesin Kif2 or the spindle pole component

Numa, as well as molecules regulating the kinetochore-microtubules interface, like the kinesin CENP-E, are indeed downregulated with ageing (Pan *et al.*, 2008).

We are confident that such a multiplicity of complementary approaches and tools will most likely allow in the very close future a better understanding of the complexity of meiotic spindle assembly and function, and will significantly contribute to the characterization of the cellular and molecular basis of meiotic chromosome missegregations and embryonic aneuploidies.

Acknowledgements

We apologize to many people whose work we were unable to cite owing to space constraints. Julien Dumont is the recipient of an EMBO long-term postdoctoral fellowship. We are grateful to J. C. Canman for critical reading of part of this manuscript.

References

Ahringer, J. (2003) Control of cell polarity and mitotic spindle positioning in animal cells. *Curr. Opin. Cell Biol.*, **15**(1), 73–81.

Albertson, D.G. and Thomson, J.N. (1993) Segregation of holocentric chromosomes at meiosis in the nematode, Caenorhabditis elegans. *Chromosome Res.*, **1**(1), 15–26.

Antonio, C., Ferby, I., Wilhelm, H. *et al.* (2000) Xkid, a chromokinesin required for chromosome alignment on the metaphase plate. *Cell*, **102**(4), 425–435.

Azimzadeh, J. and Bornens, M. (2007) Structure and duplication of the centrosome. *J. Cell Sci.*, **120** (Pt 13), 2139–2142.

Blower, M.D. and Karpen, G.H. (2001) The role of Drosophila CID in kinetochore formation, cell-cycle progression and heterochromatin interactions. *Nat. Cell Biol.*, **3**(8), 730–739.

Bringmann, H., Skiniotis, G., Spilker, A. *et al.* (2004) A kinesin-like motor inhibits microtubule dynamic instability. *Science*, **303**(5663), 1519–1522.

Brunet, S., Polanski, Z., Verlhac, M.H. *et al.* (1998) Bipolar meiotic spindle formation without chromatin. *Curr. Biol.*, **8**(22), 1231–1234.

Brunet, S., Maria, A.S., Guillaud, P. *et al.* (1999) Kinetochore fibers are not involved in the formation of the first meiotic spindle in mouse oocytes, but control the exit from the first meiotic M phase. *J. Cell Biol.*, **146**(1), 1–12.

Brunet, S., Pahlavan, G., Taylor, S. and Maro, B. (2003) Functionality of the spindle checkpoint during the first meiotic division of mammalian oocytes. *Reproduction*, **126**(4), 443–450.

Brunet, S., Dumont, J., Lee, K.W. *et al.* (2008) Meiotic regulation of TPX2 protein levels governs cell cycle progression in mouse oocytes. *PLoS ONE*, **3**(10), e3338

Buck, R.C. (1967) Mitosis and meiosis in Rhodnius prolixus: the fine structure of the spindle and diffuse kinetochore. *J. Ultrastruct. Res.*, **18**(5), 489–501.

Buendia, B., Draetta, G. and Karsenti, E. (1992) Regulation of the microtubule nucleating activity of centrosomes in Xenopus egg extracts: role of cyclin A-associated protein kinase. *J. Cell Biol.*, **116** (6), 1431–1442.

Calarco, P.G. (2000) Centrosome precursors in the acentriolar mouse oocyte. *Microsc. Res. Tech.*, **49** (5), 428–434.

Carabatsos, M.J., Combelles, C.M., Messinger, S.M. and Albertini, D.F. (2000) Sorting and reorganization of centrosomes during oocyte maturation in the mouse. *Microsc. Res. Tech.*, **49**(5), 435–444.

Carpenter, A.T. (1973) A meiotic mutant defective in distributive disjunction in Drosophila melanogaster. *Genetics*, **73**(3), 393–428.

de Carvalho, C.E., Zaaijer, S., Smolikov, S. *et al.* (2008) LAB-1 antagonizes the Aurora B kinase in C. elegans. *Genes Dev.*, **22**(20), 2869–2885.

Castoldi, M. and Vernos, I. (2006) Chromokinesin Xklp1 contributes to the regulation of microtubule density and organization during spindle assembly. *Mol. Biol. Cell*, **17**(3), 1451–1460.

Caudron, M., Bunt, G., Bastiaens, P. and Karsenti, E. (2005) Spatial coordination of spindle assembly by chromosome-mediated signaling gradients. *Science*, **309**(5739), 1373–1376.

Cheeseman, I.M. and Desai, A. (2008) Molecular architecture of the kinetochore-microtubule interface. *Nat. Rev. Mol. Cell Biol.*, **9**(1), 33–46.

Cheeseman, I.M., Niessen, S., Anderson, S. *et al.* (2004) A conserved protein network controls assembly of the outer kinetochore and its ability to sustain tension. *Genes Dev.*, **18**(18), 2255–2268.

Cheeseman, I.M., MacLeod, I., Yates, J.R. 3rd *et al.* (2005) The CENP-F-like proteins HCP-1 and HCP-2 target CLASP to kinetochores to mediate chromosome segregation. *Curr. Biol.*, **15**(8), 771–777.

Chelysheva, L., Diallo, S., Vezon, D. *et al.* (2005) AtREC8 and AtSCC3 are essential to the monopolar orientation of the kinetochores during meiosis. *J. Cell Sci.*, **118** (Pt 20), 4621–4632.

Colombie, N., Cullen, C.F., Brittle, A.L. *et al.* (2008) Dual roles of Incenp crucial to the assembly of the acentrosomal metaphase spindle in female meiosis. *Development*, **135**(19), 3239–3246.

Comings, D.E. and Okada, T.A. (1971) Fine structure of kinetochore in Indian muntjac. *Exp. Cell Res.*, **67**(1), 97–110.

Comings, D.E. and Okada, T.A. (1972) Holocentric chromosomes in Oncopeltus: kinetochore plates are present in mitosis but absent in meiosis. *Chromosoma*, **37**(2), 177–192.

Cui, W., Sproul, L.R., Gustafson, S.M. *et al.* (2005) Drosophila Nod protein binds preferentially to the plus ends of microtubules and promotes microtubule polymerization in vitro. *Mol. Biol. Cell*, **16** (11), 5400–5409.

Cytrynbaum, E.N., Scholey, J.M. and Mogilner, A. (2003) A force balance model of early spindle pole separation in Drosophila embryos. *Biophys. J.*, **84**(2 Pt 1), 757–769.

Desai, A., Rybina, S., Muller-Reichert, T. *et al.* (2003) KNL-1 directs assembly of the microtubule-binding interface of the kinetochore in C. elegans. *Genes Dev.*, **17**(19), 2421–2435.

Dumont, J., Petri, S., Pellegrin, F. *et al.* (2007) A centriole- and RanGTP-independent spindle assembly pathway in meiosis I of vertebrate oocytes. *J. Cell Biol.*, **176**(3), 295–305.

Ekwall, K. (2007) Epigenetic control of centromere behavior. *Annu. Rev. Genet.*, **41**, 63–81.

Evsikov, A.V., Graber, J.H., Brockman, J.M. *et al.* (2006) Cracking the egg: molecular dynamics and evolutionary aspects of the transition from the fully grown oocyte to embryo. *Genes Dev.*, **20**(19), 2713–2727.

Flemming, W. (1882) Beiträge zur Kenntnis der Zelle und ihrer Lebenserscheinungen. *Arch. Mikr. Anat.*, **20**, 1–86.

Funabiki, H. and Murray, A.W. (2000) The Xenopus chromokinesin Xkid is essential for metaphase chromosome alignment and must be degraded to allow anaphase chromosome movement. *Cell*, **102** (4), 411–424.

Gard, D.L. (1992) Microtubule organization during maturation of Xenopus oocytes: assembly and rotation of the meiotic spindles. *Dev. Biol.*, **151**(2), 516–530.

Gilliland, W.D., Hughes, S.E., Cotitta, J.L. *et al.* (2007) The multiple roles of Mps1 in drosophila female meiosis. *PLoS Genet.*, **3**(7), e113.

Goday, C., Ciofi-Luzzatto, A. and Pimpinelli, S. (1985) Centromere ultrastructure in germ-line chromosomes of Parascaris. *Chromosoma*, **91**(2), 121–125.

Goldstein, P. (1977) Spermatogenesis and spermiogenesis in Ascaris lumbricoides Var. suum. *J. Morphol.*, **154**(3), 317–337.

Goldstein, L.S. (1981) Kinetochore structure and its role in chromosome orientation during the first meiotic division in male D. melanogaster. *Cell*, **25**(3), 591–602.

Gueth-Hallonet, C., Antony, C., Aghion, J. *et al.* (1993) gamma-tubulin is present in acentriolar MTOCs during early mouse development. *J. Cell Sci.*, **105** (Pt 1), 157–166.

Hamant, O., Golubovskaya, I., Meeley, R. *et al.* (2005) A REC8-dependent plant Shugoshin is required for maintenance of centromeric cohesion during meiosis and has no mitotic functions. *Curr. Biol.*, **15**(10), 948–954.

Hannak, E. and Heald, R. (2006) Xorbit/CLASP links dynamic microtubules to chromosomes in the Xenopus meiotic spindle. *J. Cell Biol.*, **172**(1), 19–25.

Hassold, T., Hall, H. and Hunt, P. (2007) The origin of human aneuploidy: where we have been, where we are going. *Hum. Mol. Genet.*, **16** (Spec No 2), R203–R208.

Hauf, S. and Watanabe, Y. (2004) Kinetochore orientation in mitosis and meiosis. *Cell*, **119**(3), 317–327.

Hertig, A.T. and Adams, E.C. (1967) Studies on the human oocyte and its follicle. I. Ultrastructural and histochemical observations on the primordial follicle stage. *J. Cell Biol.*, **34**(2), 647–675.

Hodgkin, J., Horvitz, H.R. and Brenner, S. (1979) Nondisjunction mutants of the nematode Caenorhabditis elegans. *Genetics*, **91**(1), 67–94.

Howe, M., McDonald, K.L., Albertson, D.G. and Meyer, B.J. (2001) HIM-10 is required for kinetochore structure and function on Caenorhabditis elegans holocentric chromosomes. *J. Cell Biol.*, **153**(6), 1227–1238.

Howman, E.V., Fowler, K.J., Newson, A.J. *et al.* (2000) Early disruption of centromeric chromatin organization in centromere protein A (Cenpa) null mice. *Proc. Natl. Acad. Sci. USA*, **97**(3), 1148–1153.

Huchon, D., Crozet, N., Cantenot, N. and Ozon, R. (1981) Germinal vesicle breakdown in the Xenopus laevis oocyte: description of a transient microtubular structure. *Reprod. Nutr. Dev.*, **21**(1), 135–148.

Jang, J.K., Rahman, T. and McKim, K.S. (2005) The kinesinlike protein Subito contributes to central spindle assembly and organization of the meiotic spindle in Drosophila oocytes. *Mol. Biol. Cell*, **16** (10), 4684–4694.

Janson, M.E., Loughlin, R., Loiodice, I. *et al.* (2007) Crosslinkers and motors organize dynamic microtubules to form stable bipolar arrays in fission yeast. *Cell*, **128**(2), 357–368.

Januschke, J., Gervais, L., Gillet, L. *et al.* (2006) The centrosome-nucleus complex and microtubule organization in the Drosophila oocyte. *Development*, **133**(1), 129–139.

Jones, G.M., Cram, D.S., Song, B. *et al.* (2008) Gene expression profiling of human oocytes following in vivo or in vitro maturation. *Hum. Reprod.*, **23**(5), 1138–1144.

Kaitna, S., Pasierbek, P., Jantsch, M. *et al.* (2002) The aurora B kinase AIR-2 regulates kinetochores during mitosis and is required for separation of homologous chromosomes during meiosis. *Curr. Biol.*, **12**(10), 798–812.

Kallio, M., Eriksson, J.E. and Gorbsky, G.J. (2000) Differences in spindle association of the mitotic checkpoint protein Mad2 in mammalian spermatogenesis and oogenesis. *Dev. Biol.*, **225**(1), 112–123.

Karpen, G.H. and Allshire, R.C. (1997) The case for epigenetic effects on centromere identity and function. *Trends. Genet.*, **13**(12), 489–496.

Karsenti, E. and Vernos, I. (2001) The mitotic spindle: a self-made machine. *Science*, **294**(5542), 543–547.

Keefe, D., Liu, L., Wang, W. and Silva, C. (2003) Imaging meiotic spindles by polarization light microscopy: principles and applications to IVF. *Reprod. Biomed. Online*, **7**(1), 24–29.

Kerrebrock, A.W., Miyazaki, W.Y., Birnby, D. and Orr-Weaver, T.L. (1992) The Drosophila mei-S332 gene promotes sister-chromatid cohesion in meiosis following kinetochore differentiation. *Genetics*, **130**(4), 827–841.

Kirkham, W.B. and Burr, H.S. (1913) The breeding habits, maturation of eggs and ovulation of the albino rat. *Am. J. Anat.*, **15**, 291–318.

Kitajima, T.S., Kawashima, S.A. and Watanabe, Y. (2004) The conserved kinetochore protein shugoshin protects centromeric cohesion during meiosis. *Nature*, **427**(6974), 510–517.

Lee, J., Miyano, T., Dai, Y. *et al.* (2000) Specific regulation of CENP-E and kinetochores during meiosis I/meiosis II transition in pig oocytes. *Mol. Reprod. Dev.*, **56**(1), 51–62.

Lee, J., Kitajima, T.S., Tanno, Y. *et al.* (2008) Unified mode of centromeric protection by shugoshin in mammalian oocytes and somatic cells. *Nat. Cell Biol.*, **10**(1), 42–52.

Levesque, A.A., Howard, L., Gordon, M.B. and Compton, D.A. (2003) A functional relationship between NuMA and kid is involved in both spindle organization and chromosome alignment in vertebrate cells. *Mol. Biol. Cell*, **14**(9), 3541–3552.

Maddox, P.S., Oegema, K., Desai, A. and Cheeseman, I.M. (2004) 'Holo'er than thou: chromosome segregation and kinetochore function in C. elegans. *Chromosome Res.*, **12**(6), 641–653.

Mailhes, J.B., Mastromatteo, C. and Fuseler, J.W. (2004) Transient exposure to the Eg5 kinesin inhibitor monastrol leads to syntelic orientation of chromosomes and aneuploidy in mouse oocytes. *Mutat. Res.*, **559**(1–2), 153–167.

Manandhar, G., Simerly, C. and Schatten, G. (2000) Highly degenerated distal centrioles in rhesus and human spermatozoa. *Hum. Reprod.*, **15**(2), 256–263.

Manandhar, G., Schatten, H. and Sutovsky, P. (2005) Centrosome reduction during gametogenesis and its significance. *Biol. Reprod.*, **72**(1), 2–13.

Maro, B., Howlett, S.K. and Webb, M. (1985) Non-spindle microtubule organizing centers in metaphase II-arrested mouse oocytes. *J. Cell Biol.*, **101**(5 Pt 1), 1665–1672.

Marston, A.L. and Amon, A. (2004) Meiosis: cell-cycle controls shuffle and deal. *Nat. Rev. Mol. Cell Biol.*, **5**(12), 983–997.

Matos, J., Lipp, J.J., Bogdanova, A. *et al.* (2008) Dbf4-dependent CDC7 kinase links DNA replication to the segregation of homologous chromosomes in meiosis I. *Cell*, **135**(4), 662–678.

Matthies, H.J., McDonald, H.B., Goldstein, L.S. and Theurkauf, W.E. (1996) Anastral meiotic spindle morphogenesis: role of the non-claret disjunctional kinesin-like protein. *J. Cell Biol.*, **134**(2), 455–464.

Matthies, H.J., Baskin, R.J. and Hawley, R.S. (2001) Orphan kinesin NOD lacks motile properties but does possess a microtubule-stimulated ATPase activity. *Mol. Biol. Cell*, **12**(12), 4000–4012.

Mazumdar, M. and Misteli, T. (2005) Chromokinesins: multitalented players in mitosis. *Trends. Cell Biol.*, **15**(7), 349–355.

McEwen, B.F., Hsieh, C.E., Mattheyses, A.L. and Rieder, C.L. (1998) A new look at kinetochore structure in vertebrate somatic cells using high-pressure freezing and freeze substitution. *Chromosoma*, **107**(6–7), 366–375.

McGuinness, B.E., Anger, M., Kouznetsova, A. *et al.* (2009) Regulation of APC/C activity in oocytes by a Bub1-dependent spindle assembly checkpoint. *Curr. Biol.*, **19**(5), 369–380.

McNally, K., Audhya, A., Oegema, K. and McNally, F.J. (2006) Katanin controls mitotic and meiotic spindle length. *J. Cell Biol.*, **175**(6), 881–891.

Monen, J., Maddox, P.S., Hyndman, F. *et al.* (2005) Differential role of CENP-A in the segregation of holocentric C. elegans chromosomes during meiosis and mitosis. *Nat. Cell Biol.*, **7**(12), 1248–1255.

Moore, L.L., Morrison, M. and Roth, M.B. (1999) HCP-1, a protein involved in chromosome segregation, is localized to the centromere of mitotic chromosomes in Caenorhabditis elegans. *J. Cell Biol.*, **147**(3), 471–480.

Moy, G.W., Brandriff, B. and Vacquier, V.D. (1977) Cytasters from sea urchin eggs parthenogenetically activated by procaine. *J. Cell Biol.*, **73**(3), 788–793.

Nokkala, S., Laukkanen, A. and Nokkala, C. (2002) Mitotic and meiotic chromosomes in Somatochlora metallica (Cordulidae, Odonata). The absence of localized centromeres and inverted meiosis. *Hereditas*, **136**(1), 7–12.

Nomura, M., Nakajima, A. and Inaba, K. (2009) Proteomic profiles of embryonic development in the ascidian Ciona intestinalis. *Dev. Biol.*, **325**(2), 468–481.

Nordenskiold, H. (1961) Tetrad analysis and the course of meiosis in three hybrids of *Luzula campestris*. *Hereditas*, **47**, 203–238.

Nordenskiold, H. (1962) Studies of meiosis in *Luzula purpurea*. *Hereditas*, **48**, 503–519.

Oegema, K., Desai, A., Rybina, S. *et al.* (2001) Functional analysis of kinetochore assembly in Caenorhabditis elegans. *J. Cell Biol.*, **153**(6), 1209–1226.

Paliulis, L.V. and Nicklas, R.B. (2000) The reduction of chromosome number in meiosis is determined by properties built into the chromosomes. *J. Cell Biol.*, **150**(6), 1223–1232.

Pan, H., Ma, P., Zhu, W. and Schultz, R.M. (2008) Age-associated increase in aneuploidy and changes in gene expression in mouse eggs. *Dev. Biol.*, **316**(2), 397–407.

Parra, M.T., Viera, A., Gomez, R. *et al.* (2004) Involvement of the cohesin Rad21 and SCP3 in monopolar attachment of sister kinetochores during mouse meiosis I. *J. Cell Sci.*, **117** (Pt 7), 1221–1234.

Pasierbek, P., Fodermayr, M., Jantsch, V. *et al.* (2003) The Caenorhabditis elegans SCC-3 homologue is required for meiotic synapsis and for proper chromosome disjunction in mitosis and meiosis. *Exp. Cell Res.*, **289**(2), 245–255.

Perez, L.H., Antonio, C., Flament, S. *et al.* (2002) Xkid chromokinesin is required for the meiosis I to meiosis II transition in Xenopus laevis oocytes. *Nat. Cell Biol.*, **4**(10), 737–742.

Petronczki, M., Matos, J., Mori, S. *et al.* (2006) Monopolar attachment of sister kinetochores at meiosis I requires casein kinase 1. *Cell*, **126**(6), 1049–1064.

Pimpinelli, S. and Goday, C. (1989) Unusual kinetochores and chromatin diminution in Parascaris. *Trends. Genet.*, **5**(9), 310–315.

Rabitsch, K.P., Petronczki, M., Javerzat, J.P. *et al.* (2003) Kinetochore recruitment of two nucleolar proteins is required for homolog segregation in meiosis I. *Dev. Cell*, **4**(4), 535–548.

Rieder, C.L. (1982) The formation, structure, and composition of the mammalian kinetochore and kinetochore fiber. *Int. Rev. Cytol.*, **79**, 1–58.

Rieder, C.L. and Salmon, E.D. (1994) Motile kinetochores and polar ejection forces dictate chromosome position on the vertebrate mitotic spindle. *J. Cell Biol.*, **124**(3), 223–233.

Ris, H. and Witt, P.L. (1981) Structure of the mammalian kinetochore. *Chromosoma*, **82**(2), 153–170.

Rogers, E., Bishop, J.D., Waddle, J.A. *et al.* (2002) The aurora kinase AIR-2 functions in the release of chromosome cohesion in Caenorhabditis elegans meiosis. *J. Cell Biol.*, **157**(2), 219–229.

Sakuno, T., Tada, K. and Watanabe, Y. (2009) Kinetochore geometry defined by cohesion within the centromere. *Nature*, **458**(7240), 852–858.

Schatten, G. (1994) The centrosome and its mode of inheritance: the reduction of the centrosome during gametogenesis and its restoration during fertilization. *Dev. Biol.*, **165**(2), 299–335.

Schueler, M.G. and Sullivan, B.A. (2006) Structural and functional dynamics of human centromeric chromatin. *Annu. Rev. Genomics Hum. Genet.*, **7**, 301–313.

Schuh, M. and Ellenberg, J. (2007) Self organization of MTOCs replaces centrosome function during acentrosomal spindle assembly in live mouse oocytes. *Cell*, **130**(3), 484–498.

Skold, H.N., Komma, D.J. and Endow, S.A. (2005) Assembly pathway of the anastral Drosophila oocyte meiosis I spindle. *J. Cell Sci.*, **118** (Pt 8), 1745–1755.

Sluder, G., Miller, F.J., Lewis, K. *et al.* (1989) Centrosome inheritance in starfish zygotes: selective loss of the maternal centrosome after fertilization. *Dev. Biol.*, **131**(2), 567–579.

Srayko, M., O'Toole, E.T., Hyman, A.A. and Muller-Reichert, T. (2006) Katanin disrupts the microtubule lattice and increases polymer number in C. elegans meiosis. *Curr. Biol.*, **16**(19), 1944–1949.

Stoler, S., Keith, K.C., Curnick, K.E. and Fitzgerald-Hayes, M. (1995) A mutation in CSE4, an essential gene encoding a novel chromatin-associated protein in yeast, causes chromosome nondisjunction and cell cycle arrest at mitosis. *Genes Dev.*, **9**(5), 573–586.

Sullivan, B.A., Blower, M.D. and Karpen, G.H. (2001) Determining centromere identity: cyclical stories and forking paths. *Nat. Rev. Genet.*, **2**(8), 584–596.

Szollosi, D., Calarco, P. and Donahue, R.P. (1972) Absence of centrioles in the first and second meiotic spindles of mouse oocytes. *J. Cell Sci.*, **11**(2), 521–541.

Talbert, P.B., Bryson, T.D. and Henikoff, S. (2004) Adaptive evolution of centromere proteins in plants and animals. *J. Biol.*, **3**(4), 18.

Tavosanis, G., Llamazares, S., Goulielmos, G. and Gonzalez, C. (1997) Essential role for gamma-tubulin in the acentriolar female meiotic spindle of Drosophila. *EMBO J.*, **16**(8), 1809–1819.

Theurkauf, W.E. and Hawley, R.S. (1992) Meiotic spindle assembly in Drosophila females: behavior of nonexchange chromosomes and the effects of mutations in the nod kinesin-like protein. *J. Cell Biol.*, **116**(5), 1167–1180.

Tokai-Nishizumi, N., Ohsugi, M., Suzuki, E. and Yamamoto, T. (2005) The chromokinesin Kid is required for maintenance of proper metaphase spindle size. *Mol. Biol. Cell*, **16**(11), 5455–5463.

Toth, A., Rabitsch, K.P., Galova, M. *et al.* (2000) Functional genomics identifies monopolin: a kinetochore protein required for segregation of homologs during meiosis I. *Cell*, **103**(7), 1155–1168.

Van Hooser, A.A., Ouspenski, I.I., Gregson, H.C. *et al.* (2001) Specification of kinetochore-forming chromatin by the histone H3 variant CENP-A. *J. Cell Sci.*, **114** (Pt 19), 3529–3542.

Verde, F., Berrez, J.M., Antony, C. and Karsenti, E. (1991) Taxol-induced microtubule asters in mitotic extracts of Xenopus eggs: requirement for phosphorylated factors and cytoplasmic dynein. *J. Cell Biol.*, **112**(6), 1177–1187.

Walczak, C.E., Vernos, I., Mitchison, T.J. *et al.* (1998) A model for the proposed roles of different microtubule-based motor proteins in establishing spindle bipolarity. *Curr. Biol.*, **8**(16), 903–913.

Wassarman, P.M. and Fujiwara, K. (1978) Immunofluorescent anti-tubulin staining of spindles during meiotic maturation of mouse oocytes in vitro. *J. Cell Sci.*, **29**, 171–188.

Wassmann, K., Niault, T. and Maro, B. (2003) Metaphase I arrest upon activation of the Mad2-dependent spindle checkpoint in mouse oocytes. *Curr. Biol.*, **13**(18), 1596–1608.

Watanabe, Y. and Nurse, P. (1999) Cohesin Rec8 is required for reductional chromosome segregation at meiosis. *Nature*, **400**(6743), 461–464.

Wheatley, S.P., Kandels-Lewis, S.E., Adams, R.R. *et al.* (2001) INCENP binds directly to tubulin and requires dynamic microtubules to target to the cleavage furrow. *Exp. Cell Res.*, **262**(2), 122–127.

Yokobayashi, S. and Watanabe, Y. (2005) The kinetochore protein Moa1 enables cohesion-mediated monopolar attachment at meiosis I. *Cell*, **123**(5), 803–817.

Yokobayashi, S., Yamamoto, M. and Watanabe, Y. (2003) Cohesins determine the attachment manner of kinetochores to spindle microtubules at meiosis I in fission yeast. *Mol. Cell Biol.*, **23**(11), 3965–3973.

Zhang, D., Yin, S., Jiang, M.X. *et al.* (2007) Cytoplasmic dynein participates in meiotic checkpoint inactivation in mouse oocytes by transporting cytoplasmic mitotic arrest-deficient (Mad) proteins from kinetochores to spindle poles. *Reproduction*, **133**(4), 685–695.

Zheng, Y. (2004) G protein control of microtubule assembly. *Annu. Rev. Cell Dev. Biol.*, **20**, 867–894.

11

Mechanisms of asymmetric division in metazoan meiosis

Marie-Hélène Verlhac and Karen Wingman Lee

Biologie du Développement – UMR 7622, UPMC-CNRS, Paris CEDEX 05, France

The formation of female reproductive cells, called oocytes, relies on several rounds of asymmetric cell divisions. In this chapter we will not discuss the asymmetric divisions of the germline stem cells but only those occurring during meiotic divisions of the oocyte; meiotic divisions in higher eukaryotes are extremely asymmetric, leading to the formation of cells having both different genetic contents and different sizes. In most species, except insects, the first meiotic division, which enables the segregation of homologous chromosomes paired by crossovers, leads to the formation of a large cell, the oocyte, and a tiny first polar body. Without intervening DNA replication, the second meiotic division allows sister chromatid separation and gives rise to a haploid oocyte and the second polar body. In insects, these two successive meioses take place in the absence of cytokinesis. All three DNA products hence generated remain apposed to the cortex of the oocyte and will degenerate. In species which extrude two polar bodies, these tiny cells degenerate progressively. Therefore, the coupling of meiotic recombination with the elimination of all products of the two meiotic divisions except one, the oocyte, allows higher eukaryotes to produce female gametes that are all unique genetically. This reproductive strategy may help to maintain the genetic diversity of egg-producing species. The different sizes of the daughter cells during meiosis results mainly from the eccentric positioning of the chromosomes within the oocyte, which is generally huge, compared to somatic cells. We will first describe the various strategies used by oocytes from different organisms to position chromosomes eccentrically. Then we will present the mechanisms involved in the control of this eccentric positioning. Finally, we will discuss how the differentiation of a specialized cortical area of the oocyte helps to control the size of polar bodies being extruded.

Oogenesis: The Universal Process Marie-Hélène Verlhac and Anne Villeneuve

11.1 Strategies used for asymmetric positioning of chromosomes within the oocyte

Oocytes of all metazoans are arrested in prophase I of the first meiotic division in the ovary. Hormonal stimulation will result in triggering meiosis resumption, which is characterized by nuclear envelope breakdown, also called germinal vesicle breakdown (GVBD; the nucleus generally being voluminous in oocytes, formally named the germinal vesicle). Following the resumption of meiosis, the two consecutive divisions occur. A huge diversity in timing of asymmetric positioning of chromosomes in oocytes is observed through the animal kingdom: depending on the species, it takes place before, at or after GVBD. Therefore, the chromosomes are asymmetrically positioned either as a decondensed interphasic chromatin mass surrounded by an intact nuclear envelope and/or as condensed individualized metaphasic chromosomes.

11.1.1 Asymmetric positioning of the nucleus before GVBD

In the starfish *Asterina pectinifera* and the fly *Drosophila melanogaster*, the nucleus is already apposed to the cell cortex before meiosis resumes (Endow and Komma, 1997; Miyazaki, Kamitsubo and Nemoto, 2000). In flies, the localization of the GV participates in the definition of the dorsoanterior part of the future embryo (Figure 11.1a). In other species, like worms, sea urchin, some fishes and amphibians, the germinal vesicle is positioned at the animal pole of the oocyte, but the chromosomes will reach the cortex at or after GVBD. These asymmetric localizations of the GV take place during the process of oogenesis, concomitant with oocyte growth. In fully-grown sea cucumber oocytes, the GV migrates after the application of nerve extract and when meiosis is being reinitiated (Miyazaki, Kato and Nemoto, 2005). Meiosis resumption and GVBD take place in these oocytes when the GV has reached the cortex.

11.1.2 Asymmetric positioning of the condensed chromosomes at or after GVBD

In cells containing centrosomes, the spindle axis is determined by the position of the two opposing centrosomes before the nuclear envelope breaks down. Except in sea urchin (Egana, Boyle and Ernst, 2007), oocytes from most organisms are devoid of canonical centrosomes, made of centrioles surrounded by pericentriolar material. Nevertheless, when GVBD occurs, the first meiotic spindle forms where the GV was last positioned in the oocyte. In *Xenopus*, the chromosomes finish their migration to the cortex at GVBD even before the first meiotic spindle has become bipolar (Figure 11.1b). In this model system, a transient array of microtubules (microtubule transient array) forms around the condensed chromosomes and functions as an elevator, delivering the chromatin to the cortex of the animal pole (Huchon *et al.*, 1981). The first meiotic spindle then assembles parallel to the cortical layer of the oocyte and rotates through 90° to allow extrusion of the first polar body (Gard, 1992). In worms, the first meiotic spindle migrates over a

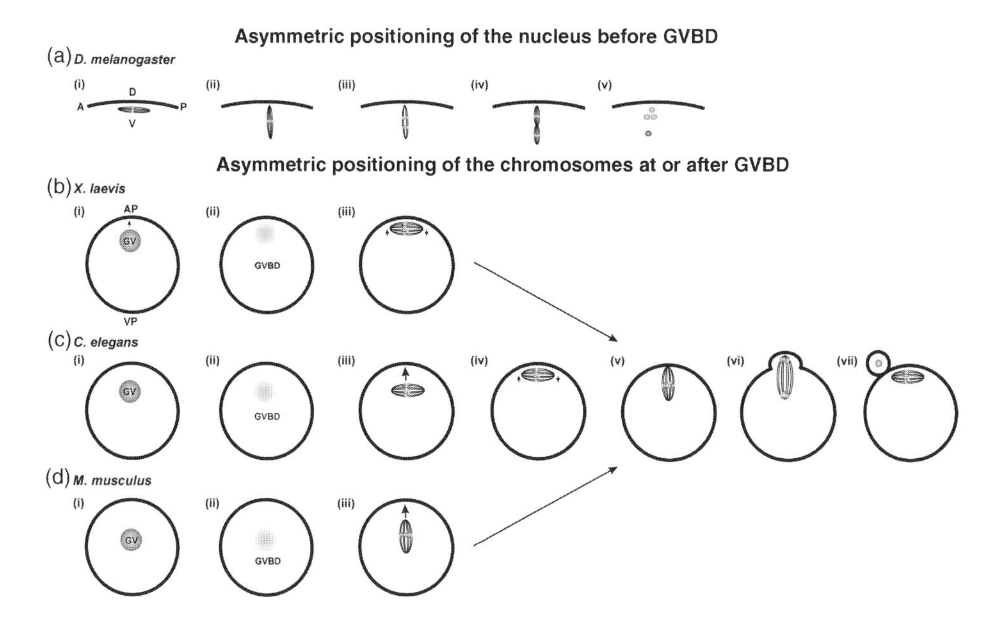

Figure 11.1 Strategies used by different organisms in asymmetric meiotic divisions. (a) In *Drosophila*, the GV is apposed to the dorsoanterior (D–A) part of the future embryo. After meiotic resumption, the first meiotic spindle forms parallel to the cortex (panel i). After fertilization (panel ii), the first meiotic spindle rotates through 90° and two successive meioses take place without cytokinesis (panels iii and iv). The innermost nucleus will remain and the other three DNA products will degenerate (panel v). (b) In *Xenopus*, the GV is located close to the animal pole (AP; panel i). After resumption of meiosis, the chromosomes migrate, via a transient array of microtubules, to the cortex at GVBD (panel ii) before the formation of the first meiotic spindle (panel iii), which will then rotate through 90° before the extrusion of the first polar body (PBE). (c) In worms, the first meiotic spindle migrates towards the cortex in a direction perpendicular to the spindle axis (panel iii). After fertilization, the spindle will shorten its axis and rotate through 90° (panel iv) before the extrusion of the polar body. (d) In mouse and ascidians, the GV is slightly off-centre (panel i). After GVBD (panel ii), the first meiotic spindle forms and migrates along its long axis towards the cortex (panel iii). In *Xenopus*, worm and mouse (b–d), after PBE, the second meiotic spindle forms parallel to the cortex (panel cvii), which will consequently rotate through 90° before the extrusion of the second polar body. GV = germinal vesicle (panels bi, ci and di); GVBD = germinal vesicle breakdown (panels bii, cii and dii); A = anterior; P = posterior; D = dorsal; V = ventral; AP = animal pole; VP = vegetal pole

short distance, following a path perpendicular to its axis and will then, at fertilization, both shorten and rotate through 90° to allow emission of the first polar body (Figure 11.1c; Albertson and Thomson, 1993; Yang, Mcnally and Mcnally, 2003). In ascidians and mice, the first meiotic spindle organizes approximately in the centre of the oocyte and migrates towards the cortex (Figure 11.1d). The spindle migrates along its long axis towards the cortex taking the shortest path, with the pole closest to the cortex leading (Verlhac *et al.*, 2000). The first meiotic spindle reaches the cortex perpendicularly. In these species, there is no rotation of the first meiotic spindle prior to the extrusion of the first polar body. However, in ascidians, amphibians, worms and mice, the second meiotic spindle forms parallel to the cortex and consequently rotates through 90° before extrusion of the second polar body occurs.

Similarly, the molecular mechanisms used for asymmetric positioning of chromosomes within the oocyte are also quite diverse.

11.2 Molecular mechanisms employed in asymmetric positioning of chromosomes within the oocyte

While extensive studies exist on the control of nuclear positioning in interphasic somatic cells, for example migrating neurons, and on spindle positioning in mitotic cells, not much is known about the control of chromosome positioning during meiosis.

11.2.1 Asymmetric GV positioning

The molecular mechanisms controlling asymmetric GV positioning have been mostly studied in flies. In *Drosophila*, the anterior–posterior axis is established during oogenesis, when the oocyte becomes polarized concomitantly with the acquisition of its fate. This polarization also leads to the asymmetric movement of the nucleus of the oocyte, which defines the dorsal side of the egg chamber as well as dorsoventral patterning of the embryo. Oocyte polarity arises from reciprocal signalling between the oocyte and surrounding follicle cells. The asymmetric positioning of the GV in *Drosophila* oocytes first requires the movement of the nucleus from a central position at the posterior pole to an asymmetric position at the anterior cortex, and also anchoring of the oocyte nucleus to the cortex, to prevent its drifting away.

In the case of somatic cells, these two processes are controlled by a combination of forces from microtubule- and actin-based networks. Directed nuclear movements depend on interactions between microtubules and cortical dynactin, via the minus-end-directed motor, dynein. Dynactin is a multisubunit complex that is required for most, if not all, types of cytoplasmic dynein activity in eukaryotes. Dynactin binds dynein directly and allows the motor to walk on the microtubule lattice over long distances. The dynactin complex contains 11 different polypeptide subunits, some of which are present in more than one copy per complex. Dynactin's largest subunit, p150^{Glued}, plays an essential role, since it binds to the dynein motor (Gill *et al.*, 1991; Holzbaur *et al.*, 1991; Vaughan and Vallee, 1995). The dynactin complex also contains the p50 dynamitin (Dmn), which constitutes the link between dynactin's two functional domains, the motor-binding domain on one side and the cargo-binding one on the other side (Gammie *et al.*, 1995; Echeverri *et al.*, 1996; Waterman-Storer and Holzbaur, 1996). The protein Lisencephaly-1 (Lis-1) binds both p150^{Glued} and dynein heavy chain, thus linking the dynactin and dynein complexes (Xiang *et al.*, 1995; Swan, Nguyen and Suter, 1999). Furthermore, Lis-1 links dynein to its cargo and enhances its processivity. Evidence from different studies suggests that the same machinery, involving interactions between plus ends of microtubules emanating from MTOCs (microtubule organizing centres) and cortical dynein/dynactin complexes, is also required for oocyte nuclear movements in *Drosophila*. First, microtubule depolymerization experiments have shown that a correct organization of the microtubule lattice is essential for nuclear migration in the oocyte (Januschke *et al.*, 2006; Theurkauf and

Hawley, 1992; Theurkauf *et al.*, 1993). Second, specific over-expression of dynamitin in the oocyte, which in mammals causes dynactin to dissociate from dynein and results in the inhibition of dynein-mediated movements, causes nuclear mislocalization (Januschke *et al.*, 2002). Third, hypomorphic alleles of *DLis-1* and of the kinesin heavy chain of Kinesin I have no influence on the complex organization of microtubules in the oocyte but affect nuclear positioning (Liu, Xie and Steward, 1999; Swan, Nguyen and Suter, 1999; Lei and Warrior, 2000; Januschke *et al.*, 2002). As a consequence of these defects in nucleus positioning, the distribution of maternal transcript is altered, resulting in defects in embryo polarity.

In starfish oocytes, the premeiotic aster, organized from a huge MTOC, is required to maintain the GV at the cortex (Miyazaki, Kamitsubo and Nemoto, 2000). However, in *Drosophila*, nuclear anchoring to the plasma membrane of the oocyte seems mainly to involve actin-based networks. The first evidence that asymmetric positioning of the *Drosophila* oocyte nucleus required, first, its migration and, second, its anchoring came from the analysis of germline clone mutants for a bZIP transcription factor, product of the gene Cap 'n' Collar: in these mutant cells, oocyte nuclear migration took place but, probably due to a lack of anchoring, the nucleus drifted away from its position (Guichet, Peri and Roth, 2001). Studies performed both in *C. elegans* and *Drosophila* have identified a protein complex, termed the LINC (*li*nkers of *n*ucleoskeleton and *c*ytoskeleton) complex, which is responsible for nuclear anchoring to the cytoskeleton (for a review see Crisp *et al.*, 2006; Crisp and Burke, 2008). This complex consists of huge KASH (*K*larsicht, *A*nc-1, *S*yne *h*omology) domain-containing proteins spanning the outer nuclear membrane, which interact, in the lumen of the nuclear envelope, with proteins of the SUN family (for *S*ad1, *Un*c-84) and with F-actin or microtubules in the cytoplasm (for a review see Starr and Han, 2003; Worman and Gundersen, 2006). Proteins of the SUN family dimerize, are embedded in the inner nuclear membrane and interact with Lamins inside the nucleus. Mutants of Anc-1, an extremely large type II membrane protein containing an actin-binding domain, show floating nuclei in syncytial hypodermal and gut cells of *C. elegans*. An Anc-1-related protein, MSP-300, seems also to be essential for nuclear anchoring in the *Drosophila* oocyte for the following reasons: first, MSP-300 localizes at nuclear envelopes of nurse cells and the oocyte; and second, egg chambers carrying a hypomorphic allele of msp-300 contain oocyte nuclei that are mislocalized to the posterior region of the oocyte (Yu *et al.*, 2006). Therefore, one can imagine that positioning of the GV in large cells such as oocytes involves similar mechanisms to those required for placing the nucleus in somatic cells.

11.2.2 Asymmetric positioning of condensed chromosomes after GVBD

Movement and orientation of mitotic spindles occurs through astral microtubules, emanating from spindle poles (for a review see Gönczy, 2002). In certain organisms, such as some annelids (Lutz, Hamaguchi and Inoué, 1988), some molluscs (Palazzo *et al.*, 1992) and starfish (Zhang *et al.*, 2004), meiotic spindles contain true centrosomes with centrioles from which astral microtubules emanate. In these species, meiotic spindle positioning may occur in much the same way as in mitotic animal cells. However, female meiotic spindles in human (Sathananthan, 1997), mice (Szöllösi,

Calarco and Donahue, 1972), *Drosophila* (Theurkauf and Hawley, 1992) and worms (Albertson and Thomson, 1993) do not have centrioles and their associated astral microtubules. Worms and mice use different mechanisms for positioning the first meiotic spindle to the cortex: the process is microtubule-dependent in *C. elegans* or mostly F-actin-dependent in mice.

A microtubule-dependent movement of the first meiotic spindle to the cortex

In *C. elegans* oocytes, the meiosis I spindle assembles several microns away from the cortex and migrates towards the cortex in an F-actin-independent but microtubule-dependent manner (Yang, Mcnally and Mcnally, 2003). It has also been shown that the ATP-dependent microtubule-severing enzyme MEI-1/katanin, a heterodimer (see Chapter 10 for more details on katanin), is required for this translocation to the cortex to occur (Yang, Mcnally and Mcnally, 2003). In addition, using an RNA interference (RNAi)-based screen, Yang, Mains and Mcnally (2005) identified a complex of proteins containing the kinesin-1 heavy chain orthologue, UNC-116, the kinesin light chain orthologues, KLC-1 and -2, and a novel cargo adaptor, KCA-1, necessary for the translocation of the meiosis I spindle. The translocation of the meiotic spindle to the cortex via this motor complex might be either direct or indirect, as a result of controlling microtubule organization.

Once the meiosis I spindle reaches the cortex, it rotates prior to polar body extrusion. In *C. elegans*, meiotic spindle rotation depends on a Lin-5/ASPM-1/Calmodulin-1 complex (van der Voet *et al.*, 2009). In cells with canonical centrosomes, spindle movements are ensured by pulling of astral microtubules via the dynein microtubule motor. Dynein interacts on one side with the minus-end of microtubules and on the other side with the Lin-5/GPR-1/2/Gα complex hooked to the cell cortex (Gotta *et al.*, 2003; Couwenbergs *et al.*, 2007; Nguyen-Ngoc, Afshar and Gönczy, 2007). It has been shown recently that meiotic spindle rotation, which depends on interactions between meiotic spindle microtubules and a meshwork of cortical microtubules, involves a Lin5/ASPM-1/Calmodulin-1 complex but not a Lin-5/GPR-1/2/Gα complex (van der Voet *et al.*, 2009).

An actin-dependent movement of the first meiotic spindle to the cortex

Spindle migration to the cortex in ascidian and mouse oocytes depends mainly on actin filaments. Oocytes treated during chromosome migration with nocodazole, a microtubule-depolymerizing agent, harbour chromosomes that can still migrate to the cortex. On the other hand, oocytes treated with cytochalasin D, an agent that induces F-actin depolymerization, display chromosomes that remain centrally located (Longo and Chen, 1985; Verlhac *et al.*, 2000; Prodon, Chenevert and Sardet, 2006). F-actin dynamics are required for chromosome positioning to the cortex in mouse oocytes, since treatment with jasplakinolide, which induces microfilament polymerization and stabilization, also prevents their movement to the cortex (Terada, Simerly and Schatten, 2000; Li *et al.*, 2008). In mouse oocytes, the spindle migrates along its long axis towards the nearest cortex late in meiosis I (from three to five hours after meiosis resumption, depending on the strain) and attaches to it. By displacing the meiotic spindle using a micro-glass needle, Schuh and Ellenberg (2008) have shown elegantly that indeed it is always the pole nearest to the cortex which takes the lead for the direction of spindle migration. There is no predetermined site in the cortex able to attract the spindle (Verlhac *et al.*, 2000). Furthermore, in these oocytes, the activation of the Mos/MAPK pathway is required for spindle migration to the cortex; however, the targets of this pathway remain largely unknown (Verlhac *et al.*, 2000).

On the other hand, spindle migration in mouse oocytes has been shown to depend unambiguously on Formin-2 nucleated actin microfilaments (Leader *et al.*, 2002; Dumont *et al.*, 2007a). Actin filaments are helical polarized polymers with a fast-growing 'barbed' plus end and a slow-growing 'pointed' minus end. New actin subunits are added to the barbed ends at approximately 10 times the rate of addition to the pointed ends (for an extensive review, see Renault, Bugyi and Carlier, 2008). Eukaryotic cells require *de novo* nucleation of F-actin from a large pool of monomeric actin in order to elicit spatial and temporal remodelling of their actin cytoskeleton. The nucleation is a rate-limiting step for actin filament polymerization. There are five classes of *de novo* actin nucleators: Formins, the Arp2/3 complex, Spire, Cordon Bleu and Leiomodin (for a review, see Liu *et al.*, 2008). Some very recently discovered nucleators are apparently tissue specific, for example Cordon Bleu and Leiomodin in the nervous system and muscle cells respectively (Ahuja *et al.*, 2007; Chereau *et al.*, 2008). Among the actin nucleators, the Arp2/3 complex binds to the side of pre-existing microfilaments initiating a branch by mimicking a new pointed end (Mullins, Heuser and Pollard, 1998). The protein Spire, which contains WASP homology (WH2) domain, nucleates straight F-actin and also binds to the pointed ends of filaments (Quinlan *et al.*, 2005). Formins constitute another class of F-actin nucleators, which function as processive assembly motors that remain bound to growing barbed ends (Romero *et al.*, 2004; Renault, Bugyi and Carlier, 2008). They are highly conserved proteins, expressed in most organisms and are usually characterized by the presence of three Formin homology (FH1, FH2, FH3) domains (for reviews see Faix and Grosse, 2006; Renault, Bugyi and Carlier, 2008). The FH1 domain binds profilin and the FH2 domain binds actin (Romero *et al.*, 2004). Unlike the Arp2/3 complex and Spire, Formins bind to the barbed end of actin filaments, protecting them from capping proteins. As a result, Formins induce the generation of relatively long straight actin filaments (for a review see Kovar and Pollard, 2004). There are two classes of Formins: proteins of the Diaphanous family, characterized by the presence of a Rho-GTPase binding domain and proteins of the Cappucino family, which lack this domain (for reviews see Evangelista, Zigmond and Boone, 2003; Liu *et al.*, 2008). The Formin-2 gene encodes a large protein containing both FH1 and FH2 domains but no obvious FH3 domain. In the mouse, it is expressed in the central nervous system and in oocytes (Leader and Leder, 2000). Oocytes from Formin-2-deficient mice present chromosomes which remain centrally located, do not extrude polar bodies and, when fertilized, lead to aneuploid embryos (Leader *et al.*, 2002). Live video-microscopy experiments showed that Fmn2$^{-/-}$ oocytes undergo anaphase I without cytokinesis (Dumont *et al.*, 2007a). Recently, using an F-actin-specific probe and live confocal microscopy, it was shown that Formin-2 organizes a highly dynamic F-actin cytoplasmic meshwork, essential for spindle migration (Azoury *et al.*, 2008, Azoury, Verlhac and Dumont, 2009; Schuh and Ellenberg, 2008). This F-actin meshwork connects the chromosomes and the spindle at GVBD and in late meiosis I; it organizes around the division spindle into a spindle-like structure connected to the cortex via straight actin filaments (Figure 11.2a and b). By FRAP (fluorescence recovery after photobleaching) analysis, it was also shown that the dynamics of this meshwork are regulated in time and space during meiotic maturation (Azoury *et al.*, 2008). It has been shown that *Drosophila* oocytes also present an F-actin meshwork required for oocyte polarization (Dahlgaard *et al.*, 2007). Indeed, oocytes from mutant flies of Spire, Cappucino or profilin show the same phenotype as oocytes treated with cytochalasin D: defaults in cortical microtubule organization resulting in early cytoplasmic streaming and loss of oocyte polarity (Manseau and Schüpbach, 1989; Cooley and Theurkauf, 1994; Theurkauf, 1994;

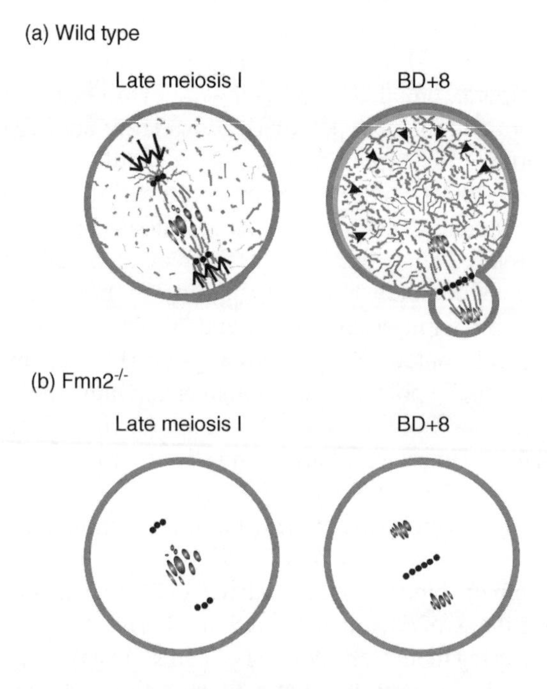

Figure 11.2 A dynamic Formin-2-nucleated F-actin meshwork is essential for spindle migration. (a) In wild-type mouse oocytes, a dynamic F-actin meshwork becomes organized around the division spindle during meiosis I into a spindle-like F-actin structure. In late meiosis I, this actin spindle is connected to the cortex via straight actin filaments. Myosin II is activated at the spindle poles and provides the pulling forces (black arrows) for the F-actin to relocate the spindle. Due to the slightly off-centring of the spindle in the oocyte, one of the spindle poles is closer to the cortex. Hence, the pulling force at this pole is more efficient and this results in the migration of the spindle in the direction of the nearest cortex. At anaphase, a global change of both cortical and cytoplasmic F-actin may ensure pushing (black arrowheads) of the spindle into the extruding polar body. (b) In oocytes from Formin-$2^{-/-}$ (Fmn$2^{-/-}$) mouse, the F-actin meshwork and the spindle-like F-actin structure are absent, correlating with the failure of spindle migration. Even though chromosome segregation occurs, the chromosomes remain in the centre of the oocyte. The grey solid lines represent F-actin. The subcortical F-actin layer is in light gray. The grey ellipses are chromosomes. The black circles represent active phospho-myosin II. BD + 8 corresponds to eight hours after germinal vesicle breakdown

Manseau, Calley and Phan, 1996; Wellington *et al.*, 1999; Rosales-Nieves *et al.*, 2006). Indeed, in wild-type oocytes, vigorous ooplasmic streaming is associated with rapid growth during stages 10b to 13 and is never observed prior to this stage. Although the ooplasmic streaming in *Drosophila* oocytes is microtubule based, actin assembly is required for its timing. A premature streaming interferes with transport mechanisms that are required for the localization of early polarity markers, resulting in disruption of dorsal–ventral and anterior–posterior body axes (Manseau and Schüpbach, 1989; Theurkauf, 1994). Oocytes from Spire mutant can be rescued by expression of constitutively active fragments of Cappucino, provided profilin is present (Dahlgaard *et al.*, 2007). By contrast, a constitutively active mutant of Spire, which rescues the Spire phenotype, cannot rescue the Cappucino mutant phenotype. This cooperation between two actin nucleators, Spire and Formin, for the formation of a cytoplasmic actin meshwork has so far only been described in *Drosophila* oocytes and it remains to be determined whether mouse oocytes also require (i) such an interaction in order to organize cytoplasmic F-actin and (ii) ooplasmic streaming to position the spindle.

Nonetheless, in mouse oocytes, the movement of the spindle to the cortex seems to depend on the activity of myosin II, since blocking Myosin IIA activity, via antibody injection or via the use of a Myosin Light Chain kinase inhibitor, prevents spindle migration to the cortex (Simerly *et al.*, 1998; Schuh and Ellenberg, 2008). Furthermore, active phospho-Myosin II is mislocalized in Formin-2-knockout oocytes (Dumont *et al.*, 2007a; Schuh and Ellenberg, 2008). Also it has been shown that active phospho-Myosin II accumulates at both poles of the meiotic spindle (Schuh and Ellenberg, 2008). Active Myosin II at the poles would pull the spindle towards the cortex (Figure 11.2a). Myosin II multimers associated with actin filaments can generate contractility by antiparallel sliding of F-actin. Therefore, published work supports the idea that movement of the first meiotic spindle to the cortex in mouse oocytes could potentially rely on a dynamic contractile actomyosin meshwork, similar to the case of starfish oocytes, where chromosome congression depends on the presence of a contractile meshwork of F-actin (Lenart *et al.*, 2005).

11.2.3 After migration, anchoring of the meiotic apparatus

The distinction between migration and anchoring has not always been made clearly and, in some cases, the two processes may actually depend on the same molecules. Yet, it is essential that either the GV or the meiotic spindle remain anchored to the cortex to ensure the asymmetry of the meiotic divisions. F-actin is required for meiotic spindle rotation in *Xenopus* and mouse (Maro *et al.*, 1984; Gard, Cha and Roeder, 1995; Zhu *et al.*, 2003; Sun and Schatten, 2006). In *Xenopus* oocytes, an unconventional myosin, Myo X, which contains a MyTH4 domain that is present in an *Arabidopsis* kinesin and has been shown to bind microtubules (Narasimhulu and Reddy, 1998), localizes on the meiotic spindle, concentrates in the region where the spindle contacts the cortex and overlaps with some actin microfilaments (Weber *et al.*, 2004). Importantly, injection of dominant-negative MyoX tail or of anti-MyoX antibodies disturbs both spindle assembly and nuclear anchoring, suggesting that this actin-based motor serves to link microtubules and cortical F-actin. *Drosophila* oocytes also use microtubules to anchor their meiotic spindles to the cortex. Indeed, feeding *Drosophila* females with colchicine, a microtubule-depolymerizing drug, induces defects in meiotic spindle positioning: spindles form deep in the cytoplasm and become oriented differently towards the cortical surface. In these oocytes, a microtubule motor of the Kif-13 family, KLP10A, also localizes at meiotic spindle poles and is required for spindle anchoring to the cortex (Zou *et al.*, 2008). It remains to be determined whether KLP10A acts in *Drosophila* spindle positioning via its motor function or via its role in the regulation of microtubule dynamics. *Drosophila* KLP10A may exert a similar function in spindle positioning as MEI1/katanin in *C. elegans* oocytes (Yang, Mcnally and Mcnally, 2003).

Although the biological significance of these observations has yet to be established, it appears that perturbation of golgi-based membrane fusion and/or vesicular traffic perturbs either positioning or anchoring of meiotic spindle to the cortex of mouse oocytes. Indeed the use of Brefeldin A, BFA (Wang *et al.*, 2008) or the injection of a dominant negative form of ARF1 (ADP-ribosylation factor 1) induces cleavage of mouse oocytes both in meiosis I and II (Wang *et al.*, 2009). The link between the effect of BFA, the over-expression of the dominant negative form of ARF1 and their potential physiological targets is still to be uncovered.

Nevertheless, in mouse oocytes, the cortical differentiated area which forms above chromosomes (see Section 11.3) seems to be involved in the anchoring of both meiosis I and II spindles. Indeed, this area is connected to the spindle both at the end of meiosis I and in meiosis II via actin microfilaments (Azoury *et al.*, 2008) and, as we will develop below, active G-proteins, which accumulate in this region, are required for spindle anchoring (Halet and Carroll, 2007).

11.3 Cortical differentiation leading to polar body extrusion

11.3.1 Peculiarities of cytokinesis during meiotic divisions

Polar body extrusion (PBE) is the final step in asymmetric meiotic division. Similar to cytokinesis, it involves the pinching of the cytoplasm into two daughter cells (a large oocyte and a small polar body). Once the spindle has migrated to the cortex, the oocyte is ready to divide asymmetrically. In mitotic cells, it is generally accepted that microtubules (either astral or midzone) determine the position of the cleavage plane, midway between the poles of the mitotic spindle (Straight and Field, 2000; Murata-Hori and Wang, 2002; Alsop and Zhang, 2003). Micromanipulation studies in marine invertebrate embryos suggest that astral microtubules play an important role in determining the positioning of the cleavage furrow (Rappaport, 1961). In a now classical 'torus' experiment with sand dollar eggs, related in many cell biology textbooks, Rappaport demonstrated that the cleavage furrow could be induced to form between two adjacent asters that are not linked by a spindle/chromosomes. He showed that the mitotic spindle of the egg could be displaced to one side of the cell by a glass bead. As a result, a cleavage furrow formed only on one side, and this resulted in a binucleate 'doughnut-shaped' egg. During the second cell-division cycle, both nuclei enter mitosis. Thus, cleavage furrows form not only between centrosomes linked by mitotic spindles, but also between adjacent centrosomes, not linked by mitotic spindles/chromosomes. It was therefore suggested that the astral microtubules are required for determining the site of cleavage. Although the asters seem to be sufficient to induce cleavage furrows in most animal cells, experiments performed in cultured cells have suggested that the midzone microtubules play a crucial role in cleavage plan specification (Cao and Wang, 1996; Eckley *et al.*, 1997; Rieder *et al.*, 1997; Dechant and Glotzer, 2003). In tissue culture cells, the asters are typically small, whereas the midzone spindle is relatively large. In addition, the presence of the asters is not sufficient to induce cleavage furrows in cultured cells. In such cells, the midzone microtubules arise from elongation and reorganization of the mitotic spindle during anaphase and telophase (Jantsch-Plunger *et al.*, 2000). These interdigitated microtubules undergo excessive bundling that results in formation of the central spindle (Dechant and Glotzer, 2003). The presence of midzone microtubules has been shown to correlate with induction of the cleavage furrow in cultured epithelial cells (Wheatley and Wang, 1996). Thus, whereas one of these populations (i.e. the astral and midzone microtubules) may be essential for cytokinesis in one system, they may be partially or even fully dispensible in another (for a review, see Glotzer, 1997). The mechanism actually employed to specify the division plane may vary from system to system in response to cell geometry and volume, and may also be related to the location of the mitotic spindle within the dividing cell. In some cases, both midzone and astral microtubules are involved in positioning the cleavage furrow (Murata-Hori and Wang, 2002).

It must be emphasized that meiotic divisions represent a very peculiar mode of cytokinesis, since meiotic spindles in most species are typically anastral, and because the spindle midzone forms deep in the cytoplasm away from the site of the future cleavage plane. Therefore, neither astral nor midzone microtubules are responsible for positioning the site of the polar body. We shall describe below the specific mechanisms that ensure polar body extrusion during meiosis.

11.3.2 Extruding the polar body

Role of chromatin in cortical differentiation

It has been known for a long time that chromatin plays an important role in defining the site of PBE (Maro *et al.*, 1986). In immature mouse oocytes, the cortex of oocytes appears to be structurally homogeneous. Concomitant with the migration of the first meiotic spindle towards the cortex, the cortical region above the spindle becomes gradually polarized (Figure 11.3a; Verlhac *et al.*, 2000). In many species (such as starfish, *Xenopus* and mouse), this polarization involves the formation of an area devoid of microvilli and an actin-enriched cap (Longo and Chen, 1985; Maro *et al.*, 1986; Heil-Chapdelaine and Otto, 1996; Ma *et al.*, 2006; Sun and Schatten, 2006; Hamaguchi, Numata and K Satoh, 2007). Treatment of mouse oocytes arrested in metaphase II (MII) with nocadazole results in chromosome dispersal. A cortical differentiated zone overlying each clump of chromosomes is induced under these conditions (Longo and Chen, 1985; Maro *et al.*, 1986; Azoury, Verlhac and Dumont, 2009). Consistent with this observation, by injecting DNA-coated beads into MII mouse oocytes, Deng *et al.* (2007) showed that this cortical differentiation is induced by chromatin and that

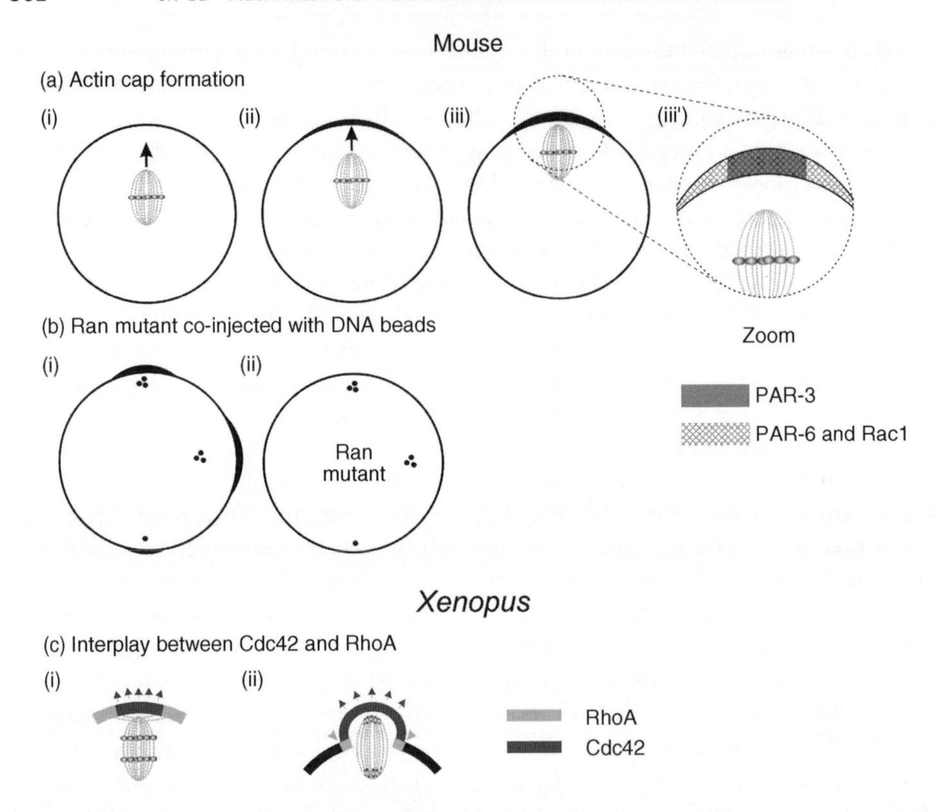

Figure 11.3 Formation of the actin cap and its signalling molecules. (a) In the mouse, as the first meiotic spindle migrates to the cortex, an actin-enriched cap progressively forms in the cortical region above the spindle (panels i–iii). This actin-enriched cap coincides with the localization of PAR-6 and active Rac1, while PAR-3 is localized in the inner region of the actin cap (panel iii'). (b) Mouse oocytes injected with DNA-coated beads form a cortical actin cap (panel i). Oocytes co-injected with DNA-coated beads and a Ran mutant fail to form the actin-enriched region in the cortex (panel ii). (c) In *Xenopus*, Cdc42 colocalizes with the actin cap in the cortex (panel i). This Cdc42 zone is circumscribed by a RhoA contractile ring (panel ii). The activated Cdc42 may facilitate the out-pocketing of the plasma membrane during polar body extrusion, while the RhoA ring plays a role in restricting Cdc42 (panel ii). Interplay between Cdc42 and RhoA facilitates the out-pocketing and pinching of the plasma membrane required during polar body extrusion. Solid black lines represent F-actin. Grey ellipses are chromosomes. Black circles represent DNA-coated beads. Dotted grey lines are spindles

the size of the actin cap is directly proportional to the amount of DNA injected (Figure 11.3b). The actin cap polarization usually coincides with the site of polar body extrusion both in meiosis I and II. The actin cap is believed to restrict the progression of the cleavage furrow to the differentiated area overlying the chromosomes. Such a mechanism might ensure a minimal size for the polar body: big enough to enclose all the chromosomes yet small enough to ensure that a sufficient amount of cytoplasm is retained in the oocyte. In addition to the actin cap, in pig and mouse oocytes, cortical granules are excluded from this region as well (cortical granule-free domain; Kim *et al.*, 1996; Deng *et al.*, 2003). It has been shown that sperm is less likely to penetrate this region during fertilization, thereby minimizing the potential damage to the spindle and the possibility of extruding chromosomes of the sperm into the second polar body (Nicosia, Wolf and Inoue, 1977; Wilson and Snell, 1998).

Role of RanGTP in inducing cortical differentiation

Although the influence of chromosomes on the cortex has been observed for a long time, it is only very recently that a factor that may be responsible for this activity has been proposed. Indeed, it has been shown that chromatin can induce the nucleation of microtubules via the activity of the small RanGTPase (for reviews see Gruss and Vernos, 2004; Zheng, 2004). Like all GTPases, the activation of Ran is favoured by its guanine nucleotide exchange factor (GEF), and its inactivation is catalyzed by GTPase-activating proteins (GAPs). The GEF of Ran, RCC1, localizes on chromosomes, whereas RanGAPs are found in the cytosol. Since opposing activities are spatially segregated, a gradient of RanGTP is produced, with higher RanGTP levels near chromatin and lower RanGTP concentrations further away from the chromosomes (Kalab, Weis and Heald, 2002; Caudron *et al.*, 2005; Kalab *et al.*, 2006). A broad RanGTP gradient centred on chromosomes is also present at all steps of meiotic maturation in mouse oocytes (Dumont *et al.*, 2007b). This local accumulation of RanGTP follows chromosome migration towards the cortex, providing a spatial indication of their position within the cell (Dumont *et al.*, 2007b). Furthermore, the co-injection of DNA-coated beads together with active/inactive mutants of the RanGTPase abolishes cortical differentiation normally induced by the beads (Figure 11.3b; Deng *et al.*, 2007). Taken together, these two sets of data suggest that the RanGTP gradient mediates the polarization of the cortex overlying the chromosomes (for a review see Verlhac and Dumont, 2008). Consistent with the idea that the cortical differentiated area defines the region where polar bodies are extruded and, as a consequence, where the progression of the cleavage furrow will be restricted, it was shown that parthenogenetic activation of metaphase II-arrested oocytes expressing a dominant-negative form of Ran cleave symmetrically instead of extruding the second polar body (Dumont *et al.*, 2007b). The effectors of RanGTP in the cortex are as yet unknown.

Role of Rac1, Cdc42 and RhoA in the formation of a cortical differentiated area

Recent works have shown that small GTPases are involved in defining the extent of the cortical differentiated area. In mouse oocytes, activated Rac1, in its GTP-bound form, covers the whole cortex in immature oocytes and is progressively restricted to the cortex above chromosomes at the end of meiosis I (Figure 11.3a; Halet and Carroll, 2007). Inhibition of Rac1 activity, using a dominant-negative Rac1 mutant, disturbs spindle organization and prevents spindle anchoring to the cortex in meiosis II. It is still not really clear whether local activation of Rac1 is important for the function of the cortical differentiated area in the process of PBE. In *Xenopus* oocytes, an interplay between Cdc42 and RhoA is involved in defining the extent of the region of the cortex that will bud-off and deliver one set of chromosomes in the polar body (Ma *et al.*, 2006; Zhang *et al.*, 2008). At anaphase, activation of Cdc42 in the region of the cortex above the contacting spindle pole would be essential for the formation of dynamic microfilaments allowing the out-pocketing of the plasma membrane (Figure 11.3ci). This region is circumscribed by a RhoA contractile ring in order to restrain the out-pocketing to the region above the spindle (Figure 11.3cii; Zhang *et al.*, 2008). It has been shown that the localized activity of Cdc42 restricts the activation of RhoA to the contractile ring. Moreover, RhoA exerts a positive feedback action on local Cdc42 activation. RhoA would act much as it does in somatic cells undergoing cytokinesis: by locally allowing Myosin II activation and by recruiting stable F-actin.

In turn, Cdc42 or Rac1 may regulate the activity of PAR (partitioning defective) proteins. The PAR proteins have been shown to control cortical polarity and asymmetric divisions in various model systems. The *par* genes were first identified in *C. elegans* by genetic screens aimed at identifying genes required for the first asymmetric division of the zygote (Kemphues *et al.*, 1988; Watts *et al.*, 1996). Two of these genes encode PDZ domain-containing proteins, PAR-3 and PAR-6, which interact with a Ser/Thr atypical protein kinase (aPKC) and constitute the PAR complex (Joberty *et al.*, 2000; Lin *et al.*, 2000). The kinase activity of aPKC is required to form a functional PAR complex and this kinase also participates in the phosphorylation of key effectors of cell polarity (for an extensive review see Goldstein and Macara, 2008). It has been shown in *Xenopus* that both aPKC and PAR-3 localize uniformly on the oocyte cortex in immature oocytes and colocalize to the animal cortex after meiosis resumption (Nakaya *et al.*, 2000). Similarly, in mouse oocytes, PAR-3 and one isoform of PAR-6 accumulate in the cortical differentiated area (Figure 11.3aiii and iii; Duncan *et al.*, 2005; Vinot *et al.*, 2004). In addition, PAR proteins could also potentially link spindle migration to extrusion of the polar body, since in mouse meiosis I, one isoform of PAR-6, mPAR-6a, localizes on the first meiotic spindle and accumulates at the pole of the spindle which is closest to the cortex, and another isoform of PAR-6, mPAR-6b, is found on spindle microtubules until MII entry and then is relocated in the cortical differentiated area during MII arrest (Vinot *et al.*, 2004).

We still need to figure out what signals trigger local activation of Cdc42 and Rac1 in *Xenopus* and mouse oocytes respectively. One potential candidate is the RanGTP activity transported by chromosomes reaching the cortex at the end of spindle migration in meiosis I mouse oocytes (Deng *et al.*, 2007; Dumont *et al.*, 2007b). In many models, the activity of the PAR complex is regulated by Cdc42 (Yamanaka *et al.*, 2001; Iden and Collard, 2008). The inhibition of Cdc42 activity in mouse oocytes perturbs the first PBE (Na and Zernicka-Goetz, 2006). Although Cdc42, Rac1 and the PAR proteins are known to interact in many model systems (for a review see Iden and Collard, 2008), the potential link between Cdc42, Rac1 and PAR proteins has not been directly addressed in the process of cortical differentiation. Nonetheless it is somewhat puzzling that active Rac1 and PARs localize in the same region of the mouse oocyte.

11.4 Conclusion

Observations from different model systems demonstrate the diversity of processes that mediate the asymmetry of meiotic divisions in oocytes: from the control of GV positioning to the control of polar body extrusion. In this chapter, we have emphasized mainly the role of the cytoskeleton and small GTPases in these processes. A great deal of work still needs to be done in order to have a more complete picture of how the networks of microtubules and microfilaments are coordinated both temporally and spatially for the progression of these peculiar meiotic divisions.

Acknowledgements

We thank Julian Smith for reading this manuscript. This work was supported by grants from the *Association pour la Recherche sur le Cancer* (ARC4968 to M.-H. Verlhac) and from the *Agence Nationale pour la Recherche* (ANR-08-BLAN-0136-01 to M.-H. Verlhac). K. Wingman Lee is supported by the *Centre Nationale pour la Recherche Scientifique*.

References

Ahuja, R., Pinyol, R., Reichenbach, N. *et al.* (2007) Cordon-bleu is an actin nucleation factor and controls neuronal morphology. *Cell*, **131**, 337–350.

Albertson, D.G. and Thomson, J.N. (1993) Segregation of holocentric chromosomes at meiosis in the nematode, Caenorhabditis elegans. *Chromosome Res.*, **1**, 15–26.

Alsop, G.B. and Zhang, D. (2003) Microtubules are the only structural constituent of the spindle apparatus required for induction of cell cleavage. *J. Cell Biol.*, **162**, 383–390.

Azoury, J., Lee, K.W., Georget, V. *et al.* (2008) Spindle positioning in mouse oocytes relies on a dynamic meshwork of actin filaments. *Curr. Biol.*, **18**, 1514–1519.

Azoury, J., Verlhac, M.-H. and Dumont, J. (2009) Actin filaments: key players in the control of asymmetric divisions in mouse oocytes. *Biol. Cell.*, **101**, 69–78.

Cao, L.G. and Wang, Y.-L. (1996) Signals from the spindle midzone are required for the stimulation of cytokinesis in cultured epithelial cells. *Mol. Biol. Cell*, **7**, 225–232.

Caudron, M., Bunt, G., Bastiaens, P. and Karsenti, E. (2005) Spatial coordination of spindle assembly by chromosome-mediated signaling gradients. *Science*, **309**, 1373–1376.

Chereau, D., Boczkowska, M., Skwarek-Maruszewska, A. *et al.* (2008) Leiomodin is an actin filament nucleator in muscle cells. *Science*, **320**, 239–243.

Cooley, L. and Theurkauf, W.E. (1994) Cytoskeletal functions during Drosophila oogenesis. *Science*, **266**, 590–596.

Couwenbergs, C., Labbe, J.C., Goulding, M. *et al.* (2007) Heterotrimeric G protein signaling functions with dynein to promote spindle positioning in C. elegans. *J. Cell Biol.*, **179**, 15–22.

Crisp, M. and Burke, B. (2008) The nuclear envelope as an integrator of nuclear and cytoplasmic architecture. *FEBS Lett.*, **582**, 2023–2032.

Crisp, M., Liu, Q., Roux, K. *et al.* (2006) Coupling of the nucleus and cytoplasm: role of the LINC complex. *J. Cell Biol.*, **172**, 41–53.

Dahlgaard, K., Raposo, A.A., Niccoli, T. and St Johnston, D. (2007) Capu and Spire assemble a cytoplasmic actin mesh that maintains microtubule organization in the drosophila oocyte. *Dev. Cell.*, **13**, 539–553.

Dechant, R. and Glotzer, M. (2003) Centrosome separation and central spindle assembly act in redundant pathways that regulate microtubule density and trigger cleavage furrow formation. *Dev. Cell*, **4**, 333–344.

Deng, M., Kishikawa, H., Yanagimachi, R. *et al.* (2003) Chromatin-mediated cortical granule redistribution is responsible for the formation of the cortical granule-free domain in mouse eggs. *Dev. Biol.*, **257**, 166–176.

Deng, M., Suraneni, P., Schultz, R.M. and Li, R. (2007) The RanGTPase mediates chromatin signaling to control cortical polarity during polar body extrusion in mouse oocytes. *Dev. Cell*, **12**, 301–308.

Dumont, J., Million, K., Sunderland, K. *et al.* (2007a) Formin-2 is required for spindle migration and for late steps of cytokinesis in mouse oocytes. *Dev. Biol.*, **301**, 254–265.

Dumont, J., Petri, S., Pellegrin, F. *et al.* (2007b) A centriole- and RanGTP-independent spindle assembly pathway in meiosis I of vertebrate oocytes. *J. Cell Biol.*, **176**, 295–305.

Duncan, F.E., Moss, S.B., Schultz, R.M. and Williams, C.J. (2005) PAR-3 defines a central subdomain of the cortical actin cap in mouse eggs. *Dev. Biol.*, **280**, 38–47.

Echeverri, C.J., Paschal, B.M., Vaughan, K.T. and Vallee, R.B. (1996) Molecular characterization of the 50-kD subunit of dynactin reveals function for the complex in chromosome alignment and spindle organization during mitosis. *J. Cell Biol.*, **132**, 617–633.

Eckley, D.M., Ainsztein, A.M., Mackay, A.M. *et al.* (1997) Chromosomal proteins and cytokinesis: patterns of cleavage furrow formation and inner centromere protein positioning in mitotic heterokaryons and mid-anaphase cells. *J. Cell Biol.*, **136**, 1169–1183.

Egana, A.L., Boyle, J.A. and Ernst, S.G. (2007) Strongylocentrotus drobachiensis oocytes maintain a microtubule organizing center throughout oogenesis: implications for the establishment of egg polarity in sea urchins. *Mol. Reprod. Dev.*, **74**, 76–87.

Endow, S.A. and Komma, D.J. (1997) Spindle dynamics during meiosis in Drosophila oocytes. *J. Cell Biol.*, **137**, 1321–1336.

Evangelista, M., Zigmond, S. and Boone, C. (2003) Formins: signaling effectors for assembly and polarization of actin filaments. *J. Cell Sci.*, **116**, 2603–2611.

Faix, J. and Grosse, R. (2006) Staying in shape with formins. *Dev. Cell*, **10**, 693–706.

Gammie, A.E., Kurihara, L.J., Vallee, R.B. and Rose, M.D. (1995) DNM1, a dynamin-related gene, participates in endosomal trafficking in yeast. *J. Cell Biol.*, **130**, 553–566.

Gard, D.L. (1992) Microtubule organization during maturation of Xenopus oocytes: Assembly and rotation of the meiotic spindles. *Dev. Biol.*, **151**, 516–530.

Gard, D.L., Cha, B.J. and Roeder, A.D. (1995) F-actin is required for spindle anchoring and rotation in Xenopus oocytes: a re-examination of the effects of cytochalasin B on oocyte maturation. *Zygote*, **3**, 17–26.

Gill, S.R., Schroer, T.A., Szilak, I. *et al.* (1991) Dynactin, a conserved, ubiquitously expressed component of an activator of vesicle motility mediated by cytoplasmic dynein. *J. Cell Biol.*, **115**, 1639–1650.

Glotzer, M. (1997) The mechanism and control of cytokinesis. *Curr. Opin. Cell Biol.*, **9**, 815–823.

Goldstein, B. and Macara, I.G. (2008) The PAR proteins: fundamental players in animal cell polarization. *Dev. Cell*, **13**, 609–622.

Gönczy, P. (2002) Mechanisms of spindle positioning: focus on flies and worms. *Trends Cell Biol.*, **12**, 332–339.

Gotta, M., Dong, Y., Peterson, Y.K. *et al.* (2003) Asymmetrically distributed C. elegans homologs of AGS3/PINS control spindle position in the early embryo. *Curr. Biol.*, **13**, 1029–1037.

Gruss, O.J. and Vernos, I. (2004) The mechanism of spindle assembly: functions of Ran and its target TPX2. *J. Cell Biol.*, **166**, 949–955.

Guichet, A., Peri, F. and Roth, S. (2001) Stable anterior anchoring of the oocyte nucleus is required to establish dorsoventral polarity of the Drosophila egg. *Dev. Biol.*, **237**, 93–106.

Halet, G. and Carroll, J. (2007) Rac activity is polarized and regulates meiotic spindle stability and anchoring in mammalian oocytes. *Dev. Cell*, **12**, 309–317.

Hamaguchi, Y., Numata, T. and Satoh, S.K. (2007) Quantitative analysis of cortical actin filaments during polar body formation in starfish oocytes. *Cell Struct. Funct.*, **32**, 29–40.

Heil-Chapdelaine, R.A. and Otto, J.J. (1996) Characterization of changes in F-actin during maturation of starfish oocytes. *Dev. Biol.*, **177**, 204–216.

Holzbaur, E.L., Hammarback, J.A., Paschal, B.M. *et al.* (1991) Homology of a 150K cytoplasmic dynein-associated polypeptide with the Drosophila gene Glued. *Nature*, **351**, 579–583.

Huchon, D., Crozet, N., Cantenot, N. and Ozon, R. (1981) Germinal vesicle breakdown in the Xenopus laevis oocyte: description of a transient microtubular structure. *Reprod. Nutr. Dev.*, **21**, 135–148.
Iden, S. and Collard, J.G. (2008) Crosstalk between small GTPases and polarity proteins in cell polarization. *Nat. Rev. Mol. Cell Biol.*, **9**, 846–859.
Jantsch-Plunger, V., Gönczy, P., Romano, A. *et al.* (2000) CYK-4: A Rho family gtpase activating protein (GAP) required for central spindle formation and cytokinesis. *J. Cell Biol.*, **149**, 1391–1404.
Januschke, J., Gervais, L., Dass, S. *et al.* (2002) Polar transport in the Drosophila oocyte requires Dynein and Kinesin I cooperation. *Curr. Biol.*, **12**, 1971–1981.
Januschke, J., Gervais, L., Gillet, L. *et al.* (2006) The centrosome-nucleus complex and microtubule organization in the Drosophila oocyte. *Development*, **133**, 129–139.
Joberty, G., Petersen, C., Gao, L. and Macara, I.G. (2000) The cell-polarity protein Par6 links Par3 and atypical protein kinase C to Cdc42. *Nat. Cell Biol.*, **2**, 531–539.
Kalab, P., Pralle, A., Isacoff, E.Y. *et al.* (2006) Analysis of a RanGTP-regulated gradient in mitotic somatic cells. *Nature*, **440**, 697–701.
Kalab, P., Weis, K. and Heald, R. (2002) Visualization of a Ran-GTP gradient in interphase and mitotic Xenopus egg extracts. *Science*, **295**, 2452–2456.
Kemphues, K.J., Priess, J.R., Morton, D.G. and Cheng, N.S. (1988) Identification of genes required for cytoplasmic localization in early C. elegans embryos. *Cell*, **52**, 311–320.
Kim, N.H., Day, B.N., Lee, H.T. and Chung, K.S. (1996) Microfilament assembly and cortical granule distribution during maturation, parthenogenetic activation and fertilisation in the porcine oocyte. *Zygote*, **4**, 145–149.
Kovar, D.R. and Pollard, T.D. (2004) Progressing actin: formin as a processive elongation machine. *Nat. Cell Biol.*, **6**, 1158–1159.
Leader, B. and Leder, P. (2000) Formin-2, a novel formin homology protein of the cappuccino subfamily, is highly expressed in the developing and adult central nervous system. *Mech. Dev.*, **93**, 221–231.
Leader, B., Lim, H., Carabatsos, M.J. *et al.* (2002) Formin-2, polyploidy, hypofertility and positioning of the meiotic spindle in mouse oocytes. *Nat. Cell Biol.*, **4**, 921–928.
Lei, Y. and Warrior, R. (2000) The Drosophila Lisencephaly1 (DLis1) is required for nuclear migration. *Dev. Biol.*, **226**, 57–72.
Lenart, P., Bacher, C.P., Daigle, N. *et al.* (2005) A contractile nuclear actin network drives chromosome congression in oocytes. *Nature*, **436**, 812–818.
Li, H., Guo, F., Rubinstein, B. and Li, R. (2008) Actin-driven chromosomal motility leads to symmetry breaking in mammalian meiotic oocytes. *Nat. Cell Biol.*, **10**, 1301–1308.
Lin, D., Edwards, A.S., Fawcett, J.P. *et al.* (2000) A mammalian PAR-3-PAR-6 complex implicated in Cdc42/Rac1 and aPKC signalling and cell polarity. *Nat. Cell Biol.*, **2**, 540–547.
Liu, R., Linardopoulou, E.V., Osborn, G.E. and Parkhurst, S.M. (2008) Formins in development: Orchestrating body plan origami. *Biochim. Biophys. Acta.* doi: 10.1016/j.bbamcr.2008.09.016
Liu, Z., Xie, T. and Steward, R. (1999) Lis1, the Drosophila homolog of a human lissencephaly disease gene, is required for germline cell division and oocyte differentiation. *Development*, **126**(20), 4477–4488.
Longo, F.J. and Chen, D.-Y. (1985) Development of cortical polarity in mouse eggs: Involvement of the meiotic apparatus. *Dev. Biol.*, **107**, 382–394.
Lutz, D.A., Hamaguchi, Y. and Inoué, S. (1988) Micromanipulation studies of the asymmetric positioning of the maturation spindle in Chaetopterus sp. oocytes. I. Anchorage of the spindle to the cortex and migration of a displaced spindle. *Cell. Mot. Cyto.*, **11**, 83–96.
Ma, C., Benink, H.A., Cheng, D. *et al.* (2006) Cdc42 activation couples spindle positioning to first polar body formation in oocyte maturation. *Curr. Biol.*, **16**, 214–220.
Manseau, L., Calley, J. and Phan, H. (1996) Profilin is required for posterior patterning of the Drosophila oocyte. *Development*, **122**, 2109–2116.

Manseau, L.J. and Schüpbach, T. (1989) cappuccino and spire: two unique maternal-effect loci required for both the anteroposterior and dorsoventral patterns of the Drosophila embryo. *Genes Dev.*, **3**, 1437–1452.

Maro, B., Johnson, M.H., Pickering, S.J. and Flach, G. (1984) Changes in the actin distribution during fertilisation of the mouse egg. *J. Embryol. Exp. Morph.*, **81**, 211–237.

Maro, B., Johnson, M.H., Webb, M. and Flach, G. (1986) Mechanism of polar body formation in the mouse oocyte: an interaction between the chromosomes, the cytoskeleton and the plasma membrane. *J. Embryol. Exp. Morph.*, **92**, 11–32.

Miyazaki, A., Kamitsubo, E. and Nemoto, S.I. (2000) Premeiotic aster as a device to anchor the germinal vesicle to the cell surface of the presumptive animal pole in starfish oocytes. *Dev. Biol.*, **218**, 161–171.

Miyazaki, A., Kato, K.H. and Nemoto, S. (2005) Role of microtubules and centrosomes in the eccentric relocation of the germinal vesicle upon meiosis reinitiation in sea-cucumber oocytes. *Dev. Biol.*, **280**, 237–247.

Mullins, R.D., Heuser, J.A. and Pollard, T.D. (1998) The interaction of Arp2/3 complex with actin: nucleation, high affinity pointed end capping, and formation of branching networks of filaments. *PNAS*, **95**, 6181–6186.

Murata-Hori, M. and Wang, Y.-L. (2002) Both midzone and astral microtubules are involved in the delivery of cytokinesis signals: insights from the mobility of aurora B. *J. Cell Biol.*, **159**, 45–53.

Na, J. and Zernicka-Goetz, M. (2006) Asymmetric positioning and organization of the meiotic spindle of mouse oocytes requires CDC42 function. *Curr. Biol.*, **16**, 1249–1254.

Nakaya, M., Fukui, A., Izumi, Y. *et al.* (2000) Meiotic maturation induces animal-vegetal asymmetric distribution of aPKC and ASIP/PAR-3 in Xenopus oocytes. *Development*, **127**, 5021–5031.

Narasimhulu, S.B. and Reddy, A.S. (1998) Characterization of microtubule binding domains in the Arabidopsis kinesin-like calmodulin binding protein. *Plant Cell*, **10**, 957–965.

Nguyen-Ngoc, T., Afshar, K. and Gönczy, P. (2007) Coupling of cortical dynein and G alpha proteins mediates spindle positioning in Caenorhabditis elegans. *Nat. Cell Biol.*, **9**, 1294–1302.

Nicosia, S.V., Wolf, D.P. and Inoue, M. (1977) Cortical granule distribution and cell surface characteristics in mouse eggs. *Dev. Biol.*, **57**, 56–74.

Palazzo, R.E., Vaisberg, E., Cole, R.W. and Rieder, C.L. (1992) Centriole duplication in lysates of Spisula solidissima oocytes. *Science*, **256**, 219–221.

Prodon, F., Chenevert, J. and Sardet, C. (2006) Establishment of animal-vegetal polarity during maturation in ascidian oocytes. *Dev. Biol.*, **290**, 297–311.

Quinlan, M.E., Heuser, J.E., Kerkhoff, E. and Mullins, R.D. (2005) Drosophila Spire is an actin nucleation factor. *Nature*, **433**, 382–388.

Rappaport, R. (1961) Experiments concerning the cleavage stimulus in sand dollar eggs. *J. Exp. Zool.*, **148**, 81–89.

Renault, L., Bugyi, B. and Carlier, M.F. (2008) Spire and Cordon-bleu: multifunctional regulators of actin dynamics. *Trends Cell Biol.*, **18**, 494–504.

Rieder, C.L., Khodjakov, A., Paliulis, L.V. *et al.* (1997) Mitosis in vertebrate somatic cells with two spindles: implications for the metaphase/anaphase transition checkpoint and cleavage. *PNAS*, **94**, 5107–5112.

Romero, S., Le Clainche, C., Didry, D. *et al.* (2004) Formin is a processive motor that requires profilin to accelerate actin assembly and associated ATP hydrolysis. *Cell*, **119**, 419–429.

Rosales-Nieves, A.E., Johndrow, J.E., Keller, L.C. *et al.* (2006) Coordination of microtubule and microfilament dynamics by Drosophila Rho1, Spire and Cappuccino. *Nat. Cell Biol.*, **8**, 367–376.

Sathananthan, A.H. (1997) Ultrastructure of the human egg. *Hum. Cell*, **10**, 21–38.

Schuh, M. and Ellenberg, J. (2008) A new model for asymmetric spindle positioning in mouse oocytes. *Curr. Biol.*, **18**(24), 1986–1992.

Simerly, C., Nowak, G., De Lanerolle, P. and Schatten, G. (1998) Differential expression and functions of cortical myosin IIA and IIB isotypes during meiotic maturation, fertilization and mitosis in mouse oocytes and embryos. *Mol. Biol. Cell*, **9**, 2509–2525.

Starr, D.A. and Han, M. (2003) ANChors away: an actin based mechanism of nuclear positioning. *J. Cell Sci.*, **116**, 211–216.
Straight, A.F. and Field, C.M. (2000) Microtubules, membranes and cytokinesis. *Curr. Biol.*, **10**, R760–R770.
Sun, Q.Y. and Schatten, H. (2006) Regulation of dynamic events by microfilaments during oocyte maturation and fertilization. *Reproduction*, **131**, 193–205.
Swan, A., Nguyen, T. and Suter, B. (1999) Drosophila Lissencephaly-1 functions with Bic-D and dynein in oocyte determination and nuclear positioning. *Nat. Cell Biol.*, **1**, 444–449.
Szöllösi, D., Calarco, P. and Donahue, R.P. (1972) Absence of centrioles in the first and second meiotic spindles of mouse oocytes. *J. Cell Sci.*, **11**, 521–541.
Terada, Y., Simerly, C. and Schatten, G. (2000) Microfilament stabilization by jasplakinolide arrests oocyte maturation, cortical granule exocytosis, sperm incorporation cone resorption, and cell-cycle progression, but not DNA replication, during fertilization in mice. *Mol. Reprod. Dev.*, **56**, 89–98.
Theurkauf, W.E. (1994) Premature microtubule-dependent cytoplasmic streaming in cappuccino and spire mutant oocytes. *Science*, **265**, 2093–2096.
Theurkauf, W.E., Alberts, B.M., Jan, Y.N. and Jongens, T.A. (1993) A central role for microtubules in the differentiation of Drosophila oocytes. *Development*, **118**, 1169–1180.
Theurkauf, W.E. and Hawley, R.S. (1992) Meiotic spindle assembly in Drosophila females: behavior of nonexchange chromosomes and the effects of mutations in the nod kinesin-like protein. *J. Cell Biol.*, **116**, 1167–1180.
Van der Voet, M., Berends, C.W., Perreault, A. *et al.* (2009) NuMA-related LIN-5, ASPM-1, calmodulin and dynein promote meiotic spindle rotation independently of cortical LIN-5/GPR/Galpha. *Nat. Cell Biol.*, **11**, 269–277.
Vaughan, K.T. and Vallee, R.B. (1995) Cytoplasmic dynein binds dynactin through a direct interaction between the intermediate chains and p150Glued. *J. Cell Biol.*, **131**, 1507–1516.
Verlhac, M.-H. and Dumont, J. (2008) Interactions between chromosomes, microfilaments and microtubules revealed by the study of small GTPases in a big cell, the vertebrate oocyte. *Mol. Cell. Endo.*, **282**, 12–17.
Verlhac, M.-H., Lefebvre, C., Guillaud, P. *et al.* (2000) Asymmetric division in mouse oocytes: with or without Mos. *Curr. Biol.*, **10**, 1303–1306.
Vinot, S., Le, T., Maro, B. and Louvet-Vallée, S. (2004) Two PAR6 proteins become asymmetrically localized during establishment of polarity in mouse oocytes. *Curr. Biol.*, **14**, 520–525.
Wang, L., Wang, Z.B., Zhang, X. *et al.* (2008) Brefeldin A disrupts asymmetric spindle positioning in mouse oocytes. *Dev. Biol.*, **313**, 155–166.
Wang, S., Hu, J., Guo, X. *et al.* (2009) ADP-ribosylation factor 1 regulates asymmetric cell division in female meiosis in the mouse. *Biol. Reprod.*, **80**, 555–562.
Waterman-Storer, C.M. and Holzbaur, E.L. (1996) The product of the Drosophila gene, Glued, is the functional homologue of the p150Glued component of the vertebrate dynactin complex. *J. Biol. Chem.*, **271**, 1153–1159.
Watts, J.L., Etemad-Moghadam, B., Guo, S. *et al.* (1996) par-6, a gene involved in the establishment of asymmetry in early C. elegans embryos, mediates the asymmetric localization of PAR-3. *Development*, **122**, 3133–3140.
Weber, K.L., Sokac, A.M., Berg, J.S. *et al.* (2004) A microtubule-binding myosin required for nuclear anchoring and spindle assembly. *Nature*, **431**, 325–329.
Wellington, A., Emmons, S., James, B. *et al.* (1999) Spire contains actin binding domains and is related to ascidian posterior end mark-5. *Development*, **126**, 5267–5274.
Wheatley, S.P. and Wang, Y.-L. (1996) Midzone microtubule bundles are continuously required for cytokinesis in cultured epithelial cells. *J. Cell Biol.*, **135**, 981–989.
Wilson, N.F. and Snell, W.J. (1998) Microvilli and cell-cell fusion during fertilization. *Trends Cell Biol.*, **8**, 93–96.
Worman, H.J. and Gundersen, G.G. (2006) Here come the SUNs: a nucleocytoskeletal missing link. *Trends Cell Biol.*, **16**, 67–69.

Xiang, X., Osmani, A.H., Osmani, S.A. *et al.* (1995) NudF, a nuclear migration gene in Aspergillus nidulans, is similar to the human LIS-1 gene required for neuronal migration. *Mol. Biol. Cell.*, **6**, 297–310.

Yamanaka, T., Horikoshi, Y., Suzuki, A. *et al.* (2001) PAR-6 regulates aPKC activity in a novel way and mediates cell-cell contact-induced formation of the epithelial junctional complex. *Genes Cells*, **6**, 721–731.

Yang, H.Y., Mains, P.E. and Mcnally, F.J. (2005) Kinesin-1 mediates translocation of the meiotic spindle to the oocyte cortex through KCA-1, a novel cargo adapter. *J. Cell Biol.*, **169**, 447–457.

Yang, H.Y., Mcnally, K. and Mcnally, F.J. (2003) MEI-1/katanin is required for translocation of the meiosis I spindle to the oocyte cortex in C. elegans. *Dev. Biol.*, **260**, 245–259.

Yu, J., Starr, D.A., Wu, X. *et al.* (2006) The KASH domain protein MSP-300 plays an essential role in nuclear anchoring during Drosophila oogenesis. *Dev. Biol.*, **289**, 336–345.

Zhang, Q.Y., Tamura, M., Uetake, Y. *et al.* (2004) Regulation of the paternal inheritance of centrosomes in starfish zygotes. *Dev. Biol.*, **266**, 190–200.

Zhang, X., Ma, C., Miller, A.L. *et al.* (2008) Polar body emission requires a RhoA contractile ring and Cdc42-mediated membrane protrusion. *Dev. Cell*, **15**, 386–400.

Zheng, Y. (2004) G protein control of microtubule assembly. *Ann. Rev. Cell and Dev. Biol.*, **20**, 867–894.

Zhu, Z.Y., Chen, D.Y., Li, J.S. *et al.* (2003) Rotation of meiotic spindle is controlled by microfilaments in mouse oocytes. *Biol. Reprod.*, **68**, 943–946.

Zou, J., Hallen, M.A., Yankel, C.D. and Endow, S.A. (2008) A microtubule-destabilizing kinesin motor regulates spindle length and anchoring in oocytes. *J. Cell Biol.*, **180**, 459–466.

Section VI

Biological clocks regulating meiotic divisions

12

The control of the metaphase-to-anaphase transition in meiosis I

M. Emilie Terret

Molecular Biology Program, Memorial Sloan-Kettering Cancer Center, New York, NY 10021, USA

12.1 Control of the metaphase-to-anaphase transition in mitosis

12.1.1 General view of the spindle assembly checkpoint (SAC)

During S phase, a single round of DNA replication produces two identical DNA copies (the sister chromatids), which undergo segregation during mitosis, leading to two identical daughter cells. Sister chromatids are generated and held together during S phase. This cohesion is mediated by a highly conserved protein complex called cohesin, which comprises four subunits: Smc1, Smc3, Scc3 and Scc1 (Uhlmann, 2004; Huang, Milutinovich and Koshland, 2005; Nasmyth, 2005). Sister chromatids attach to spindle microtubules via structures assembled on their centromeric DNA called kinetochores (Fukagawa, 2008). Cohesion is essential to create tension between the sister chromatids in order to bi-orient them to the opposite poles of the spindle, a configuration referred to as amphitelic attachment, necessary for accurate segregation of the genome. The spindle assembly checkpoint (SAC) ensures that all sister chromatids have become amphitelically attached. Once this occurs, anaphase can begin. If not, the SAC prevents a ubiquitin ligase called the anaphase-promoting complex/cyclosome (APC/C) from ubiquitinating proteins whose degradation is required for anaphase onset and mitotic exit (Peters, 2006). Substrates ubiquitinated by the APC/C are recognized and degraded by the 26S proteasome (Castro *et al.*, 2005;

Oogenesis: The Universal Process Marie-Hélène Verlhac and Anne Villeneuve

Musacchio and Salmon, 2007). In consequence, anaphase is delayed to allow the attachment or the correction of aberrant attachments of the sister chromatids.

12.1.2 The anaphase-promoting complex/cyclosome (APC/C)

Ubiquitination reactions require three types of enzyme, named E1, E2 and E3. The APC/C has an E3 ubiquitin ligase activity: it assembles polyubiquitin chains on lysine residues of substrate proteins. It is a protein complex composed of at least a dozen subunits. It can only ubiquitinate substrates with the help of three cofactors: an E1 ubiquitin-activating enzyme, an E2 ubiquitin-conjugating enzyme and a coactivator protein. All of these coactivators are characterized by the presence of specific sequences, known as the C-box (Schwab *et al.*, 2001) and the IR-tail (Passmore *et al.*, 2003; Vodermaier *et al.*, 2003), that mediate their binding to the APC/C. They also all contain a C-terminal WD40 domain that is predicted to fold into a propeller-like structure, and that is believed to recognize APC/C substrates by interacting with specific recognition elements in these substrates (Kraft *et al.*, 2005), called D-boxes (Glotzer, Murray and Kirschner, 1991) and KEN-boxes (Pfleger and Kirschner, 2000). An A-box is also required for Aurora A destruction by APC/Cdh1 during mitotic exit (Littlepage and Ruderman, 2002).

The main coactivators of APC/C are Cdc20 and Cdh1. They associate with APC/C transiently, in a tightly regulated manner. Cdc20 can only associate efficiently with APC/C in mitosis once several subunits of APC/C have been phosphorylated by mitotic kinases such as the M phase-promoting factor (MPF) and polo-like kinase-1 (PLK1) (Shteinberg *et al.*, 1999; Kramer *et al.*, 2000; Rudner and Murray, 2000; Golan, Yudkovsky and Hershko, 2002; Kraft *et al.*, 2003). By contrast, Cdh1 is prevented from efficient interaction with the APC/C as long as Cdh1 is phosphorylated by different Cdks during the S and G2 phases and in the early stages of mitosis (Zachariae *et al.*, 1998; Jaspersen, Charles and Morgan, 1999; Blanco *et al.*, 2000; Kramer *et al.*, 2000; Yamaguchi, Okayama and Nurse, 2000). As a result, APC/C–Cdc20 is active early in mitosis. Cdh1 can only activate the APC/C once APC/C–Cdc20 has decreased MPF activity by initiating cyclin B destruction. In yeast, phosphatases, such as Cdc14, have been shown to facilitate this transition by dephosphorylating Cdh1 (Visintin *et al.*, 1998). These opposing effects of phosphorylation on APC/C–Cdc20 and APC/C–Cdh1 result in the switch from high to low MPF activity which is required for mitotic exit and subsequent DNA replication. High MPF activity in mitosis leads to the assembly of APC/C–Cdc20, which initiates cyclin proteolysis and decreases MPF activity. This drop in MPF activity promotes the formation of APC/C–Cdh1, which then maintains cyclin instability in G1 and enables a new round of DNA replication. Cdc20 is a substrate of APC/C–Cdh1 at the end of mitosis (Pfleger and Kirschner, 2000).

This switch in APC/C coactivators is one way to confer selectivity to substrate recognition by APC/C. The activity of the APC/C is also regulated via phosphorylation (inhibitory or activatory), and via inhibitors such as the SAC proteins. A recent study showed that there are 71 phosphorylation sites on 9 of the APC/C subunits in HeLa cells in response to various drug treatments arresting cells in prometaphase (Steen *et al.*, 2008). Despite the common state of arrest (prometaphase), the various drug treatments result in differences in the phosphorylation patterns of the APC/C (Steen *et al.*, 2008). Furthermore, during drug arrest, the phosphorylation state of the APC/C changes, indicating that the mitotic arrest is not a static condition (Steen *et al.*, 2008). The phosphorylation of APC/C subunits is initiated at the beginning of mitosis in prophase, before spindle assembly has started (Kraft *et al.*, 2003). APC/C phosphorylation promotes binding to Cdc20, so APC/C–Cdc20 is already assembled at this stage. Indeed, the destruction of several APC/C substrates, such as cyclin A, is initiated as soon as the nuclear envelope disintegrates at the transition from prophase to prometaphase (den Elzen and Pines, 2001; Geley *et al.*, 2001). To avoid the precocious destruction of key substrates required for the metaphase-to-anaphase transition by APC/C–Cdc20, the SAC inhibits APC/C–Cdc20 until every chromosome is bioriented. Thus, the SAC can control APC/C–Cdc20 in a substrate-specific manner; it inhibits the capability of APC/C–Cdc20 to ubiquitinate cyclin B and securin but does not protect cyclin A from degradation.

12.1.3 APC/C targets

Cyclin B

Entry into mitosis is induced by the activation of the MPF. The MPF is composed of two subunits: Cdc2 (or Cdk1), the catalytic subunit, and cyclin B, the regulatory subunit (Labbé *et al.*, 1989). At the metaphase-to-anaphase transition, cyclin B is degraded. This induces a conformational change in Cdc2 that prevents both ATP hydrolysis and access of protein substrates to the active site, resulting in the complete inactivation of Cdc2 (Jeffrey *et al.*, 1995). As a result, MPF activity decreases, triggering mitosis exit. Protein phosphatases then dephosphorylate Cdc2 substrates, an essential prerequisite for disassembly of the mitotic spindle, chromosome decondensation, reformation of a nuclear envelope and formation of a cytokinetic furrow (Murray, Solomon and Kirschner, 1989). Time-lapse microscopy of cyclin B fused to GFP (green fluorescent protein) in human cells showed that degradation starts when the last chromosome is aligned on the metaphase plate (Clute and Pines, 1999). Indeed, cyclin B degradation requires ubiquitination by APC/C–Cdc20 once the SAC is satisfied (Clute and Pines, 1999). This highly regulated mechanism to degrade cyclin B ensures that mitotic exit only happens when every chromosome is aligned and the cell is ready to divide (Figure 12.1).

Securin

Securin is another target of APC/C–Cdc20, important for anaphase onset (Zou *et al.*, 1999). To allow chromosome segregation, the cohesion that holds sister chromatids together first has to be dissolved. APC/C–Cdc20 initiates this process by ubiquitinating

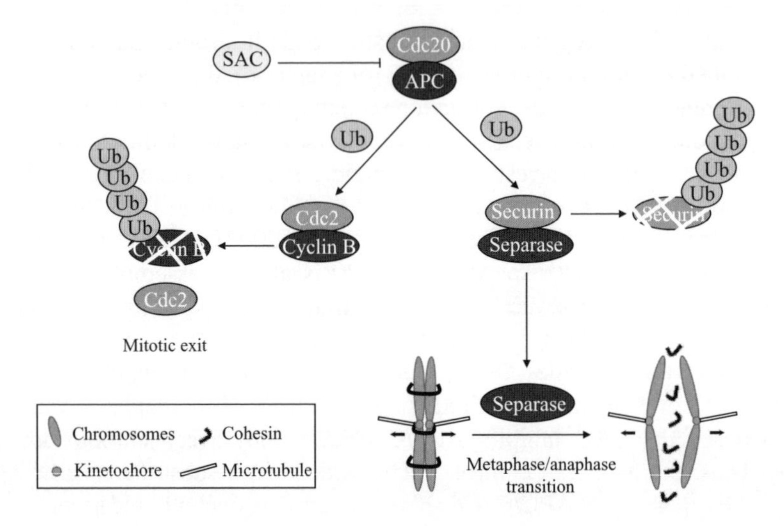

Figure 12.1 Regulation of the metaphase-to-anaphase transition in mitosis. The spindle assembly checkpoint (SAC) prevents chromosome missegregation and aneuploidy. When all the chromosomes are aligned and under tension on the metaphase plate of the spindle, APC/C is active and ubiquitinates securin and cyclin B. Separase becomes active and cleaves the cohesins, allowing sister chromatid separation between the two daughter cells. MPF activity drops, allowing mitotic exit

securin, a small protein that functions both as a chaperone and as an inhibitor of the protease, separase (Figure 12.1). Once activated, separase cleaves the Scc1 subunit of the cohesins, dissolving cohesion between sister chromatids (Nasmyth, 2001; Figure 12.1). Sister chromatids are held together by cohesins along all their length (chromosome arms and centromeres). In vertebrates, the bulk of arm cohesins is removed during prometaphase by a cleavage-independent mechanism mediated by polo-like kinase, Aurora B kinase and Wapl (Waizenegger *et al.*, 2000; Alexandru *et al.*, 2001; Sumara *et al.*, 2002; Giménez-Abián *et al.*, 2004; Gandhi, Gillespie and Hirano, 2006; Kueng *et al.*, 2006). However, complete resolution of cohesion, and hence anaphase onset, is ultimately dependant upon cleavage of Scc1 by separase, mostly at the centromere (Waizenegger *et al.*, 2000; Hauf, Waizenegger and Peters, 2001).

Securin has homologues in *Schizosaccharomyces pombe* (Pds1), *Saccharomyces cerevisiae* (Cut2), *Drosophila* (PIM) and vertebrates (PTTG), but they have almost no sequence homology, even though their roles are highly conserved. Although securin destruction is essential for the activation of separase, and securin is an essential gene in fission yeast and *Drosophila* (Funabiki, Kumada and Yanagida, 1996; Stratmann and Lehner, 1996; Jäger *et al.*, 2001), budding yeast, cultured human cells and mice can live without securin (Yamamoto, Guacci and Koshland, 1996; Jallepalli *et al.*, 2001; Mei, Huang and Zhang, 2001; Wang, Yu and Melmed, 2001). The implication is that there must be securin-independent mechanisms that control separase activity. One such mechanism might be the phosphorylation of separase, and its subsequent stoichiometric association with MPF, which is sufficient for separase inhibition in *Xenopus* egg extracts (Stemmann *et al.*, 2001; Gorr, Boos and Stemmann, 2005).

12.1.4 Components of the SAC

Most of the genes that contribute to the SAC signal transduction pathway were first described in *S. cerevisiae* mutant screens. Their mutation bypassed the ability of wild-type *S. cerevisiae* cells to arrest in mitosis in the presence of spindle poisons. These genes are the *MAD* (mitotic-arrest deficient) genes *MAD1*, *MAD2* and *MAD3* (*BUBR1* in humans), and the *BUB* (budding uninhibited by benzimidazole) gene, *BUB1* (Li and Murray, 1991; Hoyt, Totis and Roberts, 1991). Another *BUB* gene, *BUB3*, was identified later. These genes are conserved in all eukaryotes, and they are collectively involved in the SAC. They are enriched on unattached kinetochores where checkpoint signalling is initiated.

Besides the *MAD* and *BUB* genes, other SAC components include monopolar spindle-1 (MPS1; Weiss and Winey, 1996) and Aurora B (Kallio *et al.*, 2002). These proteins are required respectively to amplify the SAC signal and activate the SAC by creating unattached kinetochores (Musacchio and Salmon, 2007). Additional proteins that regulate SAC activity in higher eukaryotes include:

- Constituents of the Rough Deal (Rod), Zeste White 10 (Zw10) and Zwilch (RZZ) complex (Karess, 2005).
- p31comet (Habu *et al.*, 2002; Xia *et al.*, 2004; Mapelli *et al.*, 2006).
- Several protein kinases including mitogen-activated protein kinase (MAPK), MPF, NEK2, PLK1 (Musacchio and Salmon, 2007), TAO1 (Draviam *et al.*, 2007) and Prp4 (Montembault *et al.*, 2007).
- Microtubule motors like centromere protein E (CENP-E) (Abrieu *et al.*, 2000; Mao *et al.*, 2005; Mao, Desai and Cleveland, 2005) and dynein.
- Dynein-associated proteins such as dynactin, cytoplasmic linker protein 170 (CLIP170) and lissencephaly-1 (LIS1) (Howell *et al.*, 2001; Wojcik *et al.*, 2001; Tai *et al.*, 2002).

12.1.5 Defects detected

During prometaphase, Cdc20 and all SAC proteins concentrate at unattached kinetochores (Cleveland, Mao and Sullivan, 2003). This localization is highly dynamic, as most of SAC proteins are removed from the kinetochore after its correct attachment with spindle microtubules. This indicates that kinetochores generate the signal for the SAC to inhibit APC/C–Cdc20. Indeed the SAC detects defects of the spindle, via modifications of the kinetochores. The kinetochores have to be functional to be able to signal to the SAC (Rieder *et al.*, 1995; Tavormina and Burke, 1998). The SAC monitors the interaction between kinetochores and spindle microtubules and can detect two defects: attachment and tension. This characteristic is conserved between species. In *Drosophila*, for example, Bub1 mutants can enter anaphase before the attachment of all their kinetochores (Waters *et al.*, 1998), and Rod and Zw10 mutants can enter anaphase after attachment of all their kinetochores but before chromosome alignment (Basto, Gomes and Karess, 2000). Distinguishing the relative contributions of tension and attachment when manipulating spindles is difficult, as interfering with the creation of tension probably affects attachment.

The attachment branch of the SAC pathway is obvious, as the SAC inhibits the metaphase-to-anaphase transition until all chromosomes are bi-oriented on the spindle. For example, human cells treated with poisons depolymerizing microtubules (removing attachment) are able to induce the SAC. Furthermore, all SAC proteins are localized at unattached kinetochores in mitosis.

The tension branch of the SAC pathway is less obvious but also very important. Micromanipulation experiments in praying mantid spermatocytes (Li *et al.*, 1997) showed that the SAC is sensitive to tension at kinetochores. Human cells treated with poisons able to stabilize microtubules (removing tension at kinetochores, as measured by interkinetochore distance) are able to induce the SAC, showing that tension is also sensed by the SAC. Tension could be a way to discriminate against incorrect attachments, as a mono-oriented chromosome (syntelic attachment, or two kinetochores attached to the same pole) is under less tension than a bi-oriented chromosome. In this case, microtubule–kinetochore attachment is destabilized to correct the improper attachment. This destabilization depends on Aurora B (Tanaka *et al.*, 2002; Lampson *et al.*, 2004; Pinsky and Biggins, 2005). Aurora B is also critical for correcting merotelic attachments, which are not sensed by the SAC. Merotelic attachment occurs when one sister kinetochore attaches to microtubules from opposite poles. Bi-orientation of chromosomes with merotelic kinetochores produces sufficient occupancy and tension to turn off SAC activity. As a result, merotelic kinetochores, if left uncorrected, can produce lagging chromatids and potential chromosome missegregation in anaphase (Cimini and Degrassi, 2005).

In conclusion, once all the chromosomes are bi-oriented and under tension on the spindle (the only condition that ensures accurate segregation at anaphase), the SAC is turned off, APC/C–Cdc20 is active, and can induce the destruction of securin and cyclin B, triggering anaphase and mitotic exit (Figure 12.1). The SAC hence coordinates chromosome capture, MPF inactivation, and the segregation of sister chromatids, via its fine spatial and temporal control of APC/C-mediated degradation of cyclin B and securin. These events need to be spatially and temporally controlled to ensure a proper partitioning of the genetic material between the two daughter cells (Figure 12.1).

12.2 Control of the metaphase-to-anaphase transition in meiosis I

During sexual reproduction, two gametes fuse and combine their genomes to form the next generation. To avoid the doubling in genetic material with every new generation, genome copy number must be reduced by half before the next round of gametes is formed. This reduction in copy number to produce haploid gametes is achieved during meiosis, where two rounds of chromosome segregation occur without an intervening S phase. The main difference between meiosis and mitosis is the number of chromosome

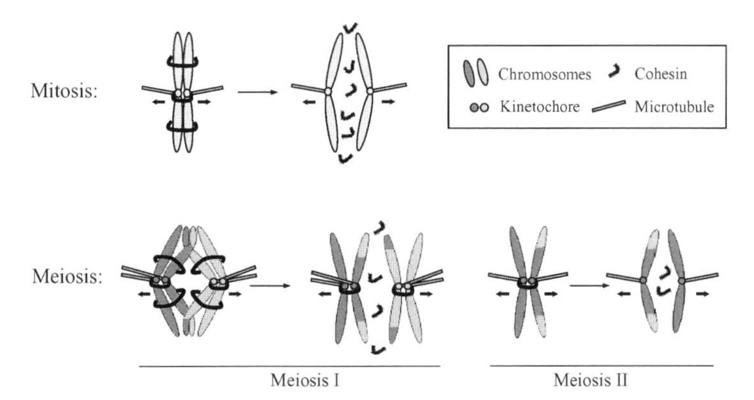

Figure 12.2 Metaphase-to-anaphase transition in mitosis and meiosis. In mitosis, sister chromatids are separated between the two daughter cells. During meiosis I, homologous chromosomes are separated between the oocyte and the first polar body. During meiosis II, sister chromatids are separated between the oocyte and the second polar body like in mitosis

separation steps that follow chromosome duplication. In mitosis, one round of chromosome segregation allows the separation of the sister chromatids (Figure 12.2). In meiosis, two rounds of chromosome segregation occur without an intervening S phase (Figures 12.2 and 12.3). The first meiotic division (meiosis I) is reductional: homologous chromosomes (from mom and dad) are separated, and as a result ploidy

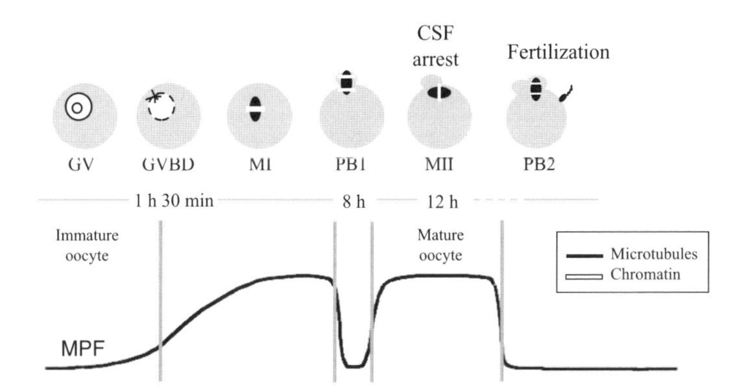

Figure 12.3 Meiotic maturation in mouse oocyte. Mouse oocytes are arrested in prophase of meiosis I (MI) in the ovaries, called the germinal vesicle (GV) stage. Hormonal signals induce meiosis resumption, marked by the germinal vesicle breakdown (GVBD). DNA forms condensed chromosomes that congress on the meiotic spindle to the metaphase plate during metaphase of meiosis I. The spindle migrates toward the cortex, the homologous chromosomes are separated during anaphase, and the first polar body is extruded (PB1). The oocyte enters into meiosis II without DNA replication. The spindle assembles under the cortex with sister chromatids aligned on the metaphase plate. The oocyte is arrested in metaphase of meiosis II (MII) by cytostatic factor (CSF) activity, until fertilization, which induces sister chromatid separation and extrusion of the second polar body (PB2). The curve represents M phase-promoting factor (MPF) activity. Timings vary between different mouse strains

is divided by two (Figure 12.2). The second meiotic division (meiosis II) is equational and very similar to mitosis: the sister chromatids are separated (Figure 12.2).

Chromosome segregation in human meiosis is surprisingly poor, resulting in aneuploid embryos (Hassold and Hunt, 2001). Aneuploid embryos usually arrest development at an early stage (Balton *et al.*, 1989), but some develop further, implant into the uterus, and undergo spontaneous abortion (Cowchock, Gibas and Jackson, 1993; Robinson, Mcfadden and Stephenson, 2001; Rubio, Simon and Vida, 2003), while others develop to term and carry genetic disorders (Tseng *et al.*, 1994; Bruyere, Rupps and Kuchinka, 2000), the most frequent being trisomies 13, 18 or 21 (Koehler *et al.*, 1996; Hunt and Hassold, 2002). Aneuploidy occurs when chromosomes or chromatids separate unequally during cell division, and is referred to as nondisjunction. As described in the previous section of this chapter, to prevent aneuploidy, mitotic cells have developed a high-fidelity surveillance system to monitor and coordinate the segregation machinery: the SAC (Nicklas, 1997; Amon, 1999; Hoyt, 2001). The SAC delays the onset of anaphase until all chromatids are correctly oriented in a bipolar position at the metaphase plate of the bipolar spindle (Amon, 1999; Nicklas, 1997; Hoyt, 2001). The existence of such a surveillance system in meiosis was debated, due to the high rate of aneuploidy in human oocytes, mostly due to errors during meiosis I, and due to the fact that meiotic segregation in wild-type female X/O mice is neither blocked, delayed nor disrupted despite the presence of an unaligned univalent chromosome at metaphase I (LeMaire-Adkins, Radke and Hunt, 1997). In this chapter, I highlight recent discoveries concerning the regulation of the separation of the homologous chromosomes during meiosis I, the most error-prone division of meiosis. The understanding of the regulatory pathways involved in coordinating meiosis I in mammalian oocytes is essential to get a better understanding of the origins of human aneuploidy.

12.2.1 Separating homologous chromosomes in meiosis I

Meiosis I is unique because homologous chromosome pairs, as opposed to sister chromatids, must be separated from each other (Figure 12.2). In mitosis, DNA replication leads to duplicated sister chromatids that are connected by sister-chromatid cohesion, mediated by the cohesins. Cohesins resist the pulling forces when microtubule fibres from opposite spindle poles attach to the kinetochores of the two sister chromatids. As a result of this resistance, sister chromatids come under mechanical tension on the spindle, which is required for their proper alignment on the metaphase plate. Once all sister chromatid pairs are aligned, cohesins are destroyed and chromatids are pulled to opposite sides, into the future daughter cells. Meiotic cells also use the establishment of tension as a mechanism to align and separate chromosomes. However, the need to separate homologous chromosomes in addition to sisters adds a number of differences.

First, similar to sister chromatids, pairs of homologous chromosomes must also be connected to allow establishment of tension between them. Linkage of homologous chromosomes occurs after meiotic DNA replication and involves two steps. First, homologous chromosomes are paired on the basis of sequence similarity. Then, in a process called crossover recombination, physical connections (chiasmata) are established by exchanging DNA strands between homologous chromosomes. Thus, to successfully separate homologous chromosomes during meiosis I, each pair of homologous chromosomes has to crossover at least once. These chiasmata link the two homologous chromosomes during prophase and metaphase of meiosis I (for a review about chromosome pairing, synapsis and recombination, see Pawlowski and Cande, 2005).

Second, unlike in mitosis, sister chromatids must move to the same spindle pole during meiosis I (Figure 12.2). A pair of homologous chromosomes consists of two pairs of sister chromatids, each of which has the potential to bind microtubules through its kinetochore. During meiosis I, to establish tension, the kinetochores of one sister pair need to bind to microtubules from one spindle pole, whereas the kinetochores of the other sister pair need to attach to microtubules from the opposite pole (Petronczki, Siomos and Nasmyth, 2003; Marston and Amon, 2004; Figure 12.2). Sister kinetochores are not normally arranged in a configuration that supports such monopolar attachment. Studies in *S. cerevisiae* have identified a group of proteins, named 'monopolins', that are required for monopolar attachment in meiosis I (Toth *et al.*, 2000; Watanabe, 2004). But the mammalian homologues remain unknown.

Third, sister chromatids have to remain linked until meiosis II. This can be done because of the stepwise loss of sister-chromatid cohesion between meiosis I and II (Klein *et al.*, 1999; Watanabe and Nurse, 1999; Lee *et al.*, 2003). During mitosis, cohesion is destroyed in one step to allow separation of sister chromatids (Figure 12.2). If cohesion were completely lost during meiosis I, sister chromatids would separate prematurely because no new cohesin complexes are loaded between meiosis I and meiosis II. Only the sister-chromatid cohesion distal to the site of crossing over is responsible for connecting homologous chromosomes, whereas cohesion close to centromeres still links sister chromatids (Carpenter, 1994; Orr-Weaver, 1996; Pawlowski and Cande, 2005; Figure 12.2). As a result, sister-chromatid cohesion must be lost in a stepwise manner, first along chromosome arms to separate homologous chromosomes during meiosis I, then at centromeres to separate sister chromatids during meiosis II (Figure 12.2). In the centromeric regions, the cohesins are protected from degradation by the Sgo1 and 2 proteins, also known as shugoshins (Tang *et al.*, 1998; Katis *et al.*, 2004; Kitajima *et al.*, 2003; Rabitsch *et al.*, 2004). Sgo1 is itself degraded prior to the second meiotic division (Salic, Waters and Mitchison, 2004). Thereby, centromeric cohesins are rendered accessible to degradation in meiosis II. *In vivo*, the loss of Sgo2 promotes a premature release of the meiosis-specific Rec8 cohesin complexes from anaphase I centromeres, inducing the complete loss of centromere cohesion at metaphase II. This leads to aneuploid gametes that give rise to infertility of the Sgo2-deficient mice (Lee *et al.*, 2008; Llano *et al.*, 2008).

Meiotic cohesins are different from their mitotic counterparts. The Scc1 subunit is replaced by the meiosis-specific Rec8 protein (Buonomo *et al.*, 2000), and the Smc1 subunit is replaced by the meiosis-specific SMC1β protein (Revenkova *et al.*, 2001). The fact that the cohesin complex holding homologous chromosomes together is established during foetal S phase and has to remain functional decades later can make it susceptible to age-related damage because it may be difficult or impossible to repair. Indeed, mutation of the meiosis-specific cohesin SMC1β makes female mice sterile due to large levels of chromosome defects (Revenkova *et al.*, 2004). Interestingly, these mice display an age-related incidence in chromosome defects (Hodges *et al.*, 2005).

12.2.2 Requirement of the APC/C in meiosis I

Among species

The loss of cohesion at the metaphase-to-anaphase I transition depends on the APC/C and separase activation in most organisms tested, including yeasts, worm and mouse (Buonomo *et al.*, 2000; Kitajima *et al.*, 2003; Salah and Nasmyth, 2000; Siomos *et al.*, 2001; Terret *et al.*, 2003; Wassmann, Niault and Maro, 2003a).

In yeast, Cdc20 is required for the degradation of securin and the inactivation of separase in both meiotic divisions (Salah and Nasmyth, 2000). In *S. pombe* and *S. cerevisiae*, noncleavable forms of Rec8 and separase invalidation cause a meiotic metaphase I arrest, as expected if cohesion release is needed to progress through meiosis I (Buonomo *et al.*, 2000; Kitajima *et al.*, 2003).

In *Caenorhabditis elegans*, mutations in, or RNA interference against several subunits of the APC/C cause a meiotic metaphase I arrest, as would be expected if separase cleavage of cohesin was needed for release of arm cohesion and separation of homologous chromosomes (Furuta *et al.*, 2000; Golden *et al.*, 2000; Davis *et al.*, 2002).

Female-sterile mutations in fzy (the homologue of Cdc20) cause both meiosis I and meiosis II arrests in *Drosophila* eggs, indicating again that the APC/C is required for the transition (Swan and Schupbach, 2007).

Several studies in mouse oocytes have demonstrated a requirement for APC/C-mediated degradation of securin and cyclin B, and activation of separase for Rec8 removal from chromosome arms and homologous disjunction in meiosis I (Herbert *et al.*, 2003; Terret *et al.*, 2003; Kudo *et al.*, 2006). I will discuss these findings in detail in the next section.

The only organism where APC/C and separase seemed to be dispensable for chromosome segregation in meiosis I was *Xenopus* (Peter *et al.*, 2001; Taieb *et al.*, 2001). Indeed, microinjection of *Xenopus* oocytes with antibodies against Cdc20 or Cdc27 (an APC/C subunit) or antisense oligonucleotides against Cdc20 does not disrupt progression through meiosis I, but only causes an arrest in meiosis II (Peter *et al.*, 2001; Taieb *et al.*, 2001). It was unclear whether these findings represented a fundamental biological difference between organisms or, rather, whether they were due to different experimental designs. Recent evidence points to the latter. First, *Xenopus* separase gets transiently activated after metaphase I (Fan, Sun and Zou, 2006). Additionally, ectopic expression of Emi2/Erp1, an APC/C inhibitor that is active later in meiosis, induces a metaphase I arrest in *Xenopus* oocytes, again suggesting that APC/C is required for the transition from metaphase I-to-anaphase I (Ohe *et al.*, 2007; Tung *et al.*, 2007). Finally, it has been shown very recently that cyclin B and securin degradation are required for the extrusion of the first polar body and the separation of the homologous chromosomes in meiosis I in *Xenopus* oocytes (Zhang *et al.*, 2008). Indeed oocytes injected with a truncated form of cyclin B1 lacking the D-box required for APC/C targeted degradation (Gross *et al.*, 2000) or with a securin D-box mutant

(Zou *et al.*, 1999) fail to extrude the first polar body and to separate the homologous chromosomes in meiosis I (Zhang *et al.*, 2008).

In mammalian oocytes

In this section I will discuss further the role of the APC/C in mammalian oocytes, as degradation of securin and cyclin B are required for anaphase I and exit from meiosis I.

The degradation of securin and cyclin B: a requirement for meiosis I exit The decline in MPF activity at the exit of meiosis I in mouse oocytes is the consequence of cyclin B degradation as shown first by pulse-chase experiments (Hampl and Eppig, 1995; Winston, 1997), and later confirmed using real-time analysis of a cyclin B GFP (Ledan *et al.*, 2001; Herbert *et al.*, 2003; Tsurumi *et al.*, 2004; Homer *et al.*, 2005b, 2005c). One unique feature of meiosis I is its extraordinary length. The length of mitosis is on average thirty minutes (Rieder *et al.*, 1994; Meraldi, Draviam and Sorger, 2004), compared with the several hours from germinal vesicle breakdown (GVBD) until the first polar body extrusion (PB1) in mammalian oocytes (Figure 12.3). The extended period of meiosis I is reflected in the dynamics of MPF activity, which rises abruptly in mitosis but gradually in meiosis, reaching a peak several hours after GVBD (Choi *et al.*, 1991; Gavin, Cavadore and Schorderet-Slatkine, 1994; Verlhac *et al.*, 1994, 1996; Figure 12.3). Loss of MPF activity is an essential requirement for completion of meiosis I, via degradation of cyclin B1. The destruction of securin is also important since it activates separase. In mouse, loss of both cyclin B1 and securin occur synchronously in a period which terminates with PB1 extrusion and is dependant on APC/C–Cdc20 (Reis *et al.*, 2007), as it is in mitosis. However, early in prometaphase I, APC/C–Cdh1 is active (made possible by low MPF activity at this time) and degrades Cdc20. This means Cdc20 has to be resynthesized in order for oocytes to complete meiosis I. Premature loss of Cdh1 brings forward the period of APC/C–Cdc20 activity and consequently the period of cyclin B1 and securin degradation (Reis *et al.*, 2007). This premature metaphase I induces high rates of nondisjunction and leads to a disruption of the integrity of the metaphase II spindle. Disruptions of the spindle have also been observed in human oocytes from older women (Battaglia, Klein and Soules, 1996). Loss of MPF activity through degradation of cyclin B1 may not be the sole mechanism for regulating MPF in oocytes. Free separase, generated by proteolysis of securin, can also bind Cdc2 and thus inhibit its kinase activity (Stemmann *et al.*, 2001; Stemmann, Gorr and Boos, 2006). The inhibition is mutual since Cdc2 binding inhibits separase proteolytic activity too (Gorr, Boos and Stemmann, 2005). The ability of separase to bind Cdc2 (independently of its protease activity) also appears essential for PB1, since PB1 extrusion is blocked when separase binding is inhibited (Gorr *et al.*, 2006). In addition, the catalytic activity of separase is not required for PB1 extrusion, as catalytically inactive separase is able to rescue PB1 extrusion in oocytes lacking separase (Kudo *et al.*, 2006).

Degradation of cyclin B and securin depends on their APC/C-specific D-box The requirement for D-box mediated destruction in meiosis I mouse oocytes has been examined using time-lapse imaging of two mutant constructs fused to GFP. One mutant is a truncated form of cyclin B lacking the D-box-containing N-terminal 90 amino acids (Δ90 cyclin B1) (Glotzer, Murray and Kirschner, 1991), while the other is a securin D-box mutant in which the D-box is mutated (RXXL to AXXA). Both of these mutants are resistant to degradation; expression of either one inhibited homologous chromosome disjunction and polar body extrusion (Herbert *et al.*, 2003), thus strongly implicating the APC/C in progression through meiosis I.

Polyubiquitination and degradation by the 26S proteasome are required for meiosis I progression Preventing the formation of a multiubiquitin chain with methylated ubiquitin induces a meiosis I arrest in rat oocytes (Dekel, 2005), indicating that progression through meiosis I is dependent upon polyubiquitination. Furthermore, inhibition of the proteasome by the proteasome inhibitor MG132 in rat and mouse arrests the oocytes in meiosis I with intact metaphase I spindles and high MPF activity, due to inhibition of cyclin B destruction (Josefsberg *et al.*, 2000; Terret *et al.*, 2003), strongly implicating again the APC/C in progression through meiosis I.

APC/C-mediated degradation and separase activity are required for meiosis I exit During mitosis, overexpression of securin induces a mitotic delay at metaphase (Hagting *et al.*, 2002) due to saturation of the APC/C. Like mitosis, overexpression of securin induces a metaphase I arrest in mouse oocytes (Terret *et al.*, 2003). Saturation of a putative destruction machinery is further corroborated by the fact that overexpression of exogenous securin also inhibits the destruction of endogenous cyclin B, resulting in high MPF activity (Terret *et al.*, 2003). Consistent with a requirement for securin destruction (and hence APC/C activity), mouse oocytes require separase activity for proper homologous chromosome disjunction. Indeed, an inhibitor of separase induces defects in homologous chromosome disjunction including failure of homologues to segregate and a 'cut' phenotype in which chromosomes become trapped between the oocyte and the first polar body (Terret *et al.*, 2003). More recently, it has been shown that oocytes lacking separase can't separate their homologous chromosomes and can't extrude the first polar body (Kudo *et al.*, 2006).

In summary, the degradation of securin and cyclin B are required for the metaphase I-to-anaphase I transition and exit from meiosis I in mammalian oocytes. Moreover, their degradation depends on their APC/C-specific D-box and is inhibited by overexpression of APC/C substrates, implying sensitivity to APC/C saturation. Finally, disrupting the polyubiquitination process or inhibiting the 26S proteasome impairs meiosis I progression. Taken together, these data strongly implicate the APC/C as an important effector of homologous chromosome disjunction and meiosis I exit in mammalian oocytes (Figure 12.4).

APC/C regulation in meiosis

The finding that cohesion between sister chromatids is dissolved in two steps in meiosis, both mediated by securin proteolysis and separase activation, implies that APC/C–Cdc20 gets activated twice during meiosis. However, APC/C needs to remain inactive during

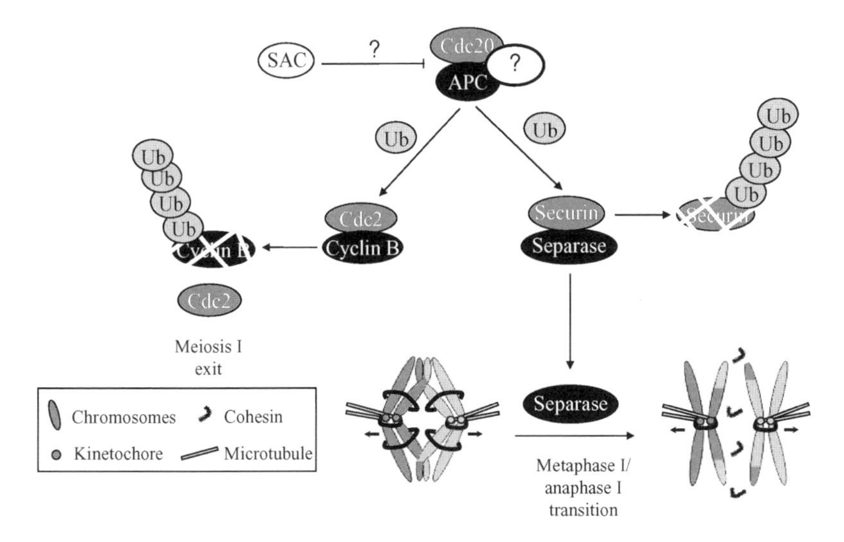

Figure 12.4 Regulation of the metaphase-to-anaphase transition in meiosis I. Like in mitosis, securin and cyclin B degradation are required for the metaphase-to-anaphase transition, to allow separase activation and meiosis II progression. However, there is no direct proof that the spindle assembly checkpoint (SAC) controls the anaphase-promoting complex/cyclosome (APC/C) in meiosis I. A specific APC/C activator remains elusive in mammalian oocytes

the long prophase arrest to prevent premature loss of cohesion, and between meiotic divisions to allow securin reaccumulation (Figure 12.3). Thus, both activation and inactivation of APC/C are key processes in meiosis.

APC/C activators During meiosis in some organisms, there are special forms of the APC/C that are activated by meiosis-specific activators. In addition to Cdc20 and Cdh1, yeast and *Drosophila* use meiosis-specific APC/C activators: Ama1 in *S. cerevisiae* (Cooper *et al.*, 2000), Mfr1/Fzr1 in *S. pombe* (Asakawa, Kitamura and Shimoda, 2001; Blanco, Pelloquin and Moreno, 2001), and Cort in *Drosophila* (Page and Orr-Weaver, 1996; Chu *et al.*, 2001). An interesting hypothesis that remains to be tested is that these meiotic activators could target the degradation of a unique, meiosis-specific set of APC/C substrates. Regulation of the APC/C during meiosis in these organisms utilizes both mitotic APC/C regulators in addition to meiosis-specific regulators, demonstrating a need for increased control of APC/C activity during the more complex meiotic cell cycle. But a few questions remain: do meiosis-specific APC/C activators target a unique set of substrates, and how is this specificity achieved? Are there meiosis-specific APC/C activators in vertebrates (Figure 12.4)?

APC/C inhibitors Just as Mes1 is required in *S. pombe* to inhibit APC/C activity between meiosis I and meiosis II, Emi2/Erp1 (Izawa *et al.*, 2005), a homologue of Emi1, is required for this role in *Xenopus* and mouse oocytes in meiosis II (for a review see Wu and Kornbluth, 2008). Emi2/Erp1 is not expressed until after GVBD, and its expression coincides with polyadenylation of Emi2/Erp1 mRNA (Ohe *et al.*, 2007;

Tung *et al.*, 2007). Inhibition of Emi2/Erp1 by injection of morpholinos or antisense oligonucleotides reduces cyclin B2 reaccumulation after meiosis I, prevents entry into meiosis II, and, in the case of morpholino injection, induces DNA replication (Ohe *et al.*, 2007; Tung *et al.*, 2007). Injection of Emi2/Erp1 morpholinos into mouse oocytes generates a very similar phenotype, suggesting that Emi2/Erp1 inhibits APC/C-mediated degradation of cyclin B after meiosis I to prevent DNA replication and to allow entry into meiosis II (Madgwick *et al.*, 2006).

12.2.3 Requirement of the SAC in meiosis I

Among species

In mitosis, the SAC inhibits APC/C–Cdc20 in the presence of improper kinetochore microtubule attachments. A limiting step in the study of SAC proteins during meiotic progression of multicellular organisms is the identification of viable alleles, since these proteins are essential for proper development. Therefore, only a few studies have addressed whether the SAC mechanism is functional during meiotic progression *in vivo*. But all the data suggest that SAC proteins appear to be required not only upon spindle damage by microtubule poisons but also for the normal mechanism of meiosis I. It remains to be determined whether this role of the SAC is mediated by inhibition of the APC/C (Figure 12.4).

In *S. cerevisiae*, mutations in *MAD1* or *MAD2* cause increased nondisjunction of homologous chromosomes in meiosis I (Shonn, McCarroll and Murray, 2000; Cheslock *et al.*, 2005). Levels of nondisjunction are restored if anaphase is artificially delayed, suggesting that Mad1 and Mad2 are important for inducing a metaphase I delay in a normal meiosis. Loss of recombination in a spo11 mutant, which causes a lack of tension on kinetochores, induces a Mad2-dependent suppression of APC/C activity, suggesting that the checkpoint also responds to spindle defects in meiosis I. The requirement for spindle checkpoint function in a normal meiosis I division may be the consequence of the increased complexity of bi-orienting homologous chromosome pairs on the spindle compared with bi-orienting sister-chromatid kinetochores in mitosis.

Control of chromosome segregation in meiosis I in *Drosophila* females is an interesting problem, because in *Drosophila* the secondary meiotic arrest occurs in metaphase I, not metaphase II, as is the case in the CSF (cytostatic factor) arrest of vertebrates (Figure 12.3). Oocytes mutant for Mps1 enter anaphase I prematurely, suggesting a role for the spindle checkpoint in mediating this arrest (Gilliland *et al.*, 2007). Reduction of Mps1 function in these oocytes causes nondisjunction of chromosomes, which is likely due, in part, to defects in bi-orientation of homologous chromosomes in meiosis I (Gilliland, Wayson and Hawley, 2005; Gilliland *et al.*, 2007). Additionally, in female meiosis of BubR1 mutants in *Drosophila*, nondisjunction of sister chromatids is elevated (Malmanche *et al.*, 2007).

There is little evidence to date as to whether the SAC functions in meiosis by inhibition of the APC/C. In *Drosophila* BubR1 mutant oocytes, cohesin is lost from chromosomes in prophase I of meiosis, but it has not been determined whether this effect occurs through loss of an inhibitory effect on APC/C activity. In *S. cerevisiae*, Mad3 (BubR1) mediates a prophase I delay that becomes essential for chromosome segregation when chromosomes do not recombine (Cheslock *et al.*, 2005). It is also unclear whether this defect is due to an uninhibited APC/C. The sole example of a link between the SAC in meiosis and the APC/C comes from studies in *C. elegans*, in which mutations in spindle checkpoint genes suppress a metaphase I arrest caused by leaky alleles of the Cdc23 gene, which encodes the APC8 subunit (Stein *et al.*, 2007).

In mammalian oocytes

SAC and spindle poisons SAC competence in mitosis is characterized by mitotic arrest and stabilization of securin and cyclin B upon spindle disruption. In response to microtubule poisons, mitotic cells arrest in a prometaphase-like stage with strong accumulation of the Mad and Bub proteins at the kinetochores. Two types of microtubule–kinetochore defects are observed following pharmacological treatment. In the first type, usually induced by high doses of nocodazole, the spindle is depolymerized, thereby depriving all kinetochores of both attachment and tension. The second type of defect is classically associated with taxol (paclitaxel), but may be produced by low doses of spindle-depolymerizing agents, and is associated with an intact spindle which lacks tension. The mitotic arrests induced by both types of spindle defects are robust and endure indefinitely in many mammalian cells in culture (Skoufias *et al.*, 2001).

Spindle depolymerization Depolymerization of the spindle in mouse oocytes using colcemid (demecolcine) reduces rates of polar body extrusion compared to untreated oocytes (Hashimoto and Kishimoto, 1988). This is accompanied by stabilization of MPF activity (Hashimoto and Kishimoto, 1988). Another indirect assay for MPF activity, based on the protein synthesis inhibitor puromycin, also suggests that four hours of spindle depolymerization stabilizes MPF activity (Brunet *et al.*, 2003). The absence of a spindle has been demonstrated to stabilize Cdc2 activity measured directly by histone H1 kinase assays (Wassmann, Niault and Maro, 2003a; Homer *et al.*, 2005c). Together these experiments demonstrate that spindle depolymerization induces a meiosis I arrest in which MPF activity is stabilized.

In mouse oocytes, Cdc2 inactivation at the meiosis I-to-meiosis II transition is due to cyclin B degradation (Hampl and Eppig, 1995; Winston, 1997; Figure 12.3). Thus, stabilization of Cdc2 activity upon spindle depolymerization suggests that this is due to inhibition of cyclin B destruction. Formal proof that cyclin B is stabilized by spindle depolymerization was provided by immunoblotting oocytes for cyclin B after a three hour incubation in nocodazole (Lefebvre *et al.*, 2002). In addition, Cdc2 and cyclin B stabilization upon spindle depolymerization were demonstrated together, thereby confirming that Cdc2 stabilization in the absence of a spindle is due to inhibition of cyclin B degradation (Homer *et al.*, 2005c).

In addition to cyclin B, the other principal downstream target of the SAC is securin. The obvious question is whether spindle depolymerization also stabilizes securin. Spindle depolymerization for 15 hours prevents homologous chromosome disjunction as assessed by chromosome spreads (Soewarto, Schmiady and Eichenlaub-Ritter, 1995). DNA staining also suggests that a four hour duration of spindle depolymerization inhibits homologous chromosome disjunction (Brunet *et al.*, 2003). Given that securin destruction and separase activity are required for homologous chromosome disjunction (Herbert *et al.*, 2003; Terret *et al.*, 2003), the extrapolation of these data is that spindle depolymerization inhibits securin destruction. Indeed, time-lapse fluorescence imaging shows that securin is stabilized by spindle depolymerization and, like cyclin B, can be stabilized for prolonged periods (Homer *et al.*, 2005c). It was also shown that prolonged periods of securin stabilization are accompanied by inhibition of homologous chromosome disjunction, as assessed by chromosome spreads (Homer *et al.*, 2005c).

But is the meiosis I arrest and protein stabilization following spindle depolymerization mediated by the SAC? This was addressed by examining the role of the SAC protein, Mad2, under conditions in which the spindle was depolymerized (Homer *et al.*, 2005c). Somatic cells depleted of the majority of Mad2 using RNA interference are unable to sustain a mitotic arrest upon spindle depolymerization, and prematurely degrade securin and cyclin B (Kops, Foltz and Cleveland, 2004; Michel *et al.*, 2004). In oocytes in which the majority of Mad2 was depleted using morpholino antisense, securin and cyclin B are unstable following spindle depolymerization; whereas mock-depleted oocytes, like wild-type oocytes, sustain high levels of securin and cyclin B for several hours (Homer *et al.*, 2005c). Although polar body extrusion is completely inhibited in control oocytes treated with nocodazole, a fraction of Mad2-depleted oocytes extrude polar bodies which, due to the absence of a spindle, are devoid of DNA (Homer *et al.*, 2005c). Moreover, protein destabilization in Mad2-depleted oocytes is the result of unrestrained APC/C activity, as Δ90 cyclin B is stable following Mad2 knockdown (Homer *et al.*, 2005c). Another recent study shows that in the oocytes from *MAD2* heterozygote mice, oocytes missegregate chromosomes at a high rate after a nocodazole-induced arrest (Niault *et al.*, 2007). From these data, we can conclude that in response to spindle depolymerization, mouse oocytes react by a sustained SAC-mediated response which arrests meiosis I by inhibiting the destruction of securin and cyclin B, likely by inhibiting APC/C activity (Figure 12.4). This is reminiscent of the response of mammalian somatic cells to spindle depolymerization and indicates that the molecular players involved in the SAC are likely conserved between mitosis and meiosis I.

Reduced tension Nanomolar concentrations of nocodazole, which leave an intact spindle in mouse oocytes, inhibit polar body extrusion and stabilize Cdc2 activity during a three hour period of drug exposure (Wassmann, Niault and Maro, 2003a). In addition, a four hour exposure to paclitaxel, which stabilizes microtubules without depolymerizing the spindle, also inhibits homologous chromosome disjunction and polar body extrusion; Cdc2 activity is stabilized, as assayed indirectly using puromycin (Brunet *et al.*, 2003). The role of Mad2 under these conditions has been addressed using a mutant form of Mad2 harbouring serine-to-aspartic acid substitutions at positions 170, 178 and 195 (3S-D Mad2) (Wassmann, Niault and Maro, 2003a). 3S-D Mad2 acts as a dominant negative in human somatic cells by impairing the ability of endogenous Mad2 to form ternary complexes with APC/C–Cdc20 (Wassmann, Liberal and Benezra, 2003b). In mouse oocytes cultured in low doses of nocodazole, it was found that expression of 3S-D Mad2 enables oocytes to transiently arrest in meiosis I: Cdc2 activity is destabilized, the homologous chromosomes are separated and polar body extrusion happens (Wassmann, Niault and Maro, 2003a). Assuming that 3S-D Mad2 also exhibits dominant negative properties in mouse oocytes, this indicates that Mad2 is required for the meiosis I delay in response to low doses of nocodazole. During mitosis, treatment with taxol or low doses of spindle depolymerizing agents is associated with the persistence of an intact spindle in which tension is reduced as determined using interkinetochore distance measurements (Waters *et al.*, 1998; Skoufias *et al.*, 2001; Pinsky and Biggins, 2005). Experiments in mouse oocytes have not formally measured interkinetochore distances. However, given the similar types of pharmacological treatments and the persistence of an intact spindle in both mammalian somatic cells (Waters *et al.*, 1998; Skoufias *et al.*, 2001; Pinsky and Biggins, 2005) and oocytes (Brunet *et al.*, 2003; Wassmann, Niault and Maro, 2003a), the assumption is that these drug treatments also induce a tension defect in female meiosis I. Therefore these data could be interpreted as evidence that tension defects activate a Mad2-dependent SAC response which protects securin and cyclin B from degradation. This would be consistent with the requirement for Mad2 in sensing tension defects in other meiosis I systems including maize (Yu, Muszynski and Dawe, 1999) and budding yeast (Shonn, McCarroll and Murray, 2000). Cultivating oocytes in low doses of nocodazole does not block meiosis I indefinitely: there are substantial rates of PBE, and Cdc2 inactivation and homologous chromosome disjunction still occur (Wassmann, Niault and Maro, 2003a; Shen *et al.*, 2005). Therefore, unlike the response to spindle depolymerization (Homer *et al.*, 2005c), drug treatment which leaves an intact spindle induces only a transient Mad2-mediated meiosis I arrest.

Experiments involving spindle poisons constitute an *in vitro* technique for demonstrating the existence or not of a SAC, and for helping to dissect the molecular details of the SAC. However, it does not define the link between SAC activity and chromosome segregation fidelity in unperturbed cells. For instance, although the SAC is required for cell survival after spindle disruption in yeast mitosis, under normal growth conditions chromosomes can be segregated reasonably accurately in the absence of *MAD2* (Li and Murray, 1991). In contrast to yeast, *MAD2* is essential in both drug-treated as well as unperturbed mammalian somatic cells (Gorbsky, Chen and Murray, 1998; Dobles *et al.*, 2000; Michel *et al.*, 2001).

SAC and unperturbed meiosis I

Role of Mad2 The most documented SAC protein so far in meiosis I is Mad2. Mad2 is endogenously expressed in unperturbed rat and mouse oocytes during meiosis I (Zhang *et al.*, 2004; Homer *et al.*, 2005b), at a similar level to the estimated Mad2 concentration of mammalian somatic cells (Fang, 2002; Homer *et al.*, 2005b). Unlike mitosis in which Mad2 levels remain stable (Fang, 2002), Mad2 levels increase during progression through meiosis I in mouse oocytes, so that relative to levels at the GV (germinal vesicle) stage, Mad2 increases about 2-fold by mid-meiosis I and about 10-fold by metaphase II (Homer *et al.*, 2005b). Thus, Mad2 is present in mammalian oocytes at levels that are consistent with SAC activity. Immunolocalization studies in mammalian mitosis indicate that Mad2 localizes primarily to unattached kinetochores (Li and Benezra, 1996; Waters *et al.*, 1998; Skoufias *et al.*, 2001), consistent with the notion that mis-attached kinetochores are the source of a Mad2-based signal important for SAC activation (Li and Benezra, 1996; Waters *et al.*, 1998; Howell *et al.*, 2000; Skoufias *et al.*, 2001; De Antoni *et al.*, 2005). Similarly, Mad2 immunostaining in mouse and rat oocytes reveals kinetochore localization during early prometaphase I which gradually declines as meiosis I progresses, becoming undetectable at metaphase I (Wassmann, Niault and Maro, 2003a; Zhang *et al.*, 2004). This suggests that Mad2 dissociates from kinetochores during meiosis I as they accumulate microtubules. Moreover, spindle depolymerization at metaphase I induced Mad2 to rebind kinetochores in mouse and rat oocytes (Wassmann, Niault and Maro, 2003a; Zhang *et al.*, 2004), providing further evidence that Mad2 responds to kinetochore attachment status in mammalian oocytes. Thus, Mad2 is expressed in mammalian oocytes and dynamically localizes to unattached kinetochores.

In somatic cells with wild-type levels of Mad2, improperly attached kinetochores act as a platform for amplifying Mad2-based Cdc20 inhibition (Howell *et al.*, 2000; De Antoni *et al.*, 2005; Nasmyth, 2005). However, following Mad2 overexpression, Cdc20 is sequestered, rendering unattached kinetochores incompetent for signal amplification and APC/C inhibition (De Antoni *et al.*, 2005). Thus, in the presence of high levels of Mad2, although chromosome congression is unaffected, mitosis does not progress beyond metaphase even though kinetochores are occupied fully by microtubules (Howell *et al.*, 2000; De Antoni *et al.*, 2005). This effect of exogenous Mad2 is dose dependent, as a 10-fold excess of Mad2 over endogenous levels arrests mitotic progression whereas a 2-fold excess has no effect on mitosis (Howell *et al.*, 2000). Like mitosis, the response of mouse oocytes to Mad2 overexpression is graded, as 4-fold excess has no effect on meiosis I, 15-fold excess induces a partial arrest, and 35-fold excess arrests meiosis I completely (Homer *et al.*, 2005b). Oocytes arrested by Mad2 overexpression have their chromosomes aligned on the bipolar spindle, high MPF activity and intact homologous chromosomes, consistent with being in metaphase I (Wassmann, Niault and Maro, 2003a; Homer *et al.*, 2005b, 2005d). Furthermore, oocytes not arrested by moderate Mad2 overexpression complete meiosis I with normal kinetics, implying that excess Mad2 does not induce a delay; meiosis I is either arrested or proceeds at a normal rate when Mad2 is overexpressed (Homer *et al.*, 2005b). Overall, in response to Mad2 overexpression, mouse oocytes exhibit a dose-dependent arrest at metaphase I reminiscent of mitosis. This suggests that Mad2 overexpression constitutively activates the SAC in mammalian oocytes. Moreover, this provides another indirect proof in support of a role for the APC/C in mammalian oocytes.

During mitosis, the role of Mad2 has been comprehensively defined using a number of different approaches, like function-blocking agents (antibodies and dominant negative-acting mutants) and reverse genetic approaches which deplete Mad2 (Gorbsky, Chen and Murray, 1998; Howell *et al.*, 2000; Canman, Salmon and Fang, 2002; Wassmann, Liberal and Benezra, 2003b; Kops, Foltz and Cleveland, 2004; Michel *et al.*, 2004; De Antoni *et al.*, 2005). Based on these studies, it has been shown that one of the essential functions of Mad2 is to give sufficient time for chromosomes to become properly aligned prior to anaphase onset (Gorbsky, Chen and Murray, 1998; Kops, Foltz and Cleveland, 2004; Meraldi, Draviam and Sorger, 2004; Michel *et al.*, 2004). In the absence of Mad2, mitosis is accelerated, inducing defects in chromosome alignment and subsequent aneuploidy. In contrast to mammalian somatic cells, it appears that the intrinsic mitotic timing machinery in mitotic yeast cells provides sufficient time for chromosome alignment to be completed, as Mad2 is not essential in normal mitosis (Dobles *et al.*, 2000). Initial attempts to define Mad2 function in unperturbed mouse oocytes used dominant negative mutants (Wassmann, Niault and Maro, 2003a; Tsurumi *et al.*, 2004). Although 3S-D Mad2 shows that Mad2 is required for the meiosis I arrest in response to nanomolar concentrations of nocodazole, no noticeable effect is observed in the absence of drug treatment (Wassmann, Niault and Maro, 2003a). Another study used a Mad2 mutant lacking its C-terminal 10 amino acids (Mad2ΔC) (Tsurumi *et al.*, 2004). Mad2ΔC produces a dominant negative effect in mitotic cells due to its inability to adopt the closed conformation required for sequestering Cdc20 while maintaining its ability to be recruited to kinetochores, thereby competing with endogenous Mad2 (Luo *et al.*, 2000; De Antoni *et al.*, 2005). In contrast to 3S-D Mad2, Mad2ΔC shortens the duration of meiosis I by about two hours (Tsurumi *et al.*, 2004). It was therefore unclear whether or not Mad2 contributed to the timing of meiosis I in mammalian oocytes. Experiments using morpholinos helped to resolve this discrepancy in dominant negative data (Homer *et al.*, 2005b, 2005c). In Mad2-depleted oocytes an increase in meiosis I nondisjunction is observed in association with a decrease in meiosis I duration and precocious degradation of cyclin B and securin (Homer *et al.*, 2005b). These data show that mouse Mad2 is required for preventing premature APC/C activation and for accurate chromosome segregation in meiosis I. These results are therefore consistent with the effects of Mad2ΔC, and together define a role for Mad2 in determining the timing of meiosis I in mammalian oocytes. Analysis of chromosome spreads shows that homologous chromosome disjunction fidelity is impaired, as aneuploidy rates increases dramatically following Mad2 depletion (Homer *et al.*, 2005b). From this we can conclude that Mad2 is indispensable for accurate homologue disjunction. More recent studies analysed oocyte maturation in *MAD2* heterozygote mice (Niault *et al.*, 2007). In mouse oocytes from females heterozygous for *MAD2*, meiosis I is shortened and anaphase I onset accelerated (Niault *et al.*, 2007). These oocytes display a large increase in aneuploidy in metaphase II, which is likely the result of the premature anaphase I onset. This shows that SAC control is impaired in *MAD2*$^{+/-}$ oocytes, leading to the generation of aneuploidies in meiosis I. In conclusion for these studies, meiosis I in mouse oocytes must be of a sufficient length to allow enough time for homologous chromosomes to orient properly on the spindle and to form stable connections with microtubules. Control of the timing of meiosis I is therefore crucial. The accelerated progression through meiosis I with Mad2 morpholino and Mad2ΔC suggests that the APC/C is prematurely activated due to deficient Mad2 inhibition of Cdc20. Consistently with a precocious APC/C activation, securin and cyclin B destruction occur two hours earlier in Mad2-depleted oocytes (Homer *et al.*, 2005b). Further evidence that premature APC/C activation is due to inefficient Cdc20 inhibition is derived from experiments involving a phosphorylation-resistant Cdc20 mutant (Tsurumi *et al.*, 2004). Cdc20 phosphorylation at residues 50, 64, 68 and 79 increases its affinity for Mad2 upon activation of the SAC (Chung and Chen, 2003). Consequently, a phosphorylation-resistant Cdc20 mutant (Cdc20-4AV) binds less avidly to Mad2, resulting in unrestrained APC/C activity and reduced SAC competence (Chung and Chen, 2003). In unperturbed mouse oocytes, Cdc20-4AV shortens the duration of meiosis I to a similar degree as Mad2 morpholino and Mad2ΔC (Tsurumi *et al.*, 2004; Homer *et al.*, 2005b). Thus, disrupting Mad2 function produces similar effects to a Mad2-resistant Cdc20, implying that when disrupting Mad2 function, the observed effects were due to defective Cdc20 inhibition.

Role of other SAC proteins Mad1 is present in mouse oocytes (Zhang *et al.*, 2005) and is observed around the nuclei at the GV stage (prophase), on kinetochores in prometaphase I, and moves to spindle poles at metaphase I and early anaphase I (Zhang *et al.*, 2005). In mouse oocytes after nocodazole treatment, Mad1 is partly relocated to the kinetochores (Zhang *et al.*, 2005). However, Mad1 localization was not altered in mouse oocytes after tension had been changed with taxol (Zhang *et al.*, 2005). These results indicate that Mad1 senses attachment of chromosomes to microtubules, but not the tension between microtubules and chromosomes. When anti-Mad1 antibodies are injected into mouse oocytes, it does not affect oocyte nuclear maturation and spindle formation, but induces chromosome misalignment (Zhang *et al.*, 2005). However, whether the oocytes with misaligned chromosomes have chromosome abnormalities, such as gain or loss of chromosomes after completion of meiosis, needs further investigation by chromosome analysis. Taken together, these data imply a role of Mad1 in the SAC during meiosis I.

Bub1 is observed on kinetochores of mouse oocytes from GVBD to early anaphase I and disappears only at late anaphase I (Brunet *et al.*, 2003). A dominant negative mutant of Bub1 (Bub1dn amino acids 1–331) accelerates progression through meiosis I (Tsurumi *et al.*, 2004). This mutant lacks the kinase domain, and disrupts the SAC in somatic cells by competing with the endogenous kinase for kinetochore localization. Furthermore, microinjection of Bub1 antibodies into mouse oocytes causes chromosome misalignment on the meiosis I spindle that is not corrected by delaying anaphase onset (Yin *et al.*, 2006). Hence Bub1 seems to have a role in the SAC during meiosis I.

A dominant negative mutant of BubR1 accelerates progression through meiosis I (Tsurumi *et al.*, 2004). BubR1d (BubR1 351–700) can bind Bub3 and Cdc20 but cannot inhibit APC/C *in vitro*, and overexpressed BubR1d acts as a dominant negative in human cells (Tang *et al.*, 2001). Furthermore, BubR1 mutant female mice contain oocytes with highly abnormal metaphase II configuration (Baker *et al.*, 2004). Therefore, BubR1 might have a role in the SAC in meiosis I.

CENP-E is a transient kinetochore component that binds to kinetochores soon after the breakdown of the nuclear envelope and remains fully bound throughout chromosome congression to the metaphase plate. In late anaphase or telophase, it is relocated to midzone microtubules of the spindles (Duesbery *et al.*, 1997; Lee *et al.*, 2000). CENP-E is localized on kinetochores during prometaphase and metaphase of meiosis I. Injection of an anti-CENP-E antibody into mouse oocytes at prophase completely prevented the oocytes from progressing to anaphase I, as all oocytes were blocked in metaphase I (Duesbery *et al.*, 1997), suggesting a role for CENP-E in the SAC during meiosis I.

From all these data, it is evident that the basic scheme of mitotic SAC signalling is conserved during meiosis I in mammalian oocytes (Figure 12.4). Thus in both systems, securin and cyclin B are important downstream targets of SAC proteins, likely via the APC/C as intermediary (Figure 12.4).

12.2.4 Concluding remarks

It is a well-established phenomenon that in humans the incidence of oocytes with an aberrant chromosome number increases with maternal age. A weakening of the spindle checkpoint is implicated in the age-dependent aneuploidies. Indeed, there is a reduced expression of *MAD2* and *BUBR1* genes in human ovaries of older women compared to younger women (Steuerwald *et al.*, 2001; Steuerwald, Steuerwald and Mailhes, 2005; Shonn, Murray and Murray, 2003; Homer *et al.*, 2005a). As a reduction of Mad2 levels in mammalian mitosis results in chromosome missegregation, a reduction of Mad2 levels in mammalian meiosis could very well result in aneuploidy (Michel *et al.*, 2001). RNAi is a feasible approach for downregulating Mad2 expression in human oocytes during meiosis I, (Homer *et al.*, 2005d). RNAi will be a useful approach to test the requirement of SAC proteins in human oocytes and to possibly mimic and analyse the consequence of lower levels of SAC proteins in older women's oocytes. Declining transcripts encoding SAC proteins in an age-dependent way in human oocytes suggests that declining SAC function is a feature of the ageing process (Steuerwald *et al.*, 2001). Indeed, in mice, ovarian BubR1 levels have been shown to decrease with age, suggesting that impairment of SAC function might be a universal feature of mammalian ageing (Baker *et al.*, 2004). Furthermore, declining levels of BubR1 in mice have not only been linked to age-related rises in aneuploidy but also with subfertility, implicating the SAC in wider aspects of mammalian reproduction (Baker *et al.*, 2004). As these studies provide a more detailed mechanistic picture of meiosis, we will be able to get a better understanding of the complexities of human meiosis and hopefully identify some of the causes and possible treatments of infertility and birth defects.

Acknowledgements

I thank John Maciejowski and Jeremie Szeftel for critical review of this manuscript. My work in the laboratory of Prasad. V. Jallepalli is supported by a grant from the National Institutes of Health (CA 107 342).

References

Abrieu, A., Kahana, J.A., Wood, K.W. and Cleveland, D.W. (2000) CENP-E as an essential component of the mitotic checkpoint *in vitro*. *Cell*, **102**, 817–826.

Acquaviva, C., Herzog, F., Kraft, C. and Pines, J. (2004) The anaphase promoting complex/cyclosome is recruited to centromeres by the spindle assembly checkpoint. *Nature Cell Biol.*, **6**, 892–898.

Alexandru, G., Uhlmann, F., Mechtler, K. *et al.* (2001) Phosphorylation of the cohesin subunit Scc1 by Polo/Cdc5 kinase regulates sister chromatid separation in yeast. *Cell*, **105**(4), 459–472.

Amon, A. (1999) The spindle checkpoint. *Curr. Opin. Cell Biol.*, **9**, 69–75.

Asakawa, H., Kitamura, K. and Shimoda, C. (2001) A novel Cdc20-related WDrepeat protein, Fzr1, is required for spore formation in *S. pombe*. *Mol. Genet. Genomics*, **265**, 424–435.

Baker, D.J., Jeganathan, K.B., Cameron, J.D. *et al.* (2004) BubR1 insufficiency causes early onset of aging-associated phenotypes and infertility in mice. *Nat. Genet.*, **36**, 744–749.

Balton, V.N., Hawes, S.M., Taylor, C.T. and Parsons, J.H. (1989) Development of spare human implantation embryos in vitro: an analysis of the correlations among gross morphology, cleavage rates, and development to blastocyst. *J. In Vitro Fertil. Embryo Transfer*, **6**, 30–35.

Basto, R., Gomes, R. and Karess, R.E. (2000) Rough deal and Zw10 are required for the metaphase checkpoint in Drosophila. *Nat. Cell Biol.*, **2**(12), 939–943.

Battaglia, D.E., Goodwin, P., Klein, N.A. and Soules, M.R. (1996) Influence of maternal age on meiotic spindle assembly in oocytes from naturally cycling women. *Hum. Reprod.*, **11**(10), 2217–22.

Blanco, M.A., Sanchez-Diaz, A., de Prada, J.M. and Moreno, S. (2000) APCSte9/Srw1 promotes degradation of mitotic cyclins in G1 and is inhibited by Cdc2 phosphorylation. *EMBO J.*, **19**, 3945–3955.

Blanco, M.A., Pelloquin, L. and Moreno, S. (2001) Fission yeast mfr1 activates APC and coordinates meiotic nuclear division with sporulation. *J. Cell Sci.*, **114**, 2135–2143.

Brunet, S., Pahlavan, G., Taylor, S.S. and Maro, B. (2003) Functionality of the spindle checkpoint during the first meiotic division of mammalian oocytes. *Reproduction*, **126**, 443–450.

Bruyere, H., Rupps, R. and Kuchinka, B.D. (2000) Recurrent trisomy 21 in a couple with a child presenting trisomy 21 mosaicism and maternal uniparental disomy for chromosome 21 in the euploid cell line. *Am. J. Hum. Genet.*, **94**, 35–41.

Buonomo, S.B., Clyne, R.K., Fuchs, J. *et al.* (2000) Disjunction of homologous chromosomes in meiosis I depends on proteolytic cleavage of the meiotic cohesin Rec8 by separin. *Cell*, **103**, 387–398.

Canman, J.C., Salmon, E.D. and Fang, G. (2002) Inducing precocious anaphase in cultured mammalian cells. *Cell Motil. Cytoskeleton*, **52**, 61–65.

Carpenter, A.T. (1994) Chiasma function. *Cell*, **77**, 957–962.

Castro, A., Bernis, C., Vigneron, S. *et al.* (2005) The anaphase-promoting complex: a key factor in the regulation of cell cycle. *Oncogene*, **24**(3), 314–325.

Cheslock, P.S., Kemp, B.J., Boumil, R.M. and Dawson, D.S. (2005) The roles of MAD1, MAD2 and MAD3 in meiotic progression and the segregation of nonexchange chromosomes. *Nat. Genet.*, **37**, 756–760.

Choi, T., Aoki, F., Mori, M. *et al.* (1991) Activation of p34cdc2 protein kinase activity in meiotic and mitotic cell cycles in mouse oocytes and embryos. *Development*, **113**, 789–795.

Chu, T., Henrion, G., Haegeli, V. and Strickland, S. (2001) *cortex*, a Drosophila gene required to complete oocyte meiosis, is a member of the Cdc20/fizzy protein family. *Genesis*, **29**, 141–152.

Chung, E. and Chen, R. (2003) Phosphorylation of Cdc20 is required for its inhibition by the spindle checkpoint. *Nat. Cell Biol.*, **5**, 748–753.

Cimini, D. and Degrassi, F. (2005) Aneuploidy: a matter of bad connections. *Trends Cell Biol.*, **15**, 442–451.

Cleveland, D.W., Mao, Y. and Sullivan, K.F. (2003) Centromeres and kinetochores: from epigenetics to mitotic checkpoint signaling. *Cell*, **112**(4), 407–421.

Clute, P. and Pines, J. (1999) Temporal and spatial control of cyclin B1 destruction in metaphase. *Nature Cell Biol.*, **1**, 82–87.

Cooper, K.F., Mallory, M.J., Egeland, D.B. *et al.* (2000) Ama1p is a meiosis-specific regulator of the anaphase promoting complex/cyclosome in yeast. *Proc. Natl. Acad. Sci. USA*, **97**, 14548–14553.

Cowchock, F.S., Gibas, Z. and Jackson, L.G. (1993) Chromosome errors as a cause of spontaneous abortion: the relative importance of maternal age and obstetric history. *Fertil. Steril.*, **59**, 1011–1014.

Davis, E.S., Wille, L., Chestnut, B.A. *et al.* (2002) Multiple subunits of the Caenorhabditis elegans anaphase-promoting complex are required for chromosome segregation during meiosis I. *Genetics*, **160**, 805–813.

De Antoni, A., Pearson, C., Cimini, D. *et al.* (2005) The Mad1/Mad2 complex as a template for Mad2 activation in the spindle assembly checkpoint. *Curr. Biol.*, **15**, 214–225.

Dekel, N. (2005) Cellular, biochemical and molecular mechanisms regulating oocyte maturation. *Mol. Cell Endocrinol.*, **234**, 19–25.

Dobles, M., Liberal, V., Scott, M.L. *et al.* (2000) Chromosome missegregation and apoptosis in mice lacking the mitotic checkpoint protein Mad2. *Cell*, **101**, 635–645.

Draviam, V.M., Stegmeier, F., Nalepa, G. *et al.* (2007) A functional genomic screen identifies a role for TAO1 kinase in spindle-checkpoint signalling. *Nat. Cell Biol.*, **9**(5), 556–564.

Duesbery, N.S., Choi, T., Brown, K.D. *et al.* (1997) CENP-E is an essential kinetochore motor in maturing oocytes and is masked during mos-dependent, cell cycle arrest at metaphase II. *Proc. Natl. Acad. Sci. USA*, **94**, 9165–9170.

den Elzen, N. and Pines, J. (2001) Cyclin A is destroyed in prometaphase and can delay chromosome alignment and anaphase. *J. Cell Biol.*, **153**, 121–136.

Fan, H.Y., Sun, Q.Y. and Zou, H. (2006) Regulation of separase in meiosis: Separase is activated at the metaphase I-II transition in Xenopus Oocytes during meiosis. *Cell Cycle*, **5**, 198–204.

Fang, G. (2002) Checkpoint protein BubR1 acts synergistically with Mad2 to inhibit anaphase-promoting complex. *Mol. Biol. Cell*, **13**, 755–766.

Fukagawa, T. (2008) The kinetochore and spindle checkpoint in vertebrate cells. *Front. Biosci.*, **13**, 2705–2713.

Funabiki, H., Kumada, K. and Yanagida, M. (1996) Fission yeast Cut1 and Cut2 are essential for sister chromatid separation, concentrate along the metaphase spindle and form large complexes. *EMBO J.*, **15**(23), 6617–6628.

Furuta, T., Tuck, S., Kirchner, J. *et al.* (2000) EMB-30: an APC4 homologue required for metaphase-to-anaphase transitions during meiosis and mitosis in Caenorhabditis elegans. *Mol. Biol. Cell*, **11**, 1401–1419.

Gandhi, R., Gillespie, P.J. and Hirano, T. (2006) Human Wapl is a cohesin-binding protein that promotes sister-chromatid resolution in mitotic prophase. *Curr. Biol.*, **16**(24), 2406–2417.

Gavin, A.C., Cavadore, J.C. and Schorderet-Slatkine, S. (1994) Histone H1 kinase activity, germinal vesicle breakdown and M phase entry in mouse oocytes. *J. Cell Sci.*, **107**, 275–283.

Geley, S., Kramer, E., Gieffers, C. *et al.* (2001) Anaphase-promoting complex/cyclosome-dependent proteolysis of human cyclin A starts at the beginning of mitosis and is not subject to the spindle assembly checkpoint. *J. Cell Biol.*, **153**, 137–148.

Gilliland, W.D., Wayson, S.M. and Hawley, R.S. (2005) The meiotic defects of mutants in the Drosophila mps1 gene reveal a critical role of Mps1 in the segregation of achiasmate homologs. *Curr. Biol.*, **15**, 672–677.

Gilliland, W.D., Hughes, S.E., Cotitta, J.L. *et al.* (2007) The multiple roles of Mps1 in Drosophila female meiosis. *PLoS Genet.*, **3**, e113.

Giménez-Abián, J.F., Sumara, I., Hirota, T. *et al.* (2004) Regulation of sister chromatid cohesion between chromosome arms. *Curr. Biol.*, **14**(13), 1187–1193.

Golan, A., Yudkovsky, Y. and Hershko, A. (2002) The cyclinubiquitin ligase activity of cyclosome/APC is jointly activated by protein kinases Cdk1–cyclin B and Plk. *J. Biol. Chem.*, **277**, 15552–15557.

Golden, A., Sadler, P.L., Wallenfang, M.R. *et al.* (2000) Metaphase to anaphase (mat) transition-defective mutants in Caenorhabditis elegans. *J. Cell Biol.*, **151**, 1469–1482.

Gorbsky, G.J., Chen, R.H. and Murray, A.W. (1998) Microinjection of antibody to Mad2 protein into mammalian cells in mitosis induces premature anaphase. *J. Cell Biol.*, **141**, 1193–1205.

Gorr, I.H., Boos, D. and Stemmann, O. (2005) Mutual inhibition of separase and Cdk1 by two-step complex formation. *Mol. Cell.*, **19**, 135–141.

Gorr, I.H., Reis, A., Boos, D. *et al.* (2006) Essential CDK1-inhibitory role for separase during meiosis I in vertebrate oocytes. *Nat. Cell Biol.*, **8**, 1035–1037.

Glotzer, M., Murray, A.W. and Kirschner, M.W. (1991) Cyclin is degraded by the ubiquitin pathway. *Nature*, **349**, 132–138.

Gross, S.D., Schwab, M.S., Taieb, F.E. *et al.* (2000) The critical role of the MAP kinase pathway in meiosis II in Xenopus oocytes is mediated by p90(Rsk). *Curr. Biol.*, **10**, 430–438.

Habu, T., Kim, S.H., Weinstein, J. and Matsumoto, T. (2002) Identification of a MAD2-binding protein, CMT2, and its role in mitosis. *EMBO J.*, **21**, 6419–6428.

Hagting, A., den Elzen, N., Vodermaier, H.C. *et al.* (2002) Human securin proteolysis is controlled by the spindle checkpoint and reveals when the APC/C switches from activation by Cdc20 to Cdh1. *J. Cell Biol.*, **157**, 1125–1127.

Hampl, A. and Eppig, J.J. (1995) Analysis of the mechanism(s) of metaphase I arrest in maturing mouse oocytes. *Development*, **121**, 925–933.

Hashimoto, N. and Kishimoto, T. (1988) Regulation of meiotic metaphase by a cytoplasmic maturation promoting factor during mouse oocyte maturation. *Dev. Biol.*, **126**, 242–252.

Hassold, T. and Hunt, P. (2001) To err (meiotically) is human: the genesis of human aneuploidy. *Nat. Rev. Genet.*, **2**, 280–291.

Hauf, S., Waizenegger, I.C. and Peters, J.M. (2001) Cohesin cleavage by separase required for anaphase and cytokinesis in human cells. *Science*, **293**(5533), 1320–1323.

Herbert, M., Levasseur, M., Homer, H. *et al.* (2003) Homologue disjunction in mouse oocytes requires proteolysis of securin and cyclin B1. *Nat. Cell Biol.*, **5**, 1023–1025.

Hodges, C.A., Revenkova, E., Jessberger, R. *et al.* (2005) SMC1beta-deficient female mice provide evidence that cohesins are a missing link in age-related nondisjunction. *Nat. Genet.*, **37**, 1351–1355.

Homer, H.A., McDougall, A., Levasseur, M. and Herbert, M. (2005a) Restaging the spindle assembly checkpoint in female mammalian meiosis I. *Cell Cycle*, **4**, 650–653.

Homer, H.A., McDougall, A., Levasseur, M. *et al.* (2005b) Mad2 prevents aneuploidy and premature proteolysis of cyclin B and securin during meiosis I in mouse oocytes. *Genes Dev.*, **19**, 202–207.

Homer, H.A., McDougall, A., Levasseur, M. *et al.* (2005c) Mad2 is required for inhibiting securin and cyclin B degradation following spindle depolymerisation in meiosis I mouse oocytes. *Reproduction*, **130**, 829–843.

Homer, H.A., McDougall, A., Levasseur, M. *et al.* (2005d) RNA interference in human oocytes: towards an understanding of human aneuploidy. *Mol. Hum. Reprod.*, **11**, 397–404.

Howell, B.J., Hoffman, D.B., Fang, G. *et al.* (2000) Visualization of Mad2 dynamics at kinetochores, along spindle fibers, and at spindle poles in living cells. *J. Cell Biol.*, **150**, 1233–1249.

Howell, B.J., McEwen, B.F., Canman, J.C. *et al.* (2001) Cytoplasmic dynein/dynactin drives kinetochore protein transport to the spindle poles and has a role in mitotic spindle checkpoint inactivation. *J. Cell Biol.*, **155**, 1159–1172.

Hoyt, M.A. (2001) A new view of the spindle checkpoint. *J. Cell Biol.*, **154**, 909–911.

Hoyt, M.A., Totis, L. and Roberts, B.T. (1991) *S. cerevisiae* genes required for cell cycle arrest in response to loss of microtubule function. *Cell*, **66**, 507–517.

Huang, C.E., Milutinovich, M. and Koshland, D. (2005) Rings, bracelet or snaps: fashionable alternatives for Smc complexes. *Philos. Trans. R. Soc. Lond. B Biol. Sci.*, **360**(1455), 537–542.

Hunt, P.A. and Hassold, T.J. (2002) Sex matters in meiosis. *Science*, **296**, 2181–2183.

Izawa, D., Goto, M., Yamashita, A. *et al.* (2005) Fission yeast Mes1p ensures the onset of meiosis II by blocking degradation of cyclin Cdc13p. *Nature*, **434**(7032), 529–533.

Jäger, H., Herzig, A., Lehner, C.F. and Heidmann, S. (2001) Drosophila separase is required for sister chromatid separation and binds to PIM and THR. *Genes Dev.*, **15**(19), 2572–2584.

Jallepalli, P.V., Waizenegger, I.C., Bunz, F. *et al.* (2001) Securin is required for chromosomal stability in human cells. *Cell*, **105**, 445–457.

Jaspersen, S.L., Charles, J.F. and Morgan, D.O. (1999) Inhibitory phosphorylation of the APC regulator Hct1 is controlled by the kinase Cdc28 and the phosphatase Cdc14. *Curr. Biol.*, **9**, 227–236.

Jeffrey, P.D., Russo, A.A., Polyak, K. *et al.* (1995) Mechanism of CDK activation revealed by the structure of a cyclinA–CDK2 complex. *Nature*, **376**, 313–320.

Josefsberg, L., Galiani, D., Dantes, A. *et al.* (2000) The proteasome is involved in the first metaphase-to-anaphase transition of meiosis in rat oocytes. *Biol. Reprod.*, **62**, 1270–1277.

Kallio, M.J., McCleland, M.L., Stukenberg, P.T. and Gorbsky, G.J. (2002) Inhibition of aurora B kinase blocks chromosome segregation, overrides the spindle checkpoint, and perturbs microtubule dynamics in mitosis. *Curr. Biol.*, **12**(11), 900–905.

Karess, R. (2005) Rod–Zw10–Zwilch: a key player in the spindle checkpoint. *Trends Cell Biol.*, **15**, 386–392.

Katis, V.L., Galova, M., Rabitsch, K.P. *et al.* (2004) Maintenance of cohesin at centromeres after meiosis I in budding yeast requires a kinetochore-associated protein related to MEI-S332. *Curr. Biol.*, **14**, 560–572.

Kitajima, T.S., Miyazaki, Y., Yamamoto, M. and Watanabe, Y. (2003) Rec8 cleavage by separase is required for meiotic nuclear divisions in fission yeast. *EMBO J.*, **22**, 5643–5653.

Klein, F., Mahr, P., Galova, M. *et al.* (1999) A central role for cohesins in sister chromatid cohesion, formation of axial elements, and recombination during yeast meiosis. *Cell*, **98**, 91–103.

Koehler, K.E., Hawley, R.S., Sherman, S. and Hassold, T. (1996) Recombination and nondisjunction in humans and flies. *Hum. Mol. Genet.*, **5**, 1495–1504.

Kops, G., Foltz, D. and Cleveland, D. (2004) Lethality to human cancer cells through massive chromosome loss by inhibition of the mitotic checkpoint. *Proc. Natl. Acad. Sci. USA*, **101**(23), 8699–8704.

Kraft, C., Herzog, F., Gieffers, C. *et al.* (2003) Mitotic regulation of the human anaphase-promoting complex by phosphorylation. *EMBO J.*, **22**, 6598–6609.

Kraft, C., Vodermaier, H.C., Maurer-Stroh, S. *et al.* (2005) The WD40 propeller domain of Cdh1 functions as a destruction box receptor for APC/C substrates. *Mol. Cell*, **18**, 543–553.

Kramer, E.R., Scheuringer, N., Podtelejnikov, A.V. *et al.* (2000) Mitotic regulation of the APC activator proteins CDC20 and CDH1. *Mol. Biol. Cell*, **11**(5), 1555–1569.

Kudo, N.R., Wassmann, K., Anger, M. *et al.* (2006) Resolution of chiasmata in oocytes requires separase-mediated proteolysis. *Cell*, **126**, 135–146.

Kueng, S., Hegemann, B., Peters, B.H. *et al.* (2006) Wapl controls the dynamic association of cohesin with chromatin. *Cell*, **127**(5), 955–967.

Labbé, J.C., Capony, J.P., Caput, D. *et al.* (1989) MPF from starfish oocytes at first meiotic metaphase is a heterodimer containing one molecule of cdc2 and one molecule of cyclin B. *EMBO J.*, **8**(10), 3053–3058.

Lampson, M.A., Renduchitala, K., Khodjakov, A. and Kapoor, T.M. (2004) Correcting improper chromosome spindle attachments during cell division. *Nat. Cell Biol.*, **6**, 232–237.

Ledan, E., Polanski, Z., Terret, M.E. and Maro, B. (2001) Meiotic maturation of the mouse oocyte requires an equilibrium between cyclin B synthesis and degradation. *Dev. Biol.*, **232**, 400–413.

Lee, J., Miyano, T., Dai, Y. *et al.* (2000) Specific regulation of CENP-E and kinetochores during meiosis I/meiosis II transition in pig oocytes. *Mol. Reprod. Dev.*, **56**, 51–62.

Lee, J., Iwai, T., Yokota, T. and Yamashita, M. (2003) Temporally and spatially selective loss of Rec8 protein from meiotic chromosomes during mammalian meiosis. *J. Cell Sci.*, **116**, 2781–2790.

Lee, J., Kitajima, T.S., Tanno, Y. *et al.* (2008) Unified mode of centromeric protection by shugoshin in mammalian oocytes and somatic cells. *Nat. Cell Biol.*, **10**(1), 42–52.

Lefebvre, C., Terret, M.E., Djiane, A. *et al.* (2002) Meiotic spindle stability depends on MAPK-interacting and spindle-stabilizing (MISS), a new MAPK substrate. *J. Cell Biol.*, **157**, 603–613.

LeMaire-Adkins, R., Radke, K. and Hunt, P.A. (1997) Lack of checkpoint control at the metaphase/anaphase transition: a mechanism of meiotic nondisjunction in mammalian females. *J. Cell Biol.*, **139**, 1611–1619.

Li, R. and Murray, A. (1991) Feedback control of mitosis in budding yeast. *Cell*, **66**, 519–531.

Li, Y. and Benezra, R. (1996) Identification of a human mitotic checkpoint gene: hsMAD2. *Science*, **274**, 246–248.

Li, Y., Gorbea, C., Mahaffey, D. *et al.* (1997) MAD2 associates with the cyclosome/anaphase-promoting complex and inhibits its activity. *Proc. Natl. Acad. Sci. USA*, **94**, 12431–12436.

Littlepage, L.E. and Ruderman, J.V. (2002) Identification of a new APC/C recognition domain, the A box, which is required for the Cdh1-dependent destruction of the kinase Aurora-A during mitotic exit. *Genes Dev.*, **16**(17), 2274–2285.

Llano, E., Gómez, R., Gutiérrez-Caballero, C. *et al.* (2008) Shugoshin-2 is essential for the completion of meiosis but not for mitotic cell division in mice. *Genes Dev.*, **22**(17), 2400–2413.

Luo, X., Fang, G., Coldiron, M. *et al.* (2000) Structure of the Mad2 spindle assembly checkpoint protein and its interaction with Cdc20. *Nat. Struct. Biol.*, **7**, 224–229.

Madgwick, S., Hansen, D.V., Levasseur, M. *et al.* (2006) Mouse Emi2 is required to enter meiosis II by reestablishing cyclin B1 during interkinesis. *J. Cell Biol.*, **174**, 791–801.

Malmanche, N., Owen, S., Gegick, S. *et al.* (2007) Drosophila BubR1 is essential for meiotic sister-chromatid cohesion and maintenance of synaptonemal complex. *Curr. Biol.*, **17**, 1489–1497.

Mao, Y., Desai, A. and Cleveland, D.W. (2005) Microtubule capture by CENP-E silences BubR1-dependent mitotic checkpoint signaling. *J. Cell Biol.*, **170**, 873–880.

Mapelli, M., Simonetta, M., Transidico, P. *et al.* (2006) Determinants of conformational dimerization of Mad2 and its inhibition by p31comet. *EMBO J.*, **25**, 1273–1284.

Marston, A.L. and Amon, A. (2004) Meiosis: Cell-cycle controls shuffle and deal. *Nat. Rev. Mol. Cell Biol.*, **5**, 983–997.

Mei, J., Huang, X. and Zhang, P. (2001) Securin is not required for cellular viability, but is required for normal growth of mouse embryonic fibroblasts. *Curr. Biol.*, **11**, 1197–1201.

Meraldi, P., Draviam, V. and Sorger, P. (2004) Timing and checkpoints in the regulation of mitotic progression. *Dev. Cell*, **7**, 45–60.

Michel, L., Liberal, V., Chatterjee, A. *et al.* (2001) MAD2 haplo-insufficiency causes premature anaphase and chromosome instability in mammalian cells. *Nature*, **409**, 355–359.

Michel, L., Diaz-Rodriguez, E., Narayan, G. *et al.* (2004) Complete loss of the tumor suppressor MAD2 causes premature cyclin B degradation and mitotic failure in human somatic cells. *Proc. Natl. Acad. Sci. USA*, **101**, 4459–4464.

Montembault, E., Dutertre, S., Prigent, C. and Giet, R. (2007) PRP4 is a spindle assembly checkpoint protein required for MPS1, MAD1, and MAD2 localization to the kinetochores. *J. Cell Biol.*, **179** (4), 601–609.

Murray, A.W., Solomon, M.J. and Kirschner, M.W. (1989) The role of cyclin synthesis and degradation in the control of maturation promoting factor activity. *Nature*, **339**, 280–286.

Musacchio, A. and Salmon, E.D. (2007) The spindle-assembly checkpoint in space and time. *Nat. Rev. Mol. Cell Biol.*, **8**(5), 379–393.

Nasmyth, K. (2001) Disseminating the genome: joining, resolving, and separating sister chromatids during mitosis and meiosis. *Annu. Rev. Genet.*, **35**, 673–745.

Nasmyth, K. (2005) How might cohesin hold sister chromatids together? *Philos. Trans. R. Soc. Lond. B Biol. Sci.*, **360**(1455), 483–496.

Niault, T., Hached, K., Sotillo, R. *et al.* (2007) Changing mad2 levels affects chromosome segregation and spindle assembly checkpoint control in female mouse meiosis I. *PLoS ONE*, **2**, e1165.

Nicklas, R.B. (1997) How cells get the right chromosomes. *Science*, **275**, 632–637.

Ohe, M., Inoue, D., Kanemori, Y. and Sagata, N. (2007) Erp1/Emi2 is essential for the meiosis I to meiosis II transition in Xenopus oocytes. *Dev. Biol.*, **303**, 157–164.

Orr-Weaver, T. (1996) Meiotic nondisjunction does the two-step. *Nat. Genet.*, **14**, 374–376.

Page, A.W. and Orr-Weaver, T.L. (1996) The Drosophila genes grauzone and cortex are necessary for proper female meiosis. *J. Cell Sci.*, **109**, 1707–1715.

Passmore, L.A., McCormack, E.A., Au, S.W. *et al.* (2003) Doc1 mediates the activity of the anaphase-promoting complex by contributing to substrate recognition. *EMBO J.*, **22**, 786–796.

Pawlowski, W.P. and Cande, W.Z. (2005) Coordinating the events of the meiotic prophase. *Trends Cell Biol.*, **15**, 674–681.

Peter, M., Castro, A., Lorca, T. *et al.* (2001) The APC is dispensable for first meiotic anaphase in Xenopus oocytes. *Nat. Cell Biol.*, **3**, 83–87.

Peters, J.M. (2006) The anaphase promoting complex/cyclosome: a machine designed to destroy. *Nat. Rev. Mol. Cell Biol.*, **7**, 644–656.

Petronczki, M., Siomos, M.F. and Nasmyth, K. (2003) Un menage a quatre: The molecular biology of chromosome segregation in meiosis. *Cell*, **112**, 423–440.

Pfleger, C.M. and Kirschner, M.W. (2000) The KEN box: an APC recognition signal distinct from the D box targeted by Cdh1. *Genes Dev.*, **14**, 655–665.

Pinsky, B.A. and Biggins, S. (2005) The spindle checkpoint: tension versus attachment. *Trends Cell Biol.*, **15**(9), 486–493.

Rabitsch, K.P., Gregan, J., Schleiffer, A. *et al.* (2004) Two fission yeast homologs of Drosophila Mei-S332 are required for chromosome segregation during meiosis I and II. *Curr. Biol.*, **14**, 287–301.

Reis, A., Madgwick, S., Chang, H.Y. *et al.* (2007) Prometaphase APCcdh1 activity prevents non-disjunction in mammalian oocytes. *Nat. Cell Biol.*, **9**, 1192–1198.

Revenkova, E., Eijpe, M., Heyting, C. *et al.* (2001) Novel meiosis-specific isoform of mammalian SMC1. *Mol. Cell Biol.*, **21**(20), 6984–6998.

Revenkova, E., Eijpe, M., Heyting, C. *et al.* (2004) Cohesin SMC1 beta is required for meiotic chromosome dynamics, sister chromatid cohesion and DNA recombination. *Nat. Cell Biol.*, **6**, 555–562.

Rieder, C.L., Schultz, A., Cole, R. and Sluder, G. (1994) Anaphase onset in vertebrate somatic cells is controlled by a checkpoint that monitors sister kinetochore attachment to the spindle. *J. Cell Biol.*, **127**, 1301–1310.

Rieder, C.L., Cole, R.W., Khodjakov, A. and Sluder, G. (1995) The checkpoint delaying anaphase in response to chromosome monoorientation is mediated by an inhibitory signal produced by unattached kinetochores. *J. Cell Biol.*, **130**(4), 941–948.

Robinson, W.P., Mcfadden, D.E. and Stephenson, M.D. (2001) The origin of abnormalities in recurrent aneuploidy/polyploidy. *Am. J. Med. Genet.*, **69**, 1245–1254.

Rubio, C., Simon, C. and Vida, T. (2003) Chromosomal abnormalities and embryo development in recurrent miscarriage couples. *Hum. Reprod.*, **18**, 182–188.

Rudner, A.D. and Murray, A.W. (2000) Phosphorylation by Cdc28 activates the Cdc20-dependent activity of the anaphase-promoting complex. *J. Cell Biol.*, **149**, 1377–1390.

Salah, S.M. and Nasmyth, K. (2000) Destruction of the securin Pds1p occurs at the onset of anaphase during both meiotic divisions in yeast. *Chromosoma*, **109**, 27–34.

Salic, A., Waters, J.C. and Mitchison, T.J. (2004) Vertebrate shugoshin links sister centromere cohesion and kinetochore microtubule stability in mitosis. *Cell*, **118**, 567–578.

Schwab, M., Neutzner, M., Mocker, D. and Seufert, W. (2001) Yeast Hct1 recognizes the mitotic cyclin Clb2 and other substrates of the ubiquitin ligase APC. *EMBO J.*, **20**, 5165–5175.

Shen, Y., Betzendahl, I., Sun, F. *et al.* (2005) Non-invasive method to assess genotoxicity of nocodazole interfering with spindle formation in mammalian oocytes. *Reprod. Toxicol.*, **19**, 459–471.

Shonn, M.A., McCarroll, R. and Murray, A.W. (2000) Requirement of the spindle checkpoint for proper chromosome segregation in budding yeast meiosis. *Science*, **289**, 300–303.

Shonn, M., Murray, A.L. and Murray, A.W. (2003) Spindle checkpoint component Mad2 contributes to biorientation of homologous chromosomes. *Curr. Biol.*, **13**, 1979–1984.

Shteinberg, M., Protopopov, Y., Listovsky, T. *et al.* (1999) Phosphorylation of the cyclosome is required for its stimulation by Fizzy/cdc20. *Biochem. Biophys. Res. Commun.*, **260**(1), 193–198.

Siomos, M.F., Badrinath, A., Pasierbek, P., Livingstone, D. *et al.* (2001) Separase is required for chromosome segregation during meiosis I in Caenorhabditis elegans. *Curr. Biol.*, **11**, 1825–1835.

Skoufias, D.A., Andreassen, P.R., Lacroix, F.B. *et al.* (2001) Mammalian mad2 and bub1/bubR1 recognize distinct spindle-attachment and kinetochore-tension checkpoints. *Proc. Natl. Acad. Sci. USA*, **98**, 4492–4497.

Soewarto, D., Schmiady, H. and Eichenlaub-Ritter, U. (1995) Consequences of non-extrusion of the first polar body and control of the sequential segregation of homologues and chromatids in mammalian oocytes. *Hum. Reprod.*, **10**, 2350–2360.

Steen, J.A.J., Steen, H., Georgi, A. *et al.* (2008) Different phosphorylation states of the anaphase promoting complex in response to antimitotic drugs: a quantitative proteomic analysis. *Proc. Natl. Acad. Sci. USA*, **105**(16), 6069–6074.

Stein, K.K., Davis, E.S., Hays, T. and Golden, A. (2007) Components of the spindle assembly checkpoint regulate the anaphase-promoting complex during meiosis in Caenorhabditis elegans. *Genetics*, **175**, 107–123.

Stemmann, O., Zou, H., Gerber, S.A. *et al.* (2001) Dual inhibition of sister chromatid separation at metaphase. *Cell*, **107**, 715–726.

Stemmann, O., Gorr, I.H. and Boos, D. (2006) Anaphase topsy-turvy: Cdk1 a securin, separase a CKI. *Cell Cycle*, **5**, 11–13.

Steuerwald, N., Cohen, J., Herrera, R.J. *et al.* (2001) Association between spindle assembly checkpoint expression and maternal age in human oocytes. *Mol. Hum. Reprod.*, **7**, 49–55.

Steuerwald, N.M., Steuerwald, M.D. and Mailhes, J.B. (2005) Post-ovulatory aging of mouse oocytes leads to decreased MAD2 transcripts and increased frequencies of premature centromere separation and anaphase. *Mol. Hum. Reprod.*, **11**(9), 623–630.

Stratmann, R. and Lehner, C.F. (1996) Separation of sister chromatids in mitosis requires the Drosophila pimples product, a protein degraded after the metaphase/anaphase transition. *Cell*, **84**(1), 25–35.

Sumara, I., Vorlaufer, E., Stukenberg, P.T. *et al.* (2002) The dissociation of cohesin from chromosomes in prophase is regulated by Polo-like kinase. *Mol. Cell*, **9**(3), 515–525.

Swan, A. and Schupbach, T. (2007) The Cdc20 (Fzy)/Cdh1-related protein, Cort, cooperates with Fzy in cyclin destruction and anaphase progression in meiosis I and II in Drosophila. *Development*, **134**, 891–899.

Tai, C.Y., Dujardin, D.L., Faulkner, N.E. and Vallee, R.B. (2002) Role of dynein, dynactin, and CLIP-170 interactions in LIS1 kinetochore function. *J. Cell Biol.*, **156**, 959–968.

Taieb, F.E., Gross, S.D., Lewellyn, A.L. and Maller, J.L. (2001) Activation of the anaphase-promoting complex and degradation of cyclin B is not required for progression from Meiosis I to II in Xenopus oocytes. *Curr. Biol.*, **11**, 508–513.

Tanaka, T.U., Rachidi, N., Janke, C. *et al.* (2002) Evidence that the Ipl1–Sli15 (Aurora kinase–INCENP) complex promotes chromosome bi-orientation by altering kinetochore spindle pole connections. *Cell*, **108**, 317–329.

Tang, T.T., Bickel, S.E., Young, L.M. and Orr-Weaver, T.L. (1998) Maintenance of sister-chromatid cohesion at the centromere by the Drosophila MEI-S332 protein. *Genes Dev.*, **12**, 3843–3856.

Tang, Z., Bharadwaj, R., Li, B. and Yu, H. (2001) Mad2-independent inhibition of APCCdc20 by the mitotic checkpoint protein BubR1. *Dev. Cell*, **1**, 227–237.

Tavormina, P.A. and Burke, D.J. (1998) Cell cycle arrest in cdc20 mutants of *S. cerevisiae* is independent of Ndc10p and kinetochore function but requires a subset of spindle checkpoint genes. *Genetics*, **148**(4), 1701–1713.

Terret, M.E., Wassmann, K., Waizenegger, I. *et al.* (2003) The meiosis I-to-meiosis II transition in mouse oocytes requires separase activity. *Curr. Biol.*, **13**, 1797–1802.

Toth, A., Rabitsch, K.P., Galova, M. *et al.* (2000) Functional genomics identifies monopolin: A kinetochore protein required for segregation of homologs during meiosis I. *Cell*, **103**, 1155–1168.

Tseng, L.H., Chung, S.M., Lee, T.Y. and Ko, T.M. (1994) Recurrent Down's syndrome due to maternal ovarian trisomy 21 mosaicism. *Arch. Gynecol. Obstet.*, **255**, 213–216.

Tsurumi, C., Hoffmann, S., Geley, S. *et al.* (2004) The spindle assembly checkpoint is not essential for CSF arrest of mouse oocytes. *J. Cell Biol.*, **167**, 1037–1050.

Tung, J.J., Padmanabhan, K., Hansen, D.V. *et al.* (2007) Translational unmasking of Emi2 directs cytostatic factor arrest in meiosis II. *Cell Cycle*, **6**, 725–731.

Uhlmann, F. (2004) The mechanism of sister chromatid cohesion. *Exp. Cell Res.*, **296**(1), 80–85.

Verlhac, M.H., Kubiak, J.Z., Clarke, H.J. and Maro, B. (1994) Microtubule and chromatin behavior follow MAP kinase activity but not MPF activity during meiosis in mouse oocytes. *Development*, **120**, 1017–1025.

Verlhac, M.H., Kubiak, J.Z., Weber, M. *et al.* (1996) Mos is required for MAP kinase activation and is involved in microtubule organization during meiotic maturation in the mouse. *Development*, **122**, 815–822.

Visintin, R., Craig, K., Hwang, E.S. *et al.* (1998) The phosphatase Cdc14 triggers mitotic exit by reversal of Cdk-dependent phosphorylation. *Mol. Cell*, **2**, 709–718.

Vodermaier, H.C., Gieffers, C., Maurer-Stroh, S. *et al.* (2003) TPR subunits of the anaphase-promoting complex mediate binding to the activator protein CDH1. *Curr. Biol.*, **13**(17), 1459–1468.
Waizenegger, I.C., Hauf, S., Meinke, A. and Peters, J.M. (2000) Two distinct pathways remove mammalian cohesin from chromosome arms in prophase and from centromeres in anaphase. *Cell*, **103**(3), 399–410.
Wang, Z., Yu, R. and Melmed, S. (2001) Mice lacking pituitary tumor transforming gene show testicular and splenic hypoplasia, thymic hyperplasia, thrombocytopenia, aberrant cell cycle progression, and premature centromere division. *Mol. Endocrinol.*, **15**, 1870–1879.
Wassmann, K., Niault, T. and Maro, B. (2003a) Metaphase I arrest upon activation of the Mad2-dependent spindle checkpoint in mouse oocytes. *Curr. Biol.*, **13**, 1596–1608.
Wassmann, K., Liberal, V. and Benezra, R. (2003b) Mad2 phosphorylation regulates its association with Mad1 and the APC/C. *EMBO J.*, **22**, 797–806.
Watanabe, Y. (2004) Modifying sister chromatid cohesion for meiosis. *J. Cell Sci.*, **117**, 4017–4023.
Watanabe, Y. and Nurse, P. (1999) Cohesin Rec8 is required for reductional chromosome segregation at meiosis. *Nature*, **400**, 461–464.
Waters, J.C., Chen, R.H., Murray, A.W. and Salmon, E.D. (1998) Localization of Mad2 to kinetochores depends on microtubule attachment, not tension. *J. Cell Biol.*, **141**(5), 1181–1191.
Weiss, E. and Winey, M. (1996) The *S. cerevisiae* spindle pole body duplication gene MPS1 is part of a mitotic checkpoint. *J. Cell Biol.*, **132**, 111–123.
Winston, N.J. (1997) Stability of cyclin B during meiotic maturation and the first meiotic cell division in mouse oocytes. *Biol. Cell*, **89**, 211–219.
Wojcik, E., Basto, R., Serr, M. *et al.* (2001) Kinetochore dynein: its dynamics and role in the transport of the Rough deal checkpoint protein. *Nat. Cell Biol.*, **3**, 1001–1007.
Wu, J.Q. and Kornbluth, S. (2008) Across the meiotic divide – CSF activity in the post-Emi2/XErp1 era. *J. Cell Sci.*, **121**(21), 3509–3514.
Xia, G., Luo, X., Habu, T. *et al.* (2004) Conformation-specific binding of p31(comet) antagonizes the function of Mad2 in the spindle checkpoint. *EMBO J.*, **23**, 3133–3143.
Yamaguchi, S., Okayama, H. and Nurse, P. (2000) Fission yeast Fizzy-related protein Srw1p is a G1-specific promoter of mitotic cyclin B degradation. *EMBO J.*, **19**, 3968–3977.
Yamamoto, A., Guacci, V. and Koshland, D. (1996) Pds1p is required for faithful execution of anaphase in the yeast, *S. cerevisiae*. *J. Cell Biol.*, **133**, 85–97.
Yin, S., Wang, Q., Liu, J.H. *et al.* (2006) Bub1 prevents chromosome misalignment and precocious anaphase during mouse oocyte meiosis. *Cell Cycle*, **5**, 2130–2137.
Yu, H., Muszynski, M.G. and Dawe, R.K. (1999) The maize homologue of the cell cycle checkpoint protein MAD2 reveals kinetochore substructure and contrasting mitotic and meiotic localization patterns. *J. Cell Biol.*, **145**, 425–435.
Zachariae, W., Schwab, M., Nasmyth, K. and Seufert, W. (1998) Control of cyclin ubiquitination by CDK-regulated binding of Hct1 to the anaphase promoting complex. *Science*, **282**, 1721–1724.
Zhang, D., Ma, W., Li, Y.H. *et al.* (2004) Intra-oocyte localization of MAD2 and its relationship with kinetochores, microtubules and chromosomes in rat oocytes during meiosis. *Biol. Reprod.*, **71**, 740–748.
Zhang, D., Li, M., Ma, W. *et al.* (2005) Localization of MAD1 in mouse oocytes during the first meiosis and its functions as a spindle checkpoint protein. *Biol. Reprod.*, **72**, 58–68.
Zhang, X., Ma, C., Miller, A.L. *et al.* (2008) Polar body emission requires a RhoA contractile ring and Cdc42-mediated membrane protrusion. *Dev. Cell*, **15**(3), 386–400.
Zou, H., McGarry, T.J., Bernal, T. and Kirschner, M.W. (1999) Identification of a vertebrate sister-chromatid separation inhibitor involved in transformation and tumorigenesis. *Science*, **285**, 418–422.

13

Mechanisms controlling maintenance and exit of the CSF arrest

Thierry Lorca and Anna Castro

CNRS UMR 5237, IFR 122, Universités Montpellier 2 et 1, Centre de Recherche de Biochimie Macromoléculaire, Montpellier CEDEX 5, France

13.1 Introduction

In the animal kingdom, oocytes arrest the cell cycle at the G2–prophase boundary of the first meiotic cycle after duplicated chromosomes have undertaken recombination. In most species, resumption of meiosis is then triggered by the steroid hormone progesterone, which primes ovulation and induces oocytes to resume meiosis I and to enter into meiosis II. Oocytes then remain arrested at metaphase for prolonged periods of time while awaiting for fertilization. Upon fertilization, eggs rapidly exit meiosis and generate a haploid pronucleus that will then fuse with the male pronucleus to form the diploid zygote.

The molecular mechanisms that drive meiosis entry and arrest have been extensively studied. Over 30 years ago, Masui and Markert (1971) described the presence of an activity in the cytoplasm of mature oocytes that, when injected into G2-arrested oocytes, induces maturation. This activity was referred to as the maturation-promoting factor (MPF), and is now identified as the cyclin B–Cdk1 complex (Lee and Nurse, 1987; Lohka, Hayes and Maller, 1998). They also described the presence of a second activity in this cytoplasm that, when injected into two-cell embryos, induces metaphase arrest. This led them to postulate the presence of a factor that inhibited meiosis at metaphase II, named the cytostatic factor or CSF. This factor should fulfil the following criteria: (i) appearing during oocyte maturation; (ii) be present and functional during metaphase

Oogenesis: The Universal Process Marie-Hélène Verlhac and Anne Villeneuve

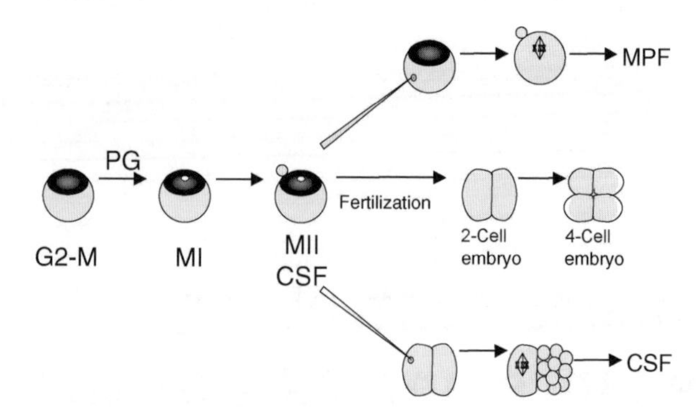

Figure 13.1 The MPF and CSF activities are essential for correct oocyte maturation and fertilization. Vertebrate oocytes are arrested at prophase of meiosis I (G2-M). Progesterone (PG) promotes the progression of the oocytes from prophase of meiosis I (MI) to metaphase of meiosis II (MII), where they will arrest until fertilization. Fertilization will finally induce metaphase II release. MPF and CSF activities were first described by Masui and Markert (1971). They showed that the injection of cytoplasm from metaphase II-arrested oocytes (CSF) to G2-M oocytes induced the meiotic progression of these oocytes up to metaphase II. They referred to this activity as the maturation-promoting factor or MPF. They also reported that the injection of a small amount of cytoplasm from metaphase II-arrested oocytes (CSF) to one blastomere of a two-cell embryo induced the arrest of the injected blastomere at metaphase II. From this observation they postulated the presence of a second activity that they named cytostatic activity or CSF, required to maintain metaphase II arrest of unfertilized oocytes

II arrest; and (iii) be rapidly inactivated after egg fertilization (Masui and Markert, 1971) (Figure 13.1). The biochemical identity of the CSF has been extensively studied; however, despite these criteria, the exact molecular mechanisms inducing the metaphase arrest are just beginning to be elucidated after 30 years of intensive study.

At the molecular level, exit of mitosis and meiosis are induced by the inactivation of the cyclin B–Cdk1 complex through the ubiquitin-dependent degradation of its cyclin B subunit. Cyclin B ubiquitination is mediated by the E3 ubiquitin ligase APC (anaphase-promoting complex). The APC is a large protein complex containing at least 11 core subunits and 2 different activators, Cdc20 and Cdh1. APC–Cdc20 targets degradation of mitotic substrates at the metaphase–anaphase transition, whereas APC–Cdh1 controls degradation during late mitosis and G1 (for a review see Castro *et al.*, 2005; Harper, Burton and Solomon, 2002). The CSF arrest is characterized by the presence of a metaphase plate and a high cyclin B–Cdk1 activity. This high MPF activity is required to maintain the meiotic state and is thought to be the result of a block on cyclin B proteolysis through the inhibition of the APC–Cdc20 ubiquitin ligase (Lorca *et al.*, 1998; Vorlaufer and Peters, 1998). In this regard recent studies have investigated the mechanisms by which the CSF could mediate cyclin B stabilization. Three different pathways have been proposed to explain APC–Cdc20 inhibition during CSF arrest: (i) Emi1-mediated pathway (Reimann *et al.*, 2001; Reimann and Jackson, 2002); (ii) spindle-checkpoint mediated pathway (review in Liu, Grimison and Maller, 2007; Tunquist and Maller, 2003); (iii) Erp1-mediated pathway (for a review see Schmidt *et al.*, 2006; Wu and Kornbluth, 2008).

The CSF release is characterized by the inactivation of the cyclin B–Cdk1 complex through the APC–Cdc20-dependent degradation of cyclin B. New insight into the molecular mechanism inducing APC–Cdc20 activation and CSF release has recently emerged by the characterization of the mechanisms regulating the Erp1 ubiquitin-dependent degradation pathway. The new findings demonstrate that Ca^{2+}/calmodulin-dependent protein kinase II (CaMKII), which is rapidly activated upon fertilization by the Ca^{2+} signal, induces a first phosphorylation of the APC–Cdc20 inhibitor Erp1, priming this protein for a subsequent phosphorylation by Plx1. Phosphorylation of Erp1 by Plx1 will then create a recognition motif for the ubiquitin ligase Skp1-Cullin-Fbox-βtrcp ($SCF^{\beta trcp}$), causing Erp1 destruction, APC–Cdc20 activation and cyclin B proteolysis (for a review see Schmidt *et al.*, 2006; Wu and Kornbluth, 2008). Besides APC–Cdc20 activation and cyclin B degradation, a new mechanism involved in CSF exit has been identified. This mechanism involves the phosphatase calcineurin, whose activation by the Ca^{2+} signal after fertilization is required to completely dephosphorylate the different mitotic substrates and to induce a rapid exit of meiosis (Mochida and Hunt, 2007; Nishiyama *et al.*, 2007a).

In this chapter, we will only briefly summarize the recent data describing the different new pathways involved in the inhibition of the APC–Cdc20 complex during CSF arrest, and we will focus on the recent advances in the mechanisms inducing CSF exit. Moreover, despite the fact that significant differences exist between frogs and mammals, the major pathways involved in CSF arrest and exit are conserved. Taking into account that the great majority of works studying metaphase II arrest are developed in *Xenopus* oocytes, we will specifically cover the results obtained in the frog model.

13.2 CSF establishment

The first candidate to fulfil the role of the CSF factor was proposed by Sagata *et al.* in 1989. These authors showed that the microinjection of c-Mos mRNA into two-cell embryos induces cleavage arrest at metaphase, and, as expected, blastomeres arrested with a high MPF activity (Sagata *et al.*, 1989). Moreover, the level of c-Mos protein increases rapidly during maturation and is maintained until egg fertilization (Sagata *et al.*, 1988). Finally the c-Mos protein disappears due to proteolytic degradation after fertilization (Castro *et al.*, 2001; Lorca *et al.*, 1991; Watanabe *et al.*, 1989). Thus, c-Mos satisfied the criteria proposed by Masui and Markert for the CSF. Subsequent studies revealed that c-Mos acts by the activation of the MEK/MAPK/Rsk pathway. Accordingly, microinjection of thiophosphorylated MAP kinase (an irreversibly activated kinase) into two-cell embryos induces metaphase arrest (Haccard *et al.*, 1993), and depletion of $p90^{Rsk}$ from egg extracts inhibits the capacity to undergo mitotic arrest, an effect that is restored by the re-addition of this kinase (Bhatt and Ferrell, 1999; Gross *et al.*, 1999). However, the same reports present data showing that Rsk is required for CSF establishment but not for CSF maintenance in *Xenopus* oocytes (Bhatt and Ferrell, 1999; Gross *et al.*, 1999), whereas in mouse oocytes, the activity of Rsk appears not to be required either for CSF establishment or for the maintenance of this activity (Dumont *et al.*, 2005). Finally, the c-Mos/MAPK/Rsk pathway is not sufficient to prevent cyclin B degradation and to stabilize MPF activity, since overexpression of c-Mos does not arrest *Xenopus* oocytes at metaphase I, although cytoplasm taken before germinal vesicle breakdown (GVBD) from these oocytes readily induces metaphase arrest when transferred into two-cell embryos (Kanki and Donoghue, 1991). This suggested the presence of another factor that could be the target of this pathway and that would finally inhibit the APC. This factor was proposed to be the early mitotic inhibitor 1 (Emi1). Emi1 is an APC inhibitor that promotes APC–Cdh1 and APC–Cdc20 inhibition at the S and G2 phases of the cell cycle, promoting cyclin A and cyclin B accumulation. Late in prometaphase Emi1 is destroyed, allowing APC–Cdc20 activation and completion of mitosis via cyclin A and cyclin B proteolysis (Reimann *et al.*, 2001; Hsu *et al.*, 2002). According to a putative role of Emi1 in CSF arrest, Reiman *et al.* showed that injection of Emi1 in two-cell embryos induced mitotic arrest with high MPF activity, and that depletion of this protein from *Xenopus* egg extracts promoted CSF exit (Reimann and Jackson, 2002). However, doubts about the involvement of Emi1 in CSF arrest were raised by the observations of Ohsumi *et al.* (2004). Contrary to Reiman *et al.* (Reimann and Jackson, 2002), Ohsumi *et al.* did not detect Emi1 protein in metaphase II-arrested oocytes. Moreover, they showed that the $SCF^{\beta trcp}$-dependent Emi1 degradation pathway was functional in these oocytes. Thus they concluded that Emi1 could not be present in these oocytes, since it would be immediately degraded. However, recent results from our laboratory show that Emi1 can be stabilized even in the presence of a functional $SCF^{\beta trcp}$. Thus, we cannot exclude the possibility that, if present, as described by Jackson and co-workers, Emi1 could participate in the CSF arrest (Reimann and Jackson, 2002).

Besides Emi1, the spindle assembly checkpoint (SAC) pathway has also been proposed as a mechanism involved in the CSF arrest (for a review see Liu, Grimison and Maller, 2007; Tunquist and Maller, 2003). During mitosis, the SAC restrains cells from entering anaphase until all replicated chromatids are correctly attached to the bipolar spindle by inducing APC–Cdc20 inhibition (for a review see Burke and Stukenberg, 2008; Musacchio and Hardwick, 2002); thus, this pathway could also participate in the inhibition of the APC–Cdc20 during CSF arrest. In these regards Maller and co-workers have shown a role of the SAC constituents Bub1, Mad2, and Mps1 in the metaphase arrest induced by c-Mos or cyclin E overexpression in interphase *Xenopus* egg extracts (Grimison *et al.*, 2006; Tunquist *et al.*, 2003; Tunquist *et al.*, 2002). However, the fact that most of these proteins are not required for the maintenance of the CSF arrest suggests that the c-Mos/MAPK/Rsk pathway induces metaphase II block by a different mechanism (Abrieu *et al.*, 2001; Sharp-Baker and Chen, 2001).

The last candidate proposed to fulfil the role of CSF is the Emi1-related protein, Erp1 (also named Emi2). Erp1 is an APC inhibitor that is present from GVBD in maturing oocytes, stable in metaphase II-arrested oocytes and rapidly degraded after fertilization (Schmidt *et al.*, 2005; Tung *et al.*, 2005). Moreover, it is required to maintain the CSF arrest, since its depletion from CSF extracts induces CSF release (Schmidt *et al.*, 2005). Thus, it completely satisfies all the criteria for the CSF, suggesting that it could be the final target of the c-Mos/MAPK/Rsk pathway. According to this hypothesis, recent data indicate that Erp1 directly binds and inhibits the APC and that this inhibition is controlled by the phosphorylation of Erp1 by $p90^{Rsk}$ (Inoue *et al.*, 2007;

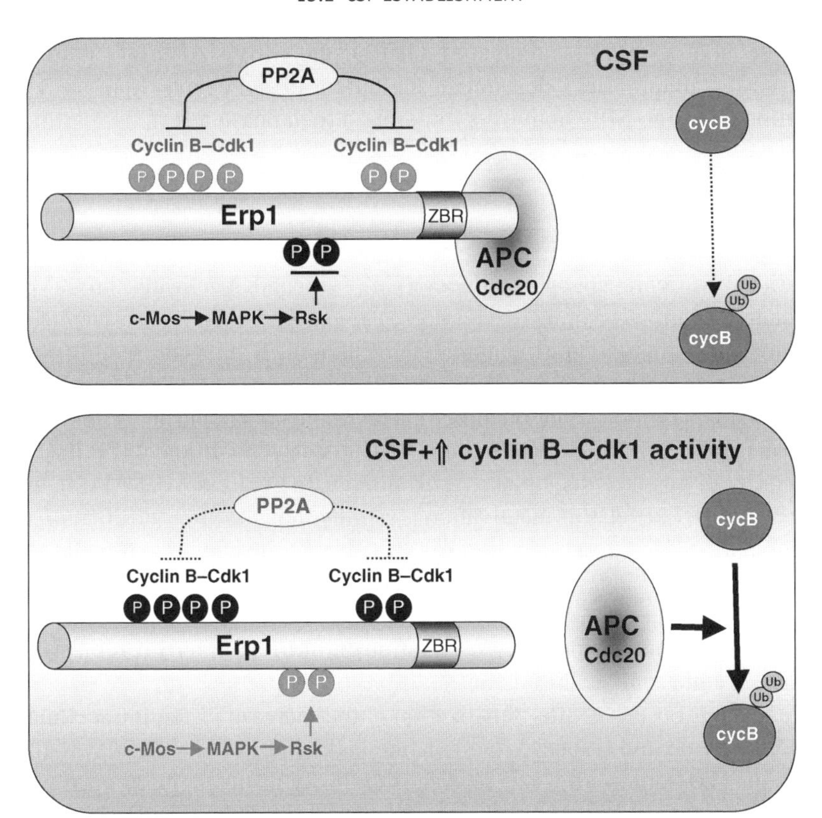

Figure 13.2 Erp1 regulation in CSF arrest. During CSF arrest, Erp1 is phosphorylated at two different clusters by cyclin B–Cdk1. This phosphorylation weakens APC^{Cdc20}–Erp1 binding, triggering APC^{Cdc20} dissociation and activation. However, this phosphorylation is antagonized by the c-Mos/MAPK/Rsk pathway. Rsk-dependent phosphorylation of Erp1 promotes PP2A binding to this protein. This association will induce dephosphorylation of the cyclin B–Cdk1-mediated phosphorylation of Erp1, and will maintain its binding to the APC^{Cdc20}, inhibiting cyclin B (cycB) ubiquitination and degradation. The addition of ectopic cyclin B to CSF extracts will increase cyclin B–Cdk1 activity, counterbalancing PP2A phosphatase activity. This increase will trigger Erp1–APC^{Cdc20} dissociation and partial degradation of cyclin B until cyclin B–Cdk1 activity reaches endogenous levels. At this moment, PP2A will be capable of again counterbalancing cyclin B–Cdk1 phosphorylation and favouring Erp1–APC^{Cdc20} binding and inhibition. ZBR = zinc-binding region

Nishiyama, Ohsumi and Kishimoto, 2007a; Wu *et al.*, 2007a). Indeed, Erp1 can be phosphorylated by both cyclin B–Cdk1 and $p90^{Rsk}$ (Figure 13.2). Phosphorylation of Erp1 by cyclin B–Cdk1 takes place at two different clusters of the Erp1 amin-oacid sequence: S213/T239/T252/T267 and T545/T551. This phosphorylation not only inhibits Erp1 activity by weakening Erp1/APC interaction, but also participates in the proteolysis of this protein (Wu *et al.*, 2007a; Wu *et al.*, 2007b). On the contrary, phosphorylation of Erp1 by $p90^{Rsk}$ reinforces Erp1–APC binding and stabilizes this protein. Erp1 phosphorylation by $p90^{Rsk}$ induces the recruitment of PP2A to Erp1 which, in turn, triggers dephosphorylation of the different cyclin B–Cdk1 phosphorylation sites, promoting, in this way, Erp1 binding to the APC (Wu *et al.*, 2007a; Wu *et al.*, 2007b).

All these findings suggest that after GVBD, the Mos/MAPK/Rsk pathway will activate and maintain Erp1-dependent inhibition of APC, resulting in cyclin B stabilization and high MPF activity, promoting a metaphase arrest.

13.3 CSF exit

Fertilization of vertebrate oocytes induces a Ca^{2+} signal that activates the CaMKII and finally induces resumption of meiosis II (Lorca *et al.*, 1993). The egg then releases from the CSF arrest, progresses from metaphase II to anaphase II, emits the second polar body and finally enters into the mitotic cell cycles. As described above, CSF arrest is mediated by Rsk-dependent phosphorylation of Erp1. This phosphorylation promotes Erp1–APC binding and inhibits APC-dependent degradation of cyclin B and MPF inactivation. Thus, in order to exit meiosis, oocytes must dissociate Erp1 from the APC to correctly degrade cyclin B. Indeed, this inactivation is rapidly induced after fertilization through Erp1 degradation. However, after meiosis exit, Erp1 is resynthesized again, and it is accumulated at about 20% of the amount found in CSF-arrested eggs. This amount is sufficient to induce metaphase arrest if the Mos/MAPK/Rsk pathway is activated (Liu *et al.*, 2006). Thus, as for Erp1, inactivation of c-Mos after fertilization is crucial to allow correct embryonic cell divisions.

The first event observed after fertilization is the increase of the intracellular Ca^{2+} levels (Cuthbertson and Cobbold, 1985; Kline and Kline, 1992; Ridgway, Gilkey and Jaffe, 1977). The initiator of Ca^{2+} release is not fully resolved, but it appears to be a protein delivered into the egg by the sperm. The most likely candidate is a sperm-specific member of the phospholipase C (PLC) family, namely PLCζ (Knott *et al.*, 2005; Saunders *et al.*, 2002; Yoda *et al.*, 2004). The recent reported data supports the hypothesis of a liberation of the PLCζ present at the sperm pronucleus after the fusion of the sperm with the egg. The liberation of this protein would catalyze production of inositol 1,4,5-triphosphate (IP3), and would finally act on type 1 IP3 receptors to initiate Ca^{2+} release from the endoplasmic reticulum (Runft, Jaffe and Mehlmann, 2002). After nuclear formation, PLCζ would then be retained at the nucleus through its nuclear localization signal and would, in this way, terminate Ca^{2+} spikes (Larman *et al.*, 2004; Marangos, FitzHarris and Carroll, 2003).

Upon fertilization, the calcium signal induces two main effects: release of cortical granules, which block polyspermy, and resumption of meiosis from metaphase II arrest. Contrary to cortical granules release, for which the Ca^{2+} -induced signalling pathway is not known (Knott *et al.*, 2006), the CSF release induced by Ca^{2+} is mediated by CaMKII. The first data describing a role of CaMKII in CSF release were reported in our laboratory by Lorca *et al.* (1993). We showed that the addition of a constitutive form of CaMKII to CSF extracts induced cyclin B degradation and MPF inactivation in the absence of Ca^{2+}. Moreover, when these extracts were supplemented with a specific inhibitor of this kinase, cyclin B remained stable even after Ca^{2+} addition. Thus, the Ca^{2+} signal induces metaphase II exit through a CaMKII-dependent pathway.

It took almost a decade to identify the downstream target of CaMKII. This target, simultaneously described by two different groups, appears to be the Erp1 protein (Schmidt *et al.*, 2005; Liu and Maller, 2005). Accordingly, these two groups showed that

Erp1 was phosphorylated upon CSF release by CaMKII. Indeed, CaMKII functions as a 'priming kinase' inducing a first phosphorylation of Erp1 at T195, which in turn promotes the interaction of Erp1 with the Polo-Box domain of Plx1. This interaction will be followed by a phosphorylation by Plx1 of Erp1 at S33 and S38 within its $DSGX_3S$ motif, a critical phosphorylation that is required for Erp1 binding to the Skp1-Cullin-Fbox (SCF) subunit, βtrcp. Finally, although it has not been formally demonstrated, the results obtained strongly suggest that Erp1 is then ubiquitinated by this ubiquitin ligase and degraded by the proteasome pathway (Schmidt *et al.*, 2005; Liu and Maller, 2005; Hansen, Tung and Jackson, 2006; Rauh *et al.*, 2005; Figure 13.3).

As reported above, Erp1 is rapidly synthesized again after meiosis exit and remains constant at least up to the early blastula, although in a higher electrophoretic mobility form, suggesting that at the first mitotic divisions it is not phosphorylated (Inoue *et al.*, 2007). Since phosphorylation and activation of Erp1 is mediated by the

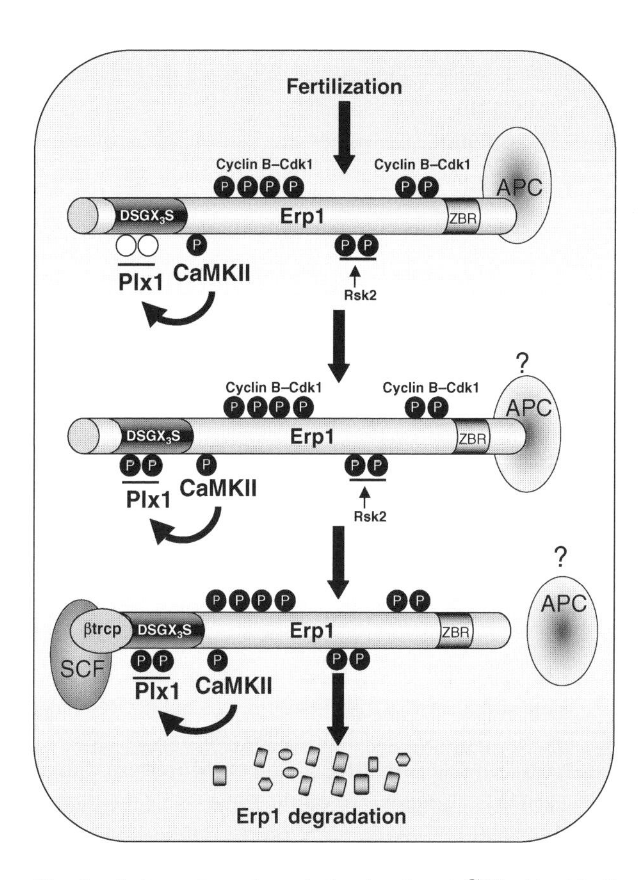

Figure 13.3 Fertilization induces Erp1 degradation by the $SCF^{\beta trcp}$ ubiquitin ligase. Fertilization triggers phosphorylation of Erp1 at T195 by CaMKII. This phosphorylation will permit the binding of Plx1 to Erp1, and its phosphorylation at the $DSGX_3S$ motif of this protein. This phosphorylation will trigger the binding of Erp1 to the SCF subunit, βtrcp, and its ubiquitination and degradation. Erp1 proteolysis will then lead to APC activation, cyclin B degradation and CSF exit. Phosphorylations of Erp1 by cyclin B–Cdk1 and Rsk2, regulating Erp1–APC association, are indicated. ZBR = zinc-binding region

c-Mos/MAPK/Rsk pathway during CSF arrest, this pathway must be inactivated in embryos in order to prevent the establishment of a new metaphase block. In fact, this is the case. Time-course experiments show that c-Mos is dephosphorylated and inactivated within 20 minutes after egg activation and is subsequently degraded by the ubiquitin-proteasome pathway. Although the exact ubiquitin ligase required for this degradation is still unknown, evidence suggests that this degradation is mediated by polyubiquitination of Lys 34 (Ishida *et al.*, 1993; Nishizawa *et al.*, 1993). However, this polyubiquitination cannot be performed when c-Mos is phosphorylated at S3. Indeed, phosphorylation at this residue not only prevents c-Mos recognition by the ubiquitin-dependent machinery, but also induces c-Mos activation. The main kinase involved in this phosphorylation is cyclin B–Cdk1. Phosphorylation of c-Mos by cyclin B–Cdk1 at GVBD increases c-Mos stability and activity, promoting the activation of the MAPK/Rsk pathway, the phosphorylation of Erp1 and the installation of the CSF arrest. However, upon fertilization, Erp1 is degraded, promoting cyclin B destruction and dephosphorylation of S3 of c-Mos. This dephosphorylation will trigger a rapid inactivation of c-Mos and a subsequent degradation of this protein by the ubiquitin-dependent pathway, preventing the establishment of a new CSF arrest at the first mitotic cell cycle of the embryo (Castro *et al.*, 2001). Thus cyclin B–Cdk1 and the c-Mos/MAPK/Rsk pathway act as a loop to install and to release the CSF arrest (Figure 13.4).

Recently, a new Ca^{2+}-dependent pathway required for CSF release has been described. This pathway involves the Ca^{2+}/calmodulin-dependent phosphatase calcineurin. The results, described simultaneously by two different laboratories, report that calcineurin is transiently activated immediately after the addition of Ca^{2+} to unfertilized *Xenopus* eggs, and that it is required for a metaphase II exit; however, different targets of this phosphatase on CSF exit were proposed by these two groups (Mochida and Hunt, 2007; Nishiyama *et al.*, 2007a). Nishiyama *et al.* showed that the addition of a phosphatase-dead form of calcineurin completely blocks cyclin B degradation and metaphase II exit after calcium addition. Interestingly, they also showed that the inhibition of calcineurin in activated CSF extracts by calcium addition reduced Erp1 binding to βtrcp and, thus, decreased Erp1 degradation and cyclin B–Cdk1 inactivation. Their results also indicated that calcineurin could dephosphorylate Erp1 protein and, in this way, promote Erp1 binding to βtrcp, suggesting that calcineurin could target Erp1 on CSF release (Nishiyama *et al.*, 2007a). On the contrary, the results obtained by Mochida *et al.* indicate that the inhibition of this phosphatase by the specific inhibitor, cyclosporin, only delayed cyclin B degradation; whereas, Erp1 destruction was identical to nonsupplemented control extracts. Moreover, they showed a major effect of the inhibition of this phosphatase on the kinetics of the global dephosphorylation of mitotic phosphoproteins. Finally, they propose that the target of calcineurin on CSF exit is the APC–Cdc20 complex, since, when they treated APC immunoprecipitates from CSF extracts with recombinant calcineurin, they observed the reversion of the mitotic mobility shift of a structural subunit of the APC, APC3, and of its activator, Cdc20. Thus, despite the fact that different results were obtained by these two groups on the mechanism controlled by calcineurin in CSF exit (probably due to the different methods used to inhibit this phosphatase), the data reported by both groups converge in

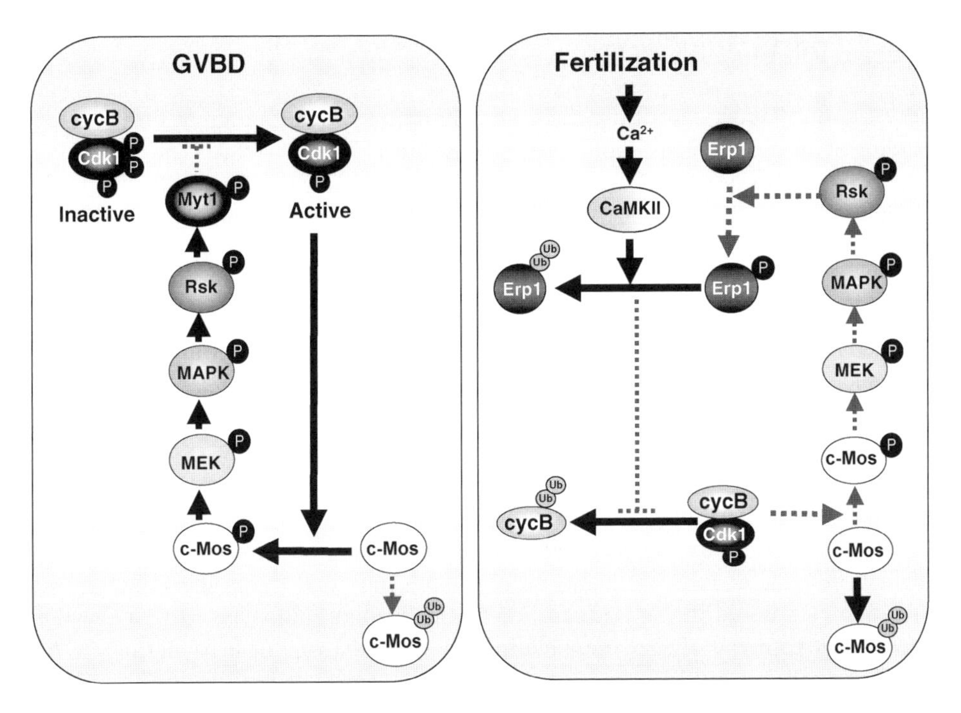

Figure 13.4 MPF and the c-Mos pathway act as a loop to install and to release the CSF. In immature oocytes, cyclin B–Cdk1 is present as a pre-MPF complex with its Cdk1 subunit phosphorylated at the threonine 14 and tyrosine 15 inhibitory residues. At GVBD, c-Mos protein is neosynthesized and the amplification loop is activated. The exact mechanism triggering the first activation of this loop is not known; however, once cyclin B–Cdk1 is partially active, it induces the c-Mos/MEK/MAPK/Rsk pathway. Rsk then promotes Myt1 phosphorylation inhibiting this kinase and, in this way, triggering Cdk1 dephosphorylation of inhibitory residues and total activation of cyclin B–Cdk1. Upon fertilization, the activation of CaMKII will induce Erp1 phosphorylation and will promote the ubiquitination and the degradation of this protein. This will raise the inhibition of the APC^{Cdc20} ubiquitin ligase by Erp1 and will promote cyclin B (cycB) degradation. Finally, as a consequence of cyclin B proteolysis and MPF inactivation, c-Mos will not be further phosphorylated in S3 and will be rapidly ubiquitinated and degraded

the fact that calcineurin is required to induce a correct exit of metaphase II after egg activation.

In summary, the reported data suggest that upon fertilization, the fusion of the male sperm pronucleous with the oocyte will induce an increase of the intracellular calcium levels, which will activate both the CaMKII and the phosphatase calcineurin. CaMKII activation will induce Erp1 phosphorylation, which will promote a second phosphorylation of this protein by the Plx1 kinase at the $DSGX_3S$ motif. Erp1, phosphorylated in this way, will bind βtrcp and will consequently be ubiquitinated by the $SCF^{\beta trcp}$ ubiquitin ligase and degraded by the proteasome. On the other hand, the activation of the phosphatase calcineurin will induce, through the dephosphorylation of either the Erp1 protein or the APC, the activation of this ubiquitin ligase, the degradation of cyclin B and the exit of metaphase II arrest.

13.4 Conclusions

Since the first report describing the CSF activity, significant progress on the understanding of the mechanisms implicated in this pathway has been made. The discovery of a role of the c-Mos/MAPK/Rsk pathway in the establishment of this activity was the first step in this research. This was followed by the identification, more than 10 years later, of the APC inhibitor Erp1, whose inhibitory activity is modulated by the c-Mos/MAPK/Rsk pathway, linking CSF arrest and cyclin B stabilization by APC inhibition. Finally, the characterization of the CaMKII-Plx1-dependent degradation of Erp1, as well as the characterization of a new calcineurin-dependent pathway on egg activation, has allowed us to understand the molecular mechanisms triggering CSF release. However, many questions remain to be answered. For example, if the APC inhibitor activity of Erp1 is maintained by Rsk-dependent phosphorylation of this protein, how could we explain the different results obtained by several laboratories that reported a requirement of the c-Mos/MAPK/Rsk pathway for CSF establishment but not for its maintenance (Bhatt and Ferrell, 1999; Gross *et al.*, 1999; Dumont *et al.*, 2005)? What is the target of calcineurin on CSF release? Is Erp1 exclusively inhibited by proteolysis upon egg activation? Or, could it also be inhibited by dephosphorylation? Is there a CaMKII-independent pathway of Erp1 degradation? What is the role of Erp1 during the first embryonic cell cycles? Why is Erp1 synthesized again after fertilization?

Further studies must be performed in order to answer these questions.

Acknowledgements

This work was supported by the Ligue Nationale Contre le Cancer (Equipe Labellisée).

References

Abrieu, A., Magnaghi-Jaulin, L., Kahana, J.A. *et al.* (2001) Mps1 is a kinetochore-associated kinase essential for the vertebrate mitotic checkpoint. *Cell*, **106**, 83–93.

Bhatt, R.R. and Ferrell, J.E. Jr (1999) The protein kinase p90 rsk as an essential mediator of cytostatic factor activity. *Science*, **286**, 1362–1365.

Burke, D.J. and Stukenberg, P.T. (2008) Linking kinetochore-microtubule binding to the spindle checkpoint. *Dev. Cell*, **14**, 474–479.

Castro, A., Peter, M., Magnaghi-Jaulin, L. *et al.* (2001) Cyclin B/cdc2 induces c-Mos stability by direct phosphorylation in Xenopus oocytes. *Mol. Biol. Cell*, **12**, 2660–2671.

Castro, A., Bernis, C., Vigneron, S. *et al.* (2005) The anaphase-promoting complex: a key factor in the regulation of cell cycle. *Oncogene*, **24**, 314–325.

Cuthbertson, K.S. and Cobbold, P.H. (1985) Phorbol ester and sperm activate mouse oocytes by inducing sustained oscillations in cell Ca2+. *Nature*, **316**, 541–542.

Dumont, J., Umbhauer, M., Rassinier, P. *et al.* (2005) p90Rsk is not involved in cytostatic factor arrest in mouse oocytes. *J. Cell Biol.*, **169**, 227–231.

Grimison, B., Liu, J., Lewellyn, A.L. and Maller, J.L. (2006) Metaphase arrest by cyclin E-Cdk2 requires the spindle-checkpoint kinase Mps1. *Curr. Biol.*, **16**, 1968–1973.

Gross, S.D., Schwab, M.S., Lewellyn, A.L. and Maller, J.L. (1999) Induction of metaphase arrest in cleaving Xenopus embryos by the protein kinase p90Rsk. *Science*, **286**, 1365–1367.

Haccard, O., Sarcevic, B., Lewellyn, A. *et al.* (1993) Induction of metaphase arrest in cleaving Xenopus embryos by MAP kinase. *Science*, **262**, 1262–1265.

Hansen, D.V., Tung, J.J. and Jackson, P.K. (2006) CaMKII and polo-like kinase 1 sequentially phosphorylate the cytostatic factor Emi2/XErp1 to trigger its destruction and meiotic exit. *Proc. Natl. Acad. Sci. USA*, **103**, 608–613.

Harper, J.W., Burton, J.L. and Solomon, M.J. (2002) The anaphase-promoting complex: it's not just for mitosis any more. *Genes Dev.*, **16**, 2179–2206.

Hsu, J.Y., Reimann, J.D., Sorensen, C.S. *et al.* (2002) E2F-dependent accumulation of hEmi1 regulates S phase entry by inhibiting APC(Cdh1). *Nat. Cell Biol.*, **4**, 358–366.

Inoue, D., Ohe, M., Kanemori, Y. *et al.* (2007) A direct link of the Mos-MAPK pathway to Erp1/Emi2 in meiotic arrest of Xenopus laevis eggs. *Nature*, **446**, 1100–1104.

Ishida, N., Tanaka, K., Tamura, T. *et al.* (1993) Mos is degraded by the 26S proteasome in a ubiquitin-dependent fashion. *FEBS Lett.*, **324**, 345–348.

Kanki, J.P. and Donoghue, D.J. (1991) Progression from meiosis I to meiosis II in Xenopus oocytes requires de novo translation of the mosxe protooncogene. *Proc. Natl. Acad. Sci. USA*, **88**, 5794–5798.

Kline, D. and Kline, J.T. (1992) Repetitive calcium transients and the role of calcium in exocytosis and cell cycle activation in the mouse egg. *Dev. Biol.*, **149**, 80–89.

Knott, J.G., Kurokawa, M., Fissore, R.A. *et al.* (2005) Transgenic RNA interference reveals role for mouse sperm phospholipase Czeta in triggering Ca2+ oscillations during fertilization. *Biol. Reprod.*, **72**, 992–996.

Knott, J.G., Gardner, A.J., Madgwick, S. *et al.* (2006) Calmodulin-dependent protein kinase II triggers mouse egg activation and embryo development in the absence of Ca2+ oscillations. *Dev. Biol.*, **296**, 388–395.

Larman, M.G., Saunders, C.M., Carroll, J. *et al.* (2004) Cell cycle-dependent Ca2+ oscillations in mouse embryos are regulated by nuclear targeting of PLCzeta. *J. Cell Sci.*, **117**, 2513–2521.

Lee, M.G. and Nurse, P. (1987) Complementation used to clone a human homologue of the fission yeast cell cycle control gene cdc2. *Nature*, **327**, 31–35.

Liu, J. and Maller, J.L. (2005) Calcium elevation at fertilization coordinates phosphorylation of XErp1/Emi2 by Plx1 and CaMK II to release metaphase arrest by cytostatic factor. *Curr. Biol.*, **15**, 1458–1468.

Liu, J., Grimison, B., Lewellyn, A.L. and Maller, J.L. (2006) The anaphase-promoting complex/cyclosome inhibitor Emi2 is essential for meiotic but not mitotic cell cycles. *J. Biol. Chem.*, **281**, 34736–34741.

Liu, J., Grimison, B. and Maller, J.L. (2007) New insight into metaphase arrest by cytostatic factor: from establishment to release. *Oncogene*, **26**, 1286–1289.

Lohka, M.J., Hayes, M.K. and Maller, J.L. (1988) Purification of maturation-promoting factor, an intracellular regulator of early mitotic events. *Proc. Natl. Acad. Sci. USA*, **85**, 3009–3013.

Lorca, T., Galas, S., Fesquet, D. *et al.* (1991) Degradation of the proto-oncogene product p39mos is not necessary for cyclin proteolysis and exit from meiotic metaphase: requirement for a Ca(2+)-calmodulin dependent event. *EMBO J.*, **10**, 2087–2093.

Lorca, T., Cruzalegui, F.H., Fesquet, D. *et al.* (1993) Calmodulin-dependent protein kinase II mediates inactivation of MPF and CSF upon fertilization of Xenopus eggs. *Nature*, **366**, 270–273.

Lorca, T., Castro, A., Martinez, A.M. *et al.* (1998) Fizzy is required for activation of the APC/cyclosome in Xenopus egg extracts. *EMBO J.*, **17**, 3565–3575.

Marangos, P., FitzHarris, G. and Carroll, J. (2003) Ca2+ oscillations at fertilization in mammals are regulated by the formation of pronuclei. *Development*, **130**, 1461–1472.

Masui, Y. and Markert, C.L. (1971) Cytoplasmic control of nuclear behavior during meiotic maturation of frog oocytes. *J. Exp. Zool.*, **177**, 129–145.

Mochida, S. and Hunt, T. (2007) Calcineurin is required to release Xenopus egg extracts from meiotic M phase. *Nature*, **449**, 336–340.

Musacchio, A. and Hardwick, K.G. (2002) The spindle checkpoint: structural insights into dynamic signalling. *Nat. Rev. Mol. Cell Biol.*, **3**, 731–741.

Nishiyama, T., Yoshizaki, N., Kishimoto, T. and Ohsumi, K. (2007a) Transient activation of calcineurin is essential to initiate embryonic development in Xenopus laevis. *Nature*, **449**, 341–345.

Nishiyama, T., Ohsumi, K. and Kishimoto, T. (2007b) Phosphorylation of Erp1 by p90rsk is required for cytostatic factor arrest in Xenopus laevis eggs. *Nature*, **446**, 1096–1099.

Nishizawa, M., Furuno, N., Okazaki, K. *et al.* (1993) Degradation of Mos by the N-terminal proline (Pro2)-dependent ubiquitin pathway on fertilization of Xenopus eggs: possible significance of natural selection for Pro2 in Mos. *EMBO J.*, **12**, 4021–4027.

Ohsumi, K., Koyanagi, A., Yamamoto, T.M. *et al.* (2004) Emi1-mediated M-phase arrest in Xenopus eggs is distinct from cytostatic factor arrest. *Proc. Natl. Acad. Sci. USA*, **101**, 12531–12536.

Rauh, N.R., Schmidt, A., Bormann, J. *et al.* (2005) Calcium triggers exit from meiosis II by targeting the APC/C inhibitor XErp1 for degradation. *Nature*, **437**, 1048–1052.

Reimann, J.D. and Jackson, P.K. (2002) Emi1 is required for cytostatic factor arrest in vertebrate eggs. *Nature*, **416**, 850–854.

Reimann, J.D., Freed, E., Hsu, J.Y. *et al.* (2001) Emi1 is a mitotic regulator that interacts with Cdc20 and inhibits the anaphase promoting complex. *Cell*, **105**, 645–655.

Ridgway, E.B., Gilkey, J.C. and Jaffe, L.F. (1977) Free calcium increases explosively in activating medaka eggs. *Proc. Natl. Acad. Sci. USA*, **74**, 623–627.

Runft, L.L., Jaffe, L.A. and Mehlmann, L.M. (2002) Egg activation at fertilization: where it all begins. *Dev. Biol.*, **245**, 237–254.

Sagata, N., Oskarsson, M., Copeland, T. *et al.* (1988) Function of c-mos proto-oncogene product in meiotic maturation in Xenopus oocytes. *Nature*, **335**, 519–525.

Sagata, N., Watanabe, N., Vande Woude, G.F. and Ikawa, Y. (1989) The c-mos proto-oncogene product is a cytostatic factor responsible for meiotic arrest in vertebrate eggs. *Nature*, **342**, 512–518.

Saunders, C.M., Larman, M.G., Parrington, J. *et al.* (2002) PLC zeta: a sperm-specific trigger of Ca(2+) oscillations in eggs and embryo development. *Development*, **129**, 3533–3544.

Schmidt, A., Duncan, P.I., Rauh, N.R. *et al.* (2005) Xenopus polo-like kinase Plx1 regulates XErp1, a novel inhibitor of APC/C activity. *Genes Dev.*, **19**, 502–513.

Schmidt, A., Rauh, N.R., Nigg, E.A. and Mayer, T.U. (2006) Cytostatic factor: an activity that puts the cell cycle on hold. *J. Cell Sci.*, **119**, 1213–1218.

Sharp-Baker, H. and Chen, R.H. (2001) Spindle checkpoint protein Bub1 is required for kinetochore localization of Mad1, Mad2, Bub3, and CENP-E, independently of its kinase activity. *J. Cell Biol.*, **153**, 1239–1250.

Tung, J.J., Hansen, D.V., Ban, K.H. *et al.* (2005) A role for the anaphase-promoting complex inhibitor Emi2/XErp1, a homolog of early mitotic inhibitor 1, in cytostatic factor arrest of Xenopus eggs. *Proc. Natl. Acad. Sci. USA*, **102**, 4318–4323.

Tunquist, B.J. and Maller, J.L. (2003) Under arrest: cytostatic factor (CSF)-mediated metaphase arrest in vertebrate eggs. *Genes Dev.*, **17**, 683–710.

Tunquist, B.J., Schwab, M.S., Chen, L.G. and Maller, J.L. (2002) The spindle checkpoint kinase bub1 and cyclin e/cdk2 both contribute to the establishment of meiotic metaphase arrest by cytostatic factor. *Curr. Biol.*, **12**, 1027–1033.

Tunquist, B.J., Eyers, P.A., Chen, L.G. *et al.* (2003) Spindle checkpoint proteins Mad1 and Mad2 are required for cytostatic factor-mediated metaphase arrest. *J. Cell Biol.*, **163**, 1231–1242.

Vorlaufer, E. and Peters, J.M. (1998) Regulation of the cyclin B degradation system by an inhibitor of mitotic proteolysis. *Mol. Biol. Cell*, **9**, 1817–1831.

Watanabe, N., Vande Woude, G.F., Ikawa, Y. and Sagata, N. (1989) Specific proteolysis of the c-mos proto-oncogene product by calpain on fertilization of Xenopus eggs. *Nature*, **342**, 505–511.

Wu, J.Q. and Kornbluth, S. (2008) Across the meiotic divide – CSF activity in the post-Emi2/XErp1 era. *J. Cell Sci.*, **121**, 3509–3514.

Wu, J.Q., Hansen, D.V., Guo, Y. *et al.* (2007a) Control of Emi2 activity and stability through Mos-mediated recruitment of PP2A. *Proc. Natl. Acad. Sci. USA*, **104**, 16564–16569.

Wu, Q., Guo, Y., Yamada, A. *et al.* (2007b) A role for Cdc2- and PP2A-mediated regulation of Emi2 in the maintenance of CSF arrest. *Curr. Biol.*, **17**, 213–224.

Yoda, A., Oda, S., Shikano, T. *et al.* (2004) Ca2+ oscillation-inducing phospholipase C zeta expressed in mouse eggs is accumulated to the pronucleus during egg activation. *Dev. Biol.*, **268**, 245–257.

14

Cytostatic arrest: post-ovulation arrest until fertilization in metazoan oocytes

Tomoko Nishiyama,[1] Kazunori Tachibana[2] and Takeo Kishimoto[2]

[1]Institute of Molecular Pathology, 1030 Vienna, Austria
[2]Graduate School of Bioscience and Biotechnology, Laboratory of Cell and Developmental Biology, Tokyo Institute of Technology, Yokohama 226-8501, Japan

14.1 Introduction

After meiotic reinitiation from the primary arrest at prophase of meiosis I (pro-I), the cell cycle of mature eggs arrests again at a particular stage of meiosis until fertilization, except in species where fertilization releases the primary arrest. The second arrest, or 'post-meiotic reinitiation arrest', enables mature eggs to prevent parthenogenesis and to begin development as a zygote only after fertilization, thus coordinating meiotic maturation and fertilization. The stage of the second arrest depends on the species: typically, metaphase of meiosis I (meta-I) in insects and ascidians, metaphase of meiosis II (meta-II) in most vertebrates, or G1 phase (pronuclear stage) after completion of meiosis II in echinoderms (Adiyodi and Adiyodi, 1983; Masui, 1985; Sagata, 1996; Kishimoto, 2003).

Amongst these, the meta-II arrest in frog eggs has been most extensively studied. The meta-II arrest is also called 'CSF arrest', as CSF stands for 'cytostatic factor', which was a hypothetical activity thought to cause meta-II arrest (Masui and Markert, 1971). Based on studies for over 30 years, we now can explain at least the principal molecular mechanism of CSF arrest. In addition, although the original CSF was described only in relation to meta-II arrest, recent elucidation of molecular components responsible for various second arrests now allows extension of the idea of CSF to the other second arrests. Here, we review first the original 'meta-II CSF' and then the other 'cytostatic arrests'. Based on these, we will propose a comprehensive view of CSF signalling.

Oogenesis: The Universal Process Marie-Hélène Verlhac and Anne Villeneuve

14.2 Frog meta-II arrest by CSF

14.2.1 Origin of the CSF

In the early twentieth century, the importance of the second arrest had already been realized, but only in terms of cell biological observations. One of the hypotheses was that the lack of an organelle for cell division, such as the centrosome, caused the arrest (Boveri, 1907). Another was that a lack of communication between the egg nucleus and the egg cytoplasm caused the arrest until fertilization (Lillie, 1912). In addition, stable cortex of unfertilized egg was also thought to cause the arrest (Loeb, 1913). From the 1930s to the 1960s, several candidates such as carbon dioxide, the polysaccharide heparin, polynucleotides and metabolic inhibitors were proposed as cell-division inhibitors, which accumulate during oogenesis (Masui, 1985). However, they did not seem to be the endogenous inhibitor, because the cell cycle of oocytes could be inhibited easily by nonspecific toxicity (Masui, 2000).

In 1971, Yoshio Masui and Clement Markert first described an activity in unfertilized mature eggs of the leopard frog *Rana pipiens*, that caused meta-II arrest (Masui and Markert, 1971). In this paper, they demonstrated that blastomeres of two-cell embryos frequently stopped cleaving when they were injected with cytoplasm of unfertilized meta-II eggs (Figure 14.1a). In the arrested blastomere, the mitotic spindle showed the same features as the meta-II arrested spindle (Masui, 1985). By contrast, when cytoplasm from a zygote was injected, cleavage was not inhibited. From these results, they concluded that some transferable cytoplasmic factor in unfertilized eggs was responsible for meta-II arrest. This hypothetical factor was named 'cytostatic factor (CSF)', and they proposed five criteria which CSF should satisfy: (1) Appear during oocyte maturation. (2) Disappear during fertilization (egg activation). (3) Be destroyed under the same physicochemical conditions as those that cause egg activation. (4) Provide the arrested zygote with the same properties as those of the unfertilized egg. (5) Inhibit mitosis of the zygote reversibly (Masui, 2000). In this cytoplasmic transfer experiment, Masui and Markert originally expected that MPF (maturation-promoting factor) activity in meta-II eggs might 'accelerate' mitosis when injected into zygotes, because MPF induced resumption of meiosis in immature oocytes. Thus, an unexpected observation led to the discovery of CSF.

14.2.2 Discovery of the Mos/MAPK pathway: the first breakthrough of CSF history

Since the discovery of CSF in 1971, researchers have struggled to purify CSF molecules from unfertilized egg. However, purification of CSF seemed to be difficult because CSF activity, which causes cleavage arrest in zygotes when the cytoplasm is injected into blastomeres (criteria 5), was readily decreased by pricking of the egg cortex. Masui and his associates succeeded in preparing egg extracts without losing the activity of CSF, and revealed that its activity was decreased by addition of Ca^{2+} to, or by removal of Mg^{2+} from, the extracts, and enhanced by phosphatase inhibitors and ATP

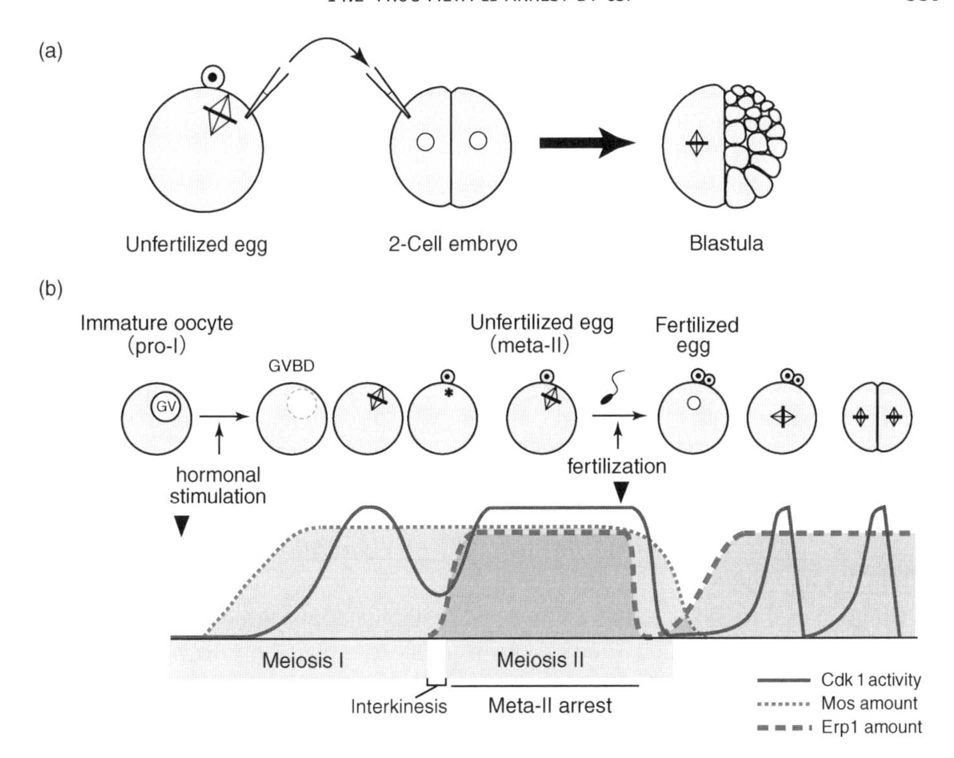

Figure 14.1 CSF and meiotic and cleavage cycles in *Xenopus* eggs. (a) Experiment of cytoplasmic transfer leading to CSF discovery. When cytoplasm of unfertilized eggs is injected into one blastomere of a two-cell embryo, the cell cycle in the recipient blastomere is arrested at metaphase. From this experiment, existence of cytostatic factor (CSF) in unfertilized eggs was proposed. (b) Meiosis progression of *Xenopus* oocytes and changes in Cdk1 activity, Mos, and Erp1 levels. The cell cycle of fully grown immature oocytes arrests at prophase of meiosis I (pro-I). Meiosis is resumed by hormonal stimulation and it is arrested again at metaphase of meiosis II (meta-II), which is released by fertilization. Although Mos and Erp1 accumulate during progression of meiosis, both are present only in meiosis II. GV = germinal vesicle; GVBD = germinal vesicle breakdown

(Masui, 1974; Meyerhof and Masui, 1977; Shibuya and Masui, 1988). In 1989, CSF was partially purified as a protease-sensitive but RNase-resistant molecule with a sedimentation coefficient of 3S (Shibuya and Masui, 1989). These findings indicated that the CSF is a small protein activated by its Mg^{2+}-dependent phosphorylation, but inactivated by Ca^{2+} ions.

In 1989, from studies done in a different context, Noriyuki Sagata and his colleagues identified Mos, the c-*mos* proto-oncogene product, as a component of CSF (Sagata *et al.*, 1989). The c-*mos* proto-oncogene was originally isolated as the cellular homologue of the v-*mos* oncogene of Moloney murine sarcoma virus (Oskarsson *et al.*, 1980). Because the mRNA of c-*mos* was found in vertebrate gonadal tissues (Propst and Vande Woude, 1985; Goldman *et al.*, 1987; Mutter and Wolgemuth, 1987; Propst *et al.*, 1987; Schmidt *et al.*, 1988), the function of Mos in oogenesis was vigorously investigated from the view of cellular transformation. However, Mos actually had a cytostatic function. Mos is a serine/threonine protein kinase and expressed specifically during oocyte maturation (Figure 14.1b). Mos not only satisfies the five criteria of CSF, but is also required for the CSF activity of the unfertilized meta-II egg's cytoplasm to induce mitotic arrest in blastomeres, since Mos-deprived or suppressed cytoplasm did not induce metaphase arrest in a cytoplasmic transfer experiment (Sagata *et al.*, 1989). This was the first breakthrough of CSF history.

Following this landmark discovery, downstream molecules were identified one after another. It was found that Mos directly phosphorylates and activates MEK1 (MAPK/ERK kinase 1), which is the immediate upstream activator of MAPK (mitogen-activated protein kinase), and, in turn, MAPK is activated in *Xenopus* oocytes (Nebreda and Hunt, 1993; Posada *et al.*, 1993; Shibuya and Ruderman, 1993). Furthermore, the introduction of constitutively active MEK or a thiophosphorylated, constitutively active form of MAPK (Haccard *et al.*, 1993) into *Xenopus* embryos or cell-free cycling extracts (Abrieu *et al.*, 1996, 1997; Bitangcol *et al.*, 1998; Murakami and Vande Woude, 1998; Walter, Guadagno and Ferrell, 1997) causes a metaphase arrest. And finally, in 1999, the 90 kDa ribosomal protein S6 kinase ($p90^{Rsk}$, RSK), which is one of the MAPK-activated proteins, was found to have CSF activity (Bhatt and Ferrell, 1999; Gross *et al.*, 1999). Thus, the Mos/MEK/MAPK/Rsk cascade was revealed to comprise CSF in *Xenopus* oocytes.

Although the Mos/MAPK/Rsk pathway was established as a key catalytic component of CSF, its target was unknown and nobody was able to explain how this pathway causes meta-II arrest. To understand this issue, we had to wait for the next breakthrough, which came from an unexpected direction.

14.2.3 Discovery of Erp1: an authentic inhibitor of APC/C

From the view of cell cycle control, meta-II arrest often might be considered to be a state similar to mitotic arrest in somatic cells by a so-called 'spindle assembly checkpoint' (Musacchio and Salmon, 2007). Indeed the basis of cell cycle regulation is common to both mitotic and meiotic arrests, where destruction of cyclin B is inhibited and Cdk1 activity is maintained at a high level. Since findings that the anaphase-promoting complex/cyclosome (APC/C) is responsible for poly-ubiquitination and destruction of cyclin B (King *et al.*, 1995; Sudakin *et al.*, 1995), APC/C has been thought to be suppressed also in meta-II arrest. Based on evidence that the APC/C coactivator Cdc20 is suppressed by the spindle assembly checkpoint protein Mad2 (Hwang *et al.*, 1998; Kim *et al.*, 1998), some of the known inhibitory proteins of APC/C–Cdc20 in spindle assembly checkpoint arrest, such as Mad1, Mad2, Bub1, and Mps1, were proposed to function in meta-II arrest (Schwab *et al.*, 2001; Tunquist *et al.*, 2002, 2003; Liu and Maller, 2005). However, because spindle assembly checkpoint arrest is essentially different from meta-II arrest in its dependency on Mos or sensitivity to Ca^{2+}, these checkpoint proteins are unlikely to act in Mos-dependent meta-II arrest, as some reports suggested in both *Xenopus* and mouse (Tsurumi *et al.*, 2004; Liu and Maller, 2005; Grimison *et al.*, 2006).

Another inhibitor of APC/C, Emi1 (Early mitotic inhibitor 1), which inhibits APC/C from G1/S to mitotic prophase in somatic cells, was also proposed to be an essential component of CSF (Reimann and Jackson, 2002). However, other evidence strongly suggested that Emi1 could not be involved in meta-II arrest (Ohsumi *et al.*, 2004):

(i) Emi1 protein is absent in unfertilized meta-II eggs of *Xenopus* due to its instability; (ii) Emi1 induces metaphase arrest independently of MAPK activity; and (iii) the Emi1-induced metaphase arrest cannot be released by Ca^{2+} (see also Chapter 13).

The confusion over identifying the APC/C inhibitor was ended by the discovery of Erp1. In 2005, Thomas Mayer and colleagues reported that Erp1 (Emi1-related protein 1, also called Emi2), a novel APC/C inhibiting protein, is crucial for meta-II arrest in *Xenopus* eggs (Schmidt *et al.*, 2005). They identified Erp1 through two-hybrid screening to seek a direct target of Plk1 (Polo-like kinase 1), which is required for CSF exit (Descombes and Nigg, 1998) (see Chapter 13). They showed that (i) Erp1 begins to accumulate significantly after interkinesis, an interphase-like stage between meiosis I and II and where Cdk1 activity is transiently decreased (Figure 14.1b); (ii) its amount reaches a maximum level at meta-II; (iii) Erp1 is degraded immediately after fertilization; (iv) immunodepletion of Erp1 from meta-II arrested egg extract causes cell cycle exit to interphase without Ca^{2+}; and (v) Erp1 can inhibit APC/C activity directly *in vitro* (Schmidt *et al.*, 2005). These pieces of evidence indicated that Erp1 is the long-sought APC/C inhibitor for meta-II arrest. This finding was the second breakthrough in CSF history.

Erp1 has several characteristic domains (Figure 14.2). An F-box in the C-terminal half is required for interaction of Erp1 with Skp1 in yeast two-hybrid analysis (P.I. Duncan and E.A. Nigg, unpublished), but might not contribute to APC/C inhibitory activity of Erp1 (A. Schmidt, N.R. Raugh and T.U. Mayer, unpublished). There is also a D-box (destruction box; see Section 14.2.4), and a Zn^{2+}-binding region (ZBR), in which the Cys583 residue is essential for APC/C inhibition (Schmidt *et al.*, 2005). Mayer and colleagues also elegantly demonstrated how Erp1 is degraded by a Ca^{2+}-dependent process (Rauh *et al.*, 2005; see also Chapter 13). Briefly, direct phosphorylation at

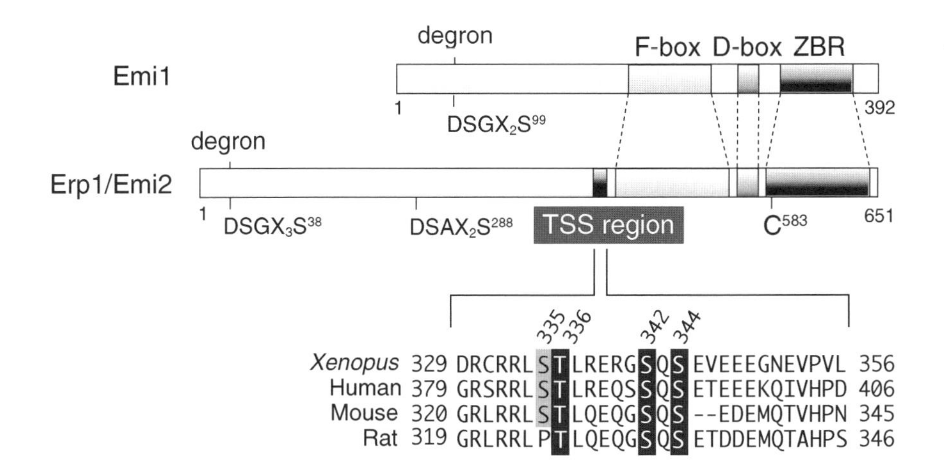

Figure 14.2 Comparison of *Xenopus* Emi1 and Erp1/Emi2 proteins. Both Emi1 and Erp1 contain the degron sequence ($DSGX_{2-3}S$) required for Ca^{2+}-dependent degradation in their N-terminus, and D-box, ZBR, and F-box in C-terminus. Cys583 in ZBR is essential for Erp1 to inhibit APC/C. The central TSS region, which is highly conserved in vertebrates and phosphorylated by Rsk, is unique to Erp1. Erp1 further contains the central degron sequence ($DSAX_2S$) which contributes to Ca^{2+}-independent instability

Thr195 of Erp1 by CaMKII (Ca^{2+}/CaM-dependent protein kinase II) generates a Plk1 binding site on Erp1 and, in turn, Plk1 binds to Erp1. Plk1 phosphorylates a 'phosphodegron sequence' in the N-terminus of Erp1, which is recognized by ubiquitin ligase $SCF^{\beta TrCP}$, and, then, Erp1 is degraded by proteasome (Rauh *et al.*, 2005; see also Chapter 13). Although the C-terminal half of Erp1, including the F-box and ZBR (the functional core for APC/C inhibition), shows 39% homology with Emi1, the N-terminal half (the Ca^{2+}-dependent regulatory domain for Erp1 destruction) is completely unique to Erp1 (Schmidt *et al.*, 2005; Figure 14.2). This unique feature of the N-terminal sequence impressed on us that Erp1 is an authentic core factor for meta-II arrest.

14.2.4 How is CSF arrest established?: cooperation of Mos and Erp1

Detailed description of changes in Erp1 levels revealed that Erp1 is degraded immediately after fertilization, accumulates until the first M phase to a level comparable to that in unfertilized eggs, and survives until midblastula (Inoue *et al.*, 2007; Nishiyama, Ohsumi and Kishimoto, 2007a; Figure 14.1b). If Erp1 alone could cause metaphase arrest, the comparable amount of Erp1 as present in unfertilized eggs should also induce metaphase arrest during the embryonic cell cycles. From this consideration, it was postulated that a regulator should be required to confine Erp1 function to unfertilized meta-II eggs. In this context, Mos protein already begins to accumulate before meiosis I, where Erp1 has not yet accumulated (Inoue *et al.*, 2007; Nishiyama, Ohsumi and Kishimoto, 2007a; Ohe *et al.*, 2007); it disappears shortly after fertilization and, thereafter, never reaccumulates (Figure 14.1b). Comparison of changes in both Mos and Erp1 levels showed us a correspondence – only when both Mos and Erp1 are present, metaphase arrest is induced. Based on the idea that the Mos/MAPK/Rsk pathway, an essential component of CSF, might function as the regulator, the relationship between Erp1 and the Mos/MAPK/Rsk pathway was tested.

Because the electrophoretic mobility of Erp1 changed in parallel with Mos/MAPK activity, three amino acids Thr336, Ser342, and Ser344 (referred to as TSS hereafter) of Erp1, which are conserved in many vertebrates (Figure 14.2), were identified as phosphorylation residues that are responsible for the mobility shift. More importantly, Rsk, which functions directly downstream of MAPK, phosphorylated TSS directly *in vitro* (Nishiyama, Ohsumi and Kishimoto, 2007a). When endogenous Erp1 was substituted by non-phosphorylatable mutant of TSS (Erp1-3A), the oocytes did not arrest at meta-II and degenerated like Erp1-deprived oocytes, indicating that phosphorylation of TSS by Rsk is required for meta-II arrest (Nishiyama, Ohsumi and Kishimoto, 2007a). Because a four to five-times higher concentration of Erp1-3A than wild-type caused meta-II arrest, phosphorylation seems to enhance the activity of Erp1 rather than generate it. Although phosphorylation of TSS also stabilizes Erp1 through suppression of the degron in the central part, a nondegradable mutant of Erp1-3A still fails to induce meta-II arrest (Nishiyama, Ohsumi and Kishimoto, 2007a). Thus, phosphorylation of Erp1 by Rsk not only enhances the activity of Erp1 but also stabilizes it, while the activity might be regulated independently of the stability. At the same time, it was also reported that phosphorylation of Ser335 and Ser336 of Erp1 by Rsk is essential for meta-II arrest and stability of Erp1 (Inoue *et al.*, 2007). These two residues were found as a consensus site for Rsk phosphorylation. Although Ser335 is less conserved in some vertebrates, it is clear that phosphorylation of these four residues, Ser335 + TSS (referred to as the TSS region hereafter: Figure 14.2), in the centre of Epr1, is essential for its stability and meta-II arresting activity in *Xenopus* eggs.

The next questions are how phosphorylation of Erp1 enhances its activity and, more importantly, how Erp1 inhibits APC/C. Both Emi1 and Erp1 have D-box and ZBR domains (Figure 14.2). The D-box is the APC/C recognition sequence (Arg-x-x-Leu), which is present in APC/C substrates and necessary for binding to APC/C or its coactivators (Yamano *et al.*, 2004; Kraft *et al.*, 2005; Peters, 2006). Emi1 is known to inhibit APC/C by direct binding to APC/C through its D-box and acting as a pseudosubstrate of APC/C (Miller *et al.*, 2006). Interestingly, when the ZBR function of Emi1 is deleted, Emi1 is ubiquitinated, presumably by APC/C, and loses its function as a pseudosubstrate (Miller *et al.*, 2006). Because the C-terminus of Erp1 containing the D-box and ZBR is similar to that of Emi1, it might be possible that both proteins inhibit APC/C in the same manner. Indeed, a D-box mutant of Erp1 fails to induce meta-II arrest, and it shows only weak binding to APC/C (Nishiyama, Ohsumi and Kishimoto, 2007a). It is noteworthy that a ZBR mutant, which also completely fails to induce meta-II arrest like the D-box mutant, associates with APC/C to the same extent as wild-type (Nishiyama, Ohsumi and Kishimoto, 2007a), as is also reported for Emi1 (Miller *et al.*, 2006). It is likely, therefore, that Erp1's ability to inhibit APC/C depends on both D-box-mediated binding to APC/C and on ZBR-mediated inhibition, each of which is inter-independent. Since Erp1 carrying non-phosphorylatable TSS shows reduction of binding to APC/C, and it is further decreased by a D-box mutation to the same extent as Erp1 with D-box mutation alone, phosphorylation of TSS seems to facilitate at least the D-box-mediated binding to APC/C (Nishiyama, Ohsumi and Kishimoto, 2007a). In addition, phosphorylation of TSS might also play a role in enhancing ZBR function, which still remains largely unknown. The question of how ZBR inhibits APC/C is a future challenge.

14.2.5 Maintenance of meta-II arrest: dynamic equilibrium of Cdk1 activity

Mature eggs of *Xenopus laevis* can be halted at meta-II for a long period: up to several days. This arrest is often thought to be a static or quiet state because fertilized eggs look more dynamic than unfertilized eggs. Actually, because, during meta-II arrest, degradation of cyclin B is inhibited and Cdk1 activity is maintained at a high level, APC/C activity is apparently inhibited completely. However, recent studies show that meta-II arrest is not quiet at all but rather a dynamic state, where synthesis and degradation of cyclin B are balanced (Yamamoto *et al.*, 2005; Wu *et al.*, 2007).

It has been suggested that synthesis of cyclin B continues in meta-II arrested eggs (Kubiak *et al.*, 1993; Thibier *et al.*, 1997), as well as egg extracts (Yamamoto *et al.*, 2005). Surprisingly, the protein synthesis rate in meta-II arrested egg extracts is approximately 60% of that in interphase extracts. Nevertheless, the total level of cyclin B amount or Cdc2 activity does not change in meta-II arrested egg extracts, suggesting that a considerable amount of cyclin B degradation, equal to the amount of new synthesis, takes place during meta-II arrest. Indeed, when exogenous cyclin B was added to meta-II arrested egg extracts, Cdk1 activity was increased proportionally with the concentration of cyclin B added; subsequently, the Cdk1 activity dropped to the initial level through cyclin B degradation, and the level of Ckd1 activity was maintained thereafter (Yamamoto *et al.*, 2005). The degradation of cyclin B induced by elevated Cdk1 activity depends on APC/C–Cdc20, indicating that a negative feedback system mediated by APC/C–Cdc20 keeps Cdk1 activity constant at the meta-II level in the presence of continuous cyclin B synthesis (Yamamoto *et al.*, 2005). The next question then is what is the sensor of the feedback system?

Recent studies suggest that Erp1 could be the sensor. Sally Kornbluth and colleagues found that Erp1 was dissociated from APC/C when exogenous cyclin B was added to meta-II arrested egg extracts (Wu *et al.*, 2007). Phosphorylation of Erp1 at two threonine residues on its C-terminus (Thr545 and Thr551) decreases its binding ability to APC/C. They also showed that, *in vitro*, the protein phosphatase 2A (PP2A) could bind directly to Erp1 and dephosphorylate the C-terminus fragment of Erp1 including Thr545 and Thr551 (Wu *et al.*, 2007). These results provide a model for the feedback system that maintains Cdk1 activity in meta-II arrest. PP2A prevents phosphorylation of Thr545 and Thr551, and keeps Erp1 associated with APC/C. Once Cdk1 activity exceeds a threshold and becomes predominant over phosphatase activity, Thr545 and Thr551 are phosphorylated, leading to dissociation of Erp1 from APC/C and, in turn, APC/C is transiently activated (Figure 14.3).

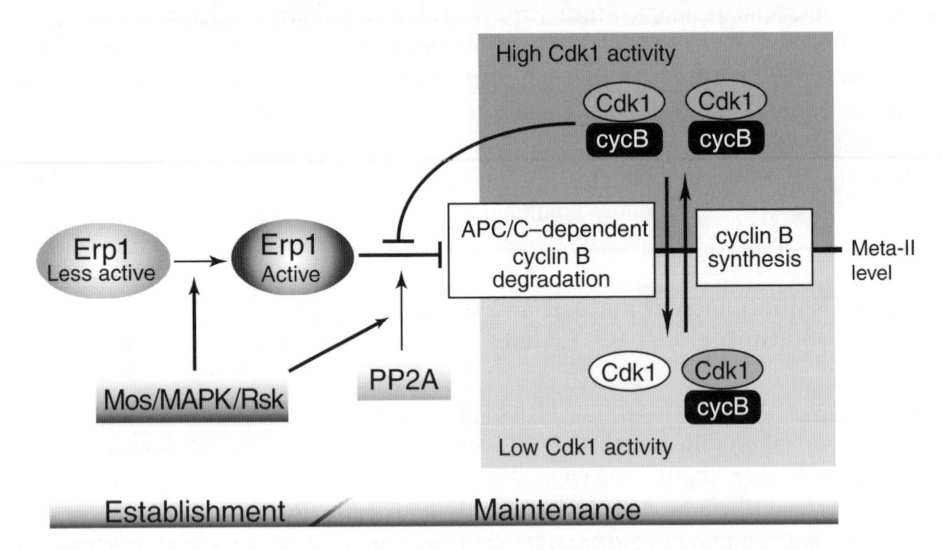

Figure 14.3 Mos/MAPK pathway and Erp1 cooperate for establishment and maintenance of meta-II arrest. Erp1 is phosphorylated at TSS region by Rsk dependently of Mos/MAPK activity. The phosphorylation increases both APC/C-inhibiting activity and stability of Erp1, thereby meta-II arrest is established and maintained. During meta-II arrest, once Cdk1 activity exceeds a threshold (meta-II level) due to continuous synthesis of cyclin B (cycB), Thr545 and Thr551 of Erp1 are phosphorylated by Cdk1. This phosphorylation causes dissociation of Erp1 from APC/C, inducing transient activation of APC/C, followed by cyclin B degradation. As a result, Cdk1 activity is dropped below meta-II level; Thr545 and Thr551 are dephosphorylated by PP2A, resulting in recovery of Erp1 activity. The binding of Erp1 to PP2A depends on Mos-dependent phosphorylation of Erp1

The Mos/MAPK pathway also seems to play an important role in this feedback system, because Mos-dependent phosphorylation of Erp1 at the TSS region is required for binding of Erp1 to PP2A, that is to say, for binding of Erp1 to APC/C (Wu *et al.*, 2007). Furthermore, during meiosis I, where the ratio of Mos/MAPK activity to Cdk1 activity is less than that in meiosis II, Erp1 protein is highly unstable due to Cdk1-mediated phosphorylation at 4 N-terminus sites (Ser213/Thr239/Thr252/Thr267) (Tang *et al.*, 2008). As is evident from these results, Mos/MAPK activity is required to maintain the level of cyclin B dynamically during meta-II arrest. Indeed, when Mos is depleted from meta-II arrested egg extracts, cyclin B is gradually degraded and finally the cell cycle shifts to interphase (Yamamoto *et al.*, 2005). Therefore, the feedback system regulated by Erp1 clearly depends on Mos/MAPK activity (Figure 14.3).

Thus, one of the most important features of CSF arrest is its dynamic stability. It might seem to be inadequate for a secure arrest because APC/C could be activated readily. However, if we think about release of the arrest, prompt activation of APC/C and degradation of cyclin B are prerequisite for precise activation of zygotes, which should not be arrested anymore. CSF might be an ingenious system not only to achieve meta-II arrest, but also to ensure the subsequent activation of fertilized zygotes.

14.3 Meta-II arrest in mouse

Mos is also required for meta-II arrest in mouse oocytes. Oocytes from mice lacking the *c-mos* gene (mos$^{-/-}$) were found to lack active MAPK. Although it has not been clearly demonstrated whether these oocytes undergo meiosis II normally, 70% of them were activated spontaneously (Colledge *et al.*, 1994; Hashimoto *et al.*, 1994; Verlhac *et al.*, 1996). In these mos$^{-/-}$ oocytes, meta-II arrest is restored by activation of MAPK (Verlhac *et al.*, 2000). However, in mouse oocytes, Rsk is unlikely to play an essential role for meta-II arrest, because oocytes from Rsk1, 2, and 3 triple-knockout mice arrest the cell cycle at meta-II normally, and constitutively active Rsk does not rescue the meta-II arrest of oocytes from mos$^{-/-}$ strain (Dumont *et al.*, 2005). As alluded to previously, some other target of MAPK, like a MISS or DOC1R, is required for spindle assembly during meta-II arrest in mouse (Lefebvre *et al.*, 2002; Terret *et al.*, 2003), and spindle assembly is necessary for release from meta-II arrest (Winston *et al.*, 1995). It seems likely that, downstream of MAPK, molecules other than Rsk somehow mediate the meta-II arrest activity in mouse oocytes.

Nevertheless, Erp1 is required for meta-II arrest in mouse oocytes as well. Erp1 is also found in mouse, and depletion of Erp1 *in situ* by RNA interference causes precocious meiotic exit in mature oocytes (Shoji *et al.*, 2006). Because Erp1 is found in protein sequence databases of many vertebrates, and its TSS region is highly conserved (Figure 14.2), the mechanism of Erp1-dependent meta-II arrest seems to be common in vertebrates. Because the TSS region does not seem to have a consensus site for MAPK phosphorylation, an alternative kinase to Rsk activated by MAPK might be responsible for Erp1 activation, or MAPK might regulate Erp1 through a region distinct from TSS. In any case, a major question that remains to be addressed is how Erp1 is regulated in mouse oocytes.

14.4 Pronuclear stage arrest in starfish

14.4.1 Dual-lock for G1 arrest

Interestingly, the Mos/MAPK pathway has also been shown to be responsible for the second arrest in unfertilized eggs of some invertebrates, which are arrested at a stage other than meta-II (see below). Among these, most extensively studied is the starfish, in which, unless fertilization occurs after ovulation (see below), mature eggs are arrested at the pronuclear stage after completion of meiosis II (Kishimoto, 2003). Particularly in the starfish *Asterina pectinifera* (renamed *Patiria pectinifera* in the 2007 NCBI Taxonomy Browser), unfertilized mature eggs are arrested at the G1 phase of the pronuclear stage (Kishimoto, 1998; see below). Fertilization releases the G1 arrest to initiate S phase, with no requirement of new protein synthesis.

Prevention of entry into S phase

During meiotic maturation in oocytes of *A. pectinifera*, Mos is initially synthesized around metaphase of meiosis I, dependently on cyclin B–Cdc2 activation (Tachibana *et al.*, 2000), and induces activation of MAPK (Tachibana *et al.*, 1997, 2000) and Rsk (Mori *et al.*, 2006). Unless fertilization occurs, Mos protein and activities of MAPK and Rsk remain elevated until G1 arrest after completion of meiosis II. Fertilization causes immediate degradation of Mos protein, resulting in inactivation of MAPK and Rsk. Suppression of either MAPK (Tachibana *et al.*, 1997) or Rsk (Mori *et al.*, 2006) in unfertilized G1 eggs is necessary and sufficient for release from G1 arrest and entry into S phase. Thus, the Mos/MAPK/Rsk pathway causes the starfish G1 arrest, in which Mos functions as an MAPK kinase kinase, and Rsk functions as a mediator immediately downstream of MAPK.

Prevention of entry into M phase

Although release from the G1 arrest by fertilization leads to S phase and the following M phase that results in embryonic cell cycling, there is an issue particular to the G1 arrest in eggs. In somatic cells, the orderly progression of the cell cycle is ensured by checkpoint controls (Hartwell and Weinert, 1989). For example, the DNA replication checkpoint allows entry into M phase only after completion of S phase, by monitoring progression of S phase. Unlike the ordinary somatic cell cycle, however, a functional cell cycle checkpoint is lacking in the early embryonic cell cycle of some organisms including starfish (Yamada *et al.*, 1985). Indeed, even when DNA replication is inhibited, the early embryonic cell cycle can progress with M phase cycling without S phase (Nagano *et al.*, 1981; Hara *et al.*, 2009). Thus, to block the start of the embryonic cell cycle at G1 phase in unfertilized starfish eggs, entry into M phase must be suppressed independently of prevention of S phase.

Entry into M phase in fertilized starfish eggs is regulated both by cyclin A–Cdk1 and cyclin B–Cdk1 (Okano-Uchida *et al.*, 1998). In G1-arrested eggs, protein levels of cyclin A and cyclin B remain low, and fertilization triggers their accumulation. Although each cyclin associates with Cdk1, cyclin B–Cdk1 remains inactive due to inhibitory phosphorylation on Cdk1 by Wee1 and Myt1. By contrast, cyclin A–Cdk1 is activated due solely to the association of both proteins. The active cyclin A–Cdk1 inactivates Wee1 and Myt1 via Plk1, resulting in activation of cyclin B–Cdk1 and entry into M phase (Okano-Uchida *et al.*, 2003; Tachibana *et al.*, 2008). Thus, accumulation of cyclin A and cyclin B is indispensable for starting the embryonic M phase.

Consistently, in unfertilized G1 eggs of the starfish, suppression of Rsk alone is not sufficient for cell cycle progression into M phase, due to lack in accumulation of cyclin A and cyclin B, even though S phase occurs (Hara *et al.*, 2009). Instead, even in the presence of active Rsk, suppression of MAPK alone causes accumulation of cyclins A and B, resulting in entry into M phase without S phase. Thus, MAPK prevents entry into M phase through a pathway that is not mediated by Rsk but that leads to repression of synthesis of cyclin A and cyclin B proteins. Synthesis of cyclins A and B proteins is repressed in a manner independent of elongation of the poly(A) tail (Hara *et al.*, 2009); in contrast, their synthesis during oocyte maturation is likely to be regulated in a poly(A) tail elongation-dependent manner (Standart *et al.*, 1987; Lapasset *et al.*, 2008). Together, to maintain G1 arrest in unfertilized starfish eggs, there is a 'dual-lock' mechanism (Hara *et al.*, 2009) in which two separate pathways are operating downstream of the Mos/MAPK pathway: one is a Rsk-dependent pathway that leads to prevention of entry into S phase, and the other is a Rsk-independent pathway that leads to prevention of entry into M phase (Figure 14.4). Such a dual-lock downstream of MAPK is necessary for G1 arrest.

So far, this is the first demonstration that the Mos/MAPK cascade separates into Rsk-dependent and Rsk-independent pathways, thereby arresting the cell cycle prior to fertilization. Although the physiological substrates of Rsk and MAPK in the respective pathways remain unclear, an effector in the Rsk pathway should be implicated in the DNA replication machinery (Kearsey and Cotterill, 2003). On the other hand, an effector in the MAPK pathway should be implicated in the protein synthesis machinery, possibly regulating 5′ cap-dependent translation (Vardy and Orr-Weaver, 2007b). A future challenge is to elucidate the components in the respective pathways downstream of Rsk and MAPK.

14.4.2 G2 arrest

Although unfertilized mature starfish eggs arrest at the pronuclear stage after completion of meiosis II, the cell cycle phase differs depending on the starfish species (Kishimoto, 1998). In contrast to the above-mentioned G1 arrest in Japanese *Asterina pectinifera* and American *Asterina miniata* (Sadler and Ruderman, 1998; also renamed *Patiria miniata*), the pronuclear arrest occurs at G2 phase after DNA replication in European *Marthasterias glacialis* and *Astropecten aranciacus* (Picard *et al.*, 1996;

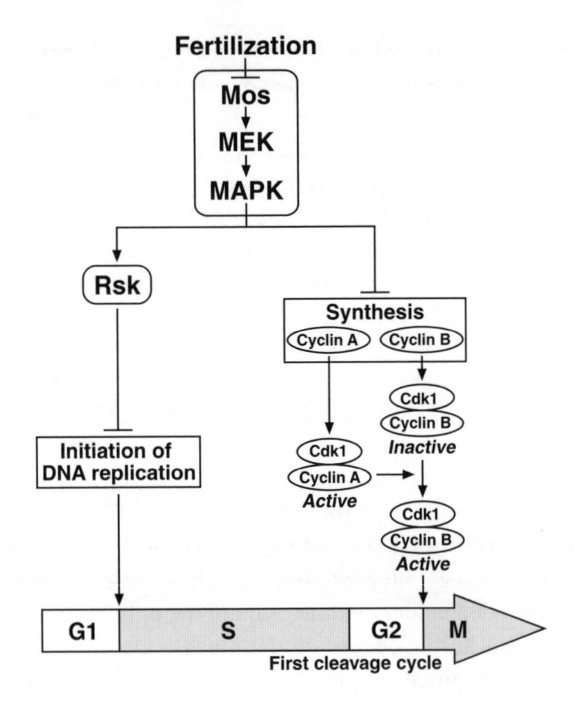

Figure 14.4 Dual-lock for G1 arrest in unfertilized mature starfish eggs. Unless fertilization occurs, mature eggs of the starfish, *Asterina pectinifera*, arrest at G1 phase of the pronuclear stage. To execute the G1 arrest, two separate pathways function downstream of the Mos/MAPK pathway. One is an Rsk-mediated pathway that leads to prevention of entry into S phase through inhibition of initiation of DNA replication. The other is an Rsk-independent pathway that prevents entry into M phase through suppression of protein synthesis of cyclin A and cyclin B. Such a dual-lock downstream of MAPK is necessary for G1 arrest, because the lack of a DNA replication checkpoint allows entry into M phase in the absence of S phase

Fisher *et al.*, 1998). Nevertheless, both G1 and G2 phase arrests depend on MAPK (Kishimoto, 1998, 2004). As for the MAPK-dependent G2 arrest, artificial elevation of MAPK activity can cause G2 arrest in the first embryonic cycle of *Xenopus* (Walter, Guadagno and Ferrell, 1997; Bitangcol *et al.*, 1998). In this arrest, accumulation of cyclin A and cyclin B occurs, but MAPK directly phosphorylates and activates Wee1 to inhibit Cdk1 (Murakami, Copeland and Vande Woude, 1999; Walter, Guadagno and Ferrell, 2000). In contrast, in European starfish eggs arrested at G2 phase, cyclin B accumulation and inhibitory phosphorylation of Cdk1 are observed, but cyclin A, though not examined directly, does not appear to accumulate, because no significant H1 kinase activity is detectable (Fisher *et al.*, 1998). These findings indicate that the G2 arrest in European starfish eggs is different from that induced by MAPK in *Xenopus* egg extracts, but rather reminiscent of that seen in Japanese starfish eggs in which Rsk is suppressed but MAPK remains active. It is most likely that in G2-arrested eggs of the European starfish, a single-lock downstream of MAPK, that is the Rsk-independent pathway alone, maintains cell cycle arrest prior to fertilization. Considering that Rsk is essential for progression into meiosis II (Mori *et al.*, 2006; see below), G2-arrested starfish eggs might be lacking not in Rsk activity but in Rsk target(s) that lead to prevention of DNA replication.

14.5 Evolution of meiotic arrest

14.5.1 Diverse stages of post-meiotic reinitiation arrest: defining of cytostatic arrest

Frog (*Xenopus laevis*), mouse and starfish (*Asterina pectinifera*) are the top three species in which cytostatic arrest prior to fertilization has been extensively studied. Looking among metazoans more generally (Adiyodi and Adiyodi, 1983; Masui, 1985; Figure 14.5), however, other aspects of cytostatic arrest are emerging. In some species of metazoans, fully grown oocytes which remain arrested at pro-I are ovulated, and fertilization releases the pro-I arrest. In most other species, the pro-I arrest is released by ovulation or by a maturation-inducing hormone, but the cell cycle arrests again after meiotic reinitiation, unless fertilization occurs. Such a postmeiotic reinitiation arrest is an almost universal phenomenon throughout Metazoa, but the arrest stage differs depending on the species: meta-I, meta-II, and pronuclear stage (G1 or G2 phase) after completion of meiosis II.

Meta-I arrest is seen in more than half of the phyla, including the large phyla of Arthropoda and Mollusca, and, hence, meta-I is the most common stage of arrest. G1 arrest occurs in some species of the phyla of Ctenophora, Cnidaria and Echinodermata. Meta-II arrest is seen only in some of phylum Chordata, including vertebrates and amphioxus but not ascidians. In fact, even within the same phylum, the stage of arrest is not always the same. For example, Arthropoda and Mollusca contain species with either pro-I or meta-I arrest. Thus, the stage of the second arrest has diverged in the course of evolution, even though all of these arrests are released by fertilization. It is intriguing nevertheless to consider whether there is a common molecular principle among the various second arrests.

Prior to this consideration, however, we should note that what is meant by 'post-meiotic reinitiation arrest' (or 'the second arrest') is not so simple. It occurs even before ovulation (i.e. inside of the ovary) in some species, and its stage is not always the stage of fertilization (i.e. sperm entry), implying that post-meiotic reinitiation arrest may not be always equal to cytostatic arrest. To clear up the confusion, we try to classify various pre-meiotic and post-meiotic reinitiation arrests into several types in Table 14.1. The primary classification is based on whether pro-I arrest is either maintained (type A) or released (types B and C) until ovulation. The secondary classification is based on whether post-meiotic reinitiation arrest occurs either only after ovulation (type B) or both before and after ovulation (type C). Subtypes are detailed as follows:

Type A: Pro-I oocytes are ovulated

- **Type A1**: Pro-I-arrested oocytes are ovulated, and pro-I arrest is maintained until fertilization. Pro-I oocytes are fertilized. (e.g. surf clam, echiuroid)
- **Type A2**: Pro-I arrest is released at/by ovulation, and meta-I arrest occurs in unfertilized oocytes. Fertilization occurs until meta-I. (e.g. polychaete, ascidian)

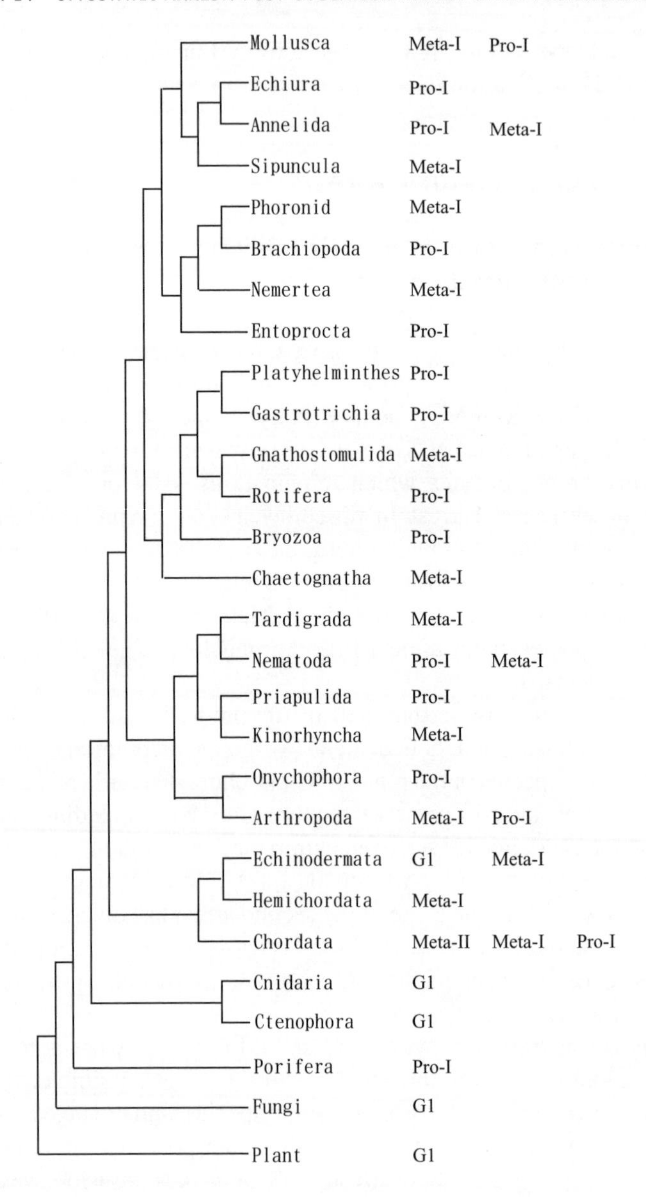

Figure 14.5 Diverse stages of sperm entry or cell cycle arrest in unfertilized eggs of metazoans. The stages of sperm entry (fertilization) or the stages of cell cycle arrest in unfertilized eggs after meiotic reinitiation are indicated in relation to each phylum of Metazoa, along with fungi and plant for reference. More than two stages in the same phylum are shown in order of higher frequency, so far reported. However, the stage of sperm entry and the stage of cell cycle arrest in unfertilized eggs are not likely to be precisely distinguished in some previous reports, despite the fact that both are not always the same. The information was obtained from Adiyodi and Adiyodi (1983) and Masui (1985). Phylogram is shown according to Dunn *et al.* (2008), except that the positions of Porifera and Ctenophora are switched

Table 14.1 Stages of meiotic arrest in metazoan oocytes

Type	Genus	Trigger of meiotic reinitiation	Stage of oocytes			
			Pre-meiotic reinitiation arrest	Post-meiotic reinitiation arrest		Sperm entry
			After ovulation (unfertilized)	Before ovulation	After ovulation (unfertilized)	
A1	Surf clam	Fertilization	GV	—	—	GV
A1	Echiuroid	Fertilization	GV	—	—	GV
A2	Polychaete	Ovulation	—	No	Meta-I	Prometa-I to meta-I
A2	Ascidian	Ovulation	—	No	Meta-I	Meta-I
B1	Nematode	MSP (*C. elegans*)	—	No	No (parthenogenesis)	Around meta-I
B2	Dog	LH	—	No	Meta-II	GV
B3	Mouse	LH	—	No	Meta-II	Meta-II
B3	Frog	Progesterone	—	No	Meta-II	Meta-II
B4	Jellyfish	Light (peptide?)	—	No	G1	G1
C1	Sawfly	?	—	Meta-I	Meta-I	Meta-I
C2	Starfish	1-methyladenine	—	Meta-I	G1 (or G2)[a]	Meta-I to G1 (G2)
C3	Fruit fly	?	—	Meta-I	G2	Meta-I
C4	Sea urchin	(1-methyladenine?)	—	G1	G1	G1

[a]G1 arrest in Japanese *Asterina pectinifera* and American *Asterina miniata*, but G2 arrest in European *Marthasterias glacialis* and *Astropecten aranciacus*.
Source: the information on various stages was obtained basically from Adiyodi and Adiyodi (1983), Masui (1985), and Sagata (1996), with slight modification.

Type B: Pro-I arrest is released and no further arrest occurs before ovulation

Type B1: Pro-I arrest is released in the ovary (by MSP in *C. elegans*). After ovulation, no arrest occurs in unfertilized oocytes. Fertilization occurs around meta-I, but no fertilization results in parthenogenesis. (e.g. nematode)

Type B2: Pro-I arrest is released in the ovary (by luteinizing hormone (LH)). After ovulation, meta-II arrest occurs in unfertilized oocytes. Fertilization occurs at germinal vesicle (GV) stage. (e.g. dog, fox)

Type B3: Pro-I arrest is released in the ovary (by LH in mouse or by progesterone in frog). After ovulation, meta-II arrest occurs in unfertilized oocytes. Meta-II oocytes are fertilized. (e.g. mouse, frog)

Type B4: Pro-I arrest is released in the ovary. After ovulation, G1 arrest occurs in unfertilized oocytes. G1 oocytes are fertilized. (e.g. jellyfish)

Type C: Pro-I arrest is released but further arrest occurs before ovulation

Type C1: After release from pro-I arrest, further arrest occurs at meta-I in the ovary. After ovulation, meta-I arrest is maintained in unfertilized oocytes. Meta-I oocytes are fertilized. (e.g. sawfly)

Type C2: After release from pro-I arrest (by 1-methyladenine in starfish), further arrest occurs at meta-I in the ovary. Meta-I arrest is released at/by ovulation, and unfertilized oocytes arrest at G1 (or G2; see Section 14.4.2). Fertilization occurs during meta-I to G1 (or G2). (e.g. starfish)

Type C3: After release from pro-I arrest, further arrest occurs at meta-I in the ovary. Meta-I arrest is released at/by ovulation, and unfertilized oocytes arrest at G2. Fertilization occurs at meta-I. (e.g. fruit fly)

Type C4: Meiosis is completed in the ovary. After ovulation, G1 arrest is maintained in unfertilized oocytes. G1 oocytes are fertilized. (e.g. sea urchin)

Among the above patterns, the most typical would be the cases of mouse (B3), frog (B3) and sea urchin (C4), in which stages of the second arrest, sperm entry and cytostatic arrest are the same. By contrast, atypical would be the cases of starfish (C2) and fruit fly (C3), in which, after meiotic reinitiation, the cell cycle arrests once again at meta-I in the ovary, restarts from meta-I at/by ovulation (Harada, Oita and Chiba (2003) for starfish; Page and Orr-Weaver (1997) for fruit fly), and finally arrests at G1 or G2 unless fertilization occurs (see below, Section 14.5.3). Looking at this case, the cytostatic arrest should be precisely defined as a stage in which oocytes are arrested after ovulation unless fertilization occurs. The meta-I arrest of these species is actually the second arrest or post-meiotic reinitiation arrest, but should not be considered as the cytostatic arrest, because its occurrence in the ovary precludes the opportunity of fertilization (see also below, Section 14.5.2). Even though sperm entry occurs after release from meta-I arrest until pronuclear stage arrest, this should be simply regarded as an inconsistency between the stages of sperm entry and cytostatic arrest. Another atypical example would be nematode (B1) which lacks in cytostatic arrest, but this lack is in good agreement with lack of *mos* in this species (see below, Section 14.5.4).

14.5.2 Meta-I arrest

Among the metazoans, meta-I is the stage at which sperm entry occurs most frequently (Figure 14.5). In a precise sense, however, it remains unclear whether the stage of sperm entry is actually the stage of cytostatic arrest (i.e. the stage at which unfertilized oocytes remain arrested after ovulation), because both stages have not been well distinguished in some of the previous literature. To consider molecular mechanisms involved in meta-I arrest, based on Table 14.1, we will classify the meta-I arrest into two types: post-ovulation meta-I arrest and pre-ovulation meta-I arrest.

Post-ovulation arrest at meta-I

Among many metazoans, regulatory mechanisms for the post-ovulation meta-I arrest have been studied in only a limited numbers of species. In ascidians and in a hymenopteran insect, the sawfly, oocytes arrest at meta-I after ovulation unless fertilization occurs. In ascidians, MAPK is likely to be involved in meta-I arrest (McDougall and Levasseur, 1998; Russo *et al.*, 1998; Sensui, unpublished) and *mos* is identified in its genome (Russo *et al.*, 2009). In the sawfly, a maternally expressed Mos has been characterized and probably acts upstream of MAPK-mediated meta-I arrest (Yamamoto *et al.*, 2008). Thus, the Mos/MAPK pathway is most likely to be involved in the post-ovulation meta-I arrest.

The molecular mechanisms underlying meta-I and meta-II arrests might be similar, because both arrests occur at the same cell cycle stage: metaphase. However, Erp1, which is the main effector for meta-II arrest, has so far not been found in genome databases of species that undergo meta-I arrest (Russo *et al.*, 2009). The post-ovulation meta-I arrest might be independent of Erp1, but dependent on Mos/MAPK.

Preovulation arrest at meta-I

In the dipteran insect *Drosophila* (Page and Orr-Weaver, 1997) and the starfish *Asterina pectinifera* (Harada, Oita and Chiba, 2003), after release from pro-I arrest, oocytes arrest again at meta-I inside the ovary (Table 14.1). Such a pre-ovulation meta-I arrest is released at ovulation. In starfish, MAPK is not yet activated at meta-I when oocytes remain inside the ovary (although MAPK is already active at meta-I when isolated oocytes undergo maturation *in vitro* (Tachibana *et al.*, 1997)), and hence, meta-I arrest does not depend on MAPK activity (Usui, Hirohashi and Chiba, 2008), contrary to a previous report (Harada, Oita and Chiba, 2003). In *Drosophila* as well, mutant females in which the *mos* gene is deleted are fertile and their oocytes are normally arrested at meta-I (Ivanovska *et al.*, 2004). It is thus most likely that pre-ovulation meta-I arrest is independent of MAPK. As sawfly oocytes arrest at meta-I both pre-ovulation and post-ovulation (Yamamoto *et al.*, 2008), it is intriguing whether the pre-ovulation arrest as well as the post-ovulation arrest require MAPK.

In summary, meta-I arrest depends on Mos/MAPK when it occurs outside the ovary, but does not when it occurs inside the ovary. If rare cases of intraovarian fertilization are excluded, cytostatic arrest can be considered as an arrest that is seen in oocytes outside the ovary, because cytostatic arrest is an arrest seen in a situation that allows fertilization, but where fertilization has not yet occurred. We can thus conclude that cytostatic meta-I arrest also requires Mos/MAPK, even though the pre-ovulation meta-I arrest does not. The pre-ovulation meta-I arrest may be a vestige of the meta-I arrest that is seen in most deuterostome animals, including hemichordates and ascidians, an evolutionarily early chordate, while it is intriguing as to how the pre-ovulation meta-I arrest is induced in the absence of MAPK.

14.5.3 G1 or G2 arrest other than in starfish

In *Asterina* starfish, unfertilized mature eggs arrest at G1, regardless of whether isolated immature oocytes undergo maturation *in vitro* or meta-I oocytes are released from the ovary (Table 14.1; see above, Section 14.4.1). Similarly to the *Asterina* starfish, unless fertilization occurs, mature eggs of sea urchin and the marine hydrozoan jellyfish (phylum Cnidaria) remain arrested at G1 of the pronuclear stage after spawning (ovulation). In their arrest as well, MAPK is necessary and sufficient for preventing release from the G1 arrest (Kumano *et al.* (2001) for sea urchin; Kondo, Tachibana and

Deguchi (2006) for the jellyfish *Cladonema pacificum*). Particularly in jellyfish, MAPK negatively regulates all of the three major postfertilization events; that is, not only the start of the embryonic cell cycle but also the cessation of sperm attraction and the expression of surface adhesive materials (Kondo, Tachibana and Deguchi, 2006). In the G1 arrest of sea urchin and jellyfish as well, Mos is most likely to function upstream of MAPK, because Mos orthologues have been recently identified in the sea urchin *Strongylocentrotus purpuratus* genome and cnidarian *Clytia hemisphaerica* EST (expressed sequence tag) database (Amiel *et al.*, 2009; see below, Section 14.5.4). Indeed, *Clytia* Mos is required for G1 arrest (Amiel *et al.*, 2009). Thus, the Mos/MAPK pathway is most likely to be conserved for G1 arrest as well. It is intriguing as to whether Rsk functions downstream of MAPK in these species.

In *Drosophila*, meta-I oocytes are normally fertilized just after ovulation. However, whether or not the oocyte is fertilized, meta-I arrest is released at ovulation; following meiosis II, all four meiotic products in the egg go through a transient interphase state, and then the chromosomes of all the products condense. Given that BrdU can be observed incorporated into these chromosomes, it is likely that S phase occurs during the interphase transition (T.L. Orr-Weaver, personal communication; Fenger *et al.*, 2000; Table 14.1). Such a G2 arrest in unfertilized *Drosophila* eggs is apparently equivalent to that in starfish eggs with the 'single-lock' downstream of MAPK (see Section 14.4.2). Interestingly, although the Mos/MAPK pathway is dispensable for meta-I and G2 arrests, MAPK is still active in unfertilized eggs and is inactivated in fertilized eggs (Ivanovska *et al.*, 2004). These behaviours of the MAPK in *Drosophila* eggs suggest that Mos/MAPK might be vestigial for cytostatic arrest (see also Section 14.5.4). In the actual *Drosophila* G2 eggs, however, the PNG (PAN GU) kinase complex positively regulates entry into M phase through promoting cyclin B synthesis (Vardy and Orr-Weaver, 2007a).

Taken together, we can conclude that Mos/MAPK is generally conserved in all of cytostatic arrests at meta-I, meta-II and G1 so far examined.

14.5.4 Conservation of Mos in metazoans

In the early studies, Mos was thought to have a major role in regulating meta-II arrest only in eggs of vertebrates such as frogs and mice. Hence, it was supposed that Mos should be present only in vertebrates. Later studies, however, revealed that MAPK regulates the cytostatic arrest at G1 in invertebrate starfish eggs as well, even though the arrest stage is not meta-II (Tachibana *et al.*, 1997). This prompted us to identify the upstream activator for starfish MAPK, and finally we isolated starfish *A. pectinifera* Mos as the first one in invertebrates (Tachibana *et al.*, 2000). Thereafter, *Drosophila* Mos was identified (Ivanovska *et al.*, 2004), followed by identification of various *mos* orthologues using genomes or EST databases (Amiel *et al.*, 2009).

Phylogenetic survey revealed that Mos genes are conserved in most of the metazoa (Amiel *et al.*, 2009). Among the metazoa, *mos* genes were found in Cnidaria, Ctenophora, Placozoa, Annelida, Arthropoda (the fruit fly *Drosophila melanogaster* and the sawfly *Athalia rosae*), Mollusca, Echinodermata and Chordata. Exceptions in the metazoa in which *mos* genes were not found are Porifera (sponges) and Nematoda

(*Caenorhabditis elegans*; see above, Section 14.5.1). So far, no *mos* gene has been found outside the metazoa, that is, in Choanoflagellata and Eumycota (fungi), including the yeast *Saccharomyces cerevisiae* and *Aspergillus niger*. Their orthology suggests that the *mos* gene may have originated in a common metazoan ancestor and then been secondarily lost in the sponges and the nematodes. Another feature is that multiple *mos* genes were found in cnidaria, while only a single intact *mos* gene was found in bilaterians, despite well-documented whole-gene duplications.

The above phylogenetic analysis, indicating that Mos is conserved but is restricted in the metazoa, implies that Mos should be seen not as a core regulator of meiosis, which is a much older process, but of a particularity of meiosis in metazoans (Amiel *et al.*, 2009). In accordance with this notion, the cytostatic arrest until fertilization is not a phenomenon that is commonly seen in meiosis in all of the eukaryotes including yeasts, but is restricted to metazoan oocyte meiosis. Thus, a general role for Mos appears to be regulation of the cytostatic arrest in metazoan oocytes (see below, Section 14.7). Consistently, *Drosophila mos*, which is not essential for meiosis (see Section 14.5.3; Ivanovska *et al.*, 2004), is highly divergent compared with other metazoan *mos* genes. And oocytes of *Caenorhabditis,* which are lacking in *mos*, do not undergo cytostatic arrest (see above, Section 14.5.1). Lack of the *mos* gene in *Amphimedon queenslandica* (sponge) may also relate to the complicated process of intraovarian fertilization in growing oocytes of sponges (Masui, 1985), which precludes the opportunity for cytostatic arrest.

14.6 Reconsideration of CSF

14.6.1 Extended CSF: core element, transducer, effector and target

Since the discovery of CSF by Yoshio Masui in 1971, our knowledge of amphibian CSF has been largely clarified through two great discoveries: Mos and Erp1. Furthermore, extensive studies on the post-meiotic reinitiation arrest (the second arrest) in various metazoans have expanded the 'CSF-like' activity into arrests other than meta-II. At this point, we can ask again 'What is CSF?' In reconsidering CSF, one way would be to restrict the concept of CSF to meta-II arrest, based on its original definition by Masui. Another way would be to extend the idea of CSF to cytostatic arrest at various stages, which occurs after meiotic reinitiation and ovulation, but before fertilization, depending on the species. Below, we will discuss each of these two views.

If we restrict 'CSF' to the original meta-II arrest, a possible consideration is as follows. In the original experiment of cytoplasmic transfer using amphibians, by which Masui found CSF (Figure 14.1a), CSF was thought not to be present in the recipient blastomere of a two-cell embryo, based on his CSF criteria. However, we now know that a comparable amount of Erp1 is present in recipient blastomeres as in unfertilized eggs. This indicates that the substance of CSF in unfertilized eggs, which caused metaphase arrest in the recipient blastomere, was Mos, not Erp1. Thus, Mos is a bona fide molecule of CSF. In this context, we can ask whether Erp1 is CSF or not. Interestingly, in mouse, if cytoplasm of unfertilized eggs (Shoji *et al.*, 2006), or active MEK or MAPK (Kashima, Kano and Naito, 2007), is injected into one- or two-cell embryos, their cleavage is not inhibited. One possibility is that, in the case of mouse, unlike *Xenopus*, because Erp1 has already disappeared in the two-cell embryo (Shoji *et al.*, 2006), Mos might not be able to induce metaphase arrest in blastomeres in the absence of Erp1. This case also clearly indicates that the ability of Mos to induce metaphase arrest in the recipient blastomere largely depends on the presence of Erp1 in the blastomere. In other words, Mos, that is to say CSF in the original definition, is insufficient for meta-II arrest. To establish and maintain meta-II arrest, the Mos/MAPK pathway has to facilitate the stability and activity of Erp1. Moreover, Erp1 is also regulated by other kinases and phosphatases, such as Cdk1, PP2A (Wu *et al.*, 2007) and even calcineurin (Nishiyama *et al.*, 2007b). Thus, an alternative to the classical definition of CSF, which does not sufficiently encompass the molecules contributing to meta-II arrest, would be a new term, 'Cytostatic system (CSS)', for the whole machinery including cooperation of Mos/MAPK, Erp1, and other regulatory molecules involving in successful meta-II arrest.

An alternative way to reconsider CSF is not to restrict the term CSF to the original meta-II arrest. Indeed, even if CSF remains in the original meta-II arrest, some inconsistencies are emerging, for example, between frog and mouse. Instead, considering the expansion of the role of Mos/MAPK for various post-meiotic reinitiation arrests in metazoans, it would be more appropriate to redefine CSF, simply based on its original main concept: CSF causes 'cytostatic arrest', that is, the post-meiotic reinitiation arrest unless fertilization occurs after ovulation. CSF can be thus considered as a common cytostatic arrest factor, irrespective of the particular stage of arrest.

How then can we reconcile CSF into a comprehensive view? Accumulating evidence strongly indicates that the Mos/MAPK pathway is a core module of CSF, which causes diverse cytostatic arrests at meta-I, meta-II and G1. The evidence further indicates that pathways downstream of MAPK are variable depending on the arrest stage and animal species. These variations can be distinguished by introducing the distinctions of 'transducer, effector and target' into each element of the pathway downstream of MAPK (Figure 14.6). Typically, in meta-II arrest, the transducer is Rsk, the effector is Erp1, and the target is APC/C in *Xenopus*, while the transducer is unknown in mouse. In starfish G1 arrest, for preventing entry into S phase, the transducer is Rsk, the effector is unknown, and the target should be a component of initiation machinery for DNA replication; for preventing entry into M phase, both the transducer and the effector are unknown, and the target should be a component of the translation initiation machinery. Thus, if we redefine Mos/MAPK (or MAPK) alone as CSF, the variety in cytostatic arrests could be reconciled by a particular transducer, effector or target downstream of CSF. We may distinguish each CSF as 'meta-I CSF' (e.g. for ascidians and sawfly), 'meta-II CSF' (e.g. for frog and mouse), or 'G1 CSF' (e.g. for starfish, sea urchin and jellyfish). It is intriguing to consider how the 'rewiring' downstream of CSF/MAPK has been introduced during metazoan evolution.

14.6.2 Release from the cytostatic arrest

In the physiological condition, fertilization generally induces an increase in intracellular free Ca^{2+}, followed by degradation of Mos to shut down its downstream pathway

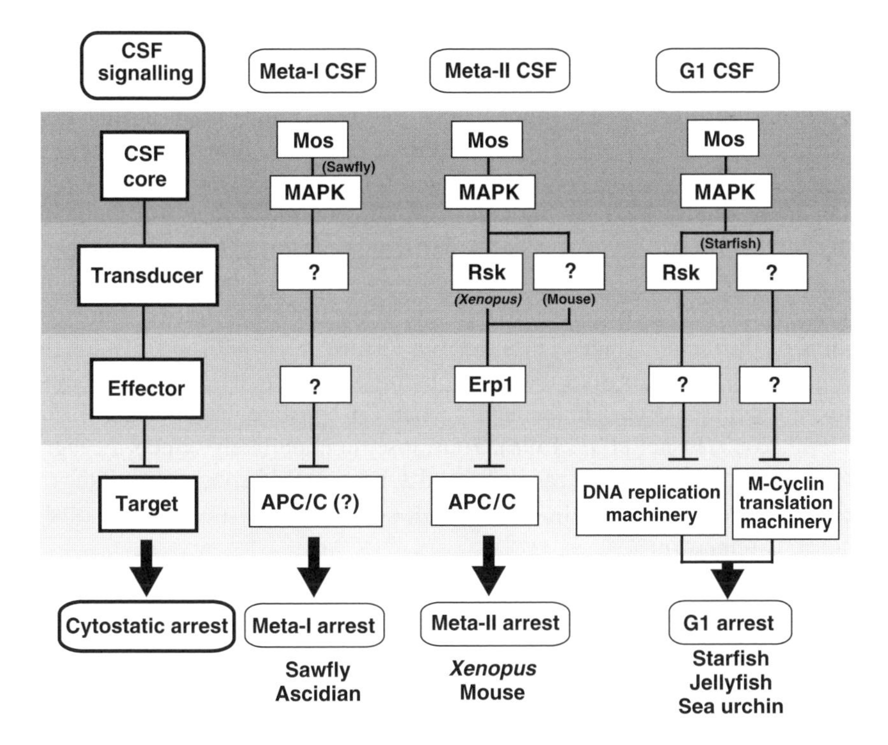

Figure 14.6 CSF signalling for cytostatic arrest in metazoans. CSF was originally defined as an activity that causes meta-II arrest in frog eggs. Based on the expansion of the role of its core component, the Mos/MAPK pathway, for various cytostatic arrests other than meta-II arrest, here we propose a conceptual extension in the definition of CSF. CSF can be considered as a common cytostatic arrest factor, irrespective of the particular stage of arrest, in metazoan eggs which are not fertilized after meiotic reinitiation and ovulation. The core of CSF is universal Mos/MAPK, downstream of which there are variable transducer, effector and target, leading to cytostatic arrest at a particular stage. The rewiring downstream of Mos/MAPK produces the variety in arrest stages, either meta-I, meta-II or G1. Thus we may distinguish 'meta-I CSF', 'meta-II CSF' or 'G1 CSF'. M-Cyclin represents cyclin A and cyclin B

(see Chapter 13). Mos is reported to be destroyed through the ubiquitin-proteasome machinery (Watanabe *et al.*, 1991; Nishizawa *et al.*, 1993; Castro *et al.*, 2001), although its E3 ligase remains unidentified. In parallel, CSF arrest is released, finally leading to embryonic cell cycle progression. More precisely, however, there is some inconsistency between degradation of Mos and exit from cytostatic arrest. In frog eggs, exit from cytostatic arrest depends on a Ca^{2+} increase, and occurs before (or even in the absence of) Mos degradation. By contrast, in starfish eggs, exit from cytostatic arrest never occurs until degradation of Mos even after an increase in free Ca^{2+} (Tachibana *et al.*, 1997, 2000; Mori *et al.*, 2006). Thus, at exit from cytostatic arrest, the Ca^{2+} signal is dominant over Mos in frog, and vice versa in starfish. In eggs of jellyfish (Kondo, Tachibana and Deguchi, 2006) and sea urchin (Kumano *et al.*, 2001) as well, G1 arrest is not released until MAPK is inactivated even after Ca^{2+} release. Such a different nature of cytostatic arrest between meta-II and G1 might be due to particular regulation of the effector molecules downstream of the Mos/MAPK pathway.

Another issue that should be noted is the lag between fertilization and Mos degradation. In G1-arrested eggs of starfish, sea urchin and jellyfish, Mos degradation or MAPK inactivation occurs immediately (within 15 min) after fertilization. By contrast, it takes more than 1 h in frog and 6 h in mouse. These differences in the lag suggest that the system for Mos degradation might not be the same among these animals. Interestingly, the lag is variable depending on the timing of insemination in starfish oocytes. These oocytes can be fertilized anytime through the GVBD (germinal vesicle breakdown) stage to G1 arrest, and an increase in intracellular Ca^{2+} occurs immediately after insemination (although the amount of Ca^{2+} release varies greatly depending on the meiotic stage of insemination). However, MAPK inactivation (hence, Mos degradation) does not start until completion of meiosis when meta-I oocytes are inseminated, while it occurs immediately when G1 eggs are inseminated (Tachibana *et al.*, 1997, 2000; Fisher *et al.*, 1998). Thus, insemination likely directs the signal for Mos degradation, but the signal can be 'stored as a memory' in oocytes until the time of its execution. What determines the timing of Mos degradation in oocytes of starfish, frog and mouse is unknown.

14.7 Concluding remarks

CSF was originally found as an activity that causes meta-II arrest in frog eggs. Further studies revealed that the core element of frog CSF is Mos/MAPK signalling, and that Mos/MAPK also shows CSF-like cytostatic activity in diverse metazoan oocytes which exhibit post-ovulation arrest at meta-I or G1, as well as meta-II, unless fertilization occurs. In fact, Mos is an ancestral kinase that has been expressed in oocytes throughout metazoan evolution, and MAPK is also highly conserved in eukaryotes. Here we propose to redefine CSF as a cytostatic activity that causes all post-ovulation arrests until fertilization, each of which is executed by a core element of Mos/MAPK and variable downstream transducer, effector and target elements that direct the specificity of the stage of the arrest.

It should be noted, however, that the role of Mos/MAPK is not restricted to cytostatic activity. Even when post-ovulation arrest does not occur after release from pro-I or meta-I arrest, MAPK remains active or is further activated during the meiosis I to II transition (for Spisula, Shibuya *et al.* (1992); for Urechis, Gould and Stephano (1999); for ascidian, Sensui and Tachibana, unpublished). Indeed, Mos/MAPK is essential for the transition from meiosis I to II, which lacks intervening S phase (Sagata, 1996; Kishimoto, 2003). Furthermore, Mos/MAPK contributes to the formation of the meiotic spindle leading to unequal cell division (polar body extrusion) (Hirao and Eppig, 1997; Gould and Stephano, 1999; Tachibana *et al.*, 2000; Brunet and Maro, 2005) and also inactivation of maternal centrioles during oocyte maturation (Tamura and Nemoto, 2001; Shirato *et al.*, 2006).

Together, these multiple roles of Mos/MAPK serve to prevent parthenogenesis, by coordinating meiotic maturation and fertilization through cytostatic arrest, meiosis I to II transition and other meiotic events. Mos/MAPK is indeed in the centre of metazoan female meiosis.

Acknowledgements

We thank Ryusaku Deguchi, Masatsugu Hatakeyama and Noburu Sensui for collaboration; Evelyn Houliston and Terry Orr-Weaver for valuable comments and suggestions; all of lab members, particularly Keita Ohsumi and Masatoshi Hara, for discussion; and Rindy Jaffe for critical reading of the manuscript. This work was supported by grants from the Ministry of Education, Science and Culture, Japan to T. Nishiyama, K. Tachibana and T. Kishimoto.

Note added in proof

In the G1 CSF signalling of the starfish *Asterina pectinifera* (see Figure 14.6), the target downstream of the Mos-MAPK-Rsk pathway was shown to be Cdc45, a component of the pre-initiation complex for DNA replication that is capable of origin-unwinding and of promoting assembly of replication forks at replication origin (Tachibana *et al.*, 2010).

References

Abrieu, A., Lorca, T., Labbe, J.C. *et al.* (1996) MAP kinase does not inactivate, but rather prevents the cyclin degradation pathway from being turned on in Xenopus egg extracts. *J. Cell Sci.*, **109**(1), 239–246.

Abrieu, A., Fisher, D., Simon, M.N. *et al.* (1997) MAPK inactivation is required for the G2 to M-phase transition of the first mitotic cell cycle. *EMBO J.*, **16**, 6407–6413.

Adiyodi, K.G. and Adiyodi, R.G. (eds.) (1983) Reproductive Biology of Invertebrates, Vol I. *Oogenesis, Oviposition, and Oosorption*, John Wiley & Sons, Ltd, Chichester.

Amiel, A., Leclere, L., Robert, L. *et al.* (2009) Conserved functions for Mos in eumetazoan oocyte maturation revealed by studies in a cnidarian. *Curr. Biol.*, **19**, 305–311.

Bhatt, R.R. and Ferrell, J.E. Jr (1999) The protein kinase p90 rsk as an essential mediator of cytostatic factor activity. *Science*, **286**, 1362–1365.

Bitangcol, J.C., Chau, A.S., Stadnick, E. *et al.* (1998) Activation of the p42 mitogen-activated protein kinase pathway inhibits Cdc2 activation and entry into M-phase in cycling Xenopus egg extracts. *Mol. Biol. Cell*, **9**, 451–467.

Boveri, T. (1907) *Zellenstudien VI. Die Entwicklung dispermer Seeigeleier. Ein Beitrag zur Befruchtungslehre und zur Theorie des Kerns*, Gustav Fischer, Jena.

Brunet, S. and Maro, B. (2005) Cytoskeleton and cell cycle control during meiotic maturation of the mouse oocyte: integrating time and space. *Reproduction*, **130**, 801–811.

Castro, A., Peter, M., Magnaghi-Jaulin, L. *et al.* (2001) Cyclin B/cdc2 induces c-Mos stability by direct phosphorylation in Xenopus oocytes. *Mol. Biol. Cell*, **12**, 2660–2671.

Colledge, W.H., Carlton, M.B., Udy, G.B. and Evans, M.J. (1994) Disruption of c-mos causes parthenogenetic development of unfertilized mouse eggs. *Nature*, **370**, 65–68.

Descombes, P. and Nigg, E.A. (1998) The polo-like kinase Plx1 is required for M phase exit and destruction of mitotic regulators in Xenopus egg extracts. *EMBO J.*, **17**, 1328–1335.

Dumont, J., Umbhauer, M., Rassinier, P. *et al.* (2005) p90Rsk is not involved in cytostatic factor arrest in mouse oocytes. *J. Cell Biol.*, **169**, 227–231.

Dunn, C.W., Hejnol, A., Matus, D.Q. *et al.* (2008) Broad phylogenomic sampling improves resolution of the animal tree of life. *Nature*, **452**, 745–749.

Fenger, D.D., Carminati, J.L., Burney-Sigman, D.L. *et al.* (2000) PAN GU: a protein kinase that inhibits S phase and promotes mitosis is early *Drosophila* development. *Development*, **127**, 4763–4774.

Fisher, D., Abrieu, A., Simon, M.N. *et al.* (1998) MAP kinase inactivation is required only for G2-M phase transition in early embryogenesis cell cycles of the starfishes Marthasterias glacialis and Astropecten aranciacus. *Dev. Biol.*, **202**, 1–13.

Goldman, D.S., Kiessling, A.A., Millette, C.F. and Cooper, G.M. (1987) Expression of c-mos RNA in germ cells of male and female mice. *Proc. Natl. Acad. Sci. USA*, **84**, 4509–4513.

Gould, M.C. and Stephano, J.L. (1999) MAP kinase, meiosis, and sperm centrosome suppression in *Urechis caupo*. *Dev. Biol.*, **216**, 348–358.

Grimison, B., Liu, J., Lewellyn, A.L. and Maller, J.L. (2006) Metaphase arrest by cyclin E-Cdk2 requires the spindle-checkpoint kinase Mps1. *Curr. Biol.*, **16**, 1968–1973.

Gross, S.D., Schwab, M.S., Lewellyn, A.L. and Maller, J.L. (1999) Induction of metaphase arrest in cleaving Xenopus embryos by the protein kinase p90Rsk. *Science*, **286**, 1365–1367.

Haccard, O., Sarcevic, B., Lewellyn, A. *et al.* (1993) Induction of metaphase arrest in cleaving Xenopus embryos by MAP kinase. *Science*, **262**, 1262–1265.

Hara, M., Mori, M., Wada, T. *et al.* (2009) Start of the embryonic cell cycle is dually locked in unfertilized starfish eggs. *Development*, **136**, 1687–1696.

Harada, K., Oita, E. and Chiba, K. (2003) Metaphase I arrest of starfish oocytes induced via the MAP kinase pathway is released by an increase of intracellular pH. *Development*, **130**, 4581–4586.

Hartwell, L.H. and Weinert, T.A. (1989) Checkpoints: controls that ensure the order of cell cycle events. *Science*, **246**, 629–634.

Hashimoto, N., Watanabe, N., Furuta, Y. *et al.* (1994) Parthenogenetic activation of oocytes in c-mos-deficient mice. *Nature*, **370**, 68–71.

Hirao, Y. and Eppig, J.J. (1997) Parthenogenetic development of Mos-deficient mouse oocytes. *Mol. Reprod. Dev.*, **48**, 391–396.

Hwang, L.H., Lau, L.F., Smith, D.L. *et al.* (1998) Budding yeast Cdc20: a target of the spindle checkpoint. *Science*, **279**, 1041–1044.

Inoue, D., Ohe, M., Kanemori, Y. *et al.* (2007) A direct link of the Mos-MAPK pathway to Erp1/Emi2 in meiotic arrest of Xenopus laevis eggs. *Nature*, **446**, 1100–1104.

Ivanovska, I., Lee, E., Kwan, K.M. *et al.* (2004) The Drosophila MOS ortholog is not essential for meiosis. *Curr. Biol.*, **14**, 75–80.

Kashima, K., Kano, K. and Naito, K. (2007) Mos and the mitogen-activated protein kinase do not show cytostatic factor activity in early mouse embryos. *J. Reprod. Dev.*, **53**, 1175–1182.

Kearsey, S.E. and Cotterill, S. (2003) Enigmatic variations: Divergent modes of regulating eukaryotic DNA replication. *Mol. Cell*, **12**, 1067–1075.

Kim, S.H., Lin, D.P., Matsumoto, S. *et al.* (1998) Fission yeast Slp1: an effector of the Mad2-dependent spindle checkpoint. *Science*, **279**, 1045–1047.

King, R.W., Peters, J.M., Tugendreich, S. *et al.* (1995) A 20S complex containing CDC27 and CDC16 catalyzes the mitosis-specific conjugation of ubiquitin to cyclin B. *Cell*, **81**, 279–288.

Kishimoto, T. (1998) Cell cycle arrest and release in starfish oocytes and eggs. *Semin. Cell Dev. Biol.*, **9**, 549–557.

Kishimoto, T. (2003) Cell-cycle control during meiotic maturation. *Curr. Opin. Cell Biol.*, **15**, 654–663.

Kishimoto, T. (2004) More than G1 or G2 arrest: useful starfish oocyte system for investigating skillful MAP kinase. *Biol. Cell*, **96**, 241–244.

Kondo, E., Tachibana, K. and Deguchi, R. (2006) Intracellular Ca^{2+} increase induces post-fertilization events via MAP kinase dephosphorylation in eggs of the hydrozoan jellyfish *Cladonema pacificum*. *Dev. Biol.*, **276**, 228–241.

Kraft, C., Vodermaier, H.C., Maurer-Stroh, S. *et al.* (2005) The WD40 propeller domain of Cdh1 functions as a destruction box receptor for APC/C substrates. *Mol. Cell*, **18**, 543–553.

Kubiak, J.Z., Weber, M., de Pennart, H. *et al.* (1993) The metaphase II arrest in mouse oocytes is controlled through microtubule-dependent destruction of cyclin B in the presence of CSF. *EMBO J.*, **12**, 3773–3778.

Kumano, M., Carroll, D.J., Denu, J.M. and Foltz, K.R. (2001) Calcium-mediated inactivation of the MAP kinase pathway in sea urchin eggs at fertilization. *Dev. Biol.*, **236**, 244–257.

Lapasset, L., Pradet-Balade, B., Verge, V. *et al.* (2008) Cyclin B synthesis and rapamycin-sensitive regulation of protein synthesis during starfish oocyte meiotic divisions. *Mol. Reprod. Dev.*, **75**, 1617–1626.

Lefebvre, C., Terret, M.E., Djiane, A. *et al.* (2002) Meiotic spindle stability depends on MAPK-interacting and spindle-stabilizing protein (MISS), a new MAPK substrate. *J. Cell Biol.*, **157**, 603–613.

Lillie, F.R. (1912) The fertilizing power of porations of the spermatozoon. *J. Exp. Zool.*, **12**, 427–476.

Liu, J. and Maller, J.L. (2005) Calcium elevation at fertilization coordinates phosphorylation of XErp1/Emi2 by Plx1 and CaMK II to release metaphase arrest by cytostatic factor. *Curr. Biol.*, **15**, 1458–1468.

Loeb, J. (1913) *Artificial Parthenogenesis and Fertilization*, Univ. Chicago Press, Chicago.

Masui, Y. (1974) A cytostatic factor in amphibian oocytes: its extraction and partial characterization. *J. Exp. Zool.*, **187**, 141–147.

Masui, Y. (1985) Meiotic arrest in animal oocytes, in *Biology of Fertilization*, Vol. **1** (eds C.B. Metz and A. Monroy), Academic Press, New York, pp. 189–219.

Masui, Y. (2000) The elusive cytostatic factor in the animal egg. *Nat. Rev. Mol. Cell Biol.*, **1**, 228–232.

Masui, Y. and Markert, C.L. (1971) Cytoplasmic control of nuclear behavior during meiotic maturation of frog oocytes. *J. Exp. Zool.*, **177**, 129–145.

McDougall, A. and Levasseur, M. (1998) Sperm-triggered calcium oscillations during meiosis in ascidian oocytes first pause, restart, then stop: correlations with cell cycle kinase activity. *Development*, **125**, 4451–4459.

Meyerhof, P.G. and Masui, Y. (1977) Ca and Mg control of cytostatic factors from Rana pipiens oocytes which cause metaphase and cleavage arrest. *Dev. Biol.*, **61**, 214–229.

Miller, J.J., Summers, M.K., Hansen, D.V. *et al.* (2006) Emi1 stably binds and inhibits the anaphase-promoting complex/cyclosome as a pseudosubstrate inhibitor. *Genes Dev.*, **20**, 2410–2420.

Mori, M., Hara, M., Tachibana, K. and Kishimoto, T. (2006) p90Rsk is required for G1 phase arrest in unfertilized starfish eggs. *Development*, **133**, 1823–1830.

Murakami, M.S. and Vande Woude, G.F. (1998) Analysis of the early embryonic cell cycles of Xenopus; regulation of cell cycle length by Xe-wee1 and Mos. *Development*, **125**, 237–248.

Murakami, M.S., Copeland, T.D. and Vande Woude, G.F. (1999) Mos positively regulates Xe-Wee1 to lengthen the first mitotic cell cycle of *Xenopus*. *Genes Dev.*, **13**, 620–631.

Musacchio, A. and Salmon, E.D. (2007) The spindle-assembly checkpoint in space and time. *Nat. Rev. Mol. Cell Biol.*, **8**, 379–393.

Mutter, G.L. and Wolgemuth, D.J. (1987) Distinct developmental patterns of c-mos protooncogene expression in female and male mouse germ cells. *Proc. Natl. Acad. Sci. USA*, **84**, 5301–5305.

Nagano, H., Hirai, S., Okano, K. and Ikegami, S. (1981) Achromosomal cleavage of fertilized starfish eggs in the presence of aphidicolin. *Dev. Biol.*, **85**, 409–415.

Nebreda, A.R. and Hunt, T. (1993) The c-mos proto-oncogene protein kinase turns on and maintains the activity of MAP kinase, but not MPF, in cell-free extracts of Xenopus oocytes and eggs. *EMBO J.*, **12**, 1979–1986.

Nishiyama, T., Ohsumi, K. and Kishimoto, T. (2007a) Phosphorylation of Erp1 by p90rsk is required for cytostatic factor arrest in Xenopus laevis eggs. *Nature*, **446**, 1096–1099.

Nishiyama, T., Yoshizaki, N., Kishimoto, T. and Ohsumi, K. (2007b) Transient activation of calcineurin is essential to initiate embryonic development in Xenopus laevis. *Nature*, **449**, 341–345.

Nishizawa, M., Furuno, N., Okazaki, K. *et al.* (1993) Degradation of Mos by the N-terminal proline (Pro2)-dependent ubiquitin pathway on fertilization of Xenopus eggs: possible significance of natural selection for Pro2 in Mos. *EMBO J.*, **12**, 4021–4027.

Ohe, M., Inoue, D., Kanemori, Y. and Sagata, N. (2007) Erp1/Emi2 is essential for the meiosis I to meiosis II transition in Xenopus oocytes. *Dev. Biol.*, **303**, 157–164.

Ohsumi, K., Koyanagi, A., Yamamoto, T.M. *et al.* (2004) Emi1-mediated M-phase arrest in Xenopus eggs is distinct from cytostatic factor arrest. *Proc. Natl. Acad. Sci. USA*, **101**, 12531–12536.

Okano-Uchida, T., Sekiai, T., Lee, K. *et al.* (1998) In vivo regulation of cyclin A/Cdc2 and cyclin B/ Cdc2 through meiotic and early cleavage cycles in starfish. *Dev. Biol.*, **197**, 39–53.

Okano-Uchida, T., Okumura, E., Iwashita, M. *et al.* (2003) Distinct regulators for Plk1 activation in starfish meiotic and early embryonic cycles. *EMBO J.*, **22**, 5633–5642.

Oskarsson, M., Mcclements, W.L., Blair, D.G. *et al.* (1980) Properties of a normal mouse cell DNA sequence (sarc) homologous to the src sequence of Moloney sarcoma virus. *Science*, **207**, 1222–1224.

Page, A.W. and Orr-Weaver, T.L. (1997) Stopping and starting the meiotic cell cycle. *Curr. Opin. Genet. Dev.*, **7**, 23–31.

Peters, J.M. (2006) The anaphase promoting complex/cyclosome: a machine designed to destroy. *Nat. Rev. Mol. Cell Biol.*, **7**, 644–656.

Picard, A., Galas, S., Peaucellier, G. and Doree, M. (1996) Newly assembled cyclin B-cdc2 kinase is required to suppress DNA replication between meiosis I and meiosis II in starfish oocytes. *EMBO J.*, **15**, 3590–3598.

Posada, J., Yew, N., Ahn, N.G. *et al.* (1993) Mos stimulates MAP kinase in Xenopus oocytes and activates a MAP kinase kinase in vitro. *Mol. Cell Biol.*, **13**, 2546–2553.

Propst, F. and Vande Woude, G.F. (1985) Expression of c-mos proto-oncogene transcripts in mouse tissues. *Nature*, **315**, 516–518.

Propst, F., Rosenberg, M.P., Iyer, A. *et al.* (1987) c-mos proto-oncogene RNA transcripts in mouse tissues: structural features, developmental regulation, and localization in specific cell types. *Mol. Cell Biol.*, **7**, 1629–1637.

Rauh, N.R., Schmidt, A., Bormann, J. *et al.* (2005) Calcium triggers exit from meiosis II by targeting the APC/C inhibitor XErp1 for degradation. *Nature*, **437**, 1048–1052.

Reimann, J.D. and Jackson, P.K. (2002) Emi1 is required for cytostatic factor arrest in vertebrate eggs. *Nature*, **416**, 850–854.

Russo, G.L., Wilding, M., Marino, M. and Dale, B. (1998) Ins and outs of meiosis in ascidians. *Semin. Cell Dev. Biol.*, **9**, 559–567.

Russo, G.L., Bilotto, S., Ciarcia, G. and Tosti, E. (2009) Phylogenetic conservation of cytostatic factor related genes in the ascidian Ciona intestinalis. *Gene*, **429**, 104–111.

Sadler, K.C. and Ruderman, J.V. (1998) Components of the signaling pathway linking the 1-methyladenine receptor to MPF activation and maturation in starfish oocytes. *Dev. Biol.*, **197**, 25–38.

Sagata, N. (1996) Meiotic metaphase arrest in animal oocytes: its mechanisms and biological significance. *Trends Cell Biol.*, **6**, 22–28.

Sagata, N., Watanabe, N., Vande Woude, G.F. and Ikawa, Y. (1989) The c-mos proto-oncogene product is a cytostatic factor responsible for meiotic arrest in vertebrate eggs. *Nature*, **342**, 512–518.

Schmidt, A., Duncan, P.I., Rauh, N.R. *et al.* (2005) Xenopus polo-like kinase Plx1 regulates XErp1, a novel inhibitor of APC/C activity. *Genes Dev.*, **19**, 502–513.

Schmidt, M., Oskarsson, M.K., Dunn, J.K. *et al.* (1988) Chicken homolog of the mos proto-oncogene. *Mol. Cell Biol.*, **8**, 923–929.

Schwab, M.S., Roberts, B.T., Gross, S.D. *et al.* (2001) Bub1 is activated by the protein kinase p90(Rsk) during Xenopus oocyte maturation. *Curr. Biol.*, **11**, 141–150.

Shibuya, E.K. and Masui, Y. (1988) Stabilization and enhancement of primary cytostatic factor (CSF) by ATP and NaF in amphibian egg cytosols. *Dev. Biol.*, **129**, 253–264.

Shibuya, E.K. and Masui, Y. (1989) Molecular characteristics of cytostatic factors in amphibian egg cytosols. *Development*, **106**, 799–808.

Shibuya, E.K. and Ruderman, J.V. (1993) Mos induces the in vitro activation of mitogen-activated protein kinases in lysates of frog oocytes and mammalian somatic cells. *Mol. Biol. Cell*, **4**, 781–790.

Shibuya, E.K., Boulton, T.G., Cobb, M.H. and Ruderman, J.V. (1992) Activation of p42 MAP kinase and the release of oocytes from cell cycle arrest. *EMBO J.*, **11**, 3963–3975.

Shirato, Y., Tamura, M., Yoneda, M. and Nemoto, S. (2006) Centrosome destined to decay in starfish oocytes. *Development*, **133**, 343–350.

Shoji, S., Yoshida, N., Amanai, M. *et al.* (2006) Mammalian Emi2 mediates cytostatic arrest and transduces the signal for meiotic exit via Cdc20. *EMBO J.*, **25**, 834–845.

Standart, N., Minshull, J., Pines, J. and Hunt, T. (1987) Cyclin synthesis, modification and destruction during meiotic maturation of the starfish oocyte. *Dev. Biol.*, **124**, 248–258.

Sudakin, V., Ganoth, D., Dahan, A. *et al.* (1995) The cyclosome, a large complex containing cyclin-selective ubiquitin ligase activity, targets cyclins for destruction at the end of mitosis. *Mol. Biol. Cell*, **6**, 185–197.

Tachibana, K., Machida, T., Nomura, Y. and Kishimoto, T. (1997) MAP kinase links the fertilization signal transduction pathway to the G1/S-phase transition in starfish eggs. *EMBO J.*, **16**, 4333–4339.

Tachibana, K., Tanaka, D., Isobe, T. and Kishimoto, T. (2000) c-Mos forces the mitotic cell cycle to undergo meiosis II to produce haploid gametes. *Proc. Natl. Acad. Sci. USA*, **97**, 14301–14306.

Tachibana, K., Hara, M., Hattori, Y. and Kishimoto, T. (2008) Cyclin B-Cdk1 controls pronuclear union in interphase. *Curr. Biol.*, **18**, 1308–1313.

Tachibana, K., Mori, M., Matsuhira, T. *et al.* (2010) Initiation of DNA replication by fertilization is regulated by p90Rsk at pre-RC/pre-IC transition in starfish eggs. *Proc. Natl. Acad. Sci. USA*, in press.

Tamura, M. and Nemoto, S. (2001) Reproductive maternal centrosomes are cast off into polar bodies during maturation division in starfish oocytes. *Exp. Cell Res.*, **269**, 130–139.

Tang, W., Wu, J.Q., Guo, Y. *et al.* (2008) Cdc2 and Mos regulate Emi2 stability to promote the meiosis I-meiosis II transition. *Mol. Biol. Cell*, **19**, 3536–3543.

Terret, M.E., Lefebvre, C., Djiane, A. *et al.* (2003) DOC1R: a MAP kinase substrate that control microtubule organization of metaphase II mouse oocytes. *Development*, **130**, 5169–5177.

Thibier, C., De Smedt, V., Poulhe, R. *et al.* (1997) In vivo regulation of cytostatic activity in Xenopus metaphase II-arrested oocytes. *Dev. Biol.*, **185**, 55–66.

Tsurumi, C., Hoffmann, S., Geley, S. *et al.* (2004) The spindle assembly checkpoint is not essential for CSF arrest of mouse oocytes. *J. Cell Biol.*, **167**, 1037–1050.

Tunquist, B.J., Eyers, P.A., Chen, L.G. *et al.* (2003) Spindle checkpoint proteins Mad1 and Mad2 are required for cytostatic factor-mediated metaphase arrest. *J. Cell Biol.*, **163**, 1231–1242.

Tunquist, B.J., Schwab, M.S., Chen, L.G. and Maller, J.L. (2002) The spindle checkpoint kinase bub1 and cyclin e/cdk2 both contribute to the establishment of meiotic metaphase arrest by cytostatic factor. *Curr. Biol.*, **12**, 1027–1033.

Usui, K., Hirohashi, N. and Chiba, K. (2008) Involvement of mitogen-activating protein kinase and intracellular pH in the duration of the metaphase I (MI) pause of starfish oocytes after spawning. *Develop. Growth Differ.*, **50**, 357–364.

Vardy, L. and Orr-Weaver, T.L. (2007a) The *Drosophila* PNG kinase complex regulates the translation of cyclin B. *Dev. Cell*, **12**, 157–166.

Vardy, L. and Orr-Weaver, T.L. (2007b) Regulating translation of maternal messages: multiple repression mechanisms. *Trends Cell Biol.*, **17**, 547–554.

Verlhac, M.H., Kubiak, J.Z., Weber, M. *et al.* (1996) Mos is required for MAP kinase activation and is involved in microtubule organization during meiotic maturation in the mouse. *Development*, **122**, 815–822.

Verlhac, M.H., Lefebvre, C., Kubiak, J.Z. *et al.* (2000) Mos activates MAP kinase in mouse oocytes through two opposite pathways. *EMBO J.*, **19**, 6065–6074.

Walter, S.A., Guadagno, T.M. and Ferrell, J.E. Jr (1997) Induction of a G2-phase arrest in Xenopus egg extracts by activation of p42 mitogen-activated protein kinase. *Mol. Biol. Cell*, **8**, 2157–2169.

Walter, S.A., Guadagno, S.N. and Ferrell, J.E. Jr (2000) Activation of Wee1 by p42 MAPK in vitro and in cycling *Xenopus* egg extracts. *Mol. Biol. Cell*, **11**, 887–896.

Watanabe, N., Hunt, T., Ikawa, Y. and Sagata, N. (1991) Independent inactivation of MPF and cytostatic factor (Mos) upon fertilization of Xenopus eggs. *Nature*, **352**, 247–248.

Winston, N.J., McGuinness, O., Johnson, M.H. and Maro, B. (1995) The exit of mouse oocytes from meiotic M-phase requires an intact spindle during intracellular calcium release. *J. Cell Sci.*, **108**, 143–151.

Wu, Q., Guo, Y., Yamada, A. *et al.* (2007) A role for Cdc2- and PP2A-mediated regulation of Emi2 in the maintenance of CSF arrest. *Curr. Biol.*, **17**, 213–224.

Yamada, H., Hirai, S., Ikegami, S. *et al.* (1985) The fate of DNA originally existing in the zygote nucleus during achromosomal cleavage of fertilized echinoderm eggs in the presence of aphidicolin: Microscopic studies with anti-DNA antibody. *J. Cell Physiol.*, **124**, 9–12.

Yamamoto, D.S., Tachibana, K., Sumitani, M. *et al.* (2008) Involvement of Mos-MEK-MAPK pathway in cytostatic factor (CSF) arrest in eggs of the parthenogenetic insect, Athalia rosae. *Mech. Dev.*, **125**, 996–1008.

Yamamoto, T.M., Iwabuchi, M., Ohsumi, K. and Kishimoto, T. (2005) APC/C-Cdc20-mediated degradation of cyclin B participates in CSF arrest in unfertilized Xenopus eggs. *Dev. Biol.*, **279**, 345–355.

Yamano, H., Gannon, J., Mahbubani, H. and Hunt, T. (2004) Cell cycle-regulated recognition of the destruction box of cyclin B by the APC/C in Xenopus egg extracts. *Mol. Cell*, **13**, 137–147.

Section VII

Oocyte ageing in mammals

15

Mammalian oocyte population throughout life

Roger Gosden,[1] Eujin Kim,[1] Bora Lee,[1] Katia Manova[2] and Malcolm Faddy[3]

[1]*Weill Medical College of Cornell University, New York, NY 10021, USA*
[2]*Memorial Sloan-Kettering Cancer Center, New York, NY 10065, USA*
[3]*School of Mathematical Sciences, Queensland University of Technology, Brisbane, QLD 4001, Australia*

15.1 Ovarian follicles as developmental units

The two principal functions of mammalian ovaries – gametogenesis and hormone production – emerge from the unitary character of follicles. None of the three specific cell types (oocyte, granulosa and theca) can maintain a normal phenotype or develop autonomously unless they interact with their follicular partners. Oocytes cannot grow without an envelope of granulosa cells to support their growth (Canipari *et al.*, 1984), but they do orchestrate follicle development and influence granulosa cell metabolism (Vanderhyden, 1996; Eppig, Wigglesworth and Pendola, 2002; Sugiura *et al.*, 2007), while the granulosa cells are, in turn, required for differentiation of theca cells (Honda *et al.*, 2007). The ability of small follicles to grow to maturity and ovulate *in vitro* has provided compelling evidence of the role of theca cells and the sufficiency of the three cell types for fulfilling both follicular functions. Theca cells are needed for development to the Graafian stage (Spears *et al.*, 1994; Liu, He and Rosenwaks, 2002), and they also have a complementary role in oestrogen biosynthesis by providing steroidal metabolites for aromatization in the granulosa cell compartment (McNatty and Sawers, 1975; Fortune and Armstrong, 1978; Armstrong, Goff and Dorrington, 1979).

The physiological implication of interdependency between follicular cell types is that the ovary ceases to function altogether in the absence of any one of them. In primates the final menses is an external manifestation of the simultaneous exhaustion of oocytes and

Oogenesis: The Universal Process Marie-Hélène Verlhac and Anne Villeneuve

their associated endocrine cells. Semantically, 'oocyte' and 'follicle' are used interchangeably here in the context of the ovarian reserve. Primordial follicles represent the major oocyte reserve in the ovary for maintaining menstrual or oestrous cycles, being the most abundant follicle stage at all ages and remaining developmentally quiescent until they are recruited for growth towards Graafian maturity or their common fate, atresia. The potential span of reproductive life therefore depends on the numbers and dynamics of these small follicles, and recent controversies about follicular renewal after birth are germane to the biology of ovarian ageing and menopause.

15.2 Follicular reserve in young ovaries

The follicle population is already established at the time of birth in human ovaries and comprises some $1–2 \times 10^6$ follicles (Baker, 1963; Forabosco and Sforza, 2007). Birth is not a major developmental transition in this organ since folliculogenesis can begin before, around or shortly after birth, depending on the species (Mossman and Duke, 1973). A few antral follicles are present at birth in humans; they become more abundant during childhood as serum gonadotrophin levels rise towards the age of puberty, but they eventually become atretic (Himelstein-Braw *et al.*, 1976; Peters, Byskov and Grinsted, 1978). The cortex and medulla of the ovary are not well defined at perinatal ages, although there are three discrete zones in humans, reflecting a developmental progression with the most mature follicles at the interior. There is an outer zone virtually devoid of follicles, an intermediate zone densely packed with primordial follicles, and an inner zone of mainly primary follicles, characterized by a growing oocyte with a single layer of cuboidal granulosa cells, and a few more advanced follicle stages. The factors responsible for creating and maintaining this gradient are unknown, though specific niches are likely to exist within the stroma for regulating the survival and fate of resident follicles.

Newborn mouse ovaries by contrast contain only $\sim 2 \times 10^4$ small oocytes, mostly at prefollicular pachytene or diplotene stages (Borum, 1967), and more-or-less evenly distributed between the paired organs (contrary to the left-sided bias in birds). The immaturity of perinatal ovaries in this altricial species reflects a short gestation period of barely three weeks. The primordial germ cells that migrated to colonize the gonadal ridges over a week earlier, at mid-gestation, formed clusters of oogonia united by intercellular bridges, called 'nests' or 'cysts'. These structures have been conserved during evolution (Pepling, de Cuevas and Spradling, 1999), though not strictly analogous to the cysts in *Drosophila* ovaries because mammalian germ cells never transform to nurse cells.

However, large numbers of germ cells undergo apoptosis when the nests break down and follicles emerge shortly after birth (Ratts *et al.*, 1995; McGee *et al.*, 1998), but it is unclear whether cells with defective genotypes are being selectively eliminated or if a random process is operating to control the size of the nascent follicle population. Targeted disruption of *bcl2* reduces germ cell survival (Ratts *et al.*, 1995), whereas ectopic expression of the same gene increases the number of germ cells (Flaws *et al.*, 2001). Moreover, in the absence of the proapoptotic gene, *bax*, oocyte survival increases threefold (Perez *et al.*, 1999), and without caspase-2, a component of the cell death pathway, there is a comparable effect (Bergeron *et al.*, 1998). Genetic deletion of the Cdk inhibitor 1B ($p27^{kip1-/-}$) likewise doubles the oocyte endowment at birth, as well as having downstream effects on follicle dynamics (Rajareddy *et al.*, 2007). These data imply that the oocyte population size is not tightly regulated.

Primordial follicles first emerge in the ovigerous cords (Byskov *et al.*, 1977; Byskov, 1986; Hirshfield, 1991) and, at least in species with a proximal mesonephros, the pregranulosa cells may be derived from the rete ovarii (Byskov, 1975). These follicles in mouse ovaries consist of an oocyte surrounded by a single layer of ~6 squamous, pregranulosa cells; in humans they are twice the diameter (35 μm) and have ~30 cells, also changing in morphology, size and cell number with age (Westergaard, Byskov and Andersen, 2007). Follicle growth begins as soon as the population is established, but the first recruits reach multilaminar stages and become gonadotrophin dependent long before puberty when the hormonal milieu can support maturation, and hence they undergo atresia. Proportionately more follicles are lost before puberty than afterwards, reflecting the high rate of recruitment from a large reserve. In adult ovaries, most follicles only become atretic after starting to grow, although in CBA mice, and possibly other strains, small follicles continue to disappear at a high rate, causing sterility as early as ~12–14 months of age (Jones and Krohn, 1961a; Faddy, Gosden and Edwards, 1983).

Apoptosis is not the only mechanism responsible for eliminating germ cells in newborn and infant ovaries. Some oogonial or prophase I oocytes are found in the surface epithelium, actively propelling themselves into the peritoneal cavity, or the periovarial sac in rodents, where they presumably die. This migration occurs during infancy in mice but is mainly prenatal in human ovaries, reflecting differences between species in timetables for folliculogenesis (Motta and Makabe, 1986).

In general, follicles contain a single oocyte, but binovular and polyovular follicles exist, sometimes abundantly, and have been recorded since Von Baer's seminal paper (1827) describing the mammalian ovum. These variants have been historically attributed to dividing oocytes, despite obvious conceptual difficulties with a meiotic cell, or to concrescence, even though they occur at all follicle stages. Most probably, they formed when a nest of prefollicular oocytes failed to separate completely but became enclosed in a common envelope of pregranulosa cells. Their frequency varies greatly between individual women or breeds of animals, independently of age (Gougeon, 1981), and there are dramatic differences between species, in the following order of relative abundance: rabbits < rhesus monkeys < humans < cats < dogs (Telfer and Gosden, 1987). Uniovular follicles are always the most common type, and the frequency of polyovular follicles varies inversely with their number of oocytes (up to ten or even more). Despite the high frequency in some animals, such as puppies (68%) (Payan-Carreira and Pires, 2008), polyovular follicles probably affect the ovulation rate only slightly, if at all, even in naturally superovulating species, such as tenrecs, elephant shrews and the plains viscacha (Weir, 1971). One reason is that the growth of oocytes in a cluster is often desynchronized, which probably impairs developmental competence. Follicles with the largest number of oocytes have disappeared almost completely by mid-life, at least in dogs, and the overall frequency of polyovular follicles has diminished with age from 14% at 1–2 years to 5% at 7–11 years of age (Telfer and Gosden, 1987).

15.3 Initiation of primordial follicle growth

Recruitment of follicles to the primary follicle stage is the fundamental and irreversible transition that fixes a limit on fecundity as well as the onset of ovarian senescence (Figure 15.1). The first signs of oocyte activation are increased cytoplasmic, nuclear and nucleolar volumes, upregulated RNA synthesis (Lintern-Moore and Moore, 1979) and secretion of zona pellucida components. Granulosa cells change from a squamous to cuboidal morphology at the same time, and begin mitosis (Hirshfield, 1991). A major increase in gene expression occurs at this time, especially affecting growth and immune factors and hormones, their receptors and corresponding signalling pathways (Park *et al.*, 2005; Dharma, Modi and Nandedkar, 2009), although carbohydrate metabolism is unchanged per unit volume (Harris *et al.*, 2009).

Although deletion of either *Foxo3* or *PTEN* has no effect on prenatal oogenesis, after birth global activation of growth is triggered, with many oocytes enlarging without corresponding changes in their granulosa cells (Castrillon *et al.*, 2003; John *et al.*, 2007; Reddy *et al.*, 2008). In consequence, the ovaries become prematurely sterile, demonstrating the key importance of regulated follicle recruitment and the PI-3-kinase signalling pathway for a normal reproductive lifespan. These findings may have direct

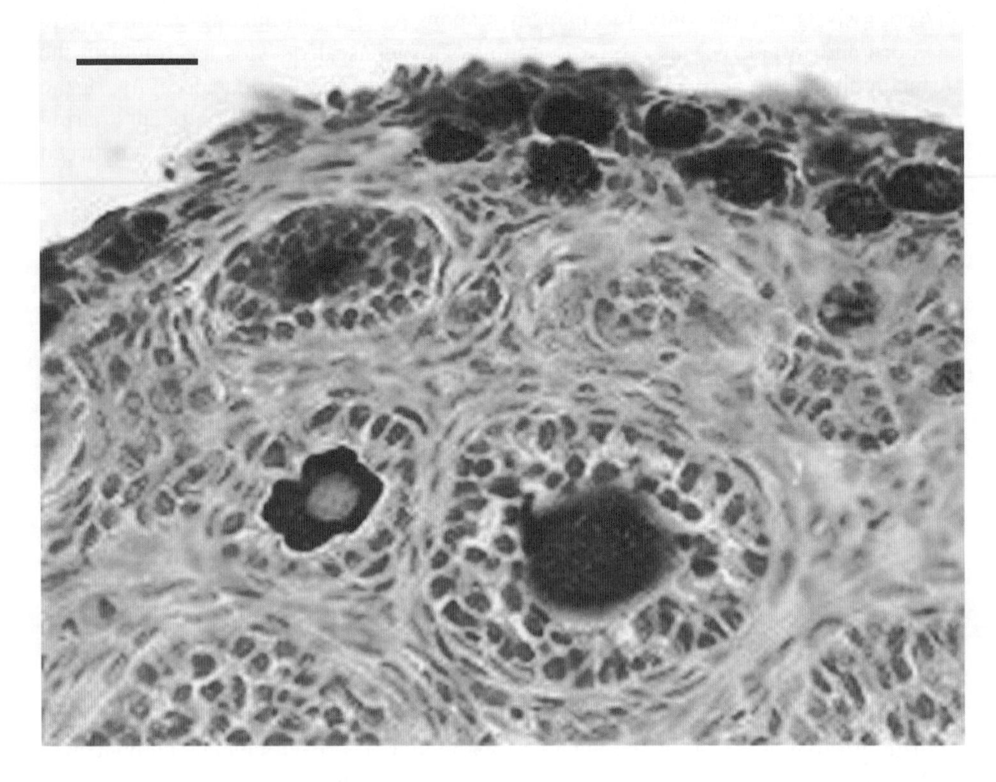

Figure 15.1 Juvenile mouse ovary showing abundant primordial follicles close to the surface epithelium, whereas the growing stages are deeper in the stroma. The oocyte cytoplasm is stained for Vasa and nuclei counterstained with methylene blue. Scale bar = 50 μm. A full colour version of this figure appears in the colour plate section.

relevance to human follicles which also express PTEN and phosphorylated Akt (Goto *et al.*, 2007). Considering the limited follicle reserve and the risks of explosive follicle growth, illustrated by ovarian hyperstimulation syndrome in women with polycystic ovaries (Brinsden *et al.*, 1995), follicle recruitment is likely to be under robust control, and to be evolutionarily conserved (Skinner, 2005). When fragments of ovarian tissue are cultured, follicle recruitment is spontaneously upregulated in all species investigated so far. This finding suggests the existence of an inhibitory factor(s), and the experimental phenomenon has been used to test candidate molecules.

Members of the TGF-β superfamily are evidently key players in follicle recruitment. Anti-Müllerian hormone (AMH), which is specifically expressed in granulosa cells of growing and small antral follicles, is a strong candidate inhibitor since gene deletion increases the recruitment of growing follicles (excess follicles subsequently undergoing atresia), whereas AMH supplementation of culture medium correspondingly inhibits follicle growth (Durlinger *et al.*, 1999, 2002; Visser *et al.*, 2007). The AMH molecule may play a paracrine role in human ovaries because it can suppress follicle recruitment *in vitro* without affecting survival (Carlsson *et al.*, 2006).

Bone morphogenetic proteins (BMPs) exert multiple effects during oogenesis and follicle development. BMP-4 from theca-stroma cells promotes both recruitment and survival (Nilsson and Skinner, 2003), whereas BMP-15, specifically expressed by oocytes, is involved in a local regulatory feedback loop with Kit ligand (KL) from granulosa cells (Su *et al.*, 2004). The growing follicle fraction is dramatically increased by incubating ovarian tissue with KL, whose cognate receptor (c-Kit) is on the surface of the oocyte (Parrott and Skinner, 1999). The pattern of KL expression is strongly suggestive of a physiological role, not only for growth initiation but at later stages (Manova *et al.*, 1990, 1993; Huang *et al.*, 1993). Growth differentiation factor-9, like its homologue BMP-15, is expressed by the oocyte but is not required for growth commencement, although necessary for progression beyond the primary follicle stage (Dong *et al.*, 1996). TGF-β-related molecules are unlikely to act alone since follicle recruitment, at least under culture conditions, is increased by leukaemia inhibiting factor (LIF) and keratinocyte growth factor (KGF), as well as by insulin and bFGF (basic fibroblast growth factor). Neither IGF-1 (insulin-like growth factor 1) nor EGF (epidermal growth factor) has any significant effect (Skinner, 2005).

Taken together, these data suggest that follicle initiation is regulated by many factors, both negatively and positively, and involves the PI-3-kinase signalling pathway in oocytes and paracrine secretion from neighbouring somatic cells. No single factor appears to have overwhelming control, instead a balance probably exists between inhibitory and stimulatory 'tone', with local, perhaps stochastic, variations responsible for the unpredictable order of follicle recruitment.

Practical benefits for animal production and reproductive medicine could emerge from further discoveries in this field if it becomes possible to control follicular fate, especially if recruitment can be inhibited in women at risk of premature ovarian failure. According to one model of the human ovary, ~37 follicles are recruited from the reserve population every day between the ages of 24 and 25 years, far exceeding the need for one ovulable follicle per month to sustain menstrual cycles. Since most of the surplus follicles undergoing atresia are believed to have potentially competent oocytes, more parsimonious utilization of the follicle reserve could extend fertility and postpone the menopause, even in women at late reproductive ages, since there are still three follicles starting to grow per day at 45–46 years of age (Faddy and Gosden, 1995).

15.4 Germline stem cells in postnatal ovaries

The German anatomist, Heinrich Wilhelm Gottfried von Waldeyer-Hartz ('Waldeyer') originally proposed that the number of oocytes formed in the ovary is fixed by the time of birth or shortly afterwards (Waldeyer, 1870). Although his hypothesis was challenged by authorities in the next century (Allen, 1923; Pincus, 1936; Vermande-van Eck, 1956), the controversy seemed to be settled in his favour in 1950 by Zuckerman's seminal review at the Laurentian Hormone Conference in New Hampshire. Zuckerman assembled a large body of data from ovarian morphology, experimental surgery and follicle counting in rat ovaries showing that the reserve declines inexorably with age (Zuckerman, 1951). As the hypothesis predicted, no variation in primordial follicle numbers occurred during the oestrous cycle: neither were any early stages of meiotic prophase present in adult ovaries, nor was there evidence of follicle regeneration after removing or destroying tissue. His arguments and confirmatory studies that followed were so compelling that Waldeyer's hypothesis became an entrenched dogma of ovarian biology. The only corollary, and a minor one, was from reports of oogonia persisting in adult prosimian ovaries, yet even this was not in serious conflict because none of the germ cells entered meiosis (Telfer, 2004).

Some recent challenges to the hypothesis were therefore unexpected. In one paper, there was evidence of follicular renewal from germline stem cells resident in adult mouse ovaries (Johnson *et al.*, 2004), and, in the second from the same laboratory, an even more revolutionary conclusion was drawn, namely that oocytes forming *de novo* after birth are derived from extraovarian precursors (Johnson *et al.*, 2005). Although these claims have not found wide acceptance, they deserve examination.

Johnson *et al.* questioned Waldeyer's hypothesis based on a similar rationale to critics of more than 50 years earlier. They doubted whether a follicle population of fixed size is sufficiently large to account for a full reproductive lifespan of up to a year in mice or 35 years in humans, especially given the high incidence of follicular atresia. They inferred that new oocytes must be forming continuously, and estimated by a rather circular argument that an additional 77 are formed per day.

Assumptions about the continuous decline in follicle numbers have not only been questioned by Johnson *et al.* (2004) but by Kerr *et al.* (2006), who claimed to have obtained 'qualified support for an as yet unknown mechanism of follicle renewal'. No significant differences were found in mean numbers of follicles between 7 and 100 days of age in mouse ovaries, but a lack of statistical significance does not necessarily mean that the null hypothesis of no change is true. Interval estimation is more informative about 'effect sizes' than significance testing, and simple regression analysis of the same data gave 95% confidence intervals for the daily change in mean numbers of (−6.5, 7.6) for primordial follicles and (−10.4, 5.5) for total follicles (Faddy and Gosden, 2007) (Figure 15.2). In other words, these data are consistent with either a fall or no change or an increase in follicle numbers, and are therefore uninformative about ageing, for which a larger sample would be required for a more accurate determination.

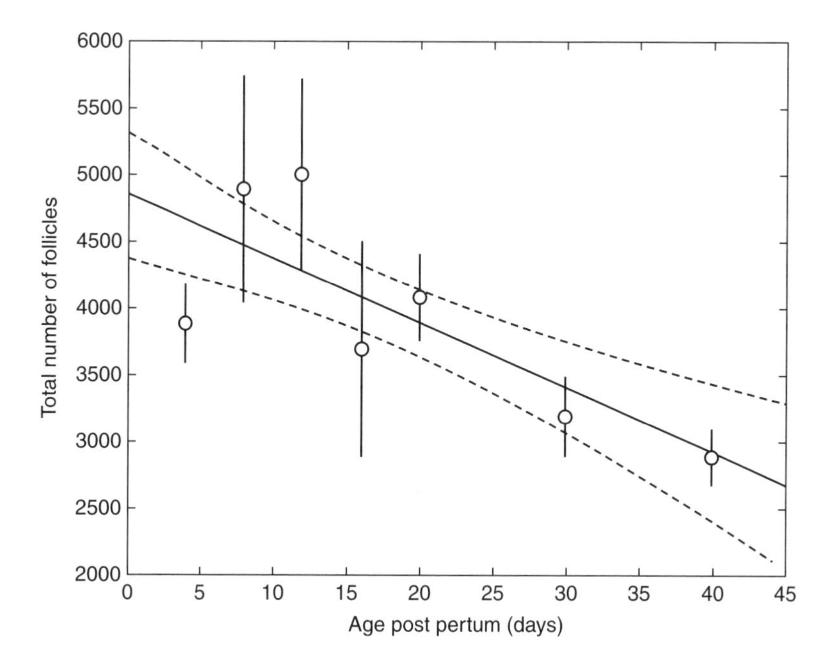

Figure 15.2 Variation in total numbers of follicles in mouse ovaries during the first 100 days of life. Reanalyzed from Kerr *et al.* (2006) by Faddy and Gosden (2007), by fitting linear regression (___) ± standard error (- - -); observed means (○) ± standard error of the mean (vertical bars)

The oogonia (germline stem cells) responsible for new oocytes were attributed at first by Johnson *et al.* (2004) to cells observed in the ovarian surface epithelium although, as mentioned above, they are generally regarded as exiting the ovary and disappear by puberty. Nevertheless, based on expression of certain mRNAs, presumptively specific for early meiosis (e.g. Scp3), and more compelling evidence of chimaerism in follicles after fusing pairs of ovaries, a prima facie case seemed to exist for neoformation of follicles. However, alternative experimental interpretations were quickly offered by other groups (Gosden, 2004; Telfer *et al.*, 2005; Byskov *et al.*, 2006). Johnson *et al.* also presented data indicating follicular regeneration after acute ovotoxicity (Johnson *et al.*, 2005), contrary to numerous studies that had not been previously challenged, showing that irradiation or chemically induced sterilization of the ovary is irreversible (e.g. Peters and Borum, 1961; Peters and Levy, 1964; Krarup, 1970; Haas, Christian and Hoyer, 2007).

Most astonishingly, they concluded that oocytes are being 'seeded' from circulating cells derived from the bone marrow (Johnson *et al.*, 2005). This radical hypothesis emerged from expression of molecular markers of germ cells in the marrow and peripheral blood, with cyclical variations that indicated a feedback loop for replenishing follicles lost by atresia. The most convincing studies involved bone marrow transplants to hosts that had been sterilized or semi-sterilized by irradiation and alkylating agents. Donor cells expressing green fluorescent protein (GFP) appeared to restore follicles in host animals, most convincingly in those presumed to be completely sterilized by deletion of the *Atm* gene (Johnson *et al.*, 2005). Yet, despite these findings they were unable to generate any offspring expressing GFP (Lee *et al.*, 2007). Nor have other groups been able to demonstrate formation of oocytes originating from transplanted GFP-labelled bone marrow or ovaries, or using parabiosis between a transgenic and wild-type partner (Eggan *et al.*, 2006; Begum, Papaioannou and Gosden, 2008). Not surprisingly, therefore, a young woman with residual ovarian function who, despite potentially sterilizing effects of Fanconi anaemia and chemotherapy from an allogeneic transplant, became pregnant with a child who was genetically related to herself rather than to her bone marrow donor (Veitia *et al.*, 2007).

Bukovsky *et al.* have also challenged the conventional wisdom of the Waldeyer hypothesis, specifically for human ovaries, but they concluded that a different mechanism was operating compared with mice. The scope for studying human ovaries is much more limited, being mainly immunohistochemistry and tissue culture, but they nevertheless claimed oocytes and granulosa cells were being generated from bipotential stem cells of mesenchymal origin in adult organs (Bukovsky *et al.*, 2004, 2008). These studies addressed the need for granulosa cell precursors in folliculogenesis, a crucial point that had been overlooked in the murine studies, even though it can be tested by transplanting primordial germ cells or small, naked oocytes. Bukovsky, Svetlikova and Caudle (2005) reported oocyte-like cells formed *in vitro*, and Virant-Klun *et al.* (2008) found cells resembling pluripotent ES cells (embryonic stem cells) from cultured scrapings of the surface epithelium of sterile human ovaries, some of which grew to the size of oocytes. The provenance of these ‘oocyte-like’ cells requires further investigation, but the evidence to date for neoformation of oocytes in human ovaries is unconvincing, and is contradicted by some molecular screens using specific molecular markers of oogonia and early meiosis (Liu *et al.*, 2007).

Waldeyer’s hypothesis still provides the best interpretation for the bulk of published experimental and observational data. It also offers a plausible explanation for the dynamics of the follicle population during ageing, and predicts that fecundity is irreversibly lost after exposure to ovotoxins. On the other hand, controversy has been beneficial in stimulating efforts to experimentally validate the dogma using modern technologies. Had the existence of ovarian regeneration been confirmed, concepts about ovarian senescence, menopause and oocyte ageing would have been turned upside down, and new opportunities might have emerged for generating fresh oocytes for clinical treatment and research. Perhaps reluctantly, we find no convincing basis for accepting such an optimistic view of ovarian physiology.

15.5 Human ovarian ageing

The total number of oocytes in human ovaries falls from one million at birth to a quarter of a million by the age of menarche (Figure 15.3), but age-specific numbers for individual women vary up to an order of magnitude (Block, 1952; Charleston *et al.*, 2007), which is probably a major factor determining the variable onset of menopause (Goswami and Conway, 2005). Lifestyle and environmental factors, apart from certain types of chemotherapy and irradiation, have a minor impact (Van Noord *et al.*, 1997) compared to the heritability of menopausal age, according to familial studies and data from twins (Snieder, MacGregor and Spector, 1998; de Bruin *et al.*, 2001). Likewise, mouse ovaries demonstrate significant variation between strains

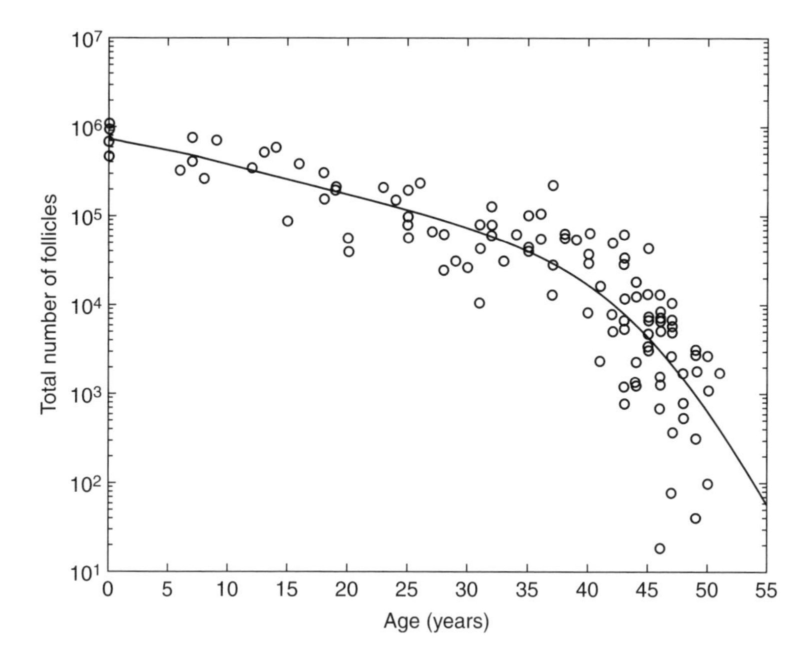

Figure 15.3 Observed (○) and estimated decline in mean (____) numbers of follicles in human ovaries, showing accelerating loss after 35–40 years of age. (Faddy and Gosden, 2003.)

in the initial endowment and the rate at which follicles are being lost (Jones and Krohn, 1961a; Faddy, Gosden and Edwards, 1983).

Considering the wide spectrum of life-history strategies and longevity in mammals, follicle numbers are expected to vary greatly between species for a comparable stage of development, such as puberty. Larger animals live longer and have a follicle endowment of corresponding magnitude, the numbers being scaled allometrically according to adult body mass ($27\,700M^{0.47}$, where M is in kg) (Gosden and Telfer, 1987). However, the rate at which follicles disappear with age is potentially at least as influential for timing follicle exhaustion as the initial numbers. Unfortunately, few data are available from which the dynamics can be estimated, the most reliable being from rodents and primates, including humans. In general, follicle loss occurs at a relatively constant rate across the adult lifespan, the half-life of the follicle population in mice being ~100 days and in humans 7 years, tentatively suggesting a scaling relationship enabling species to eke out their limited reserve according to life expectancy.

The human ovary differs not only in respect of uniovular cycles compared to mice, but because the follicles disappear faster during the two decades preceding the menopause, causing the reserve to be virtually exhausted by about 50 years of age compared to about 70 years if the rate had remained unchanged (Figure 15.3). This acceleration appears to be a biological phenomenon rather than due to counting or selection bias (Hansen *et al.*, 2008) and is reflected in the numbers of sonographically detectable antral follicles (Scheffer *et al.*, 1999). The physiological basis for accelerated ovarian ageing is, however, unknown. Serum FSH (follicle-stimulating hormone) levels start rising, variably, at approximately the same time (te Velde and Pearson, 2002), but they are unlikely to be responsible for this phenomenon because the FSH receptor is not expressed by primordial follicles (Oktay, Briggs and Gosden, 1997) and, furthermore, patients receiving repeated superstimulation with gonadotrophins for *in vitro* fertilization therapy do not appear to have an earlier menopause (Elder *et al.*, 2008).

We have modelled human ovarian ageing using differential equations for follicle counts of sectioned organs 19–50 years old based on three stages of growth: primordial, primary and subsequent follicle stages (Faddy and Gosden, 1995, 2000). The best fit to the data was obtained when the net rates of exponential loss varied, consistent with accelerated decline at each stage and similar to total follicle numbers. Rather perversely, the accelerated loss of primordial follicles was accompanied by proportionately fewer being recruited to the next stage. The occurrence of atresia at the primordial stage is virtually impossible to record from immunohistochemistry for apoptosis because follicle turnover is sluggish, but modelling suggests an absence of atresia at this stage in young, well-endowed ovaries (Faddy and Gosden, 1995), although contrary data exist (Gougeon, Ecochard and Thalabard, 1994).

The question of whether the follicle reserve is depleted only by recruitment or additionally by atresia is important for considering any strategy that might extend the functional lifespan of the ovary. Slowing the rate of recruitment is a more plausible and potentially safer option than inhibiting apoptosis. This appears to be the mechanism by which hypophysectomy or feeding mice on alternate days significantly slowed the disappearance of follicles (Jones and Krohn, 1961b; Nelson, Gosden and Felicio, 1985), potentially extending the fertile lifespan of dietary-restricted animals restored to a normal diet (Selesniemi, Lee and Tilly, 2008). The effects of dietary restriction are evidently independent of hypogonadotrophism, and possibly involve the sirtuins which have a known role in lifespan extension in invertebrates. Their interactions with the insulin/IGF-1 signalling pathway indicate a possibly fruitful avenue for investigation in mammals (Wenzel, 2006).

The menopause is conventionally defined retrospectively by 12 months of amenorrhoea, and hence the precise number of follicles remaining at the last menses is unknown. However, if the follicle decay curve is extrapolated to the median menopausal age of 51 years, $\sim 10^3$ follicles are predicted (Faddy *et al.*, 1992), consistent with sporadic reports of follicles in postmenopausal ovaries. It is unclear why maturation of this residue fails, although the elevated gonadotrophins or accumulated cellular age changes could be responsible, or perhaps in combination. At any rate, there is a close, and probably causal, temporal association between follicle exhaustion and the menopause in humans (Richardson, Senikas and Nelson, 1987; Faddy *et al.*, 1992), and probably chimpanzees also (Jones *et al.*, 2007). In rodents, hypothalamic–pituitary function plays a more important role in timing the cessation of ovulation (Felicio *et al.*, 1983). Primary ovarian ageing in humans is also in harmony with a model based on follicle dynamics and assumptions about the threshold number for sustaining cycles, because there was close concordance between the predicted and recorded distributions of menopausal age (Faddy and Gosden, 1996). It now seems almost unassailable that ovarian ageing is due to formation of a limited follicle population during prenatal ages followed by a program of continuous recruitment for ovulation and atresia, leading to inevitable exhaustion and menopause in surviving individuals. Behind the apparent simplicity of this model, theoretical opportunities exist to experimentally modulate fecundity, with implications for reproductive medicine.

Acknowledgement

We thank Ellen Kutner for help in preparing the manuscript.

References

Allen, E. (1923) Ovogenesis during sexual maturity. *Am. J. Anat.*, **31**, 439–481.

Armstrong, D.T., Goff, A.K. and Dorrington, J.H. (1979) Regulation of follicular estrogen biosynthesis, in *Ovarian Follicular Development and Function* (eds A.R. Midgley and W.A. Sadler), Raven Press, New York, pp. 169–181.

Baker, T.G. (1963) A quantitative and cytological study of germ cells in human ovaries. *Proc. R. Soc. London Ser. B*, **158**, 417–433.

Begum, S., Papaioannou, V.E. and Gosden, R.G. (2008) The oocyte population is not renewed in transplanted or irradiated adult ovaries. *Hum. Reprod.*, **23**, 2326–2330.

Bergeron, L., Perez, G.I., Macdonald, G. *et al.* (1998) Defects in regulation of apoptosis in caspase-2-deficient mice. *Genes Dev.*, **12**, 1304–1314.

Block, E. (1952) Quantitative morphological investigations of the follicular system in women: variations at different ages. *Acta Anat.*, **14**, 108–123.

Borum, K. (1967) Oogenesis in the mouse: a study of the origin of the mature ova. *Exp. Cell Res.*, **45**, 39–47.

Brinsden, P.R., Wada, I., Tan, S.-L. *et al.* (1995) Diagnosis, prevention and management of ovarian hyperstimulation syndrome. *Br. J. Obstet. Gynaecol.*, **102**, 767–772.

Bukovsky, A., Caudle, M.R., Svetlikova, M. and Upadhyaya, N.B. (2004) Origin of germ cells and formation of new primary follicles in adult human ovaries. *Reprod. Biol. Endocrinol.*, **2**, 20

Bukovsky, A., Svetlikova, M. and Caudle, M.R. (2005) Oogenesis in cultures derived from adult human ovaries. *Reprod. Biol. Endocrinol.*, **3**, 17.

Bukovsky, A., Gupta, S.K., Virant-Klun, I. *et al.* (2008) Study origin of germ cells and formation of new primary follicles in adult human and rat ovaries. *Methods Mol. Biol.*, **450**, 233–265.

Byskov, A.G. (1975) The role of the rete ovarii in meiosis and follicle formation in the cat, mink and ferret. *J. Reprod. Fertil.*, **45**, 201–209.

Byskov, A.G. (1986) Differentiation of mammalian embryonic gonad. *Physiol. Rev.*, **66**, 71–117.

Byskov, A.G., Faddy, M.J., Lemmen, J.G. and Anderson, C.Y. (2006) Eggs forever? *Differentiation*, **73**, 438–446.

Byskov, A.G., Skakkebaek, N.E., Stafanger, G. and Peters, H. (1977) Influence of ovarian surface epithelium and rete ovarii on follicle formation. *J. Anat.*, **123**, 77–86.

Canipari, R., Palombi, F., Riminucci, M. and Mangia, F. (1984) Early programming of maturation competence in mouse oogenesis. *Dev. Biol.*, **102**, 519–524.

Carlsson, I.B., Scott, J.E., Visser, J.A. *et al.* (2006) Anti-Müllerian hormone inhibits initiation of growth of human primordial ovarian follicles in vitro. *Hum. Reprod.*, **21**, 2223–2227.

Castrillon, D.H., Miao, L., Kollipara, R. *et al.* (2003) Suppression of ovarian follicle activation in mice by the transcription factor Foxo3a. *Science*, **301**, 215–218.

Charleston, J.S., Hansen, K.R., Thyer, A.C. *et al.* (2007) Estimating human ovarian non-growing follicle number: the application of modern stereology techniques to an old problem. *Hum. Reprod.*, **22**, 2103–2110.

Dharma, S.J., Modi, D.N. and Nandedkar, T.D. (2009) Gene expression profiling during early folliculogenesis in the mouse ovary. *Fertil. Steril.*, **91**(5 Suppl.), 2025–2036.

de Bruin, J.P., Bovenhuis, H., van Noord, P.A. *et al.* (2001) The role of genetic factors in age at natural menopause. *Hum. Reprod.*, **16**, 2014–2018.

Dong, J., Albertini, D.F., Nishimori, K. *et al.* (1996) Growth differentiation factor-9 is required during early ovarian folliculogenesis. *Nature*, **383**, 531–535.

Durlinger, A.L., Kramer, P., Karels, B. *et al.* (1999) Control of primordial follicle recruitment by anti-Müllerian hormone in the mouse ovary. *Endocrinology*, **140**, 5789–5796.

Durlinger, A.L., Gruijters, M.J., Kramer, P. *et al.* (2002) Anti-Müllerian hormone inhibits initiation of primordial follicle growth in the mouse ovary. *Endocrinology*, **143**, 1076–1084.

Eggan, K., Jurga, S., Gosden, R. *et al.* (2006) Ovulated oocytes in adult mice derive from non-circulating germ cells. *Nature*, **441**, 1109–1114.

Elder, K., Mathews, T., Kutner, E. *et al.* (2008) Impact of gonadotrophin stimulation for assisted reproductive technology on ovarian ageing and menopause. *Reprod. Biomed. Online*, **16**, 611–616.

Eppig, J.J., Wigglesworth, K. and Pendola, F.L. (2002) The mammalian oocyte orchestrates the rate of ovarian follicular development. *Proc. Natl. Acad. Sci. USA*, **99**, 2890–2894.

Faddy, M.J., Gosden, R.G., Gougeon, A. *et al.* (1992) Accelerated disappearance of ovarian follicles in mid-life – implications for forecasting menopause. *Hum. Reprod.*, **7**, 1342–1346.

Faddy, M.J. and Gosden, R.G. (1996) A model conforming the decline in follicle numbers to the age of menopause in women. *Hum. Reprod.*, **11**, 1484–1486.

Faddy, M.J. and Gosden, R.G. (2003) Modelling the dynamics of ovarian follicle utilization throughout life, in *Biology and Pathology of the Oocyte* (eds A.L. Trounson and R.G. Gosden), Cambridge University Press, Cambridge, pp. 44–52.

Faddy, M.J., Gosden, R.G. and Edwards, R.G. (1983) Ovarian follicle dynamics in mice: a comparative study of three inbred strains and an F1 hybrid. *J. Endocr.*, **96**, 23–33.

Faddy, M.J. and Gosden, R.G. (1995) A mathematical model for follicle dynamics in human ovaries. *Hum. Reprod.*, **10**, 770–775.

Faddy, M.J. and Gosden, R.G. (2000) Mathematical models for follicle development and depletion, in *Female Reproductive Ageing* (eds E.R. te Velde, P.L. Pearson and F.J. Broekmans), Parthenon Publishing, New York & London, pp. 71–78.

Faddy, M. and Gosden, R. (2007) Numbers of ovarian follicles and testing germ line renewal in the postnatal ovary. *Cell Cycle*, **6**, 1951–1952.

Felicio, L.S., Nelson, J.F., Gosden, R.G. and Finch, C.E. (1983) Restoration of ovulatory cycles by young ovarian grafts in aging mice: potentiation by long term ovariectomy decreases with age. *Proc. Natl. Acad. Sci. USA*, **80**, 6076–6080.

Flaws, J.A., Hirshfield, A.N., Hewitt, J.A. *et al.* (2001) Effect of bcl-2 on the primordial follicle endowment in the mouse ovary. *Biol. Reprod.*, **64**, 1153–1159.

Forabosco, A. and Sforza, C. (2007) Establishment of ovarian reserve: a quantitative morphometric study of the developing human ovary. *Fertil. Steril.*, **88**, 675–683.

Fortune, J.E. and Armstrong, D.T. (1978) Hormonal control of 17 beta-estradiol biosynthesis in proestrous rat follicles: estradiol production by isolated theca versus granulosa. *Endocrinology*, **102**, 227–235.

Gosden, R.G. and Telfer, E. (1987) Numbers of follicles in mammalian ovaries and their allometric relationships. *J. Zool.*, **211**, 169–175.

Gosden, R.G. (2004) Germline stem cells in the postnatal ovary: is the ovary more like a testis? *Hum. Reprod. Update*, **10**, 193–195.

Goswami, D. and Conway, G.S. (2005) Premature ovarian failure. *Hum. Reprod. Update*, **11**, 391–410.

Goto, M., Iwase, A., Ando, H. *et al.* (2007) PTEN and Akt expression during growth of human ovarian follicles. *J. Assist. Reprod. Genet.*, **24**, 541–546.

Gougeon, A. (1981) Frequent occurrence of multiovular follicles and multinuclear oocytes in the adult human ovary. *Fertil. Steril.*, **35**, 417–422.

Gougeon, A., Ecochard, R. and Thalabard, J.C. (1994) Age-related changes of the population of human ovarian follicles: increase in the disappearance rate of non-growing and early-growing follicles in aging women. *Biol. Reprod.*, **50**, 653–663.

Haas, J.R., Christian, P.J. and Hoyer, P.B. (2007) Effects of impending ovarian failure induced by 4-vinylcyclohexene diepoxide on fertility in C57BL/6 female mice. *Comp. Med.*, **57**, 443–449.

Hansen, K.R., Knowlton, N.S., Thyer, A.C. *et al.* (2008) A new model of reproductive aging: the decline in ovarian non-growing follicle number from birth to menopause. *Hum. Reprod.*, **23**, 699–708.

Harris, S.E., Leese, H.J., Gosden, R.G. and Picton, H.M. (2009) Pyruvate and oxygen consumption throughout the growth and development of murine oocytes. *Mol. Reprod. Dev.*, **76**(3), 231–238.

Himelstein-Braw, R., Byskov, A.G., Peters, H. and Faber, M. (1976) Follicular atresia in the infant human ovary. *J. Reprod. Fertil.*, **46**, 55–59.

Hirshfield, A.N. (1991) Development of follicles in the mammalian ovary. *Int. Rev. Cytol.*, **124**, 43–101.

Honda, A., Hirose, M., Hara, K. *et al.* (2007) Isolation, characterization, and *in vitro* and *in vivo* differentiation of putative thecal stem cells. *Proc. Natl. Acad. Sci. USA*, **104**, 12389–12394.

Huang, E.J., Manova, K., Packer, A.I. *et al.* (1993) The murine steel panda mutation affects kit ligand expression and growth of early ovarian follicles. *Dev. Biol.*, **157**, 100–109.

John, G.B., Shirley, L.J., Gallardo, T.D. and Castrillon, D.H. (2007) Specificity of the requirement for Foxo3 in primordial follicle activation. *Reproduction*, **133**, 855–863.

Johnson, J., Canning, J., Kaneko, T. *et al.* (2004) Germline stem cells and follicular renewal in the postnatal mammalian ovary. *Nature*, **428**, 145–150.

Johnson, J., Bagley, J., Skaznik-Wikiel, M. *et al.* (2005) Oocyte generation in adult mammalian ovaries by putative germ cells in bone marrow and peripheral blood. *Cell*, **122**, 303–315.

Jones, K.P., Walker, L.C., Anderson, D. *et al.* (2007) Depletion of ovarian follicles with age in chimpanzees: similarities to humans. *Biol. Reprod.*, **77**, 247–251.

Jones, E.C. and Krohn, P.L. (1961a) The relationships between age, numbers of oocytes and fertility in virgin and multiparous mice. *J. Endocrinol.*, **21**, 469–495.

Jones, E.C. and Krohn, P.L. (1961b) The effect of hypophysectomy on age changes in the ovaries of mice. *J. Endocrinol.*, **21**, 497–509.

Kerr, J.B., Duckett, R., Myers, M. *et al.* (2006) Quantification of healthy follicles in the neonatal and adult mouse ovary: evidence for maintenance of primordial follicle supply. *Reproduction*, **132**, 95–109.

Krarup, T. (1970) Oocyte survival in the mouse ovary after treatment with 9,10-dimethyl-1,2-benzanthracene. *J. Endocrinol.*, **46**, 483–495.

Lee, H.J., Selesniemi, K., Niikura, Y. *et al.* (2007) Bone marrow transplantation generates immature oocytes and rescues long-term fertility in a preclinical mouse model of chemotherapy-induced premature ovarian failure. *J. Clin. Oncol.*, **25**, 3198–3204.

Lintern-Moore, S. and Moore, G.P. (1979) The initiation of follicle and oocyte growth in the mouse ovary. *Biol. Reprod.*, **20**, 773–778.

Liu, H.C., He, Z. and Rosenwaks, Z. (2002) *In vitro* culture and *in vitro* maturation of mouse preantral follicles with recombinant gonadotropins. *Fertil. Steril.*, **77**, 373–383.

Liu, Y., Wu, C., Lyu, Q. *et al.* (2007) Germline stem cells and neo-oogenesis in the adult human ovary. *Dev. Biol.*, **306**, 112–120.

Manova, K., Nocka, K., Besmer, P. and Bachvarova, R.F. (1990) Gonadal expression of c-kit encoded at the W locus of the mouse. *Development*, **110**, 1057–1069.

Manova, K., Huang, E.J., Angeles, M. *et al.* (1993) The expression pattern of the c-kit ligand in gonads of mice supports a role for the c-kit receptor in oocyte growth and in proliferation of spermatogonia. *Dev. Biol.*, **157**, 85–99.

McGee, E.A., Hsu, S.Y., Kaipia, A. and Hsueh, A.J. (1998) Cell death and survival during ovarian follicle development. *Mol. Cell. Endocrinol.*, **140**, 15–18.

McNatty, K.P. and Sawers, R.S. (1975) Relationship between the endocrine environment within the Graafian follicle and the subsequent rate of progesterone secretion by human granulosa cells. *J. Endocrinol.*, **66**, 391–400.

Mossman, H.W. and Duke, K.L. (1973) *Comparative Morphology of the Mammalian Ovary*, Univ. Wisconsin Press, Madison.

Motta, P.M. and Makabe, S. (1986) Elimination of germ cells during differentiation of the human ovary: an electron microscopic study. *Eur. J. Obstet. Gynecol. Reprod. Biol.*, **22**, 271–286.

Nelson, J.F., Gosden, R.G. and Felicio, L.S. (1985) Effect of dietary restriction on estrous cyclicity and follicular reserves in aging C57BL/6J mice. *Biol. Reprod.*, **32**, 515–522.

Nilsson, E.E. and Skinner, M.K. (2003) Bone morphogenetic protein-4 acts as an ovarian follicle survival factor and promotes primordial follicle development. *Biol. Reprod.*, **69**, 1265–1272.

Oktay, K., Briggs, D. and Gosden, R.G. (1997) Ontogeny of follicle-stimulating hormone receptor gene expression in isolated human ovarian follicles. *J. Clin. Endocrinol. Metab.*, **82**, 3748–3751.

Park, C.E., Cha, K.Y., Kim, K. and Lee, K.A. (2005) Expression of cell cycle regulatory genes during primordial-primary follicle transition in the mouse ovary. *Fertil. Steril.*, **83**, 410–418.

Parrott, J.A. and Skinner, M.K. (1999) Kit-ligand/stem cell factor induces primordial follicle development and initiates folliculogenesis. *Endocrinology*, **140**, 4262–4271.

Payan-Carreira, R. and Pires, M.A. (2008) Multioocyte follicles in domestic dogs: a survey of frequency of occurrence. *Theriogenology*, **69**, 977–982.

Pepling, M.E., de Cuevas, M. and Spradling, A.C. (1999) Germline cysts: a conserved phase of germ cell development? *Trends Cell. Biol.*, **9**, 257–262.

Perez, G.I., Robles, R., Knudson, C.M. *et al.* (1999) Prolongation of ovarian lifespan into advanced chronological age by Bax-deficiency. *Nat. Genet.*, **21**, 200–203.

Peters, H. and Borum, K. (1961) The development of mouse ovaries after low-dose irradiation at birth. *Int. J. Radiat. Biol.*, **3**, 1–16.

Peters, H., Byskov, A.G. and Grinsted, J. (1978) Follicular growth in fetal and prepubertal ovaries of humans and other primates. *Clin. Endocrinol. Metab.*, **7**, 469–485.

Peters, H. and Levy, E. (1964) Effect of irradiation in infancy on the mouse ovary. A quantitative study of oocyte sensitivity. *J. Reprod. Fertil.*, **7**, 37–45.

Pincus, G. (1936) *The Eggs of Mammals*, MacMillan, New York, p. 14.

Rajareddy, S., Reddy, P., Du, C. *et al.* (2007) p27kip1 (cyclin-dependent kinase inhibitor 1B) controls ovarian development by suppressing follicle endowment and activation and promoting follicle atresia in mice. *Mol. Endocrinol.*, **21**, 2189–2202.

Ratts, V.S., Flaws, J.A., Kolp, R. *et al.* (1995) Ablation of bcl-2 gene expression decreases the numbers of oocytes and primordial follicles established in the post-natal female mouse gonad. *Endocrino logy*, **136**, 3665–3668.

Reddy, P., Liu, L., Adhikari, D. *et al.* (2008) Oocyte-specific deletion of Pten causes premature activation of the primordial follicle pool. *Science*, **319**, 611–613.

Richardson, S.J., Senikas, V. and Nelson, J.F. (1987) Follicular depletion during the menopausal transition: evidence for accelerated loss and ultimate exhaustion. *J. Clin. Endocrinol. Metab.*, **65**, 1231–1237.

Scheffer, G.J., Broekmans, F.J., Dorland, M. *et al.* (1999) Antral follicle counts by transvaginal ultrasonography are related to age in women with proven natural fertility. *Fertil. Steril.*, **72**, 845–851.

Selesniemi, K., Lee, H.J. and Tilly, J.L. (2008) Moderate caloric restriction initiated in rodents during adulthood sustains function of the female reproductive axis into advanced chronological age. *Aging Cell*, **7**, 622–629.

Skinner, M.K. (2005) Regulation of primordial follicle assembly and development. *Hum. Reprod. Update*, **11**, 461–471.

Snieder, H., MacGregor, A.J. and Spector, T.D. (1998) Genes control the cessation of a woman's reproductive life: a twin study of hysterectomy and age at menopause. *J. Clin. Endocrinol. Metab.*, **83**, 1875–1880.

Spears, N., Boland, N.I., Murray, A.A. and Gosden, R.G. (1994) Mouse oocytes derived from in vitro grown primary ovarian follicles are fertile. *Hum. Reprod.*, **9**, 527–532.

Su, Y.-Q., Wu, X., O'Brien, M.J. *et al.* (2004) Synergistic roles of BMP15 and GDF9 in the development and function of the oocyte–cumulus cell complex in mice: genetic evidence for an oocyte–granulosa cell regulatory loop. *Dev. Biol.*, **276**, 64–73.

Sugiura, K., Su, Y.Q., Diaz, F.J. *et al.* (2007) Oocyte-derived BMP15 and FGFs cooperate to promote glycolysis in cumulus cells. *Development*, **134**, 2593–2603.

Telfer, E.E. (2004) Germline stem cells in the postnatal mammalian ovary: a phenomenon of prosimian primates and mice? *Reprod. Biol. Endocrinol.*, **18**, 2–24.

Telfer, E. and Gosden, R.G. (1987) A quantitative cytological study of polyovular follicles in mammalian ovaries with particular reference to the domestic bitch (*Canis familiaris*). *J. Reprod. Fertil.*, **81**, 137–147.

Telfer, E.E., Gosden, R.G., Byskov, A.G. *et al.* (2005) On regenerating ovaries and generating controversy. *Cell*, **122**, 821–822.

Vanderhyden, B.C. (1996) Oocyte-secreted factors regulate granulosa cell steroidogenesis. *Zygote*, **4**, 317–321.

Van Noord, P.A., Dubas, J.S., Dorland, M. *et al.* (1997) Age at natural menopause in a population-based screening cohort: the role of menarche, fecundity, and lifestyle factors. *Fertil. Steril.*, **68**, 95–102.

Veitia, R.A., Gluckman, E., Fellous, M. *et al.* (2007) Recovery of fertility after chemotherapy, irradiation and bone marrow allograft: further evidence against massive oocyte regeneration by bone marrow-derived germline stem cells. *Stem Cells*, **25**, 1334–1335.

te Velde, E.R. and Pearson, P.L. (2002) The variability of female reproductive ageing. *Hum. Reprod. Update*, **8**, 141–154.

Vermande-van Eck, G.J. (1956) Neo-ovogenesis in the adult monkey. *Anat. Rec.*, **125**, 207–224.

Virant-Klun, I., Zech, N., Rozman, P. *et al.* (2008) Putative stem cells with an embryonic character isolated from the ovarian surface epithelium of women with no naturally present follicles and oocytes. *Differentiation*, **76**, 843–856.

Visser, J.A., Durlinger, A.L., Peters, I.J. *et al.* (2007) Increased oocyte degeneration and follicular atresia during the estrous cycle in anti-Müllerian hormone null mice. *Endocrinology*, **148**, 2301–2308.

Von Baer, K.E. (1827) *De ovi mammalium et hominis genesi. Epistola ad Academiam Imperialem Scientiarum Petropolitanam*, L. Vossius, Leipzig.

Waldeyer, W. (1870) *Eierstock und Ei*, Engelmann, Leipzig.

Weir, B.J. (1971) The reproductive organs of the female plains viscacha, Lagostomus maximus. *J. Reprod. Fertil.*, **25**, 365–373.

Wenzel, U. (2006) Nutrition, sirtuins and aging. *Genes Nutr.*, **1**, 85–93.

Westergaard, C.G., Byskov, A.G. and Andersen, C.Y. (2007) Morphometric characteristics of the primordial to primary follicle transition in the human ovary in relation to age. *Hum. Reprod.*, **22**, 2225–2231.

Zuckerman, S. (1951) The number of oocytes in the mature ovary. *Rec. Prog. Horm. Res.*, **6**, 63–108.

Section VIII

From oocyte to embryo

16

Fertilization and the evolution of animal gamete proteins

Julian L. Wong and Gary M. Wessel

Department of Molecular Biology, Cell Biology, and Biochemistry, Brown University, Providence, RI 02912, USA

16.1 Introduction

The success of various developmental mechanisms is measured by an organism's reproductive capabilities. Yet, once the germline has been determined and the gametes have developed, many parameters still affect the probability of two gametes successfully interacting. These include geographic isolation, timing of gamete release and reproductive anatomy (Wong and Wessel, 2006; Vieira and Miller, 2006). Macroscopic contributions such as habitat ecology, nutrient availability and mating behaviour also indirectly influence molecular participants (reviewed in Wong and Wessel, 2006; Marshall, Steinberg and Evans, 2004; O'Rand, 1988). Since successful fertilization ultimately depends on the productive interaction of two gametes, an understanding of how these haploid cells come together is of particular interest. Two complementary approaches are currently being used to understand the process of gamete binding on the molecular level: a reductive perspective that aims to identify the least common collection of proteins necessary to achieve fertilization (Wong and Wessel, 2006), and a divergence perspective that aims to understand how molecules evolve to establish barriers to interspecies hybridization (Vieira and Miller, 2006). These two approaches have intersected many times through the identification of common candidates, suggesting that both purifying and diversifying selection affect how fertilization evolves within a species.

In animals, female and male gametes experience different selective pressures that balance diversity and adaptation with fitness (Gavrilets, 2000; Parker and Partridge, 1998). Females usually produce the larger gamete, requiring major nutritional and

Oogenesis: The Universal Process Marie-Hélène Verlhac and Anne Villeneuve

temporal commitment from the individual. During oogenesis, oocyte precursors retain most of the cytoplasm during asymmetric divisions, and further expand their size through *de novo* synthesis and import of nutrients made by somatic cells. A female's energy expenditure per egg can be extremely high, sometimes serving also as a bank for nutritional storage that can be recycled if the female confronts dietary constraints. Further, only one-quarter of the haploid cells produced from a primary oocyte become fertilization competent, thereby increasing the investment on a per-cell basis. Sperm, on the other hand, cost each male less on a per-cell basis. Each meiotic division yields four sperm per primary spermatocyte, and the majority of cytoplasm and organelles is recycled in the testis prior to release of each mature gamete. The total yield of viable sperm is orders of magnitude greater than oocytes, which could negatively impact male fitness since the overall cost per fertilization is higher in males (Parker and Partridge, 1998). The quantitative imbalance does, however, allow for sperm diversity, which favours males in a 'sexual arms race' amongst the competing adaptation of gametes (Gavrilets, 2000; Parker and Partridge, 1998). Conversely, female success in such a race is linked to minimizing the frequency of change, to conserve resources while continuing to adapt to survive (Gavrilets, 2000). This struggle between the sexes is strongly linked to adaptive evolution, and fosters diversity of reproductive proteins amongst all organisms (Clark, Aagaard and Swanson, 2006; Swanson and Vacquier, 2002).

Most of the changes that drive the sexual arms race are thought to involve proteins found at the cell surface, since these are the sites of gamete recognition. This model is consistent with the observation that extracellular proteins in general are evolving faster than their intracellular counterparts (Julenius and Pedersen, 2006; Luz and Vingron, 2006). The origin of sperm–egg interactions is proposed to derive from ancient cell–cell recognition mechanisms (O'Rand, 1988), because the process fulfils similar requirements, such as low-affinity binding and the more complex assessment of compatibility that results in high-affinity associations; indeed, like immune response factors, reproductive proteins are classified as the most diverse gene set amongst animals (Swanson and Vacquier, 2002). The molecules involved in two phases of gamete binding are of particular interest since they represent the final barrier against interspecific fertilization, an event that is usually a reproductive dead end (Vieira and Miller, 2006; Parker and Partridge, 1998). Modifications to these cell surface proteins can dramatically affect the compatibility of sperm and egg, and with enough alteration may result in the reproductive isolation necessary for speciation (Vieira and Miller, 2006; Parker and Partridge, 1998; Palumbi and Metz, 1991). Maintenance of a new species depends on minimal alteration of the new surface proteins within the population, while they continue to diverge from their orthologues in sister species. This constant pull between neutralization and diversification generally results in positive selection on the proteins involved in the process (Swanson and Vacquier, 2002). Such adaptive evolution is recognized through identification of many, low-frequency polymorphic alleles at a genetic locus relative to a neutrally maintained locus, and/or as a ratio among orthologues of nonsynonymous-to-synonymous amino-acid changes greater than 1 ($d_N/d_S > 1$) (Clark, Aagaard and Swanson, 2006; Yang and Bielawski, 2000).

Although fertilization is an essential event in the lifecycle of all sexually reproducing organisms, few molecules have been found that function in specific animal gamete binding (Swanson and Vacquier, 2002). Such limited insight is likely due to the specialized events that gametes experience in their lifetime. First, gamete recognition molecules are only transiently present on the cell surface. The sperm, for example, exposes most egg-binding molecules only after activation; for example, the sperm has exocytosed its secretory vesicle, thereby completing the acrosome reaction (reviewed in Wong and Wessel, 2006; Töpfer-Petersen *et al.*, 1990; Tulsiani *et al.*, 1998). Secretion of the acrosomal vesicle is relatively destructive to the cell, and is sufficient to completely alter the sperm's surface protein profile (Tulsiani *et al.*, 1998; Bleil *et al.*, 1988; Usui, 1987; Dan, Ohori and Kushida, 1964), thereby altering affinity of the individual sperm for its complementary gamete. Importantly, the female receptors that recognized the sperm surface are often rendered nonfunctional following fertilization (Bauskin *et al.*, 1999; Moller *et al.*, 1990; Rossignol *et al.*, 1984; Carroll and Epel, 1975; Carroll and Epel, 1975). This prevents further fusion events by additional sperm (e.g. polyspermy), but also means the protein is harder to track. Second, gametes are amongst the select group of cells in the body that normally fuse with each other. The molecular mechanisms used by fusion-competent cells rarely overlap (reviewed in Chen and Olson, 2005), implying that the cell–cell recognition machinery and fusion catalysts are composed of a complex, complementary set of proteins, lipids, and potentially glycosylated residues that together can mediate cell–cell fusion.

One criterion used to look for protein-coding genes involved in gamete recognition is their evolution under positive selection (Clark, Aagaard and Swanson, 2006; Swanson and Vacquier, 2002; Gavrilets, 2000). Here, we review the function and coevolution of some protein pairs used by mollusc, echinoderm and eutherian gametes with respect to gamete binding and fertilization within animals. Collectively, we find that all pairings exhibit limited regions of sequence diversity within the primary sequence, but the overall structure or fold is conserved in a manner that could optimize the function of this protein pair.

16.2 Lysin: VERL

Aquatic gastropods, such as abalone and marine snails, reproduce by spawning. Their gametes are released into the water where they mix with complementary gametes as a result of signalling between adults to maximize gamete interaction. The primary barriers to interspecific hybrids include geography and/or nonoverlapping gravid seasons (Swanson and Vacquier, 2002; Parker and Partridge, 1998; O'Rand, 1988), but variations in climate can trigger overlapping spawning of species that may coexist. This tenuous situation has selected for a rigorous system to minimize interspecific fertilization events (reviewed in Vacquier and Lee, 1993). Most events cited here pertain to abalone, but some of the molecular players appear to be conserved in other molluscs (Springer and Crespi, 2007; Hellberg and Vacquier, 1999).

The cascade of fertilization events is initiated following gamete release. Eggs are spawned with an extracellular matrix, called the vitelline envelope, and with a coat of glycoproteins called the jelly layer (Figure 16.1). In the case of the abalone *Haliotis rufescens*, the chemoattractant L-tryptophan is released from the egg and establishes a gradient detected by sperm (Riffell, Krug and Zimmer, 2002, 2004). As a motile sperm

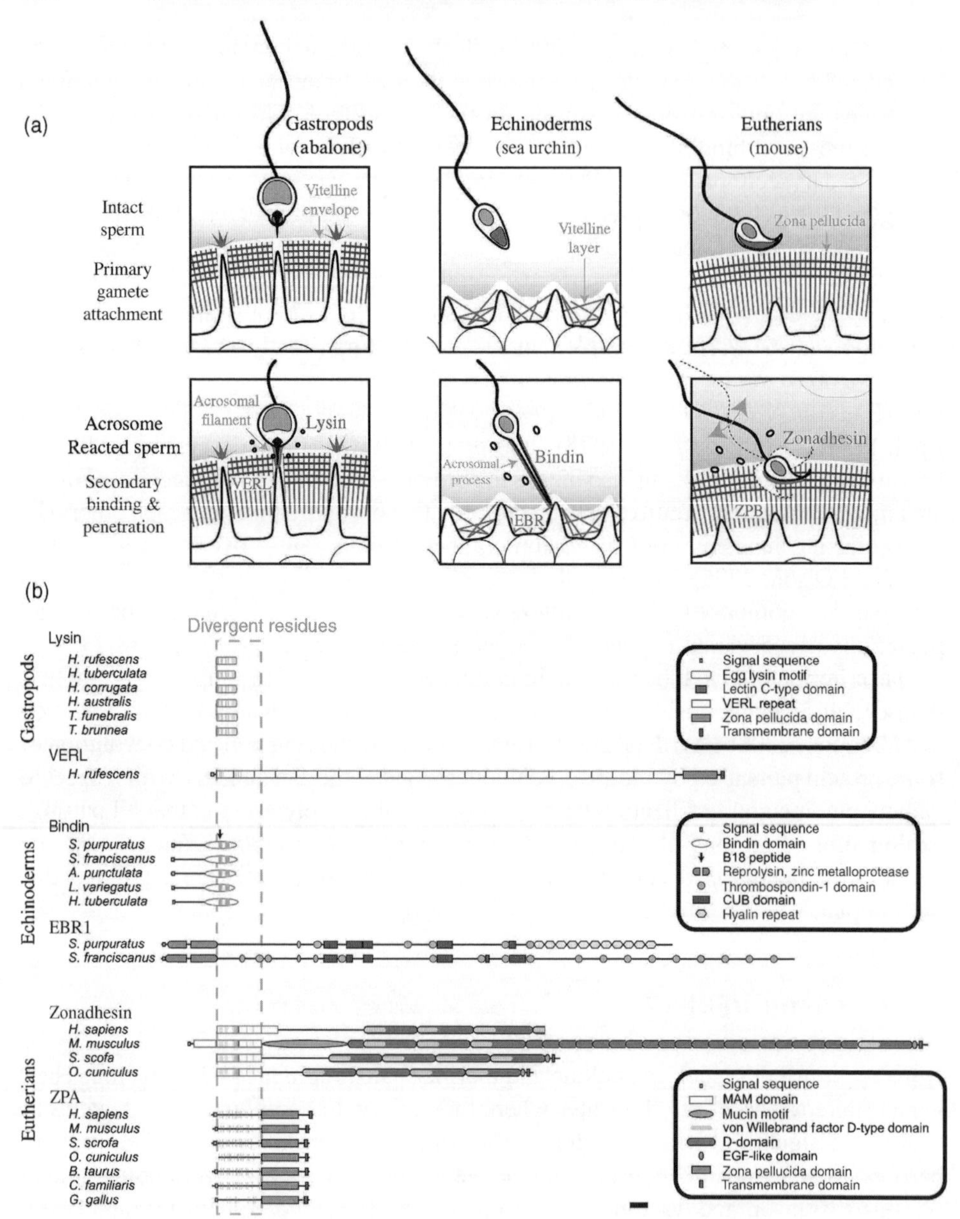

Figure 16.1 (a) Schematic of the major gamete interaction in three classes of animals. Details of egg cortex at the site of sperm binding are shown. Chemoattractant layer (yellow) covers the egg extracellular matrix (blue). The major sperm proteins (red) thought to contribute to the species-specific events are found first in the acrosome, but following exocytosis are relocated to the sperm surface. Basic images are modified from Wong and Wessel (2006). (b) Primary sequence maps of coevolving gamete-binding proteins from each class of animals. Domains specific to each orthologue are detailed in the legends. Most diverse residues (green) are clustered in select regions. Accession numbers include: [lysin] *Haliotis rufescens* (AAA29 196), *H. tuberculata* (AAB59 168), *H. corrugata* (P19 448), *H. australis* (AAA21 517), *Tegula funebralis* (AAD28 265), *T. brunnea* (AAD28 264); [VERL]

approaches within 100 microns of a conspecific egg, its behaviour changes from random kinetic to directed motility. The sperm passes through the looser jelly layer, and upon contacting the vitelline envelope, releases its acrosomal vesicle (Mozingo, Vacquier and Chandler, 1995), the main secretory organelle found in the head of the sperm (Usui, 1987). Coordinate with exocytosis comes the assembly of a cylindrical extension from the apical tip of the reacted sperm head (Usui, 1987), a structure that speeds the penetration of the sperm through the vitelline envelope. When the sperm contacts the egg plasma membrane, fusion can occur and fertilization is complete.

The cylindrical extension from the sperm head is a conduit for the protein lysin to access its receptor in the egg vitelline envelope (Figure 16.1). Abalone lysin is a 16 000 M_r protein stored in the sperm acrosome until vesicle exocytosis, when it then coats the surface of the sperm extension (Usui, 1987). This increased surface area provides maximal exposure of the vitelline envelope proteins to the population of lysin molecules, facilitating the nonenzymatic disassembly of this egg structure (Lewis, Talbot and Vacquier, 1982). The crystal structure of lysin suggests that it homodimerizes along a hydrophobic patch exposed on the surface of a monomer (Kresge, Vacquier and Stout, 2000a; Shaw *et al.*, 1995), while the interaction of these dimers with the sperm membrane may occur via patches of positive charge found along the protein surface (Kresge, Vacquier and Stout, 2000a; Shaw *et al.*, 1995). In the presence of the vitelline envelope, however, the association of lysin with the sperm's cylindrical extension is lost in favour of its egg receptor (Shaw *et al.*, 1995), the vitelline envelope receptor for lysin (VERL). VERL is a 400 000 M_r glycoprotein with about the same mass in carbohydrates decorating its surface (Galindo *et al.*, 2002; Swanson and Vacquier, 1997). Higher-affinity binding sites on VERL cause the lysin homodimers to separate, leaving active monomers to associate with its egg receptor along the hydrophobic patch (Shaw *et al.*, 1995). This lysin receptor contains a transmembrane domain, a ZP domain (common to many egg extracellular proteins (Jovine *et al.*, 2002), including its namesake the mammalian zona pellucida (see below)), and a concatenation of similar sequences designated VERL repeats (Galindo *et al.*, 2002). Each VERL repeat can bind two lysin monomers (Galindo *et al.*, 2002), which may derive from the same or different homodimers (Kresge, Vacquier and Stout, 2000a; Shaw *et al.*, 1995). All the VERL repeats are encoded within a single exon (Figure 16.1; Galindo *et al.*, 2002) and most intramolecular repeats utilized by a species exhibit a high percentage of sequence identity consistent with the concerted evolution of these domains (Galindo *et al.*, 2002; Swanson and Vacquier, 1998).

The overall homogenization of VERL repeats implies that neutral or purifying selection influences the evolution of this egg protein; yet both lysin and VERL are

Haliotis rufescens (AAL50 827); [bindin] *Strongylocentrotus purpuratus* (AAA30 038), *S. franciscanus* (AAA30 037), *Arbacia punctulata* (CAA38 094), *Lytechinus variegatus* (AAA29 997), *Heliocidaris tuberculata* (AAQ09 975); [EBR1] *S. purpuratus* (AAR03 494), *S. franciscanus* (AAP44 488); [zonadhesin] *Homo sapiens* (AAC78 790), *Mus musculus* (AAC26 680), *Sus scrofa* (Q28 983), *Oryctolagus cuniculus* (P57 999); [ZPA] *H. sapiens* (AAA61 335), *M. musculus* (P20 239), *S. scrofa* (P42 099), *O. cuniculus* (P48 829), *Bos taurus* (Q9BH10), *Canus familiaris* (P47 983), and *Gallus gallus* (NP_001 034 187). Bar represents 100 residues. A full colour version of this figure appears in the colour plate section.

diversifying (Swanson and Vacquier, 2002). In both primary sequence and structure, lysin is generally conserved (Figure 16.1); only in small regions of solvent-exposed surface near the dimerization cleft does sequence diversity occur in abalone (Yang, Swanson and Vacquier, 2000; Kresge, Vacquier and Stout, 2000a, 2000b; Lee, Ota and Vacquier, 1995; Shaw *et al.*, 1995) and in marine snails (Hellberg and Vacquier, 1999). Within abalone, this variation is still restricted by amino-acid properties, favouring basic and highly charged residues over neutral or acidic ones (Vacquier, Carner and Stout, 1990). These constraints might be associated with how lysin interacts with VERL, and the diversity in lysin has been proposed to represent drift rather than true positive selection (Swanson, Aquadro and Vacquier, 2001a). This alternative hypothesis derives from the observation that, overall, VERL is under neutral evolution; only the last two VERL repeats exhibit diversity amongst these repeats and amongst species (Galindo *et al.*, 2002; Swanson, Aquadro and Vacquier, 2001a). Given these constraints on the VERL protein sequence, maintenance of lysin variance within a population appears to have minimal benefits. Yet lysin continually changes, and some sympatric subpopulations of *Haliotis tuberculata* have even retained two lysin paralogues that differ from one another in a hypervariable region observed across species (Clark *et al.*, 2007). An allopatric subpopulation, however, only retained a single, fixed lysin allele (Clark *et al.*, 2007), suggesting that the geographic overlap and differentiation of alleles in the sympatric populations is a method of distinguishing individuals from a subpopulation. This might also represent early steps in reproductive isolation, which eventually may end in speciation. Based on the apparent coevolution of VERL, it will be useful to test if the divergent VERL repeats are also changing among individuals from these sympatric *H. tuberculata* subpopulations.

Despite subtle species-dependent variations in the coevolving lysin–VERL pair, the effects these changes have on function are clear. While the most efficient dissolution of the vitelline envelope is mediated by conspecific lysin, interspecific activity has been observed in some directional crosses between *Haliotis corrugata* lysin onto *H. rufescens* eggs (Vacquier, Carner and Stout, 1990). Based on the high degree of conservation between VERL orthologues, and the homogenization of VERL repeats within a species (Galindo *et al.*, 2002; Swanson, Aquadro and Vacquier, 2001a), it is not surprising that the vitelline envelope can be dissolved by a heterospecific lysin if there is moderate binding affinity between the structurally conserved lysin and the VERL repeats. The primary barrier would appear to be how effectively the heterospecific lysin binds the two most divergent VERL repeats (Galindo *et al.*, 2002; Swanson, Aquadro and Vacquier, 2001a). Surprisingly, a reciprocal cross of *H. rufescens* lysin to *H. corrugata* eggs fails to produce the dissolution phenotype (Vacquier, Carner and Stout, 1990). How these pockets of sequence diversity influence binding selectivity remains a mystery.

16.3 Bindin: EBR1

Echinoderms such as sea urchins also spawn freely into the ocean, so intergamete specificity and timing of release are under similar constraints as molluscs. Gamete courtship also involves chemoattraction of motile sperm to individual eggs, but the source of this attractant is a layer of jelly coating the outer surface of each spawned egg (Hirohashi and Vacquier, 2002). Species-specific peptides that most effectively trigger the acrosome reaction in echinoderm sperm are found in and released from the egg jelly coat (Hirohashi and Vacquier, 2002; Suzuki *et al.*, 1988; Ward *et al.*, 1985; Hansbrough and Garbers, 1981). The acrosome reaction in sea urchins results in the extension of an acrosomal process (Figure 16.1) assembled from polymerizing actin anchored at the tip of the nucleus (Dan *et al.*, 1964).

An essential consequence of acrosome exocytosis is the release of bindin, a small molecule that mediates cell agglutination (Vacquier and Moy, 1977; Glabe and Vacquier, 1977). Bindin is conserved throughout echinoderms, but its presence is also restricted to this class of animals (Zigler and Lessios, 2003; Biermann, 1998; Minor *et al.*, 1991). The proteins are translated in a pro-form, which is subsequently cleaved in half to produce the active bindin molecule (Minor *et al.*, 1991). The central sequence of bindin is highly conserved, enriched in nonpolar and acidic residues scattered amongst eight positionally locked cysteine residues (Biermann, 1998; Metz and Palumbi, 1996; Minor *et al.*, 1991). Within the core sequence is a hydrophobic peptide of about 18 residues that is freed following proteolysis of the pro-form (Zigler and Lessios, 2003). In the presence of zinc ions, this B18 peptide morphs from an unordered to a helical structure (Glaser *et al.*, 1999; Ulrich *et al.*, 1998); this is altered again to a beta-sheet organization in the presence of phospholipid membranes (Barre *et al.*, 2003). This plasticity in peptide configuration is consistent with the two catalytic functions of bindin in egg agglutination (Vacquier and Moy, 1977) and in liposome fusion (Ulrich *et al.*, 1998). The agglutination of eggs by bindin may occur through its association with the egg bindin receptor (EBR), a ~300 000 M_r glycoprotein composed of about 30% protein and 70% carbohydrates (Foltz, Partin and Lennarz, 1993; Ohlendieck *et al.*, 1993; Ruiz-Bravo *et al.*, 1986). EBR contains a metalloprotease domain, a cysteine-rich domain, and a series of thrombospondin 1 domains (Adams, 1997) paired with CUB domains (Bork and Beckmann, 1993) clustered at the carboxyl terminus (Kamei and Glabe, 2003). Surprisingly, no obvious transmembrane domain is present within the EBR open reading frame (Kamei and Glabe, 2003), suggesting that it is retained on the egg surface or within the vitelline layer by a tethering protein.

The interaction of bindin with EBR is species selective (Vacquier and Moy, 1977; Glabe and Vacquier, 1977). Unlike lysin and VERL (see above), both participating sea urchin proteins exhibit extensive regions of diversity. Two regions exhibit hypervariable sequence in bindin (Figure 16.1): a small patch located just amino-terminal of the conserved hydrophobic core (Zigler and Lessios, 2003; Zigler *et al.*, 2003; Biermann, 1998) and a larger region at the carboxy terminus encoding short repeats whose variability and repetitiveness is thought to predispose bindin to divergence amongst species within the same genus (Zigler *et al.*, 2003; Zigler and Lessios, 2003; Biermann, 1998; Metz and Palumbi, 1996; Minor *et al.*, 1991). Both the sequence and the total number of carboxyl repeats account for differences in the mass of bindin observed between species (Minor *et al.*, 1991). The cognate receptor EBR contains surprising sequence diversity, with alternative domains (Kamei and Glabe, 2003) and different carbohydrates used in sister species (Hirohashi and Lennarz, 2001; Rossignol *et al.*, 1984). *Strongylocentrotus purpuratus* EBR, for example, contains hyalin repeats in place of the last 10 thrombospondin 1/CUB repeats found in *S. fransiscanus* (Kamei and Glabe, 2003).

The functional consequences of these large variable regions in both bindin and EBR are evident in gamete choice assays. Interspecific hybrids are minimal in sea urchins (Palumbi, 1999; Palumbi and Metz, 1991; Minor *et al.*, 1991). While this may be due to differences in spawning periods, cellular and biochemical evidence suggests that behaviour alone is insufficient to maintain species barriers. Bindin, for example, agglutinates eggs in a species-selective fashion (Vacquier and Moy, 1977; Glabe and Vacquier, 1977), while eggs are preferentially fertilized by sperm from males with the same bindin allele as the female (Palumbi, 1999). Together, these molecular preferences suggest that EBR and bindin loci are coevolving within a population, and that their loci may be selected by similar mechanisms.

16.4 Zonadhesin: ZP

Mammalian fertilization occurs internally, so both mating and anatomical behaviour are thought to be the main barriers against interspecific fertilization (reviewed in Wong and Wessel, 2006; Shur, Rodeheffer and Ensslin, 2004; Wassarman, 1999). Ejaculated sperm enter the uterus and travel into the oviduct, where they meet an ovulated oocyte. During their travels through the uterus, small percentages of sperm continuously undergo capacitation, the final stage of sperm maturation that reorganizes cell surface proteins and primes the sperm for fertilization (Wassarman, 1999; Tulsiani *et al.*, 1998). Since the lifespan of a capacitated sperm is short as a result of low nutritional stores, waves of capacitation in the sperm population means that a freshly activated cohort of sperm are present for much longer periods in the reproductive tract than any one sperm. These primed sperm first contact a layer of granulosa cells still surrounding the oocyte. After passing through the hyaluronin-rich extracellular matrix of these somatic cells, the primed sperm contact the zona pellucida, an extracellular glycoprotein matrix surrounding the oocyte. Binding to the outer surface of the zona pellucida triggers the sperm acrosome reaction (Figure 16.1), releasing many different proteins and enzymes thought to participate in subsequent stages of mammalian fertilization (Tulsiani *et al.*, 1998). Some of the acrosome content proteins destabilize the zona pellucida, allowing the motile sperm to penetrate the extracellular matrix (Tulsiani *et al.*, 1998; Töpfer-Petersen *et al.*, 1990).

Exocytosis of the mammalian acrosome exposes many different proteins near the zona pellucida (Tulsiani *et al.*, 1998). Acrosin is a serine protease that binds to a constituent of the zona pellucida (Howes *et al.*, 2001; Jansen *et al.*, 1998; Urch and Patel, 1991). This soluble protease was originally hypothesized to digest a path through the zona pellucida (Honda, Siruntawineti and Baba, 2002; Tulsiani *et al.*, 1998), but the absence of any ultrastructural evidence for this tunnelling (Bedford, 1998) suggests that it could play an alternative role such as dispersing other acrosome proteins (Yamagata *et al.*, 1998; Takano, Yanagimachi and Urch, 1993). Zonadhesin, on the other hand, remains associated with the postacrosomal membrane (Bi *et al.*, 2003; Lea, Sivashanmugam and O'Rand, 2001). This integral transmembrane protein contains MAM domains, a mucin motif, von Willebrand factor D-type domains, an EGF-like domain, and a basic carboxy-terminus (Lea, Sivashanmugam and O'Rand, 2001; Gao and Garbers, 1998; Hardy and Garbers, 1995). It is expressed and processed during spermatogenesis, and is protected on the luminal surface of the outer matrix until acrosome exocytosis (Bi *et al.*, 2003; Lea, Sivashanmugam and O'Rand, 2001). Like lysin and bindin (see above), the fraction of zonadhesin remaining with the sperm head can bind to the zona pellucida (Hardy and Garbers, 1995). The receptor for zonadhesin is not known, however.

The zona pellucida is composed primarily of three major proteins, all named after their respective parent matrix. Each zona pellucida (ZP) protein shares a cysteine-containing module that folds into a beta-sheet configuration able to polymerize with other ZP domains to form extracellular protofilaments (Wassarman, 2008; Jovine *et al.*, 2002; Sinowatz, Kolle and Töpfer-Petersen, 2001; Bork and Sander, 1992;). Of the three major zona pellucida proteins – ZPA, ZPB and ZPC (Spargo and Hope, 2003) – ZPA is believed to be the major component that acrosome-reacted sperm binds to (Kerr *et al.*, 2002; Howes *et al.*, 2001; Tsubamoto *et al.*, 1999; Bleil *et al.*, 1988). Both protein sequence (Tsubamoto *et al.*, 1999) and conjugated carbohydrates (Rankin *et al.*, 2003; Doren *et al.*, 1999) contribute to the sperm binding function of ZPA. This could occur through the mucin domain of zonadhesin (Hardy and Garbers, 1995), although there are many additional candidates with lectin-like properties found at the surface of the acrosome-reacted sperm (reviewed in Mengerink and Vacquier, 2001), including relocated β1,3-galactosyltransferase (Larson and Miller, 1997; Lopez and Shur, 1987) and acrosome-derived acrosin (Howes *et al.*, 2001) and β-*N*-acetyl glucosaminidase (Miller, Gong and Shur, 1993).

Despite their functional conservation amongst mammals, the primary sequence of sperm acrosin and zonadhesin, and of egg ZPA are diverging (Berlin, Qu and Ellegren, 2008). Acrosin is expressed in males of hemichordates (Kodama *et al.*, 2001) and chordates ranging from ancient fish (Dabrowski, Glogowski and Ciereszko, 2004; Ciereszko *et al.*, 2000; Ciereszko, Dabrowski and Ochkur, 1996; Ciereszko *et al.*, 1994) to placental animals (Raterman and Springer, 2008; Berlin, Qu and Ellegren, 2008; Howes *et al.*, 2001; Jansen *et al.*, 1998; Richardson and O'Rand, 1996; Urch and Patel, 1991). Its participation in species-selective interactions is predicted by the sequence diversity found in pockets outside of the catalytic core (Berlin, Qu and Ellegren, 2008). A similar pattern is observed with mammalian zonadhesin (Herlyn and Zischler, 2005; Lea, Sivashanmugam and O'Rand, 2001; Gao and Garbers, 1998; Hardy and Garbers, 1995). At the molecular level, zonadhesin exhibits variability in quantity of repetitive motifs and in sequence within these motifs (Herlyn and Zischler, 2005; Lea, Sivashanmugam and O'Rand, 2001; Gao and Garbers, 1998) – a trend also observed in sea urchin bindin (see above). The most variability is described in the MAM and von Willebrand factor D-like domains (Figure 16.1): murine zonadhesin contains an additional MAM domain and an extended concatamer of partial von Willebrand factor D-like domains not present in other mammalian orthologues (Herlyn and Zischler, 2005; Lea, Sivashanmugam and O'Rand, 2001; Hardy and Garbers, 1995); the sequence of the shared primate MAM domains is diverging (Herlyn and Zischler, 2005); and, further, each of the partial von Willebrand factor D-like domains specific to the mouse orthologue is under positive selection (Herlyn and Zischler, 2006). Finally, the species cross-reactivity of antibodies generated against the zona pellucida alludes to the conservation of the ZP proteins (Moller *et al.*, 1990). Yet, there are still regions of sequence divergence within this family (Berlin, Qu and Ellegren, 2008; Swanson *et al.*, 2001b), suggesting that ZPA participates in species-selective events during gamete binding.

Evidence for the participation of the acrosin, zonadhesin and/or ZPA as coevolving pairs remains weak, however. Mutations to the evolutionarily conserved acrosin result in reduced binding affinity to homospecific zona pellucidae (Jansen *et al.*, 1998; Richardson and O'Rand, 1996), enforcing the possibility that the positive selection of this enzyme is important for species-selective gamete interactions. But, acrosin is clearly not essential for mediating these gamete interactions since acrosin-null male mice are still fertile (Nayernia *et al.*, 2002; Adham, Nayernia and Engel, 1997; Baba *et al.*, 1994). Zonadhesin similarly exhibits higher affinity for homospecific zona pellucidae (Hardy and Garbers, 1995), but the absence of data regarding the fertility of zonadhesin-null animals leaves the significance of its divergence untested. ZPA, on the other hand, is required in the zona pellucida for murine fertilization (Rankin *et al.*, 2001). The fertility defect can be rescued by proper expression of a frog (Doren *et al.*, 1999) or human (Rankin *et al.*, 2003) orthologue in the ZPA-null mice, consistent with a high level of functional conservation amongst the ZPA orthologues. Yet, the incorporation of heterospecific ZPA into a mouse zona pellucida does not alter the oocyte's selectivity for sperm: murine sperm are still preferred, implying that differences accrued due to homospecific post-translational processing, such as glycosylation, trump relatively minor changes in the primary protein sequence (Rankin *et al.*, 2003).

16.5 Glycosylation in speciation

Carbohydrate moieties associated with gamete binding proteins are likely candidates for mediating the selectivity of gamete interactions (reviewed in Wong and Wessel, 2006; Mengerink and Vacquier, 2001; Shalgi and Raz, 1997; Vacquier and Lee, 1993). Just as co-associating proteins on the cell surface may have been co-opted for species-selective gamete interactions, carbohydrate recognition can serve as a barrier to gamete interactions in the same manner that proteins mediate self-recognition (Ohtsubo and Marth, 2006). All of the diversifying receptors for sperm described above are heavily glycosylated with various combinations of basic sugar monomers: abalone VERL contains branch glycans with fucose, glucose, mannose, galactosamine and glucosamine (Swanson and Vacquier, 1997); EBR is conjugated to sulfated carbohydrates rich in mannose, galactosamine, fucose and uronic acid (Ohlendieck *et al.*, 1993; Foltz and Lennarz, 1990; DeAngelis and Glabe, 1987; Ruiz-Bravo *et al.*, 1986; Rossignol *et al.*, 1984; Glabe *et al.*, 1982); and ZPA is decorated with sialylated chains enriched in mannose, fucose and β-galactosyl glycans (Tulsiani, 2000; Easton *et al.*, 2000). Thus, the ability of an individual sperm to discriminate subtle differences in carbohydrate chemistry could be essential for its success. Consistent with this hypothesis is the observation that sperm proteins complementary to the female sperm receptor may at least partly bind to glycans: no published evidence for abalone lysin binding to carbohydrates exists, although mussel lysins contain a C-type lectin fold with species-divergent residues that regulate sugar specificity (Springer and Crespi, 2007;

Takagi *et al.*, 1994); bindin association to EBR is competed by fucoidin, a sulfated polymer of fucose (DeAngelis and Glabe, 1987; Glabe *et al.*, 1982); and zonadhesin contains a mucin domain, presumably allowing it to bind carbohydrates (Lea, Sivashanmugam and O'Rand, 2001; Gao and Garbers, 1998; Hardy and Garbers, 1995). Indeed, replacement of the genes encoding the zona pellucida in mice with the genes encoding the human zona pellucida still result in preferential binding and activation of mouse sperm (Rankin *et al.*, 2003). Thus, conspecific sperm binding may be directed more by post-translational modifications than protein sequence. It would also argue that diversification in sperm binding is a consequence of variable carbohydrates displayed on the oocyte or egg surface, which may occur through differential expression of glycosyltransferases or available sugar modifications.

Given the natural variance of products synthesized during post-translational glycosylation (Ohtsubo and Marth, 2006; Rudd and Dwek, 1997), it should be expected that a single egg would carry a range of branched glycans on a single protein. If the complementary sperm proteins preferentially bind to carbohydrates decorating the egg receptors, and the affinities for different glycans are regulated by differences in surface charge, then diversity in male protein sequence should compensate for the range of glycosylation patterns displayed on the female gamete. Examples of this change at the amino acid level are evident in the regional positive selection observed in acrosome-derived lysin (Springer and Crespi, 2007; Kresge, Vacquier and Stout, 2000a, 2000b; Yang, Swanson and Vacquier, 2000; Hellberg and Vacquier, 1999; Lee, Ota and Vacquier, 1995), bindin (Zigler *et al.*, 2003; Zigler and Lessios, 2003; Biermann, 1998; Metz and Palumbi, 1996; Minor *et al.*, 1991) and zonadhesin (Herlyn and Zischler, 2005, 2006; Lea, Sivashanmugam and O'Rand, 2001; Gao and Garbers, 1998). The functional consequences of these changes to glycan affinities, however, are not clear.

If both peptide sequence diversity and glycan synthesis control gamete binding, then subtle changes in the dominant adhesion proteins may favour adaptive evolution and even speciation. Mutation of a single amino acid, for example, could dramatically affect the affinity of either male or female protein. Within orthologues, glycosylated serine and threonine residues of extracellular proteins are less likely to be conserved than nonglycosylated residues due to their position on the protein surface (Julenius *et al.*, 2005). Regardless of the relationship between this diversity in glycosylated residues and the rapid evolution of extracellular protein sequences (Julenius and Pedersen, 2006; Luz and Vingron, 2006), the implication for gamete recognition protein evolution is clear: a mutated residue can be the source of extreme epitope variation that may drive speciation. If a key glycosylation site is lost on the female glycoprotein, then the carbohydrate that the majority of sperm require to confer selective binding is lost. This mutant female glycoprotein would then favour only those sperm that are able to identify a different region of the unadulterated version, or sperm that prefer the charged protein surface in the same position. A reciprocal scenario may be described for mutations on the male side, leading to changes in female glycoprotein configurations – hence, a sexual arms race (Clark, Aagaard and Swanson, 2006; Parker and Partridge, 1998).

An additional factor with significant impact, but never identified in this coevolutionary race, is variation in gamete-specific glycosyltransferases in the Golgi (Ohtsubo and Marth, 2006; de Graffenried and Bertozzi, 2004; Rudd and Dwek, 1997). A change in specificity of one enzyme could dramatically affect the final female glycoprotein, rendering the receptor foreign to the usual sperm population; only those sperm with a complementary mutation will then be able to fertilize that aberrant oocyte or egg. If the complementary sperm protein was also modified by the same enzyme, then allele preferences observed in some animals (Palumbi, 1999; Vacquier, Carner and Stout, 1990) could reflect an increasing role for carbohydrate–carbohydrate affinities between gametes, as already documented during fertilization of rainbow trout (Kodama *et al.*, 2001).

16.6 Coevolution of interacting gamete proteins vs. speciation

The stringency of protein–protein interactions is governed by the conformation of each participant upon their association. This 'induced fit' model (Koshland, 1958) assumes that the conformational changes induced by the proteins binding each other increases their mutual affinity. If this protein pair proved to be suitable for an essential interaction or recognition function, then the individual proteins would be expected to coevolve, perhaps under neutral or purifying selection to maintain or to enhance their affinities (Luz and Vingron, 2006). Over time, however, individual pairs drift apart if there is no selection to maintain them (Swanson and Vacquier, 2002). Minor changes associated with drift can wreak havoc on the coevolution of protein pairs when subtle changes to surface residues, repeat lengths, and/or glycosylation regulate binding affinities. The more rapid diversification of extracellular proteins compared to proteins residing inside the cell (Julenius and Pedersen, 2006; Luz and Vingron, 2006) only accelerates the incompatibility of gamete recognition proteins. It is through such subtle change that speciation can rapidly occur: the slightest modification to an individual's gamete binding protein could fundamentally alter the population it is fit to mate with (Swanson and Vacquier, 2002). Finding a compatible mate, however, is the first step towards reproductive isolation since only these founding individuals would possess the alleles necessary for the coevolution of this alternative protein pair.

References

Adams, J.C. (1997) Thrombospondin-1. *Int. J. Biochem. Cell. Biol.*, **29**, 861–865.

Adham, I.M., Nayernia, K. and Engel, W. (1997) Spermatozoa lacking acrosin protein show delayed fertilization. *Mol. Reprod. Dev.*, **46**, 370–376.

Baba, T., Azuma, S., Kashiwabara, S. and Toyoda, Y. (1994) Sperm from mice carrying a targeted mutation of the acrosin gene can penetrate the oocyte zona pellucida and effect fertilization. *J. Biol. Chem.*, **269**, 31845–31849.

Barre, P., Zschornig, O., Arnold, K. and Huster, D. (2003) Structural and dynamical changes of the bindin B18 peptide upon binding to lipid membranes. A solid-state NMR study. *Biochemistry*, **42**, 8377–8386.

Bauskin, A.R., Franken, D.R., Eberspaecher, U. and Donner, P. (1999) Characterization of human zona pellucida glycoproteins. *Mol. Hum. Reprod.*, **5**, 534–540.

Bedford, J.M. (1998) Mammalian fertilization misread? Sperm penetration of the eutherian zona pellucida is unlikely to be a lytic event. *Biol. Reprod.*, **59**, 1275–1287.

Berlin, S., Qu, L. and Ellegren, H. (2008) Adaptive evolution of gamete-recognition proteins in birds. *J. Mol. Evol.*, **67**, 488–496.

Bi, M., Hickox, J.R., Winfrey, V.P. *et al.* (2003) Processing, localization and binding activity of zonadhesin suggest a function in sperm adhesion to the zona pellucida during exocytosis of the acrosome. *Biochem. J.*, **375**, 477–488.

Biermann, C.H. (1998) The molecular evolution of sperm bindin in six species of sea urchins (Echinoida: Strongylocentrotidae). *Mol. Biol. Evol.*, **15**, 1761–1771.

Bleil, J.D., Greve, J.M. and Wassarman, P.M. (1988) Identification of a secondary sperm receptor in the mouse egg zona pellucida: role in maintenance of binding of acrosome-reacted sperm to eggs. *Dev. Biol.*, **128**, 376–385.

Bork, P. and Beckmann, G. (1993) The CUB domain. A widespread module in developmentally regulated proteins. *J. Mol. Biol.*, **231**, 539–545.

Bork, P. and Sander, C. (1992) A large domain common to sperm receptors (Zp2 and Zp3) and TGF-beta type III receptor. *FEBS Lett.*, **300**, 237–240.

Carroll, E.J. Jr and Epel, D. (1975) Elevation and hardening of the fertilization membrane in sea urchin eggs. Role of the soluble fertilization product. *Exp. Cell Res.*, **90**, 429–432.

Carroll, E.J. Jr and Epel, D. (1975) Isolation and biological activity of the proteases released by sea urchin eggs following fertilization. *Dev. Biol.*, **44**, 22–32.

Chen, E.H. and Olson, E.N. (2005) Unveiling the mechanisms of cell-cell fusion. *Science*, **308**, 369–373.

Ciereszko, A., Dabrowski, K., Lin, F. and Doroshov, S.I. (1994) Identification of a trypsin-like activity in sturgeon spermatozoa. *J. Exp. Zool.*, **268**, 486–491.

Ciereszko, A., Dabrowski, K. and Ochkur, S.I. (1996) Characterization of acrosin-like activity of lake sturgeon (Acipenser fulvescens) spermatozoa. *Mol. Reprod. Dev.*, **45**, 72–77.

Ciereszko, A., Dabrowski, K., Mims, S.D. and Glogowski, J. (2000) Characteristics of sperm acrosin-like activity of paddlefish (Polyodon spathula Walbaum). *Comp. Biochem. Physiol. B. Biochem. Mol. Biol.*, **125**, 197–203.

Clark, N.L., Aagaard, J.E. and Swanson, W.J. (2006) Evolution of reproductive proteins from animals and plants. *Reproduction*, **131**, 11–22.

Clark, N.L., Findlay, G.D., Yi, X. *et al.* (2007) Duplication and selection on abalone sperm lysin in an allopatric population. *Mol. Biol. Evol.*, **24**, 2081–2090.

Dabrowski, K., Glogowski, J. and Ciereszko, A. (2004) Effects of proteinase inhibitors on fertilization in sea lamprey (Petromyzon marinus). *Comp. Biochem. Physiol. B. Biochem. Mol. Biol.*, **139**, 157–162.

Dan, J., Ohori, Y. and Kushida, H. (1964) Studies on the acrosome. vii. Formation of the acrosomal process in sea urchin spermatozoa. *J. Ultrastruct. Res.*, **11**, 508–524.

DeAngelis, P.L. and Glabe, C.G. (1987) Polysaccharide structural features that are critical for the binding of sulfated fucans to bindin, the adhesive protein from sea urchin sperm. *J. Biol. Chem.*, **262**, 13946–13952.

Doren, S., Landsberger, N., Dwyer, N. *et al.* (1999) Incorporation of mouse zona pellucida proteins into the envelope of Xenopus laevis oocytes. *Dev. Genes. Evol.*, **209**, 330–339.

Easton, R.L., Patankar, M.S., Clark, G.F. *et al.* (2000) Pregnancy-associated changes in the glycosylation of tamm-horsfall glycoprotein. Expression of sialyl Lewis(x) sequences on core 2 type O-glycans derived from uromodulin. *J. Biol. Chem.*, **275**, 21928–21938.

Foltz, K.R. and Lennarz, W.J. (1990) Purification and characterization of an extracellular fragment of the sea urchin egg receptor for sperm. *J. Cell Biol.*, **111**, 2951–2959.

Foltz, K.R., Partin, J.S. and Lennarz, W.J. (1993) Sea urchin egg receptor for sperm: sequence similarity of binding domain and hsp70. *Science*, **259**, 1421–1425.

Galindo, B.E., Moy, G.W., Swanson, W.J. and Vacquier, V.D. (2002) Full-length sequence of VERL, the egg vitelline envelope receptor for abalone sperm lysin. *Gene*, **288**, 111–117.

Gao, Z. and Garbers, D.L. (1998) Species diversity in the structure of zonadhesin, a sperm-specific membrane protein containing multiple cell adhesion molecule-like domains. *J. Biol. Chem.*, **273**, 3415–3421.

Gavrilets, S. (2000) Rapid evolution of reproductive barriers driven by sexual conflict. *Nature*, **403**, 886–889.

Glabe, C.G. and Vacquier, V.D. (1977) Species specific agglutination of eggs by bindin isolated from sea urchin sperm. *Nature*, **267**, 836–838.

Glabe, C.G., Grabel, L.B., Vacquier, V.D. and Rosen, S.D. (1982) Carbohydrate specificity of sea urchin sperm bindin: a cell surface lectin mediating sperm-egg adhesion. *J. Cell Biol.*, **94**, 123–128.

Glaser, R.W., Grune, M., Wandelt, C. and Ulrich, A.S. (1999) Structure analysis of a fusogenic peptide sequence from the sea urchin fertilization protein bindin. *Biochemistry*, **38**, 2560–2569.

de Graffenried, C.L. and Bertozzi, C.R. (2004) The roles of enzyme localisation and complex formation in glycan assembly within the Golgi apparatus. *Curr. Opin. Cell Biol.*, **16**, 356–363.

Hansbrough, J.R. and Garbers, D.L. (1981) Speract. Purification and characterization of a peptide associated with eggs that activates spermatozoa. *J. Biol. Chem.*, **256**, 1447–1452.

Hardy, D.M. and Garbers, D.L. (1995) A sperm membrane protein that binds in a species-specific manner to the egg extracellular matrix is homologous to von Willebrand factor. *J. Biol. Chem.*, **270**, 26025–26028.

Hellberg, M.E. and Vacquier, V.D. (1999) Rapid evolution of fertilization selectivity and lysin cDNA sequences in teguline gastropods. *Mol. Biol. Evol.*, **16**, 839–848.

Herlyn, H. and Zischler, H. (2005) Identification of a positively evolving putative binding region with increased variability in posttranslational motifs in zonadhesin MAM domain 2. *Mol. Phylogenet. Evol.*, **37**, 62–72.

Herlyn, H. and Zischler, H. (2006) Tandem repetitive D domains of the sperm ligand zonadhesin evolve faster in the paralogue than in the orthologue comparison. *J. Mol. Evol.*, **63**, 602–611.

Hirohashi, N. and Lennarz, W.J. (2001) Role of a vitelline layer-associated 350 kDa glycoprotein in controlling species-specific gamete interaction in the sea urchin. *Dev. Growth Differ.*, **43**, 247–255.

Hirohashi, N. and Vacquier, V.D. (2002) Egg fucose sulfate polymer, sialoglycan, and speract all trigger the sea urchin sperm acrosome reaction. *Biochem. Biophys. Res. Commun.*, **296**, 833–839.

Honda, A., Siruntawineti, J. and Baba, T. (2002) Role of acrosomal matrix proteases in sperm-zona pellucida interactions. *Hum. Reprod. Update*, **8**, 405–412.

Howes, E., Pascall, J.C., Engel, W. and Jones, R. (2001) Interactions between mouse ZP2 glycoprotein and proacrosin; a mechanism for secondary binding of sperm to the zona pellucida during fertilization. *J. Cell Sci.*, **114**, 4127–4136.

Jansen, S., Jones, R., Jenneckens, I. *et al.* (1998) Site-directed mutagenesis of boar proacrosin reveals residues involved in binding of zona pellucida glycoproteins. *Mol. Reprod. Dev.*, **51**, 184–192.

Jovine, L., Qi, H., Williams, Z. *et al.* (2002) The ZP domain is a conserved module for polymerization of extracellular proteins. *Nat. Cell Biol.*, **4**, 457–461.

Julenius, K. and Pedersen, A.G. (2006) Protein evolution is faster outside the cell. *Mol. Biol. Evol.*, **23**, 2039–2048.

Julenius, K., Molgaard, A., Gupta, R. and Brunak, S. (2005) Prediction, conservation analysis, and structural characterization of mammalian mucin-type O-glycosylation sites. *Glycobiology*, **15**, 153–164.

Kamei, N. and Glabe, C.G. (2003) The species-specific egg receptor for sea urchin sperm adhesion is EBR1,a novel ADAMTS protein. *Genes. Dev.*, **17**, 2502–2507.

Kerr, C.L., Hanna, W.F., Shaper, J.H. and Wright, W.W. (2002) Characterization of zona pellucida glycoprotein 3 (ZP3) and ZP2 binding sites on acrosome-intact mouse sperm. *Biol. Reprod.*, **66**, 1585–1595.

Kodama, E., Baba, T., Yokosawa, H. and Sawada, H. (2001) cDNA cloning and functional analysis of ascidian sperm proacrosin. *J. Biol. Chem.*, **276**, 24594–24600.

Koshland, D.E. (1958) Application of a theory of enzyme specificity to protein synthesis. *Proc. Natl. Acad. Sci. USA*, **44**, 98–104.

Kresge, N., Vacquier, V.D. and Stout, C.D. (2000a) The high resolution crystal structure of green abalone sperm lysin: implications for species-specific binding of the egg receptor. *J. Mol. Biol.*, **296**, 1225–1234.

Kresge, N., Vacquier, V.D. and Stout, C.D. (2000b) 1.35 and 2.07 A resolution structures of the red abalone sperm lysin monomer and dimer reveal features involved in receptor binding. *Acta Crystallogr. D. Biol. Crystallogr.*, **56** (1),34–41.

Larson, J.L. and Miller, D.J. (1997) Sperm from a variety of mammalian species express beta1,4-galactosyltransferase on their surface. *Biol. Reprod.*, **57**, 442–453.

Lea, I.A., Sivashanmugam, P. and O'Rand, M.G. (2001) Zonadhesin: characterization, localization, and zona pellucida binding. *Biol. Reprod.*, **65**, 1691–1700.

Lee, Y.H., Ota, T. and Vacquier, V.D. (1995) Positive selection is a general phenomenon in the evolution of abalone sperm lysin. *Mol. Biol. Evol.*, **12**, 231–238.

Lewis, C.A., Talbot, C.F. and Vacquier, V.D. (1982) A protein from abalone sperm dissolves the egg vitelline layer by a nonenzymatic mechanism. *Dev. Biol.*, **92**, 227–239.

Lopez, L.C. and Shur, B.D. (1987) Redistribution of mouse sperm surface galactosyltransferase after the acrosome reaction. *J. Cell Biol.*, **105**, 1663–1670.

Luz, H. and Vingron, M. (2006) Family specific rates of protein evolution. *Bioinformatics*, **22**, 1166–1171.

Marshall, D.J., Steinberg, P.D. and Evans, J.P. (2004) The early sperm gets the good egg: mating order effects in free spawners. *Proc. R. Soc. Lond. B. Biol. Sci.*, **271**, 1585–1589.

Mengerink, K.J. and Vacquier, V.D. (2001) Glycobiology of sperm-egg interactions in deuterostomes. *Glycobiology*, **11**, 37R–43R.

Metz, E.C. and Palumbi, S.R. (1996) Positive selection and sequence rearrangements generate extensive polymorphism in the gamete recognition protein bindin. *Mol. Biol. Evol.*, **13**, 397–406.

Miller, D.J., Gong, X. and Shur, B.D. (1993) Sperm require beta-N-acetylglucosaminidase to penetrate through the egg zona pellucida. *Development*, **118**, 1279–1289.

Minor, J.E., Fromson, D.R., Britten, R.J. and Davidson, E.H. (1991) Comparison of the bindin proteins of Strongylocentrotus franciscanus, S. purpuratus, and Lytechinus variegatus: sequences involved in the species specificity of fertilization. *Mol. Biol. Evol.*, **8**, 781–795.

Moller, C.C., Bleil, J.D., Kinloch, R.A. and Wassarman, P.M. (1990) Structural and functional relationships between mouse and hamster zona pellucida glycoproteins. *Dev. Biol.*, **137**, 276–286.

Mozingo, N.M., Vacquier, V.D. and Chandler, D.E. (1995) Structural features of the abalone egg extracellular matrix and its role in gamete interaction during fertilization. *Mol. Reprod. Dev.*, **41**, 493–502.

Nayernia, K., Adham, I.M., Shamsadin, R. *et al.* (2002) Proacrosin-deficient mice and zona pellucida modifications in an experimental model of multifactorial infertility. *Mol. Hum. Reprod.*, **8**, 434–440.

O'Rand, M.G. (1988) Sperm-egg recognition and barriers to interspecies fertilization. *Gamete. Res.*, **19**, 315–328.

Ohlendieck, K., Dhume, S.T., Partin, J.S. and Lennarz, W.J. (1993) The sea urchin egg receptor for sperm: isolation and characterization of the intact, biologically active receptor. *J. Cell Biol.*, **122**, 887–895.

Ohtsubo, K. and Marth, J.D. (2006) Glycosylation in cellular mechanisms of health and disease. *Cell*, **126**, 855–867.

Palumbi, S.R. (1999) All males are not created equal: fertility differences depend on gamete recognition polymorphisms in sea urchins. *Proc. Natl. Acad. Sci. USA*, **96**, 12632–12637.

Palumbi, S.R. and Metz, E.C. (1991) Strong reproductive isolation between closely related tropical sea urchins (genus Echinometra). *Mol. Biol. Evol.*, **8**, 227–239.

Parker, G.A. and Partridge, L. (1998) Sexual conflict and speciation. *Philos. Trans. R. Soc. Lond. B. Biol. Sci.*, **353**, 261–274.

Rankin, T.L., O'Brien, M., Lee, E. *et al.* (2001) Defective zonae pellucidae in Zp2-null mice disrupt folliculogenesis, fertility and development. *Development*, **128**, 1119–1126.

Rankin, T.L., Coleman, J.S., Epifano, O. *et al.* (2003) Fertility and taxon-specific sperm binding persist after replacement of mouse sperm receptors with human homologs. *Dev. Cell*, **5**, 33–43.

Raterman, D. and Springer, M.S. (2008) The molecular evolution of acrosin in placental mammals. *Mol. Reprod. Dev.*, **75**, 1196–1207.

Richardson, R.T. and O'Rand, M.G. (1996) Site-directed mutagenesis of rabbit proacrosin. Identification of residues involved in zona pellucida binding. *J. Biol. Chem.*, **271**, 24069–24074.

Riffell, J.A., Krug, P.J. and Zimmer, R.K. (2002) Fertilization in the sea: the chemical identity of an abalone sperm attractant. *J. Exp. Biol.*, **205**, 1439–1450.

Riffell, J.A., Krug, P.J. and Zimmer, R.K. (2004) The ecological and evolutionary consequences of sperm chemoattraction. *Proc. Natl. Acad. Sci. USA*, **101**, 4501–4506.

Rossignol, D.P., Earles, B.J., Decker, G.L. and Lennarz, W.J. (1984) Characterization of the sperm receptor on the surface of eggs of Strongylocentrotus purpuratus. *Dev. Biol.*, **104**, 308–321.

Rudd, P.M. and Dwek, R.A. (1997) Glycosylation: heterogeneity and the 3D structure of proteins. *Crit. Rev. Biochem. Mol. Biol.*, **32**, 1–100.

Ruiz-Bravo, N., Rossignol, D.P., Decker, G. *et al.* (1986) Characterization of the Strongylocentrotus purpuratus egg cell surface receptor for sperm. *Adv. Exp. Med. Biol.*, **207**, 293–313.

Shalgi, R. and Raz, T. (1997) The role of carbohydrate residues in mammalian fertilization. *Histol. Histopathol.*, **12**, 813–822.

Shaw, A., Fortes, P.A., Stout, C.D. and Vacquier, V.D. (1995) Crystal structure and subunit dynamics of the abalone sperm lysin dimer: egg envelopes dissociate dimers, the monomer is the active species. *J. Cell Biol.*, **130**, 1117–1125.

Shur, B.D., Rodeheffer, C. and Ensslin, M.A. (2004) Mammalian fertilization. *Curr. Biol.*, **14**, R691–R692.

Sinowatz, F., Kolle, S. and Töpfer-Petersen, E. (2001) Biosynthesis and expression of zona pellucida glycoproteins in mammals. *Cells Tissues Organs*, **168**, 24–35.

Spargo, S.C. and Hope, R.M. (2003) Evolution and nomenclature of the zona pellucida gene family. *Biol. Reprod.*, **68**, 358–362.

Springer, S.A. and Crespi, B.J. (2007) Adaptive gamete-recognition divergence in a hybridizing Mytilus population. *Evolution*, **61**, 772–783.

Suzuki, N., Kajiura, H., Nomura, K. *et al.* (1988) Some more speract derivatives associated with eggs of sea urchins, Pseudocentrotus depressus, Strongylocentrotus purpuratus, Hemicentrotus pulcherrimus and Anthocidaris crassispina. *Comp. Biochem. Physiol. B*, **89**, 687–693.

Swanson, W.J. and Vacquier, V.D. (1997) The abalone egg vitelline envelope receptor for sperm lysin is a giant multivalent molecule. *Proc. Natl. Acad. Sci. USA*, **94**, 6724–6729.

Swanson, W.J. and Vacquier, V.D. (1998) Concerted evolution in an egg receptor for a rapidly evolving abalone sperm protein. *Science*, **281**, 710–712.

Swanson, W.J. and Vacquier, V.D. (2002) The rapid evolution of reproductive proteins. *Nat. Rev. Genet.*, **3**, 137–144.

Swanson, W.J., Aquadro, C.F. and Vacquier, V.D. (2001a) Polymorphism in abalone fertilization proteins is consistent with the neutral evolution of the egg's receptor for lysin (VERL) and positive Darwinian selection of sperm lysin. *Mol. Biol. Evol.*, **18**, 376–383.

Swanson, W.J., Yang, Z., Wolfner, M.F. and Aquadro, C.F. (2001b) Positive Darwinian selection drives the evolution of several female reproductive proteins in mammals. *Proc. Natl. Acad. Sci. USA*, **98**, 2509–2514.

Takagi, T., Nakamura, A., Deguchi, R. and Kyozuka, K. (1994) Isolation, characterization, and primary structure of three major proteins obtained from Mytilus edulis sperm. *J. Biochem.*, **116**, 598–605.

Takano, H., Yanagimachi, R. and Urch, U.A. (1993) Evidence that acrosin activity is important for the development of fusibility of mammalian spermatozoa with the oolemma: inhibitor studies using the golden hamster. *Zygote*, **1**, 79–91.

Töpfer-Petersen, E., Cechová, D., Henschen, A. *et al.* (1990) Cell biology of acrosomal proteins. *Andrologia*, **22** (Suppl 1), 110–121.

Tsubamoto, H., Hasegawa, A., Nakata, Y. *et al.* (1999) Expression of recombinant human zona pellucida protein 2 and its binding capacity to spermatozoa. *Biol. Reprod.*, **61**, 1649–1654.

Tulsiani, D.R. (2000) Structural analysis of the asparagine-linked glycan units of the ZP2 and ZP3 glycoproteins from mouse zona pellucida. *Arch. Biochem. Biophys.*, **382**, 275–283.

Tulsiani, D.R., Abou-Haila, A., Loeser, C.R. and Pereira, B.M. (1998) The biological and functional significance of the sperm acrosome and acrosomal enzymes in mammalian fertilization. *Exp. Cell Res.*, **240**, 151–164.

Ulrich, A.S., Otter, M., Glabe, C.G. and Hoekstra, D. (1998) Membrane fusion is induced by a distinct peptide sequence of the sea urchin fertilization protein bindin. *J. Biol. Chem.*, **273**, 16748–16755.

Urch, U.A. and Patel, H. (1991) The interaction of boar sperm proacrosin with its natural substrate, the zona pellucida, and with polysulfated polysaccharides. *Development*, **111**, 1165–1172.

Usui, N. (1987) Formation of the cylindrical structure during the acrosome reaction of abalone spermatozoa. *Gamete Res.*, **16**, 37–45.

Vacquier, V.D. and Lee, Y.H. (1993) Abalone sperm lysin: unusual mode of evolution of a gamete recognition protein. *Zygote*, **1**, 181–196.

Vacquier, V.D. and Moy, G.W. (1977) Isolation of bindin: the protein responsible for adhesion of sperm to sea urchin eggs. *Proc. Natl. Acad. Sci. USA*, **74**, 2456–2460.

Vacquier, V.D., Carner, K.R. and Stout, C.D. (1990) Species-specific sequences of abalone lysin, the sperm protein that creates a hole in the egg envelope. *Proc. Natl. Acad. Sci. USA*, **87**, 5792–5796.

Vieira, A. and Miller, D.J. (2006) Gamete interaction: is it species-specific? *Mol. Reprod. Dev.*, **73**, 1422–1429.

Ward, G.E., Brokaw, C.J., Garbers, D.L. and Vacquier, V.D. (1985) Chemotaxis of Arbacia punctulata spermatozoa to resact, a peptide from the egg jelly layer. *J. Cell Biol.*, **101**, 2324–2329.

Wassarman, P.M. (1999) Mammalian fertilization: molecular aspects of gamete adhesion, exocytosis, and fusion. *Cell*, **96**, 175–183.

Wassarman, P.M. (2008) Zona pellucida glycoproteins. *J. Biol. Chem.*, **283**, 24285–24289.

Wong, J.L. and Wessel, G.M. (2006) Defending the zygote: Search for the ancestral animal block to polyspermy. *Curr. Top. Dev. Biol.*, **72**, 1–151.

Yamagata, K., Murayama, K., Okabe, M. *et al.* (1998) Acrosin accelerates the dispersal of sperm acrosomal proteins during acrosome reaction. *J. Biol. Chem.*, **273**, 10470–10474.

Yang, Z. and Bielawski, J.P. (2000) Statistical methods for detecting molecular adaptation. *Trends Ecol. Evol.*, **15**, 496–503.

Yang, Z., Swanson, W.J. and Vacquier, V.D. (2000) Maximum-likelihood analysis of molecular adaptation in abalone sperm lysin reveals variable selective pressures among lineages and sites. *Mol. Biol. Evol.*, **17**, 1446–1455.

Zigler, K.S. and Lessios, H.A. (2003) 250 million years of bindin evolution. *Biol. Bull.*, **205**, 8–15.

Zigler, K.S., Raff, E.C., Popodi, E. *et al.* (2003) Adaptive evolution of bindin in the genus Heliocidaris is correlated with the shift to direct development. *Evolution Int. J. Org. Evolution*, **57**, 2293–2302.

17

Remodelling the oocyte into a totipotent zygote: degradation of maternal products

José-Eduardo Gomes, Jorge Merlet, Julien Burger and Lionel Pintard
Institut Jacques Monod, CNRS and Université Paris Diderot, Paris CEDEX 13, France

17.1 Introduction

In metazoans dramatic changes take place upon fertilization. The dormant oocyte is transformed into a fast-developing embryo, able to differentiate and generate the various cell types of the adult organism. Once fertilized, the oocyte – typically arrested at an intermediate stage of meiosis – resumes meiosis, thereby producing a haploid pronucleus. Subsequently, maternal and paternal pronuclei fuse, resulting in the formation of a diploid embryo, which then undergoes rapid cell cycle divisions, producing a large number of cells in a short period of time. Cell proliferation at this early stage is concomitant with the establishment of embryonic cell fate patterning. Importantly, the oocyte-to-embryo transition and early embryonic events rely largely on maternal products (mRNA and protein) loaded onto the oocyte, as zygotic gene expression typically does not start until after a few rounds of cell division, depending on the species. Once the maternal RNAs and proteins are no longer needed they must be promptly eliminated from the embryo in a regulated manner.

A considerable amount of research conducted in various model organisms, and particularly in the nematode *Caenorhabditis elegans*, has revealed the critical importance of the ubiquitin-proteolytic system (UPS) in the selective degradation of maternal proteins. In this chapter we will review these major findings taking *C. elegans* as the paradigm, and comparing it with other relevant model organisms.

Oogenesis: The Universal Process Marie-Hélène Verlhac and Anne Villeneuve

17.2 Ubiquitin-proteolytic system

The most common mechanism for regulated protein degradation, among eukaryotes, is the ubiquitin-proteolytic system in which proteins are targeted for rapid proteolysis upon conjugation to ubiquitin, a conserved 76-residue protein (Hershko *et al.*, 1980; Ciechanover *et al.*, 1980). Covalent attachment of ubiquitin to lysine residues of the substrate requires the coordination of three classes of enzymes: E1, E2, and E3 (Figure 17.1a). The ubiquitin-activating (E1) and the ubiquitin-conjugating (E2) enzymes are involved in activating and transferring ubiquitin to the substrate through thioester bond formation (Ciechanover *et al.*, 1981; Haas *et al.*, 1982; Pickart and Rose, 1985; Hershko *et al.*, 1983). The E3 ubiquitin ligases serve a dual function: they recruit the ubiquitin-loaded E2 enzyme and position it in close proximity to the substrate, thereby promoting substrate polyubiquitinylation and subsequent degradation by the 26S proteasome (reviewed in Hershko and Ciechanover, 1992; Pickart and Cohen, 2004).

Selective recognition of the substrate is fundamental for proper function of the ubiquitin/proteasome system. Indeed, rather than a promiscuous system for protein degradation, substrate recognition by the ubiquitin pathway must be highly selective and tightly regulated. Substrate specificity is determined by the E3 ligases. There are two main categories of E3 ligases: the HECT (homologous to E6-AP C-terminus) types and the RING (Really Interesting New Gene) types. In contrast to HECT-type E3

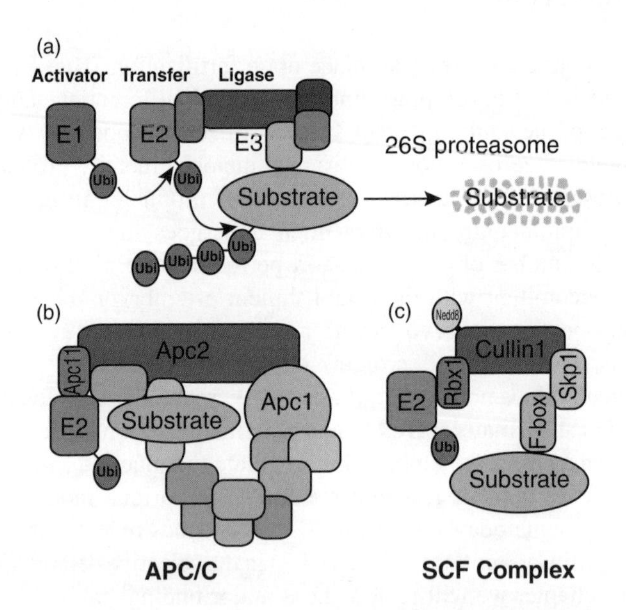

Figure 17.1 Ubiquitination and protein degradation. (a) Ubiquitin is sequentially conjugated onto the E1 activating enzyme, then onto E2 transfer enzyme, and finally E3 ligase brings the E2 close to the target protein, enabling its ubiquitination. Reiteration of these processes leads to the assembly of a polyubiquitin chain, and polyubiquitinated proteins are subsequently degraded by the 26S proteasome. (b) Basic structure of the SCF complex, the archetype of the CRL E3 type ligases. (c) A possible structure for the APC complex (adapted from Peters, 2006). A full colour version of this figure appears in the colour plate section.

enzymes, which contain an active cysteine residue and form an obligate thioester bond with ubiquitin prior to transfer to the substrate, RING-type E3s serve as a docking site for the ubiquitin-activated E2, which provides catalytic activity. The RING domain was first linked to the ubiquitin-proteolytic system through its discovery in subunits of two cell cycle-regulated E3 enzymes: APC/C (anaphase-promoting complex/cyclosome) (Zachariae *et al.*, 1998; Yu *et al.*, 1998) and SCF (Skp1 – Cullin–F-box) (Yu *et al.*, 1998; Ohta *et al.*, 1999; Tan *et al.*, 1999; Skowyra *et al.*, 1999; Seol *et al.*, 1999; Tyers and Jorgensen, 2000).

APC/C is a highly conserved RING-type E3 ligase of 13 subunits, involved in cell cycle progression, both in mitosis and meiosis (Castro *et al.*, 2005; van Leuken, Clijsters and Wolthuis, 2008). APC/C recognizes its substrates through Cdc20/Fzy and Cdh1/Hct1/Fzr, two evolutionarily conserved subunits containing WD40 propeller domains (Yu, 2007; Thornton and Toczyski, 2006; Figure 17.1b).

The SCF core complex is composed of a small BTB (Bric à Brac, Tramtrack, Broad complex)-fold containing protein called Skp1, the scaffold protein Cul1/Cdc53 (generically called a cullin) and the RING domain protein Rbx1 (also known as Roc1 or Hrt1), which provides the docking site for the ubiquitin-loaded E2 enzyme. Skp1 interacts with the ~40-residue F-box motif that defines the F-box protein family. F-box proteins recognize their substrates via protein interaction domains, such as WD40, or Leucine Rich Repeat (LRR) domains (Patton, Willems and Tyers, 1998; Skowyra *et al.*, 1997; Figure 17.1c). The SCF complex, simultaneously discovered in budding yeast and in *C. elegans*, through genetic studies on cell division, is the founding member of the most prominent family of ubiquitin ligases: the cullin-RING E3 ligases (CRL) (reviewed in Petroski and Deshaies, 2005). The eukaryotic genomes encode at least five cullins, termed Cul1 through 5 and, as with Cul1 in the SCF complex, each of these cullins assembles multisubunit E3 ubiquitin ligases containing the same catalytic core, but distinct substrate-recognition modules that are specific to each cullin. In CRL2 (Cul2-based) and CRL5 complexes, the BTB protein, elongin C, bridges the interaction between the cullins and substrate-recruitment factors, called BC-VHL box for CRL2 and BC-SOCS box for CRL5 complexes. In CRL3 complexes, a single polypeptide, containing a BTB domain and a substrate interaction motif, functions as a substrate-recognition module, merging the functions of the Skp1/F-box and elongin C/BC box heterodimers (Pintard, Willems and Peter, 2004). Finally, the large DDB1 protein, which does not contain a BTB domain but three seven-bladed β-propellers, recruits specificity factors containing WD40 repeats termed DCAF (Ddb1, and CUL4 associated factors) (Jin *et al.*, 2006; Angers *et al.*, 2006).

All CRLs are regulated by similar mechanisms. In particular, the covalent addition of the ubiquitin-like protein Nedd8 to the cullin subunit stimulates CRL activity *in vivo* and *in vitro* (Figure 17.1c). Nedd8 conjugation (also called neddylation) to the cullin induces a drastic conformational change of its C-terminal part, which exposes the RING domain of Rbx1, and allows it to adopt a variable conformation, thus stimulating substrate ubiquitination (Duda *et al.*, 2008; Saha and Deshaies, 2008).

CRLs regulate a multitude of cellular and biological processes by targeting specific proteins for degradation (reviewed in Petroski and Deshaies, 2005). In particular, CRL complexes have emerged in recent years as key components of the oocyte-to-embryo transition by coordinating the spatiotemporal degradation of various maternal factors.

17.3 Oocyte-to-embryo transition and early embryonic development in *C. elegans*

C. elegans oogenesis takes place in the syncytial gonad of the hermaphrodite adult. At the distal end of the gonad a subset of nuclei proliferates continuously. As these nuclei move proximally they enter meiosis. Cellularization takes place at the proximal end of the female gonad, where an oocyte arrested in prophase of meiosis I is ovulated (Hirsh, Oppenheim and Klass, 1976). The oocyte then passes through the spermatheca, where fertilization occurs (reviewed in Greenstein, 2005). Due to the constraint of the adult hermaphrodite anatomy, sperm usually enters the oocyte at the opposite side of the maternal nucleus (Goldstein and Hird, 1996). Sperm entry triggers both the resumption of the oocyte meiosis and the establishment of the embryonic anterior–posterior (a–p) axis (Ward and Carrel, 1979; Goldstein and Hird, 1996). As the oocyte completes meiosis I, the haploid maternal pronucleus forms, and both male and female pronuclei decondense their DNA and enter S phase (Edgar and McGhee, 1988). DNA replication is achieved before the maternal pronucleus migrates towards the paternal pronucleus, in a microtubule-dependent manner (Edgar and McGhee, 1988; Strome and Wood, 1983). Pronuclei then meet near the posterior end of the embryo, move to the centre, fuse, and the nucleocentrosomal complex undergoes a 90° rotation. After nuclear envelope breakdown, the mitotic spindle aligns along the a–p axis of the embryo (Albertson, 1984). Due to unequal pulling forces acting on both centrosomes during metaphase, the mitotic spindle is displaced towards the posterior end of the embryo (Grill *et al.*, 2001). The cytokinetic furrow then bisects the spindle, leading to an asymmetric cell division. The resulting anterior AB and posterior P1 blastomeres are unequal both in size and developmental potential. Whereas AB generates most of the somatic cells, P1 gives rise to the precursors of the germline and the remaining somatic cells (Sulston *et al.*, 1983). Asymmetric cell division is thus critical for proper cell fate specification.

Embryonic polarity contrasts with the anisotropy of the oocyte, and sperm entry is the initial event breaking the oocyte symmetry and eventually leading to the establishment of the a–p axis of the embryo. The cornerstone of a–p axis establishment is the asymmetric distribution of the evolutionarily conserved PAR (partitioning defective) proteins. PAR-3/PAR-6/aPKC form a complex localized at the anterior cortex and PAR-1 and PAR-2 localize at the opposite in the posterior cortex; the PAR's polarization in turn controls all the subsequent embryonic asymmetries, such as the asymmetric spindle positioning, the asynchronous cell cycle and the segregation of cell fate determinants (Kemphues *et al.*, 1988; Kirby, Kusch and Kemphues, 1990; Levitan *et al.*, 1994; Etemad-Moghadam, Guo and Kemphues, 1995; Guo and Kemphues, 1995; Boyd *et al.*, 1996; Watts *et al.*, 1996; Tabuse *et al.*, 1998; Hung and Kemphues, 1999; reviewed in Gonczy, 2008). Proper embryonic development requires accurate coordination of these events, in a narrow time window. Importantly, early development relies on maternal products provided by the mother in the oocyte (reviewed in Stitzel and Seydoux, 2007). The ubiquitin-proteolytic system has recently emerged as a critical pathway remodelling the oocyte and transforming it into a totipotent embryo.

17.4 Degradation of maternal proteins and progression through meiosis

At fertilization, the oocyte is arrested in prophase of meiosis I (Ward and Carrel, 1979). Completion of meiosis is therefore the first developmental task to accomplish on the oocyte-to-embryo transition. Sperm entry triggers the resumption of meiosis, such that the oocyte proceeds to anaphase I, extrudes the first (diploid) and the second (haploid) polar bodies, and completes meiosis II (reviewed in Greenstein, 2005).

Two ubiquitin ligases regulate meiotic progression: APC/C and a CRL2-based E3 ligase.

APC/C's role in cell cycle progression is widely conserved, particularly its function in the separation of sister chromatids, in both mitosis and meiosis. Strong loss of function, in the context of RNA interference (RNAi) or mutations in genes coding for APC/C subunits, causes meiotic arrest in metaphase I, as well as other meiosis defects, or abnormalities during germline proliferation (Davis *et al.*, 2002; Furuta *et al.*, 2000; Golden *et al.*, 2000; Rappleye *et al.*, 2002; Shakes *et al.*, 2003; Kitagawa *et al.*, 2002). Interestingly, upon partial loss of function in some APC/C mutants, defects can be observed in both meiosis I and meiosis II, indicating that APC/C plays a role not only in the first, but rather in both meiotic divisions. At the molecular level, one of APC/C's best-documented functions is to target the protein securin for degradation, triggering a cascade of events leading to chromosome segregation. Securin is an inhibitor of the separase protease, which in turn promotes chromosome segregation by cleaving Scc1, a subunit of the cohesin complex that holds chromosomes together at metaphase. By promoting securin degradation, APC/C releases separase from its inhibition, allowing cohesin cleavage, and thus chromosome segregation (Figure 17.2). Consistent with this

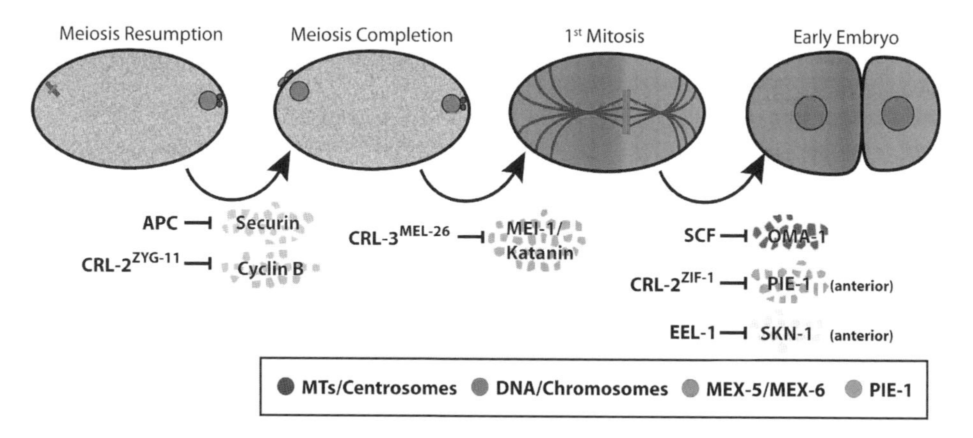

Figure 17.2 Main protein degradation events in the *C. elegans* oocyte-to-embryo transition. Triggering of degradation of specific proteins by E3 ligases is crucial for the main steps to take place. APC and CRL-2^{ZYG-11} ubiquitinate securin and cyclin B, respectively, targeting them for degradation, thus enabling completion of meiosis. CRL-3^{MEL-26} targets MEI-1/Katanin for degradation upon completion of meiosis, allowing the formation of a proper mitotic spindle. SCF targets OMA-1 for degradation, ensuring the start of zygotic transcription. In the anterior of the embryo, as part of cell fate patterning, CRL-2^{ZIF-1} targets PIE-1 (as well as POS-1 and MEX-1, not shown on the figure) for degradation in a MEX-5/6-dependent manner, and EEL-1 targets SKN-1 for degradation. A full colour version of this figure appears in the colour plate section.

role, APC/C mutants exhibit defects on chromosome segregation and spindle formation in meiosis.

Like other E3 ligases, APC/C has multiple targets. However, other APC/C targets are yet to be identified. Nonetheless, phenotypic analysis points to other roles for APC/C besides chromosome segregation. Indeed, when CUL-2 function is compromised, APC/C mutant phenotypes are enhanced, suggesting that a CRL2 complex may be redundant with APC/C (Sonneville and Gonczy, 2004). However, no defects in meiosis I are observed upon loss of CUL-2, indicating that redundancy between APC/C and CUL-2 is only partial, and likely restricted to meiosis II.

CUL-2 is involved in meiotic exit, as well as in S phase progression during germline proliferation (Feng *et al.*, 1999). Loss of CUL-2 delays progression through meiosis II and impairs meiosis exit. The same phenotypes are observed upon the loss of other components of the CRL2 complex, that is, RBX-1 and ELC-1 (Liu, Vasudevan and Kipreos, 2004). Interestingly, inactivation of another locus termed *zyg-11* phenocopies *cul-2* inactivation on meiosis II, but not on germline proliferation (Liu, Vasudevan and Kipreos, 2004; Sonneville and Gonczy, 2004). ZYG-11 contains a BC-box and binds CRL2 components *in vivo* and *in vitro* (Vasudevan, Starostina and Kipreos, 2007), and thus fulfils the criteria to act as substrate-recruitment factor of a CRL2 complex, specifically involved in meiosis progression. Nonetheless, the critical targets of this putative $CRL2^{ZYG\text{-}11}$ ligase remain elusive. Cyclins B1 and B3 accumulate at higher levels in both *zyg-11(-)* and *cul-2(-)* embryos, making them good candidates for $CRL2^{ZYG\text{-}11}$ targets (Liu, Vasudevan and Kipreos, 2004; Sonneville and Gonczy, 2004; Figure 17.2). However, it has not been shown biochemically that cyclins B1/B3 are direct targets of $CRL2^{ZYG\text{-}11}$, and therefore they may not be the relevant targets of the complex, and their regulation by $CRL2^{ZYG\text{-}11}$ could be indirect.

Although more investigation is needed to identify the relevant targets of the $CRL2^{ZYG\text{-}11}$ complex, it is clear that ubiquitin-mediated degradation of specific targets is essential to meiosis resumption and completion, the first step towards the oocyte-to-embryo transition.

17.5 Meiosis-to-mitosis transition: redundant pathways for timely degradation of the microtubule-severing protein MEI-1/Katanin

In the early embryo, two very different spindle structures form in the same cytoplasm within a 20 min interval: the meiotic spindle and the mitotic spindle (Figure 17.2). In sharp contrast to the acentrosomal meiotic spindle, the assembly of a functional mitotic spindle requires the nucleation of long arrays of astral microtubules from the centrosomes. These mitotic microtubules are crucial for asymmetric spindle positioning and spindle elongation along the a–p axis of the embryo.

The heterodimer composed of MEI-1/MEI-2 subunits (also known as Katanin) is required for the assembly of the meiotic spindle. Loss-of-function mutations in either *mei-1* or *mei-2* impair meiosis without affecting mitosis, revealing a specific function of

Katanin in meiotic divisions (Clark-Maguire and Mains, 1994a, 1994b; McNally *et al.*, 2006; Srayko *et al.*, 2000). Katanin is an evolutionarily conserved AAA-ATPase that exhibits ATP-dependent microtubule-severing activity *in vitro* (Hartman *et al.*, 1998). Presumably Katanin severs the microtubule protofilament upon conformational change induced by ATP hydrolysis. The activity of the MEI-1/MEI-2 complex has been shown to increase tubulin polymer numbers from a relatively inefficient chromatin-based microtubule nucleation pathway (McNally *et al.*, 2006; Srayko *et al.*, 2006).

MEI-1/2 proteins accumulate on the meiotic spindle, but MEI-1 rapidly disappears after completion of meiosis and is undetectable in mitosis. Importantly, a specific mutation that disrupts a PEST motif of MEI-1 causes active MEI-1/MEI-2 complex to accumulate to high levels in mitosis (Clark-Maguire and Mains, 1994a, 1994b). Likewise, mutations in a second locus, called *mel-26*, also result in high MEI-1 levels in mitosis. In *mel-26(-)* mutant embryos, MEI-1 accumulates on centrosomes and prevents the assembly of a functional mitotic spindle (Pintard *et al.*, 2003; Dow and Mains, 1998). These observations suggested a requirement of the UPS for MEI-1 degradation after meiosis.

Importantly, the first indication that a CRL complex was involved in MEI-1 degradation stems from the discovery of *rfl-1*, another locus required for MEI-1 degradation (Kurz *et al.*, 2002). *rfl-1* encodes a subunit of the E1 enzyme of the neddylation pathway, which activates CRLs by conjugating the ubiquitin-like protein Nedd8 on the cullin subunit (see above). Through RNAi-based screening of the five cullins, CUL-3 was identified as the critical target of the neddylation machinery responsible for MEI-1 degradation. Upon loss of *rfl-1* gene function, non-neddylated forms of CUL-3 accumulate, causing phenotypes similar to those observed in *cul-3(-)* or *mel-26(-)* embryos, due to the persistence of MEI-1/MEI-2 function in mitosis (Pintard *et al.*, 2003; Kurz *et al.*, 2002).

A wealth of evidence then revealed that MEL-26 functions as a substrate-recruitment factor of this CRL3 complex, and actively recruits MEI-1, but not MEI-2. MEL-26 contains a canonical BTB domain required for binding of the N-terminal part of CUL-3, and a MATH domain required for MEI-1 binding. Specific mutation of the MATH domain prevents MEL-26 interaction with MEI-1 *in vivo* and *in vitro*. Likewise, the product of the gain-of-function mutation of MEI-1, which blocks MEI-1 degradation *in vivo*, failed to bind MEL-26 *in vitro* (Pintard *et al.*, 2003). Finally, immunopurified $CRL3^{MEL-26}$ complex from HEK293T cells readily ubiquitinates MEI-1 *in vitro* (Xu *et al.*, 2003). Interestingly, lethality resulting from MEI-1 accumulation in mitosis, due to *mei-1* gain-of-function mutation, can be rescued by tubulin mutations that are partially resistant to MEI-1/MEI-2 severing (Lu, Srayko and Mains, 2004). Together these observations clearly indicate that the $CRL3^{MEL-26}$ complex is required to inactivate Katanin via MEI-1 degradation (Figure 17.2). However, a low percentage of *mel-26*-null mutant animals are viable at 15 °C, suggesting that other E3 ligases, or degradation-independent mechanisms, may contribute to MEI-1 inactivation after meiosis.

Although the role of the $CRL\text{-}3^{MEL-26}$ complex in MEI-1 degradation is well established, the signal triggering MEI-1 degradation after completion of meiosis remains to be identified. In many instances, phosphorylation of the substrate targets its recognition and consequent ubiquitination by the corresponding E3 ligase. Importantly, several groups have identified a *C. elegans* homologue of the minibrain-kinase, called MBK-2, as a regulator of MEI-1 degradation (Pellettieri *et al.*, 2003; Quintin *et al.*, 2003; Lu and Mains, 2007; Pang *et al.*, 2004). The phenotypes resulting from *mbk-2* inactivation are complex, but include specific defects in spindle positioning, that are rescued by *mei-1* inactivation. Moreover, GFP::MEI-1 accumulates to high levels in mitosis upon *mbk-2* inactivation, and MBK-2 phosphorylates MEI-1 *in vitro* (Stitzel, Pellettieri and Seydoux, 2006). Taken together, these observations suggest that MBK-2-dependent phosphorylation of MEI-1 might target its recognition by the CRL3$^{\text{MEL-26}}$ complex (Stitzel, Pellettieri and Seydoux, 2006). However, a partial loss-of-function mutation of *mbk-2* enhances the penetrance of a *mel-26* null allele, indicating that *mbk-2* and *mel-26* might not be in the same pathway, but rather in parallel and partially redundant pathways converging on MEI-1 degradation (Lu and Mains, 2007). While the signal that triggers MEI-1 degradation at the end of meiosis is unknown, MEL-26 and MEI-1 expressed in bacteria bind weakly to each other, and CUL-3 promotes MEI-1 ubiquitination *in vitro*, suggesting that CRL-3$^{\text{MEL-26}}$ may be capable of recognizing and ubiquitinating MEI-1 without occurrence of post-translational modifications (Furukawa *et al.*, 2003; Pintard *et al.*, 2003). Actually, there seems to be a considerable degree of redundancy in MEI-1 regulation. While the CUL-3$^{\text{MEL-26}}$ complex is the main E3 ligase controlling MEI-1 degradation, APC/C also contributes to MEI-1 elimination after meiosis. Partial inactivation of APC/C function significantly decreased the low percentage viability of *mel-26* null allele, and this effect correlates directly with MEI-1 accumulation (Lu and Mains, 2007).

An intriguing question remains: why is MEI-1 stable in meiosis but rapidly degraded in mitosis? The answer to this question is currently unclear, however interplay of kinases and phosphatases, and/or other binding partners not necessarily causing protein modifications, may impinge on MEI-1 degradation. Alternatively, the protein levels of MEL-26 – low in meiosis and increased in mitosis – may contribute to the regulation of MEI-1 stability (Johnson *et al.*, 2009).

The sharp contrast between a small meiotic spindle and a robust mitotic spindle separated by a short period of time is common to most animals, and is particularly evident among vertebrates. The Katanin complex is conserved in vertebrates – where it is composed of p60 and p80 subunits, homologues of MEI-1 and MEI-2 respectively – as is its function in severing microtubules (Hartman *et al.*, 1998). However, to our knowledge no study has yet been performed in other systems on whether the Katanin complex is involved in meiotic spindle formation, or whether its degradation by CUL-3-based CRLs is required during oocyte-to-embryo transition.

17.6 Embryonic transcriptional reactivation relies on an SCF complex

Silencing of zygotic transcription during early embryonic development is a general trait of metazoans, and correlates with rapid cell cycles (reviewed in Schier, 2007). Transcription silencing persists until as early as the four-cell stage or as late as gastrulation, depending on the species. In any case, specific mechanisms must exist to ensure zygotic transcription is kept silent during early stages of development. In *C. elegans*, transcription is silenced until the four-cell stage in the somatic precursors, and until after the 100-cell stage in the germline precursors (Seydoux and Fire, 1994; Seydoux *et al.*, 1996; Blackwell and Walker, 2006). This transcriptional silencing in somatic cells depends, at least in part, on the cytoplasmic sequestration of TAF-4, a subunit of the pol II transcriptional complex, TFIID. TFIID is a multicomponent transcription factor, composed of the TATA-binding protein (TBP) and TBP-associated factor (TAF), which recognizes and binds the promoter DNA and establishes the transcription start site. TFIID is unstable in the absence of TAF-4, and nuclear retention of TAF-4 requires TAF-12 binding via a histone fold-like domain (Walker *et al.*, 2001; Guven-Ozkan *et al.*, 2008).

Although TAF-4 is found in the nucleus in oocytes and later embryos, it is excluded from the nucleus in one- and two-celled embryos (Walker *et al.*, 2001), suggesting that transcription silencing in the early embryo might rely on its subcellular localization (Figure 17.3). TAF-4 was recently identified as an OMA-1-interactor in a yeast two-hybrid screen (Guven-Ozkan *et al.*, 2008). OMA-1 (oocyte maturation) and OMA-2 are two very similar maternal proteins that are essential for oocyte maturation in the

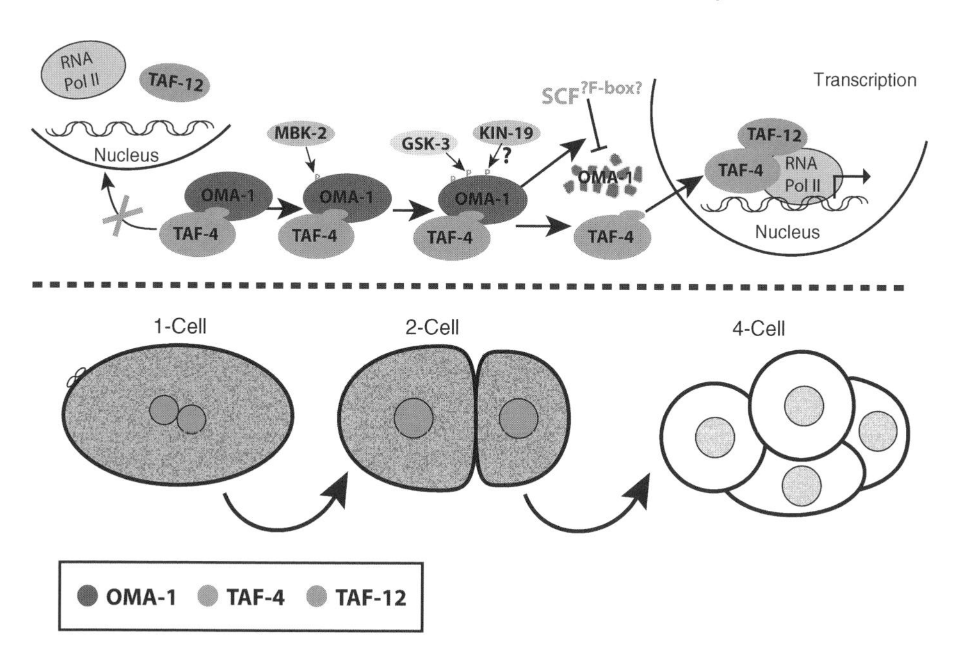

Figure 17.3 Activation of zygotic transcription. In one- and two-cell stage embryos, OMA-1 sequesters TAF-4 in the cytoplasm, precluding its translocation to the nucleus, interaction with TAF-12 and RNA Pol II, thus repressing transcription. Both OMA-1 and TAF-12 bind TAF-4 via its histone fold, thus competing for binding through this domain. OMA-1 is phosphorylated by MBK-2, priming it for further phosphorylation by GSK-3, and probably KIN-19 as well. Phosphorylated OMA-1 is ubiquitinated by a SCF complex and degraded, releasing repression of TAF-4. TAF-4 is then free to be translocated to the nucleus, bind TAF-12 and RNA Pol II, leading to transcription activation in somatic cells. Note that although TAF-4/TAF-12 are present in the nucleus of germline precursors, transcription is repressed by PIE-1 (not represented in the figure). A full colour version of this figure appears in the colour plate section.

female gonad. Besides containing a zinc finger domain, they contain a histone fold. Interestingly, upon reduction of OMA-1/2 levels, transcription is no longer repressed in one- and two-cell embryos, suggesting that OMA-1/2 interaction with TAF-4 may be crucial for transcriptional silencing (Guven-Ozkan *et al.*, 2008). OMA-1 interacts with TAF-4 via its histone-fold domain, like TAF-12 does, hence competing with TAF-12 for TAF-4-binding by means of molecular mimicking (Guven-Ozkan *et al.*, 2008). Importantly, upon SCF-mediated degradation of OMA-1 in somatic cells, TAF-4 is released, allowing for interaction with TAF-12 (Guven-Ozkan *et al.*, 2008). Therefore, OMA-1 represses transcription by sequestering TAF-4 in the cytoplasm, while OMA-1 degradation enables activation of zygotic transcription by releasing TAF-4 (Figure 17.3).

Consistent with this model, timely and rapid degradation of OMA-1/2 is necessary for normal embryo development, as protein persistence after the first mitotic division causes misexpression of crucial cell fate determinants, and consequent embryonic lethality (Lin, 2003). The next question is then to understand temporal regulation of OMA-1 degradation. The answer lies in the phosphorylation by kinases MBK-2 and GSK-3, shown to be required for timely degradation of OMA-1 (Nishi and Lin, 2005; Stitzel, Pellettieri and Seydoux, 2006). MBK-2 and GSK-3 phosphorylate sequentially two specific threonine residues of OMA-1. First MBK-2 phosphorylates T239, triggering a second phosphorylation mediated by GSK-3 on T339, after which OMA-1 is targeted for degradation (Nishi and Lin, 2005). Additionally, MBK-2 can also phosphorylate OMA-1 on S302, *in vitro* (Shirayama *et al.*, 2006; Figure 17.3). However, these are not the only kinases involved: mutations in genes encoding KIN-19 – homologue of a casein kinase 1 α, or CK1α – as well as cell cycle kinase CDK-1, also cause a failure in OMA-1 degradation (Shirayama *et al.*, 2006). Importantly, reducing *oma-1* activity partially suppresses the phenotypes caused by the mutations in *kin-19* and *cdk-1*, as well as in *mbk-2* and *gsk-3*, confirming that OMA-1 down-regulation is one important function of these kinases *in vivo*. It is unlikely, however, that OMA-1 itself is the phosphorylation target of all five kinases. Rather it is plausible that, in addition to MBK-2 and GSK-3, OMA-1 is a direct target of KIN-19, but not of CDK-1 and CKS-1. Human CK1α, close homologue of KIN-19, displays some affinity for OMA-1 *in vitro*, and this affinity is enhanced by pretreatment with MBK-2, suggesting that KIN-19, like GSK-3, may phosphorylate OMA-1 primed by MBK-2 (Shirayama *et al.*, 2006). On the other hand, no evidence of OMA-1 phosphorylation by CDK-1 was found, indicating that they function indirectly by phosphorylating other proteins in the pathway. The relevant targets of CDK-1 and CKS-1 impinging on OMA-1 regulation are yet to be determined. Nonetheless, the identification of *cdk-1(ne236* and *ne2257)* and *cks-1(ne549)* mutant alleles as being required for OMA-1 degradation highlights the link to cell cycle progression (Shirayama *et al.*, 2006). Interestingly, these mutants are not null alleles, they carry point mutations leading to single amino acid substitutions and do not display all the phenotypes detected in null alleles, but seem to be somewhat specific to OMA-1 degradation. Therefore, these mutant alleles allow genetic separation of the roles of CDK-1 and CKS-1 in OMA-1 degradation from other requirements in cell cycle progression. Wild-type CDK-1 and CKS-1 bind together to form a complex, while CKS-1(Y10F) protein, the product of the *cks-1(ne549)* allele, fails to bind CDK-1, indicating that the CDK-1/CSK-1 complex formation is required for

OMA-1 degradation (Shirayama *et al.*, 2006). As to CDK-1, both *cdk-1(ne236)* and *cdk-1(ne2257)* affect a T loop, presumably involved in interaction with cyclins. Further research is required to understand how these mutations affect CDK-1 function, as no effect on the interaction with known binding partners was detected with the CDK-1 (I173F) mutant form. Nonetheless, it is worth noting that embryos depleted of cyclin B3, but not cyclin A or cyclin B, also exhibit OMA-1 protein stabilization, suggesting that cyclin B3 is the relevant cyclin for CDK-1/CKS-1 function, although the mechanism remains elusive (Shirayama *et al.*, 2006).

A complex pathway thus converges on OMA-1 phosphorylation to target this protein for degradation. It becomes relevant then to understand the mechanism responsible for recognition and degradation of phosphorylated OMA-1. Genetic evidence points to CUL-1-based SCF E3 ligases. Depletion of CUL-1 by RNA interference in early embryos causes persistence of OMA-1/2, repressing transcription in somatic cells beyond the four-cell stage (Guven-Ozkan *et al.*, 2008; Shirayama *et al.*, 2006). Likewise, Skp1 homologues SKR-1 and SKR-2 are also required for OMA-1 degradation (Shirayama *et al.*, 2006), strengthening the hypothesis that a canonical SCF complex, nucleated by CUL-1, is responsible for OMA-1 degradation (Figure 17.3). Whereas the role of the SCF in OMA-1 degradation is well established, the identity of the F-box protein specifically targeting OMA-1 degradation remains unknown. However, one puzzling observation has no obvious explanation: loss of function of *zyg-11* also impairs timely degradation of OMA-1, while *cul-2* has little or no effect (Shirayama *et al.*, 2006). ZYG-11 is known to be an adaptor protein for CRL2 complexes (see above), but seems to have a CUL-2-independent role in OMA-1/2 degradation.

17.7 Reprogramming a germ cell into a totipotent zygote: role of the CRL2^{ZIF-1} complex in the degradation of the germ plasm proteins

During the early stages of development, the oocyte, which is a germ cell, must acquire the ability to produce both germ and somatic cell fates. Several lines of evidence, from various species, indicate that localized degradation of germ-plasm proteins in the somatic lineage contributes to the specification of the somatic cell identity. In *C. elegans*, early embryonic development proceeds through a series of asymmetric divisions, which in most cases give birth to a germline precursor cell and a somatic precursor cell (reviewed in Gonczy and Rose, 2005). These asymmetric divisions cause dramatic differences in protein and mRNA contents between sister cells, eventually leading them to adopt strikingly different developmental fates. Establishment of the asymmetric protein content relies to a large extent on selective protein degradation. At least three CCCH fingers containing proteins POS-1, MEX-1 and PIE-1 are specifically removed from the somatic cell by the UPS system, while persisting in the germline cell. POS-1, MEX-1 and PIE-1 are putative RNA-binding proteins that accumulate in the germ cell precursor. This specific localization is due not only to localized degradation, but also results from their active segregation into the germline precursors upon asymmetric division (and in the case of POS-1, of its mRNA as well) (Mello *et al.*, 1996; Reese *et al.*, 2000; Tabara *et al.*, 1999; Guedes and Priess, 1997). Selective degradation of these proteins is dependent on ZIF-1, a BC-box-containing protein. PIE-1's first CCCH zinc finger (ZF1), but not the second (ZF2), is sufficient to target protein degradation in somatic cells, and the same is true for ZF1 of MEX-1 and POS-1 (DeRenzo, Reese and Seydoux, 2003). ZIF-1 was identified in a yeast two-hybrid screen as a binding partner of PIE-1's ZF1, and was found to be required both for exclusion of PIE-1, POS-1 and MEX-1 from somatic cells and for degradation of GFP fused to the ZF1 of PIE-1 (DeRenzo, Reese and Seydoux, 2003). ZIF-1 forms a CRL complex with elongin C, elongin B and CUL-2, and is likely the adaptor protein that binds the substrates PIE-1, POS-1 and MEX-1, enabling their ubiquitination (Figure 17.2). This $CRL2^{ZIF\text{-}1}$ complex uses UBC-5 as E2 enzyme, given that inactivation of UBC-5, ZIF-1, ELC-1 or CUL-2 blocks the selective degradation of the CCCH proteins in the somatic precursor (DeRenzo, Reese and Seydoux, 2003). Two other very similar and largely redundant CCCH proteins, MEX-5 and MEX-6, are also involved in somatic degradation of PIE-1. PIE-1, POS-1 and MEX-1 are no longer degraded in the somatic blastomeres upon loss of MEX-5/6 function, and conversely upon MEX-5 misexpression in the germline are no longer detected in germline precursors, indicating that MEX-5/6 trigger degradation of those germline proteins (Schubert *et al.*, 2000). Moreover, MEX-5/6 function requires phosphorylation by MBK-2 and Polo kinases PLK-1 and PLK-2. MBK-2 phosphorylates MEX-5 on T186, priming it for PLK-1/2 phosphorylation, and mutating T186 to A impairs severely, although not completely, MEX-5 function *in vivo* (Nishi *et al.*, 2008). Importantly, in this case, and unlike other functions of MBK-2, the phosphorylated protein is not targeted for degradation by the UPS. Phosphorylation by MBK-2 does not regulate MEX-5 localization either (Pellettieri *et al.*, 2003), rather it seems to be involved specifically in activating MEX-5 function in PIE-1, POS-1 and MEX-1 degradation. In summary, the genetic data indicate that MEX-5/6 are required for ZIF-1-mediated degradation of germline proteins in somatic cells, although the molecular link between $CRL2^{ZIF\text{-}1}$ and MEX-5/6 has yet to be found (Figure 17.2). Intriguingly, MEX-5/6 contain two CCCH fingers as well, and are degraded in somatic cells, but one round of cell division later than PIE-1, POS-1 and MEX-1, once these have been degraded themselves. MEX-5/6 degradation, in contrast with the other CCCH proteins, is dependent on its second CCCH zinc finger, rather than on the first one, but is dependent on $CRL2^{ZIF\text{-}1}$ as well (DeRenzo, Reese and Seydoux, 2003).

The CCCH proteins are specifically degraded in the anterior cell, AB, but are rock stable in its sister cell, P1, raising the question of how the activity of the $CRL2^{ZIF\text{-}1}$ complex is restricted to the somatic lineage. PAR-1 kinase plays a critical role in this process. PAR-1 is itself localized asymmetrically to the posterior, starting at the early one-cell stage embryo, and is required for multiple aspects of the zygote's asymmetry. Consistent with their role in somatic precursors, MEX-5 and MEX-6 are themselves asymmetrically distributed, being enriched in the anterior cytoplasm (Schubert *et al.*, 2000). This asymmetry depends on PAR-1, as *par-1* loss of function causes MEX-5/6 to extend to the posterior (Schubert *et al.*, 2000). Importantly, both PIE-1 and GFP fused to the ZF-1 of PIE-1 are degraded in the posterior of *par-1(-)* embryos in a *mex-5/6*-dependent manner, clearly indicating that loss of PIE-1 observed in *par-1(-)* is due to protein degradation relying on MEX-5/6 (DeRenzo and Seydoux, 2004; Schubert *et al.*, 2000). Additionally, it has been shown that MEX-5 needs to be phosphorylated in a critical C-terminal Ser residue, S458, in order to become asymmetric. This phosphorylation is dependent not only on PAR-1, but also on PAR-4, another kinase involved in early asymmetry (Tenlen *et al.*, 2008). Hence, $CRL2^{ZIF-1}$ activity seems to be restricted to somatic cells by localizing MEX-5/6 to the anterior cytoplasm in a process relying on the activity of the asymmetry kinases PAR-1 and PAR-4 (Figure 17.2).

In parallel to PIE-1, other cell fate determinants become restricted to the posterior of the embryo during early development, namely the transcription factors SKN-1 and the homologue of *Drosophila* Caudal, PAL-1 (Bowerman *et al.*, 1993; Bowerman, Eaton and Priess, 1992; Hunter and Kenyon, 1996). Unlike PIE-1, however, these transcription factors are also present in somatic cells, suggesting that their asymmetry is achieved by different means. Furthermore, SKN-1 and PAL-1 asymmetries require RNA-binding proteins MEX-1 (for SKN-1), MEX-3 (for PAL-1) and SPN-4 (for both), suggesting translation regulation is the main mechanism responsible for unequal distribution of those determinants (Bowerman *et al.*, 1993; Hunter and Kenyon, 1996; Gomes *et al.*, 2001; Huang *et al.*, 2002). Nonetheless, protein degradation, in addition to translational control, ensures unequal distribution of SKN-1 in the early embryo. The HECT-type E3 ubiquitin ligase EEL-1 was found to participate in the asymmetry of SKN-1 (Page *et al.*, 2007). The *eel-1* gene was identified as an enhancer of *efl-1* mutant, in an RNAi-based screen. EFL-1 is the *C. elegans* homologue of vertebrate E2F, and mutations in *efl-1* indirectly affect the levels of SKN-1 (Page *et al.*, 2001). *eel-1(RNAi)* enhanced the phenotype of *efl-1* hypomorph mutant (Page *et al.*, 2007). The N-terminus of EEL-1 interacts with the C-terminus of SKN-1, and C-terminus of SKN-1 fused to GFP is degraded in an EEL-1-dependent manner *in vivo.* Loss of EEL-1 partly impairs the asymmetry of endogenous SKN-1, but not of other asymmetrically distributed proteins (like MEX-5 or PIE-1), indicating that EEL-1 specifically degrades SKN-1 (Page *et al.*, 2007). Therefore, in addition to $CRL2^{ZIF-1}$, EEL-1, a HECT-type E3 ligase, participates in a–p cell fate patterning by specifically regulating degradation of the transcription factor SKN-1 (Figure 17.2).

17.8 Orchestrating maternal protein degradation: role of the MBK-2/DYRK kinase

As mentioned above, MBK-2 kinase phosphorylates both MEI-1 and OMA-1, playing an instrumental role in targeting these proteins for degradation. MBK-2 seems to be central to the targeting of multiple proteins for degradation during early embryonic events (Figure 17.4). The *mbk-2* gene was originally identified in an RNAi-based screen as being required for asymmetry of PIE-1 protein (Pellettieri *et al.*, 2003). *mbk-2(RNAi)* embryos display PIE-1 distribution throughout the embryo, indicating that MBK-2 is required for PIE-1 degradation in the anterior, presumptive somatic, portion of the embryo. Analysis of the phenotype revealed, however, other phenotypes of *mbk-2* loss of function, independent of PIE-1, namely microtubule-dependent phenotypes which turned out to be due, at least in part, to MEI-1 regulation (see above). Nonetheless, still

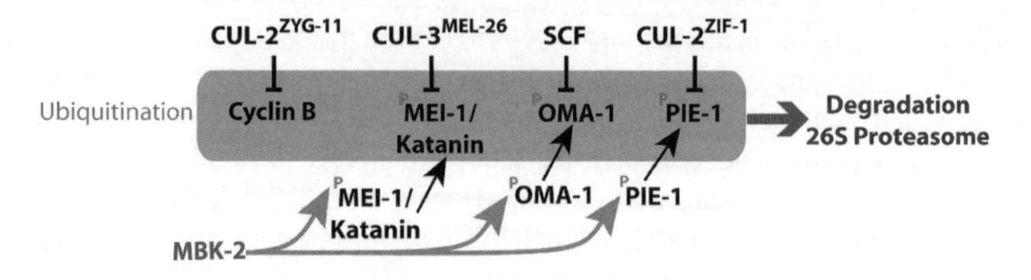

Figure 17.4 Cullin-based E3 ligases and phosphorylation. Four CRLs are involved in degradation of maternal products: CRL-2^{ZYG-11} ubiquitinates cyclin B, CRL-3^{MEL-26} ubiquitinates MEI-1/Katanin, SCF ubiquitinates OMA-1, and CRL-2^{ZIF-1} ubiquitinates PIE-1. MEI-1, PIE-1 and OMA-1 are phosphorylated by MBK-2, enabling recognition by the respective E3 ligase. A full colour version of this figure appears in the colour plate section.

other phenotypes were observed that could not be explained by lack of MEI-1 or PIE-1 degradation, such as cytokinesis and failure to localize POS-1 protein or the germline-specific P granules to the posterior of the embryo. Although MBK-2 seems to be required for the final asymmetries of PIE-1, POS-1 and P granules, it is not required for embryonic asymmetry *per se*, as unequal distribution of the PAR proteins, or MEX-5, is not affected in *mbk-2(-)* embryos (Pellettieri *et al.*, 2003). Furthermore, MBK-2 was shown to be required for PIE-1 degradation in the anterior of the embryo (Pellettieri *et al.*, 2003). Taking these findings together, MBK-2 stands out as being a core element in coordinating and temporally regulating degradation of specific protein targets during oocyte-to-embryo transition (Figure 17.4). Accordingly, MBK-2 localization and activity are highly dynamic during the early steps of embryonic development (Pellettieri *et al.*, 2003). During meiosis, MBK-2 is anchored to the cortex, thus precluding phosphorylation of MEI-1 and consequent degradation. MBK-2 is tethered at the cortex by the protein EGG-3, which upon completion of meiosis is ubiquitinated by the APC/C E3 ligase and targeted for degradation, releasing MBK-2 (Stitzel, Cheng and Seydoux, 2007). In the absence of EGG-3, degradation of MEI-1 occurs prematurely, indicating that EGG-3 is required for the timely release of MBK-2 from the cortex, enabling MEI-1 phosphorylation and degradation (see above).

17.9 Degradation of germline proteins in *Drosophila*

The establishment of soma–germline asymmetry downstream of the PAR proteins seems to be conserved, particularly in *Drosophila*, and, like in *C. elegans*, involves protein degradation by the UPS. The Oskar protein is an essential determinant of germline fate. It is provided maternally in the oocyte in the form of mRNA, is locally translated and accumulates in the posterior of the embryo (Rongo, Gavis and Lehmann, 1995; Kim-Ha, Kerr and Macdonald, 1995). Oskar is phosphorylated in a Par1 (homologue of *C. elegans* PAR-1)-dependent manner *in vivo*, and Par1 phosphorylates Oskar *in vitro* (Riechmann *et al.*, 2002). Importantly, Par1 localizes at the posterior of the embryo, and phosphorylation of Oskar protects it from degradation. Upon Par1 loss of function, Oskar levels are reduced in the posterior of the embryo, whereas Oskar protein is stabilized in embryonic extracts upon Par1 overexpression, revealing that Par1 phosphorylation protects Oskar from degradation (Riechmann *et al.*, 2002). It remains to be determined if, and which, E3 ligases are involved in Oskar degradation (reviewed in DeRenzo and Seydoux, 2004).

Oskar recruits Vasa, another protein essential for germline fate, to the posterior of the embryo (Hay, Jan and Jan, 1990). Vasa stability also seems to be locally regulated, similarly to Oskar, and evidence suggests the involvement of the UPS. Deubiquitinating enzyme Fat facets protects Vasa from degradation, as upon *fat facets* loss of function Vasa is polyubiquitinated and protein levels are reduced (Liu, Dansereau and Lasko, 2003). In addition, Fat facets and Vasa co-immunoprecipitate *in vivo*. Further supporting the involvement of the UPS, the SOCS-box protein Gustavus interacts with Vasa. However, while one would expect a SOCS-box protein to act as substrate adaptor and promote ubiquitination, Gustavus promotes stabilization of Vasa in the posterior of the embryo (Styhler, Nakamura and Lasko, 2002). Paradoxically, Gustavus seems to cooperate with Fat facets to localize Vasa and protect it from ubiquitination and degradation (reviewed in DeRenzo and Seydoux, 2004). This apparent paradox could be explained if Gustavus is actually working as a kind of dominant-negative – thus preventing Vasa degradation, rather than promoting it – and not as a bona fide substrate adaptor. It is worth pointing out that Gustavus has not been shown to be part of any CRL, despite possessing a SOCS-box. Gustavus could also be working indirectly, on the degradation of an unknown target, in which case its interaction with Vasa would not be relevant.

17.10 SCF$^{\beta\text{-TRCP}}$-mediated CPEB degradation promotes progression through meiosis in *Xenopus*

Translational control of maternal mRNAs is another critical level of regulation during early development. mRNA-binding proteins – which usually recognize 3′UTR (untranslated region) elements and influence the recruitment of the small ribosomal subunit to the 5′ cap – play a central role in these mechanisms. One of the best-studied mRNA-binding proteins is CPEB (cytoplasmic polyadenylation element binding protein), which recognizes cytoplasmic polyadenylation elements (CPEs) on certain mRNA's 3′ UTR. CPEB is essential for oocyte maturation in *Xenopus*, where it has a dual function since it represses cap-dependent translation in the oocyte but activates translation, via cytoplasmic polyadenylation in meiotically maturing eggs (Hake and Richter, 1994; Stebbins-Boaz *et al.*, 1999; reviewed in Mendez and Richter, 2001). During *Xenopus* oocyte maturation, different types of mRNA are regulated by CPEB, and mRNA selection largely depends on changes in the CPEB/CPE ratio. Upon its phosphorylation on residue Ser174, CPEB promotes translation of dormant mRNAs, such as that encoding the MAP kinase, Mos, promoting the release from prophase I arrest (Mendez *et al.*, 2000a, 2000b). However, during meiosis I, a large fraction of CPEB is actively degraded, dramatically changing the CPEB/CPE ratio (Reverte, Ahearn and Hake, 2001). Consequently, translation of another class of maternal mRNAs is activated, including Erp1, cyclin A1, B1 and B2, which are required for entry into meiosis II (Stebbins-Boaz, Hake and Richter, 1996). Modulation of the CPEB/CPE ratio is thus achieved through selective and timely degradation of CPEB, which involves the $SCF^{\beta\text{-TRCP}}$ E3 ligase. β-TRCP recognizes the degron $T_{190}SGFSS_{195}$ triply phosphorylated on Thr190, Ser191 and Ser195, triggering CPEB ubiquitination and degradation by the UPS (Setoyama, Yamashita and Sagata, 2007). It is noteworthy that the role of CPEB in oocyte maturation is highly conserved, and CPEB degradation at this stage has been observed in several other vertebrates (Thom *et al.*, 2003; Lapasset *et al.*, 2005; Uzbekova *et al.*, 2008). Therefore, it is tempting to speculate that degradation of CPEB might be a conserved mechanism to regulate the timely translation of maternal mRNAs.

17.11 Degradation of maternal mRNAs

In addition to regulation at the level of the RNA-binding proteins, maternally provided mRNAs may be themselves modified. These mRNAs can be kept dormant, translated when activation is triggered, or degraded. By default most maternal mRNAs are dormant until translation activation takes place. Activation usually starts with polyadenlyation, that is, the addition of multiple adenyl nucleotides to the 3′ end of the mRNA, tagging it for translation (reviewed in Seydoux, 1996). However, some maternal mRNAs are degraded in a temporally and spatially regulated manner to ensure proper patterns of protein expression. In *C. elegans* it has been shown that a subset of maternal mRNAs persists in the germline, suggesting specific mechanisms for maternal mRNA degradation in somatic cells (Seydoux and Fire, 1994). Additionally, in the early embryo a number of maternal mRNAs are essential, particularly for cell fate patterning, and have been shown to be differentially regulated across the embryo as a result of the early asymmetries. This is the case of the transcription factors SKN-1 and PAL-1, expressed only in posterior blastomeres starting at the four-cell stage, and GLP-1, the Notch homologue transmembrane receptor, detected only in anterior blastomeres (Bowerman *et al.*, 1993; Hunter and Kenyon, 1996; Evans *et al.*, 1994). Asymmetric expression of SKN-1, PAL-1 and GLP-1 requires the function of the RNA-binding proteins SPN-4, MEX-3, GLD-1 and POS-1 (Gomes *et al.*, 2001; Huang *et al.*, 2002; Ogura *et al.*, 2003; Marin and Evans, 2003; reviewed in Evans and Hunter, 2005). These proteins act mostly as translation repressors, and in the absence of SPN-4, mRNAs encoding SKN-1 and PAL-1 are translated in all cells, at the four-cell stage, while GLP-1 is lost (Gomes *et al.*, 2001; Huang *et al.*, 2002; Ogura *et al.*, 2003). However, in the absence of both POS-1 and SPN-4, GLP-1 protein is detected in all cells, suggesting that SPN-4 is not required for translation, and acts rather as a repressor of POS-1 (Ogura *et al.*, 2003). Similarly, MEX-3 and SPN-4 repress the translation of PAL-1 in anterior cells (Gomes *et al.*, 2001; Huang *et al.*, 2002), and both POS-1 and GLD-1 repress the translation of GLP-1 in posterior cells (Ogura *et al.*, 2003; Marin and Evans, 2003). In all these cases, RNA-binding proteins bind to the 3′UTRs of the target mRNAs and repress translation (Huang *et al.*, 2002; Ogura *et al.*, 2003; Marin and Evans, 2003). This suggests that the asymmetric translation of maternal mRNAs coding for cell fate determinants SKN-1, PAL-1 and GLP-1 is achieved by repression of translation, possibly coupled to destabilization and degradation of the mRNA in specific cells.

Nonetheless, the specific mechanisms of mRNA destabilization remain to be investigated.

In *Drosophila*, more comprehensive analysis of maternally provided mRNAs, using genomic approaches, has been performed. The amount of mRNAs provided by the mother to the oocyte has been estimated to be as high as 50% of the protein-coding genes in the fly's genome (De Renzis *et al.*, 2007; Tadros *et al.*, 2007a; Tadros, Westwood and Lipshitz, 2007b). The early embryo proceeds through 13 rounds of cell division in a common cytoplasm, producing ~5000 nuclei, before zygotic transcription begins. During this phase some mRNAs are actively transported or locally degraded, in some parts of the embryo, as part of cell fate patterning. After cellularization, zygotic transcription starts, around the time of gastrulation, coinciding with massive degradation of maternal mRNAs. Based on the few examples studied in detail, this massive degradation, like most examples of regulated mRNA destabilization, predominantly starts by the removal of the poly(A) tail, followed by decapping and finally 3′-to-5′ exonucleolytic degradation (reviewed in Semotok and Lipshitz, 2007). SMG RNA-binding protein plays a major role in the degradation of maternal mRNAs. SMG was initially found as a repressor of *nanos* mRNA through binding to an SMG responsive element (SRE) on its 3′ UTR (Dahanukar, Walker and Wharton, 1999; Smibert *et al.*, 1999). Later, SMG was found to target Hsp80 mRNA for degradation (Semotok *et al.*, 2005). SMG was found to recruit CCR4/POP2/NOT deadenylase complex in an SRE-independent manner. More recently, through microarray profiling, SMG was shown to be involved in degradation of the majority of maternal mRNAs that become unstable upon zygotic activation (Tadros *et al.*, 2007a). Furthermore, unstable mRNAs are enriched in SREs, supporting the role of SMG as a main regulator of mRNA stability.

MicroRNAs (miRNAs) are also emerging as important players in degradation of maternal mRNAs. In *Drosophila*, a cluster comprising eight miRNAs, the miR-309 cluster, was shown to destabilize a large number of maternal mRNAs. Again using microarray profiling, 410 maternal mRNAs were shown to be upregulated in the absence of the miR-309 cluster (Bushati *et al.*, 2008). Interestingly, these miRNAs are expressed zygotically, and their temporal expression correlates well with destabilization of the putative target mRNAs. A similar finding, suggesting that the role of miRNA in degradation of maternal mRNAs is conserved, was reported in zebrafish: miRNA miR-430 is expressed zygotically to destabilize a large number of maternal mRNAs (Giraldez *et al.*, 2006). Furthermore, the authors show that miRNAs promote the deadenylation of the target mRNAs, establishing the link with the RNA degradation machinery.

17.12 Concluding remarks

At every step of the way in the transformation of the oocyte into an embryo, degradation of specific maternal products, proteins and mRNAs, is instrumental for normal development. The ubiquitin proteolytic system is the main pathway controlling protein degradation, and the E3 ligases are the essential piece in the puzzle with regards to target specificity, ensuring both temporal and spatial regulation of protein degradation. In the last decade, the cullin-RING based E3 ligases, CRLs, have been taking central stage, mostly through the study of the *C. elegans* early embryo. CRL2s regulate meiosis completion, as well as cell fate patterning, SCF (or CRL1) triggers timely onset of zygotic transcription, and CRL3 controls the assembly of the mitotic spindle. Phosphorylation is known to prime proteins for interaction and ubiquitination by E3 ligases, and it seems that timely phosphorylation is the main mechanism for temporal regulation of protein degradation. In the *C. elegans* oocyte-to-embryo transition, the highly conserved MBK-2 kinase has emerged as a core regulator of degradation. MBK-2 phosphorylates and targets for degradation a surprisingly large number of key proteins. In spite of all the knowledge accumulated in recent years, a considerable number of mechanistic details on the function of the UPS in the oocyte-to-embryo transition remain to be elucidated, and more research is required to shed light on them in the years ahead. It also remains to be investigated how much the discoveries made in *C. elegans* hold true in other systems. The components of the UPS are highly conserved, as are most of their targets and other key players, and studies in other model organisms are needed to test whether the mechanisms are conserved.

A considerable amount of work on regulation of maternal mRNAs has also been done, in several model organisms, mainly *Xenopus* and *Drosophila*. This work has highlighted the mechanisms of the degradation of maternal mRNAs, as zygotic transcription is activated. Recent work in zebrafish and *Drosophila* has unravelled the importance of microRNAs in regulating the degradation of maternal mRNAs, and further work should elucidate their mechanisms of action.

In summary, for normal development, and particularly the oocyte-to-embryo transition, there is as much need for the organism to discard old products as to synthesize new ones. This topic has attracted much attention over the last decade, and more discoveries are expected in the years to come.

References

Albertson, D.G. (1984) Formation of the first cleavage spindle in nematode embryos. *Dev. Biol.*, **101**, 61–72.

Angers, S., Li, T., Yi, X. *et al.* (2006) Molecular architecture and assembly of the DDB1-CUL4A ubiquitin ligase machinery. *Nature*, **443**, 590–593.

Blackwell, T.K. and Walker, A.K. (September 5, 2006) Transcription mechanisms, in *WormBook* (ed. The *C. elegans* Research Community). www.wormbook.org. doi: 10.1895/wormbook.1.121.1

Bowerman, B., Draper, B.W., Mello, C.C. and Priess, J.R. (1993) The maternal gene skn-1 encodes a protein that is distributed unequally in early C. elegans embryos. *Cell*, **74**, 443–452.

Bowerman, B., Eaton, B.A. and Priess, J.R. (1992) skn-1, a maternally expressed gene required to specify the fate of ventral blastomeres in the early C. elegans embryo. *Cell*, **68**, 1061–1075.

Boyd, L., Guo, S., Levitan, D. *et al.* (1996) PAR-2 is asymmetrically distributed and promotes association of P granules and PAR-1 with the cortex in C. elegans embryos. *Development*, **122**, 3075–3084.

Bushati, N., Stark, A., Brennecke, J. and Cohen, S.M. (2008) Temporal reciprocity of miRNAs and their targets during the maternal-to-zygotic transition in Drosophila. *Curr. Biol.*, **18**, 501–506.

Castro, A., Bernis, C., Vigneron, S. *et al.* (2005) The anaphase-promoting complex: a key factor in the regulation of cell cycle. *Oncogene*, **24**, 314–325.

Ciechanover, A., Heller, H., Elias, S. *et al.* (1980) ATP-dependent conjugation of reticulocyte proteins with the polypeptide required for protein degradation. *Proc. Natl. Acad. Sci. USA*, **77**, 1365–1368.

Ciechanover, A., Heller, H., Katz-Etzion, R. and Hershko, A. (1981) Activation of the heat-stable polypeptide of the ATP-dependent proteolytic system. *Proc. Natl. Acad. Sci. USA*, **78**, 761–765.

Clark-Maguire, S. and Mains, P.E. (1994a) mei-1, a gene required for meiotic spindle formation in Caenorhabditis elegans, is a member of a family of ATPases. *Genetics*, **136**, 533–546.

Clark-Maguire, S. and Mains, P.E. (1994b) Localization of the mei-1 gene product of Caenorhabditis elegans, a meiotic-specific spindle component. *J. Cell. Biol.*, **126**, 199–209.

Dahanukar, A., Walker, J.A. and Wharton, R.P. (1999) Smaug, a novel RNA-binding protein that operates a translational switch in Drosophila. *Mol. Cell.*, **4**, 209–218.

Davis, E.S., Wille, L., Chestnut, B.A. *et al.* (2002) Multiple subunits of the Caenorhabditis elegans anaphase-promoting complex are required for chromosome segregation during meiosis I. *Genetics*, **160**, 805–813.

De Renzis, S., Elemento, O., Tavazoie, S. and Wieschaus, E.F. (2007) Unmasking activation of the zygotic genome using chromosomal deletions in the Drosophila embryo. *PLoS Biol.*, **5**, e117

DeRenzo, C., Reese, K.J. and Seydoux, G. (2003) Exclusion of germ plasm proteins from somatic lineages by cullin-dependent degradation. *Nature*, **424**, 685–689.

DeRenzo, C. and Seydoux, G. (2004) A clean start: degradation of maternal proteins at the oocyte-to-embryo transition. *Trends Cell. Biol.*, **14**, 420–426.

Dow, M.R. and Mains, P.E. (1998) Genetic and molecular characterization of the Caenorhabditis elegans gene, mel-26, a postmeiotic negative regulator of mei-1, a meiotic-specific spindle component. *Genetics*, **150**, 119–128.

Duda, D.M., Borg, L.A. Scott, D.C. *et al.* (2008) Structural insights into NEDD8 activation of cullin-RING ligases: conformational control of conjugation. *Cell*, **134**, 995–1006.

Edgar, L.G. and McGhee, J.D. (1988) DNA synthesis and the control of embryonic gene expression in C. elegans. *Cell*, **53**, 589–599.

Etemad-Moghadam, B., Guo, S. and Kemphues, K.J. (1995) Asymmetrically distributed PAR-3 protein contributes to cell polarity and spindle alignment in early C. elegans embryos. *Cell*, **83**, 743–752.

Evans, T.C., Crittenden, S.L., Kodoyianni, V. and Kimble, J. (1994) Translational control of maternal *glp-1* mRNA establishes an asymmetry in the C. elegans embryo. *Cell*, **77**, 183–194.

Evans, T.C. and Hunter, C.P. (November 10, 2005) Translational control of maternal RNAs, in *WormBook* (ed. The *C. elegans* Research Community). www.wormbook.org. doi: 10.1895/wormbook.1.34.1

Feng, H., Zhong, W., Punkosdy, G. *et al.* (1999) CUL-2 is required for the G1-to-S-phase transition and mitotic chromosome condensation in *Caenorhabditis elegans*. *Nat. Cell. Biol.*, **1**, 486–492.

Furukawa, M., He, Y.J., Borchers, C. and Xiong, Y. (2003) Targeting of protein ubiquitination by BTB-Cullin 3-Roc1 ubiquitin ligases. *Nat. Cell. Biol.*, **5**, 1001–1007.

Furuta, T., Tuck, S., Kirchner, J. *et al.* (2000) EMB-30: an APC4 homologue required for metaphase-to-anaphase transitions during meiosis and mitosis in Caenorhabditis elegans. *Mol. Biol. Cell.*, **11**, 1401–1419.

Giraldez, A.J., Mishima, Y., Rihel, J. *et al.* (2006) Zebrafish MiR-430 promotes deadenylation and clearance of maternal mRNAs. *Science*, **312**, 75–79.

Golden, A., Sadler, P.L., Wallenfang, M.R. *et al.* (2000) Metaphase to anaphase (mat) transition-defective mutants in Caenorhabditis elegans. *J. Cell. Biol.*, **151**, 1469–1482.

Goldstein, B. and Hird, S.N. (1996) Specification of the anteroposterior axis in Caenorhabditis elegans. *Development*, **122**, 1467–1474.

Gomes, J.E., Encalada, S.E., Swan, K.A. *et al.* (2001) The maternal gene spn-4 encodes a predicted RRM protein required for mitotic spindle orientation and cell fate patterning in early C. elegans embryos. *Development*, **128**, 4301–4314.

Gonczy, P. (2008) Mechanisms of asymmetric cell division: flies and worms pave the way. *Nat. Rev. Mol. Cell. Biol.*, **9**, 355–366.

Gonczy, P. and Rose, L.S. (October 15, 2005) Asymmetric cell division and axis formation in the embryo, in *WormBook* (ed. The *C. elegans* Research Community). www.wormbook.org. doi: 10.1895/wormbook.1.30.1

Greenstein, D. (December 28, 2005) Control of oocyte meiotic maturation and fertilization, in *WormBook* (ed. The *C. elegans* Research Community). www.wormbook.org. doi: 10.1895/wormbook.1.53.1

Grill, S.W., Gonczy, P., Stelzer, E.H. and Hyman, A.A. (2001) Polarity controls forces governing asymmetric spindle positioning in the Caenorhabditis elegans embryo. *Nature*, **409**, 630–633.

Guedes, S. and Priess, J.R. (1997) The C. elegans MEX-1 protein is present in germline blastomeres and is a P granule component. *Development*, **124**, 731–739.

Guo, S. and Kemphues, K.J. (1995) par-1, a gene required for establishing polarity in C. elegans embryos, encodes a putative Ser/Thr kinase that is asymmetrically distributed. *Cell*, **81**, 611–620.

Guven-Ozkan, T., Nishi, Y., Robertson, S.M. and Lin, R. (2008) Global transcriptional repression in C. elegans germline precursors by regulated sequestration of TAF-4. *Cell*, **135**, 149–160.

Haas, A.L., Warms, J.V., Hershko, A. and Rose, I.A. (1982) Ubiquitin-activating enzyme. Mechanism and role in protein-ubiquitin conjugation. *J. Biol. Chem.*, **257**, 2543–2548.

Hake, L.E. and Richter, J.D. (1994) CPEB is a specificity factor that mediates cytoplasmic polyadenylation during Xenopus oocyte maturation. *Cell*, **79**, 617–627.

Hartman, J.J., Mahr, J., McNally, K. *et al.* (1998) Katanin, a microtubule-severing protein, is a novel AAA ATPase that targets to the centrosome using a WD40-containing subunit. *Cell*, **93**, 277–287.

Hay, B., Jan, L.Y. and Jan, Y.N. (1990) Localization of vasa, a component of Drosophila polar granules, in maternal-effect mutants that alter embryonic anteroposterior polarity. *Development*, **109**, 425–433.

Hershko, A. and Ciechanover, A. (1992) The ubiquitin system for protein degradation. *Annu. Rev. Biochem.*, **61**, 761–807.

Hershko, A., Ciechanover, A., Heller, H. *et al.* (1980) Proposed role of ATP in protein breakdown: conjugation of protein with multiple chains of the polypeptide of ATP-dependent proteolysis. *Proc. Natl. Acad. Sci. USA*, **77**, 1783–1786.

Hershko, A., Heller, H., Elias, S. and Ciechanover, A. (1983) Components of ubiquitin-protein ligase system. Resolution, affinity purification, and role in protein breakdown. *J. Biol. Chem.*, **258**, 8206–8214.

Hirsh, D., Oppenheim, D. and Klass, M. (1976) Development of the reproductive system of Caenorhabditis elegans. *Dev. Biol.*, **49**, 200–219.

Huang, N.N., Mootz, D.E., Walhout, A.J. *et al.* (2002) MEX-3 interacting proteins link cell polarity to asymmetric gene expression in Caenorhabditis elegans. *Development*, **129**, 747–759.

Hung, T.J. and Kemphues, K.J. (1999) PAR-6 is a conserved PDZ domain-containing protein that colocalizes with PAR-3 in Caenorhabditis elegans embryos. *Development*, **126**, 127–135.

Hunter, C.P. and Kenyon, C. (1996) Spatial and temporal controls target pal-1 blastomere-specification activity to a single blastomere lineage in C. elegans embryos. *Cell*, **87**, 217–226.

Johnson, J.L., Lu, C., Raharjo, E. *et al.* (2009) Levels of the ubiquitin ligase substrate adaptor MEL-26 are inversely correlated with MEI-1/katanin microtubule-severing activity during both meiosis and mitosis. *Dev. Biol.*, **330**, 349–357.

Jin, J., Arias, E.E., Chen, J. *et al.* (2006) A family of diverse Cul4-Ddb1-interacting proteins includes Cdt2, which is required for S phase destruction of the replication factor Cdt1. *Mol. Cell.*, **23**, 709–721.

Kemphues, K.J., Priess, J.R., Morton, D.G. and Cheng, N.S. (1988) Identification of genes required for cytoplasmic localization in early C. elegans embryos. *Cell*, **52**, 311–320.

Kim-Ha, J., Kerr, K. and Macdonald, P.M. (1995) Translational regulation of oskar mRNA by bruno, an ovarian RNA-binding protein, is essential. *Cell*, **81**, 403–412.

Kirby, C., Kusch, M. and Kemphues, K. (1990) Mutations in the par genes of Caenorhabditis elegans affect cytoplasmic reorganization during the first cell cycle. *Dev. Biol.*, **142**, 203–215.

Kitagawa, R., Law, E., Tang, L. and Rose, A.M. (2002) The Cdc20 homolog, FZY-1, and its interacting protein, IFY-1, are required for proper chromosome segregation in Caenorhabditis elegans. *Curr. Biol.*, **12**, 2118–2123.

Kurz, T., Pintard, L., Willis, J.H. *et al.* (2002) Cytoskeletal regulation by the Nedd8 ubiquitin-like protein modification pathway. *Science*, **295**, 1294–1298.

Lapasset, L., Pradet-Balade, B., Lozano, J.C. *et al.* (2005) Nuclear envelope breakdown may deliver an inhibitor of protein phosphatase 1 which triggers cyclin B translation in starfish oocytes. *Dev. Biol.*, **285**, 200–210.

van Leuken, R., Clijsters, L. and Wolthuis, R. (2008) To cell cycle, swing the APC/C. *Biochim. Biophys. Acta*, **1786** (1),49–59.

Levitan, D.J., Boyd, L., Mello, C.C. *et al.* (1994) par-2, a gene required for blastomere asymmetry in Caenorhabditis elegans, encodes zinc-finger and ATP-binding motifs. *Proc. Natl. Acad. Sci. USA*, **91**, 6108–6112.

Lin, R. (2003) A gain-of-function mutation in oma-1, a C. elegans gene required for oocyte maturation, results in delayed degradation of maternal proteins and embryonic lethality. *Dev. Biol.*, **258**, 226–239.

Liu, J., Vasudevan, S. and Kipreos, E.T. (2004) CUL-2 and ZYG-11 promote meiotic anaphase II and the proper placement of the anterior-posterior axis in C. elegans. *Development*, **131**, 3513–3525.

Liu, N., Dansereau, D.A. and Lasko, P. (2003) Fat facets interacts with vasa in the Drosophila pole plasm and protects it from degradation. *Curr. Biol.*, **13**, 1905–1909.

Lu, C. and Mains, P.E. (2007) The C. elegans anaphase promoting complex and MBK-2/DYRK kinase act redundantly with CUL-3/MEL-26 ubiquitin ligase to degrade MEI-1 microtubule-severing activity after meiosis. *Dev. Biol.*, **302**, 438–447.

Lu, C., Srayko, M. and Mains, P.E. (2004) The Caenorhabditis elegans microtubule-severing complex MEI-1/MEI-2 katanin interacts differently with two superficially redundant beta-tubulin isotypes. *Mol. Biol. Cell.*, **15**, 142–150.

Marin, V.A. and Evans, T.C. (2003) Translational repression of a C. elegans Notch mRNA by the STAR/KH domain protein GLD-1. *Development*, **130**, 2623–2632.

McNally, K., Audhya, A., Oegema, K. and McNally, F.J. (2006) Katanin controls mitotic and meiotic spindle length. *J. Cell. Biol.*, **175**, 881–891.

Mello, C.C., Schubert, C., Draper, B. *et al.* (1996) The PIE-1 protein and germline specification in C. elegans embryos. *Nature*, **382** (6593),710–712.

Mendez, R., Hake, L.E., Andresson, T. *et al.* (2000a) Phosphorylation of CPE binding factor by Eg2 regulates translation of c-mos mRNA. *Nature*, **404**, 302–307.

Mendez, R., Murthy, K.G., Ryan, K. *et al.* (2000b) Phosphorylation of CPEB by Eg2 mediates the recruitment of CPSF into an active cytoplasmic polyadenylation complex. *Mol. Cell.*, **6**, 1253–1259.

Mendez, R. and Richter, J.D. (2001) Translational control by CPEB: a means to the end. *Nat. Rev. Mol. Cell. Biol.*, **2**, 521–529.

Nishi, Y. and Lin, R. (2005) DYRK2 and GSK-3 phosphorylate and promote the timely degradation of OMA-1, a key regulator of the oocyte-to-embryo transition in C. elegans. *Dev. Biol.*, **288**, 139–149.

Nishi, Y., Rogers, E., Robertson, S.M. and Lin, R. (2008) Polo kinases regulate C. elegans embryonic polarity via binding to DYRK2-primed MEX-5 and MEX-6. *Development*, **135**, 687–697.

Ogura, K., Kishimoto, N., Mitani, S. *et al.* (2003) Translational control of maternal glp-1 mRNA by POS-1 and its interacting protein SPN-4 in Caenorhabditis elegans. *Development*, **130**, 2495–2503.

Ohta, T., Michel, J.J., Schottelius, A.J. and Xiong, Y. (1999) ROC1, a homolog of APC11, represents a family of cullin partners with an associated ubiquitin ligase activity. *Mol. Cell.*, **3**, 535–541.

Page, B.D., Diede, S.J., Tenlen, J.R. and Ferguson, E.L. (2007) EEL-1, a Hect E3 ubiquitin ligase, controls asymmetry and persistence of the SKN-1 transcription factor in the early C. elegans embryo. *Development*, **134**, 2303–2314.

Page, B.D., Guedes, S., Waring, D. and Priess, J.R. (2001) The C. elegans E2F- and DP-related proteins are required for embryonic asymmetry and negatively regulate Ras/MAPK signaling. *Mol. Cell.*, **7**, 451–460.

Pang, K.M., Ishidate, T., Nakamura, K. *et al.* (2004) The minibrain kinase homolog, mbk-2, is required for spindle positioning and asymmetric cell division in early C. elegans embryos. *Dev. Biol.*, **265**, 127–139.

Patton, E.E., Willems, A.R. and Tyers, M. (1998) Combinatorial control in ubiquitin-dependent proteolysis: don't Skp the F-box hypothesis. *Trends Genet.*, **14**, 236–243.

Pellettieri, J., Reinke, V., Kim, S.K. and Seydoux, G. (2003) Coordinate activation of maternal protein degradation during the egg-to-embryo transition in C. elegans. *Dev. Cell.*, **5**, 451–462.

Peters, J.M. (2006) The anaphase promoting complex/cyclosome: a machine designed to destroy. *Nat. Rev. Mol. Cell. Biol.*, **7**, 644–656.

Petroski, M.D. and Deshaies, R.J. (2005) Function and regulation of cullin-RING ubiquitin ligases. *Nat. Rev. Mol. Cell. Biol.*, **6**, 9–20.

Pickart, C.M. and Cohen, R.E. (2004) Proteasomes and their kin: proteases in the machine age. *Nat. Rev. Mol. Cell. Biol.*, **5**, 177–187.

Pickart, C.M. and Rose, I.A. (1985) Functional heterogeneity of ubiquitin carrier proteins. *J. Biol. Chem.*, **260**, 1573–1581.

Pintard, L., Willems, A. and Peter, M. (2004) Cullin-based ubiquitin ligases: Cul3-BTB complexes join the family. *EMBO J.*, **23**, 1681–1687.

Pintard, L., Willis, J.H., Willems, A. *et al.* (2003) The BTB protein MEL-26 is a substrate-specific adaptor of the CUL-3 ubiquitin-ligase. *Nature*, **425**, 311–316.

Quintin, S., Mains, P.E., Zinke, A. and Hyman, A.A. (2003) The mbk-2 kinase is required for inactivation of MEI-1/katanin in the one-cell Caenorhabditis elegans embryo. *EMBO Rep.*, **4**, 1175–1181.

Rappleye, C.A., Tagawa, A., Lyczak, R. *et al.* (2002) The anaphase-promoting complex and separin are required for embryonic anterior-posterior axis formation. *Dev. Cell.*, **2**, 195–206.

Reese, K.J., Dunn, M.A., Waddle, J.A. and Seydoux, G. (2000) Asymmetric segregation of PIE-1 in C. elegans is mediated by two complementary mechanisms that act through separate PIE-1 protein domains. *Mol. Cell.*, **6**, 445–455.

Reverte, C.G., Ahearn, M.D. and Hake, L.E. (2001) CPEB degradation during Xenopus oocyte maturation requires a PEST domain and the 26S proteasome. *Dev. Biol.*, **231**, 447–458.

Riechmann, V., Gutierrez, G.J., Filardo, P. *et al.* (2002) Par-1 regulates stability of the posterior determinant Oskar by phosphorylation. *Nat. Cell. Biol.*, **4**, 337–342.

Rongo, C., Gavis, E.R. and Lehmann, R. (1995) Localization of oskar RNA regulates oskar translation and requires Oskar protein. *Development*, **121**, 2737–2746.

Saha, A. and Deshaies, R.J. (2008) Multimodal activation of the ubiquitin ligase SCF by Nedd8 conjugation. *Mol. Cell.*, **32**, 21–31.

Schier, A.F. (2007) The maternal-zygotic transition: death and birth of RNAs. *Science*, **316**, 406–407.

Schubert, C.M., Lin, R., de Vries, C.J. *et al.* (2000) MEX-5 and MEX-6 function to establish soma/germline asymmetry in early C. elegans embryos. *Mol. Cell.*, **5**, 671–682.

Semotok, J.L., Cooperstock, R.L., Pinder, B.D. *et al.* (2005) Smaug recruits the CCR4/POP2/NOT deadenylase complex to trigger maternal transcript localization in the early Drosophila embryo. *Curr. Biol.*, **15**, 284–294.

Semotok, J.L. and Lipshitz, H.D. (2007) Regulation and function of maternal mRNA destabilization during early Drosophila development. *Differentiation*, **75**, 482–506.

Seol, J.H., Feldman, R.M., Zachariae, W. *et al.* (1999) Cdc53/cullin and the essential Hrt1 RING-H2 subunit of SCF define a ubiquitin ligase module that activates the E2 enzyme Cdc34. *Genes Dev.*, **13**, 1614–1626.

Setoyama, D., Yamashita, M. and Sagata, N. (2007) Mechanism of degradation of CPEB during Xenopus oocyte maturation. *Proc. Natl. Acad. Sci. USA*, **104**, 18001–18006.

Seydoux, G. (1996) Mechanisms of translational control in early development. *Curr. Opin. Genet. Dev.*, **6**, 555–561.

Seydoux, G. and Fire, A. (1994) Soma-germline asymmetry in the distributions of embryonic RNAs in Caenorhabditis elegans. *Development*, **120**, 2823–2834.

Seydoux, G., Mello, C.C., Pettitt, J. *et al.* (1996) Repression of gene expression in the embryonic germ lineage of C. elegans. *Nature*, **382**, 713–716.

Shakes, D.C., Sadler, P.L., Schumacher, J.M. *et al.* (2003) Developmental defects observed in hypomorphic anaphase-promoting complex mutants are linked to cell cycle abnormalities. *Development*, **130**, 1605–1620.

Shirayama, M., Soto, M.C., Ishidate, T. *et al.* (2006) The conserved kinases CDK-1, GSK-3, KIN-19, and MBK-2 promote OMA-1 destruction to regulate the oocyte-to-embryo transition in C. elegans. *Curr. Biol.*, **16**, 47–55.

Skowyra, D., Craig, K.L., Tyers, M. *et al.* (1997) F-box proteins are receptors that recruit phosphorylated substrates to the SCF ubiquitin-ligase complex. *Cell*, **91**, 209–219.

Skowyra, D., Koepp, D.M., Kamura, T. *et al.* (1999) Reconstitution of G1 cyclin ubiquitination with complexes containing SCF^{Grr1} and Rbx1. *Science*, **284**, 662–665.

Smibert, C.A., Lie, Y.S., Shillinglaw, W. *et al.* (1999) Smaug, a novel and conserved protein, contributes to repression of nanos mRNA translation in vitro. *RNA*, **5**, 1535–1547.

Sonneville, R. and Gonczy, P. (2004) *zyg-11* and *cul-2* regulate progression through meiosis II and polarity establishment in *C. elegans. Development*, **131** (15),3527–3543.

Srayko, M., Buster, D.W., Bazirgan, O.A. *et al.* (2000) MEI-1/MEI-2 katanin-like microtubule severing activity is required for Caenorhabditis elegans meiosis. *Genes Dev.*, **14**, 1072–1084.

Srayko, M., O'toole, E.T., Hyman, A.A. and Müller-Reichert, T. (2006) Katanin disrupts the microtubule lattice and increases polymer number in C. elegans meiosis. *Curr. Biol.*, **16**, 1944–1949.

Stebbins-Boaz, B., Cao, Q., de Moor, C.H. *et al.* (1999) Maskin is a CPEB-associated factor that transiently interacts with elF-4E. *Mol. Cell.*, **4**, 1017–1027.

Stebbins-Boaz, B., Hake, L.E. and Richter, J.D. (1996) CPEB controls the cytoplasmic polyadenylation of cyclin, Cdk2 and c-mos mRNAs and is necessary for oocyte maturation in Xenopus. *EMBO J.*, **15**, 2582–2592.

Stitzel, M.L., Cheng, K.C. and Seydoux, G. (2007) Regulation of MBK-2/Dyrk kinase by dynamic cortical anchoring during the oocyte-to-zygote transition. *Curr. Biol.*, **17**, 1545–1554.

Stitzel, M.L., Pellettieri, J. and Seydoux, G. (2006) The C. elegans DYRK kinase MBK-2 marks oocyte proteins for degradation in response to meiotic maturation. *Curr. Biol.*, **16**, 56–62.

Stitzel, M.L. and Seydoux, G. (2007) Regulation of the oocyte-to-zygote transition. *Science*, **316**, 407–408.

Strome, S. and Wood, W.B. (1983) Generation of asymmetry and segregation of germ-line granules in early C. elegans embryos. *Cell*, **35**, 15–25.

Styhler, S., Nakamura, A. and Lasko, P. (2002) VASA localization requires the SPRY-domain and SOCS-box containing protein, GUSTAVUS. *Dev. Cell.*, **3**, 865–876.

Sulston, J.E., Schierenberg, E., White, J.G. and Thomson, J.N. (1983) The embryonic cell lineage of the nematode Caenorhabditis elegans. *Dev. Biol.*, **100**, 64–119.

Tabara, H., Hill, R.J., Mello, C.C. *et al.* (1999) pos-1 encodes a cytoplasmic zinc-finger protein essential for germline specification in C. elegans. *Development*, **126**, 1–11.

Tabuse, Y., Izumi, Y., Piano, F. *et al.* (1998) Atypical protein kinase C cooperates with PAR-3 to establish embryonic polarity in Caenorhabditis elegans. *Development*, **125**, 3607–3614.

Tadros, W., Goldman, A.L., Babak, T. *et al.* (2007a) SMAUG is a major regulator of maternal mRNA destabilization in Drosophila and its translation is activated by the PAN GU kinase. *Dev. Cell.*, **12**, 143–155.

Tadros, W., Westwood, J.T. and Lipshitz, H.D. (2007b) The mother-to-child transition. *Dev. Cell.*, **12**, 847–849.

Tan, P., Fuchs, S.Y., Chen, A. *et al.* (1999) Recruitment of a ROC1-CUL1 ubiquitin ligase by Skp1 and HOS to catalyze the ubiquitination of I kappa B alpha. *Mol. Cell.*, **3**, 527–533.

Tenlen, J.R., Molk, J.N., London, N.,*et al.* (2008) MEX-5 asymmetry in one-cell C. elegans embryos requires PAR-4- and PAR-1-dependent phosphorylation. *Development*, **135** (22), 3665–3675.

Thom, G., Minshall, N., Git, A. *et al.* (2003) Role of cdc2 kinase phosphorylation and conserved N-terminal proteolysis motifs in cytoplasmic polyadenylation-element-binding protein (CPEB) complex dissociation and degradation. *Biochem. J.*, **370**, 91–100.

Thornton, B.R. and Toczyski, D.P. (2006) Precise destruction: an emerging picture of the APC. *Genes Dev.*, **20**, 3069–3078.

Tyers, M. and Jorgensen, P. (2000) Proteolysis and the cell cycle: with this RING I do thee destroy. *Curr. Opin. Genet. Dev.*, **10**, 54–64.

Uzbekova, S., Arlot-Bonnemains, Y., Dupont, J. *et al.* (2008) Spatio-temporal expression patterns of aurora kinases A, B, and C and cytoplasmic polyadenylation-element-binding protein in bovine oocytes during meiotic maturation. *Biol. Reprod.*, **78**, 218–233.

Vasudevan, S., Starostina, N.G. and Kipreos, E.T. (2007) The Caenorhabditis elegans cell-cycle regulator ZYG-11 defines a conserved family of CUL-2 complex components. *EMBO Rep.*, **8**, 279–286.

Walker, A.K., Rothman, J.H., Shi, Y. and Blackwell, T.K. (2001) Distinct requirements for C.elegans TAF(II)s in early embryonic transcription. *EMBO J.*, **20**, 5269–5279.

Ward, S. and Carrel, J.S. (1979) Fertilization and sperm competition in the nematode Caenorhabditis elegans. *Dev. Biol.*, **73**, 304–321.

Watts, J.L., Etemad-Moghadam, B., Guo, S. *et al.* (1996) par-6, a gene involved in the establishment of asymmetry in early C. elegans embryos, mediates the asymmetric localization of PAR-3. *Development*, **122**, 3133–3140.

Xu, L., Wei, Y., Reboul, J.,*et al.* (2003) BTB proteins are substrate-specific adaptors in an SCF-like modular ubiquitin ligase containing CUL-3. *Nature*, **425**, 316–321.

Yu, H. (2007) Cdc20: a WD40 activator for a cell cycle degradation machine. *Mol. Cell.*, **27**, 3–16.

Yu, H., Peters, J.M., King, R.W. *et al.* (1998) Identification of a cullin homology region in a subunit of the anaphase-promoting complex. *Science*, **279**, 1219–1222.

Zachariae, W., Shevchenko, A., Andrews, P.D. *et al.* (1998) Mass spectrometric analysis of the anaphase-promoting complex from yeast: identification of a subunit related to cullins. *Science*, **279**, 1216–1219.

18

Chromatin remodelling in mammalian oocytes

Rabindranath De La Fuente,[1] Claudia Baumann,[1] Feikun Yang,[1] and Maria M. Viveiros[2]

[1]*Department of Clinical Studies, Center for Animal Transgenesis and Germ Cell Research, New Bolton Center, University of Pennsylvania, School of Veterinary Medicine, Kennett Square, PA 19348, USA*
[2]*Department of Animal Biology, Center for Animal Transgenesis and Germ Cell Research, New Bolton Center, University of Pennsylvania, School of Veterinary Medicine, Kennett Square, PA 19348, USA*

18.1 Introduction

Under the classical definition, the field of epigenetics has been described as the study of chromatin-based mechanisms leading towards stable and heritable changes in gene expression that are brought about without any changes in the underlying DNA sequence of a gene or gene cluster (Jablonka *et al.*, 2002; Mager and Bartolomei, 2005; Goldberg, Allis and Bernstein, 2007). However, an alternative definition has recently been put forward which bears particular relevance to thc study of epigenetics in the germline. This view places less emphasis on the heritability of chromatin changes and includes in the definition a series of critical, albeit transient chromosomal marks that occur during meiosis and/or mitosis. These chromosome marks have been identified in various nuclear processes including DNA repair or changes in chromosome architecture during different stages of the cell cycle. Epigenetics has, therefore, been recently redefined as: 'The structural adaptation of chromosomal regions so as to register, signal or perpetuate altered activity states' (Bird, 2007).

In mammalian cells, epigenetic phenomena include a wide range of fundamental biological processes such as cell differentiation, DNA replication, repair and recombination, X-chromosome inactivation, and genomic imprinting (Jenuwein and Allis, 2001;

Oogenesis: The Universal Process Marie-Hélène Verlhac and Anne Villeneuve

Heard, 2004; Lucchesi, Kelly and Panning, 2005). Notably, recent studies indicate that epigenetic modifications in the mammalian genome are also critical for many aspects of chromosome biology and, as such, play an important role in the maintenance of genomic stability during meiosis (Celeste *et al.*, 2002; Peters *et al.*, 2001; Webster *et al.*, 2005; Bourc'his and Bestor, 2004; De La Fuente *et al.*, 2006). The range of biological mechanisms under epigenetic control is diverse, and, hence, mammalian cells employ several key molecular strategies, including DNA methylation, the coordinated expression of regulatory noncoding or structural RNAs, as well as the modification of chromatin structure through chromatin remodelling and/or histone post-translational modifications. These modifications occur synergistically, in order to elicit the kind of changes in gene expression that are required in response to a wide range of differentiation or environmental stimuli (Hashimshony *et al.*, 2003; Jaenisch and Bird, 2003; Delaval and Feil, 2004; Jeffery and Nakielny, 2004; Goldberg, Allis and Bernstein, 2007).

DNA methylation is one of the most widely studied types of epigenetic modification. In the mammalian genome, DNA methylation occurs at cytosine-phosphate-guanine (CpG) dinucleotides, where such chemical modification is associated with transcriptional repression (Bird and Wolffe, 1999; Chen and Li, 2006). Importantly, recent strategies designed to determine the patterns of DNA methylation throughout the entire genome indicate that most cytosine methylation takes place at intergenic or nonregulatory regions as well as repetitive elements of the mammalian genome. These results are consistent with a prominent role for DNA methylation in epigenetic gene silencing and maintenance of genome stability in somatic cells (Weber and Schubeler, 2007; Rollins, Haghighi and Edwards, 2006). Cytosine methylation patterns are maintained during cell division by the action of the maintenance DNA methyltransferase (DNMT-1), whereas the establishment of *de novo* methylation patterns is under the control of DNMT3a and DNMt3b (Bird and Wolffe, 1999; Chen and Li, 2006). During gametogenesis, establishment of parental-specific DNA methylation patterns by the action of DNMT3a/DNMT3b as well as DNMT3L proteins confers the mammalian genome with a sex-specific mark or genomic imprinting that is essential for embryonic development (Barton, Surani and Norris, 1984; Surani, Barton and Norris, 1984; McGrath and Solter, 1984; Obata *et al.*, 1998; Bourc'his *et al.*, 2001; Kaneda *et al.*, 2004). Differences in the patterns of DNA methylation between the paternal and maternal genomes are critical to regulate allele-specific gene expression, and thus constitute the basis of genomic imprinting in mammals (Bestor and Bourc'his, 2004).

To date, approximately 80 genes have been identified as imprinted; for a complete list see: www.mousebook.org/catalog.php?catalog=imprinting. Importantly, a growing body of evidence indicates that imprinted genes have a direct involvement in placental growth and differentiation, regulation of foetal growth, postnatal development and maternal behaviour in mice. Furthermore, abnormal imprinting in human patients is known to be associated with several syndromes of foetal overgrowth, gestational abnormalities such as the formation of hydatidiform moles, and several types of cancer (Tilghman, 1999; Moore, 2001; Kelly and Trasler, 2004). The primary signal responsible for the establishment of differential methylation patterns during gametogenesis is not known. However, most imprinted genes identified to date in the mouse seem to be differentially methylated on a locus-by-locus basis at different times during oogenesis and/or spermatogenesis (Ferguson-Smith and Surani, 2001; Reik, Dean and Walter, 2001; Obata and Kono, 2002; Lucifero *et al.*, 2004; Hiura *et al.*, 2006; Bourc'his and Bestor, 2006).

Notably, structural or noncoding RNAs have recently emerged as key regulators of chromatin structure through a direct role in the formation of centromeric heterochromatin domains, transcriptional silencing of intragenomic parasites such as transposons and other repetitive elements, as well as X chromosome inactivation in several model organisms including the mouse (Matzke, Matzke and Kooter, 2001; Volpe *et al.*, 2002; Fukagawa *et al.*, 2004). For example, elegant experiments using a conditional loss-of-function mutation for the ribonuclease protein Dicer in mammalian cells demonstrate that noncoding RNAs, known to be involved in the RNA interference (RNAi) pathway, also have a critical role in the epigenetic regulation of heterochromatin function through the induction of specific histone modifications (Fukagawa *et al.*, 2004; Kanellopoulou *et al.*, 2005). The picture emerging so far is that of a significant crosstalk between the different epigenetic pathways identified up to now. For example, noncoding RNAs such as Dicer are known to induce histone methylation in somatic and embryonic stem cells; in turn this post-translational modification may directly or indirectly affect the patterns of DNA methylation by affecting the binding of DNA methyltransferase enzymes. Importantly, DNA methylation also exhibits a functional interaction with histone modifications by affecting chromatin structure through the induction of changes in chromatin remodelling and or the establishment of chromatin modifications in the form of histone post-translational modifications (Fukagawa *et al.*, 2004; Kanellopoulou *et al.*, 2005; Goldberg, Allis and Bernstein, 2007).

18.2 Chromatin remodelling vs. chromatin modifications

Chromatin remodelling has been broadly defined as the differential regulation of chromatin structure and function in response to an environmental or differentiation stimulus (Aalfs and Kingston, 2000). In mammalian cells, chromatin structure and function can be regulated by three primary mechanisms; namely (i) the action of ATP-dependent chromatin remodelling proteins (Varga-Weisz, 2001; Davis and Brackmann, 2003; Fry and Peterson, 2001), (ii) the incorporation of several histone variants such as histone H3.3, H2A.Z and CENP-A (Polo and Almouzni, 2006; Sarma and Reinberg, 2005), and/or (iii) the induction of histone post-translational modifications (Bannister, Schneider and Kouzarides, 2002; Kouzarides, 2007).

ATP-dependent chromatin remodelling proteins comprise a large protein family with a staggering 1300 members, subdivided into 24 subfamilies based on a recent structural characterization (Flaus *et al.*, 2006). These chromatin-remodelling enzymes are capable of altering nucleosome structure and function. The nucleosome is the structural unit of eukaryotic chromosomes, and is comprised of a string of 146 base pairs of DNA wound around a histone octamer containing two molecules each of histones H2A, H2B, H3 and H4 (Langst and Becker, 2001; Tsukiyama, 2002; Luger, 2003). Chromatin remodelling proteins use the energy of ATP hydrolysis to induce a series of noncovalent modifications that interfere with specific histone and DNA interactions, alter the position of nucleosomes on the DNA molecule inducing 'nucleosome sliding' and, thereby, can facilitate or prevent chromatin accessibility to cognate transcription factors. Hence, this family of proteins is involved in multiple cellular processes including chromatin assembly, transcriptional regulation, cell differentiation and carcinogenesis (Smith and Peterson, 2005).

18.2.1 Nuclear architecture

The nucleus of mammalian cells is organized into several functional compartments that are essential for the control of gene expression, chromosome segregation and maintenance of genome stability (Dundr and Misteli, 2001; Dillon and Festenstein, 2002). For example, heterochromatin domains are highly condensed during interphase, actively maintain a transcriptionally repressive environment, and replicate late during the cell cycle (Dillon and Festenstein, 2002). These properties confer heterochromatin domains with an essential role in the control of transcription (Festenstein *et al.*, 1999), modulation of nuclear architecture, and chromosome segregation (Bernard and Allshire, 2002; Bernard *et al.*, 2001). Covalent modifications on core histones modulate the formation of heterochromatin, and several histone modifications may occur simultaneously to determine the 'context' of a functional response to the cellular environment or the transcriptional status (Cleveland, Mao and Sullivan, 2003). Global chromatin modifications are nontargeted changes in histone acetylation or histone methylation taking place on a genome-wide scale or at nuclear domains such as centromeric heterochromatin. Large-scale chromatin remodelling refers to genome-wide changes in nuclear structure (at the chromosomal level).

Depending on their mechanism of association with DNA, histone variants have been classified into replicative and replacement forms (Polo and Almouzni, 2006). In somatic cells, deposition of histone variants shortly after DNA replication or DNA repair during the S phase of the cell cycle is essential for the regulation of nuclear architecture. Histone variants may participate in the basic 'block building' of chromatin by facilitating the assembly of individual nucleosomes or nucleosome arrays, or may even contribute to the formation of entire nuclear domains in the cell. For example, specialized histone H3 variants such as CENPA are critical for the formation of a functional centromere during mitosis (Sarma and Reinberg, 2005; Polo and Almouzni, 2006). Direct histone variant deposition at the nucleosome is regulated by the activity of dedicated histone chaperone molecules such as chromatin assembly factor 1 (CAF-1), which has been recently shown to be essential for heterochromatin formation in mouse embryos and pluripotent embryonic stem cells (Polo and Almouzni, 2006; Houlard *et al.*, 2006). Notably, replacement histone variants are expressed constitutively and can be incorporated into chromatin independently of DNA replication. For instance, germ cell-specific histone variants such as the testis-specific TH2A/B and the linker histone H1t are highly expressed in mouse testis (Govin *et al.*, 2004; Rousseaux *et al.*, 2005; Kimmins and Sassone-Corsi, 2005), and recent studies suggest that TH2B may cooperate with two newly discovered histone variants named H2AL1 and H2AL2 in the formation of pericentric heterochromatin domains in condensing spermatids (Govin *et al.*, 2007). Additional replacement histone variants such as H3.3 participate in remodelling chromatin at the sex body during sex chromosome inactivation in pachytene stage spermatocytes (Van der Heijden *et al.*, 2007). Furthermore, the linker histone H1Foo, an oocyte-specific histone variant, is highly expressed in the nucleus of primordial oocytes, concurrently with follicular activation as well as the initiation of oocyte growth in the mouse ovary (Tanaka *et al.*, 2001; Tanaka *et al.*, 2005). Notably, the histone variant H3.3 has been shown to associate with the male pronucleus after fertilization, where it may be involved in chromatin remodelling and the early onset of global transcription in the male pronucleus (Loppin *et al.*, 2005; Torres-Padilla *et al.*, 2006; Van der Heijden *et al.*, 2005). In contrast, the histone variant macroH2A is found exclusively in the female pronucleus before being displaced from the maternal genome during syngamy (Chang *et al.*, 2005). Therefore, histone variants contribute to the establishment of epigenetic asymmetry between the paternal and maternal genomes during early embryonic development.

Importantly, chromatin structure can also be regulated by histone post-translational modifications such as acetylation (Grunstein, 1997), phosphorylation (Peterson and Laniel, 2004), methylation (Bannister, Schneider and Kouzarides, 2002), deimination (Cuthbert *et al.*, 2004; Wright *et al.*, 2003), ADP ribosylation (Faraone-Mennella, 2005), ubiquitylation (Zhang, 2003) and sumoylation (Gill, 2004). The list of both histone-modifying enzymes as well as the multiple post-translational modifications taking place at specific amino-acid residues (i.e. lysine, arginine, serine, proline), of histones H2A, H2B, H3 and H4 in the somatic cells of several model organisms including the mouse, has been dramatically expanded over the past decade (Kouzarides, 2007). This is partially attributed to the combination of sensitive chromatin immunoprecipitation methods with DNA microarrays and high-throughput sequencing technology for the analysis of genome-wide epigenetic modifications in organisms such as the yeast (*Saccharomyces cerevisiae*) (Vogelauer *et al.*, 2000; Rando, 2007a, 2007b), at whole chromosome regions or throughout the human genome (Brinkman *et al.*, 2006; Barski *et al.*, 2007). These chromatin modifications are for the most part dynamic, provide topological information and can be associated with a transcriptionally permissive or repressive chromatin environment, and hence play a critical role in the control of nuclear architecture as well as gene expression in response to extracellular signalling pathways during development and differentiation (Cheung, Allis and Sassone-Corsi, 2000; Margueron, Trojer and Reinberg, 2005). For example, formation of transcriptionally repressed and highly condensed heterochromatin domains in the mammalian cell nucleus is regulated by a hierarchy of multiple and complex histone post-translational modifications taking place during the cell cycle and at different times in different tissues (Taddei *et al.*, 1999; Briggs and Strahl, 2002; Dillon and Festenstein, 2002; Richards and Elgin, 2002). Covalent histone modifications such as acetylation and phosphorylation can physically alter the chromatin fibre and thus lead to changes in higher-order chromatin structure (Grunstein, 1997; Goldberg, Allis and Bernstein, 2007). Moreover, methylation of histone H3 at lysine 9 (H3K9me) can be observed carrying one or more methylated groups, leading to the formation of mono-, di- or tri-methylated forms that are essential to recruit chromatin-binding proteins such as heterochromatin-protein 1 (HP1) to specific nuclear domains (Rea *et al.*, 2000; Lachner *et al.*, 2001; Schotta *et al.*, 2004).

It is important to emphasize that all of the mechanisms described above may be operating at a 'local' or nucleosomal level by directly inducing modifications in the organization and molecular composition of the nucleosome, the structural unit of eukaryotic chromosomes (Langst and Becker, 2001; Tsukiyama, 2002; Luger, 2003). However, as we gain a better understanding of the multiple kinds and complexity of the myriad potential interactions between post-translational modifications of core histone proteins *in vitro* and *in vivo*, it is becoming increasingly evident that these mechanisms may also be operating on a genome-wide basis in order to contribute to the formation of specific nuclear domains or to regulate large-scale chromatin structure even at the chromosomal level (Robinett *et al.*, 1996; Tumbar, Sudlow and Belmont, 1999; Vazquez, Belmont and Sedat, 2001; Ye *et al.*, 2001; Berger and Felsenfeld, 2001; Cremer, 2001).

18.2.2 Large-scale chromatin remodelling in the mammalian germline

Mammalian meiosis is a specialized type of cell division, whereby a set of homologous chromosomes is extruded into the first polar body followed by the subsequent separation of sister chromatids shortly after fertilization of a metaphase II-arrested mature egg (Eppig *et al.*, 2004; Hassold and Hunt, 2001). As such, meiotic chromosomes exhibit unique structural and functional properties that are essential to coordinate a highly dynamic interaction with the microtubular spindle apparatus during chromosome segregation (Petronczki, Siomos and Nasmyth, 2003; Page and Hawley, 2003). Importantly, a growing body of evidence indicates that differentiation of chromatin structure and function during oogenesis is essential to confer the mature egg with meiotic and developmental competence (De La Fuente, 2006; De La Fuente *et al.*, 2004a, 2004b). The evidence obtained in several model organisms, including the mouse, seems to indicate that diverse and crucial mechanisms including primordial germ cell determination, and regulation of global transcriptional activity, as well as the unique functional properties of meiotic chromosomes, are under the control of germline-specific chromatin modifications (Sassone-Corsi, 2002; Kimmins and Sassone-Corsi, 2005).

18.3 Epigenetic mechanisms in primordial germ cell (PGC) formation

In all sexually reproducing organisms, gametes have the enormous task of transmitting an organism's genetic information through subsequent generations. Primordial germ cells (PGCs) are uniquely suited to differentiate into gametes, undergo genetic reprogramming, meiotic recombination and two subsequent chromosomal divisions in order to give rise to mature haploid sperm or eggs. It is well established that genome reprogramming in primordial germ cells is subject to developmentally regulated mechanisms that affect the DNA methylation status of both imprinted and nonimprinted genes in order to erase the methylation marks from the previous generation and to reestablish genomic imprints *de novo* during gametogenesis according to the sex of the offspring (Reik, Dean and Walter, 2001; Morgan *et al.*, 2005). Notably, recent studies demonstrate that epigenetic mechanisms, including extensive changes in chromatin modifications, are also critical from the earliest stages of primordial germ cell differentiation (Ohinata *et al.*, 2005, 2006; Ancelin *et al.*, 2006; Hajkova *et al.*, 2008).

18.3.1 Role of Blimp1 in primordial germ cell (PGC) specification and chromatin modifications in the mammalian germline

In mammals, PGCs originate from the epiblast (embryonic ectoderm). At the time of embryonic gastrulation (day 7.25 *post coitum* (dpc) in the mouse) PGCs are identified by their high alkaline phosphatase content at the posterior end of the primitive streak near the base of the allantois. Germ cell fate is determined at this stage from approximately 45 progenitor cells. Once committed to the germline, PGC precursors undergo dramatic changes in cell cycle regulation and patterns of gene expression (McLaren, 2003). Importantly, the expression of pluripotency-associated factors such as alkaline phosphatase, *OCT-4*, the recently identified fragilis (a member of the interferon inducible gene family), *Smad 1* (a gene involved in signal transduction) as well as *Stella* (a germ cell-specific factor) are significantly increased, while the expression of genes required for somatic cell function is drastically reduced (Saitou, Barton and Surani, 2002).

Blimp 1 (B-lymphocyte maturation-induced protein 1), also known as *Prdm 1*, is a transcriptional repressor molecule of a histone methyltransferase subfamily that has been recently identified as the critical factor regulating primordial germ cell specification and epigenetic reprogramming in the germline (Ohinata *et al.*, 2005; Surani, Hayashi and Hajkova, 2007). This transcriptional regulator is initially detected in a small number of primordial germ cell precursors in mice on day 6.5 of embryonic development (E6.5) (Ohinata *et al.*, 2005). Importantly, recent studies indicate that the role of Blimp1 in primordial germ cell formation is mediated through an epigenetic mechanism that induces the establishment of transcriptionally repressive chromatin modifications in the nucleus of PGCs. More specifically, Blimp1 forms a large protein complex with the histone methyltransferase Prmt5 at around day E7.5 when it is found colocalized with Prmt5 in the nucleus of primordial germ cells, resulting in the subsequent di-methylation of arginine residues on histones H2A and H4 in mouse germ cells by E8.5 (Ancelin *et al.*, 2006). Interestingly, the Blimp1–Prmt5 complex translocates from the nucleus to the cytoplasm on day E10.5, a time at which extensive genome reprogramming occurs in the germline (Hajkova *et al.*, 2002; Ancelin *et al.*, 2006). Recent experiments also demonstrate that targeted deletion of an additional member belonging to the same protein family (Prdm14) results in both male and female sterility due to abnormal germ cell formation and severe defects in the progression of epigenetic reprogramming in mutant germ cells, including abnormal patterns of H3K9me2 and H3K27me3 (Yamaji *et al.*, 2008). Thus, by inducing the transcriptional repression of myriad genes involved in somatic cell differentiation, the expression of Blimp1 (Prdm1) and the structurally related factor Prdm14 in primordial germ cell precursors has been directly linked to the specification, maintenance and epigenetic reprogramming of the mammalian germline (Ohinata *et al.*, 2005; Ancelin *et al.*, 2006; Surani, Hayashi and Hajkova, 2007; Yamaji *et al.*, 2008).

18.3.2 Primordial germ cell migration and epigenetic reprogramming

By day 8.5 *post coitum*, PGCs migrate from their initial location at the base of the allantois through the dorsal portion of the hindgut and finally reach the nascent genital ridges (primitive gonads) by 10.5–11.5 dpc. Expression of germ cell-specific genes such as Vasa and germ cell nuclear antigen (GCNA) increases upon colonization of the embryonic gonad (McLaren, 2003). At this time, PGCs begin the process of epigenetic reprogramming that involves the erasure of DNA methylation patterns for both imprinted and nonimprinted genes, while maintaining some DNA methylation at repetitive elements and centromeric heterochromatin domains in order to prevent the potential reactivation of genomic parasitic elements such as transposons in the germline (Hajkova *et al.*, 2002; Lane *et al.*, 2003; Bestor and Bourc'his, 2004). Genome-wide epigenetic reprogramming is also reflected by changes in global histone modifications as well as the reactivation of the inactive X chromosome in females (Ohinata *et al.*, 2006; Kimmins and Sassone-Corsi, 2005). For example, the loss of DNA methyltransferases in the nucleus of migrating PGCs is followed by a striking reduction in both global DNA methylation (as detected by 5-methylcytosine staining), and histone H3 di-methylation at lysine 9 (H3K9me2) on E8. In contrast, the level of H3K27 methylation, a modification that is associated with transcriptional repression, is increased in the nuclei of PGCs, where it may play a role in suppressing the somatic cell programme for gene expression (Seki *et al.*, 2005; Surani, Hayashi and Hajkova, 2007). Interestingly, the levels of transcriptionally permissive histone modifications such as H3K4me2 and acetylation of histone H3 at lysine 9 have been shown to increase during colonization of the primitive gonad, a process that may be of importance to reset the epigenetic signature during germline development (Seki *et al.*, 2005). Notably, PGCs also undergo striking changes in nuclear architecture as well as the patterns of chromatin modifications that might be essential to restore the totipotential state of PGCs (Surani, Hayashi and Hajkova, 2007; Hajkova *et al.*, 2008). For example, in addition to the extensive global DNA demethylation, PGCs undergo dramatic changes in the type of chromatin configuration, with transient loss of chromocentres or heterochromatin domains as well as erasure of several chromatin modifications, most likely mediated by exchange of histone variants. Importantly, it has been suggested that this process of extensive genome reprogramming may be dependent on signalling molecules from the surrounding somatic cells of the embryonic gonad (Surani, Hayashi and Hajkova, 2007; Hajkova *et al.*, 2008).

The number of PGCs within the genital ridge increases after two or three mitotic divisions, and, by 12.5 dpc, PGCs enter a premeiotic stage (McLaren, 2003) with upregulation of meiosis-specific genes such as the synaptonemal complex protein (SYCP3). In the male genital ridge, meiosis proceeds no further. This meiotic arrest is potentially mediated by secretion of prostaglandin D2 from Sertoli cells, as well as metabolic degradation of retinoic acid by somatic cells (McLaren, 2003; Bowles *et al.*, 2006; Koubova *et al.*, 2006) until after birth. In contrast, female PGCs within the genital ridge, now referred to as oogonia, enter the prophase I stage of meiosis (McLaren, 2003). Following the initiation of meiosis, oogonial precursors are designated as oocytes and proceed through different stages that are recognized cytologically by striking changes in chromosome configuration (i.e. leptotene, zygotene and pachytene stages) before arresting at the diplotene stage at the time of birth. Mouse oocytes enter prophase I synchronously on day 13 of foetal development and reach the pachytene or early diplotene stage by day E17.5 or day E18.

18.4 Epigenetic control of meiosis

Progression of meiosis in mammalian oocytes requires a complex and dynamic interaction between homologous chromosomes of maternal and paternal origin. Although the mechanisms controlling homologous chromosome search and synapsis are not fully understood, a growing body of evidence suggests that chromatin remodelling in the germline is essential to modulate chromosome structure and, hence, for the establishment of proper synapsis between homologous chromosomes (Hayashi, Yoshida and Matsui, 2005; Matsui and Hayashi, 2007; Surani, Hayashi and Hajkova, 2007). Notably, evidence obtained through biochemical analysis as well as several genetic mouse models suggests the existence of histone methyltransferases that might be exclusively involved in the control of germline-specific epigenetic modifications. For example, Meisetz (Meiosis-induced factor containing PR/SET domain and zinc finger motif) is the first meiosis-specific histone methyltransferase found in mammals, and targeted deletion studies have demonstrated that Meisetz is required for proper meiotic prophase I progression in both male and female germlines (Hayashi, Yoshida and Matsui, 2005). Meisetz specifically induces the tri-methylation of lysine 4 on histone H3; however the protein motifs responsible for this histone modification have not been identified. Meisetz transcripts are expressed at high levels in the female foetal gonad shortly before meiosis onset, and subsequently decrease at the time when oocytes reach the late pachytene or early diplotene stage on embryonic day E17.5. While homozygous null mice are viable, both males and females are sterile due to abnormal gametogenesis associated with abnormal chromosome synapsis and inefficient repair of double-strand DNA breaks (DSBs) (Hayashi, Yoshida and Matsui, 2005).

In addition to the requirement for Meisetz-dependent tri-methylation of histone H4 during meiosis, establishment of mono- (H3K9me) and di-methylation (H3K9me2) of histone H3 at lysine 9 (Tachibana *et al.*, 2007), as well as tri-methylation (H3K9me3) of histone H3 at lysine 9 (Peters *et al.*, 2001), are essential for the successful completion of meiotic prophase I in the female as well as the male germline. For example, germline-specific deletion of the histone methyltransferase G9a results in both male and female sterility due to dramatic germ cell loss associated with a global reduction in the levels of H3K9me and H3K9me2, as well as meiotic defects including abnormal synapsis of homologous chromosomes and disruption of the mechanism of epigenetic gene silencing, with subsequent abnormal patterns of gene expression.

Interestingly, this study also revealed a sexual dimorphism in the patterns of nuclear localization of H3K9me and H3K9me2, with active removal of both epigenetic marks in male germ cells at the pachytene stage, and their maintenance up to the diplotene stage in female germ cells (Tachibana *et al.*, 2007).

Due to their prominent role in the regulation of centromeric heterochromatin formation, the histone methyltransferases SUV39h1/SUV39h2 are also essential for the control of meiosis (Peters *et al.*, 2001). Targeted deletion of the two isoforms of Suv39h, a histone methyltransferase specifically involved in tri-methylation of histone H3 at lysine 9 (H3K9me3), results in altered DNA methylation of tandem repeats at pericentric heterochromatin in mouse embryonic stem cells, and disrupts chromosome synapsis in meiotic spermatocytes (Peters *et al.*, 2001; Lehnertz *et al.*, 2003; Richards and Elgin, 2002). In this model, most of the evidence obtained so far points towards a role for SUV39h1/SUV39h2 during male meiosis, in which double-knockout mice exhibit severe abnormalities in pericentric heterochromatin formation with lack of H3K9me3 as well as the establishment of abnormal nonhomologous chromosome interactions specifically at centromeric domains. Seemingly, double-null females also revealed chromosomal defects, although the nature of which remains to be characterized (Peters *et al.*, 2001). Importantly, H3K9me3 is critical for heterochromatin formation and the maintenance of a transcriptionally repressive environment due to its function as a docking site for additional chromatin binding proteins, such as heterochromatin protein 1 (HP1) and the chromatin-remodelling factor ATRX (Rea *et al.*, 2000; Lachner *et al.*, 2001; Bannister *et al.*, 2001; Kourmouli *et al.*, 2005; Schotta *et al.*, 2004).

Although the molecular mechanisms involved in abnormal meiotic chromosome synapsis and recombination in the different models discussed above are not fully understood, it is tempting to speculate that abnormal chromatin modifications during prophase I of meiosis might affect the process of homologous chromosome search and/or synapsis and thus result in severe meiotic defects. Collectively, these studies indicate that epigenetic modifications brought about by dedicated histone methyltransferases specific to the germline play a critical role in mammalian prophase I progression. The study by Peters *et al.*, was also the first to demonstrate a direct role for heterochromatin formation in the normal progression of meiosis in mammals (Peters *et al.*, 2001).

18.4.1 Epigenetic control of heterochromatin formation

In mammalian cells, repetitive DNA sequences at centric heterochromatin are necessary, albeit not sufficient, for centromere formation, suggesting an important epigenetic component in the regulation of centromere function (Karpen and Allshire, 1997; Dillon and Festenstein, 2002). Mammalian centromere structure is complex and strikingly dynamic; it requires the binding of the histone H3 variant CENP-A with large protein complexes to induce the formation of a higher order chromatin structure through histone deacetylation and large-scale chromatin remodelling (Pluta *et al.*, 1995; Karpen and Allshire, 1997; Murphy and Karpen, 1998; Henikoff, Ahmad and Malik, 2001; Wiens and Sorger, 1998). In addition, a growing body of evidence indicates that short heterochromatic RNAs (shRNAs) are also a critical epigenetic component in the mechanisms of heterochromatin formation in several model organisms. For example, evidence obtained from *Drosophila* (Akhtar, Zink and Becker, 2000) and, more recently, mammalian cells indicates that shRNAs play a structural role in heterochromatin formation by mediating the binding of large chromatin remodelling complexes to centromeric DNA sequences (Maison *et al.*, 2002; Jenuwein, 2002; Bouzinba-Segard, Guais and Francastel, 2006). Interestingly, an increasing number of RNA functions are now being uncovered that directly link structural RNA molecules with the formation of centromeric chromatin structure and with histone post-translational modifications at specific regions of condensed chromosomes (Rasmussen *et al.*, 2001; Csankovszki, Nagy and Jaenisch, 2001; Bouzinba-Segard, Guais and Francastel, 2006). Consistent with these observations, Dicer-deficient mouse oocytes show a severe meiotic arrest at metaphase I with aberrant meiotic spindles and abnormal turnover of maternal mRNA stores, suggesting that noncoding RNAs are critical for meiotic progression in the female germline (Murchison *et al.*, 2007; Tang *et al.*, 2007).

In the mouse genome, constitutive heterochromatin corresponding to centromeric domains is comprised of two closely related chromosomal subdomains with distinct structure and function (Guenatri *et al.*, 2004; Sullivan and Karpen, 2004). Pericentric heterochromatin is formed by major satellite sequences containing several megabases of a 234 bp repeat, and is marked by large chromatin remodelling complexes comprising DNA binding proteins such as heterochromatin protein 1 (HP1), chromatin assembly factor 1 (CAF-1) and helicases of the switch/sucrose non-fermenting (SWI/SNF2) family such as ATRX and LSH (Maison and Almouzni, 2004; McDowell *et al.*, 1999; De La Fuente *et al.*, 2004a; Yan *et al.*, 2003a). In turn, HP1 exhibits a functional interaction with repressive histone post-translational modifications such as tri-methylation of histone H3 at lysine 9 (H3K9me3) and tri-methylation of histone H4 at lysine 20 (H4K20me3), two of the most stable chromatin marks that constitute a hallmark of pericentric heterochromatin (Peters *et al.*, 2003; Schotta *et al.*, 2004). Alternatively, the centric heterochromatin subdomain is the site of kinetochore formation and is formed by several hundred kilobases of the 120 bp repeat unit of the minor satellite sequence as well as by centromere-specific proteins such as the histone variant CENP-A (Maison and Almouzni, 2004; Karpen and Allshire, 1997). The formation of both centric and pericentric heterochromatin domains is required for the modulation of homologous chromosome interactions during male meiosis in mice (Peters *et al.*, 2001) and for proper chromosome segregation (Bernard and Allshire, 2002; Bernard *et al.*, 2001). Importantly, in mammalian cells, pericentric heterochromatin has been shown to be an active participant in the complex mechanisms regulating centromere cohesion and the timely separation of individual chromatids during mitosis (Guenatri *et al.*, 2004).

18.4.2 Heterochromatin: functional significance

Heterochromatin formation in eukaryotes has important roles in: (i) nuclear architecture; (ii) chromosome segregation; and (iii) gene silencing. A transcriptionally repressive heterochromatin environment is essential to silence tandem repeats, to repress retrotransposons and for X-chromosome inactivation. Heterochromatin is also essential to regulate the growth and differentiation of the mammalian embryo (Houlard *et al.*, 2006). For example, several conditions such as the ATRX and ICF (immunodeficiency, centromere instability and facial anomalies) syndromes in humans are associated with mutations in pericentric heterochromatin proteins involved in DNA methylation of major satellite sequences such as ATRX, and DNA methyltransferase 3b (DNMT3b), respectively. Mutations in DNMT3b result in facial anomalies, centromeric instability and aneuploidy (Xu *et al.*, 1999; Robertson and Wolffe, 2000). Spontaneous mutations in the ATRX gene cause the X-linked α-thalassaemia/mental retardation (ATRX) syndrome in males, which in addition to facial dysmorphism may also exhibit gonadal dysgenesis (Gibbons *et al.*, 2000; Muers *et al.*, 2007).

18.4.3 Heterochromatin formation in the mammalian germline

Chromatin in the female germline exhibits unique structural and functional properties that are essential to coordinate the complex events of meiosis and epigenetic reprogramming during foetal development, with subsequent changes leading towards nuclear and epigenetic maturation during oocyte growth and differentiation. Compelling evidence indicates that large-scale chromatin remodelling during gametogenesis is under the control of germline-specific chromatin modifications (Sassone-Corsi, 2002; Kimmins and Sassone-Corsi, 2005). However, the mechanisms involved in the regulation of heterochromatin formation and its impact in the control of gene expression in the mammalian germline are only beginning to be unravelled. The evidence obtained so far indicates that primordial germ cells are subject to a developmentally regulated mechanism of genome reprogramming that efficiently erases the DNA methylation patterns of both imprinted and nonimprinted or repetitive sequences in order to remove the epigenetic marks from the previous generation and to reestablish genomic imprints *de novo*, at different times during gametogenesis according to the sex of the offspring (Reik, Dean and Walter, 2001; Morgan *et al.*, 2005). Notably, interspersed repeat elements such as retrotransposons of the intracisternal A particle (IAP) class evade global demethylation and hence retain specific methylation patterns after genome reprogramming in the germline (Walsh, Chaillet and Bestor, 1998; Hajkova *et al.*, 2002; Lane *et al.*, 2003; Lees-Murdock, De Felici and Walsh, 2003; Kato *et al.*, 2007). These studies provide strong experimental evidence consistent with the hypothesis that the transmission of transcriptionally repressive chromatin modifications at repetitive elements in the germline constitutes an important mechanism of epigenetic inheritance, which is required to prevent the reactivation of intragenomic parasites in the mammalian germline (Hajkova *et al.*, 2002; Lane *et al.*, 2003; Lees-Murdock, De Felici and Walsh, 2003; Bestor and Bourc'his, 2004).

The mechanism(s) involved in protecting repetitive sequences from undergoing DNA demethylation after global epigenetic reprogramming in germ cells is not known. However, recent studies have demonstrated the existence of sex-specific epigenetic modifications that regulate DNA methylation of IAP elements as well as tandem repeats at centric and pericentric heterochromatin in the germline (Bourc'his *et al.*, 2001; Bourc'his and Bestor, 2004; Kaneda *et al.*, 2004; Webster *et al.*, 2005; De La Fuente *et al.*, 2006). For example, targeted deletion of two members of the DNA methyltransferase protein family (DNMT3a and DNMT3L) revealed a direct role in the establishment of *de novo* methylation patterns at imprinted genes during mouse oogenesis, but had no effect on the methylation of IAP elements or repetitive sequences at centromeric domains in the female germline (Bourc'his *et al.*, 2001; Kaneda *et al.*, 2004; Bestor and Bourc'his, 2004). In contrast, DNMT3L is essential for transcriptional silencing of IAP elements through DNA methylation during spermatogenesis. However, these studies also suggest that DNMT3L has only a minor role in the establishment of DNA methylation patterns on select paternally imprinted genes (Bourc'his and Bestor, 2004; Webster *et al.*, 2005). Therefore, the normal methylation patterns observed in tandem repeats at major and minor satellite sequences in both DNMT3L-deficient female and male germ cells (Bourc'his *et al.*, 2001; Bourc'his and Bestor, 2004; Webster *et al.*, 2005) suggests that methylation of tandem repeats at centromeric heterochromatin is regulated by an independent mechanism compared with the establishment of imprinted genes, and that additional factors might be set in place for the establishment of DNA methylation patterns and transcriptional silencing of major and minor satellite sequences in the germline (Bourc'his *et al.*, 2001; Kaneda *et al.*, 2004).

The lymphoid-specific helicase (LSH), also known as helicase, lymphoid-specific or Hells is a member of the SWI/SNF2 family of helicases with chromatin remodelling activity (Jarvis *et al.*, 1996; Geiman, Durum and Muegge, 1998; Sun *et al.*, 2004; Flaus *et al.*, 2006). The mouse *Lsh* gene has been mapped to region C3-D1 on chromosome 19, and its human homologue to chromosome 10q23-q24; the gene contains an open reading frame encoding 821 amino acids containing an ATPase domain as well as seven helicase domains sharing structural similarity with Rad54, a gene involved in DNA repair and recombination (Geiman, Durum and Muegge, 1998; Meehan, Pennings and Stancheva, 2001). In mouse embryonic fibroblasts, the LSH protein is localized to pericentric heterochromatin, where it is required for DNA methylation of tandem repeats and regulation of histone methylation (Yan *et al.*, 2003a, 2003b). Studies using *Lsh*-null mice demonstrate that LSH is a major epigenetic regulator, which participates in the maintenance of DNA methylation and transcriptional silencing of repeat elements including retroviral long terminal repeats in the mouse genome (Geiman *et al.*, 2001; Muegge, 2005; Dennis *et al.*, 2001; Sun *et al.*, 2004; Yan *et al.*, 2003b; Huang *et al.*, 2004a; Fan *et al.*, 2005).

18.4.4 Mammalian helicases involved in the establishment and/or maintenance of DNA methylation

Two chromatin-remodelling proteins of the SWI/SNF2 family (LSH and ATRX) have been recently implicated in the control of DNA methylation of repetitive sequences in the mammalian genome (Muegge, 2005). The lymphoid-specific helicase (LSH) protein is required for DNA methylation at tandem repeats of centromeric heterochromatin and dispersed retroviral elements in female germ cells, where it also plays a critical role in mediating homologous chromosome synapsis (De La Fuente *et al.*, 2006).

The α-thalassaemia/mental retardation X-linked (ATRX) protein is localized to centromeric heterochromatin, where it plays a critical role in mediating proper chromosome alignment to the meiotic spindle (De La Fuente *et al.*, 2004a). In addition, ATRX is a marker for the inactive X chromosome in somatic cells, and during imprinted X-chromosome inactivation in trophoblast stem cells (Baumann and De La Fuente, 2009).

Importantly, recent studies demonstrate that LSH is also involved in the regulation of chromatin-mediated processes during female meiosis (De La Fuente *et al.*, 2006). Double immunostaining of wild-type oocytes with an antibody against the C-terminal domain of LSH and the synaptonemal complex protein (SYCP3) revealed that LSH exhibits a dynamic nuclear localization during prophase I of meiosis. For example, at the leptotene stage, LSH exhibits a diffuse nuclear localization. However, during the zygotene stage, LSH shows a transient accumulation at pericentric heterochromatin domains. Analysis of meiotic configuration in pachytene-stage oocytes obtained from LSH-null females revealed an essential role in mediating homologous chromosome synapsis. At this stage, the incomplete synapsis observed in mutant oocytes is marked with persistent γH2AX phosphorylation as well as RAD51 foci; chromatin modifications associated with double-strand DNA breaks and meiotic recombination, respectively (Figure 18.1). Furthermore, analysis of homologous chromosome interactions by fluorescence *in situ* hybridization with an X-chromosome DNA probe showed that impaired synapsis was also associated with nonhomologous interactions and the presence of a univalent X chromosome in a high proportion of oocytes. Importantly, analysis of DNA methylation at major and

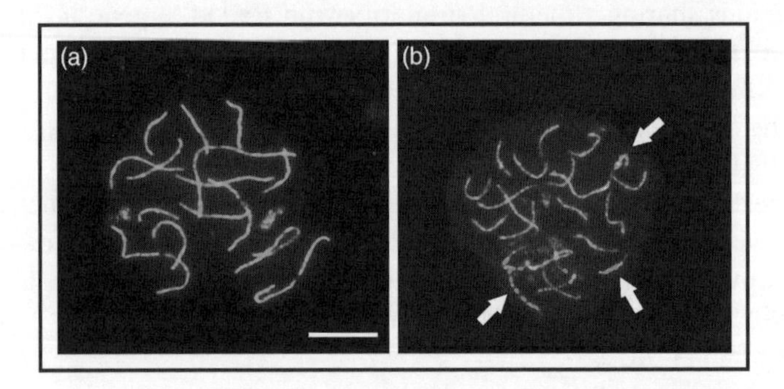

Figure 18.1 Incomplete homologous chromosome synapsis in *Lsh*$^{(-/-)}$ oocytes at the pachytene stage. (a) Control wild-type oocyte at the pachytene stage stained with SYCP3 antibody (green). SYCP3 is a component of the lateral elements of the synaptonemal complex. At this stage, control oocytes exhibit full synapsis of homologous chromosomes as indicated by the presence of 20 bivalents. (b) In contrast, following SYCP3 staining (green), *Lsh*-null oocytes exhibit incomplete homologous chromosome synapsis and persistence of double-strand DNA breaks as indicated by the colocalization of RAD51 foci (red) with asynapsed chromosomes (arrows). Note the absence of RAD51 foci in control wild-type oocytes. These results indicate that chromatin remodelling during prophase I of meiosis is required for proper chromosome synapsis in the female germline. Scale bar = 10 μM. A full colour version of this figure appears in the colour plate section.

minor satellite sequences as well as IAP elements, in oocytes obtained from mutant females, revealed a striking demethylation pattern at such repetitive elements. These results suggest that, through its role in maintaining DNA methylation at tandem repeats of centromeric heterochromatin and transcriptional repression of transposable elements in the germline, LSH might be important to ensure proper synapsis between homologous chromosomes (De La Fuente *et al.*, 2006). These results provided the initial evidence indicating that LSH plays an essential role in the epigenetic silencing of repetitive elements in the female germline. Additional studies previously demonstrated that silencing of transposable elements mediated by DNMT3L is essential for meiotic progression and viability of the male germline (Bourc'his and Bestor, 2004; Webster *et al.*, 2005). However, neither global DNA methylation nor meiotic progression was affected in DNMT3L-knockout oocytes (Bourc'his *et al.*, 2001).

The LSH protein has no methylase activity; however, previous studies suggest that the two mammalian helicases known to be involved in DNA methylation (i.e. LSH and ATRX) might regulate cytosine methylation by inducing changes in chromatin remodelling. These helicases are proposed to promote the recruitment of a protein complex, including DNA methyltransferases, that results in the accumulation of repressive chromatin modifications at centromeric heterochromatin (Meehan, Pennings and Stancheva, 2001; Zhu *et al.*, 2006). For example, transfection of episomal vectors into LSH-deficient mouse embryonic fibroblasts demonstrated that this protein is required for DNA methylation and that LSH functionally interacts with DNMT3a and DNMT3b, most likely as a protein complex, to induce *de novo* methylation of endogenous genes in embryonic stem cells (Zhu *et al.*, 2006). Interestingly, both DNMT3a and DNMT3b localize to pericentric heterochromatin in embryonic stem cells, where they function as transcriptional repressors through a plant homeodomain amino acid motif that is shared with the ATRX protein (Bachman, Rountree and Baylin, 2001; Chen *et al.*, 2003). However, in the mammalian germline, expression and nuclear localization of DNMT3L, DNMT3a and DNMT3b is restricted to embryonic day 17.5 in male germ cells, and shortly after birth in the growing oocyte (La Salle *et al.*, 2004; Lees-Murdock *et al.*, 2005; Sakai *et al.*, 2004). These results suggest that the nature of transcriptionally repressive complexes at pericentric heterochromatin differs between the male and the female germline.

18.4.5 Chromatin modifications and large-scale chromatin remodelling during oocyte growth

Following the completion of meiotic prophase I, mammalian oocytes enter a protracted meiotic arrest, known as the dictyate or diplotene stage, immediately before birth. Meiotic arrest at the diplotene stage is maintained until puberty, when luteinizing hormone (LH) stimulates the resumption of meiosis by triggering the onset of reductional meiotic divisions. Thus, in female mammals, oocytes are maintained in meiotic arrest (diplotene) for a long period of postnatal development, during which significant oocyte and follicle growth take place. Oocyte growth and differentiation take place within the context of a developing ovarian follicle and require a complex bidirectional communication between germ cells and somatic cells of the ovary (Matzuk *et al.*, 2002). Upon activation of primordial follicles, oocytes engage in a prolonged phase of intensive RNA synthesis. Such high levels of transcriptional activity in the oocyte genome are required to ensure that maternal message, ribosomes and cytoplasmic organelles accumulate in growing oocytes. Notably, in order to sustain oocyte growth while accumulating molecules essential for embryogenesis, complex mechanisms coordinate transcriptional activation in growing oocytes with a selective translational repression and accumulation of dormant maternal transcripts (Richter, 2001). The timely synthesis and accumulation of maternal products, such as cell cycle-related molecules, is thus essential for the oocyte's acquisition of meiotic and developmental competence (Evsikov *et al.*, 2006). This is also a critical window during which oocyte-specific epigenetic modifications take place, including the establishment of maternal methylation patterns for several imprinted genes (Morgan *et al.*, 2005; Lucifero *et al.*, 2004; Fedoriw *et al.*, 2004). The primary mechanisms responsible for the establishment and maintenance of maternal imprints are not fully understood; however, the role of several factors including DNMT3L (Bourc'his and Bestor, 2006; Bourc'his *et al.*, 2001) and the KRAB zinc finger protein Zfp57 (Li *et al.*, 2008) are beginning to be unravelled. The mechanisms involved in the establishment of maternal-specific methylation patterns during oogenesis are the subject of intense investigation and have been reviewed elsewhere (Surani, 2001; Surani, Hayashi and Hajkova, 2007; Ferguson-Smith and Surani, 2001; Morgan *et al.*, 2005).

Differentiation of chromatin structure and function during postnatal oocyte growth is also critical for the acquisition of meiotic and developmental competence (Zuccotti *et al.*, 1995; De La Fuente and Eppig, 2001; De La Fuente *et al.*, 2004b; De La Fuente, 2006). However, in contrast with the mechanistic studies conducted after functional ablation of Blimp1 or Prdm5 during primordial germ cell formation (Ohinata *et al.*, 2005; Ohinata *et al.*, 2006), little is known concerning the dynamics of histone and chromatin modifications during postnatal oocyte growth. For example, previous studies have been limited to determining the patterns of expression of several histone post-translational modifications as well as global DNA methylation patterns through the analysis of 5-methylcytosine staining (5-mC) during oocyte growth (Kageyama *et al.*, 2006, 2007). Although a great deal of information has been obtained on the type of histone and chromatin modifications present in the oocyte genome (Table 18.1), little is known concerning the mechanisms involved in the establishment of global epigenetic modifications or the functional significance for the myriad combinations of potential acetylation or methylation changes at specific lysine residues during oogenesis. The patterns that seem to emerge so far indicate that most histone modifications increase with oocyte growth. Importantly, similar to the situation observed in embryonic stem cells, histone modifications associated with transcriptional activity in somatic cells are not necessarily associated with a transcriptionally permissive chromatin template in the germline (Adenot *et al.*, 1997; Kimura *et al.*, 2004; De La Fuente *et al.*, 2004b; Spinaci, Seren and Mattioli, 2004; Kageyama *et al.*, 2007). Thus, alterations in the effects of well-characterized histone modifications in somatic cells might have completely different functional

Table 18.1 Chromatin modifications during meiotic progression in mouse oocytes according to the Brno nomenclature

Histone	Type of histone modification	GV	MI	MII	Reference
Variants	H1foo	+	n.d.	+	Tanaka *et al.*, 2001; Fu *et al.*, 2003
	macroH2A	+	n.d.	+	Chang *et al.*, 2005
	H3.1	n.d.	n.d.	–	Van der Heijden *et al.*, 2005
	H3.3	+	n.d.	–	Torres-Padilla *et al.*, 2006
H4/H2A	H4/H2AS1ph	+	n.d.	+	Sarmento *et al.*, 2004
H3	H3R17me	+	n.d.	–	Sarmento *et al.*, 2004
	H3K4me	+	n.d.	+	Sarmento *et al.*, 2004; Van der Heijden *et al.*, 2005
	H3K4me2	+	+	+	Wang *et al.*, 2006; Van der Heijden *et al.*, 2005; De La Fuente, 2006
	H3K4me3	+	n.d.	+	Kageyama *et al.*, 2007; Van der Heijden *et al.*, 2005
	H3K9me	+	+	+	Arney *et al.*, 2002; Liu, Kim and Aoki, 2004; De La Fuente *et al.*, 2004a
	H3K9me3	+	n.d.	+	Kageyama *et al.*, 2007; Cowell *et al.*, 2002; De La Fuente *et al.*, 2004a
	H3K27me	n.d.	n.d.	+	Van der Heijden *et al.*, 2005
	H3K27me2	n.d.	n.d.	+	Van der Heijden *et al.*, 2005
	H3K27me3	n.d.	n.d.	+	Van der Heijden *et al.*, 2005
	H3K79me2	+	n.d.	+	Ooga *et al.*, 2008
	H3K79me3	+	n.d.	+	Ooga *et al.*, 2008
	H3K9ac	+	–	–	Kim *et al.*, 2003; Wang *et al.*, 2006
	H3K9/K14ac	+	–	–	Meglicki, Zientarski and Borsuk, 2008; Wang *et al.*, 2006
	H3K14ac	n.d.	n.d.	–	Akiyama, Nagata and Aoki, 2006
		–	–	–	Kim *et al.*, 2003
		+	–	–	Meglicki *et al.*, 2008
	H3K18ac	+	n.d.	–	Kageyama *et al.*, 2007; Van der Heijden *et al.*, 2005
	H3K23ac	n.d.	n.d.	+	Van der Heijden *et al.*, 2005
	H3S10ph	–	+	+	Swain *et al.*, 2007
		n.d.	+	n.d.	Hodges and Hunt, 2002
		+	+	+	Wang *et al.*, 2006
	H3S28ph	+	+	+	Swain *et al.*, 2007; Van der Heijden *et al.*, 2005
		–	+	+	Wang *et al.*, 2006
H4	H4R3me	–	n.d.	–	Sarmento *et al.*, 2004
	H4K20me	n.d.	n.d.	+	Van der Heijden *et al.*, 2005
	H4K20me2	n.d.	n.d.	–	Van der Heijden *et al.*, 2005
	H4K20me3	n.d.	n.d.	+	Van der Heijden *et al.*, 2005; Kourmouli *et al.*, 2004

(continued)

Table 18.1 (*Continued*)

Histone	Type of histone modification	GV	MI	MII	Reference
	H4K5ac	+	–	–	Sarmento *et al.*, 2004; Kim *et al.*, 2003; Adenot *et al.*, 1997; De La Fuente *et al.*, 2004b
	H4K8ac	+	+	+	Kim *et al.*, 2003; Huang *et al.*, 2007
		n.d.	–	–	Wang *et al.*, 2006; Akiyama, Nagata and Aoki, 2006
	H4K12ac	+	–	–	Kim *et al.*, 2003; Akiyama *et al.*, 2004
	H4K16ac	+	–	–	Kim *et al.*, 2003; Wang *et al.*, 2006; Akiyama *et al.*, 2006

GV = germinal vesicle; MI = metaphase I; MII = metaphase II.
+ = detectable; – = not detectable; n.d. = not determined.

consequences in the germline. Clearly this is a critical area requiring further investigation.

Beginning on day 16 of postnatal development, the mouse oocyte genome undergoes striking changes in large-scale chromatin structure (at the chromosomal level), in which a decondensed, transcriptionally active nucleus exhibiting a configuration termed non-surrounded nucleolus (NSN), progressively acquires a condensed and transcriptionally inactive configuration called surrounded nucleolus (SN), when chromatin in the germinal vesicle of fully grown, preovulatory oocytes forms a prominent heterochromatic rim in close apposition with the nucleolus (revicwed in De La Fuente, 2006). Chromatin remodelling into the SN configuration, and the concomitant global transcriptional quiescence, occurs in a high proportion (>87%) of oocytes immediately before meiotic resumption. This is a complex process that seems to be modulated, at least in part by paracrine signals, the nature of which remain unknown, from ovarian granulosa cells immediately surrounding the oocyte (De La Fuente and Eppig, 2001; Liu and Aoki, 2002). The underlying mechanisms involved in this critical developmental transition are not fully understood. However, use of a transgenic mouse model deficient for the nuclear chaperone nucleoplasmin 2 (Npm2) provides a unique experimental paradigm to determine the relationship between chromatin remodelling and the transcriptional status of the mammalian oocyte. In this model, the transition into the SN configuration doest not occur; instead chromatin in preovulatory oocytes remains decondensed and at least morphologically resembles the NSN configuration typical of growing wild-type oocytes. However, simultaneous analysis of chromatin configuration and synthesis of nascent transcripts using transcription run-on assays revealed that, upon gonadotrophin stimulation of Npm2 mutant oocytes, nascent transcripts were no longer detectable in the nucleoplasm, indicating that transcriptional repression may occur even in the absence of chromatin remodelling into the SN configuration (De La Fuente *et al.*, 2004b). In addition, pharmacological manipulation of chromatin structure with the histone deacetylase inhibitor TSA revealed that, upon short exposure to TSA, chromatin in the germinal vesicle becomes highly decondensed and as a consequence the structure of the karyosphere is affected. However, exposure to TSA failed to restore transcriptional activity as determined by transcription run-on assays. Thus, these studies revealed for the first time that, although temporally linked in wild-type oocytes, the dramatic changes in genome-wide chromatin remodelling and global transcriptional silencing can be experimentally dissociated and thus might be under the control of distinct cellular pathways (De La Fuente *et al.*, 2004b).

Notably, the recent identification of a histone H2A kinase (NHK-1) required for the control of nuclear architecture in *Drosophila* oocytes (Ivanovska *et al.*, 2005) has provided valuable insight that might contribute to dissecting the potential mechanisms involved in the transition into the SN configuration in mammalian oocytes. For example, although the karyosome, a spherical nuclear structure formed as a result of chromosome coalescence during prophase I arrest in *Drosophila* oocytes is morphologically distinct from the karyosphere observed in mammalian oocytes, there might be intriguing functional similarities (Ivanovska and Orr-Weaver, 2006), in that both structures are required to maintain chromosomes in close proximity and in a particular configuration that is potentially relevant for meiotic progression (Gruzova and Parfenov, 1993). *Drosophila* oocytes obtained from *nhk-1* mutant females fail to form a karyosome, leading to complete sterility due to chromosomal defects including abnormal polar body formation associated with absence of histone H2A phosphorylation, lack of acetylation of histone H3 at lysine 14 (H3K14ac) and acetylation of histone H4 at lysine 5 (H4K5ac), as well as lack of condensin loading into chromosomes (Ivanovska *et al.*, 2005; Ivanovska and Orr-Weaver, 2006). Based on these observations, the hypothesis that condensing might be involved in karyosphere formation in mammalian oocytes has been put forward (Ivanovska and Orr-Weaver, 2006). However, this proposal remains to be formally tested.

18.4.6 Large-scale chromatin remodelling and chromosome segregation in mammalian oocytes

Use of fluorescence *in situ* hybridization with a pan-centromeric DNA probe to detect major satellite sequences revealed that, upon chromatin remodelling into the SN configuration, pericentric heterochromatin domains become associated with, and, hence, are an important component of, the perinucleolar heterochromatin rim or karyosphere (De La Fuente *et al.*, 2004b). Therefore, remodelling chromatin into the SN configuration may confer centromeric domains with a functional configuration essential to recruit heterochromatin-binding proteins such as ATRX that are in turn required to mediate proper chromosome alignment on the meiotic spindle (De La Fuente *et al.*, 2004b). Both the lack of karyosphere formation in Npm2 mutant oocytes, as well as the experimental evidence obtained following pharmacological manipulation of chromatin structure with the histone deacetylase inhibitor trichostatin A (TSA), indicate that the unique nuclear architecture acquired following the transition into the SN configuration is essential for proper chromosome segregation and hence of critical importance to confer the mammalian oocyte with full meiotic and developmental competence (De La Fuente *et al.*, 2004a, 2004b). Consistent with this notion, recent studies have also demonstrated that the lack of developmental potential associated with oocytes that exhibit the NSN configuration is also due to the presence of abnormal patterns of gene expression, including downregulation of maternal stores for the transcription factor OCT-4 and the pluripotency-associated factor Stella, as well as an upregulation of up to 23 genes whose transcriptional activity is affected by the patterns of OCT-4 gene expression in the early preimplantation embryo (Zuccotti *et al.*, 2008).

Although multiple chromatin modifications have been described during the process of meiotic chromosome segregation in mammalian oocytes (Table 18.1), little is known concerning the enzymatic activities that are directly responsible for inducing post-translational modifications at specific lysine residues on histone proteins during the different stages of meiotic maturation. Moreover, our understanding of the critical relationship between chromatin modifications and proper chromosome segregation remains incomplete. The evidence obtained so far indicates that some epigenetic marks such as histone H3/H4 methylation are established during oocyte growth and remain stably associated with either pericentric or interstitial segments of meiotic chromosomes throughout meiosis (Arney *et al.*, 2002; Cowell *et al.*, 2002; Fu *et al.*, 2003; Kourmouli *et al.*, 2004; Liu, Kim and Aoki, 2004; Ooga *et al.*, 2008; Meglicki, Zientarski and Borsuk, 2008; Hodges and Hunt, 2002; Swain *et al.*, 2007; Wang *et al.*, 2006). In contrast, additional chromosomal marks such as acetylation of histone H3/H4 at several lysine residues or the methylation of arginine 3 on histone H4 are highly dynamic and exhibit dramatic changes during the resumption of meiotic cell division (Adenot *et al.*, 1997; Kim *et al.*, 2003; De La Fuente *et al.*, 2004a; Sarmento *et al.*, 2004; Huang *et al.*, 2007).

Perhaps one of the most important concepts emerging from the careful analysis of the information summarized in Table 18.1 is that developmental transitions in chromatin modifications during meiosis are an essential epigenetic mechanism for the maintenance of chromosome stability in the female gamete. Changes in histone modifications may indeed be important for a fast and efficient response to a developmental transition, such as the onset of chromosome condensation upon germinal vesicle breakdown, or to subtle changes in the metabolic state of the oocyte as a response to extracellular signals provided by different hormonal or physical environments. Importantly, it is becoming increasingly clear that the normal progression of meiotic transitions, as well as the establishment of epigenetic marks during gametogenesis, can be adversely influenced by several environmental factors (Susiarjo *et al.*, 2007; Dolinoy *et al.*, 2006; Jirtle and Skinner, 2007). The initial evidence suggesting the presence of genome-wide chromatin modifications during the resumption of meiosis was obtained after analysis of histone acetylation patterns in mouse oocytes (Kim *et al.*, 2003; De La Fuente *et al.*, 2004a). These studies reported the presence of acetylated histones H3 and H4 in the germinal vesicle of preovulatory oocytes, as well as the onset of a wave of global histone deacetylation coincident with germinal vesicle breakdown. Importantly, these modifications included the extensive deacetylation of histone H4 at lysine 12 (H4K12ac), a process that is exclusively found during meiotic chromosome condensation (Kim *et al.*, 2003), as indicated by the persistence of this epigenetic mark in the chromosomes of somatic cells during mitosis (Kruhlak *et al.*, 2001). Similarly, genome-wide histone deacetylation during meiotic resumption was also observed for histone H4 at lysine 5 (H4K5ac) (De La Fuente *et al.*, 2004a), a chromatin modification associated with histone hyperacetylation in somatic cells (Kruhlak *et al.*, 2001). The mechanisms and specific factors responsible for global histone deacetylation during meiosis are still not fully understood. However, several lines of evidence indicate that histone deacetylases play a critical role in this process. For example, exposure of maturing oocytes to roscovitine, an inhibitor of cdc2 kinase, interferes with deacetylation of H4K12, suggesting that histone deacetylases become activated by an increase in the levels of cdc2 kinase following germinal vesicle breakdown (Akiyama *et al.*, 2004). Moreover, pharmacological inhibition of histone deacetylases with TSA prevents the onset of global deacetylation upon meiotic resumption and results in the formation of hyperacetylated chromosomes (Kim *et al.*, 2003; De La Fuente *et al.*, 2004a).

Importantly, studies also demonstrate that global histone deacetylation during meiosis is of functional significance to recruit heterochromatin-binding proteins such as ATRX to centromeric domains, in order to mediate proper chromosome alignment at the meiotic spindle. Following exposure to TSA, hyperacetylated chromosomes no longer exhibit centromeric ATRX foci. Instead, faint ATRX signals are redistributed throughout the length of the chromatids at the metaphase II stage (De La Fuente *et al.*, 2004a). Moreover, analysis of meiotic spindle configuration using laser scanning confocal microscopy indicates that TSA exposure induced the formation of elongated chromosomes and highly abnormal meiotic figures, including misaligned chromosomes and chromosome lagging, in 56–76% of oocytes, depending on the time of exposure to TSA (Figure 18.2; De La Fuente *et al.*, 2004b). Similar experiments

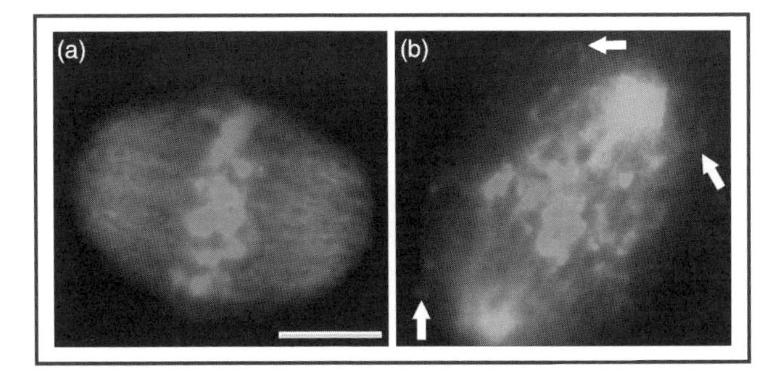

Figure 18.2 Inhibition of histone deacetylases (HDACs) disrupts meiotic progression and induces aberrant chromosome segregation. (a) Meiotic metaphase II spindle in control oocytes showing proper alignment of chromosomes (red) to the equatorial region. β-Tubulin staining (green) confirms the formation of a bipolar spindle. (b) Inhibition of HDACs with trichostatin A (TSA) results in the formation of abnormal meiotic spindles, elongated chromatids and a high incidence of chromosome lagging. Scale bar = 10 μM. A full colour version of this figure appears in the colour plate section.

later confirmed that these abnormal meiotic figures result in severe oocyte aneuploidy and early demise in 50% of embryos derived from TSA-treated oocytes (Akiyama, Nagata and Aoki, 2006)

18.4.7 Role of ATRX in chromosome segregation

The α-thalassaemia/mental retardation X-linked (ATRX) protein is another member of the SWI/SNF2 family of chromatin remodelling proteins. ATRX has been shown to bind pericentric heterochromatin domains in human and mouse somatic cells, and is essential to establish DNA methylation at repetitive sequences of the human genome (McDowell *et al.*, 1999; Gibbons *et al.*, 2000; Picketts *et al.*, 1998; Gibbons *et al.*, 1997). The ATRX gene has been mapped to the long arm of the human X chromosome (Xq13.3), and contains an open reading frame encoding a 280 kDa protein with a plant homeodomain (PHD) region at the amino terminus responsible for interactions with HP1, and a helicase domain at the carboxyl terminal region essential for interaction with the human methyl-CpG binding protein, MeCP2 (Picketts *et al.*, 1998; Bérubé, Smeenk and Picketts, 2000; Nan *et al.*, 2007). The gene also encodes a truncated isoform of approximately 200 kDa lacking the helicase domain (Garrick *et al.*, 2004). Due to the presence of several helicase domains, a stretch of several glutamic acid residues and a coil-coil domain, the ATRX protein has the potential to interact with many protein partners and may indeed acquire different functions by assembling with different protein complexes according to the cell type or stage of the cell cycle (Tang *et al.*, 2004; Ishov, Vladimirova and Maul, 2004).

Previous studies indicate that ATRX binds to pericentric heterochromatin in the chromosomes of mouse oocytes at the metaphase II stage, where it is involved in mediating chromosome alignment at the meiotic spindle (De La Fuente *et al.*, 2004a). The role of ATRX in female meiosis is not fully understood. However, in several organisms including mammals, pericentric heterochromatin formation is required to regulate homologous chromosome interactions during meiosis, and for proper chromosome segregation during meiotic and mitotic cell division (Peters *et al.*, 2001; Bernard and Allshire, 2002; Bernard *et al.*, 2001). Importantly, a growing body of evidence indicates that chromatin remodelling proteins of the SWI/SNF2 family specifically recruited to pericentric heterochromatin are essential to maintain sister chromatid cohesion until the onset of anaphase in order to ensure accurate chromatid segregation during mitosis. For example, initial evidence obtained in the yeast *Saccharomyces cerevisiae* indicated that the HP1 orthologue Swi6 is required for the specific binding of the cohesin subunit Rad21 to centromeric domains (Bernard *et al.*, 2001). Moreover, an ATP-dependent chromatin remodelling complex belonging to the SWI/SNF family, the RSC complex, also plays a critical role in the differential loading of cohesin subunits to chromosome arms in the budding yeast (Bernard and Allshire, 2002; Huang, Hsu and Laurent, 2004b). In mammalian cells, pericentric heterochromatin is essential to coordinate sister centromere cohesion and the timely separation of individual chromatids during mitosis (Guenatri *et al.*, 2004; Maison and Almouzni, 2004). Studies have shown that chromatin-remodelling complexes such as SNF2h are essential to load Rad21 at the centromeres of human mitotic cells (Hakimi *et al.*, 2002). Moreover, loss of HP1 from pericentric heterochromatin in mouse somatic cells deficient for the Suv39 histone methyltransferase protein severely affects centromeric cohesion (Peters *et al.*, 2001; Maison *et al.*, 2002). Collectively, these studies indicate that pericentric heterochromatin formation has a direct impact on centromere cohesion in mitotic cells. However, the potential role of pericentric heterochromatin formation in kinetochore function during female mammalian meiosis remains to be determined.

18.4.8 Prospects and potential for the prevention of aneuploidy

During the past few years we have witnessed the emergence of a key concept in epigenetic regulation, namely the existence of a 'histone code' for the control of nuclear structure and function that attempted to explain chromatin modifications in simple and predictable terms of activation or repression of transcription at single copy genes (Jenuwein and Allis, 2001; Nightingale, O'Neill and Turner, 2006). However, the concept of an underlying 'code' for histone post-translational modifications has rapidly evolved into what seems to be a complex and exquisitely regulated 'language' of hierarchical and combinatorial histone modifications that affect gene expression at multiple levels, ranging from single copy genes to whole chromosomes and even nuclear domains (Nightingale, O'Neill and Turner, 2006; Berger and Felsenfeld, 2001) that affect fundamental biological processes including meiosis. It is also becoming evident that environmental factors such as diet and age have a direct impact on the mammalian epigenome, a process that is of particular importance for the culture of oocytes and embryos of several species. Epigenetic reprogramming in the germline demonstrates that epigenetic modifications including DNA methylation are reversible during meiosis. This experimental paradigm provides tantalizing evidence that may pave the way for the application of 'epigenetic therapy' strategies to reestablish proper genomic imprints in cancer cells, to prevent any potential environmental disruption of imprinting or chromosome stability during assisted reproductive technologies, and perhaps to identify reliable early detection markers for induced epigenetic changes with potential application to prevent aneuploidy in the female gamete.

Acknowledgements

Funding is provided by a grant from the National Institutes of Health RO1-HD 042 740 to R. De La Fuente. Funding for M. Viveiros is provided by a grant from the McCabe Fund and the University of Pennsylvania Research Foundation. Correspondence should be directed to: R. De La Fuente DVM., Ph.D. Department of Clinical Studies, University of Pennsylvania, School of Veterinary Medicine. E-mail: rfuente@vet.upenn.edu.

References

Aalfs, J.D. and Kingston, R.E. (2000) What does 'chromatin remodeling' mean? *Trends Biochem. Sci.*, **25**, 548–555.

Adenot, P.G., Mercier, Y., Renard, J.P. and Thompson, E.M. (1997) Differential H4 acetylation of paternal and maternal chromatin precedes DNA replication and differential transcriptional activity in pronuclei of 1-cell mouse embryos. *Development*, **124**, 4615–4625.

Akhtar, A., Zink, D. and Becker, P.B. (2000) Chromodomains are protein-RNA interaction modules. *Nature*, **407**, 405–409.

Akiyama, T., Kim, J.M., Nagata, M. and Aoki, F. (2004) Regulation of histone acetylation during meiotic maturation in mouse oocytes. *Mol. Reprod. Dev.*, **69**, 222–227.

Akiyama, T., Nagata, M. and Aoki, F. (2006) Inadequate histone deacetylation during oocyte meiosis causes aneuploidy and embryo death in mice. *Proc. Natl. Acad. Sci. USA*, **103**, 7339–7344.

Ancelin, K., Lange, U.C., Hajkova, P. *et al.* (2006) Blimp1 associates with Prmt5 and directs histone arginine methylation in mouse germ cells. *Nat. Cell Biol.*, **8**, 623–630.

Arney, K.L., Bao, S., Bannister, A.J. *et al.* (2002) Histone methylation defines epigenetic asymmetry in the mouse zygote. *Int. J. Dev. Biol.*, **46**, 317–320.

Bachman, K.E., Rountree, M.R. and Baylin, S.B. (2001) DNMT3a and DNMT3b are transcriptional repressors that exhibit unique localization properties to heterochromatin. *J. Biol. Chem.*, **276**, 32282–32287.

Bannister, A.J., Schneider, R. and Kouzarides, T. (2002) Histone methylation: dynamic or static? *Cell*, **109**, 801–806.

Bannister, A.J., Zegerman, P., Partridge, J.F. *et al.* (2001) Selective recognition of methylated lysine 9 on histone H3 by the HP1 chromo domain. *Nature*, **410**, 120–124.

Barski, A., Cuddapah, S., Cui, K. *et al.* (2007) High-resolution profiling of histone methylations in the human genome. *Cell*, **129**, 823–837.

Barton, S.C., Surani, M.A.H. and Norris, M.L. (1984) Role of paternal and maternal genomes in mouse development. *Nature*, **311**, 374–376.

Baumann, C. and De La Fuente, R. (2009) ATRX marks the inactive X chromosome (Xi) in somatic cells and during imprinted X chromosome inactivation in trophoblast stem cells. *Chromosoma*, **118** (2),209–222.

Berger, S. and Felsenfeld, G. (2001) Chromatin goes global. *Mol. Cell*, **8**, 263–268.

Bernard, P. and Allshire, R. (2002) Centromeres become unstuck without heterochromatin. *Trends Cell Biol.*, **12**, 419–424.

Bernard, P., Maure, J., Partridge, J. *et al.* (2001) Requirement of heterochromatin for cohesion at centromeres. *Science*, **294**, 2539–2542.

Bérubé, N., Smeenk, C. and Picketts, D. (2000) Cell cycle-dependent phosphorylation of the ATRX protein correlates with changes in nuclear matrix and chromatin association. *Hum. Mol. Genet.*, **9**, 539–547.

Bestor, T.H. and Bourc'his, D. (2004) Transposon silencing and imprint establishment in mammalian germ cells. *Cold Spring Harb. Symp. Quant. Biol.*, **69**, 381–387.

Bird, A. (2007) Perceptions of epigenetics. *Nature*, **447**, 396–398.

Bird, A. and Wolffe, A.P. (1999) Methylation-induced repression - belts, braces, and chromatin. *Cell*, **99**, 451–454.

Bourc'his, D. and Bestor, T.H. (2004) Meiotic catastrophe and retrotransposon reactivation in male germ cells lacking DNMT3L. *Nature*, **431**, 96–99.

Bourc'his, D. and Bestor, T.H. (2006) Origins of extreme sexual dimorphism in genomic imprinting. *Cytogenet. Genome Res.*, **113**, 36–40.

Bourc'his, D., Xu, G.L., Lin, C.S. *et al.* (2001) DNMT3L and the establishment of maternal genomic imprints. *Science*, **294**, 2536–2539.

Bouzinba-Segard, H., Guais, A. and Francastel, C. (2006) Accumulation of small murine minor satellite transcripts leads to impaired centromeric architecture and function. *Proc. Natl. Acad. Sci. USA*, **103**, 8709–8714.

Bowles, J., Knight, D., Smith, C. *et al.* (2006) Retinoid signaling determines germ cell fate in mice. *Science*, **312**, 596–600.

Briggs, S.D. and Strahl, B.D. (2002) Unraveling heterochromatin. *Nat. Genet.*, **30**, 241–242.

Brinkman, A.B., Roelofsen, T., Pennings, S.W. *et al.* (2006) Histone modification patterns associated with the human X chromosome. *EMBO Rep.*, **7**, 628–634.

Celeste, A., Petersen, S., Romanienko, P.J. *et al.* (2002) Genomic instability in mice lacking histone H2AX. *Science*, **296**, 922–927.

Chang, C.C., Ma, Y., Jacobs, S. *et al.* (2005) A maternal store of macroH2A is removed from pronuclei prior to onset of somatic macroH2A expression in preimplantation embryos. *Dev. Biol.*, **278**, 367–380.

Chen, T. and Li, E. (2006) Establishment and maintenance of DNA methylation patterns in mammals. *Curr. Top. Microbiol. Immunol.*, **301**, 179–201.

Chen, T., Ueda, Y., Dodge, J.E. *et al.* (2003) Establishment and maintenance of genomic methylation patterns in mouse embryonic stem cells by DNMT3a and DNMT3b. *Mol. Cell Biol.*, **23**, 5594–5605.

Cheung, P., Allis, C.D. and Sassone-Corsi, P. (2000) Signaling to chromatin through histone modifications. *Cell*, **103**, 263–271.

Cleveland, D.W., Mao, Y. and Sullivan, K.F. (2003) Centromeres and kinetochores: from epigenetics to mitotic checkpoint signaling. *Cell*, **112**, 407–421.

Cowell, I.G., Aucott, R., Mahadevaiah, S.K. *et al.* (2002) Heterochromatin, HP1 and methylation at lysine 9 of histone H3 in animals. *Chromosoma*, **111**, 22–36.

Cremer, T.C.C. (2001) Chromosome territories, nuclear architecture and gene regulation in mammalian cells. *Nat. Rev. Genet.*, **2**, 292–301.

Csankovszki, G., Nagy, A. and Jaenisch, R. (2001) Synergism of Xist RNA, DNA methylation, and histone hypoacetylation in maintaining X chromosome inactivation. *J. Cell Biol.*, **153**, 773–784.

Cuthbert, G.L., Daujat, S., Snowden, A.W. *et al.* (2004) Histone deimination antagonizes arginine methylation. *Cell*, **118**, 545–553.

Davis, P.K. and Brackmann, R.K. (2003) Chromatin remodeling and cancer. *Cancer Biol. Ther.*, **2**, 22–29.

De La Fuente, R. (2006) Chromatin modifications in the germinal vesicle (GV) of mammalian oocytes. *Dev. Biol.*, **292**, 1–12.

De La Fuente, R., Baumann, C., Fan, T. *et al.* (2006) Lsh is required for meiotic chromosome synapsis and retrotransposon silencing in female germ cells. *Nat. Cell Biol.*, **8**, 1448–1454.

De La Fuente, R. and Eppig, J.J. (2001) Transcriptional activity of the mouse oocyte genome: Companion granulosa cells modulate transcription and chromatin remodeling. *Dev. Biol.*, **229**, 224–236.

De La Fuente, R., Viveiros, M., Wigglesworth, K. and Jj, E. (2004a) ATRX, a member of the SNF2 family of helicase/ATPases, is required for chromosome alignment and meiotic spindle organization in metaphase II stage mouse oocytes. *Dev. Biol.*, **272**, 1–14.

De La Fuente, R., Viveiros, M., Burns, K. *et al.* (2004b) Major chromatin remodeling in the germinal vesicle (GV) of mammalian oocytes is dispensable for global transcriptional silencing but required for centromeric heterochromatin function. *Dev. Biol.*, **275**, 447–458.

Delaval, K. and Feil, R. (2004) Epigenetic regulation of mammalian genomic imprinting. *Curr. Opin. Genet. Dev.*, **14**, 188–195.

Dennis, K., Fan, T., Geiman, T. *et al.* (2001) Lsh, a member of the SNF2 family, is required for genome-wide methylation. *Genes. Dev.*, **15**, 2940–2944.

Dillon, N. and Festenstein, R. (2002) Unravelling heterochromatin: competition between positive and negative factors regulates accessibility. *Trends Genet.*, **18**, 252–258.

Dolinoy, D.C., Weidman, J.R., Waterland, R.A. and Jirtle, R.L. (2006) Maternal genistein alters coat color and protects Avy mouse offspring from obesity by modifying the fetal epigenome. *Environ. Health Perspect.*, **114**, 567–572.

Dundr, M. and Misteli, T. (2001) Functional architecture in the cell nucleus. *Biochem. J.*, **356**, 297–310.

Eppig, J.J., Viveiros, M.M., Marin-Bivens, C. and De La Fuente, R. (2004) Regulation of mammalian oocyte maturation, in *The Ovary* (eds P.C.K. Leung and E.Y. Adashi), Eslevier, Amsterdam, pp. 113–129.

Evsikov, A.V., Graber, J.H., Brockman, J.M. *et al.* (2006) Cracking the egg: molecular dynamics and evolutionary aspects of the transition from the fully grown oocyte to embryo. *Genes Dev.*, **20**, 2713–2727.

Fan, T., Hagan, J.P., Kozlov, S.V. *et al.* (2005) Lsh controls silencing of the imprinted Cdkn1c gene. *Development*, **132**, 635–644.

Faraone-Mennella, M.R. (2005) Chromatin architecture and functions: the role(s) of poly(ADP-RIBOSE) polymerase and poly(ADPribosyl)ation of nuclear proteins. *Biochem. Cell Biol.*, **83**, 396–404.

Fedoriw, A.M., Stein, P., Svoboda, P. *et al.* (2004) Transgenic RNAi reveals essential function for CTCF in H19 gene imprinting. *Science*, **303**, 238–240.

Ferguson-Smith, A.C. and Surani, M.A. (2001) Imprinting and the epigenetic asymmetry between parental genomes. *Science*, **293**, 1086–1089.

Festenstein, R., Sharghi-Namini, S., Fox, M. *et al.* (1999) Heterochromatin protein 1 modifies mammalian PEV in a dose- and chromosomal-context-dependent manner. *Nat. Genet.*, **23**, 457–461.

Flaus, A., Martin, D.M., Barton, G.J. and Owen-Hughes, T. (2006) Identification of multiple distinct Snf2 subfamilies with conserved structural motifs. *Nucleic Acids Res.*, **34**, 2887–2905.

Fry, C.J. and Peterson, C.L. (2001) Chromatin remodeling enzymes: who's on first? *Curr. Biol.*, **11**, R185–R197.

Fu, G., Ghadam, P., Sirotkin, A. *et al.* (2003) Mouse oocytes and early embryos express multiple histone H1 subtypes. *Biol. Reprod.*, **68**, 1569–1576.

Fukagawa, T., Nogami, M., Yoshikawa, M. *et al.* (2004) Dicer is essential for formation of the heterochromatin structure in vertebrate cells. *Nat. Cell Biol.*, **6**, 784–791.

Garrick, D., Samara, V., Mcdowell, T.L. *et al.* (2004) A conserved truncated isoform of the ATR-X syndrome protein lacking the SWI/SNF-homology domain. *Gene*, **326**, 23–34.

Geiman, T., Durum, S. and Muegge, K. (1998) Characterization of gene expression, genomic structure, and chromosomal localization of Hells (Lsh). *Genomics*, **54**, 477–483.

Geiman, T., Tessarollo, L., Anver, M. *et al.* (2001) Lsh, a SNF2 family member, is required for normal murine development. *Biochim. Biophys. Acta*, **1526**, 211–220.

Gibbons, R., Bachoo, S., Picketts, D. *et al.* (1997) Mutations in transcriptional regulator ATRX establish the functional significance of a PHD-like domain. *Nat. Genet.*, **17**, 146–148.

Gibbons, R.J., Mcdowell, T.L., Raman, S. *et al.* (2000) Mutations in ATRX, encoding a SWI/SNF-like protein, cause diverse changes in the pattern of DNA methylation. *Nat. Genet.*, **24**, 368–371.

Gill, G. (2004) SUMO and ubiquitin in the nucleus: different functions, similar mechanisms? *Genes Dev.*, **18**, 2046–2059.

Goldberg, A.D., Allis, C.D. and Bernstein, E. (2007) Epigenetics: a landscape takes shape. *Cell*, **128**, 635–638.

Govin, J., Caron, C., Lestrat, C. *et al.* (2004) The role of histones in chromatin remodelling during mammalian spermiogenesis. *Eur. J. Biochem.*, **271**, 3459–3469.

Govin, J., Escoffier, E., Rousseaux, S. *et al.* (2007) Pericentric heterochromatin reprogramming by new histone variants during mouse spermiogenesis. *J. Cell Biol.*, **176**, 283–294.

Grunstein, M. (1997) Histone acetylation in chromatin structure and transcription. *Nature*, **389**, 349–352.

Gruzova, M.N. and Parfenov, V.N. (1993) Karyosphere in oogenesis and intranuclear morphogenesis. *Int. Rev. Cytol.*, **144**, 1–52.

Guenatri, M., Bailly, D., Maison, C. and Almouzni, G. (2004) Mouse centric and pericentric satellite repeats form distinct functional heterochromatin. *J. Cell Biol.*, **166**, 493–505.

Hajkova, P., Ancelin, K., Waldmann, T. *et al.* (2008) Chromatin dynamics during epigenetic reprogramming in the mouse germ line. *Nature*, **452**, 877–881.

Hajkova, P., Erhardt, S., Lane, N. *et al.* (2002) Epigenetic reprogramming in mouse primordial germ cells. *Mech. Dev.*, **117**, 15–23.

Hakimi, M.A., Bochar, D.A., Schmiesing, J.A. *et al.* (2002) A chromatin remodelling complex that loads cohesin onto human chromosomes. *Nature*, **418**, 994–998.

Hashimshony, T., Zhang, J., Keshet, I. *et al.* (2003) The role of DNA methylation in setting up chromatin structure during development. *Nat. Genet.*, **34**, 187–192.

Hassold, T. and Hunt, P. (2001) To err (meiotically) is human: the genesis of human aneuploidy. *Nat. Rev. Genet.*, **2**, 280–291.

Hayashi, K., Yoshida, K. and Matsui, Y. (2005) A histone H3 methyltransferase controls epigenetic events required for meiotic prophase. *Nature*, **438**, 374–378.

Heard, E. (2004) Recent advances in X-chromosome inactivation. *Curr. Opin. Cell Biol.*, **16**, 247–255.

Henikoff, S., Ahmad, K. and Malik, H.S. (2001) The centromere paradox: stable inheritance with rapidly evolving DNA. *Science*, **293**, 1098–1102.

Hiura, H., Obata, Y., Komiyama, J. *et al.* (2006) Oocyte growth-dependent progression of maternal imprinting in mice. *Genes Cells*, **11**, 353–361.

Hodges, C.A. and Hunt, P. (2002) Simultaneous analysis of chromosomes and chromosome-associated proteins in mammalian oocytes and embryos. *Chromosoma*, **111**, 165–169.

Houlard, M., Berlivet, S., Probst, A.V. *et al.* (2006) CAF-1 is essential for heterochromatin organization in pluripotent embryonic cells. *PLoS Genet.*, **2** (11),e181

Huang, J., Fan, T., Yan, Q. *et al.* (2004a) Lsh, an epigenetic guardian of repetitive elements. *Nucleic Acids Res.*, **32**, 5019–5028.

Huang, J., Hsu, J.M. and Laurent, B.C. (2004b) The RSC nucleosome-remodeling complex is required for Cohesin's association with chromosome arms. *Mol. Cell*, **13**, 739–750.

Huang, J.C., Yan, L.Y., Lei, Z.L. *et al.* (2007) Changes in histone acetylation during postovulatory aging of mouse oocyte. *Biol. Reprod.*, **77**, 666–670.

Ishov, A.M., Vladimirova, O.V. and Maul, G.G. (2004) Heterochromatin and ND10 are cell-cycle regulated and phosphorylation-dependent alternate nuclear sites of the transcription repressor Daxx and SWI/SNF protein ATRX. *J. Cell Sci.*, **117**, 3807–3820.

Ivanovska, I., Khandan, T., Ito, T. and Orr-Weaver, T.L. (2005) A histone code in meiosis: the histone kinase, NHK-1, is required for proper chromosomal architecture in Drosophila oocytes. *Genes Dev.*, **19**, 2571–2582.

Ivanovska, I. and Orr-Weaver, T.L. (2006) Histone modifications and the chromatin scaffold for meiotic chromosome architecture. *Cell Cycle*, **5**, 2064–2071.

Jablonka, E., Matzke, M., Thieffry, D. and Van Speybroeck, L. (2002) The genome in context: biologists and philosophers on epigenetics. *Bioessays*, **24**, 392–394.

Jaenisch, R. and Bird, A. (2003) Epigenetic regulation of gene expression: how the genome integrates intrinsic and environmental signals. *Nat. Genet.*, **33** (Suppl), 245–254.

Jarvis, C., Geiman, T., Vila-Storm, M. *et al.* (1996) A novel putative helicase produced in early murine lymphocytes. *Gene*, **169**, 203–207.

Jeffery, L. and Nakielny, S. (2004) Components of the DNA methylation system of chromatin control are RNA-binding proteins. *J. Biol. Chem.*, **279**, 49479–49487.

Jenuwein, T. (2002) An RNA-guided pathway for the epigenome. *Science*, **297**, 2215–2218.

Jenuwein, T. and Allis, C.D. (2001) Translating the histone code. *Science*, **293**, 1074–1080.

Jirtle, R.L. and Skinner, M.K. (2007) Environmental epigenomics and disease susceptibility. *Nat. Rev. Genet.*, **8**, 253–262.

Kageyama, S., Liu, H., Nagata, M. and Aoki, F. (2006) Stage specific expression of histone deacetylase 4 (HDAC4) during oogenesis and early preimplantation development in mice. *J. Reprod. Dev.*, **52**, 99–106.

Kageyama, S.-I., Liu, H., Kaneko, N. *et al.* (2007) Alterations in epigenetic modifications during oocyte growth in mice. *Reproduction*, **133**, 85–94.

Kaneda, M., Okano, M., Hata, K. *et al.* (2004) Essential role for de novo DNA methyltransferase DNMT3a in paternal and maternal imprinting. *Nature*, **429**, 900–903.

Kanellopoulou, C., Muljo, S.A., Kung, A.L. *et al.* (2005) Dicer-deficient mouse embryonic stem cells are defective in differentiation and centromeric silencing. *Genes Dev.*, **19**, 489–501.

Karpen, G.H. and Allshire, R.C. (1997) The case for epigenetic effects on centromere identity and function. *Trends Genet.*, **13**, 489–496.

Kato, Y., Kaneda, M., Hata, K. *et al.* (2007) Role of the Dnmt3 family in *de novo* methylation of imprinted and repetitive sequences during male germ cell development in the mouse. *Hum. Mol. Genet.*, **16** (19),2272–2280.

Kelly, T.L. and Trasler, J.M. (2004) Reproductive epigenetics. *Clin. Genet.*, **65**, 247–260.

Kim, J., Liu, H., Tazaki, M. *et al.* (2003) Changes in histone acetylation during mouse oocyte meiosis. *J. Cell Biol.*, **162**, 37–46.

Kimmins, S. and Sassone-Corsi, P. (2005) Chromatin remodelling and epigenetic features of germ cells. *Nature*, **434**, 583–589.

Kimura, H., Tada, M., Nakatsuji, N. and Tada, T. (2004) Histone code modifications on pluripotential nuclei of reprogrammed somatic cells. *Mol. Cell Biol.*, **24**, 5710–5720.

Koubova, J., Menke, D.B., Zhou, Q. *et al.* (2006) Retinoic acid regulates sex-specific timing of meiotic initiation in mice. *Proc. Natl. Acad. Sci. USA*, **103**, 2474–2479.

Kourmouli, N., Jeppesen, P., Mahadevhaiah, S. *et al.* (2004) Heterochromatin and tri-methylated lysine 20 of histone H4 in animals. *J. Cell Sci.*, **117**, 2491–2501.

Kourmouli, N., Sun, Y.M., Van Der Sar, S. *et al.* (2005) Epigenetic regulation of mammalian pericentric heterochromatin in vivo by HP1. *Biochem. Biophys. Res. Commun.*, **337**, 901–907.

Kouzarides, T. (2007) Chromatin modifications and their function. *Cell*, **128**, 693–705.

Kruhlak, M.J., Hendzel, M.J., Fischle, W. *et al.* (2001) Regulation of global acetylation in mitosis through loss of histone acetyltransferases and deacetylases from chromatin. *J. Biol. Chem.*, **276**, 38307–38319.

La Salle, S., Mertineit, C., Taketo, T. *et al.* (2004) Windows for sex-specific methylation marked by DNA methyltransferase expression profiles in mouse germ cells. *Dev. Biol.*, **268**, 403–415.

Lachner, M., O'carroll, D., Rea, S. *et al.* (2001) Methylation of histone H3 lysine 9 creates a binding site for HP1 proteins. *Nature*, **410**, 116–120.

Lane, N., Dean, W., Erhardt, S. *et al.* (2003) Resistance of IAPs to methylation reprogramming may provide a mechanism for epigenetic inheritance in the mouse. *Genesis*, **35**, 88–93.

Langst, G. and Becker, P.B. (2001) Nucleosome mobilization and positioning by ISWI-containing chromatin-remodeling factors. *J. Cell Sci.*, **114**, 2561–2568.

Lees-Murdock, D.J., De Felici, M. and Walsh, C.P. (2003) Methylation dynamics of repetitive DNA elements in the mouse germ cell lineage. *Genomics*, **82**, 230–237.

Lees-Murdock, D.J., Shovlin, T.C., Gardiner, T. *et al.* (2005) DNA methyltransferase expression in the mouse germ line during periods of de novo methylation. *Dev. Dyn.*, **232**, 992–1002.

Lehnertz, B., Ueda, Y., Derijck, A.A. *et al.* (2003) Suv39h-mediated histone H3 lysine 9 methylation directs DNA methylation to major satellite repeats at pericentric heterochromatin. *Curr. Biol.*, **13**, 1192–1200.

Li, X., Ito, M., Zhou, F. *et al.* (2008) A maternal-zygotic effect gene, Zfp57, maintains both maternal and paternal imprints. *Dev. Cell*, **15**, 547–557.

Liu, H. and Aoki, F. (2002) Transcriptional activity associated with meiotic competence in fully grown mouse GV oocytes. *Zygote*, **10**, 327–332.

Liu, H., Kim, J. and Aoki, F. (2004) Regulation of histone H3 lysine 9 methylation in oocytes and early pre-implantation embryos. *Development*, **131**, 2269–2280.

Loppin, B., Bonnefoy, E., Anselme, C. *et al.* (2005) The histone H3.3 chaperone HIRA is essential for chromatin assembly in the male pronucleus. *Nature*, **437**, 1386–1390.

Lucchesi, J.C., Kelly, W.G. and Panning, B. (2005) Chromatin remodeling in dosage compensation. *Annu. Rev. Genet.*, **39**, 615–651.

Lucifero, D., Mann, M.R., Bartolomei, M.S. and Trasler, J.M. (2004) Gene-specific timing and epigenetic memory in oocyte imprinting. *Hum. Mol. Genet.*, **13**, 839–849.

Luger, K. (2003) Structure and dynamic behavior of nucleosomes. *Curr. Opin. Genet. Dev.*, **13**, 127–135.

Mager, J. and Bartolomei, M.S. (2005) Strategies for dissecting epigenetic mechanisms in the mouse. *Nat. Genet.*, **37**, 1194–1200.

Maison, C. and Almouzni, G. (2004) HP1 and the dynamics of heterochromatin maintenance. *Nat. Rev. Mol. Cell Biol.*, **5**, 296–304.

Maison, C., Bailly, D., Peters, A.H.F.M. *et al.* (2002) Higher-order structure in pericentric heterochromatin involves a distinct pattern of histone modification and an RNA component. *Nat. Genet.*, **30**, 329–334.

Margueron, R., Trojer, P. and Reinberg, D. (2005) The key to development: interpreting the histone code? *Curr. Opin. Genet. Dev.*, **15**, 163–176.

Matsui, Y. and Hayashi, K. (2007) Epigenetic regulation for the induction of meiosis. *Cell Mol. Life Sci.*, **64**, 257–262.

Matzke, M., Matzke, A.J. and Kooter, J.M. (2001) RNA: guiding gene silencing. *Science*, **293**, 1080–1083.

Matzuk, M.M., Burns, K.H., Viveiros, M.M. and Eppig, J.J. (2002) Intercellular communication in the mammalian ovary: oocytes carry the conversation. *Science*, **296**, 2178–2180.

McDowell, T.L., Gibbons, R.J., Sutherland, H. *et al.* (1999) Localization of a putative transcriptional regulator (ATRX) at pericentromeric heterochromatin and the short arms of acrocentric chromosomes. *Proc. Natl. Acad. Sci. USA*, **96**, 13983–13988.

McGrath, J. and Solter, D. (1984) Completion of mouse embryogenesis requires both the maternal and paternal genomes. *Cell*, **37**, 179–183.

McLaren, A. (2003) Primordial germ cells in the mouse. *Dev. Biol.*, **262**, 1–15.

Meehan, R.R., Pennings, S. and Stancheva, I. (2001) Lashings of DNA methylation, forkfuls of chromatin remodeling. *Genes. Dev.*, **15**, 3231–3236.

Meglicki, M., Zientarski, M. and Borsuk, E. (2008) Constitutive heterochromatin during mouse oogenesis: the pattern of histone H3 modifications and localization of HP1alpha and HP1beta proteins. *Mol. Reprod. Dev.*, **75**, 414–428.

Moore, T. (2001) Genetic conflict, genomic imprinting and establishment of the epigenotype in relation to growth. *Reproduction*, **122**, 185–193.

Morgan, H.D., Santos, F., Green, K. *et al.* (2005) Epigenetic reprogramming in mammals. *Hum. Mol. Genet.*, **14** (Rev. Issue 1), R47–R48.

Muegge, K. (2005) Lsh, a guardian of heterochromatin at repeat elements. *Biochem. Cell Biol.*, **83**, 548–554.

Muers, M.R., Sharpe, J.A., Garrick, D. *et al.* (2007) Defining the cause of skewed X-chromosome inactivation in X-linked mental retardation by use of a mouse model. *Am. J. Hum. Genet.*, **80**, 1138–1149.

Murchison, E.P., Stein, P., Xuan, Z. *et al.* (2007) Critical roles for Dicer in the female germline. *Genes. Dev.*, **21**, 682–693.

Murphy, T.D. and Karpen, G.H. (1998) Centromeres take flight: alpha satellite and the quest for the human centromere. *Cell*, **93**, 317–320.

Nan, X., Hou, J., Maclean, A. *et al.* (2007) Interaction between chromatin proteins MECP2 and ATRX is disrupted by mutations that cause inherited mental retardation. *Proc. Natl. Acad. Sci. USA*, **104**, 2709–2714.

Nightingale, K.P., O'Neill, L.P. and Turner, B.M. (2006) Histone modifications: signalling receptors and potential elements of a heritable epigenetic code. *Curr. Opin. Genet. Dev.*, **16**, 125–136.

Obata, Y., Kaneko-Ishino, T., Koide, T. *et al.* (1998) Disruption of primary imprinting during oocyte growth leads to the modified expression of imprinted genes during embryogenesis. *Development*, **125**, 1553–1560.

Obata, Y. and Kono, K. (2002) Maternal primary imprinting is established at a specific time for each gene throughout oocyte growth. *J. Biol. Chem.*, **277**, 5285–5289.

Ohinata, Y., Payer, B., O'Carroll, D. *et al.* (2005) Blimp1 is a critical determinant of the germ cell lineage in mice. *Nature*, **436**, 207–213.

Ohinata, Y., Seki, Y., Payer, B. *et al.* (2006) Germline recruitment in mice: a genetic program for epigenetic reprogramming. *Ernst Schering Res. Found. Workshop*, **60**, 143–174.

Ooga, M., Inoue, A., Kageyama, S. *et al.* (2008) Changes in H3K79 methylation during preimplantation development in mice. *Biol. Reprod.*, **78**, 413–424.

Page, S. and Hawley, R. (2003) Chromosome choreography: the meiotic ballet. *Science*, **301**, 785–789.

Peters, A.H., Kubicek, S., Mechtler, K. *et al.* (2003) Partitioning and plasticity of repressive histone methylation states in mammalian chromatin. *Mol. Cell*, **12**, 1577–1589.

Peters, A.H., O'Carroll, D., Scherthan, H. *et al.* (2001) Loss of the Suv39h histone methyltransferases impairs mammalian heterochromatin and genome stability. *Cell*, **107**, 323–337.

Peterson, C.L. and Laniel, M.A. (2004) Histones and histone modifications. *Curr. Biol.*, **14**, R546–R551.

Petronczki, M., Siomos, M. and Nasmyth, K. (2003) Un ménage à quatre: the molecular biology of chromosome segregation in meiosis. *Cell*, **112**, 423–440.

Picketts, D.J., Tastan, A.O., Higgs, D.R. and Gibbons, R.J. (1998) Comparison of the human and murine ATRX gene identifies highly conserved, functionally important domains. *Mamm. Genome*, **9**, 400–403.

Pluta, A.F., Mackay, A.M., Ainsztein, A.M. *et al.* (1995) The centromere: hub of chromosomal activities. *Science*, **270**, 1591–1594.

Polo, S.E. and Almouzni, G. (2006) Chromatin assembly: a basic recipe with various flavours. *Curr. Opin. Genet. Dev.*, **16**, 104–111.

Rando, O.J. (2007a) Chromatin structure in the genomics era. *Trends Genet.*, **23**, 67–73.

Rando, O.J. (2007b) Global patterns of histone modifications. *Curr. Opin. Genet. Dev.*, **17**, 94–99.

Rasmussen, T.P., Wutz, A., Pehrson, J.R. and Jaenisch, R. (2001) Expression of Xist RNA is sufficient to initiate machrochromatin body formation. *Chromosoma*, **110**, 411–420.

Rea, S., Eisenhaber, F., O'Carroll, D. *et al.* (2000) Regulation of chromatin structure by site-specific histone H3 methyltransferases. *Nature*, **406**, 593–599.

Reik, W., Dean, W. and Walter, J. (2001) Epigenetic reprogramming in mammalian development. *Science*, **293**, 1089–1093.

Richards, E.J. and Elgin, S.C. (2002) Epigenetic codes for heterochromatin formation and silencing: rounding up the usual suspects. *Cell*, **108**, 489–500.

Richter, J. (2001) Think globally, translate locally: what mitotic spindles and neuronal synapses have in common. *Proc. Natl. Acad. Sci. USA*, **98**, 7069–7071.

Robertson, K.D. and Wolffe, A.P. (2000) DNA methylation in health and disease. *Nat. Rev. Genet.*, **1**, 11–19.

Robinett, C.C., Straight, A., Li, G. *et al.* (1996) In vivo localization of DNA sequences and visualization of large-scale chromatin organization using Lac operator/repressor recognition. *J. Cell Biol.*, **135**, 1685–1700.

Rollins, R.A., Haghighi, F. and Edwards, J.R. (2006) Large-scale structure of genomic methylation patterns. *Genome Res.*, **16**, 157–163.

Rousseaux, S., Caron, C., Govin, J. *et al.* (2005) Establishment of male-specific epigenetic information. *Gene*, **345**, 139–153.

Saitou, M., Barton, S.C. and Surani, M.A. (2002) A molecular programme for the specification of germ cell fate in mice. *Nature*, **418**, 293–300.

Sakai, Y., Suetake, I., Shinozaki, F. *et al.* (2004) Co-expression of de novo DNA methyltransferases DNMT3a2 and DNMT3L in gonocytes of mouse embryos. *Gene. Expr. Patterns*, **5**, 231–237.

Sarma, K. and Reinberg, D. (2005) Histone variants meet their match. *Nat. Rev. Mol. Cell Biol.*, **6**, 139–149.

Sarmento, O.F., Digilio, L.C., Wang, Y. *et al.* (2004) Dynamic alterations of specific histone modifications during early murine development. *J. Cell Sci.*, **117**, 4449–4459.

Sassone-Corsi, P. (2002) Unique chromatin remodeling and transcriptional regulation in spermatogenesis. *Science*, **296**, 2176–2178.

Schotta, G., Lachner, M., Sarma, K. *et al.* (2004) A silencing pathway to induce H3-K9 and H4-K20 trimethylation at constitutive heterochromatin. *Genes Dev.*, **18**, 1251–1262.

Seki, Y., Hayashi, K., Itoh, K. *et al.* (2005) Extensive and orderly reprogramming of genome-wide chromatin modifications associated with specification and early development of germ cells in mice. *Dev. Biol.*, **278**, 440–458.

Smith, C.L. and Peterson, C.L. (2005) ATP-dependent chromatin remodeling. *Curr. Top. Dev. Biol.*, **65**, 115–148.

Spinaci, M., Seren, E. and Mattioli, M. (2004) Maternal chromatin remodeling during maturation and after fertilization in mouse oocytes. *Mol. Reprod. Dev.*, **69**, 215–221.

Sullivan, B.A. and Karpen, G.H. (2004) Centromeric chromatin exhibits a histone modification pattern that is distinct from both euchromatin and heterochromatin. *Nat. Struct. Mol. Biol.*, **11**, 1076–1083.

Sun, L., Lee, D., Zhang, Q. *et al.* (2004) Growth retardation and premature aging phenotypes in mice with disruption of the SNF2-like gene. *PASG Genes. Dev.*, **18**, 1035–1046.

Surani, M. (2001) Reprogramming of genome function through epigenetic inheritance. *Nature*, **414**, 122–128.

Surani, M., Barton, S.C. and Norris, M.L. (1984) Development of reconstituted mouse eggs suggests imprinting of the genome during gametogenesis. *Nature*, **308**, 548–550.

Surani, M.A., Hayashi, K. and Hajkova, P. (2007) Genetic and epigenetic regulators of pluripotency. *Cell*, **128**, 747–762.

Susiarjo, M., Hassold, T.J., Freeman, E. and Hunt, P.A. (2007) Bisphenol A exposure in utero disrupts early oogenesis in the mouse. *PLoS Genet.*, **3**, e5

Swain, J.E., Ding, J., Brautigan, D.L. *et al.* (2007) Proper chromatin condensation and maintenance of histone H3 phosphorylation during mouse oocyte meiosis requires protein phosphatase activity. *Biol. Reprod.*, **76**, 628–638.

Tachibana, M., Nozaki, M., Takeda, N. and Shinkai, Y. (2007) Functional dynamics of H3K9 methylation during meiotic prophase progression. *Embo. J.*, **26**, 3346–3359.

Taddei, A., Roche, D., Sibarita, J. *et al.* (1999) Duplication and maintenance of heterochromatin domains. *J. Cell Biol.*, **147**, 1153–1166.

Tanaka, M., Hennebold, J.D., Macfarlane, J. and Adashi, E.Y. (2001) A mammalian oocyte-specific linker histone gene H1oo: homology with the genes for the oocyte-specific cleavage stage histone (cs-H1) of sea urchin and the B4/H1M histone of the frog. *Development*, **128**, 655–664.

Tanaka, M., Kihara, M., Hennebold, J.D. *et al.* (2005) H1FOO is coupled to the initiation of oocytic growth. *Biol. Reprod.*, **72**, 135–142.

Tang, F., Kaneda, M., O'Carroll, D. *et al.* (2007) Maternal microRNAs are essential for mouse zygotic development. *Genes Dev.*, **21**, 644–648.

Tang, J., Wu, S., Liu, H. *et al.* (2004) A novel transcription regulatory complex containing death domain-associated protein and the ATR-X syndrome protein. *J. Biol. Chem.*, **279**, 20369–20377.

Tilghman, S. (1999) The sins of the fathers and mothers: genomic imprinting in mammalian development. *Cell*, **96**, 185–193.

Torres-Padilla, M.E., Bannister, A.J., Hurd, P.J. *et al.* (2006) Dynamic distribution of the replacement histone variant H3.3 in the mouse oocyte and preimplantation embryos. *Int. J. Dev. Biol.*, **50**, 455–461.

Tsukiyama, T. (2002) The in vivo functions of ATP-dependent chromatin-remodelling factors. *Nat. Rev. Mol. Cell Biol.*, **3**, 422–429.

Tumbar, T., Sudlow, G. and Belmont, A.S. (1999) Large-scale chromatin unfolding and remodeling induced by VP16 acidic activation domain. *J. Cell Biol.*, **145**, 1341–1354.

Van der Heijden, G.W., Derijck, A.A., Posfai, E. *et al.* (2007) Chromosome-wide nucleosome replacement and H3.3 incorporation during mammalian meiotic sex chromosome inactivation. *Nat. Genet.*, **39**, 251–258.

Van der Heijden, G.W., Dieker, J.W., Derijck, A.A. *et al.* (2005) Asymmetry in histone H3 variants and lysine methylation between paternal and maternal chromatin of the early mouse zygote. *Mech. Dev.*, **122**, 1008–1022.

Varga-Weisz, P. (2001) ATP-dependent chromatin remodeling factors: nucleosome shufflers with many missions. *Oncogene*, **20**, 3076–3085.

Vazquez, J., Belmont, A.S. and Sedat, J.W. (2001) Multiple regimes of constrained chromosome motion are regulated in the interphase Drosophila nucleus. *Curr. Biol.*, **11**, 1227–1239.

Vogelauer, M., Wu, J., Suka, N. and Grunstein, M. (2000) Global histone acetylation and deacetylation in yeast. *Nature*, **408**, 495–498.

Volpe, T.A., Kidner, C., Hall, I.M. *et al.* (2002) Regulation of heterochromatic silencing and histone H3 lysine-9 methylation by RNAi. *Science*, **297**, 1833–1837.

Walsh, C.P., Chaillet, J.R. and Bestor, T.H. (1998) Transcription of IAP endogenous retroviruses is constrained by cytosine methylation. *Nat. Genet.*, **20**, 116–117.

Wang, Q., Wang, C.M., Ai, J.S. *et al.* (2006) Histone phosphorylation and pericentromeric histone modifications in oocyte meiosis. *Cell Cycle*, **5**, 1974–1982.

Weber, M. and Schubeler, D. (2007) Genomic patterns of DNA methylation: targets and function of an epigenetic mark. *Curr. Opin. Cell Biol.*, **19**, 273–280.

Webster, K.E., O'Bryan, M.K., Fletcher, S. *et al.* (2005) Meiotic and epigenetic defects in DNMT3L-knockout mouse spermatogenesis. *Proc. Natl. Acad. Sci. USA*, **102**, 4068–4073.

Wiens, G.R. and Sorger, P.K. (1998) Centromeric chromatin and epigenetic effects in kinetochore assembly. *Cell*, **93**, 313–316.

Wright, P.W., Bolling, L.C., Calvert, M.E. *et al.* (2003) ePAD, an oocyte and early embryo-abundant peptidylarginine deiminase-like protein that localizes to egg cytoplasmic sheets. *Dev. Biol.*, **256**, 73–88.

Xu, G.L., Bestor, T.H., Bourc'his, D. *et al.* (1999) Chromosome instability and immunodeficiency syndrome caused by mutations in a DNA methyltransferase gene. *Nature*, **402**, 187–191.

Yamaji, M., Seki, Y., Kurimoto, K. *et al.* (2008) Critical function of Prdm14 for the establishment of the germ cell lineage in mice. *Nat. Genet.*, **40**, 1016–1022.

Yan, Q., Cho, E., Lockett, S. and Muegge, K. (2003a) Association of Lsh, a regulator of DNA methylation, with pericentromeric heterochromatin is dependent on intact heterochromatin. *Mol. Cell Biol.*, **23**, 8416–8428.

Yan, Q., Huang, J., Fan, T. *et al.* (2003b) Lsh, a modulator of CpG methylation, is crucial for normal histone methylation. *EMBO J.*, **22**, 5154–5162.

Ye, Q., Hu, Y.F., Zhong, H. *et al.* (2001) BRCA1-induced large-scale chromatin unfolding and allele-specific effects of cancer-predisposing mutations. *J. Cell. Biol.*, **155**, 911–921.

Zhang, Y. (2003) Transcriptional regulation by histone ubiquitination and deubiquitination. *Genes. Dev.*, **17**, 2733–2740.

Zhu, H., Geiman, T.M., Xi, S. *et al.* (2006) Lsh is involved in de novo methylation of DNA. *EMBO J.*, **25**, 335–345.

Zuccotti, M., Merico, V., Sacchi, L. *et al.* (2008) Maternal OCT-4 is a potential key regulator of the developmental competence of mouse oocytes. *BMC Dev. Biol.*, **8**, 97

Zuccotti, M., Piccinelli, A., Rossi, P.G. *et al.* (1995) Chromatin organization during mouse oocyte growth. *Mol. Reprod. Dev.*, **41** (4),479–485.

19

Follicles and medically assisted reproduction

Susan L. Barrett[1,2,3] **and Teresa K. Woodruff**[1,2,3]

[1]*Department of Obstetrics and Gynecology, Feinberg School of Medicine Northwestern University, Chicago, IL 60611, USA*
[2]*Center for Reproductive Science, Northwestern University, Evanston, IL 60208, USA*
[3]*The Oncofertility Consortium, Northwestern University, Chicago, IL 60611, USA*

19.1 Introduction

Oncofertility is a new and exciting field of medicine incorporating reproductive biology, reproductive endocrinology and oncology, in hopes of giving patients with a diagnosis of cancer the best possible options for preserving their fertility. Because current cancer regimens run the risk of infertility following treatment, many challenges facing fertility preservation stem from diagnosis and the state of the ovarian tissue to the development of fertility preservation techniques.

There have been several prior chapters discussing the present understanding of follicle growth and development. Though scientists have a grasp on the underlying genetics and cell signalling pathways that regulate follicle growth, we still face challenges in adapting this knowledge for fertility preservation. Every follicle stage has unique properties and must be treated differently for fertility preservation. For example, primordial and primary follicles have shown to be difficult to grow in culture, but can be easily cryopreserved in thin cortical tissue strips for ovarian tissue transplant (Figure 19.1a; Kagawa *et al.*, 2007; Silber *et al.*, 2008), whereas early, secondary follicles are difficult to cryopreserve in cortical strips, however are capable of growing in culture (Figure 19.1b; Kreeger *et al.*, 2005; Telfer *et al.*, 2008; Tempone, 2008;

Oogenesis: The Universal Process Marie-Hélène Verlhac and Anne Villeneuve

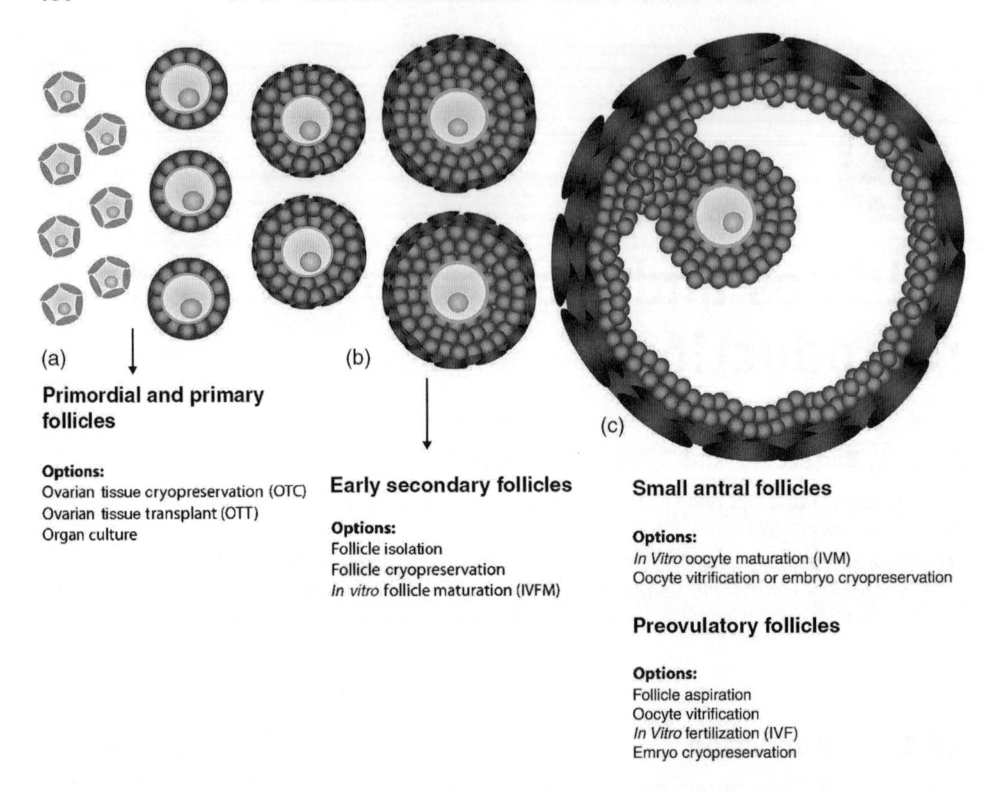

Figure 19.1 Fertility preservation options according to follicle stage. Key: granulosa cells (green); oocyte (tan); theca cells (purple). A full colour version of this figure appears in the colour plate section.

Xu *et al.*, 2006). Antral follicles do not survive the freezing process and are too large for proper nutrient exchange in culture, though, depending on the follicle, oocytes can be retrieved and stored for later use (Figure 19.1c). Though many of these techniques are experimental, we hope to both continue the discovery process as well as expand the options for young women with a fertility-threatening medical diagnosis such as cancer or autoimmune disease. Here, we review treatments and the effects on the female reproductive system as well as current and experimental options in fertility preservation.

19.2 Cancer and reproduction

One-third of women diagnosed with breast cancer are premenopausal. Although the percentages of women surviving breast cancer have increased over the years due to improved chemotherapy and more aggressive use of radiation, these life-preserving treatments can leave women infertile. Depending on the type of cancer diagnosis, the treatments may result in loss of fertility, either temporarily or permanently. These sections will discuss anticancer agents and their effects on the ovarian follicular reserve.

19.2.1 Chemotherapeutics

Several factors may affect the extent of follicle loss related to chemotherapeutics. This can range from the disruption of growing follicles or reduction of the follicular reserve, to its complete elimination, which results in premature ovarian failure (POF). The type of chemotherapeutic agent, the dose, as well as the age at treatment can be excellent predictors of the risk of POF (Agarwal and Chang, 2007; Georgescu *et al.*, 2008; Tempone, 2008).

It is well documented that alkylating agents such as cyclophosphamide and ifosfamide are toxic to germ cells (Gracia and Ginsberg, 2007; Oktay and Sonmezer, 2008). They target cancer by damaging the tumour cells' DNA, directing them to undergo apoptosis. These agents are commonly used for treating cancers such as leukaemia and lymphoma, as well as sarcomas. Despite their efficacy relative to cancer control, alkylating agents are extremely toxic to follicles – particularly follicles that have been recruited into the growing pool. Importantly, the effect of alkylating agents on the gonads depends on a variety of parameters including patient age (the younger the patient, the bigger the ovarian reserve). The five-year survival rates for prepubescent girls with cancer are greater than 70%; however, when treated with alkylating agents these girls are at risk of premature ovarian failure (POF). Most girls who do not lose ovarian function post-treatment will, however, experience premature menopause prior to 40 years. It has been shown that girls have an overall 6.3% chance of developing POF (children with Hodgkin's lymphoma being at the greatest risk of POF at 30.7%), whereas 60% of women below 30 and 100% of women over 30 run the risk of POF (Chapman, Sutcliffe and Malpas, 1979a, 1979b; Chemaitilly *et al.*, 2006).

It had been thought that preventing follicles from entering the growing pool, by suppressing ovulation by treating patients with birth control or gonadotrophin-releasing hormone (GnRH) analogues, would prevent the loss of follicles during chemotherapy. However, data in rats have shown that cyclophosphamide stimulates the ovary, resulting in a larger number of growing follicles, and treatment with Lupron (leuprorelin), to suppress ovulation, did not protect the ovary from POF (Letterie, 2004). Research in humans has been conflicting. Data from Blumenfeld (Blumenfeld, 2007; Blumenfeld *et al.*, 2008) have shown that women with Hodgkin's lymphoma that were exposed to GnRH agonists resumed cyclicity in 96.9% of cases, compared to nontreated controls (63%). Currently GnRH analogues are not suggested as a strategy for fertility preservation due to lack of randomized trials and its ineffectiveness in males (Oktay and Sonmezer, 2008).

Other chemotherapeutics, such as doxorubicin (known as Adriamycin), target tumours by intercolating into DNA and inhibiting topoisomerase II, thereby stopping transcription. These agents are used to treat lymphomas and leukaemias, as well as solid tumours of the breast, lung and ovaries. Though doxorubicin is considered to be mildly toxic to the ovary, many women treated with this chemotherapeutic experience irreversible amenorrhoea or early menopause (Faddy *et al.*, 1992).

Knowing that chemotherapeutics may either affect the overall follicle reserve or may cause genetic defects in oocytes stresses the importance of discovering drugs that are fertoprotective. Ideally, creating drugs with targeted delivery mechanisms, such as compounds targeted to receptors that are overexpressed on the surface of cancer cells, would keep ovaries as well as hormone-producing tissues free from toxicity.

19.2.2 Radiation

Just as certain chemotherapeutics affect the follicular reserve, radiation can also have detrimental effects on fertility. Radiation causes DNA damage in cancer cells as well as normal cells of the body. This causes cells to initiate their own cell cycle checkpoint control and activate DNA repair mechanisms, or results in cell death. Women who are exposed to pelvic, abdominal and spinal radiation are at risk for developing ovarian failure. Depending on the woman's level of exposure, radiation exposure can result in damage to ovaries as well as the hormone-producing regions of the brain (hypothalamus and pituitary). It is found that young girls exposed to 10–20 Gy of pelvic radiation often do not complete or fail to undergo normal puberty. Treatment with 4–6 Gy of pelvic radiation in adult women, and as little as 600 cGy in women over the age of 40, is enough radiation exposure to result in POF (Wallace and Thomson, 2003; Wallace, Thomson and Kelsey, 2003).

Not only are the ovarian follicles affected by radiation; many different tissues can be affected, leading to infertility. The uterus, for example, is greatly affected. It must develop a lining thick enough for implantation to occur (~7 mm); however, with radiation the uterus can become scarred, reducing blood flow and resulting in a thin endometrium that cannot support implantation (Larsen *et al.*, 2004). The concept of a uterus transplant has existed since 2000, when the first human uterus transplant was made. Though the surgery was a success, the transplant failed after three months (Fageeh *et al.*, 2002). Interestingly, uterus transplants in rats have proven to be successful (>70%); however, much research will have to be completed before it is attempted in humans again (Wranning *et al.*, 2008).

Young girls and women who have been treated with radiation for tumours of the head and neck run the risk of developing hypogonadotropic hypogonadism. This affects hypothalamic-pituitary function, resulting in reduced secretion or a failure to secrete gonadotrophin-releasing hormone (GnRH). Recently, it has been found that puberty can be initiated in young girls by administering conjugated oestrogens; however, to maintain menstrual cyclicity, exogenous hormones must continuously be given (Ascoli and Cavagnini, 2006). Fertility may also be restored in adults with hypogonadotropic hypogonadism by giving exogenous pulsatile GnRH (Ascoli and Cavagnini, 2006; Hall *et al.*, 1994).

19.2.3 Other fertility-threatening diseases

Although the term 'oncofertility' centres around cancer, fertility preservation methods and assisted reproductive technologies (ARTs) used for oncofertility apply to all diseases and/or therapies that threaten fertility. Here we discuss two diseases that result in loss of fertility in which preservation techniques can be used.

systemic lupus erythematosus (SLE) is a chronic autoimmune disease that can affect any part of the body, particularly the nervous system, heart, lungs, skin, joints, liver and kidneys. The disease used to be fatal; however, with advancing medicine approximately 80% of patients survive 20 years from the time of diagnosis. It is unclear whether SLE directly affects fertility; however, treatments for SLE, namely NSAIDS and corticosteroids, as well as cyclophosphamide, are all implicated in infertility (Gracia and Ginsberg, 2007; Ostensen *et al.*, 2006). The risk of amenorrhoea is correlated with the age of the patient as well as the cumulative dose of cyclophosphamide; because of this, SLE patients would make excellent candidates for reproductive intervention early in treatment if fertility preservation is desired. Lupus patients not only have an increased risk for amenorrhoea, they are also at increased risk for early pregnancy loss due to a tendency for their symptoms to flare due to elevated oestradiol (Le Thi Huong *et al.*, 1994; Macut *et al.*, 2000). Because of the risks to SLE patients as well as potential foetal risk, many SLE women opt for controlled low-dose ovarian stimulation, followed by embryo transfer and careful monitoring as a high-risk pregnancy (Costa and Colia, 2008), or the use of gestational surrogates.

Turner syndrome is a genetic disorder, occurring in 50 per 100 000 births, caused by chromosomal aneuploidy, most commonly (45,X), with at least half of the cases being mosaic (Hjerrild, Mortensen and Gravholt, 2008). There are several health-related issues in classical Turner patients, such as infertility, hypogonadism, short stature, cardiovascular malformations, liver abnormalities, type-2 diabetes, and low bone density. Interestingly, congenital malformations are frequent amongst the 45,X karyotype, where other Turner karyotypes frequently see an increase in endocrine disorders.

Classical Turner patients lose all germ cells at the 18th week of gestation, and the lack of oestradiol and testosterone during the teen years results in the failure to develop secondary sexual characteristics (hypogonadism) (Hjerrild, Mortensen and Gravholt, 2008). It has been found that some girls do spontaneously initiate puberty (15–30%), with 2–5% reaching menarche, indicating that ovarian follicles may survive in some Turner cases, suggesting potential for fertility preservation (Birgit *et al.*, 2009). For other patients, it is advised that they have endocrine therapy to initiate puberty for bone mineralization and socialization. Follicles have been found in Turner patients with both 45,X karyotype and mosaic phenotypes, in girls with and without spontaneous puberty. It is possible to use ovarian tissue cryopreservation to preserve fertility for these patients if ovarian tissue can be isolated at a young enough age, prior to POF. The ovarian tissue could then be transplanted back into the abdomen after endocrine therapy and pubertal onset to achieve regular menses (Birgit *et al.*, 2009; Hjerrild, Mortensen and Gravholt, 2008). Due to the high probability of abnormalities in oocytes from Turner patients, preimplantation genetic diagnosis, chorionic villus sampling and amniocentesis is advised in patients who choose to have children (Birgit *et al.*, 2009; Verschraegen-Spae *et al.*, 1992).

19.3 Options for oncofertility

There are several fertility-sparing options for women and men who are facing a diagnosis of cancer. However, even today many patients as well as physicians concentrate on the diagnosis at hand and forget about the post-treatment effects of the anticancer regimen.

19.3.1 Hormone stimulation

The most successful technology to ensure the possibility of having biological children in the future is hormone stimulation followed by *in vitro* fertilization or intracytoplasmic sperm injection (ICSI) and embryo cryopreservation. This is 'standard of care' for women with cancers that are hormone insensitive. If the woman has the ability to postpone treatment, she can be stimulated with exogenous gonadotrophins to produce a large number of growing follicles that can be aspirated to collect mature oocytes that will be fertilized and stored for her use. Depending on the case, the woman is usually stimulated every day for approximately 10 days with 225 IU of recombinant human FSH, followed by hCG to induce oocyte maturation. Cumulus–oocyte complexes are aspirated out of the ovary prior to ovulation, where they are mixed with sperm for fertilization. After 24 hours, eggs that show two pronuclei are cryopreserved or vitrified for the patient's later use. In the event that the woman does not have a sperm donor, several clinics are currently using protocols to cryopreserve and vitrify MII oocytes. Currently, it is thought that vitrification is superior to slow-freeze cryopreservation of MII oocytes, with development to blastocyst rates of 33.1 and 12%, respectively (Cao *et al.*, 2009; Cobo *et al.*, 2008a, 2008b).

19.3.2 Ovarian tissue cryopreservation (OTC)

For patients in whom hormone stimulation is not an option due to time constraints, or for patients that want to pursue more avenues of fertility preservation, ovarian tissue cryopreservation is another option. As detailed earlier in this chapter, depending on a woman's age, the ovary is filled with thousands of primordial follicles and numerous growing follicles. Interestingly, the outer 1 mm of ovarian tissue consists mainly of primordial follicles. Primordial follicles are very hardy and can easily survive the cryopreservation (slow freeze) or vitrification (fast freeze, no ice crystal formation) process due to their size and limited fluid volume.

The cryopreservation process consists of two critical steps: freeze and thaw. Both steps are equally important in order to minimize damage to the tissue and cells from ice crystal formation. To prevent ice crystal formation the pieces of tissue are incubated in a cryoprotectant, namely ethylene glycol, at a concentration that is cryoprotective yet is not toxic to the tissue. For an excellent review on cryopreservation and vitrification please see Mullen and Critser (2007). Currently several institutions have generated protocols for removing tissue from 1–2 mm × 1–2 mm × 1 cm to 1 mm × 1 cm × 1 cm of cortex from an ovary prior to treatment, and cryopreserved it for patients' later use. Cryopreserved and vitrified tissue has been transplanted successfully back into the abdomen of patients. These cases will be discussed later in this chapter.

19.3.3 *In vitro* maturation and oocyte vitrification

Women who opt to store ovarian tissue for later use also have an option to vitrify mature oocytes if they are present. Currently, women who opt to cryopreserve their ovarian tissue usually have a large portion of their ovary or one of their ovaries removed. The ovary is halved and the inner medulla or vasculature is removed. The outer 3–5 mm of cortex is cut into thin sections for cryopreservation. As the ovarian tissue is sectioned, several large follicles are often punctured. Immature oocytes can be collected at each stage of the tissue removal process. Oocytes are found at all stages from naked, completely immature to mature MII oocytes with expanded cumulus. Immature oocytes are placed in human *in vitro* maturation medium (Sage) containing hCG and FSH for up to 48 hours. Oocytes that reach MII stage are denuded and vitrified for the patient's later use.

19.4 Frontiers in oncofertility

There is exciting work going on in the field of oncofertility. Scientists are examining new ways to ensure that patients undergoing fertility-threatening therapies will have options post-treatment.

19.4.1 3D follicle culture

An exciting new direction in oncofertility research is the isolation and growth of small secondary follicles in culture in the hopes of producing fertilizable oocytes. Research has already shown that individual secondary follicles isolated from 12 (Kreeger *et al.*, 2005) and 16 day-old (Xu, Woodruff and Shea, 2007) mice can be enveloped in a biomaterial called alginate and grown in culture. Alginate, a product of seaweed, acts as an inert matrix allowing for the 3D growth of the follicle (West *et al.*, 2007). Mouse follicles grow to form an antrum, and by day 4 of culture secrete oestradiol and progesterone as well as androstenedione, which is representative of theca cell differentiation (Xu, Woodruff and Shea, 2007).

When the oocyte is stimulated to undergo meiotic resumption by adding LH, EGF and FSH at day 12 of culture, the cumulus–oocyte complex fully expands and breaks through the boundary of the follicle. Metaphase II oocytes retrieved from these follicles have been fertilized, and the resulting embryos have been transferred into pseudopregnant mice, leading to the birth of live pups (Xu *et al.*, 2006).

This experimental technique is now being adapted to nonhuman primates, and humans. To date, researchers are able to encapsulate early secondary follicles in alginate and culture them for a period of 30 days. Human follicles grow up to be approximately 1 mm in diameter and contain large antral cavities. Hormone profiles of these follicles are similar to what is measured in mice.

There are definite differences between follicles in mice and follicles in primates, particularly the length of the follicular cycle. Nonhuman primates and humans take much longer to grow a preovulatory follicle (~90 days) compared to mice (~20 days), and they more than likely will take much longer to grow in culture. Determining the right concentrations of hormones as well as the ideal biomaterial for growth will be key to the production of a fertilizable oocyte.

19.4.2 Organ cultures

The majority of follicles in ovaries at any one time are of the primordial and primary stage (Gosden and Telfer, 1987). It has been extremely difficult, particularly in primates, to achieve the transition process from primordial follicle stage into the growing pool of follicles in culture. It has been shown that primordial follicles from newborn mouse ovaries are capable of being grown in organ cultures to the preantral stage. After removing cumulus–oocyte complexes from growing follicles, they were cultured for 12–14 days, resulting in oocytes that were competent to resume meiosis, to be fertilized and give rise to the birth of live pups (Eppig, O'Brien and Wigglesworth, 1996; Eppig and O'Brien, 1996; O'Brien, Pendola and Eppig, 2003). Recently, it has been shown that biopsies of human tissue, containing mostly primordial follicles, can be cultured for six days, at which point secondary follicles can be isolated. During an additional four days of individual follicle culture, it has been found that media supplemented with activin can induce follicle growth and small antrum formation (Telfer *et al.*, 2008). The question still remains as to what is the best way to culture a secondary follicle in order to obtain a mature oocyte capable of being fertilized and forming a healthy embryo.

19.4.3 Ovarian tissue transplant

Though ovarian tissue transplant is still considered to be an experimental procedure, it has proven to be a successful fertility-restoring technique. Scientists have shown the successes of ovarian transplant studies beginning in animals with allografts in rabbits (Knauer, 1986), autografts in sheep (Gosden *et al.*, 1994) and xenotransplants from wombats to rats (Wolvekamp *et al.*, 2001). Since then, several human cases have shown that not only can fresh tissue be transplanted between monozygotic twins (Silber *et al.*, 2008; Silber and Gosden, 2007; Silber *et al.*, 2005), ovarian tissue that has been cryopreserved and stored for several years can be transplanted back to reinitiate follicular growth and cyclicity (Camboni *et al.*, 2008; Donnez *et al.*, 2008).

For women using ovarian tissue transplant for fertility restoration, cryopreserved or vitrified ovarian tissue removed prior to chemotherapy and/or radiation would be transplanted orthotopically (near the infundibulopelvic ligament) if the fallopian tubes were kept in place. In the instance of a heterotopic transplant, sites such as the forearm are used. Depending on the size of the cryopreserved ovarian tissue, cortical strips are thawed and sutured together to form a patchwork, which is then connected to supportive vasculature. Presently, there are over 20 successful cases of fresh and frozen ovarian transplant cases (Gosden, 2008). On average it takes approximately three to four months (approximately the same time as follicular growth) for cyclicity to begin after transplant, and women have become pregnant as early as the third cycle (Silber *et al.*, 2008; Silber and Gosden, 2007; Silber *et al.*, 2005). Still, techniques of tissue cryopreservation and vitrification are being perfected, which will only make ovarian tissue transplant a more effective and common method of fertility preservation.

19.5 Conclusions

Much has been learned in the last three decades about the regulation of ovarian follicle development and activation. The development of *in vitro* systems that faithfully recapitulate the *in vivo* environment are useful not only to our fund of knowledge about reproductive systems but may also be applied to young women with a diagnosis that results in medically induced infertility. There are still many hurdles in both the basic and applied reproductive fields, but there are rapid advancements that make this an exciting time to be in reproductive biology and medicine.

Acknowledgements

This research was supported by grants from the National Institutes of Health to the Oncofertility Consortium, grant numbers: 1UL1RR024 926 and 1RL1HD0 582 951. The content is solely the responsibilities of the authors and does not necessarily reflect the official views of the National Institutes of Health.

References

Agarwal, S.K. and Chang, R.J. (2007) Fertility management for women with cancer, in *Oncofertility: Fertility Preservations for Cancer Survivors* (eds T.K. Woodruff and K.A. Snyder), Springer, New York, pp. 15–26.

Ascoli, P. and Cavagnini, F. (2006) Hypopituitarism. *Pituitary*, **9**(4), 335–342.

Birgit, B., Julius, H., Carsten, R. *et al.* (2009) Fertility preservation in girls with Turner syndrome: prognostic signs of the presence of ovarian follicles. *J. Clin. Endocrinol. Metab.*, **94**(1), 74–80.

Blumenfeld, Z. (2007) How to preserve fertility in young women exposed to chemotherapy? The role of GnRH agonist cotreatment in addition to cryopreservation of embrya, oocytes, or ovaries. *Oncologist*, **12**(9), 1044–1054.

Blumenfeld, Z., Avivi, I., Eckman, A. *et al.* (2008) Gonadotropin-releasing hormone agonist decreases chemotherapy-induced gonadotoxicity and premature ovarian failure in young female patients with Hodgkin lymphoma. *Fertil. Steril.*, **89**(1), 166–173.

Camboni, A., Martinez-Madrid, B., Dolmans, M.M. *et al.* (2008) Autotransplantation of frozen-thawed ovarian tissue in a young woman: ultrastructure and viability of grafted tissue. *Fertil. Steril.*, **90**(4), 1215–1218.

Cao, Y.X., Xing, Q., Li, L. *et al.* (2009) Comparison of survival and embryonic development in human oocytes cryopreserved by slow-freezing and vitrification. *Fertil. Steril.* **92**(4), 1306–1311.

Chapman, R.M., Sutcliffe, S.B. and Malpas, J.S. (1979a) Cytotoxic-induced ovarian failure in women with Hodgkin's disease. I. Hormone function. *JAMA*, **242**(17), 1877–1881.

Chapman, R.M., Sutcliffe, S.B. and Malpas, J.S. (1979b) Cytotoxic-induced ovarian failure in Hodgkin's disease. II. Effects on sexual function. *JAMA*, **242**(17), 1882–1884.

Chemaitilly, W., Mertens, A.C., Mitby, P. *et al.* (2006) Acute ovarian failure in the childhood cancer survivor study. *J. Clin. Endocrinol. Metab.*, **91**(5), 1723–1728.

Cobo, A., Domingo, J., Perez, S. *et al.* (2008a) Vitrification: an effective new approach to oocyte banking and preserving fertility in cancer patients. *Clin. Transl. Oncol.*, **10**(5), 268–273.

Cobo, A., Kuwayama, M., Perez, S. *et al.* (2008b) Comparison of concomitant outcome achieved with fresh and cryopreserved donor oocytes vitrified by the Cryotop method. *Fertil. Steril.*, **89**(6), 1657–1664.

Costa, M. and Colia, D. (2008) Treating infertility in autoimmune patients. *Rheumatology (Oxford)*, **47**(Suppl 3), iii38–iii41.

Donnez, J., Squifflet, J., Van Eyck, A.S. *et al.* (2008) Restoration of ovarian function in orthotopically transplanted cryopreserved ovarian tissue: a pilot experience. *Reprod. Biomed. Online*, **16**(5), 694–704.

Eppig, J.J. and O'Brien, M.J. (1996) Development in vitro of mouse oocytes from primordial follicles. *Biol. Reprod.*, **54**(1), 197–207.

Eppig, J.J., O'Brien, M. and Wigglesworth, K. (1996) Mammalian oocyte growth and development in vitro. *Mol. Reprod. Dev.*, **44**(2), 260–273.

Faddy, M.J., Gosden, R.G., Gougeon, A. *et al.* (1992) Accelerated disappearance of ovarian follicles in mid-life: implications for forecasting menopause. *Hum. Reprod.*, **7**(10), 1342–1346.

Fageeh, W., Raffa, H., Jabbad, H. and Marzouki, A. (2002) Transplantation of the human uterus. *Int. J. Gynaecol. Obstet.*, **76**(3), 245–251.

Georgescu, E.S., Goldberg, J.M., du Plessis, S.S. and Agarwal, A. (2008) Present and future fertility preservation strategies for female cancer patients. *Obstet. Gynecol. Surv.*, **63**(11), 725–732.

Gosden, R.G. (2008) Ovary and uterus transplantation. *Reproduction*, **136**(6), 671–680.

Gosden, R.G. and Telfer, E. (1987) Number of follicles and oocytes in mammalian ovaries and their allometric relationships. *J. Zool.*, **211**,169–175.

Gosden, R.G., Baird, D.T., Wade, J.C. and Webb, R. (1994) Restoration of fertility to oophorectomized sheep by ovarian autografts stored at −196 degrees C. *Hum. Reprod.*, **9**(4), 597–603.

Gracia, C.R. and Ginsberg, J.P. (2007) Fertility risk in pediatric and adolescent cancers, in *Oncofertility: Fertility Preservation fro Cancer Survivors* (eds T.K. Woodruff and K.A. Snyder), Springer, New York, pp. 57–68.

Hall, J.E., Martin, K.A., Whitney, H.A. *et al.* (1994) Potential for fertility with replacement of hypothalamic gonadotropin-releasing hormone in long term female survivors of cranial tumors. *J. Clin. Endocrinol. Metab.*, **79**(4), 1166–1172.

Hjerrild, B.E., Mortensen, K.H. and Gravholt, C.H. (2008) Turner syndrome and clinical treatment. *Br. Med. Bull.*, **86**,77–93.

Kagawa, N., Kuwayama, M., Nakata, K. *et al.* (2007) Production of the first offspring from oocytes derived from fresh and cryopreserved pre-antral follicles of adult mice. *Reprod. Biomed. Online*, **14**(6), 693–699.

Knauer, E. (1986) Einige Versuche uber Ovarientransplantation bei Kaninchen. *Zentralbl. Gynäkol.*, **20**,524–528.

Kreeger, P.K., Fernandes, N.N., Woodruff, T.K. and Shea, L.D. (2005) Regulation of mouse follicle development by follicle-stimulating hormone in a three-dimensional in vitro culture system is dependent on follicle stage and dose. *Biol. Reprod.*, **73**(5), 942–950.

Larsen, E.C., Schmiegelow, K., Rechnitzer, C. *et al.* (2004) Radiotherapy at a young age reduces uterine volume of childhood cancer survivors. *Acta Obstet. Gynecol. Scand.*, **83**(1), 96–102.

Le Thi Huong, D., Wechsler, B., Piette, J.C. *et al.* (1994) Pregnancy and its outcome in systemic lupus erythematosus. *QJM*, **87**(12), 721–729.

Letterie, G.S. (2004) Anovulation in the prevention of cytotoxic-induced follicular attrition and ovarian failure. *Hum. Reprod.*, **19**(4), 831–837.

Macut, D., Micic, D., Suvajdzic, N. *et al.* (2000) Ovulation induction and early pregnancy loss in a woman susceptible to autoimmune diseases: a possible interrelationship. *Gynecol. Endocrinol.*, **14**(3), 153–157.

Mullen, S.F. and Critser, J.K. (2007) The science of cryobiology, in *Oncofertility: Fertility Preservation for Cancer Survivors* (eds T.K. Woodruff and K.A. Snyder), Springer, New York, pp. 83–103.

O'Brien, M.J., Pendola, J.K. and Eppig, J.J. (2003) A revised protocol for in vitro development of mouse oocytes from primordial follicles dramatically improves their developmental competence. *Biol. Reprod.*, **68**(5), 1682–1686.

Oktay, K. and Sonmezer, M. (2008) Chemotherapy and amenorrhea: risks and treatment options. *Curr. Opin. Obstet. Gynecol.*, **20**(4), 408–415.

Ostensen, M., Khamashta, M., Lockshin, M. *et al.* (2006) Anti-inflammatory and immunosuppressive drugs and reproduction. *Arthritis Res. Ther.*, **8**(3), 209

Silber, S.J. and Gosden, R.G. (2007) Ovarian transplantation in a series of monozygotic twins discordant for ovarian failure. *N. Engl. J. Med.*, **356**(13), 1382–1384.

Silber, S.J., Lenahan, K.M., Levine, D.J. *et al.* (2005) Ovarian transplantation between monozygotic twins discordant for premature ovarian failure. *N. Engl. J. Med.*, **353**(1), 58–63.

Silber, S.J., DeRosa, M., Pineda, J. *et al.* (2008) A series of monozygotic twins discordant for ovarian failure: ovary transplantation (cortical versus microvascular) and cryopreservation. *Hum. Reprod.*, **23**(7), 1531–1537.

Telfer, E.E., McLaughlin, M., Ding, C. and Thong, K.J. (2008) A two-step serum-free culture system supports development of human oocytes from primordial follicles in the presence of activin. *Hum. Reprod.*, **23**(5), 1151–1158.

Tempone, A. (2008) Ovarian function preservation in patients under chemotherapy treatment. *Gynecol. Endocrinol.*, **24**(9), 481–482.

Verschraegen-Spae, M.R., Depypere, H., Speleman, F. *et al.* (1992) Familial Turner syndrome. *Clin. Genet.*, **41**(4), 218–220.

Wallace, W.H. and Thomson, A.B. (2003) Preservation of fertility in children treated for cancer. *Arch. Dis. Child.*, **88**(6), 493–496.

Wallace, W.H., Thomson, A.B. and Kelsey, T.W. (2003) The radiosensitivity of the human oocyte. *Hum. Reprod.*, **18**(1), 117–121.

West, E.R., Xu, M., Woodruff, T.K. and Shea, L.D. (2007) Physical properties of alginate hydrogels and their effects on in vitro follicle development. *Biomaterials*, **28**(30), 4439–4448.

Wolvekamp, M.C., Cleary, M.L., Cox, S.L. *et al.* (2001) Follicular development in cryopreserved Common Wombat ovarian tissue xenografted to Nude rats. *Anim. Reprod. Sci.*, **65**(1-2), 135–147.

Wranning, C.A., Akhi, S.N., Kurlberg, G. and Brannstrom, M. (2008) Uterus transplantation in the rat: model development, surgical learning and morphological evaluation of healing. *Acta Obstet. Gynecol. Scand.*, **87**(11), 1239–1247.

Xu, M., Kreeger, P.K., Shea, L.D. and Woodruff, T.K. (2006) Tissue-engineered follicles produce live, fertile offspring. *Tissue Eng.*, **12**(10), 2739–2746.

Xu, M., Woodruff, T.K. and Shea, L.D. (2007) Bioengineering and the ovarian follicle. *Cancer Treat. Res.*, **138**,75–82.

Index

Page numbers in *italics* refer to figures and tables.

Oogenesis: The Universal Process Marie-Hélène Verlhac and Anne Villeneuve